BIOLOGY OF SHARKS AND THEIR RELATIVES

CRC
MARINE BIOLOGY
SERIES
Peter L. Lutz, Editor

BIOLOGY OF SHARKS AND THEIR RELATIVES

EDITED BY
Jeffrey C. Carrier, John A. Musick,
and Michael R. Heithaus

CRC PRESS

Boca Raton   London   New York   Washington, D.C.

COVER PHOTOGRAPHS — Front Cover: Blue shark (*Prionace glauca*) photographed off Montauk, New York. Back Cover: Nurse shark (*Ginglymostoma cirratum*) carrying a CritterCam™ (a self-contained, animal-borne video and data-recording system) for remote observations of reproductive behavior. (Photographs copyright Harold L. Pratt, Jr. With permission.)

Senior Editor: John Sulzycki
Production Editor: Christine Andreasen
Project Coordinator: Pat Roberson
Marketing Manager: Carolyn Spence

Library of Congress Cataloging-in-Publication Data

Biology of sharks and their relatives / [edited by] Jeffrey C. Carrier, John A. Musick, and
   Michael R. Heithaus.
            p. cm. — (CRC marine biology series)
      Includes bibliographical references and index.
      ISBN 0-8493-1514-X
         1. Chondrichthyes. I. Carrier, Jeffrey C. II. Musick, John A. III. Heithaus, Michael R.
      IV. Series.

   QL638.6.B56 2004
   597.3—dc22   2003069759

Visit the CRC Press Web site at www.crcpress.com

Dedication

To Perry Gilbert, Shelley Applegate, and Sonny Gruber —

for your inspiration, leadership, and perseverance

Preface

In 1963 and again in 1967, Perry Gilbert edited two collections of research papers that provided a comprehensive examination of shark research and served as points of departure for future studies. *Sharks and Survival* and *Sharks, Skates, and Rays* provided an extensive review of the understanding of elasmobranch fishes at the time and are still cited frequently. The most recent summary of the biology of sharks, skates, and rays, *Elasmobranchs as Living Resources,* was edited by Wes Pratt, Sonny Gruber, and Toru Taniuchi in 1990, but much has changed in the world of elasmobranch biology since then. When we first considered developing a modern synthesis of the biology of sharks and their relatives, we were forced to look at what major changes have occurred in our world, how those changes have influenced the worldwide status of sharks and their relatives, and how advances in technology and analytical techniques have changed, not only how we approach problem solving and scientific investigations, but also how we formulate questions.

At least three major influences can be identified for their profound influence on our approach to studies of these animals. Foremost among them is the tremendous interest in sharks and their relatives by the public, perhaps influenced by their main-character roles in movies and popular literature. Such an interest has resulted in the development of public displays and public encounter exhibits that cater to our curiosity about these animals. Captive facilities have increased their basic research components in order to develop better ways to maintain these animals in captive environments and to create aquatic "petting zoos" where rays are stroked and fed by hand. Shark and ray dive adventures have proliferated and the public interest continues to help drive basic research into aspects of shark behavior. In some regards, we have seen a shift from blind fascination with shark attacks to a greater interest in the intricacies of their lives. The media fascination never seems to dwindle, however, and stories of predators driven to maniacal attacks on humans still sell newspapers and television shows, testifying to our more morbid interests in these animals, despite our emerging understanding of the natural behavior of these predators.

A second factor is the significant commercial value of these animals and the resultant worldwide threat to populations that is a result of commercial overexploitation. We have seen areas where populations of sharks and rays have been so reduced that encounters are now almost non-existent. This has forced biologists studying these animals to dramatically increase their focus on studies of life histories. As we developed a better understanding of age, growth, and reproduction, we discovered that these animals, which have survived so well for 400 million years, do not possess high rates of natural replenishment. Partially as a result of this low reproductive rate, the past decade has seen a tremendous increase in conservation and management initiatives around the world that hope to recover depleted populations.

Finally, virtually every area of research associated with these animals has been strongly affected by the revolutionary growth in technology, and the questions we can now ask are very different than those reported in Perry Gilbert's work not so long ago. A careful reading of the chapters presented in this book will show conclusions based on emergent technologies that have revealed some long-hidden secrets of these animals. Modern immunological and genetic techniques, satellite telemetry and archival tagging, modern phylogenetic analysis, geographic information systems (GIS), and bomb dating are just a few of the techniques and procedures that have become a part of our investigative lexicon.

Recently, Bill Hamlett (1999) published an extensive review of the anatomy and fine structure of elasmobranch fishes. In this volume we have taken a different approach, and present a broad survey of the evolution, ecology, behavior, and physiology of sharks and their relatives. Our chapters have been contributed by some of the most eminent chondrichthyan biologists in the field, as well as by some of its most promising "rising stars." We hope that these efforts will not only provide a synopsis of our current understanding of elasmobranchs, but also show the gaps in our knowledge and help to stimulate further studies.

Jeffrey C. Carrier
John A. Musick
Michael R. Heithaus

The Editors

Jeffrey C. Carrier, Ph.D., holds the W.W. Diehl Trustees Professorship in Biology at Albion College, Albion, Michigan, where he has been a faculty member since 1979. He earned a B.S. in biology in 1970 from the University of Miami, and completed a Ph.D. in biology from the University of Miami in 1974. While at Albion College, Dr. Carrier has received awards for teaching and scholarship. His primary research interests have centered on various aspects of the physiology and ecology of nurse sharks in the Florida Keys. His most recent work focuses on the reproductive biology and mating behaviors of this species in a long-term study in an isolated region of south Florida. Dr. Carrier is a member of the American Elasmobranch Society, the American Society of Ichthyologists and Herpetologists, Sigma Xi, and the Council on Undergraduate Research. He has served as Secretary, Editor, and President of the American Elasmobranch Society.

John A. (Jack) Musick, Ph.D., holds the Marshall Acuff Chair in Marine Science at the Virginia Institute of Marine Science (VIMS), The College of William and Mary, Gloucester Point, Virginia, where he has served on the faculty since 1967. He earned his B.A. in biology from Rutgers University in 1962 and his M.A. and Ph.D. in biology from Harvard University in 1964 and 1969, respectively. While at VIMS he has successfully mentored 32 Master's and 39 Ph.D. students. Dr. Musick has been awarded the Thomas Ashley Graves Award for Sustained Excellence in Teaching from The College of William and Mary and the Outstanding Faculty Award from the State Council on Higher Education in Virginia. He has published more than 100 scientific papers and seven books focused on the ecology and conservation of sharks, marine fishes, and sea turtles. In 1985, he was elected a Fellow by the American Association for the Advancement of Science. He has received Distinguished Service Awards from both the American Fisheries Society and the American Elasmobranch Society, for which he has served as president and chair of the Conservation Committee. Dr. Musick also has served as president of the Annual Sea Turtle Symposium (now the International Sea Turtle Society), and as member of the World Conservation Union (IUCN) Marine Turtle Specialist Group. Dr. Musick currently serves as co-chair of the IUCN Shark Specialist Group and on two national, five regional, and five state scientific advisory committees concerned with marine resource management and conservation.

Michael R. Heithaus, Ph.D., is Assistant Professor of Marine Biology at Florida International University, Miami, Florida. He received his B.A. in biology from Oberlin College, Oberlin, Ohio and his Ph.D. from Simon Fraser University in British Columbia, Canada. He was a Postdoctoral Scientist and Staff Scientist in the Center for Shark Research at Mote Marine Laboratory in Sarasota, Florida (2001–2003) and also served as a research fellow at the National Geographic Society (2002–2003). Dr. Heithaus' main research interests are in predator–prey interactions and the factors influencing behavioral decisions, especially of large marine taxa including marine mammals, sharks and rays, and sea turtles. Currently, he is also investigating how behavioral decisions, especially antipredator behavior, may influence behavioral decisions of other individuals and community dynamics. The majority of Dr. Heithaus' previous fieldwork has focused on tiger sharks and their prey species in Western Australia. Dr. Heithaus is a member of the Ecological Society of America, Animal Behavior Society, International Society for Behavioral Ecology, Society for Marine Mammalogy, and Sigma Xi.

Contributors

Neil Aschliman
Department of Biology
Texas A&M University
College Station, Texas

Ashby B. Bodine
Department of Animal
    and Veterinary Sciences
Clemson University
Clemson, South Carolina

Gregor M. Cailliet
Moss Landing Marine Laboratories
Moss Landing, California

Janine N. Caira
Department of Ecology
    and Evolutionary Biology
University of Connecticut
Storrs, Connecticut

John K. Carlson
NOAA/National Marine Fisheries Service
Southeast Fisheries Science Center
Panama City Laboratory
Panama City, Florida

Jeffrey C. Carrier
Department of Biology
Albion College
Albion, Michigan

José I. Castro
Center for Shark Research
Mote Marine Laboratory
Sarasota, Florida
and
NOAA/National Marine Fisheries Service
Southeast Fisheries Science Center
Miami, Florida

Keith P. Choe
Department of Zoology
University of Florida
Gainesville, Florida

Leonard J.V. Compagno
Shark Research Centre
Department of Marine Biology
South African Museum
Cape Town, South Africa

Enric Cortés
NOAA/National Marine Fisheries Service
Southeast Fisheries Science Center
Panama City Laboratory
Panama City, Florida

Leo S. Demski
Department of Biology
New College of Florida
Sarasota, Florida

Dominique A. Didier
Department of Ichthyology
The Academy of Natural Sciences
Philadelphia, Pennsylvania

David H. Evans
Department of Zoology
University of Florida
Gainesville, Florida

James Gelsleichter
Elasmobranch Physiology and
    Environmental Biology Program
Center for Shark Research
Mote Marine Laboratory
Sarasota, Florida

Kenneth J. Goldman
Department of Biology
Jackson State University
Jackson, Mississippi

Eileen D. Grogan
Department of Biology
Saint Joseph's University
Philadelphia, Pennsylvania

Melanie M. Harbin
Virginia Institute of Marine Science
The College of William and Mary
Gloucester Point, Virginia

Claire J. Healy
Department of Ecology and Evolutionary
	Biology
University of Connecticut
Storrs, Connecticut

Edward J. Heist
Fisheries and Illinois Aquaculture Center
Southern Illinois University
Carbondale, Illinois

Michael R. Heithaus
Department of Biological Sciences
Marine Biology Program
Florida International University
Miami, Florida

Michelle R. Heupel
Elasmobranch Behavioral Ecology
	Program
Center for Shark Research
Mote Marine Laboratory
Sarasota, Florida

Robert E. Hueter
Shark Biology Program
Center for Shark Research
Mote Marine Laboratory
Sarasota, Florida

George V. Lauder
Organismic and Evolutionary Biology
Museum of Comparative Zoology
Harvard University
Cambridge, Massachusetts

Christopher G. Lowe
Department of Biological Sciences
California State University
Long Beach, California

Carl A. Luer
Marine Biomedical Research Program
Center for Shark Research
Mote Marine Laboratory
Sarasota, Florida

Richard Lund
Department of Vertebrate Paleontology
Carnegie Museum of Natural History
Philadelphia, Pennsylvania

David A. Mann
College of Marine Science
University of South Florida
St. Petersburg, Florida

Karen P. Maruska
Department of Zoology
University of Hawaii at Manoa
Honolulu, Hawaii

John D. McEachran
Department of Wildlife and Fisheries
	Sciences
Texas A&M University
College Station, Texas

Philip J. Motta
Department of Biology
University of South Florida
Tampa, Florida

John A. Musick
Virginia Institute of Marine Science
The College of William and Mary
Gloucester Point, Virginia

Peter M. Piermarini
Department of Zoology
University of Florida
Gainesville, Florida

Harold L. Pratt, Jr.
Center for Shark Research
Mote Marine Laboratory
Summerland Key, Florida

Colin A. Simpfendorfer
Elasmobranch Fisheries and Conservation
	Biology Program
Center for Shark Research
Mote Marine Laboratory
Sarasota, Florida

Joseph A. Sisneros
Department of Psychology
University of Washington
Seattle, Washington

Catherine J. Walsh
Marine Immunology Program
Center for Shark Research
Mote Marine Laboratory
Sarasota, Florida

Bradley M. Wetherbee
Department of Biological Sciences
University of Rhode Island
Kingston, Rhode Island

Cheryl A. D. Wilga
Department of Biological Sciences
University of Rhode Island
Kingston, Rhode Island

Contents

Part I

Phylogeny and Zoogeography

1

The Origin and Relationships of Early Chondrichthyes

Eileen D. Grogan and Richard Lund

CONTENTS

1.1   Introduction

Chondrichthyan fishes are probably the most successful of all fishes if success is measured in terms of historical endurance, based on ability to survive the mass extinctions of the last 400 million years or so. They are essentially defined by a cartilaginous skeleton that is superficially mineralized by prismatic calcifications (tesserae) and by the modification, within males, of mixopterygia (claspers) for the purpose of internal fertilization. It is generally accepted that the Class Chondrichthyes is a monophyletic group divisible into two sister taxa, the Elasmobranchii and Holocephali, and that extant chondrichthyans (sharks, skates, rays, and chimaeras) are derivable from Mesozoic forms. Yet, how these forms relate to the distinctly more diverse Paleozoic forms and even the relationship of the Chondrichthyes to all other fishes are poorly resolved issues. Unquestionable evidence of cartilaginous fishes extends to the Lower Devonian or even to the Silurian if isolated scales (putatively shark in origin) can be considered with any confidence. However, in reviewing such information it is imperative to consider what features or characters ensure that these earliest preserved vestiges are remains of a chondrichthyan form rather than a non-chondrichthyan. The purpose of this chapter is to discuss the evidence for the origin, diversification, and life histories of the Chondrichthyes, to address trends in their morphological divergence and innovation, and to explore the possible relationships between fossil and modern forms. In a general discussion of relationships, we adopt the classification scheme for shark and sharklike fishes put forth by Compagno (2001), as a consensus of the analyses of Compagno (1984), Shirai (1996), and de Carvalho (1996). The classification scheme used to describe the relationships of all Chondrichthyes is that developed in Lund and Grogan (1997a,b, in press, a,b) and Grogan and Lund (2000).

1.2   On the Synapomorphic Chondrichthyan Characters: Tesserate Mineralization and Internal Fertilization by Male Claspers

The interrelationships of the gnathostome classes is beyond the scope of this study. Outgroup comparison of characters used in this work, however, strongly supports the thesis that the Chondrichthyes are a monophyletic group (e.g., Maisey, 1984, 2001; Lund and Grogan, 1997a; Grogan and Lund, 2000). Although a variety of characters have been proposed to define this monophyletic group, two synapomorphies are generally accepted to define these fish: the prismatic endoskeletal calcification and pelvic claspers.

1.2.1   Tesserate Mineralization

The tesserate mode of mineralizing endoskeletal tissues peripherally is *the* critical defining character of the group (see Coates and Sequeira, 1998; Lund and Grogan, 1997a; Maisey, 1984, 2000). It is therefore unfortunate that the term *tesserae* has been applied to both the chondrichthyan and placoderm conditions (Ørvig, 1951; Applegate, 1967; Denison, 1978) for they represent two different phenomena. At best, they share an extremely remote relationship derived from common vertebrate patterns of skeletal tissue determination, regulation, and, therefore, development. Chondrichthyan tesserae represent a developmental deviation from the pattern of endoskeletal tissue formation that characterizes primitive gnathostomes. Previous studies and current work in progress address the difficult question of the transition from the primitive gnathostome condition of perichondral bone (Janvier, 1996; Basden et al., 2000) to the chondrichthyan states of perichondral mineralized cartilage (see Ørvig, 1951; Applegate, 1967; Kemp and Westrin, 1979; Rosenberg, 1998; Yucha, 1998; Grogan and Yucha, 1999; Grogan et al., in prep.). All data generally support the idea that the endoskeletal mineralization of chondrichthyans represents an autapomorphic condition relative to other gnathostomes. Developmental responses to mechanical and growth parameters (possibly even including regulatory features associated with the pituitary:gonadal axis) lead to variants of the common mineralized plan.

"Prismatic" calcification, as used here, refers to the macroscopically visible state of separate, peripheral mineralized units (Ørvig, 1951). Thin sections typically reveal either the more primitive state of globular calcified cartilage or a highly ordered, star-shaped architecture parallel to the cartilage surface and an hourglass microstructure perpendicular to that surface when viewed under crossed polarizer and analyzer (Ørvig, 1951; Applegate, 1967; Yucha et al., pers. obs.). The latter configuration is due to the mineralized unit having two subunits, Kemp and Westrin's (1979) cap and body components. Subsequent studies further indicated that these tessera subunits are distinct in their origin and the extent of their development, thus offering an explanation for apparent differences in tesserate appearance within and across taxa (Yucha, 1998; Rosenberg, 1998; Fluharty and Grogan, 1999; Grogan and Yucha, 1999; Grogan et al., in prep.). In keeping with these observations and with observations of fossil forms, then, we use "continuous" calcified cartilage, in the sense of Ørvig (1951), to refer to a modified tesserate condition wherein adjacent tesserae undergo an early ontogenetic fusion and, therefore, do not exhibit the more typical prismatic microstructure.

The primitive gnathostome condition, in contrast, is most likely to be that of endoskeletal elements having a cartilaginous core covered by perichondral bone (Ørvig, 1951; Janvier, 1996).

1.2.2 Modification of Pelvic Girdle in Males to Generate Claspers

Claspers (mixopterygia) are extensions of the endoskeletal axis of the pelvic fin of a male chondrichthyan that form sperm-conducting structures (copulatory organs) to facilitate internal fertilization of a female. These axial modifications may or may not be accompanied by modifications of the fin radials or modifications of the adjacent squamation. The development of claspers, however, also involves the coordinated development of the musculature necessary to pump sperm and the musculature necessary to maneuver the claspers. A model of clasper development and its morphoclinal transition within Chondrichthyes has been presented in Lund and Grogan (1997a).

All mature male chondrichthyans display intromittent organs. The Upper Devonian *Cladoselache* has often been claimed to be the exception, yet such arguments ignore the high probability that the recovered forms are strictly female. The fossil deposits from which clasper-lacking specimens are derived are shallow, epicontinental, and marine and appear to indicate a paleoenvironment that was like that of a coastal margin/shelf or contiguous bay (Ettensohn and Barron, 1981; Hansen, 1999). This combination of evidence is consistent with the interpretation that evidence of females rather than males may be due to life history styles, including sexual segregation, that are reminiscent of extant forms. It is also clearly established that other Upper Devonian male elasmobranchs, including the comparatively smaller *Diademodus* from the same deposit as *Cladoselache*, displayed pelvic claspers. We also maintain that the preponderance of evidence supports the view that all mature male chondrichthyans are identifiable by their possession of claspers since all other members of the cladodont group have claspers to identify males.

It has been proposed that at least some male placoderms exhibited pelvic claspers akin to those of chondrichthyans (Denison, 1978), thereby possibly rendering this a synapomorphy of Chondrichthyes plus Placodermi (Miles and Young, 1977; Young, 1986). Yet, there is no convincing evidence that the placoderm structures (which are principally dermal in nature) are other than analogous rather than developmentally homologous.

Claspers are primary sexual characters of males (not secondary). Alar scales or plates, as found in some skates and in iniopterygians (Zangerl and Case, 1973) as well as the prepelvic tenaculae of chimaeroids and some cochliodonts are secondary sexual characters that are accessory to the reproductive (sperm transfer) function of claspers. The pelvic plates of ptyctodont arthrodires (e.g., Young, 1986) show no morphology indicative of either formation from a pelvic endoskeletal axis or a sperm transfer function, and are thus not homologous with chondrichthyan claspers. The elongate pelvic basal interpreted for *Ctenurella* (Ørvig, 1960) has not been confirmed as endoskeletal and shows no obvious indication of being reproductive in nature.

1.3 Historic Evidence of Early Chondrichthyans

The first scales and spines attributed to chondrichthyans range from the Lower Silurian, and more diverse forms of these scales are generally abundant within the Devonian (Goujet, 1976; Karatajute-Talimaa, 1992; Cappetta et al., 1993; Karatajute-Talimaa and Predtechenskyj, 1995; Zhu, 1998; Rodina, 2002). It has been argued on the basis of putative scale morphotypes that chondrichthyan origins may reach as far back as the Ordovician or Cambrian and that a maximal adaptive radiation in the Early Devonian led to the rise of ctenacanthid, hybodontid, and protacrodontid forms (Karatajute-Talimaa, 1992), lineages which undeniably extend into the Carboniferous and beyond. However, the nature and affinity of the earliest of these elements will remain subject to skepticism without further developmental and histological studies of these tissues across vertebrates and as long as whole-organism fossils bearing such scales are wanting. Of particular concern is an apparent ontogenetic and morphological continuum between thelodont, acanthodian, and chondrichthyan scales and buccopharyngeal denticles (Rodina, 2002; Lund and Grogan, pers. obs.). The complexity of this problem is further accentuated by additional observations: (1) Upper Silurian *Elegestolepis*-type scales and Devonian *Ctenacanthus*-type scales are both found within the Carboniferous elasmobranchian *Falcatus* as cranial and buccopharyngeal denticles, respectively, and (2) both *Elegestolepis* and Devonian *Protacrodus*-type scales correlate, respectively, with the generalized and specialized cranial scales of the Carboniferous euchondrocephalan *Venustodus argutus* St. John and Worthen 1875 (Bear Gulch specimen, CM 41097).

A more reliable indicator of the presence of chondrichthyan(s) would logically be in the form of a diagnostic feature, that of the tesserate mode of cartilage mineralization. To our knowledge, the first reports of chondrichthyan-type calcified cartilage (i.e., *sensu* Ørvig, 1951, and Applegate, 1967; Lund et al., in prep.), are from the marine Devonian deposits of Bolivia. The frequency with which these calcified cartilage fragments occur suggests that the chondrichthyans were the most abundant of all vertebrates in a marine environment in which agnathan thelodonts, actinopterygians (e.g., *Moythomasia*), acanthodians (e.g., *Sinacanthus*), and placoderms are also indicated (Janvier and Suarez-Riglos, 1986; Gagnier et al., 1989). It has also been deduced that the three chondrichthyan species of *Zamponiopteron*, organ taxa originally established on the basis of fin elements and those recently identified as possessing prismatic calcified cartilage (Lund et al., in prep.), were possibly endemic to this site.

Other data from this deposit confirm not only the presence of chondrichthyans by the middle Devonian but have provided vital morphological evidence of the one of the earliest chondrichthyans known from articulated endoskeletal material. A cranium, identified as *Pucapampella* (Janvier and Suarez-Riglos, 1986), attests not only to the retention of stem gnathostome features in an Early Devonian chondrichthyan but also emphasizes, by comparison with the other cold-water pucapampellids — from the slightly older Emsian of South Africa (Anderson et al., 1999; Maisey and Anderson, 2001) and from the younger Late Eifelian-Givetian of Bolivia (Maisey, 2001) — that distinct cranial morphs with holocephalan affinities on the one hand and selachian affinities on the other were already established by this time (Grogan, pers. obs.).

In hindsight, recovery of such data reinforces what was previously known, in general, about the Devonian. The radiations associated with this "Age of Fishes" were supported by an increasingly diverse suite of estuarine, brackish to freshwater, and marine continental margin environments in equatorial Euramerica and in the southern continent of Gondwana. Especially in Euramerica, shallow seas supported extensive reef-building by stromatoporoids and corals and, so, were likely to have favored the retention and diversification of many fishes along or near the continental margins as a consequence of high primary productivity. In keeping with this, there is significant evidence to document that, by the middle to late Devonian, the chondrichthyans were represented by a number of strikingly different forms that inhabited environments ranging from fresh and brackish water to continental margins and oceans (Ivanov and Rodina, 2002a,b). Yet, perhaps because of their propensity for poor holomorphic preservation, they apparently remained relatively scarce compared to the placoderms and actinopterygians.

The freshwater xenacanthids included *Leonodus, Antarctilamna, Portalodus*, and *Aztecodus* (Young, 1982). Elasmobranchs and euchondrocephalans, including *Zamponiopteron, Diademodus, Siamodus, Ctenacanthus, Plesioselachus, Phoebodus, Thrinacodus, Orodus, Protacrodus, Stethacanthus,* and hybontids, are reported from marine, estuarine, and coastal lagoonal environments at some point in the Devonian (Janvier and Suarez-Riglos, 1986; Gagnier et al., 1989; Ginter, 1999; Lelievre and Derycke, 1998; Anderson et al., 1999; Ivanov and Rodina, 2002a,b). Chondrichthyan microremains (teeth, scales) from Germany suggest a wide distributional range for the group at the end of the Devonian (Famennian-Tournasian) but with a progressive partitioning of forms according to an environmental gradient (Ginter, 1999; Ivanov and Rodina, 2002a). Protacrodonts primarily occupied shallow epicontinental seas and the proximal aspect of continental margins, the tooth-taxon *Jalodus* was principally associated with deep marine waters, and the cladodonts reflected a more cosmopolitan distribution as they exhibited more of an ocean-roaming habit. The broad, blunt, durophagous teeth of orodonts, helodonts, and *Psephodus*-like forms, as well as those of *Ageleodus*-like forms (whose teeth are closest to those of the Debeeriidae), are found in what are probable estuarine to freshwater deposits toward the Upper Devonian (Downs and Daeschler, 2001). It is also important to note that no evidence currently exists for cochliodont tooth plates in Devonian deposits, possibly indicating that the Holocephali *sensu stricto* had yet to evolve or to diversify. It is true that, if the Holocephali *sensu stricto* had evolved by the Devonian and if they inhabited deep waters, then any fossilized remains of them would be the least likely of all forms to be recovered. Yet, the morphological, chronological, and developmental data all support the view that, at best, paraselachian-type holocephalan ancestors existed during this phase of vertebrate life.

In terms of community structure, arthrodiran placoderms were the apex predators of the Devonian. The known elasmobranchs and protacrodonts were lower-trophic-level forms; the former predominantly bore piercing teeth whereas the latter possessed lower, blunt-crowned teeth. On the other hand, the various and well-established xenacanths may have vied for the apex predator level with Crossopterygii in freshwater environments.

1.3.1 Evidence from the Carboniferous

1.3.1.1 *Carboniferous Communities and Chondrichthyan Adaptations* — The Lower Carboniferous witnessed the extinction of the Placodermi, the reduction of the formerly diverse Acanthodii to one or two toothless genera, and slow diversification of freshwater Amphibia. Crossopterygii were limited to very few freshwater and marine species. Coelacanths, however, diversified to the extent that correlated with highly specialized habitat preferences. Small actinopterygians broadly diversified within their primary consumer trophic-level specializations in the marine environment. Chondrichthyans, in contrast, radiated rapidly and expansively in all available aquatic regimes. Marine waters included the stethacanthids, protacrodonts, petalodonts, "helodonts," and a host of other forms known only from teeth (Rodina and Ivanov, 2002a,b). The freshwater environs, on the other hand, were inhabited by forms such as *Hybodus* and *Helodus*, in addition to various xenacanths (Romer, 1952; Lund, 1976). Carboniferous marine deposits that offer information beyond isolated chondrichthyan remains are those of Glencartholm, Scotland (Traquair, 1888a,b; Moy-Thomas and Dyne, 1938), and Bearsden, Scotland (Coates, 1988). The Lower Carboniferous tooth and spine faunas of Armagh, Ireland (Davis, 1883) and the upper Mississippi Valley of the United States (e.g., Newberry and Worthen, 1866, 1870; St. John and Worthen, 1875, 1883) are either small and limited deposits or organ-taxon deposits. The fish fauna of the penecontemporaneous Upper Mississippian Bear Gulch Limestone includes forms comparable to those uncovered in these deposits but has additional advantages. The biota of the Bear Gulch and the dimensions and conditions of the shallow, tropical, marine bay from which they came are sufficiently detailed by lagerstätten-type preservation to permit what is likely to be the most comprehensive and reliable documentation of community structure and ecology reported for Upper Paleozoic fish to date.

We believe that the community structure of the adjacent epicontinental and open waters was not inconsistent with that of the Bear Gulch to the extent that the Paleozoic Bear Gulch bay was obviously accessible to migratory forms and provided breeding and nursery grounds for those not endemic to the bay. Given this and the continuity between the other Carboniferous deposits noted above (i.e., select genera or even species in common), it is likely that the diversity of the Bear Gulch fauna may potentially be representative of Upper Mississippian marine faunas. Yet, detailed faunal analyses of the Glencartholm and Bearsden deposits would be required to evaluate this possibility further. It is known that the Bear Gulch fauna exhibits a higher diversity and species richness than later Pennsylvanian deposits, which are characterized by both freshwater and marine fishes (Lund and Poplin, 1999; Schultze and Maples, 1992). Yet, the latter situation may simply reflect the correlation between lower diversity in newly developing ecosystems or in areas of recent disturbance and invasion by generalists or ecological opportunists (e.g., Downs and Daeschler, 2001). By contrast, analyses of the paleoenvironment at the time of the Bear Gulch do not indicate catastrophic or revolutionary change (Grogan and Lund, 2002). The conditions prevailing during the deposition of this deposit suggest periodic disturbances, but not of a magnitude that would dramatically reduce diversity and richness and lead to a permanent shift in community composition. In any event, it is certain that the preserved remains of Bear Gulch chondrichthyans provide a rare view and index of the range of chondrichthyan diversity evident at this early stage in the evolution of the group.

1.3.1.2 *Bear Gulch Limestone* — The data that follow are based on both published and nonpublished material. Specimen abbreviation codes are as follows: CM, Carnegie Museum of Natural History, Section of Vertebrate Fossils; MV, University of Montana Geological Museum.

1.3.1.2.1   Taxonomic Diversity and Species Richness.   Of the 126 Bear Gulch taxa currently identified, 60% of the species are chondrichthyan (compared with 32.8% actinopterygian and 4.8% coela-
canth). This pattern of taxonomic composition is radically different from the general pattern evidenced
by extant fishes. Given the conservative estimate of 24,618 species of extant fish (Nelson, 1994), today's
chondrichthyans represent approximately 3.4%, actinopterygians make up approximately 96%, and the
coelacanths, 0.004%. Likewise, Compagno (2001) indicates that the diversity of living chondrichthyans
(~1200 identified forms) is strikingly low compared to that indicated in the record of fossil chondrichthyans (2500). He further approximates that, of the identified living and valid species of chondrichthyans,
about 4.2% are chimaeroids, 41.7% are sharks, and 50% are represented by the highly modified batoids.
Again, these percentages are radically different from the Bear Gulch numbers. Elasmobranchs (sharklike
and related forms) represent only 25% of the chondrichthyans. The majority of the chondrichthyans are
euchondrocephalans, 45.3% of which are holocephalimorphs and 29.3% are paraselachii (forms intermediate between the selachian and holocephalan morphological plans).

There is also an inverse relationship between the number of Bear Gulch taxonomic units and numbers
of individuals representing these units (Lund and Poplin, 1999), such that the coelacanths with low species
richness are the most abundant fish numerically, the actinopterygians follow next, and the chondrichthyans, with the greatest species richness, are typically known from fewer individuals (Figure 1.1). Within
the Chondrichthyes of the Bear Gulch community there is only one abundant species in each of the three
major adaptive divisions of the class: the small stethacanthid elasmobranch *Falcatus falcatus*; the small
debeeriid euchondrocephalan *Heteropetalus elegantulus*; and the small chondrenchelyid holocephalan
Harpagofututor volsellorhinus. The overall ecological and morphological aspects of the chondrichthyan
community show low numbers of individuals in each species, commensurate with a *k*-selected strategy,
and fine niche partitioning on the basis of both feeding and propulsive specializations.

Dominance analyses confirm that the Bear Gulch community shows low dominance indices and
moderate equitability and evenness statistics, as do both Chondrichthyan and Osteichthyan fishes (Table
1.1). However, chondrichthyans have considerably higher richness statistics with both Menhinick's index
and Margaleff's index than the osteichthyans. Chondrichthyan rarefaction data also display much greater
retrieval of taxa in relation to specimen numbers than those of the Osteichthyes of equal sample size
(Table 1.2). Such analyses of the Bear Gulch Limestone community provide a measure of the fundamental
adaptive radiation of the class during the Permo-Carboniferous.

1.3.1.2.2   The Bear Gulch and Devonian-Permian Chondrichthyan Groups.   Some major types
of Devonian-Permian chondrichthyans are listed in Figure 1.2. Illustrations of some of these and other
relevant chondrichthyans are presented in Figure 1.3 through Figure 1.5. The Devonian-Permian Elasmobranchii (Figure 1.3) divide into a large clade, herein designated as the Paleoselachii (including the
marine cladodonts, the freshwater xenacanths, *Squatinactis*, and the undescribed fish code-named
"Snipe") and a smaller group with closer ties to the Euselachii (early hybodonts, and, possibly, ctenacanths). The predatory Paleoselachii had amphistylic suspensoriums, elasmobranchian branchial structure
and function, and multicuspid piercing teeth and were amply represented in the Bear Gulch Limestone
by the several taxa of Stethacanthidae, the eel-shaped "Snipe," and the angel-sharklike *S. montanus*.

The nonholostylic Euchondrocephali (Figure 1.4: 2, 3A and 3B, 6; Figure 1.5: 1A and 1B, 2) have
autodiastylic suspensoriums that have been variously modified from the primitive gnathostome condition.
They also had primitively operculate branchial chambers with branchial baskets that varied toward the
snugly nested condition of the modern chimaeroids. Dentitions were variable and frequently heterodont,
with anterior nipping or plucking teeth and posterior comminuting teeth. There were also frequent
specializations of the premandibular parasymphysial/symphysial dentition that seem to have originated
as a primitive character complex for gnathostomes. Included here are whole-bodied forms that display
protacrodont and orodontiform teeth, which are basal forms and likely to have a close relation to the
Devonian tooth-taxa. Most of the nonholostylic Bear Gulch Euchondrocephali are less than 1 m in length.

The Holocephalimorpha were a guild of holostylic fishes (Figure 1.4: 4, 5A; Figure 1.5: 3 to 6). Most
combined small low, flat, anterior teeth or tooth plates with large posterior tooth plates that indicate a
primarily durophagous diet. The genus *Echinochimaera* developed cutting edges on their tooth plates.
Numerous specializations of cranial plates and spines, dorsal fins, and anterior tenacular secondary sexual

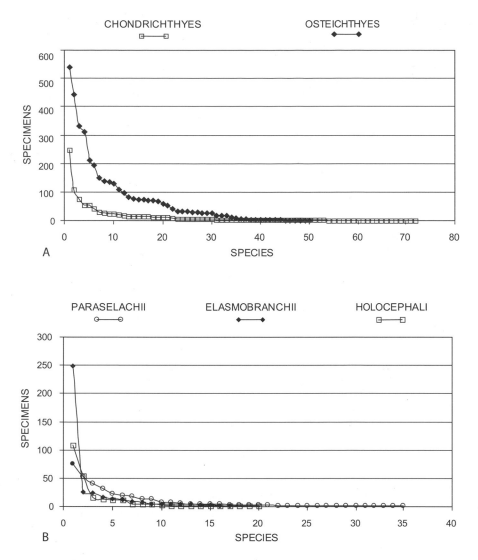

FIGURE 1.1 Distribution of specimen numbers by species for the Upper Mississippian Bear Gulch Limestone fish fauna (for collections from 1968–2002; one species each of Acanthodii and Agnatha excluded). (A) Distribution of Chondrichthyes and Osteichthyes. (B) Distribution of Elasmobranchii, Holocephali, and non-holocephalan Euchondrocephali (Paraselachii).

structures further characterize this speciose group. Most of these taxa are also less than 1 m in length in the Bear Gulch Limestone bay.

1.3.1.2.3 Community Structure and Population Dynamics. Akin to today's sharks (reviewed by Camhi et al., 1998), most Bear Gulch and other Carboniferous chondrichthyans appear to have been restricted in their distribution, with the majority confined to continental shelf and slope waters and only a small percentage large-scale migrants. The Paleozoic forms reflect a range of habitats and exhibit feeding-based specializations.

The predatory stethacanthids (3 to 3.5 m) represent some of the largest of the Bear Gulch fishes. Their size, the nature of their preserved remains (principally as disarticulated remains of bloated fish), and the proximity of these fossils to the deeper aspects of the bay support the interpretation of these as migratory or opportunistic vagrants from adjacent waters. Also, the geographic distribution of stethacanthid sharks ranges as far as current Scotland and Moscow, Russia. Most other paleoselachians (e.g., *Falcatus*, *Damocles*, *Squatinactis*, code-name Tristy) are smaller (~150 mm adults). Because of their size, they

TABLE 1.1

Diversity Indexes (D) and Bootstrap 95% Confidence Intervals (lower, upper) for
Bear Gulch Limestone Chondrichthyes and Osteichthyes Collected from 1968
through 2002

	Chondrichthyes			Osteichthyes		
	D	**Lower**	**Upper**	**D**	**Lower**	**Upper**
Taxa	72	58	68	50	44	50
Individuals	988	988	988	3791	3791	3791
Dominance	0.094	0.083	0.108	0.064	0.061	0.067
Shannon	3.119	2.983	3.172	3.119	3.078	3.142
Simpson	0.906	0.892	0.917	0.936	0.933	0.939
Menhinick	2.291	1.845	2.163	0.812	0.715	0.812
Margaleff	10.3	8.266	9.716	5.946	5.218	5.946
Equitability	0.729	0.72	0.762	0.797	0.793	0.821
Fisher's alpha	17.86	13.46	16.57	8.135	6.986	8.135
Berger-Parker	0.251	0.22	0.276	0.142	0.129	0.154

Note: Indexes presented by the PAST software program of Hammer et al. (2001).

TABLE 1.2

Rarefaction Data for Specimen Numbers Equal
to the Maximum Number of Specimens of Bear
Gulch Limestone Chondrichthyes Collected
from 1968 through 2002 (bootstrap estimate;
Hammer et al., 2001)

	n	**Taxa**	**SD**
Osteichthyes	980	42.3	1.78
Chondrichthyes	980	71.85	0.388
All fish	980	84.45	3.73

were likely to have been persistent inhabitants of, or spent the majority of their time in, the more
protective environs of the bay even though they had the ability to extend into the epicontinental sea
(Lund, 1985, 1986a). The highly specialized, eel-shaped form with *Thrinacodus*-style teeth (code-named
Snipe) is likely to have ranged from open water to the reef environment. The only elasmobranch in the
fauna known to have a hyostylic suspensorium, code-named Tristy, possessed diminutive teeth in small
mobile jaws and, so, was probably a microphagous suction feeder.

Euchondrocephalans, including the crown group Holocephalimorpha, make up the bulk of the consti-
tutive bay inhabitants, but not all were restricted to the bay. They range from a benthic habitat (two
predatory iniopterygian taxa, Iniop 1 and 2, with shrimp-filled guts), to near bottom (the cochliodonts,
Debeerius), to reef dwellers (petalodonts, Elweir), to mid- and upper-water-column swimmers/flyers
(*Heteropetalus* and taxa code-named Little 2 Spine and Iniop 3). The dentitions of this speciose group
suggest varied feeding tactics and diets. Fossil data reflect that diets ranged from worms, shrimp, and
mollusks to amorphous bituminous (plant) remains that were likely to have been ingested as the fish
scavenged or sifted through bottom sediments. The taxon code-named Elweir lacked marginal teeth,
bore greatly enlarged labial cartilages, and, with an oral rim mechanism resembling that of modern
Clupeid teleosts, was most probably a suction feeder (Grogan, 1993).

Overall, the data indicate that the comparatively smaller chondrichthyans, which also represent the
majority of the chondrichthyan forms, reveal a more restricted distribution, that of the epicontinental
sea margins and adjacent shallow bodies of water. Like the smaller shark species of the coastal to
inshore environs (Smith et al., 1998), it is probable that these smaller Bear Gulch chondrichthyans
matured earlier and were shorter lived compared to the larger, apex predatory stethacanthids. By virtue
of the range of morphological designs that had become possible earlier in their history and, apparently,
by retaining considerable developmental plasticity in cranial and feeding design, the smaller

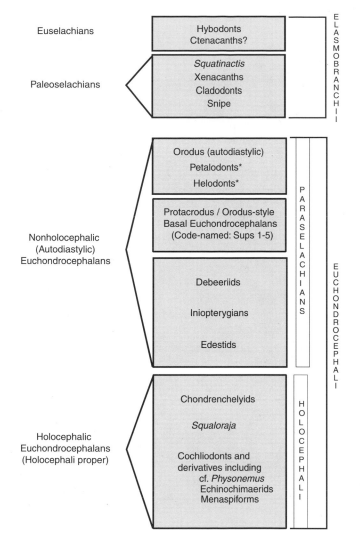

FIGURE 1.2 Devonian-Permian chondrichthyans and their assignments to higher taxonomic units or groupings. The listing of representative taxa and units for Devonian-Permian forms (paleoselachians, hybodontids, ctenacanthids, holocephalic euchondrocephalans, and non-holocephalic euchondrocephalans) generally corresponds to the arrangement in the cladogram of Figure 1.6. * Specifies derived euchondrocephalans that are convergently holostylic and have no close relationship to the Holocephali *sensu* Lund and Grogan, 1997a. ? Designates the commonly presumed assignment of ctenacanths.

chondrichthyans were able to expand quickly into a variety of habitats and niches, and to outnumber non-chondrichthyans in terms of taxonomic diversity (e.g., as specifically demonstrated by the species richness of holocephalimorph forms). Rapid vertebrate diversification such as this can be explained by duplication of body pattern–determining genes, such as Hox-gene homologues, which permit rapid diversification in form, and by promoting heterochronic manipulation of a common developmental plan through neoteny, progenesis, and/or peramorphosis. Over the subsequent evolution of the group, however, as the majority of surviving forms became increasingly predatory and/or large (with an oceanic lifestyle and/or ability to seek refuge in deep waters during cataclysmic periods), the attributes of continued growth and increase in size were likely favored (heterochronically) at the expense of a timely progression to reproductive maturity.

1.3.1.2.4 Segregation According to Age, Sex, and Reproductive Stage. The male-to-female ratio, the isolation of individuals by sex and sexual maturity, and the sex-associated size difference of

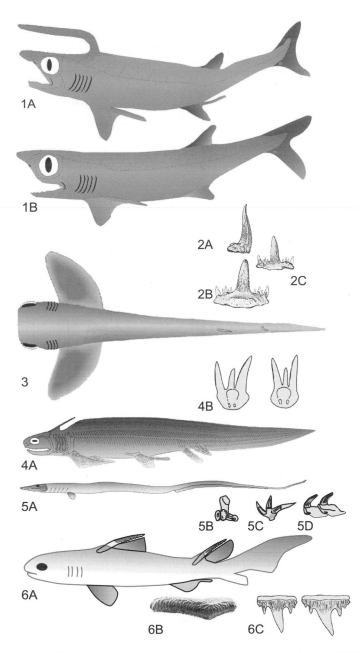

FIGURE 1.3 Examples of Fossil Elasmobranchii. (1A) Restoration of male *Falcatus falcatus,* MV5385; (1B) restoration of female *F. falcatus,* MV5386. (2) General cladodont tooth morphology as represented by *Cladodus springeri.* (Modified from St. John and Worthen, 1875, Plate II.) (2A) proximal view; (2B) lingual view; (2C) labial view. (3) Restoration of *Squatinactis caudispinatus* CM62701 (immature specimen). (4A) *Xenacanthus* sp. (Modified from Carroll, 1988.) (4B) *Xenacanthus parallelus* teeth. (Modified from Schneider and Zajic, 1994.) (5A) Restoration of undescribed fish code-named Snipe, CM62724; (5B) Snipe tooth in occlusal view, CM62724; (5C) Snipe tooth in labial view, MV7699; (5D) two successional Snipe teeth in lateral view, CM62724. (6A) Reconstruction of the Jurassic *Hybodus* sp. (Modified from Maisey, 1982a.) (6B) *Hamitonichthys mapesi* tooth. (Modified from Maisey, 1989.) (6C) Teeth of the Jurassic *Hybodus basanus.* (Modified from Maisey, 1983.) *Note:* Different genera or species are not scaled to one another.

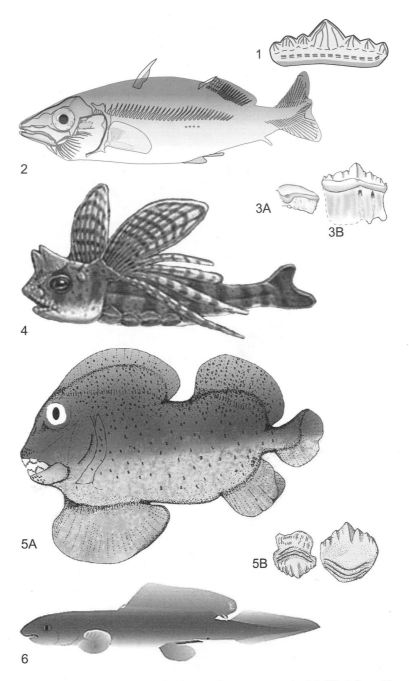

FIGURE 1.4 Examples of euchondrocephalans. (1) *Protacrodus vetustus* tooth. (Modified from Obruchev, 1967.) (2) Composite reconstruction of Sup1. (3A and 3B) Parasymphysial and mandibular tooth, respectively, of Sup3. (4) Iniopterygian species 1. (Courtesy of R. Troll, troll-art.com.) (5A) Petalodont, restoration of *Belantsea montana*, MV7698. (Modified from Lund, 1989.) (5B) Teeth of *Belantsea montana*, MV7698. (6) Reconstruction of male *Heteropetalus elegantulus. Note:* Different genera or species are not scaled to one another.

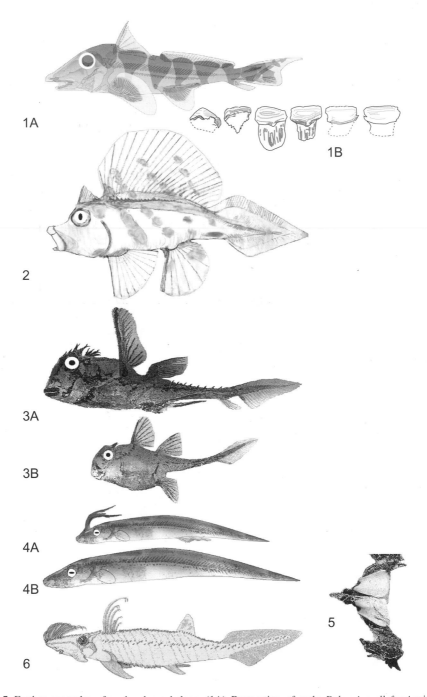

FIGURE 1.5 Further examples of euchondrocephalans. (1A) Restoration of male *Debeerius ellefseni* with preserved pigment pattern, ROM41073; (1B) teeth of *Debeerius ellefseni,* ROM41073. (Both modified from Grogan and Lund, 2000.) (2) Restoration of undescribed female fish code-named Elweir (CM41033) with pigment pattern as preserved. (3A and 3B) Restoration of male (CM30630) and female (CM25588) *Echinchimaera meltoni*, respectively, depicting the relative size of male to female. (4A and 4B) Reconstruction of male (MV7700) and restoration of female (MV5370) *Harpagofututor volsellorhinus*, respectively, depicting the relative male to female size. (5) *Traquairius nudus* (CM46196) lower jaw and tooth plates in dorsal view. (6) *Traquairius agkistrocephalus* (CM48662), reconstructed male. *Note:* Different genera or species are not scaled to one another.

individuals are characteristics of extant populations that are also exhibited by some, but not all, Paleozoic Chondrichthyes. Within the Bear Gulch, the newly described (Lund and Grogan, in press) euchondrocephalan previously referred to as Sup 1 has been preserved in one instance as a school of young to subadult specimens (both sexes) and a probable mature female. *Falcatus* has been preserved *en masse* and, in the most spectacular case, record a ratio of nine to ten mature males to one immature male and one (supposedly adult) female. The form code-named Orochom has been recovered as a group kill of four to six very immature (neonatal?) individuals. Yet, other forms (e.g., *Debeerius ellefseni*) are principally known from a single sex (males) and show no evidence of group or school-based distribution. Similarly, the holocephalic *Harpagofututor volsellorhinus* is typically found individually, and data across the fossil deposit demonstrate a seven-to-one ratio of female and male specimens.

Taxa show great variation in size range. Male and female specimens of *Harpagofututor* demonstrate sexual maturity in individuals of 110 mm total length (Grogan and Lund, 1997), with 155 mm as the longest specimen recovered to date. By contrast, specimens of *Stethacanthus productus* are generally large, with maturity indicated for those attaining 1.5 to 2 m in total length. (The largest *S. "productus"* recovered to date was of an estimated 3 to 3.5 m total length.) Although appreciable evidence is lacking thus far for subadult to adult specimens of smaller size, one partial specimen of *S. altonensis* that shows only the earliest stages of neurocranial mineralization but significant mineralization of vertebral elements was approximately 38 cm. Unfortunately, pelvic information was lacking, due to preservational conditions.

1.3.1.2.5 Reproductive Strategies. Paleozoic chondrichthyan reproductive strategies have purportedly included oviparity but, more often than not, the precise nature or the ownership of supposed fossil egg cases is difficult to demonstrate. The Devonian egg cases reported by Crookall (1928) are sufficiently different from any known chondrichthyan egg case morphology as to question whether ownership is just as likely for another coexisting vertebrate with internal fertilization. There is morphological similarity between a chimaeroid-type egg case and material reported from the Devonian Bokkeveld Series of South Africa (Chaloner et al., 1980). Yet, the question remains whether morphological similarity alone correlates with an accurate identification. Also, the identification of a Carboniferous fossil as a chondrichthyan egg case (Zidek, 1976) remains debatable and likely to be plant remains (Lund, pers. obs.). Our 35 years of excavation across all areas and sections of the Bear Gulch deposit have resulted in the discovery of only *one* possible egg case (Grogan and Lund, in prep.). Although we cannot absolutely confirm the nature of this fossil, it does show similarity to some euselachian egg cases in its general appearance and in details, including a series of respiratory-type apertures. This leads us to consider that this fossil may be an egg case, and, if so, it is more likely to have belonged to a chondrichthyan than a non-chondrichthyan. If it is eventually confirmed as an egg case, then the low recovery rate of this material may simply indicate that these egg cases do not fossilize well or that egg cases were not typically deposited in the confines of the Paleozoic Bear Gulch bay. Alternatively, it could be interpreted as an indication that oviparity with an egg case was a rarity at best for the resident chondrichthyan fishes within this bay. All other accepted reports of chondrichthyan egg cases are Cretaceous in age or later and the degree of confidence in identifying the nature of these is generally a direct result of extant chondrichthyan lineages being reliably traced to Mesozoic forms.

There is evidence of viviparity in the Bear Gulch. Direct evidence includes the demonstration of both viviparity and intrauterine feeding for the holocephalan *Delphyodontos* (Lund, 1980). Indirect evidence is indicated in a proportionate number of Bear Gulch euchondrocephalans. There is evidence of an aborted cochliodont fetus (MV 6207) and other neonatal-sized cochliodonts that show evidence of post-embryonic tooth plate development and proportionate total body growth (CM 62713, CM 43164, CM 30625). Tooth plate wear has been observed in one of the latter. Other euchondrocephalans, *Heteropetalus, Debeerius,* Elweir, and the orochoms reveal a size range through to sexual maturity. There is no evidence of a yolk sac in even the smallest of these fossils and no indication of the type of allometric variation as expected between embryonic forms and young or adult forms. Ovoviviparity or viviparity is equally plausible as a reproductive mode for these forms while oviparity is not. We conclude, then, that none of these particular forms was likely to have undergone any extended (and distinctly extra-oviductal) embryonic developmental phase such as that correlated with egg-case development.

All available data seem to suggest that the small chondrichthyans in the bay are less likely to be characterized by an externally deposited egg case stage of development. It is possible, however, to conceive of a smaller-sized chondrichthyan as having this reproductive strategy and depositing encapsulated fertilized eggs or encapsulated embryos outside the bay, but this places both the adult and the young at a high risk for predation. [Unfortunately, we cannot currently discriminate between pregnant vs. nonpregnant chondrichthyan females. At best, we have direct evidence of female reproductive maturity vs. immaturity for only one form, *Harpagofututor volsellorhinus* (Grogan and Lund, 1997). Factors of preservation (disposition of the body) and fossil recovery (number and age and size distribution of female specimens) are not sufficient to permit the sort of morphological analysis that may allow us to identify whether abdominal girth or some other body feature may be an indicator of pregnancy.] Thus, (1) given the size of the Bear Gulch "egg case" (~40 mm in total length, ~30 mm minus tendrils/horns), (2) the paucity of fossilized evidence of egg capsules in the bay following more than three decades of widespread excavation, and, this, in light of the fact that this deposit is acknowledged for fossil preservation that ranges to the superb (Grogan and Lund, 2002), we believe that these fossilized remains probably belonged to a larger, opportunistic visitor to the Bear Gulch bay than to one of its endemic forms.

In the larger scope, it generally remains most plesiomorphous to consider external fertilization as primitive for all vertebrates (and gnathostomes in particular) and internal fertilization as apomorphic. Live birth and the extended development of a meso- to macrolecithal fertilized egg in an encapsulating case are, then, alternative and further derived states. Data thus far available for Paleozoic chondrichthyans, albeit limited, then lead us to speculate that oviparity with an intrinsic (egg-derived) hyaline coat may be primitive for the earliest chondrichthyans and loosely associated with early euchondrocephalans. Oviparity with an extrinsic (maternally derived) egg capsule may be an apomorphic reproductive mode and loosely associated with the Paleozoic elasmobranchians. (This also agrees with the interpretation of Paleozoic Elasmobranchii, in contrast to the Euchondrocephalans, as heterochronically derived, neotenic forms rather than as representative of the primitive chondrichthyan condition.) Chondrichthyan viviparity may be secondarily derived from either condition.

We note that these interpretations are not necessarily in conflict with the cladistical conclusion of Dulvy and Reynolds (1997, p. 1310: "This phylogeny suggests that egg-laying is ancestral in chondrichthyans") if we consider that their analysis is principally based on neoselachians and extant chimaeroids. (The fossil data are restricted to two taxonomic units, the viviparous Bear Gulch *Delphyodontos* and, as an outgroup, the Placodermi.) We agree with Dulvy and Reynolds that egg-laying may be the plesiomorphous and primitive condition for non-Paleozoic forms. However, this may not apply to the ancestral chondrichthyan stock. Additional paleontological finds beyond the Bear Gulch and more in-depth analyses will be necessary to confirm or deny these propositions.

1.3.2 Upper Carboniferous and Permian Record

The Pennsylvanian (Upper Carboniferous) and Permian chondrichthyan faunas are diminished in diversity above the period boundary, but continue the major lineages of the Mississippian adaptive radiation. Edestoid euchondrocephalans extended into the Permian, giving rise to forms like *Helicoprion* and *Parahelicoprion* (Karpinski, 1899, 1925) and achieving a wide geographic distribution by the latter part of Early Permian (Nassichuk, 1971; Chorn, 1978). Their trend was toward a greatly increased size, highly specialized dentition, and an increasingly oceanic distribution. Similarly, the xenacanths continued to flourish in freshwater environments. The hybodonts and helodonts extended into a variety of aquatic environments (Romer, 1952) as the cochliodontomorphs continued as morphologically radical forms like the Menaspiformes (Schaumberg, 1992; Lund and Grogan, 1997b). Whereas the Carboniferous holocephalimorphs were shallow-water forms that exhibited high morphological diversity, data suggest that a limited number of these lineages later survived the Permo-Triassic extinction, probably by having the ability to extend to deeper waters (see below).

1.4 Theorized Relationships between Recent and Fossilized Forms

1.4.1 On Holocephalan Origins

The chimaeroid/holocephalimorph grouping apparently achieved its greatest diversity during the Carboniferous and most of the descendant forms appear to have become extinct by the end of the Permian. Thereafter, the holocephalimorphs are represented by the Jurassic *Squaloraja*, *Acanthorhina*, the myriacanthoids, and the chimaeroids. All extant forms are believed to be traced to the last, with *Eomanodon* (Ward and Duffin, 1989) purported to be the oldest. That there are no chimaeroid/holocephalimorph data spanning the hundreds of millions of years from the Permian to the Jurassic has been used to argue that today's chimaeroid fishes are not likely to share a direct ancestry with the Paleozoic forms because it is unlikely for lineages to persist for such an extended period of time (Stahl, 1999). Yet, there is paleontological evidence of chondrichthyan lineages persisting from the Devonian to the Triassic (the xenacanths, ctenacanths), and from the Carboniferous to the Mesozoic (the hybodonts).

Stahl (1999) also argues that there are no Paleozoic holocephalans after the Permian, that a decline in the number of fossil finds (which are principally tooth plates) reflects a holocephalan decline after the Carboniferous, and that the Mesozoic forms are of two groups that diverged from some yet-to-be-discovered Permian basal group. We disagree. We find the logic posed in the first part of this argument to be faulty, because a lack of evidence does not equate as evidence of extinction or loss. Rather, it is plausible that some holocephalimorphs survived the Permian by having sought refuge in or having adopted a deeper water lifestyle, as is evidenced by the cochliodonts of the Permian Phosphoria Formation. Any remains of these forms would, necessarily, have a very low probability of preservation and recovery due to inaccessibility, lower potential of fossilization, and loss due to subductive forces acting on the ocean floor. More importantly, however, and in response to the second half of the argument, the "unusual" or odd morphologies of Jurassic myriacanthids, Squalorajidae, and chimaeroids show such confluence with Carboniferous taxa that it is difficult to dismiss direct developmental links between the Permo-Carboniferous and Mesozoic forms (Lund and Grogan, 1997b, in press). Repeated cladistic analyses (Figure 1.6) using revised and expanded character matrices consistently associate these forms in a highly stable, robust topology, which traces both modern chimaeroid and the other Mesozoic holocephalan lineages from the Carboniferous cochliodonts, a derived group of euchondrocephalans (Lund and Grogan, 1997b). Thus, the Cochliodontomorpha comprise [Squalorajiformes + (Chondrenchelyiformes + Menaspiformes)] and [cochliodonts + Chimaeriformes].

1.4.2 On Elasmobranch Origins

Neoselachians (*sensu* de Carvalho, 1996) are monophyletically defined as a common ancestor of all living forms plus all of its descendants. Gaudin (1991), Shirai (1996), and de Carvalho (1996) have provided cladistic models for the relationships between the Recent chondrichthyans (principally elasmobranchs) and Mesozoic forms. (Paleozoic forms were included in the analyses but the treatment of these select chondrichthyans is so limited as to essentially render them into a Hennegian ladder of distant sister group associations.) The latter analyses resolve neoselachians into two major divisions (de Carvalho's Galeomorphi and Squalea vs. Shirai's Galea and Squalea), with origins in the Jurassic and an appreciable diversification in the Cretaceous. The early neoselachians, like many Paleozoic forms, were principally near-shore predators, but offshore predators by the mid-Cretaceous. So, the Jurassic–Cretaceous neoselachian radiation is attributed to an increased availability of basal neopterygians as prey (Thies and Reif, 1985).

Euselachians are neoselachians plus those Paleozoic and Mesozoic forms deemed as the closest allies to neoselachians. Data indicate that cladodonts, the stethacanthids and their allies (*Squatinactis*, taxon code-named Snipe), and the xenacanths are all amphistylic and represent uniquely derived offshoot(s) of basal elasmobranchs (gill-plate chondrichthyans). Therefore, these paleoselachians have no direct relationship to any recent form generally referred to as a shark. Ctenacanths and hybodonts, however, are often invoked as likely neoselachian allies. Zangerl (1973) proposed that recent forms, the Paleozoic

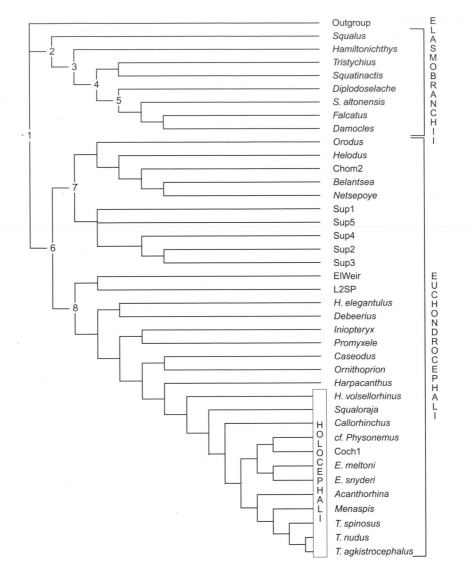

FIGURE 1.6 Relationships of the Paleozoic Chondrichthyes. The cladistic diagram is based on the character table of Appendix 1.1 (Chon46). It is compiled from two analytical treatments; the first includes the Elasmobranchii to best explore basal relationships whereas the second excludes the Elasmobranchii but includes the holocephalan *Acanthorhina* to best explore the relationships of the higher Euchondrocephali.

ctenacanths and the Paleozoic and Mesozoic hybodonts (his phalacanthous sharks), are monophyletic on the basis of dorsal fin spine structure and a tribasal pectoral fin. Maisey (1975) subsequently specified that dorsal fin spine morphology supports a closer relationship between modern elasmobranchs and the ctenacanths rather than hybodonts but later demonstrated paraphyly for the ctenacanth assemblage (and, consequently, the paraphyly of Zangerl's phalacanthous grouping). No significant new ctenacanth evidence has been presented since. As such, then, the Paleozoic ctenacanth information is too incomplete for further cladistical discussion with euselachians at this time. Although ctenacanth spine and scale evidence clearly extends at least to the Devonian, overall the evidence of the group is scanty and generally ranges from slightly informative to uninformative. The Bear Gulch ctenacanth(s) have preserved very poorly and, so, provide little information beyond occurrence. Consequently, where and how these forms lived and any qualitative indication of either their numbers or diversity remains essentially elusive (Maisey, 1981, 1982b, 1983).

As for the early hybodont record, microremains are purported to exist as far back as the Devonian (Lelievre and Derycke, 1998) and the Mesozoic *Hybodus* originally appeared to provide a possible link to the neoselachians through *Heterodontus* (reviewed in Maisey, 1982a, 1989). In an attempt to link neoselachians to Paleozoic and Mesozoic hybodonts, Young (1982) proposed *Hybodus*, *Tristychius*, and *Onychoselache* as the sister group to living elasmobranchs. Maisey refined this further, ultimately arguing that hybodonts are a monophyletic sister group to the neoselachians, with the latter grouping comprising modern elasmobranchs plus Mesozoic forms including *Synechodus* and *Paleospinax* (Maisey, 1982a, 1984, 1985, 1987, 1989, in press). This particular paradigm of relationships was especially strengthened by the discovery of the first appreciably detailed and whole-body evidence of the Upper Carboniferous *Hamiltonichthys* (Maisey, 1989). These specimens permitted the first qualitative morphological comparison of a Paleozoic hybodont with the Mesozoic forms and, so, helped to firmly establish the phylogenetic position that is most likely for hybodonts relative to neoselachians, i.e., as the monophyletic sister group.

1.5   Cladistical Evaluation of Paleozoic Chondrichthyan Relationships and Comments on the Higher Systematic Groupings of Chondrichthyans

Cladistic analyses were performed with Hennig86 (Farris, 1988) and Winclada (Nixon, 1999). The data matrix used in these analyses (Appendix 1.1) consists of 99 characters (50 of which are cranial). Character states are treated non-additively (unordered) and a zero-based hypothetical taxon serves as the outgroup. This reveals the Paleozoic chondrichthyans to be divided into three main assemblages or tribes, as described above and indicated in Figure 1.2 and Figure 1.6. These are (1) the Paleozoic elasmobranchians (the paleoselachians plus a smaller group with closer ties to the Euselachii), (2) the non-holostylic Euchondrocephali, and (3) the Holocephalimorpha.

1.5.1   Trends in the Basal Diversification of the Chondrichthyes (Figure 1.6, Appendix 1.1)

The derived characters and character states of our matrix that identify the Class Chondrichthyes are tesserate mineralization of endoskeletal elements (characters 0, 21, 75); multiple in-line cusps on (protacrodont-like?) teeth (character 36); a ventral braincase that is wide both anteriorly and posteriorly (characters 9, 10); a pelvic fin with an elongate basipterygium, and the majority of radials originating on the basipterygium; and, although we did not use this complex in our matrix, internal fertilization with copulation by means of claspers developed by extension of the male pelvic fin basal axis.

One of the primary derived adaptive character complexes of the Subclass Elasmobranchii centers around the elaboration of the primitive electrosensory system of vertebrates into the ampullary system as a rostral dome complex. This dome is supported by a specialized, and reduced, set of extravisceral cephalic cartilages (*sensu* Grogan and Lund, 2000); labials are reduced, and the primitive and plesiomorphous upper parasymphysial and lower symphysial cartilages are suppressed (node 2, characters 1, 17, 22, 23). Modifications of the neurocranium, suspensorium, and dentition for a well-braced and leveraged predatory feeding mechanism include characters 8, 12, 14, 16, and 25. Neotenic retention of separate vertebrate gill chambers into adulthood, and significant extension of the branchial basket behind the head with the consequent posterior positioning of the pectoral girdle (characters 24 and 28), are probably both correlated with the development of ram-gill ventilation; this in turn is most likely also related to the highly specialized Elasmobranch feeding system. (This specialized feeding mechanism is presumably derived from an early chordate design as the feeding mechanism becomes increasingly independent of, and disjunct from, the respiratory mechanism.) The propulsive system is modified away from fine degrees of maneuverability and toward controlled cruising by characters 73, 82, and 92.

The Euselachii (node 3) develop palatoquadrates that meet in a median symphysis (character 18), an aspect of further feeding system refinement. Specializations of the propulsive system involve the evolution of a tribasal pectoral fin and elimination of a post-metapterygial axis (characters 73, 74), the development of a puboischisadic bar, and anal and dorsal fin modifications (characters 78, 80, 92, 97).

The functional significance of the puboischiadic bar is uncertain. It may be associated with the development of a more finely tuned control within the fin or of select fin elements (i.e., by providing attachment sites for specialized muscles controlling distal vs. proximal elements; Daniel, 1934; Zalisko and Kardong, 1998). This interpretation provides a plausible explanation for why, in the earliest evidence of a chondrichthyan puboischiadic bar (Upper Pennsylvanian *Hamiltonichthys*; Maisey, 1989), that it is restricted to males of the species. Subsequently, selective pressures may have favored the retention of a puboischiadic bar in both sexes of descendant lineages if it provided increased abdominal support and protection during copulation or for bearing embryos of significant size and/or number.

Nodes 4 and 5 concern the common tendency to develop secondary sexually dimorphic structures, including patterns of scale expression, among lineages of the paleoselachians. The development of significant lingual expansion of tooth bases, which interlock successional teeth (cladodont and xenacanth teeth) together, and the co-occurrence of this with those amphistylic suspensoriums that articulate the palatoquadrate to the rear of the postorbital process delineate a unique set of feeding specializations for this group. According to the Hennig86 and Winclada programs, there are either retentions of or reversals to more plesiomorphous states of pelvic and dorsal fin structures at this level.

The Euchondrocephali have less obvious elaborations of a rostral ampullary concentration (character 1). Also, the focal point of support in the primitively autodiastylic suspensorium (i.e., the basitrabecular process) shifts forward and laterally (characters 6, 7, 15, 20), and the ethmoid region of the neurocranium is extended anteriorly for jaw support. A recurrent trend among euchondrocephalan taxa involves the reduction in numbers of tooth positions on the jaws (character 29), reduction in tooth cusp height, and the evolution of heterodonty in jaw teeth. Feeding adaptations are varied. Branchial denticles are rare, first dorsal fin spines occur only sporadically through the euchondrocephalans, and second dorsal fins tend toward elongation, changes that are reflected in characters 26 and 85 to 90.

The orodont–helodont–petalodont clade (node 7; characters 18, 33, 36, 42, 43) further accentuates the common trend among the Euchondrocephali to reduce jaw tooth cusp size, reflecting less of a piercing dental action, and to shift tooth histology to compensate for the changes in the tooth organ aspect ratio and function (see Reif, 1978, 1979). At this node, flexure of the basicranium becomes marked (character 5), the basitrabecular process is flared anteroventrolaterally thus widening but shortening the buccal cavity, and the common trend among the euchondrocephalans to broaden, reduce, and even suppress tooth cusps becomes more conspicuous (character 37). Members of this clade reveal significantly variable development of symphysial/parasymphysial dentitions. Symphysial dentitions are distinct from the dentition of the jaws when present. There is holostylic fusion of palatoquadrates to the ethmoid neurocranium in the helodont–petalodont group (character 16). Holostyly also develops convergently elsewhere among the euchondrocephalans, but in a morphologically distinct manner (see below). Thus, feeding adaptations are varied and branchial denticles (character 26) are rare in the euchondrocephalans. The presence of the first dorsal fin spine (character 85) occurs very sporadically.

Although the character state changes from node 8 and higher reflect the characteristics of a "Hennig's ladder" of successive adaptive branches, several important trends can be extracted from the data. Squamation is both variable in occurrence among taxa and subject to secondary sexual modifications. The ethmoid region of the neurocranium is elongate, narrow, and closed above, the otic region is short and high, and the orbital region is somewhat reduced in size over the plesiomorphous condition. The jaw articulation shifts to a preorbital position and reduction in numbers of jaw tooth positions continues. The branchial basket becomes tightly nested under the otic braincase. Pectoral fins become unibasal and stenobasal, as in modern Chimaeriformes. Collectively, the evidence records that these changes in feeding and respiratory mechanism accompany changes in pectoral fin and girdle design. This suite of characters would appear to equate with a shift in feeding and swimming patterns from a predominantly cruising habitus to specializations for more geometrically complex niches (Webb, 1984).

It is difficult, in holocephalan fish, to identify directly the fate of parasymphysial and symphysial skeletal elements and any dental tissues that they supported. However, the separate origins and distinct types of holostyly (indicated above node 8 in the helodont–petalodont–"chom" clade and in the Holocephali at node 15) help to resolve this issue by highlighting those dental fields that are developmentally derivable from the parasymphysial/symphysial complex. In the more basal taxa of Euchondrocephali, when symphysial and parasymphysial dentitions are present, their tooth morphologies are distinctly different from those found on the jaws. Thereafter, the trends in character states indicate an early ontogenetic fusion of parasymphysials, symphysials, and palatoquadrates to the neurocranium and yet the persistence of their separate inductive fields. Ultimately, the persistence of these separate fields in members of the Holocephali not only reflects but further amplifies the basic euchondrocephalan propensity for distinctly different patterns of tooth occurrences, morphologies, and histologies among the variety of anterior, middle, and median teeth, tooth whorls, or tooth plates in the holocephalimorph and holocephalan taxa (characters 18, 22, 23, 44 to 49).

The Holocephalimorpha reveal two other noteworthy tendencies. The first is a propensity for fusion and elaboration of areas of cranial squamation into large spikes, spines, or plates (Lund and Grogan, 1997b, in review). The second is the repeated evolution of anterior cranial tenaculae of broadly different designs (Lund and Grogan, in press, b) among males.

1.5.2 Other Concluding Remarks on the Origins of Chondrichthyans, Trends in Chondrichthyan Evolution, and on Characters of the Class

The Chondrichthyes are a monophyletic clade and are principally distinguished on the basis of two unique autapomorphous character sets. These are the development of tesserate endoskeletal mineralization and the presence of intromittent organs (claspers and their supporting structures) developed from extensions of the pelvic axis of mature males. The Chondrichthyes share the basic patterns of their scale development with several agnathan scale types, but there is as yet no morphological evidence to support the assignment of some Ordovician and Silurian scale types to the Chondrichthyes.

As chondrichthyan teeth are absent from the earlier scale-bearing deposits, a toothless condition or one in which teeth are not differentiated from scales may still be considered a viable state for the earliest chondrichthyans. However, all evidence suggests a progressive transition from individual, homodont teeth derived from single odontodes to partially or entirely fused organs derived from multiple odontodes. Beyond this, the types and positions of dental organs vary depending on the status of premandibular and mandibular arch elements and the presence of appropriate preoral, palatal, mandibular, and buccopharyngeal inductive fields. The holocephalan dental developmental trend culminates in expansive dental-organ fusions, which are identified as tooth plates. A symphysial tooth organ or complex is apparently plesiomorphous for all gnathostomes minus Placodermi, but is absent in Elasmobranchii.

Outgroup comparisons lead to the conclusion that the pelvic fin radials plesiomorphously articulated with a horizontally oriented, paired pelvic plate. There is a notable trend to shift the articulation of the radials onto the basipterygium in both Elasmobranchii and Euchondrocephali. In the hybodont Elasmobranchii, a further change in pelvic structure results in the development of a puboischiadic bar that separates the pelvic fins widely across the midline. Holocephalans maintain the primitive and plesiomorphic condition of separate halves to the pelvic girdle. Some Holocephali, however, adopt an alternative mode of pelvic muscular support, by the elaboration of an iliac process, and the differentiation of a prepelvic tenaculum from a disjunct anterior fin radial.

The Elasmobranchii and Euchondrocephali express divergent trends in the proportions of the neurocranium, the structure, and the support of the visceral skeleton, which reflect on their inhabitation of distinctly different feeding styles and modes. The scheme of chondrichthyan relationships presented here reflects a diversity of the chondrichthyans during the Carboniferous and, possibly, into the Permian, which was a least an order of magnitude greater than that of osteichthyans. Moreover, the nature of the diversity is striking. The morphological diversity of the actinopterygians reflect what are, fundamentally, transitions in *individual* characters (e.g., cranial bone shape) while the diversity of the chondrichthyans reflects significant morphological modification and coordination in *suites* of characters. How or what this may say about the evolutionary history, reproductive history, and/or even the developmental genetics of these distinct vertebrate groupings has yet to be fully resolved. Yet, it is clear that the differences between these early vertebrates were already emerging by the end of the Devonian and that any common evolutionary history was already distant by this time.

For the chondrichthyan fishes to have capitalized on the emerging diversity of ecological and environmental settings of the Devonian would have required, *a priori,* some degree of genetically inherent adaptiveness (e.g., due to the generation of Hox gene paralogues) that might best be correlated with an earlier (pre-Devonian) radiation event. Alternatively, it would have depended on the environmental/evolutionary selection for those Devonian forms that, by that point in time, were phenotypically expressing the consequences of such duplication. In this context, then, it is conceivable that a Silurian basal radiation would have fundamentally supported the range of chondrichthyan diversity that is identified by the Lower Devonian (and that fueled the subsequent apex of Paleozoic diversity) while also allowing the possibility for retention of gnathostome stem group features in some members. For example, some Lower to Mid-Devonian chondrichthyans display variation in a cranial feature that, until recently, was strictly associated with acanthodians and osteichthyans (Janvier, 1997). An oticooccipital fissure, albeit variable in the extent of its development, has now been confirmed within a range of elasmobranchs — *Orthacanthus* and *Tamiobatis* (Schaeffer, 1981); the Bearsden *Stethacanthus* (Coates and Sequeira, 1998); *Cladodoides gutturensis* (Cladodus) *nielseni*, and *Cobelodus* (Maisey, 2000) — from the Devonian and Carboniferous. Of the chondrichthyans noted to date, only the Devonian pucapampellids retain the primitive gnathostome cranial fissure, generated by the confluence between an oticooccipital fissure and a ventral otic fissure (Maisey, 2001). Furthermore, the morphology of articulated endoskeletal, tooth, and scale information from Carboniferous protacrodonts, orodonts, and Sups, and that of the protacrodontid organ (scale)-species that extends to the Devonian collectively suggests that these are basal forms, not far removed from stem chondrichthyans.

Acknowledgments

The authors are indebted to the Montana families and friends who have made the Bear Gulch fieldwork possible. We thank them and our field crew for their special contributions to this and other Bear Gulch research. We also acknowledge the contributions of the reviewers and thank many colleagues, including Jack Musick, for numerous enthusiastic discussions of both fossil and extant chondrichthyans.

Appendix 1.1: Characters and States for the Cladogram of Figure 1.6

No.	Character	State 0	State 1	State 2	State 3	State 4
0	Mineralized endoskeletal tissue	Perichondral bone	Prismatic calcified cartilage	Superficially continuous	Trabecular	
1	Ampullary system	Not present	Skin tight/thin if present	Rostral ampullary dome		
2	Supraorbital cartilage	Present	Absent			
3	Frontal clasper	Absent	Median	Single pair	Median elongate	Multiple pairs
4	Anterior braincase opening	Closed/small opening	Large/ precerebral fontanelle			
5	Rostrum:basal plane	180°	>80°			
6	Ethmoid% of neurocranium	<25%	25–40	40–50	>50	
7	Orbital% of neurocranium	35–46	>46	<35%		
8	Postorbital% of neurocranium	25–40	15–25	>40	<15%	
9	Anterior ventral braincase (X-section)	Narrow V-shaped	Platybasic, narrow	Platybasic, wide		
10	Posterior ventral braincase (X-section)	Narrow V-shaped	Narrow shelf	Wide shelf		
11	Palatine area shape	Generalized	Medial shelf			
12	Palatoquadrate: basitrabecular articulation	Present	Absent	Fused		
13	pq-Postorbital articulation	Absent	Under postorbital process	Posterior side of postorbital process	Otic-postotic	
14	Basitrabecular process attitude	In line with postorbital	Flared ventrolateral to postorbital	Flared anteroventro-laterally	Absent	
15	Basitrabecular/ pq articular position	Postorbital	Orbital	Antorbital	Absent	
16	Suspensorium	Autodiastyly	Hyostyly	Amphistyly	Holostyly	Methyostyly
17	Extravisceral cephalic cartilages	Prominent labials	Few/reduced labials	None	Pre-mandibular jaw bones	
18	Palatoquadrate anteriad	Separate	Median symphysis	Parallel (parasymphysial) extension		
19	Palatoquadrate-otic process shape	Slight/ postorbital	Dorsall recurved	Posteriorly extended	Dorsally expanded	
20	Meckel's-quadate articulation	Postorbital	Orbital	Preorbital		

No.	Character	State 0	State 1	State 2	State 3	State 4
21	Mandibular mineralization	Bone	Prismatic calcified cartilage	Solid fibrocartilage	Other mineralization	
22	Upper parasymphysial cartilage(s)	Small	Anteriorly extended	Absent		
23	Lower symphysial element	Present between mandibles	Extended anteriad	Absent		
24	Gill openings	Single opercular valve	Separate gill openings			
25	Epihyal	Opercular support	Mandibular support			
26	Branchial denticles	Present	Absent			
27	Ventral hyoid and branchials	Normal	Anteriorly elongate			
28	Branchial basket	Subotic to postcranial	Subcranial	Principally postcranial		
29	Tooth families per jaw	~10–15	<9	<4		
30	Extended neurocranial rostrum	Absent	Present			
31	Functional jaw tooth families	1–2 per family	Pavement occlusion	Reduced or absent		
32	Tooth shapes on jaw	Homodont	Heterodont	Teeth and tooth plates	Plates alone	
33	Occlusion	Cutting/piercing	Crushing/grinding	Both	Absent/NA	
34	Lower symphysial family	Generalized	Prominent	Whorl	Fused plate	Absent
35	Upper parasymphysial family	Generalized	Prominent	Whorl	Fused plate	Absent
36	Cusps on teeth	Single	Multiple in-line	Reduced/suppressed	Absent	
37	Cusp shape	High, sharp	Low, rounded	Bladed	Heterodont	Absent
38	Crown base	Generalized	Lingual, labial ridges	Lingual basin	Basin and ridges	Absent
39	Tooth root	Short below crown	Long below crown	Extended lingual	Fused	Below lingual edge
40	Root shape	Straight below crown	Lingual S-shape	Lingual shelf	Fused	
41	Enameloid layers	Radial crystallite	Complex	Absent		
42	Orthodentine	Present	Absent			
43	Osteodentine	Absent	Present			
44	Paired, upper dental positions	>3 (teeth)	2 anterior, 1 posterior	1 anterior, 1 posterior	1 posterior	Absent
45	Paired anterior lowers	>3 (teeth)	2 anterior	1 anterior	Absent	
46	Middle, lower condition	Tooth	Whorl	Plate	Absent	
47	Posterior dental histology	Coronal tooth tissues	Tritoral dentine	Pleromin tritors	Other	Absent

No.	Character	State 0	State 1	State 2	State 3	State 4
48	Anterior and middle dental surfaces	Coronal tooth tissues	Limited tritors	Complete coverage	Bone, no tritor	Absent
49	Anterior dental histology	Coronal tooth tissues	Tritoral dentine	Pleromin tritors	Bone, no tritor	Absent
50	Scale type	Placoid	Zonal growth	Fused placoid	Absent	Ganoid
51	Head scale coverage	Generalized	Both generalized and specialized	Specialized areas only	Absent	
52	Head scale modifications	Generalized	Enlarged scales/spines	Plates	Absent	
53	Head dermal bones	Absent	Present			
54	Head lateral line scales	Small, simple, oriented	Thin rings	Bones	Canal(s) enclosed in plates	Absent
55	Ethmorostral scales	Generalized	Spikes	Enlarged denticles	Few plates	Absent
56	Supraorbital scales	Generalized	Enlarged scales/spines	Plates	Absent	
57	Otic scales	Generalized	Enlarged scales/spines	Plates	Absent	
58	Occipital scales	Generalized	Enlarged scales/spines	Absent		
59	Mandibular squamation	Generalized	Posterior specialization	Longitudinal specialization (plates)	Absent	
60	Mandibular bones	Present	Absent			
61	Posterior mandibular scales	Generalized	Small sharp spine	Broad spine	Buttressed plate	Absent
62	Basitrabecular rim scales	Generalized	Few large denticles	Plate	Plate and spine	Absent
63	Body scales	Generalized	Generalized and specialized areas	Only specialized areas	Absent	
64	Enlarged paired dorsal body scales	Generalized	First dorsal to caudal	Between fins	Past second dorsal	Absent
65	Dorsal fin squamation	Complete covering	Crest of fin only	Upon radials	Absent	
66	Squamation/sex	Monomorphic	Sexually dimorphic			
67	Lateral line scales of body	Small, simple, oriented	Rings	Canal through scales	None	
68	Pectoral fin position	Ventrolateral	Mid-lank	Nape of the neck		
69	Pectoral girdle	Principally endoskeletal	Principally exoskeletal			
70	Coracoid length	Normal	Extended anteriorly	Truncated anteriorly		
71	Pectoral fin radial support	>50% on anterior basals	>50% on metapterygium	>50% on postmetapterygial axis		
72	Pectoral fin	Uniserial	Partially biserial	Entirely biserial		
73	Pectoral fin base	Multibasal, eurybasal	Unibasal, stenobasal	Tribasal		

No.	Character	State 0	State 1	State 2	State 3	State 4
74	Pectoral post-metapterygial axis	Absent	1–4 small elements	>4 small elements		
75	Pelvic girdle mineralization	Bone	Prismatic calcified cartilage	Solid, three-dimensional		
76	Pelvic dorsal process	Absent	Short	Tall		
77	Pelvic basipterygium	Minor or absent	Elongate	Triangular		
78	Pelvic radials	Majority on girdle	~50% on basipterygium	Majority on basipterygium		
79	Prepelvic tenaculum	Absent	Adjacent to fin	Anterior to fin		
80	Anal fin and/or anal plate	Absent	Present			
81	Dorsal fin numbers	Two fins	One fin			
82	Dorsal spines	Anterior fin only	Two spines	No spines		
83	Anterior dorsal fin	Large	Small flap	Absent	Rod	
84	Anterior dorsal fin/spine support	Basal plate/radials	Synarcuum	On head	Shoulder girdle	Absent
85	Anterior dorsal spine	Deeply fixed	Long, mobile	Superficial insertion	Absent	
86	Anterior dorsal presence	Found in both sexes	Absent in both sexes	Sexually dimorphic		
87	Anterior dorsal fin/spine development	At birth	At puberty			
88	Anterior dorsal spine shape	Posteriorly directed narrow	Triangular	Forwardly curved	Absent	
89	Anterior dorsal spine enameloid, den	Present	Absent	No spine		
90	Posterior dorsal fin	Short	Elongate	Absent		
91	Posterior dorsal fin base	All radials	Basal plate and radials	Basal plate	Other	
92	Posterior dorsal spine	Absent	Superficial insertion	Deeply fixed		
93	Caudal fin	Heterocercal	Extended heterocercal	Homocercal	Diphycercal	
94	Caudal endoskeleton	Serial hypochordal	Epichordal component	Fusions/expansions	Homocercal	
95	Notochordal calcification	Uncalcified	Chordacentra	Complete centra		
96	Vertebral arcual mineralization	Uncalcified	Regionalized	Entire column		
97	Pelvic girdles	Separate across midline	Puboischiadic bar in males	Puboischiadic bar both sexes		
98	Tooth crown linguo/labial buttresses	Absent	Crenellated	Buttressed		

References

Anderson, M. E., J. E. Almond, F. J. Evans, and J. A. Long. 1999. Devonian (Emsian-Eifelian) fish from the Lower Bokkeveld Group (Ceres Subgroup), South Africa. *J. Afr. Earth Sci.* 29(1):179–184.

Applegate, S. P. 1967. A survey of shark hard parts, in *Sharks, Skates, and Rays*. P.W. Gilbert, R.F. Matthewson, and D.P. Rall, Eds., Johns Hopkins University Press, Baltimore, 37–67.

Basden, A. M., G. C. Young, M. I. Coates, and A. Ritchie. 2000. The most primitive osteichthyan braincase? *Nature* 403: 185–188.

Camhi, M., S. Fowler, J. Musick, A. Bräutigam, and S. Fordham. 1998. The biology of the Chondrichthyan fishes, in *Sharks and Their Relatives. Ecology and Conservation*. IUCN Species Survival Commission 20, 39 pp.

Cappetta, H., C. J. Duffin, and J. Zidek. 1993. Chondrichthyes, in *The Fossil Record*, Vol. 2. M. J. Benton, Ed., Chapman & Hall, London, 593–609.

Carroll, R. L. 1988. *Vertebrate Paleontology and Evolution*. W. H. Freeman, New York, 648 pp.

Chaloner, W. G., P. L. Forey, B. G. Gardiner, A. J. Hill, and V. T. Young. 1980. Devonian fish and plants from the Bokkeveld Series of South Africa. *Ann. S. Afr. Mus.* 81(3):127–156.

Chorn, J. 1978. *Helicoprion* (Elasmobranchii, Edestidae) from the Bone Spring formation (Lower Permian of West Texas). *Univ. Kansas Paleontol. Contrib.* 89(1):2–4.

Coates, M. I. 1988. A New Fauna of Namurian (Upper Carboniferous) Fish from Bearsden, Glasgow. Ph.D. thesis, University of Newcastle upon Tyne, England.

Coates, M. I. and S. E. K. Sequeira. 1998. The braincase of a primitive shark. *Trans. R. Soc. Edinb. Earth Sci.* 89:63–85.

Compagno, L. J. V. 1984. *Sharks of the World. An Annotated and Illustrated Catalogue of Shark Species Known to Date*. Part 1. *Hexanchiformes to Lamniformes*. FAO Species Catalogue, Vol. 4. FAO Fish. Synop. 125(4):1–250.

Compagno, L. J. V. 2001. *Sharks of the World. An Annotated and Illustrated Catalogue of Shark Species Known to Date*. Vol. 2. *Bullhead, Mackerel and Carpet Sharks (Heterodontiformes, Lamniformes, and Orectolobiformes). FAO Species Catalogue for Fisheries Purposes*. FAO, Rome, 269 pp.

Crookall, R. 1928. Paleozoic species of *Vetacapsula* and *Palaeoxyris*, in *Report Geol. Surv. Great Britain, Summary of Progress for 1927*, 2:87–107.

Daniel, J. F. 1934. *The Elasmobranch Fishes*. University of California Press, Berkeley, 332 pp.

Davis, J. W. 1883. On the fossil fishes of the Carboniferous Limestone Series of Great Britain. *Sci. Trans. R. Dublin Soc.* 2(1):327–600.

de Carvalho, M. R. 1996. Higher-level elasmobranch phylogeny, basal squaleans, and paraphyly, in *Interrelationships of Fishes*. M. Stiassny, L. R. Parenti, and G. D. Johnson, Eds., Academic Press, New York, 35–62.

Denison, R. 1978. *Handbook of Paleoichthyology*, Vol. 2, *Placodermi*. H.-P. Schultze, Ed., Gustav Fischer Verlag, Stuttgart, 128 pp.

Downs, J. P. and E. B. Daeschler. 2001. Variation within a large sample of *Ageleodus pectinatus* teeth (Chondrichthyes) from the late Devonian of Pennsylvania, U.S.A. *J. Vertebr. Paleontol.* 21(4):811–814.

Dulvy, N. K. and J. D. Reynolds. 1997. Evolutionary transitions among egg-laying, live-bearing and maternal inputs in sharks and rays. *Proc. R. Soc. Lond. B* 264:1309–1315.

Ettensohn, F. R. and L. S. Barron. 1981. Depositional model for the Devonian-Mississippian black shales of North America: a paleoclimatic-paleogeographic approach, in *GSA Cincinnati'81 Field Trip Guidebooks*, Vol. 2. T. G. Roberts, Ed., American Geological Institute, Alexandria, VA, 344–357.

Farris, J. S. 1988. Hennig86. Stony Brook, New York.

Fluharty, C. and E. D. Grogan. 1999. Chondrichthyan calcified cartilage: chimaeroid cranial tissue with reference to *Squalus*, in *Annual Joint Meeting Abstracts of the American Elasmobranch Society/American Society of Ichthyologists and Herpetologists*, Pennsylvania State University, University Park, 105.

Gagnier, P. Y., F. Paris, P. Racheboeuf, P. Janvier, and M. Suarez-Riglos. 1989. Les vertebres de Bolivie: Sonnées biostratigraphiques et anatomiques complémentaires. *Bull. Inst. Fr. Etud. Andines* 18(1):75–93.

Gaudin, T. J. 1991. A re-examination of elasmobranch monophyly and chondrichthyan phylogeny. *N. Jahrb. Geol. Paleontol. Abh.* 182(2):133–160.

Ginter, M. 1999. Famennian-Tournaisian chondrichthyan microremains from the Eastern Thuringian Slate Mountains. *Abh. Ber. Naturkundemus.* 21:25–47.

Ginter, M. 2000. Chondrichthyan biofacies in the Upper Famennian of Western USA, in *Abstracts of the 9th Annual International Symposium, Early Vertebrates/Lower Vertebrates*, Flagstaff, 8–9.

Ginter, M. 2001. Chondrichthyan biofaces in the Late Famennian of Utah and Nevada. *J. Vertebr. Paleontol.* 21(4):714–729.

Goujet, D. 1976. Les poissons in les schistes et calcaires éodévoniens de Saint-Céneré (Massif Amouricain, France). *Mem. Soc. Geol. Mineral. Bretagne* 19:313–323.

Grogan, E. D. 1993. The Structure of the Holocephalan Head and the Relationships of the Chondrichthyes. Ph.D. thesis. College of William and Mary, School of Marine Sciences, Gloucester Point, VA, 240 pp.

Grogan, E. D. and R. Lund. 1997. Soft tissue pigments of the Upper Mississippian chondrenchelyid, *Harpagofututor volsellorhinus* (Chondrichthyes, Holocephali) from the Bear Gulch Limestone, Montana, USA. *J. Paleontol.* 71(2):337–342.

Grogan, E. D. and R. Lund. 2000. *Debeerius ellefseni* (Fam. Nov., Gen. Nov., Spec. Nov.), an autodiastylic chondrichthyan from the Mississippian Bear Gulch Limestone of Montana (USA), the relationships of the Chondrichthyes, and comments on Gnathostome evolution. *J. Morphol.* 243:219–245.

Grogan, E. D. and R. Lund. 2002. The geological and biological environment of the Bear Gulch Limestone (Mississippian of Montana, USA) and a model for its deposition. *Geodiversitas* 24(2):295–315.

Grogan, E. D. and D. T. Yucha. 1999. Endoskeletal mineralization in *Squalus*: types, development, and evolutionary implications, in *Annual Joint Meeting Abstracts of the American Elasmobranch Society/American Society of Ichthyologists and Herpetologists*. Pennsylvania State University, College Park, 108.

Grogan, E. D., R. Lund, and D.T. Yucha. In prep. Selachian tesserate mineralization: ontogenetic development and modeling in response to mechanical stress.

Hammer, Ø., Harper, D.A.T., and P. D. Ryan, 2001. Past: Paleontological Statistics Software Package for Education and Data Analysis. Palaeontologia Electronica 4(1):1–9. Available at http://palaeo_ electronica.org/2001_1/past/issue1_01.htm.

Hansen, M. 1999. The Geology of Ohio — The Devonian. Ohio. *Geology* 1:1–7. Available at www.dnr.state.oh.us/geosurvey/pdf/99_no_1.pdf.

Ivanov, A. O. and O. A. Rodina. 2002a. Givetian-Famennian phoebodontid zones and their distribution, in *The Fifth Baltic Stratigraphical Conference,* September 22–27, Vilnius, Lithuania, 66–68.

Ivanov, A. O. and O. A. Rodina. 2002b. Change of chondrichthyan assemblages in the Frasnian/Famennian boundary of Kuznetsk Basin, in *Proceedings of the International Symposium on the Geology of the Devonian System, II,* Syktyvkar, Komi Republic, July 9–12, 2002, 84–87.

Janvier, P. 1996. *Early Vertebrates*. Clarendon Press, Oxford, 393 pp.

Janvier, P. and M. Suarez-Riglos. 1986. The Silurian and Devonian vertebrates of Bolivia. *Bull. Inst. Fr. Etud. Andines* 15(3–4):73–114.

Karatajute-Talimaa, V. 1992. The early stages of the dermal skeleton formation in chondrichthyans, in *Fossil Fishes as Living Animals*. E. Mark-Kurik, Ed., Academy of Sciences of Estonia, Tallinn, 223–232.

Karatajute-Talimaa, V. and N. Predtechenskyj. 1995. La repartition des vertébrés dans l'Ordovicien terminal et le Silurien inférieur des paléobassins de la Plateforme sibérienne. *Bull. Mus. Natl. Hist. Nat. C* 17:39–55.

Karpinsky, A. 1899. Über die Reste von Edestiden und die neue Gattung *Helicoprion. Verh. Russ. Mineral. Ges.* 36(2): 361–475.

Karpinsky, A. 1925. Sur une nouvelle troouvaille de restes de *Parahelicoprion* et sur relations de ce genre avec *Campodus. Livre Jubil. Soc. Géol. Belg.* 1(1):127–137.

Kemp, N. E. and S. K. Westrin. 1979. Ultrastructure of calcified cartilage in the endoskeletal tesserae of sharks. *J. Morphol.* 160:75–102.

Lelievre, H. and C. Derycke. 1998. Microremains of vertebrates near the Devonian-Carboniferous boundary of southern China (Hunan Province) and their biostratigraphical significance. *Rev. Micropaléontol.* 41(4):297–320.

Lund, R. 1976. General geology and vertebrate biostratigraphy of the Dunkard basin, in *The Continental Permian in Central, West, and South Europe*. H. Falke, Ed., NATO A.S.I. Symposium.

Lund, R. 1980. Viviparity and intrauterine feeding in a new holocephalan fish from the Lower Carboniferous of Montana. *Science* (209):697–699.

Lund, R. 1985. The morphology of *Falcatus falcatus* (St. John and Worthen), a Mississippian stethacanthid chondrichthyan from the Bear Gulch Limestone of Montana. *J. Vertebr. Paleontol.* 5:1–19.

Lund, R. 1986a. On *Damocles serratus* nov. gen. et sp. (Elasmobranchii: Cladodontida) from the Upper Mississippian Bear Gulch Limestone of Montana. *J. Vertebr. Paleontol.* 6:12–19.

Lund, R. 1986b. The diversity and relationships of the Holocephali, in *Indo Pacific Fish Biology: Proceedings of the Second International Conference on Indo-Pacific Fishes.* T. Uyeno, R. Arai, T. Taniuchi, and K. Matsuura, Eds., Ichthyological Society of Japan, Tokyo.

Lund, R. 1989. New Petalodonts (Chondrichthyes) from the Upper Mississippian Bear Gulch Limestone (Namurian E2b) of Montana. *J. Vertebr. Paleontol.* 9(3):350–368.

Lund, R. 1990. Chondrichthyan life history styles as revealed by the 320 million years old Mississippian of Montana. *Environ. Biol. Fishes* 27:1–19.

Lund, R. and E. D. Grogan. 1997a. Relationships of the Chimaeriformes and the basal radiation of the Chondrichthyes. *Rev. Fish Biol. Fisheries* 7:65–123.

Lund, R. and E. D. Grogan. 1997b. Cochliodonts from the Bear Gulch Limestone (Mississippian, Montana, USA) and the evolution of the Holocephali, in *Dinofest International. Proc. Symp. Spons. by Ariz. St. Univ.* D. L. Wolberg, E. Stump, and G. D. Rosenberg, Eds., Academy of Natural Sciences, Philadelphia, 603 pp.

Lund, R. and E. D. Grogan. In press, a. Five new euchondrocephalan Chondrichthyes from the Bear Gulch Limestone (Serpukhovian, Namurian E2b) of Montana and their impact on the Class Chondrichthyes, in *Recent Advances in the Origin and Early Radiation of Vertebrates.* G. Arratia, R. Cloutier, and M. V. H. Wilson, Eds.

Lund, R. and E. D. Grogan. In press, b. Two tenaculum-bearing *Holocephalimorpha* (Chondrichthyes) from the Bear Gulch Limestone (Chesterian, Serpukhovian) of Montana, USA, and their impact on the evolution of the Holocephali, in *Recent Advances in the Origin and Early Radiation of Vertebrates.* G. Arratia, R. Cloutier, and M. V. H. Wilson, Eds.

Lund, R. and C. Poplin. 1999. Fish diversity of the Bear Gulch Limestone, Namurian, Lower Carboniferous of Montana, USA. *Geobios* 32(2):285–295.

Lund, R., P. Janvier, E. D. Grogan, and L. Mulvey. In prep. Tensegrity structures and the Chondrichthyes: zamponiopteron from the Middle Devonian of Bolivia.

Maisey, J. G. 1975. The interrelationships of phalacanthous selachians. *Neues Jahrb. Geol. Paläont.* 9:553–567.

Maisey, J. G. 1981. Studies on the Paleozoic Selachian genus *Ctenacanthus* Agassiz No. 1. Historical review and revised diagnosis of *Ctenacanthus*, with a list of referred taxa. *Am. Mus. Novit.* 2718:1–22.

Maisey, J. G. 1982a. The anatomy and interrelationships of Mesozoic hybodont sharks. *Am. Mus. Novit.* 2724:1–48.

Maisey, J. G. 1982b. Studies on the Paleozoic Selachian genus *Ctenacanthus* Agassiz: No. 2. *Bythiacanthus* St. John and Worthen, *Amelacanthus*, new genus, *Eunemacanthus* St. John and Worthen, *Sphenacanthus* Agassiz, and *Wodnika* Münster. *Am. Mus. Novit.* 2722:1024.

Maisey, J. G. 1983. Studies on the Paleozoic Selachian genus *Ctenacanthus* Agassiz: No. 3. Nominal species referred to *Ctenacanthus. Am. Mus. Novit.* 2774:1–20.

Maisey, J. G. 1984. Chondrichthyan phylogeny: a look at the evidence. *J. Vertebr. Paleontol.* 4(3):359–371.

Maisey, J. G. 1985. Cranial morphology of the fossil elasmobranch *Synechyodus dubrisiensis. Am. Mus. Novit.* 2804:1–28.

Maisey, J. G. 1987. Cranial anatomy of the Lower Jurassic Shark *Hybodus reticulatus* (Chondrichthyes: Elasmobranchii), with comments on Hybodontid Systematics. *Am. Hus. Novit.* 2878:1–39.

Maisey, J. G. 1989. *Hamiltonichthys mapesi*, g. & sp. Nov. (Chondrichthyes; Elasmobranchii), from the Upper Pennsylvanian of Kansas. *Am. Mus. Novit.* 2931:1–42.

Maisey, J. G. 2000. CT-scan reveals new cranial features in Devonian chondrichthyan *"Cladodus" wildungensis. J. Vertebr. Paleontol.* 21(4):807–810.

Maisey, J. G. 2001. A primitive chondrichthyan braincase from the Middle Devonian of Bolivia, in *Major Events in Early Vertebrate Evolution: Paleontology, Phylogeny, Genetics and Development.* P. E. Ahlberg, Ed., Taylor & Francis, London, 263–288.

Maisey, J. G. In press. Endocranial morphology in fossil and recent chondrichthyans, in *Recent Advances in the Origin and Early Radiation of Vertebrates.* G. Arratia, R. Cloutier, and M. V. H. Wilson, Eds.

Maisey, J. G. and M. E. Anderson. 2001. A primitive chondrichthyan braincase from the Early Devonian of South Africa. *J. Vertebr. Paleontol.* 21(4):702–713.

Miles, R. S. and G. C. Young. 1977. Placoderm interrelationships reconsidered in the light of new ptyctodontids from Gogo, Western Australia, in *Problems in Vertebrate Evolution.* S. M. Andrews, R. S. Miles and A. D. Walker, Eds., Academic Press, London, 123–198.

Moy-Thomas, J. A. and M. B. Dyne. 1938. The actinopterygian fishes from the Lower Carboniferous of Glencartholm, Eskdale, Dumfriesshire. *Trans. R. Soc. Edinb.* 59:437–480.

Nassichuk, W. W. 1971. *Helicoprion* and *Physonemus* Permian vertebrates from the Assistance Formation Canadian Arctic archipelago. *Can. Geol. Surv. Bull.* 192:83–93.

Nelson, J. S. 1994. *Fishes of the World*, 3rd ed. John Wiley & Sons, New York.

Newberry, J. S. and A. H. Worthen. 1866. Descriptions of new genera and species of vertebrates mainly from the sub-Carboniferous limestone and Coal Measures of Illinois. *Geol. Surv. Ill.* 2:9–134.

Newberry, J. S. and A. H. Worthen. 1870. Descriptions of fossil vertebrates. *Geol. Surv. Ill.* 4:347–374.

Nixon, K. C. 1999. Winclada (Beta) ver. 0.9.9, published by the author, Ithaca, NY.

Orbruchev, D. V. 1967. *Agnatha, Pisces. Fundamentals of Paleontology,* Vol. II. Israel Program for Scientific Translations, 825 pp.

Ørvig, T. 1951. Histologic studies of Placoderms and fossil Elasmobranchs. I. The endoskeleton, with remarks on the hard tissues of lower vertebrates in general. *Ark. Zool.* 2:321–456.

Ørvig, T. 1960. New finds of acanthodians, arthrodires, crossopterygians, ganoids and dipnoans in the Upper Middle Devonian Calcareous Flage (Oberer Plattenkalk) of the Bergisch-Paffrath Trough. Part I. *Palä-ontol. Z.* 34:295–355.

Reif, W.-E. 1978. Bending-resistant enameloid in carnivorous teleosts. *N. Jahrb. Geol. Paläontol. Abh.* 157:173–175.

Reif, W.-E. 1979. Structural convergences between enameloid of actinopterygian teeth and of shark teeth. *Scanning Electr. Microsc.* 1979(2):546–554.

Rodina, O. A. and A. O. Ivanov. 2002. Chondrichthyans from the Lower Carboniferous of Kuznetsk Basin. Russian Academy of Sciences International Symposium, Geology of the Devonian System, July 9–12, Syktyvkar, Russia, 263–268.

Romer, A. S. 1952. Late Pennsylvanian and Early Permian vertebrates from the Pittsburgh-West Virginia region. *Ann. Carnegie Mus.* 33:47–112.

Rosenberg, L. 1998. A Study of the Mineralized Tissues of Select Fishes of the Bear Gulch Limestone and Recent Fishes. M.S. thesis, Adelphi University, Garden City, NY.

Schaeffer, B. 1981. The xenacanth shark neurocranium, with comments on elasmobranch monophyly. *Bull. Am. Mus. Natl. Hist.* 169:3–66.

Schaumberg, G. 1992. Neue Informationen zu *Menaspis armata* Ewald. *Paläont. Z.* 66:311–329.

Schneider, J. W. and J. Zajic. 1994. Xenacanthiden (Pisces, Chondrichthyes) des mitteleuropäischen Oberkar-bon und perm. Revision der Originale zu Goldfuss 1847, Beyrich 1848, Kner 1867, und Fritsch 1879–1890. *Freib. Forschungsh.* 452:101–151.

Schultze, H.-P. and C. Maples. 1992. Comparison of the late Pennsylvanian faunal assemblage of Kinney Brick Company quarry, New Mexico, with other Late Pennsylvanian lägerstatten. *N.M. Bur. Mines Miner. Resour. Bull.* 138:231–235.

Shirai, S. 1996. Phylogenetic interrelationships of Neoselachians (Chondrichthyes: Euselachii), in *Interrela-tionships of Fishes*. M. Stiassny, L. R. Parenti, and G. D. Johnson, Eds., Academic Press, New York, 9–34.

Smith, S. E., D. W. Au, and C. Snow. 1998. Intrinsic rebound potentials of 26 species of Pacific sharks. *Mar. Freshwater Res.* 49(7):663–678.

St. John, O. H. and A. H. Worthen. 1875. Descriptions of fossil fishes. *Geol. Surv. Ill.* 6:245–488.

St. John, O. H. and A. H. Worthen. 1883. Descriptions of fossil fishes; a partial revision of the Cochliodonts and Psammodonts. *Geol. Surv. Ill.* 7:55–264.

Stahl, B. J. 1999. Mesozoic holocephalians, in *Mesozoic Fishes*, Vol. 2. *Systematics and Fossil Record.* G. Arratia and H. P. Schultze, Eds., Verlag Dr. Friedrich Pfeil, Munich, Germany.

Thies, D. and W. E. Reif. 1985. Phylogeny and evolutionary ecology of Mesozoic Neoselachii. *N. Jahrb. Geol. Paläeontol. Abh.* 3:333–361.

Traquair, R. H. 1888a. Notes on Carboniferous Selachii. *Geol. Mag.*, Ser. 3, 5:81–86.

Traquair, R. H. 1888b. Further notes on Carboniferous Selachii. *Geol. Mag.*, Ser. 3, 5:101–104.

Walker, T. I. 1998. Can shark resources be harvested sustainably? A question revisited with a review of shark fisheries. *Mar. Freshwater Res.* 49(7):553–572.

Ward, D. and C. J. Duffin. 1989. Mesozoic chimaeroids. 1. A new chimaeroid from the Early Jurassic of Gloucestershire, England. *Mesozoic Res.* 2(2):45–51.

Webb, P. W. 1984. Body form, locomotion and foraging in aquatic vertebrates. *Am. Zool.* 24:107–120.

Young, G. C. 1982. Devonian sharks from south-eastern Australia and Antarctica. *Paleontology* 25:817–850.

Young, G. C. 1986. The relationships of placoderm fishes. *Zool. J. Linn. Soc.* 88:1–57.

Yucha, D. T. 1998. A Qualitative Histological Analysis of Calcified Cartilage in *Squalus acanthias.* M.S. thesis. Saint Joseph's University, Philadelphia, 49 pp.

Zalisko, E. J. and K. Kardong. 1998. *Comparative Vertebrate Anatomy.* WCB/McGraw-Hill, New York, 214 pp.

Zangerl, R. and G. R. Case. 1973. Iniopterygia, A New Order of Chondrichthyan Fishes from the Pennsylvanian of North America. Fieldiana Geol. Mem. Field Museum of Natural History Publ. 1167, 67 pp.

Zhu, M. 1998. Early Silurian sinacanths (Chondrichthyes) from China. *Paleontology* 41(1):157–171.

Zidek, J. 1976. Oklahoma Paleoichthyology, Pt. V: Chondrichthyes. *Oklahoma Geol. Notes* 36:175–192.

2

Historical Zoogeography of the Selachii

John A. Musick, Melanie M. Harbin, and Leonard J.V. Compagno

CONTENTS

2.1 Introduction

2.1.1 Zoogeographic Patterns

Zoogeography is the study of patterns of distribution of animals on earth and the biological, geological and climatic processes that influence these patterns (Lieberman, 1999; Mooi and Gill, 2002). Historically, two major fields of scientific inquiry have developed relative to zoogeography: historical zoogeography and ecological zoogeography (Brown and Lomolino, 1998; Mooi and Gill, 2002). Historical zoogeography examines distributions of animals over large spatial scales, often at various taxonomic levels, and involves zoogeographic mechanisms over long temporal scales (Briggs, 1995). Ecological zoogeography focuses on short-term ecological and evolutionary processes that influence the distribution, abundance, and diversity of animals, usually at lower taxonomic levels and small spatial scales (MacArthur and Wilson, 1967). This chapter presents a review of the historical zoogeography of sharks (Selachii).

Some shark species may be euryhaline and capable of residence in freshwater, but they are members of marine families. Thus, to understand the zoogeography of sharks it is necessary to study their distribution in the marine environment. Zoogeographers have found it helpful to divide the Earth's seas

TABLE 2.1

Major Marine Zoogeographic Zones, Regions, and Subregions

A. Tropical Zone
  1. Indo-West Pacific
     a. Western Indian Ocean/Red Sea
     b. Indo-Malayan
     c. Pacific Plate
     d. North Australia
  2. Eastern Pacific
  3. Western Atlantic
  4. Eastern Atlantic
B. Warm Temperate Zone
  1. Western North Pacific (Japan)
  2. South, Western and Southeastern Australia, Northern New Zealand
  3. Southern Africa (Agulhas)
  4. Eastern North Pacific (California)
  5. Eastern South Pacific (Chilean)
  6. Western North Atlantic (Carolina and Gulf of Mexico)
  7. Western South Atlantic (Patagonian)
  8. Eastern North Atlantic and Mediterranean
  9. Eastern South Atlantic (Benguela)
C. Cold Temperate Zone
  1. Western Pacific boreal
  2. Tasmanian–south Australia
  3. Southern New Zealand
  4. Magellan (southern South America)
  5. Eastern Pacific boreal
  6. Western Atlantic boreal
  7. Eastern Atlantic boreal
D. Arctic Polar Zone
E. Antarctic Polar Zone

into regions and provinces defined primarily by sea surface temperature, latitude, and depth (Ekman, 1953; Briggs, 1974, 1995; Zezina, 1997). Zoogeographic regions were first recognized because they contained taxa that shared common distributions (Ekman, 1953). The historical reality of such regions is suspect because they may contain faunal elements with very different evolutionary histories (Mooi and Gill, 2002). Regardless, organisms co-occur in a given area because the environmental conditions there suit them. Among environmental parameters that influence biotic distributions in the sea, temperature is the most important (Briggs, 1995). Consequently, marine regions of the world and the faunas that occupy them have been divided first into tropical, temperate, and polar, then subdivided by ocean basins and their adjacent landmasses (Parenti, 1991) (Table 2.1, Figure 2.1).

The tropical zone coincides approximately with 20 to 21°C winter sea surface temperatures (SST), which generally define the poleward limits of reef coral development (Ekman, 1953) (Figure 2.1). The flow of major ocean currents dictates the extent and shape of the tropical zone. The north and south equatorial current systems are driven from east to west, turning toward higher latitudes as they approach continents to the west. Thus, warm currents flow north and south from the equator extending the tropical zone on the western sides of ocean basins. The converse is true on the eastern sides of the ocean basins where the predominant currents flow from higher latitudes toward the equator bringing cooler water and pinching in the tropical zone (Longhurst, 1998). The vast Indo-West Pacific region, stretching from East Africa through the Indo-Malayan Archipelago out to the islands of the Pacific tectonic plate, has the highest marine biodiversity on Earth with its core in the Indo-Malayan subregion (Ekman, 1953; Briggs, 1995). This appears to hold for the Selachii as well (Compagno, 2002). Springer (1982) presented a convincing and extensively documented argument that the Pacific tectonic plate itself should be recognized as a distinct zoogeographic subregion because all of the landmasses and associated neritic

FIGURE 2.1 Marine climatic zones: TR = tropical, WT = warm temperate, CT = cold temperate, C = polar. (From Briggs, J.C. 1995. *Global Biogeography. Developments in Paleontology and Stratigraphy.* Elsevier, New York. With permission.)

habitats there are oceanic islands that are separated from other coastal Indo-West Pacific faunas by deep-sea trenches.

The temperate zones are not as well defined as the tropical zones because seasonality becomes pronounced at midlatitudes and zoogeographic boundaries shift with the seasons (Parr, 1933; Musick et al., 1986). For mobile animals such as fishes this often results in mass seasonal migrations. In temperate areas like the middle-Atlantic coast of the United States with very strong seasonal temperature differences boreal species dominate in the winter and warm temperate and subtropical species dominate during the summer (Murdy et al., 1997). As with the tropical zone, the shapes of the temperate zones are strongly influenced by ocean currents, but also by dominant air masses at midlatitudes. As air masses move from west to east they are influenced by the surface of the Earth beneath. Thus, air masses that pass over continents tend to be warmer in summer and cooler in winter than those that pass over large moderating ocean expanses. These factors result in broader temperate zones on the eastern sides of ocean basins where neritic habitats are bathed by extensions of the same tropical currents that warm low latitudes on the western side of the basins. These same currents become diverted to the east at higher latitudes, slowly cooling on their way. Seasonality is much less pronounced on the eastern side of ocean basins as well because of the influence of oceanic air masses. In comparison, neritic habitats at midlatitudes on the western sides of ocean basins (as in the middle-Atlantic example) exhibit very strong seasonality because of the influence of continental air masses (Sverdrup et al., 1942).

At the outer margins of the continental shelves, the 200-m isobath is usually taken as the arbitrary boundary between coastal and bathyal habitats (Marshall, 1979). In most regions 200 m marks the shelf break, beyond which the continental slope steeply drops away to the continental rise and abyssal plain offshore. In addition, the shelf break often marks the oceanographic front between coastal and oceanic water masses (Longhurst, 1998). Temperature, light, and pressure change rapidly with depth on the continental slope (Gage and Tyler, 1991). The deepest penetration of biologically detectable sunlight is about 1000 m (Marshall, 1979), and the bottom of the permanent thermocline (~4°C) is located in most localities somewhere between 1000 and 2000 m (with notable exceptions in the Mediterranean and Red

TABLE 2.2

Benthic Bathymetric Zones

Zone	Depth
Coastal	0–200 m
Upper bathyal	200–500 m
Middle bathyal	500–1000 m
Lower bathyal	1000–2000 m
Upper abyssal	2000–2500 m
Abyssal	>2500 m

Seas) (Gage and Tyler, 1991). Fish faunas also change rapidly with depth; eurytopic coastal species are replaced in the most part by stenotopic bathyal forms between 200 and 1000 m (Musick et al., 1996).

The nature of the change is coenoclinic (Musick et al., 1996). That is, species' bathymetric distributions overlap, but the rate of faunal change is greatest where environmental gradients are steepest (Musick et al., 1996). The border between the bathyal and abyssal faunas has been variously designated by different authors, some preferring the 4°C isotherm (Gage and Tyler, 1991), others the 3000-m isobath (Zezina, 1997). For purposes of defining elasmobranch distribution in this chapter we designate the following bathymetric zones: coastal, upper bathyal, middle bathyal, lower bathyal, upper abyssal, and abyssal (Table 2.2).

Bathyal and abyssal habitats are inhabited by cold-adapted animals. Thus, the thermal barrier between shelf and bathyal faunas becomes less distinct at higher latitudes (Zezina, 1997). Likewise, cold bathyal temperatures have allowed elements of high-latitude shelf faunas to colonize tropical latitudes through submergence (Gage and Tyler, 1991; Zezina, 1997; Stuart et al., 2003).

2.1.2 Zoogeographic Mechanisms

Two principal mechanisms have been proposed to explain zoogeographic patterns: dispersal and vicariance (Mooi and Gill, 2002). The dispersal hypothesis in its simplest form explains the allopatric distribution of closely related taxa by the movement of some members of an ancestral taxon from a center of origin across an existing barrier to new areas where they subsequently evolve into distinct taxa (Briggs, 1995). In this scenario the barrier is older than the allopatric taxa. The vicariance hypothesis explains the allopatric distribution of closely related taxa by invoking the erection of a new barrier, which subsequently divides a widely distributed ancestral taxon into allopatric descendants (Platnick and Nelson, 1978). In this scenario the barrier and the allopatric taxa are of equal age.

Other factors in addition to dispersal and vicariance may influence the zoogeographic distribution of taxa. These factors are mostly ecological and are most important in trying to understand the absence of a given taxon from a region where it might be expected. This absence might be due to competitive exclusion, unsuitable habitat, or extinction. Knowledge about the ecology (food habits, habitat requirements) of the taxon in question will aid in evaluating the former two factors, whereas the fossil record might provide clues to the latter.

2.1.3 Zoogeographic Methods; Brief History

A plethora of methods for analyzing zoogeographic patterns and testing hypotheses have proliferated over the last 30 years. Among these are dispersalism (Simpson, 1952; Bremer, 1992), phylogenetic biogeography (Wiley, 1981), panbiogeography (Craw, 1988; Craw et al., 1999), and cladistic biogeography (Morone and Crisci, 1995; Humphries and Parenti, 1999). Limited space here does not permit us to examine all of these methods in detail, but excellent reviews may be found in Morone and Crisci (1995) and in Brown and Lomolino (1998). The advent of cladistic methods revolutionized taxonomy and led to inferences regarding the sequence and relative timing of speciation and assumed allopatry of taxa (Humphries and Parenti, 1999). This information then has been incorporated into zoogeographic analyses by looking for congruent geographic patterns in different taxa and corresponding patterns in potential vicariant events (Brown and Lomolino, 1998). Cladistic methods were originally based on analyses of morphological characters, but genetic analyses have been used with great success in recent years to produce phylogenies that also can be subjected to zoogeographic analyses (Avise, 2000). Phylogenetics uses genetic divergence among taxa and molecular clocks to hypothesize taxonomic relationships, dispersal direction, vicariance, etc. (Bowen et al., 2001).

Another very useful source of information in historical zoogeography is the fossil record, which can provide insights into areas of origin, climatic, evolutionary and zoogeographic history, and taxon age, all of which can contribute to understanding current zoogeographic patterns. The use of fossils in zoogeography has evolved into the discipline of paleobiogeography (Lieberman, 1999). Paleontological data can be valuable but must be viewed with caution, particularly when a taxon of interest is absent from a given place and time. The fossil record depends on sediment deposition and erosion rates, which can vary widely in space and time, and on whether organisms are soft or hard bodied. Elasmobranchs, having cartilaginous rather than bony skeletons, do not fossilize as readily as do actinopteryigians. Indeed, the elasmobranch fossil record is dominated by teeth and spines, denticles and vertebrae, but entire skeletons are relatively rare (Applegate, 1967; Case, 1972).

Brown and Lomolino (1998) have summarized recent developments in zoogeographic methodology: "The history of lineages is being advanced rapidly by research in two areas. On the one hand, systematists have been using the techniques of cladistics and molecular genetics to reconstruct phylogenetic trees. On the other hand, paleontologists and paleobiologists have been using new finds and better interpretations of fossil remains to document what kinds of organisms lived in particular places at different times in the history of the Earth. Unfortunately, there has been too little communication between these two groups of scientists, and too little synthesis of their findings." The objective of this chapter is to provide such a synthesis to better understand the zoogeography of the Selachii.

2.2 Methods

The following methods were used in this chapter in assessing zoogeographic patterns.

2.2.1 Vagility

Because elasmobranchs bear live young or deposit large benthic eggs in horny capsules, they cannot disperse through pelagic eggs and larvae as do many teleosts. Therefore, dispersal requires active swimming, which depends on the vagility of the taxon in question. Rosenblatt (1963) and Rosenblatt and Waples (1986) noted that in tropical reef fishes vagility is lowest in small benthic species without pelagic eggs or larvae. Also, vagility is inversely correlated with speciation, intrataxon diversity, and endemism. In elasmobranchs, we hypothesize that vagility increases with body size, vagility is lowest in benthic species, higher in benthopelagic species, and highest in pelagic species. Also, vagility tends to be lower in coastal species than in bathyal and oceanic species.

As a test of these hypotheses we examined range size, as defined by the number of FAO Fishing Areas occupied by each species (Compagno, 1984b, 2001, in press) within the most speciose orders (Orectolobiformes, Carcharhiniformes, Squaliformes) and for the Selachii as a whole. Then we calculated the mean range size for small (<100 cm total length, or TL), medium (100 to 150 cm TL), large (150 to 300 cm TL), and very large (>300 cm TL) species. Similar comparisons were made among species groups with coastal, bathyal, and oceanic habitats and among benthic, benothopelagic, and pelagic habits. Benthic species spend most of their time resting on the substrate using branchial ventilation, benthopelagic sharks spend most of their time actively swimming above the bottom and up into the water column using ram ventilation, and pelagic sharks spend most of their time actively swimming in the upper water column using ram ventilation. Ecological species designations were derived from the literature (Bigelow and Schroeder, 1948; Castro, 1983; Compagno, 1984b, 2001; Last and Stevens, 1994). Last, we calculated the number of species by size, habit, and habitat groups for the Selachii as a whole.

2.2.2 Geographic Distribution

Present geographic distribution patterns of families within each order were defined by pooling individual species distributions summarized in Compagno (1984a,b, 2001, in press), Last and Stevens (1994), Nakabo (2002), and elsewhere (documented below).

2.2.3 Zoogeographic Synthesis

Most recently proposed phylogenetic relationships of euselachians (Maisey et al., in press), fossil history, and vagility of major taxa were compared in order to erect zoogeographic hypotheses.

Euselachians are very conservative in form and function (Musick et al., 1990), and appear to evolve at slower rates than other vertebrates (Martin et al., 1992). Therefore, the following assumptions were made in interpreting paleontological information: (1) Fossil forms had ecological attributes similar to closely related modern taxa with well-defined thermal (tropical, boreal, etc.) and habitat (coastal, bathyal, etc.) preferences. Corroboration of this assumption sometimes may be available from other accompanying fossils and nature of the deposits. (2) Vagility (dispersal ability) of fossil forms was similar to that of closely related living forms with similar morphology and habitat preferences. Information on fossil distributions was taken from reviews, primarily by Cappetta (1987), Cappetta et al. (1993), and elsewhere (documented below).

2.3 Results and Discussion

2.3.1 Vagility Analyses

2.3.1.1 Range Size — There was a strong negative relationship between body size and range size when all orders were included in the analysis (Figure 2.2A) with small species (<100 cm TL) having very small ranges (<2 FAO regions) and very large species (>300 cm TL) often having circumglobal distributions. When the most speciose orders were analyzed (Orectolobiformes, Carcharhiniformes, and Squaliformes) the same patterns were evident. The Squaliformes in general had much larger ranges at given body sizes than the Orectolobiformes, with the exception of the "very large" category where the circumtropical *Rhincodon typus,* a pelagic, oceanic species, was the sole orectolobiform and the anti-tropical species of *Somniosus* were the only Squaliformes (Figure 2.3A and C). This pattern was related to the almost universally benthopelagic habit of the Squaliformes and the benthic habit of the Orectolo-biformes. The Carcharhiniformes also had larger ranges than the Orectolobiformes except in the smallest size category where the benthic carcharhiniform family Scyliorhinidae dominated (Figure 2.3B).

Benthic habit had a strong effect on range size apparently regardless of body size (Figure 2.2B) with pelagic sharks having ranges more than five times larger than benthic sharks. Habitat was also important with strictly coastal sharks having on average smaller ranges than those of species that occurred primarily in bathyal or oceanic habitats (Figure 2.2C).

2.3.1.2 Species Diversity — There was a strong relationship between species diversity and body size with more than twice as many small species as those of medium or large size, and around eight times as many small species as those of very large size (Figure 2.4A). This suggests that smaller species, which have smaller range sizes (Figure 2.2A), tend to become more easily isolated, thus leading to higher rates of speciation than in larger more widely ranging species. Other factors such as trophic level (small sharks tend to occupy lower trophic levels) (Cortés, 1999) and niche size (smaller species probably have smaller niches) undoubtedly play a role in species diversity, but are beyond the scope of the current discussion, which is focused on vagility. Patterns of regional diversity are discussed below.

An examination of species diversity among habitats showed the highest number of species in coastal habitats (<200 m). However, contrary to the common conception that sharks have radiated primarily in warm shallow habitats, diversity was high among those species that occurred both in outer coastal and upper bathyal habitats (with centers of distribution ~200 m) and in those primarily found in bathyal

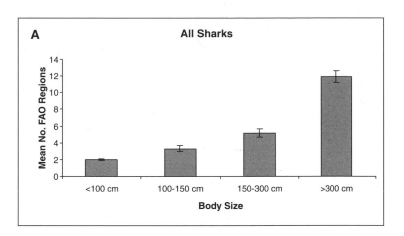

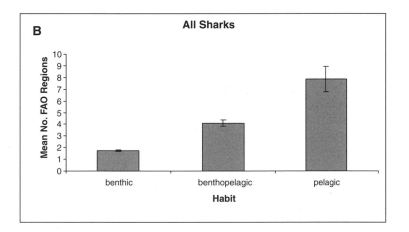

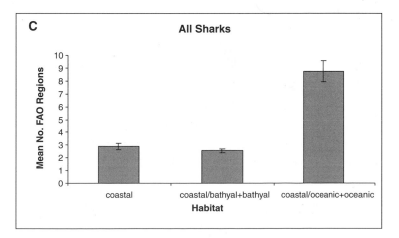

FIGURE 2.2 Average range sizes compared to (A) body size (TL), (B) habit, and (C) habitat.

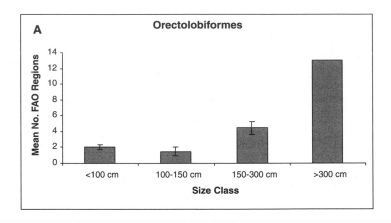

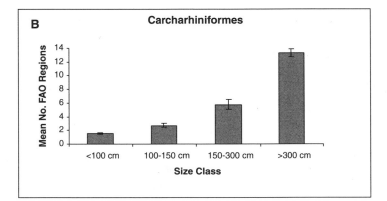

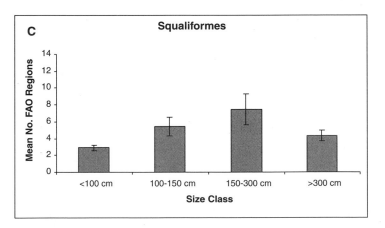

FIGURE 2.3 Average range size compared to body size among (A) Orectolobiformes, (B) Carcharhiniformes, and (C) Squaliformes.

habitats. When these two habitat categories were pooled, there was higher shark diversity in cool outer shelf and slope environments than in shallow warm coastal environments (Figure 2.4C). Last and Séret (1999) found a similar pattern in their analysis of the chondrichthyan faunas of the tropical southeast Indian and southwest Pacific Oceans. The continental slope is a region of habitat diversity with a wide variety of substrates and rapid changes in temperature, light, and pressure (Marshall, 1979; Gage and Tyler, 1991). Much of the species diversity in these deep coastal and bathyal habitats is attributable to small species within the Scyliorhinidae and Squaloidea (Appendix 2.1).

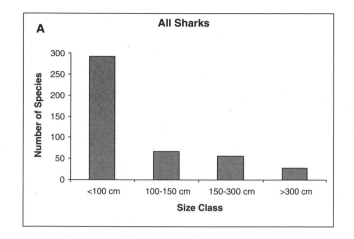

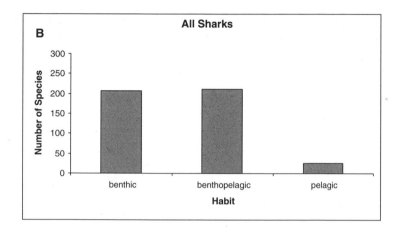

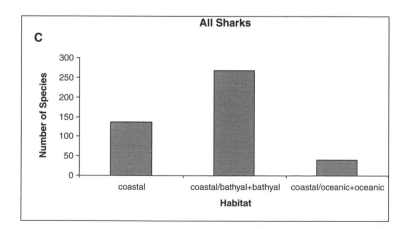

FIGURE 2.4 Number of shark species compared to (A) body size, (B) habit and (C) habitat.

Diversity is much lower among the mostly large species that occur pelagically along the edge of the continental shelf (coastal/oceanic), and even lower among mostly oceanic species (Figure 2.4C, Appendix 2.1). A plot of diversity by habit (Figure 2.4B) showed that there are slightly more benthopelagic species than strictly benthic species and a very small number of pelagic species. The small number of pelagic species is in keeping with their high vagility (Figure 2.2B) and trophic position as most are large or very large species. The high number of benthopelagic species is a bit surprising but is attributable to high diversity among the small squaloids and small to moderate sized triakids with low vagility (Figure 2.3B and C, Appendix 2.1), and to the enigmatic speciose genus *Carcharhinus* with mostly medium to large species (see below).

2.3.2   Zoogeographic Patterns by Taxon

2.3.2.1   *Galeomorphii (Division Galeomorphi sensu de Carvalho, 1996, Superorder Galea sensu Shirai, 1996)* — According to a consensus among most recent studies (de Carvalho, 1996; Shirai, 1996, Compagno, 1999; Maisey et al., in press), the Galeomorphii contains four orders: Heterodontiformes, Orectolobiformes, Lamniformes, and Carcharhiniformes.

2.3.2.1.1   *Heterodontiformes.*   The Heterodontiformes are basal to all other galeomorphs (Figure 2.5), and comprise one living family and genus and nine regional or local species (Compagno, 2001). These species are small to medium in size, benthic, and coastal. The family is absent from the Pacific Plate and oceanic islands in general except for *Heterodontus quoyi* from the Peruvian coast and nearby Galapagos Islands (Compagno, 2001). The vagility of the Heterodontiformes is apparently very limited. Modern species are distributed from East Africa, the Arabian peninsula, Southeast Asia to Japan, Australia, and in the eastern Pacific. The order is notably absent from the Atlantic. This is an ancient

FIGURE 2.5 Recent elasmobranch phylogenies (left) based on molecular genetics (after Maisey et al., in press) and (right) based on morphology. (From Shirai, S. 1996. In *Interrelationships of Fishes.* M.L.J. Stiaseny, L.R. Parenti, and G.D. Johnson, Eds., Academic Press, New York. With permission.)

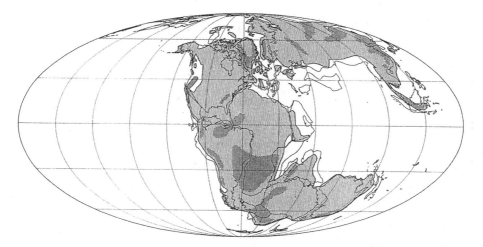

FIGURE 2.6 Lower Jurassic world map ~180 mya. (From Smith, A.G. et al. 1994. *Atlas of Mesozoic and Cenozoic Coastlines.* Cambridge University Press, Cambridge, U.K. With permission.)

group with the earliest fossils known from the lower Jurassic of northern Germany (Cappetta, 1987; Cappetta et al., 1993). Europe at that time appears to have been an archipelago nestled in the western Tethys Sea against the eastern margin of Pangaea (Smith et al., 1994) (Figure 2.6). Climate during the early and middle Jurassic was subtropical to tropical around the Tethyan margin from the Indo-Malayan region in the northeast, to the central European–north African region in the west, to the Antarctic–Australian region in the southeast. Only in the northern polar basin were temperatures estimated to be warm temperate (Vakhrameev and Hughes, 1991). Jurassic Heterdontiformes were very similar to modern species and are congeneric, suggesting that the group arose even earlier (Shirai, 1996) and may have been widespread over the entire eastern neritic margin of Pangaea. Thus, the present day distribution of Heterodontiformes may be due to the vicariant break up of Pangaea. The occurrence of the group in the eastern Pacific along with its absence in the Atlantic appears to contradict the classic zoogeographic track from the Tethys into the emerging eastern Atlantic to the western Atlantic and subsequently into the eastern Pacific via the seaway between North and South America that predated the rising of the Panamanian Isthmus 3 million years ago (mya) (Rosen, 1975; Helfman et al., 1997). However, the group has been reported from the Cretaceous of Patagonia, Argentina, the Eocene of Georgia, U.S.A. (Capetta, 1987), and the Miocene of Virginia (Applegate, pers. comm.) thus showing that it has become extinct in the Atlantic (for whatever reason). Phylogenetic analyses of this small group are needed to test this and alternative zoogeographic hypotheses.

2.3.2.1.2 Orectolobiformes. The Orectolobiformes are a diverse group of primitive galeoid sharks basal to the remainder of the galeoids (Shirai, 1996; de Carvalho, 1996; Maisey et al., in press). According to Goto (2001) the order is comprised of two suborders, the Parascylloidei with one family, and the Orectoloboidei with two well-defined clades (superfamilies?) each containing two families, the Orectolobidae and Brachaeluridae, and the Hemiscylliidae and Rhincodontidae, respectively (Figure 2.7). Most of the species are tropical, coastal, and benthic (with the notable exception of the circumtropical pelagic, oceanic species *Rhincodon typus*).

The Parascylliidae comprises two genera, *Parascyllium* with four species confined to temperate coastal waters of Australia, and *Cirrhoscyllium* with three little-known species; one from the Gulf of Tonkin, one from Formosa, and one from southern Japan, all from cool outer shelf benthic habitats (Compagno, 2001). The Parascylliidae are a deeply rooted clade, a sister group to all other Orectolobiformes, in both morphological and genetic analyses (Goto, 2001; Maisey et al., in press). Regardless, the family does not appear in the fossil record until 98 mya in the mid-Cretaceous of Lebanon (Cappetta, 1987). Other Orectolobiform families appear much earlier: Brachaeluridae, early Jurassic (180 mya); Orectolobidae, mid Jurassic (160 mya); and Ginglymostomatidae (*sensu* Compagno, 1999), early Cretaceous (105 mya) (Cappetta, 1987; Cappetta et al., 1993). Parascylliidae are small sharks with tiny teeth that can be easily

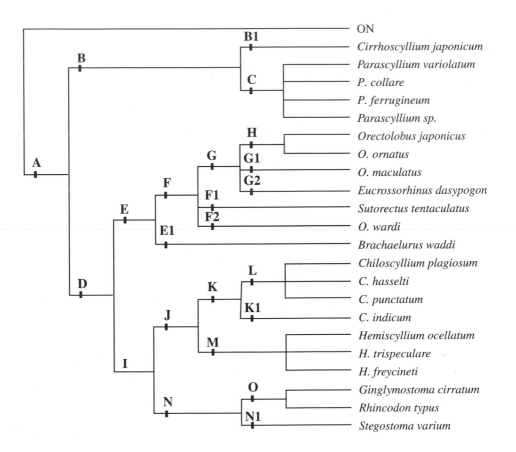

FIGURE 2.7 Phylogeny of Orectolobiformes. (From Goto, T., 2001. *Mem. Grad. Sch. Fish. Sci. Hokkaido Univ.* 48(1):1–100. With permission.)

overlooked in fossil deposits. Thus, earlier fossils may yet be found. Fossil records from the Cretaceous to the lower Eocene of the Parascylliidae are based on the genus *Pararhincodon*, mostly represented by teeth except for one partial skeleton (Cappetta, 1987). Fossils from this family have also been reported from the Miocene of North Carolina (Purdy et al., 2001). Thus, the family appears to have been widespread in the Tethys Sea. The extant genus, *Parascyllium,* endemic to Australia, may have arisen there after Australia separated from Gondwanaland (Figure 2.8). Conversely, the other living genus, *Cirrhoscyllium,* appears to be relict off east Asia.

Brachaeluridae (two species) is the sister group of Orectolobidae, which comprises three genera and seven described species. Both groups are mostly restricted to Australia–New Guinea with the exception of one large species of *Orectolobus,* which occurs from Japan and the Philippines to Viet Nam (other undescribed species of *Orectolobus* may occur in the Indo-Malayan subregion; J. Stevens, CSIRO Marine Research, Hobart, Australia, pers. comm.). The high Australian diversity and endemicity in these groups would suggest an Australian origin for this clade; however, these are ancient groups with fossils from the lower and middle Jurassic of Europe (*Paleobrachaelurus*) and from the middle Jurassic of Europe (*Orectoloboides*). Other fossils attributed to these families occur from the lower Cretaceous to the lower Eocene of Europe, North and West Africa, and Montana, U.S.A. (Cappetta, 1987; Cappetta et al., 1993) (Figure 2.8). Thus, the Brachaelurid–Orectolobid clade was widely distributed along the Tethyan margins and appears to be relict today in Australia. This would imply widespread extinction of these families during and after the early Eocene except around Australia, an island refuge slowly gliding northward to its meeting with the Indo-Malayan Archipelago in the early Miocene (Figure 2.9).

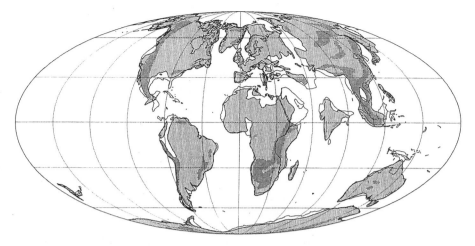

FIGURE 2.8 Lower Eocene world map ~53 mya. (From Smith, A.G. et al. 1994. *Atlas of Mesozoic and Cenozoic Coastlines*. Cambridge University Press, Cambridge, U.K. With permission.)

The family Hemiscyllidae comprises two extant genera. One, *Hemiscyllium* (five species), is endemic to Australia and New Guinea with one species reaching the nearby Solomon Islands. The other genus, *Chiloscylium,* is widespread with seven regional species from the Arabian Sea, India, and the Indo-Malayan Archipelago, to Southern Japan. One species occurs both in the Indo-Malayan Archipelago and northern Australia (Compagno, 2001). All members of the family are tropical, benthic, and neritic, of small size and apparently very low vagility. The earliest known hemiscylliid fossil (*Mesiteia*) appears to be from the Upper Cretaceous (95 mya) of Lebanon. Other fossil hemiscyllids are known from the Upper Cretaceous of Texas and *Chiloscyllium* occurs from the Upper Cretaceous of Trinidad, South Dakota, France, Belgium, and Morocco, and in the Lower and Middle Eocene of Togo. In contrast, *Hemiscyllium* has been reported only from the Upper Paleocene and lower Eocene of Belgium (Cappetta, 1987; Cappetta et al., 1993). Whereas the widespread distribution of *Chiloscyllium* in Tethys neritic habitats is well in keeping with the genus's present continental Indo-West Pacific distribution, the Belgium *Hemiscyllium* fossils are enigmatic given the present Australian endemism of that genus. The Belgium fossils, which apparently consist only of small teeth, bear reexamining. If *Hemiscyllium* arose and radiated in Australian waters, it may have done so in one of two ways: (1) Its ancestor was present in Australia previous to the

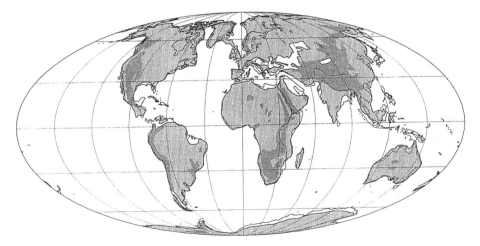

FIGURE 2.9 Early Miocene world map ~20 mya. (From Smith, A.G. et al. 1994. *Atlas of Mesozoic and Cenozoic Coastlines*. Cambridge University Press, Cambridge, U.K. With permission.)

break up of Gondwanaland (much as hypothesized for *Parascyllium*) and *Hemiscyllium* evolved in isolation from *Chiloscyllium*, (this hypothesis would require a much earlier origin for the family (Upper Jurassic, Lower Cretaceous) than suggested by the fossil record). (2) *Hemiscyllium* may have arisen in Australia after it drifted sufficiently close to the Indo-Malayan Archipelago to allow a hemiscyllid ancestor to cross the deep ocean gap from Indo-Malaya to Australia. Given the low vagility of the family this probably could not happen until the Oligocene (30 mya) or the Miocene (20 mya) (Figure 2.9). The cladistic separation between the two genera appears to be fairly deep (Goto, 2001) (Figure 2.7), i.e., much deeper than the Paleocene separation between *Rhincodon* and *Ginglymostoma*; consequently, the relatively recent origin of *Hemiscyllium* is unlikely. Neither hypothesis is satisfying.

The last clade within the Orectolobiformes recognized by Goto (2001) is the Rhincodontidae, comprising *Ginglymostoma* and *Rhincodon* as sister groups and *Stegostoma* as a primitive sister group to the other two. Goto (2001) did not examine *Paraginglymostoma* or *Nebrius*. Compagno (1999) preferred to recognize separate families for the planktivorous Rhincodontidae (monotypic), Ginglymostomatidae (comprising the monotypic genera *Ginglymostoma, Nebrius,* and *Paraginglymostoma*), and the Stegostomatidae (monotypic). The position of *Paraginglymostoma* is problematic and it may be the primitive sister group of the entire clade (Compagno, 2001). In terms of zoogeography and biology, *Rhincodon* is unique among the Orectilobiformes because it is pelagic, oceanic, and the largest living fish. It is highly mobile and circumtropical in distribution (Eckert and Stewart, 2001). With the exception of *Paraginglymostoma,* a small species endemic to Madagascar and a short stretch of East African coast, the remaining members of this clade may have higher vagility than other Orectolobiformes. *Nebrius,* in particular, is widely distributed from South Africa, around the Indian Ocean, to Southeast Asia and Australia. It is the only orectolobiform widely distributed among the oceanic islands of the Pacific Plate (Compagno, 2001). In addition, its morphology (terete body, falcate pectoral fins, etc.) suggests it may be more benthopelagic than benthic in its habits. *Ginglymostoma* is the only orectolobiform (other than *Rhincodon*) that occurs in the eastern and western Atlantic and eastern Pacific. All other orectolobiform clades have become extinct there. This distribution may be due to vicariant forces, as the genus first occurred in the fossil record in the Lower Cretaceous of Lithuania with no fewer than 12 fossil species recorded from the eastern United States, the Caribbean, western Europe, and North and West Africa up into the Miocene (Cappetta, 1987). The relationships of these fossil forms to the living *G. cirratum* remain to be determined. In addition, the interrelationships of nominal *G. cirratum* from the eastern Atlantic, western Atlantic, and eastern Pacific remain to be studied in detail. Castro et al. (2003) recently suggested the amphi-American populations differ at the species level. The Orectolobiformes have their highest diversity and endemicity in coastal habitats of the North Australian tropical subregion, followed closely by the Indo-Malayan subregion (Table 2.3).

2.3.2.1.3 Lamniformes.   The Lamniformes are a moderate-sized group of sharks with seven recognized families, the interrelationships of which are not entirely resolved. Shirai (1996) recognized Mitsukurinidae as the primitive sister group to all of the rest of the order followed by two clades, one with the Odontaspidae and the Pseudocarchariidae and the other with Megachasmidae as a sister group to Alopiidae and Lamnidae and Cetorhinidae (Figure 2.5). Conversely, a recent molecular study (Martin and Naylor, 1997) recognized two very different higher clades: one with *Carcharias* at its base with Cetorhinidae and Lamnidae above; the other with *Megachasma, Pseudocarcharias, Alopias,* and *Odontaspis* (Figure 2.10).

The division of the Odontaspidae into two major clades (families?), *Carcharias* and *Odontaspis,* with different phyletic affinities may not be far-fetched based on dentition, cranial morphology, and the fossil record (S. Applegate, Department of Vertebrate Paleontology, University of Mexico, Mexico City, pers. comm.). *Carcharias* first occurred in the lower Cretaceous of Albania (Cappetta et al., 1993) and *Odontaspis* in the Lower Cretaceous of Japan (Goto et al., 1996); thus, the two clades have been evolving separately for at least 115 million years (Cappetta, 1987). The distribution and ecology of extant members of the two clades are very different. *Carcharias* is represented by one species, *C. taurus,* which has a mostly warm temperate, antitropical distribution in shallow coastal waters. The species is circumglobal with the exception of the eastern Pacific from which Oligocene and Miocene fossils are known (Applegate, 1986). Conversely, *Odontaspis* includes two bathyal species, which are circumglobal in warm

TABLE 2.3

Distribution of Orectolobiform Species by Zoogeographic Region and Subregions from Table 2.1

Regions and Subregions	Total	Total Endemics	Coastal	Coastal Endemics	Coastal/ Bathyal + Bathyal	Coastal/ Bathyal + Bathyal Endemics	Coastal/ Oceanic + Oceanic	Coastal/ Oceanic + Oceanic Endemics
A1a	8	2	7	2	0	0	1	0
A1b	13	2	10	2	2	0	1	0
A1c	2	0	0	0	1	0	1	0
A1d	15	6	12	6	2	0	1	0
A2	2	0	1	0	0	0	1	0
A3	2	0	1	0	0	0	1	0
A4	2	0	1	0	0	0	1	0
B1	13	2	8	0	4	2	1	0
B2	10	3	9	3	0	0	1	0
B3	1	0	0	0	0	0	1	0
B4	1	0	0	0	0	0	1	0
B5	2	0	1	0	0	0	1	0
B6	2	0	1	0	0	0	1	0
B7	0	0	0	0	0	0	0	0
B8	2	0	1	0	0	0	1	0
B9	0	0	0	0	0	0	0	0
C1	0	0	0	0	0	0	0	0
C2	7	0	5	0	1	0	1	0
C3	0	0	0	0	0	0	0	0
C4	0	0	0	0	0	0	0	0
C5	0	0	0	0	0	0	0	0
C6	2	0	1	0	0	0	1	0
C7	1	0	1	0	0	0	0	0
D	0	0	0	0	0	0	0	0
E	0	0	0	0	0	0	0	0

Note: Species numbers are approximate and may change as additional information accrues.

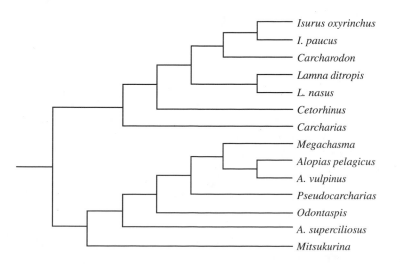

Isurus oxyrinchus
I. paucus
Carcharodon
Lamna ditropis
L. nasus
Cetorhinus
Carcharias
Megachasma
Alopias pelagicus
A. vulpinus
Pseudocarcharias
Odontaspis
A. superciliosus
Mitsukurina

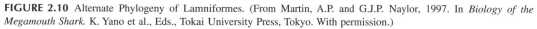

FIGURE 2.10 Alternate Phylogeny of Lamniformes. (From Martin, A.P. and G.J.P. Naylor, 1997. In *Biology of the Megamouth Shark*. K. Yano et al., Eds., Tokai University Press, Tokyo. With permission.)

temperate to tropical habitats (Compagno, 2001). The resolution of lamniform interrelationships may ultimately lie in morphological and paleontological studies because as Martin and Naylor (1997) conclude, "lack of [molecular] resolution among many of the species most likely reflects ancient origination of lineage: coupled with speciation events that happened over a relatively brief period of time." Regardless of the absence of a precise phylogeny of the Lamniformes, taxonomy may provide little insight into the current zoogeography of the group because most species are large and benthopleagic or pelagic, and most are bathyal or oceanic with circumglobal distributions and very high vagility. Indeed, as tagging and genetic studies accrue, evidence for current or very recent interpopulation movements has been documented for many species (Kohler et al., 1998; Heist, Chapter 16 in this volume).

The origin of the Lamniformes apparently lies in the mid-Jurassic with the earliest fossil *Palaeocarcharias* known from the Upper Jurassic of Europe. This genus has benthic orectoloboid body features but clearly lamniform teeth (Duffin, 1988). The earliest record of extant lamniform families, Odontaspidae and Mitsukurinidae, are from Early Cretaceous (115 mya) with the lamnids appearing in the Paleocene, the alopiids most likely in the Early Eocene, and the cetorhinids in the Oligocene (Cappetta, 1987; Cappetta et al., 1993).

The Mitsukurinidae is represented by one extant bathyal species found circumglobally in warm temperate and tropical latitudes. Its early Cretaceous origin agrees with its basal position within the order (Cappetta, 1987).

The Lamnidae comprises three genera, *Carcharodon* (monotypic), *Isurus* (two species),. and *Lamna* (two species), all species of which are large or very large, pelagic, and at least in part oceanic. All lamnids are endothermic and can maintain their body temperature above that of ambient seawater (Carlson et al., Chapter 7 in this volume). *Carcharodon* occupies mostly temperate latitudes. *Lamna* occupies cold temperate latitudes and has an antitropical distribution with one species, *L. ditropis,* endemic to the North Pacific and the other, *L. nasus,* found everywhere else in boreal and austral latitudes. The antitropical distribution of the latter species may have developed during the Pleistocene glaciation when the tropical zone became more constricted allowing *L. nasus* to pass from the Northern to the Southern Hemisphere through submergence. *Lamna* fossils first appear in the Oligocene of Belgium, and *L. nasus* is known from there in the Pliocene. Reif and Saure (1987) postulated that *L. ditropis* diverged from *L. nasus* after the Arctic Seaway was closed by formation of the Arctic ice sheet in the late Cenozoic. *Isurus* occurs in temperate and tropical latitudes, where *I. oxyrinchus* tends to prefer water of 17 to 22°C (Compagno, 2001). The thermal preferences of *I. paucus,* a relatively rare species, are unknown but its large eyes and oceanic habitat suggest it lives in cooler, deeper water.

The Cetorhinidae is monotypic with one very large pelagic planktivorous species circumglobal in cold temperate latitudes of both hemispheres (Compagno, 2001). Recent tracking data that show that the species may feed in the thermocline (Sims et al., 2002) suggest that, as with *L. nasus, Cetorhinus maximus* may have passed through the tropics through submergence during glacial periods when the tropical zones, particularly on the eastern sides of ocean basins, became more constricted than at present. The genus is known from the Oligocene of Europe and was widespread in the Northern Hemisphere by the Miocene. *Cetorhinus maximus* first appeared in the Pliocene.

The Alopiidae comprises three large to very large recognized species. All are pelagic, in part oceanic (Compagno, 2001), and endothermic to some extent (Carlson et al., Chapter 7 in this volume). *Alopias pelagicus* occurs in subtropical and tropical latitudes, but its preferred temperature and bathymetric ranges are poorly known. *Alopias superciliosus* occurs in temperate and tropical latitudes, but is mostly a deep-living oceanic species. *Alopias vulpinus* occurs from tropical into cold temperate latitudes in summer (Compagno, 2001). The temperate distributions of *A. superciliosus* and *A. vulpinus* could explain their circumglobal distributions with, at least, the Cape of Good Hope acting as a conduit between the Atlantic and the Indo-Pacific. Conversely, the tropical distribution of *A. pelagicus* could limit its distribution there and certainly at Cape Horn. Its absence from the Atlantic suggests that the species might have evolved after the Isthmus of Panama arose (~3 mya). This notion is contradicted by fossils of all these species from El Cien (Baja California, Mexico) from the lower Miocene (30 mya) (Applegate, 1986.). Undoubtedly, both *A. superciliosus* and *A. vulpinus* transited the Panamanian seaway as well as another Miocene seaway (10 mya) across the Isthmus of Tehuantepec (Banford, 1998). Therefore, the Atlantic absence of *A. pelagicus* remains enigmatic. A more complete search of Neogene Atlantic fossil records may confirm its presence (and subsequent extinction) there. Two remaining lamniform families are both monotypic, pelagic, and oceanic and distributed circumtropically. *Pseudocarcharias* is a medium-sized active predator, whereas *Megachasma* is a very large planktivore, frequenting both epipelagic and mesopelagic habitats.

2.3.2.1.4   Carcharhiniformes.   The Carcharhiniformes is the sister group to the Lamniformes although its evolutionary history and radiation are quite different with eight extant families and ~197 species (Appendix 2.1).

The family Scyliorhinidae is basal to the rest of the Carcharhiniformes and the oldest group with earliest known fossils from the Upper Jurassic of Germany (Cappetta, 1987; Cappetta et al., 1993). With few exceptions, the scyliorhinids have radiated in cool-water coastal and bathyal habitats. Most species are small and all are benthic with apparently low vagility. This family is the most diverse among the Selachii with 17 genera and about 100 species (Last and Stevens, 1994). The genus *Apristurus* is the most speciose with at least 25 species, many of which have very small ranges (Last and Stevens, 1994; Compagno, in press; Appendix 2.1). Some of the deeper-living species are widespread at lower bathyal depths, and some occur on the slopes of oceanic islands. *Apristurus* has not been noted in the fossil record, but because the teeth of extant scyliorhinids have not been adequately described, most scyliorhinid fossils have been lumped in the genus *Scyliorhinus* (Cappetta, 1987). As with *Apristurus*, the genera *Galeus* (12 species) and *Parmaturus* (6 species) occur circumglobally. *Galeus* appears in the fossil record from the early Miocene of France. *Halaelurus* (12 species) and *Cephaloscylium* (12 species) are distributed widely in the Indo-Pacific including Australia, with one species each in the eastern Pacific. Both are absent from the Atlantic.

The genus *Scyliorhinus* contains ~13 species from the Atlantic, South Africa, and the Indo-Malayan Archipelago. If the oldest *Scyliorhinus* fossils are truly in that genus (from the Lower Cretaceous of France) it could have been widespread in the Cretaceous Tethys, but absent from Australia, which had already broken away from Pangaea. This would explain the absence of *Scyliorhinus* from Australia today. Its absence from the eastern Pacific is enigmatic.

Two scyliorhinid genera are endemic to Australia, *Asymbolus* (eight species) and *Aulohalaelurus* (monotypic) (Last and Stevens, 1994). The genus *Haploblepharus* (three species) is endemic to South Africa, whereas *Holohalaelurus* (two species) ranges more widely from South Africa to East Africa, and *Poroderma* (three species) from South Africa to Madagascar. These genera with their distribution centered off South Africa suggest a second center of scyliorhinid endemism (in addition to Australia). *Schroederichthys* (four species) is amphi-American in distribution and along with *Cephalurus* (monotypic), which is endemic to the Gulf of California represents a third center of scyliorhinid endemism, the New World. Scyliorhinids have reached their highest diversity and endemicity in the tropical northern Australian region followed by the temperate western North Pacific (Table 2.4).

Two factors are apparent from the geographic distribution of the scyliorhinids: (1) the large numbers of species suggest that small size and benthic habit are conducive to isolation and speciation; (2) the widespread distribution of some of the deeper-living species and their presence around oceanic islands suggest that some species may have moderate vagility and may have served as a vehicle for the circumglobal distribution of some genera. Resolution of the generic positions of fossil scyliorhinids and detailed cladistic analysis of the live forms must be accomplished before a cogent zoogeographic analysis may be attempted for the family.

Three small carcharhiniform families (approximately seven species), the Proscyllidae, Pseudotriakidae, and Leptochariidae, are bathyal and little known with no recognized fossil record. These are not be discussed further.

The family Triakidae is diverse with nine extant genera. *Furgaleus* is monotypic and endemic to temperate Australian waters. *Galeorhinus* in contrast is also monotypic, but is distributed in all temperate continental shelves save the western north Atlantic and western north Pacific. The genus is known from the Upper Cretaceous (90 mya) of northern France and from numerous other European, North and West African, eastern U.S. and Mexican Pacific fossils up into the Oligocene, Miocene, and Lower Pliocene (Applegate, 1986; Cappetta, 1987; Welton and Farrish, 1993; Kent, 1994; Purdy et al., 2001). It first appears in California in the Pleistocene (Fitch, 1964). This small benthopelagic shark appears to have

TABLE 2.4

Distribution of Scyliorhinid Species by Zoogeographic Region and Subregions from Table 2.1

Regions and Subregions	Total	Total Endemics	Coastal	Coastal Endemics	Coastal/ Bathyal + Bathyal	Coastal/ Bathyal + Bathyal Endemics	Coastal/ Oceanic + Oceanic	Coastal/ Oceanic + Oceanic Endemics
A1a	12	7	2	0	10	7	0	0
A1b	20	10	5	1	15	9	0	0
A1c	1	0	0	0	1	0	0	0
A1d	25	12	7	4	18	12	0	0
A2	7	1	0	0	7	1	0	0
A3	11	5	0	0	11	5	0	0
A4	7	0	0	0	7	0	0	0
B1	22	15	5	2	17	13	0	0
B2	22	5	3	0	19	5	0	0
B3	13	4	3	2	10	2	0	0
B4	7	1	0	0	7	1	0	0
B5	4	0	0	0	4	0	0	0
B6	11	3	0	0	11	3	0	0
B7	2	1	0	0	2	1	0	0
B8	9	4	0	0	9	4	0	0
B9	7	0	1	0	6	0	0	0
C1	0	0	0	0	0	0	0	0
C2	13	0	1	0	12	0	0	0
C3	3	1	1	1	2	0	0	0
C4	3	0	0	0	3	0	0	0
C5	2	0	0	0	2	0	0	0
C6	5	0	0	0	5	0	0	0
C7	7	2	0	0	7	2	0	0
D	0	0	0	0	0	0	0	0
E	0	0	0	0	0	0	0	0

Note: Species numbers are approximate and may change as additional information accrues, particularly for bathyal sharks.

relatively high vagility as recent tagging studies have recorded some specimens migrating between Australia and New Zealand (West and Stevens, 2001). The origin of the genus in late Cretaceous seas off northern Europe where a boreal climate had developed by the Early Cretaceous (Middlemiss, 1984) suggests that the ancient congeners of extant cold temperate *G. galeus* had similar thermal preferences. The fossil distributions suggest that the genus may have been restricted to North Atlantic and eastern North Pacific waters until the Pleistocene when it apparently radiated widely in cold temperate coastal habitats. Its present absence in both the western North Atlantic and western North Pacific may be mitigated by the extreme seasonal and continental nature of the present climates there. In contrast, other regions where the species is common are dominated by maritime climates.

The genera *Gogolia* (one sp.), *Hemitriakis* (five spp.), *Hypogaleus* (one sp.), *Iago* (two spp.), and *Scylliogaleus* (one sp.) comprise small species distributed variously in the Indo-West Pacific and with no apparent fossil record (Cappetta, 1987; Compagno, in press).

Mustelus is a diverse genus with 22 described species (Compagno, 1999) distributed circumglobally in mostly temperate seas. Of 20 species for which we have ecological information, 18 have cool-water temperate, antitropical, or bathyal distributions, whereas only 2 occur mostly in tropical coastal seas. Many species make seasonal migrations apparently to remain in preferred temperate thermal ranges. Heemstra (1973) reported *M. norrisi* to move inshore in the Gulf of Mexico in winter and offshore to cooler, deeper water in summer. *Mustelus canis* winters in the south Atlantic bight of the United States and migrates north into the northern middle Atlantic Bight in summer (Colvocoresses and Musick, 1984) and has a preferred temperature range of 14 to 20°C (Musick, unpublished). The genus comprises small to medium-sized species, most with limited distributions (Appendix 2.1). The earliest fossils of *Mustelus* are from the Lower Eocene of Belgium with later fossils also occurring in Europe and eastern North America (Cappetta, 1987). A few species of *Mustelus* are widely distributed, with one (*M. mento*) occurring in the eastern Pacific including the Galapagos and Juan Fernandez Islands (both oceanic). Also there is an endemic species *(M. lenticulatus)* in New Zealand, which has existed as an oceanic island since before the evolution of the genus. Therefore, the widespread temporal distribution of local and regional species of *Mustelus* has probably been mitigated by occasional dispersal of some species with higher vagility, then subsequent local speciation. Further speculation on the subject awaits more complete cladistic analysis and review of the fossil record.

The genus *Triakis* comprises five small to large species with three in the eastern Pacific, one in the northwestern Pacific and one off South Africa. All are temperate in distribution. The oldest fossils are from the Paleocene of Trinidad with other fossils from the Lower Eocene to the Upper Miocene of Western Europe (Cappetta, 1987). Clearly, the genus has become extinct in the Atlantic.

The family Hemigaleidae comprises four genera of mostly small to medium-sized tropical sharks. The genus *Chaenogaleus* is monotypic and is distributed from the western Indian Ocean to China, and the Indo-Malayan archipelago. Earliest fossils are known from the Lower Miocene of southern Germany and the Middle and Upper Miocene of Portugal and southern France, and the Middle Miocene of Poland. Thus, the genus was distributed through the Tethyan corridor. *Hemigaleus* (one sp.) is restricted to the heart of the Indo-Pacific region, from southern India to China, the Philippines, the Indo-Malayan Archipelago, and northern Australia. There are no fossil records of *Hemigaleus*. *Hemipristis* includes one large tropical coastal species at present distributed from South Africa around the Indian Ocean to Southeast Asia and China, the Philippines, and northern Australia. The earliest known fossils are from the Middle Eocene of Egypt, but the genus was widespread by the Miocene with fossils known from the United States, Europe, Java, India, Australia, West Africa, and Peru. The genus persisted in Europe and West Africa until the Pliocene. Its subsequent widespread extinction and present relict distribution in the Indo-West Pacific is reminiscent of other groups of tropical coastal sharks, typified by the Orectolobidae. The genus *Paragaleus* comprises four species of small to medium-sized tropical coastal sharks. Regional species occur off West Africa, South Africa, and the northern Indian Ocean and Southeast Asia. The genus is known from the Miocene of Portugal and southern France. It probably followed the coastal Tethyan corridor in the Miocene into the eastern Atlantic but appears to be ill-adapted to crossing large ocean expanses, thus its absence in the western Atlantic, Pacific Plate, and eastern Pacific.

The very large family Carcharhinidae includes 12 genera (8 are monotypic) and >49 species (Figure 2.11). Most species are medium to large in size with greatest diversity in coastal, tropical habitats

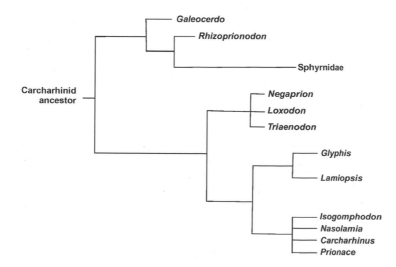

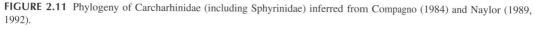

FIGURE 2.11 Phylogeny of Carcharhinidae (including Sphyrinidae) inferred from Compagno (1984) and Naylor (1989, 1992).

TABLE 2.5

Distribution of Carcharhinid Species by Zoogeographic Region and Subregions from Table 2.1

Regions and Subregions	Total	Total Endemics	Coastal	Coastal Endemics	Coastal/ Bathyal + Bathyal	Coastal/ Bathyal + Bathyal Endemics	Coastal/ Oceanic + Oceanic	Coastal/ Oceanic + Oceanic Endemics
A1a	32	2	19	2	0	0	13	0
A1b	29	2	17	2	0	0	15	0
A1c	15	0	4	0	0	0	11	0
A1d	36	2	19	2	0	0	17	0
A2	24	0	11	0	0	0	13	0
A3	29	1	16	1	0	0	13	0
A4	19	0	7	0	0	0	12	0
B1	18	0	8	0	0	0	9	0
B2	21	0	10	0	0	0	11	0
B3	22	0	10	0	0	0	12	0
B4	20	0	8	0	0	0	11	0
B5	14	0	6	0	0	0	7	0
B6	24	0	12	0	0	0	12	0
B7	21	0	11	0	0	0	9	0
B8	18	0	6	0	0	0	12	0
B9	13	0	5	0	0	0	8	0
C1	1	0	0	0	0	0	1	0
C2	10	0	5	0	0	0	5	0
C3	5	0	1	0	0	0	4	0
C4	5	0	2	0	0	0	3	0
C5	1	0	1	0	0	0	0	0
C6	13	0	5	0	0	0	8	0
C7	13	0	2	0	0	0	11	0
D	0	0	0	0	0	0	0	0
E	0	0	0	0	0	0	0	0

Note: Species numbers are approximate and may change as additional information accrues, particularly for bathyal sharks.

(Compagno, 1984, 1999) (Table 2.5, Appendix 2.1). All are benthopelagic or pelagic with high vagility. In accord with the phylogeny of Compagno (1988) based on morphology and modifications suggested by the molecular analysis of Naylor (1989, 1992), the following phylogenetic arrangement seems to be reasonable: *Galeocerdo* and *Rhizoprionodon* and the closely related *Scoliodon* may be placed as a basal clade to the rest of the family, followed by a group comprising *Negaprion, Loxodon,* and *Triaenodon,* with *Carcharhinus* (including very derived *Prionace*) being most derived. Closely allied with *Carcharhinus* and sometimes placed in its synonymy are *Isogomphodon* and *Nasolamia* (which may be the sister taxon to *C. acronotus*) and *Glyphis* and *Lamiopsis* (which appear to be closely related to one another) (Figure 2.10).

The earliest fossils in the family have been found in the Lower Eocene (53 mya) with *Galeocerdo* and *Rhizoprionodon* already present then (agreeing with their basal phylogenetic position). *Negaprion, Isogomphodon,* and *Carcharhinus* appeared by the Middle Eocene (45 mya), but *Prionace* did not appear until the Pliocene (Cappetta, 1987), supporting Naylor's (1989) suggestion that *Prionace* is a derived *Carcharhinus. Galeocerdo cuvier* is a very large benthopelagic circumtropical coastal shark, which regularly crosses large expanses of open ocean (Kohler et al., 1998; Holland et al., 1999) and visits temperate latitudes in summer (Musick et al., 1993). Fossils of *Galeocerdo* are widespread in Cenozoic deposits in Europe, North and South America, the Celebes, India, and Japan (Cappetta, 1987). *Rhizoprionodon* includes five small and two medium-sized species, all of which are tropical and neritic. The largest species, *R. acutus,* has the broadest range with an isolated population off West Africa, but widely distributed around the tropical Indian Ocean to Indo-Malaya and northern Australia (Compagno, 1984, in press). The species may have been widespread in the Oligocene Tethys, becoming isolated into Atlantic and Indo-Pacific groups with the Miocene collision of Africa and Asia that led to the formation of the Suez (Old World) land barrier, cutting off the Indian Ocean from the Mediterranean and Atlantic (Briggs, 1995). Many extant carcharhinid species are known from Oligocene and Miocene deposits (Applegate, 1986; Cappetta, 1987; Naylor and Marcus, 1994; Purdy et al., 2001), and *Rhizoprionodon fischeuri* is known from the Middle Miocene of southern France and Portugal (Cappetta, 1987). Its relationship to *R. acutus* has not been examined and they may be conspecific. Subsequent cooling after the Miocene eliminated the tropical fauna in the Mediterranean (including *Rhizoprionodon*) (Briggs, 1995). Three species of *Rhizoprionodon* are restricted to the western Atlantic: one, *R. terraenovae*, with a parapatric subtropical/warm temperate North American distribution; and two, *R. lalandei* and *R. porosus* with mostly sympatric more southerly distributions from the Caribbean to southern Brazil. An amphi-American congener *R. longurio* occurs in the eastern Pacific. In the Indo-West Pacific, *R. oligolinx* ranges from the northwest Indian Ocean through Indo-Malaya to Palau, whereas parapatric *R. taylori* is restricted to northern Australia and nearby Indonesian waters. *Rhizoprionodon* includes small species with relatively high vagility (Heist et al., 1996). Their present zoogeography is most likely a result of widespread distribution along Tethyan coasts in the mid-Cenozoic, occasional dispersal across open ocean barriers, and vicariant isolation mitigated by the rise of the Suez and Panamanian Isthmuses.

Negaprion comprises two large, sluggish coastal sharks with complementary tropical distributions. *Negaprion acutidens* is distributed throughout the Indian Ocean, through Indo-Malaya and northern Australia to Tahiti and the Marshall Islands, but not across the East Pacific oceanic barrier (Compagno, 1984; Briggs, 1995). *Negaprion brevirostris* occurs in both the eastern and western tropical Atlantic and the tropical eastern Pacific. These distributions may be another example of isolation of a widespread Tethyan taxon in the Early Miocene by tectonic creation of the Suez ("Old World") barrier; with subsequent speciation in the Indo-Pacific and Atlantic, followed by isolation of the East Pacific *N. brevirostris* by the Pliocene rise of the Panamanian Isthmus.

Loxodon macrorhinus is a small coastal species widely distributed in the tropical Indo-Pacific. *Triaenodon obesus* is a medium-sized tropical shark very widely distributed from the western Indian Ocean to the western Pacific, Pacific Plate, and eastern Pacific. Fossils of *T. obesus* are known from the Miocene of North Carolina. Thus, the species has become extinct in the Atlantic (Purdy et al., 2001).

The genus *Carcharhinus* is the most speciose in the family (31 recognized species) and the phylogeny of the genus is far from resolved, despite detailed morphological and molecular studies (Naylor, 1992). Molecular data (Naylor, 1992) suggest that the large, ridge-backed members of the genus (*C. altimus, C. falciformis, C. galapagensis, C. longimanus, C. obscurus, C. perezi,* and *C. plumbeus*) are members of a monophyletic group that should also include *Prionace glauca*. Most members within this group were also considered to be closely related on morphological grounds by Garrick (1982) and Compagno (1988). Resolution of the interrelationships among other members of the genus has not been achieved, and in many cases molecular and morphological analyses have led to different conclusions (Naylor, 1992). Consequently, the discussion of the zoogeography of other members of the genus (in so far as possible) will be based on body size (Appendix 2.1). Small species include *C. borneensis, C. leiodon, C. macloti,* and *C. sealei*. Species of moderate size include *C. acronotus, C. cautus, C. dussumieri, C. fitzroyensis, C. hemiodon, C. porosus,* and *C. isodon*. Large species include *C. albimarginatus, C. amblyrhynchoides, C. amblyrhychos, C. brachyurus, C. brevipinna, C. limbatus, C. melanopterus, C. signatus, C. sorrah, C. tilstoni,* and *C. wheeleri*. The only very large *Carcharhinus* not already included in the ridge-backed group above is *C. leucas*. The very large ridge-backed species all have circumglobal distributions, but appear to exhibit habitat or range differences. *Carcharhinus falciformis* is a coastal/oceanic species occurring pelagically in both outer coastal and oceanic water masses near the continental margin in warm temperate and tropical seas. *Carcharhinus obscurus* and *C. galapagensis* are sibling species, one *C. obscurus* dominating in coastal, warm temperate to subtropical habitats, the other associated with subtropical and tropical oceanic islands. *Carcharhinus longimanus* is mostly tropical and confined to oceanic water masses. *Prionace glauca* is widespread in the oceanic realm, but migrates into temperate coastal areas in summer. It submerges in the tropics to stay within its preferred cool temperature range (Compagno, 1984). Two of the large species within the ridge-backed group of *Carcharhinus* have wide distributions in warm temperate to tropical coastal habitats, but have mostly complementary bathymetric distributions with *C. plumbeus* found in shallower water than *C. altimus*.

In contrast, *C. perezi* is confined to the Caribbean and adjacent tropical seas mostly associated with coral reefs. All of the small species of *Carcharhinus* are from the Indo-Pacific region; two have restricted distributions, *C. borneensis* to the Indo-Malayan subregion, and *C. leidon* to the Red Sea. The other two species, *C. macloti* and *C. sealei,* are widespread.

Of the moderate-sized species, *C. acronotus* and *C. isodon* are restricted to tropical and warm temperate coastal seas of the western Atlantic, and *C. porosus* occurs both in the western Atlantic and eastern Pacific. *Carcharhinus cautus* is confined to northern Australia and *C. fitzroyensis* occurs in northern Australia out onto the Pacific Plate islands, and *C. hemiodon* to northern Australia and Indo-Malaya. The widespread *C. dussumieri* ranges from the western Indian Ocean, through Indo-Malaya to Japan. Among the remaining large species of *Carcharhinus*, *C. tilsoni* is restricted to northern Australia, *C. amblyrhynchoides* and *C. sorrah* occur from the northwestern Indian Ocean to Indo-Malaya and northern Australia, and *C. amblyrhynchos* and *C. melanopterus* range from the western Indian Ocean to the western Pacific and out to the islands of the Pacific Plate. *Carcharhinus albimarginatus* has a wider distribution, and extends all the way to the eastern Pacific. In contrast, *C. wheeleri* is known only from the western Indian Ocean. However, this species may prove to be conspecific with *C. amblyrhynchos* (Compagno, in press). The rest of the large species of *Carcharhinus*, with the exception of *C. signatus* (a pelagic oceanic species that occurs in the eastern and western Atlantic and eastern Pacific), are all widespread and coastal: *C. amboinensis* is widespread in the Indo-West Pacific and in the Gulf of Guinea in the eastern Atlantic; *C. brachyurus* is cosmopolitan in warm temperate and subtropical areas and is at present absent from the western North Atlantic although it was abundant there in the Miocene (Purdy et al., 2001); *C. brevipinna* is circumglobal in warm temperate to tropical areas except in the eastern Pacific; *C. limbatus* is circumtropical and mostly coastal (specimens from the islands of the Pacific Plate and oceanic eastern Pacific need to be reexamined). *Carcharhinus leucas*, a very large species, is circumtropical in shallow, coastal and estuarine habitats and regularly penetrates far up into freshwater rivers.

Two closely related genera to *Carcharhinus* have restricted distributions: *Nasolamia* in the tropical eastern Pacific and *Isogomphodon* in the tropical western Atlantic. Miocene fossils are known for *Isogomphodon* for the eastern Atlantic and it appears to be extinct there. The remaining two genera of carcharhinids to be discussed have Indo-West Pacific distributions: *Lamiopsis temmincki* has a scattered coastal distribution from the northern Indian Ocean to China; *Glyphis* includes a group of riverine and estuarine sharks that are poorly known with populations known from India (*G. gangeticus*), Burma (*G. siamensis*), Borneo (*G. glyphis?*), and northern Australia (two little-known and perhaps undescribed species).

Carcharhinus and related genera probably evolved in the Eocene Tethys and radiated widely in coastal tropical habitats. Some species evolved the ability to invade freshwater habitats whereas others became distributed widely in oceanic habitats. The rise of the Carcharhinidae in time parallels the rise of higher teleosts with both groups having their greatest radiation and highest modern diversity in coastal tropical habitats of the Indo-West Pacific (Briggs, 1995; Randall, 1998; Table 2.5). Radiation of higher teleosts into diverse niches may have fostered an evolutionary response in the predatory carcharhinids. Although members of the family are conservative in body form, species have evolved a diversity of dentitions (Naylor and Marcus, 1994). Detailed studies of regional carcharhinid communities using ecomorphological analyses comparing body size and form and dentition to habitat parameters such as preferred depth, temperature, and salinity and food habits are needed to better understand the ecological interactions and evolution of the carcharhinid sharks.

The Sphyrnidae has historically been considered a family unto itself, but recent morphological and molecular evidence suggests that the group would better be considered a clade (tribe? subfamily?) within the Carcharhinidae closest to *Rhizoprionodon* (Figure 2.10) (Compagno, 1988; Naylor, 1992). The clade includes two genera and eight species of tropical, warm temperate, benthopelagic, coastal sharks (Appendix 2.1). *Eusphyra* contains one species, *E. blochii*, a medium-sized shark distributed from the northwest Indian Ocean to China, Indo-Malaya, and northern Australia. *Sphyrna* includes seven species, one of which (*S. corona*) is small, three of which are medium in size (*S. media, S. tiburo, S. tudes*), and three of which are very large (*S. lewini, S. mokarran, S. zygaena*). *Sphyrna corona* is confined to the tropical eastern Pacific, and *S. media* and *S. tiburo* are confined to the New World in the tropical western Atlantic and eastern Pacific, although *S. tiburo* ranges more widely at least seasonally into warm temperate latitudes. *Sphyrna tudes* is confined to the western Atlantic from Venezuela to Uruguay. All of the very large species of Sphyrnidae are circumglobal, coastal, and tropical; although *S. zygaena* ranges more widely into temperate waters, and *S. lewini* is more likely to move into oceanic habitats adjacent to landmasses at least seasonally (Compagno, 1984). Sphyrnid fossils are known from as early as the Lower Miocene and *S. zygaena* has been recorded from the Miocene of Portugal and southern France and the Lower Pleistocene of Japan. This suggests that the sphyrnids diverged from some *Rhizoprionodon*-like ancestor in the Oligocene and dispersed through Tethyan corridors. After the rise of the Suez Isthmus, *Eusphyra* appears to have evolved in the Indo-West Pacific, whereas there was a radiation of small and medium-sized *Sphyrna* in the New World. The small and medium-sized sphyrnids appear to be more strictly coastal and have lower vagility than the very large species.

2.3.2.2 Squalomorphii (Division Squalea sensu de Carvalho, 1996, Superorder Squalea sensu Shirai, 1996) —

As recognized here the squalomorphs comprise five orders: Hexanchiformes, Echinorhiniformes, Pristiophoriformes, Squatiniformes, and Squaliformes. The derived squalomorph superorder Hypnosqualea putatively comprising the Pristiophoriformes, Squatini- formes, and the Batoidea (Shirai, 1996; de Carvalho, 1996) has recently been rejected by strong molecular evidence, which suggests that the batoids evolved separately from the remainder of the Neoselachians (Douady et al., 2003; Maisey et al., in press). The early appearance of batoids in the fossil record (Lower Jurassic; Thies, 1983) supports this assertion. Most authors agree that Pristiophoriformes and Squatin- iformes are closely related although morphological studies have them as derived squalomorphs (Shirai, 1996; de Carvalho, 1996), whereas recent molecular work suggests that they are basal to Squaliformes (Douady et al., 2003; Maisey et al., in press) (Figure 2.5). A basal position for at least Squatiniformes is supported by its early fossil appearance in the Late Jurassic (Thies, 1983).

2.3.2.2.1 Hexanchiformes. The Hexanchiformes comprise two families: the Chlamydoselachidae and the Hexanchidae. The order is basal to the rest of the squalomorphs in keeping with its Lower Jurassic fossil record (Cappetta et al., 1993) (Figure 2.5). The Hexanchidae includes three genera and four mostly large species, which are benthopelagic with well-developed musculature and apparently high vagility found in bathyal or cold temperate coastal habitats. Fossil evidence suggests that the large Cretaceous hexanchid, *Notidanodon*, also had a cold-water "bipolar" distribution (Cioni, 1996). *Hep- tranchias* (one sp.), a medium-sized species, occurs spottily in all tropical and temperate seas except the eastern North Pacific. The genus is known from the Early Eocene of the eastern United States and Morroco (Cappetta et al., 1993). *Hexanchus* includes two species, one of which (*H. griseus*) is very large and distributed circumglobally. The other (*H. nakamurai*) is distributed widely in deep waters in warm temperate and tropical seas, but is apparently absent from the eastern Pacific (Compagno, in press). Although this genus is ancient with fossils known from the Lower Jurassic (Cappetta, 1987; Cappetta et al., 1993), its present distribution is most likely due to its vagility. Unlike some other groups of Jurassic origin such as *Squatina* or *Heterodontus,* which have low vagility and have evolved local and regional coastal species, Pangaean vicariance is highly unlikely for bathyal *Hexanchus* with only two very widely distributed species.

The genus *Notorynchus* (one sp.) is neritic at cool temperate latitudes worldwide except in the North Atlantic. The genus is first known from the lower Cretaceous of southern France with subsequent Cenozoic fossils known from other European and North American localities. Its present absence in the North Atlantic is apparently due to extinction there.

The family Chlamydoselachidae includes one large bathyal species distributed in temperate and tropical seas, often occurring around oceanic islands and seamounts (Compagno, in press). Fossils of *Chlamydoselachus* first appear in the Upper Cretaceous (Cappetta, 1987). Its present distribution is undoubtedly due to deep-water dispersal.

2.3.2.2.2 Echinorhiniformes. The Echinorhiniformes are sometimes included as a family within the Squaliformes (Compagno, 1999), but Shirai (1996) recognized it as the sister group to the rest of the squalomorphs (Figure 2.5). Maisey et al. (in press) have it basal to the squalomorphs but close to the Squatiniformes and Pristiophoriformes. Echinorhiniformes includes one extant family Echinorhinidae comprised of one genus and two large benthopelagic bathyal species. *Echinorhinus brucus* occurs in the Atlantic, Indian, and Western Pacific Oceans, whereas *E. cookei* is confirmed only from the Pacific (Compagno, 1984, in press). The earliest fossils of *Echinorhinus* are known from the Upper Cretaceous of Angola (Cappetta, 1987) and subsequent records are from North Atlantic (rather than Tethyan) localities. The genus appeared by the lower Miocene in the eastern Pacific. Understanding the almost complementary present distribution of the two species awaits reevaluation of their relationship to several nominal Cenozoic species (Cappetta, 1987).

2.3.2.2.3 Pristiophoriformes.   The order Pristiophoriformes includes one family with two genera, *Pliotrema* (one sp.) confined to temperate neritic and upper bathyal habitats off South Africa, southern Mozambique, and southeastern Madagascar, and *Pristiophorus* with four recognized and four undescribed species (Last and Stevens, 1994). Compagno (in press) recognizes four species groups within the genus, a neritic temperate group of large, robust species (*P. cirratus, P. japonicus,* and *P.* sp. *A*) from the Western Pacific; a bathyal, tropical group of small slender species (*P. schroederi, P.* sp. *B,* and *P.* sp. *C*) with a wide disjunct distribution from the western North Atlantic and the west-central Pacific; the large *P. nudipinnis* from southern Australia; and the distinctive *P.* sp. *D* from the western Indian Ocean. The interrelationships among these species groups remain to be studied. The oldest pristiophoriform fossils (*Pristiophorus*) occur in the Upper Cretaceous deposits from Lebanon and are very similar to recent species (Cappetta, 1987). Other fossils show the genus to have been widely distributed from the Upper Cretaceous to Upper Oligocene of Japan, Upper Paleocene of Morocco, Upper Eocene of Oregon, Oligocene of Belgium, Holland, and California, Miocene of Southwest Germany, Portugal, and southern France, and into the Pliocene of New Zealand. The genus had a wide Tethyan distribution with species also present in the eastern Pacific by the Oligocene. At present, it is extinct in the eastern Pacific and apparently relict in the Atlantic with highest diversity in Australia.

2.3.2.2.4 Squatiniformes.   The Squatiniformes consist of one family and genus with 16 described species from mostly temperate coastal, and tropical upper bathyal habitats (Compagno, 1984, in press). Squatiniformes are strongly benthic and appear to have limited vagility even though some species are large. They are absent from oceanic islands. This is an ancient group and the genus *Squatina* is known from the Upper Jurassic of Germany (Cappetta, 1987). It has an extensive widespread fossil record from the Cretaceous and throughout the Cenozoic. It may have been widespread in Pangaean seas and its present distribution could be due to the vicariant breakup of that ancient supercontinent. Further speculation about the zoogeography of this group requires a detailed study of interrelationship among species in the genus *Squatina*.

2.3.2.2.5 Squaliformes.   The Squaliformes are a large group of small to medium-sized cold-water sharks. Most species are bathyal and benthopelagic. Six families (excluding Echinorhinidae), Squalidae, Centrophoridae, Etmopteridae, Somniosidae, Oxynotidae, and Dalatiidae, are at present recognized in the order (Compagno, 1999). There are major differences in the topology of the phyletic trees for Squaliformes based on morphology (Shirai, 1996) and molecular evidence (Maisey et al., in press) (Figure 2.5). Whereas Shirai (1996) has Centrophoridae basal to the squaliformes, with Squalidae derived and basal to the Hypnosqualeans, Maisey et al. has Squalidae basal and Centrophoridae derived. The fossil record supports the latter topology as the earliest squalid fossil may be *Squalogaleus*, a poorly preserved specimen from the Upper Jurassic of Europe. The next oldest fossil in the family, *Protosqualeus*, is from the Lower Cretaceous (125 mya) of northern Germany and northern France (Cappetta, 1987; Cappetta et al., 1993). The Centrophoridae apparently did not appear until the Upper Cretaceous (90 mya) (Cappetta, 1987).

The family Squalidae includes two genera: *Cirrhigaleus* (two sp.) and *Squalus* (eight recognized sp.). *Cirrhigaleus asper* is widespread at upper bathyal depths in temperate and tropical seas including oceanic islands, but is absent from the eastern Pacific. *Cirrhigaleus barbifer* occurs in the western Pacific from Japan to New Caledonia, New Zealand, and Australia (Compagno, in press). The genus *Squalus* may be divided into three species groups (Compagno, in press):

1. The *S. acanthias* group comprising one neritic species with an antitropical distribution in all cold temperate coastal seas

2. The *S. mitsukurii* group with at least ten species including *S. mitsukurii,* which is widespread in upper bathyal habitats and circumglobal except for the eastern North Pacific, *S. blainvillei* (*sensu* Compagno, in press) and *S. japonicus* from the western Pacific, *S. melanurus* and *S. rancureli* known only from New Caledonia, and five undescribed species from Australia (Last and Stevens, 1994)

3. The *megalops* group with *S. megalops* widely distributed from the eastern North Atlantic to South Africa, the Indian Ocean to the western Pacific, but absent in the eastern Pacific and western Atlantic, *S. cubensis* from the western Atlantic, and an undescribed species from Australia and New Guinea (Last and Stevens, 1994; Compagno, in press)

Although the Squalidae are known from fossils at least as far back as the Lower Cretaceous, the genus *Squalus* first appeared in the Upper Cretaceous with numerous fossil records during the Cenozoic from Europe, the former U.S.S.R., North Africa, North America, and Asia (Cappetta, 1987). The Squalidae are muscular, active sharks with high vagility. The taxonomy of the genus *Squalus* is replete with problems with many undescribed species and confusing species complexes (Compagno, in press). Therefore, further speculation on their zoogeography will require a more complete knowledge of their interrelationships. Regardless, dispersal through cold-water bathyal and oceanic corridors has been important in shaping the current zoogeography of the group.

The Somniosidae is a benthopelagic family of moderate size with seven genera currently recognized (Compagno, 1999, in press): *Centroscymnus, Centroselachus, Proscymnodon, Scymnodalatias, Scymnodon, Somniosus,* and *Zameus. Centroscymnus* includes two species both of which are circumglobal at middle bathyal to upper abyssal depths except in the eastern Pacific. *Centroselachus* (monotypic) is circumglobal in upper and middle bathyal depths except in the eastern North Pacific. *Proscymnodon* comprises two species, *P. macracanthus,* known only from the type specimen taken in the Straits of Magellan, and *P. plunketi* endemic at upper and middle bathyal depths in Australia and New Zealand. *Scymnodalatias* includes four little-known species, which appear to be oceanic and pelagic. *Scymnodalatias garricki* is known from the North Atlantic but records for the other three species are restricted to the Southern Hemisphere. *Scymnodon* is monotypic (*S. ringens*) and endemic at bathyal depths to the northeast Atlantic.

The genus *Somniosus* comprises two species groups (Compagno, in press). The first group contains one species, *S. rostratus,* which is a medium-sized bathyal shark from temperate latitudes in the eastern Atlantic and western Pacific off Japan and New Zealand (Compagno, 1984, in press; Francis et al., 1988). The second group comprises three very large species: *S. microcephalus* and *S. pacificus,* which are sister taxa from the North Atlantic and North Pacific basins, respectively, and *S. antarcticus* from cold temperate Australia and Southern Ocean habitats. *Somniosus microcephalus* and *S. pacificus* occur from Arctic waters where they may occur at the surface, south into temperate or even subtropical latitudes at bathyal or even abyssal depths — 2000 m off California for *S. pacificus* (Anderson et al., 1979); 2200 m on the Blake Plateau for *S. microcephalus* (Herdendorf and Berra, 1995). Records of *S. microcephalus* from the South Atlantic off South Africa and from the Southern Ocean at Kerguelen and Macquarie Islands (Cherel and Duhamel, 2004) and of *S. pacificus* from Tasmania and New Zealand (Last and Stevens, 1994) are based apparently on a closely related form *S. antarcticus* (J. Stevens, pers. comm.). The relationship of *S. antarcticus* is at present equivocal (Compagno, in press), but it appears that either species could have passed through the tropics through submergence during the Pleistocene and been ancestral to the Southern Hemisphere species. The southernmost limit of *S. antarcticus* into polar seas is unknown. Reif and Saure (1987) have suggested that *S. microcephalus* and *S. pacificus* may have diverged in the Pleistocene with the formation of Arctic sea ice. However, whether Arctic sea ice presents a barrier to these truly cryophilic sharks is debatable because *S. microcephalus,* at least, lives commonly under ice in the eastern Arctic (Skomal, 2001). The present northern limits of distribution for either species in the Arctic Ocean are unknown. Conversely, initial isolation of these species may have been caused by the Pliocene rise of the Panamanian Isthmus (Briggs, 1995), which may have separated tropical bathyal populations into Atlantic and Pacific forms.

The genus *Zameus* comprises two small, mid-bathyal to lower bathyal sharks. *Zameus squamulosus* is widespread in temperate and tropical localities except the eastern Pacific, whereas *Z. ichiharai* is known only from Japan. The family Somniosidae is known from the Upper Cretaceous of Lebanon (*Cretascymnus*), but fossils of recent genera are rare. *Scymnodon* has been reported from the Upper Oligocene and Late Pleistocene of California (Cappetta, 1987), but is noticeably absent along with most other bathyal Squaliformes from the northeast Pacific today. *Somniosus* has been reported from the Miocene of Japan and Pliocene of Belgium.

The Dalatiidae includes seven genera, five of which are monotypic — *Dalatias, Euprotomicroides, Euprotomicrus, Heteroscymnoides,* and *Mollisquama* — and *Isistius* and *Squaliolus,* each with two species. *Dalatias licha* stands out as the only medium-sized, mostly benthopelagic, bathyal shark in this family of diminutive, pelagic, oceanic sharks. *Dalatias* is widespread in tropical and temperate latitudes including around seamounts and oceanic islands, but is absent from the eastern Pacific (Compagno, 1984, in press). *Euprotomicroides zantedeschia* is known only from two specimens from the South Atlantic, and *M. parini* likewise is rare and reported only from the eastern South Pacific. *Euprotomicrus bispinatus* has an antitropical distribution and is widely distributed in central water masses in the Pacific, Indian, and South Atlantic Oceans, but so far not reported for the North Atlantic (Compagno, in press). *Squaliolus laticaudus* has a distribution that is mostly complementary to that of *Euprotomicrus* and avoids central ocean basins, occurring near the edges of landmasses and islands where productivity is relatively higher (Compagno, in press). It appears to be mesopelagic with diel vertical migrations. *Squaliolus aliae* is apparently known only off Japan. *Heteroscymnoides marleyi* is probably circumglobal in cold subantarctic waters of the Southern Hemisphere (Stehman et al., 1999). *Isistius brasiliensis* is circumtropical and mostly mesopelagic, migrating up to the surface (presumably at night). Its congener *I. plutodus* is less well known but also widely distributed (western North Atlantic, and both western North and South Pacific). Compagno (in press) has suggested that this species occurs deeper than *I. brasiliensis,* thus accounting for its uncommon capture. Vertical segregation among mesopelagic teleost congeners is a well-known zoogeographic pattern (Gibbs et al., 1971; Marshall, 1979), and thus its appearance within sharks is not unexpected.

The Dalatiids are apparently all mesopelagic with the exception of *Dalatias,* which is larger and associated with bathyal habitats, but often captured well off the bottom (Compagno, 1984). For species that are reasonably known, zoogeographic patterns follow those that have been described for mesopelagic teleosts, which have been shown to be associated with specific water masses (both vertical and horizontal) (Ebeling, 1967; Backus and Craddock, 1977; Marshall, 1979); i.e., an antitropical central water mass pattern (*E. bispinatus*), "pseudo-oceanic" pattern (*S. laticaudus*), and a sub-Antarctic water mass pattern (*H. marleyi*).

The earliest known fossils for both *Dalatias* (as *Scymnorhinus*) and *Isistius* are from Upper Paleocene deposits. Whereas *Dalatias* fossils from the Tertiary are widespread (Europe, the former U.S.S.R., Asia, New Zealand, West India) those of *Isistius* are mostly from Europe and adjacent Morroco with one outlying record from Ecuador (Cappetta, 1987). Different fossil availability between the genera is probably due to the bathyal distributions of *Dalatias* as opposed to the oceanic distribution of *Isistius.* *Squaliolus,* similarly, is known only from deep water deposits from the Miocene of southern France and Italy, and the other oceanic genera of dalatiids are as yet unknown as fossils.

The Etmopteridae includes five genera of small, benthopelagic bathyal sharks of which three genera are monotypic. *Aculeola nigra* is common, but restricted to the eastern South Pacific where other bathyal squaliformes are depauperate. *Miroscyllium sheikoi* is apparently known only from Japan, and *Trigonognathus kabeyi* occurs in the western North Pacific off Japan, Hawaii, and the Emperor seamounts. *Centroscyllium* comprises seven species, six of which are very small (<50 cm TL) and restricted to local or regional distributions with species off Japan, the northern Indian Ocean, eastern Pacific, and southwest Atlantic; many of those species have been recorded from seamounts and oceanic islands. The largest species (50 to 100 cm TL) *C. fabricii* has an antitropical distribution in the North and South Atlantic. This is primarily a high-latitude bathyal shark that submerges at lower latitudes. The genus *Etmopterus* is the most speciose within the Squalomorphii and the Selachii as whole and, with 32 recognized species, is currently ahead of *Carcharhinus* (Carcharhinidae) with 31 species and *Apristurus* (Scyliorhinidae) with 30 species (Appendix 2.1). *Etmopterus* shares with *Apristurus* the distinction of having small or even very small (<50 cm) species all confined to bathyal habitats (although *Apristurus* is basically benthic and *Etmopterus* benthopelagic).

Of the 32 species of *Etmopterus* recognized, 26 are Lilliputian in size (<50 cm TL), one species *E. perryi* maturing at <20 cm. These species have local or regional distributions around the world in tropical and temperate latitudes. Some of the larger species (50 to 100 cm TL) are more widespread. *Etmopterus bigelowi* is distributed from the western and eastern Atlantic, to the western Indian Ocean, western Pacific across the Pacific Plate to the Nazca and Sala y Gomez Ridges in the eastern Pacific. *Etmopterus princeps* occurs around the North Atlantic rim, and *E. pusillus* occurs in the western and eastern Atlantic and the Indian Ocean to the western Pacific. The western North Atlantic appears to be a center of endemism for *Etmopterus* with six recognized endemics: *E. bullisi, E. carteri, E. hillianus, E. robinsi, E. schultzi,* and *E. virens.* Conversely, the diversity of *Etmopterus* there may more likely be the result of more intense ichthyological scrutiny (Bigelow and Schroeder, 1948, 1957; Springer and Burgess, 1985). Recent studies off Australia (Last and Stevens, 1994) revealed six new *Etmopterus* species, which were very recently described (Last et al., 2002). There may be at least five more undescribed from other areas (Compagno, unpubl. data). *Etmopterus* is apparently the only genus in the family known from the fossil record and it is known from deep-water Miocene deposits in France and Italy (Cappetta, 1987). Further study of the zoogeography of *Etmopterus* must await resolution of taxonomic problems and detailed analysis of this large group.

The small family Oxynotidae comprises one extant genus and five species of small to medium-sized benthic sharks from deep coastal and bathyal habitats. Two species, *O. centrina* and *O. paradoxus*, are known from the eastern Atlantic, and one each from the southern Caribbean (*O. caribbaeus*), Australia and New Zealand (*O. bruniensis*), and Japan (*O. japonicus*). The family first appeared in the fossil record (*Protoxynotus*) in the Late Cretaceous (75 mya), and the genus *Oxynotus* is known from Lower Miocene fossils from California, and Pliocene fossils from Belgium (*O. centrina*) (Cappetta, 1987; Cappetta et al., 1994). It is difficult to believe that these awkward-looking little sharks have anything but low vagility, although there are apparently no published observations of their behavior in the aquarium or from nature. The scarcity of fossils in this bathyal group mirrors that of other bathyal groups of Squaliformes. Shirai's (1996) morphological analysis places them close to the somniosids and dalatiids, whereas Maisey et al. (in press) have them basal to the centrophorids. Perhaps they are basal to both groups and thus older than the fossil record would indicate. This would suggest a Tethyan dispersal of the family in the Lower Cretaceous with subsequent continental dispersal to the eastern North Pacific by the Lower Miocene. The present distribution of the family may be relict with late Tertiary extinction in the eastern North Pacific.

The Centrophoridae are small to medium-sized benthopelagic sharks found in upper to middle bathyal depths. The family includes two genera, *Centrophorus* with ten recognized species and *Deania* with four species. The Centrophoridae represent three extreme ecomorphotypes (Compagno, 1990). *Deania* with it small size and very long flat snout at one extreme, and a group of large, short-snouted *Centrophorus* (*C. acus, C. niaukang,* and *C. squamosus*) at the other. *Centrophorus harrissoni, C. isodon,* and *C. lusitanicus* appear to be intermediate between these two types. *Centrophorus moluccensis* represents a third morphotype with a narrow head and conical snout, whereas *C. atromarginatus, C. granulosus,* and *C. tesselatus* appear to be between *C. moluccensis* and the large short-snouted *Centrophorus* (Compagno, in press). The family is absent from the eastern North Pacific and represented in the eastern South Pacific by only one species, *Deania calcea. Centrophorus acus* occurs in the western North Pacific in tropical and warm temperate latitudes. Nominal *C. acus* from the western North Atlantic could be attributed to an undescribed species (G. Burgess, pers. comm.). *Centrophorus niaukang* and *C. squamosus* are both widespread at temperate and tropical latitudes from the eastern Pacific to the western Indian Ocean, and western and eastern Atlantic. *Centrophorus harrissoni* is widespread from the western Pacific to South Africa and perhaps the western North Atlantic (Compagno, in press), and *C. lusitanicus* is also widespread from the western Pacific to the western Indian Ocean and eastern North Atlantic. In contrast, *C. isodon* is restricted to the western Pacific and eastern Indian Ocean. *Centrophorus moluccensis* and *C. atromarginatus* are both widely distributed in the Indo-West Pacific, whereas *C. granulosus* is very widespread from the western and eastern Atlantic and Mediterranean, western Indian Ocean, and western Pacific out onto the Pacific Plate. *Centrophorus tesselatus* is rare, but widespread from the western North

Atlantic, western Indian Ocean, and western Pacific and the Pacific Plate. Distributions of most of the *Centrophorus* species may be more widespread than at present recognized. Conversely, distributions of some apparently widespread species may be substantially modified after taxonomic and species identification problems in the genus are solved. *Centrophorus* first appeared in the fossil record in the Upper Cretaceous of Lithuania and other fossils recorded in the Cenozoic are mostly from Europe with Oligocene records from California and Pliocene records from New Zealand (Cappetta, 1987).

The four recognized species of *Deania* (*D. calcea, D. hystricosum, D. profundorum, D. quadrispinosum*) are all widespread with records from the Atlantic, Indian Ocean, and western Pacific. The last species appears to be restricted mostly to the Southern Hemisphere. As with *Centrophorus* the distributions of species of *Deania* will be subject to revision as additional information accrues on these poorly known bathyal forms. Fossils of *Deania* are known from the Miocene of Europe and the West Indies.

The Squaliformes have radiated widely in bathyal habitats from tropical to cold temperate latitudes circumglobally. Bathyal Squaliformes are largely absent from the northeastern Pacific (approximately three species) and depauperate in the southeastern Pacific (seven species) (Table 2.6). This pattern cannot be explained by a lack of vagility or lack of suitable habitat. The group is very diverse in the eastern Atlantic (approximately 26 species) where similar cool coastal currents and upwelling occur. Although the dearth of bathyal squaloids in the southeastern Pacific could be explained by lack of collections from there, the bathyal fish fauna of the northeastern Pacific has been well studied (Pearcy et al., 1982). Perhaps the great diversity of large bathyal sebastine and other large scorpaeniform species there may be acting to competitively exclude the small and medium-sized bathyal squaloids. Squaliforms are also depauperate in the western Pacific boreal region (Table 2.6) where sebastines also are diverse and abundant (Nakabo, 2002). Competitive exclusion may explain the depauperate squaliform fauna for temperate latitudes in the eastern Pacific, but not for the tropics where the sebastines become rare or absent. The bathyal fish fauna in the tropical eastern Pacific has not been well studied and accurate estimates of the diversity of squaloids there and in the southeastern Pacific await further research.

2.4 Summary

The zoogeography of the Selachii is complex and diverse. In considering vicariant and dispersive zoogeographic mechanisms, an evaluation of taxon dispersal ability or vagility is very important. We found that vagility is directly proportional to body size, and that benthic sharks have very low vagility compared to benthopelagic sharks, which in turn have lower vagility than pelagic sharks. Likewise, there is an increase in vagility from coastal to bathyal to oceanic species.

Ancient vicariant events such as the breakup of Pangaea may have been important in the evolution of current zoogeographic patterns of higher benthic taxa such as Heterodontiformes, Squantiniformes, and Orectolobiformes. The data suggest a Tethyan origin for these and most other older groups with subsequent widespread coastal dispersal, and later regional extinction in the Atlantic for Heterodontiformes and all but two genera of Orectolobiformes. Conversely, fossil evidence suggests the Hexanchiformes and Squaliformes may have arisen in cool waters of semienclosed boreal basins to the north of Tethys. An alternative hypothesis is that these groups, particularly the Squaliformes, may also have had a Tethyan origin, but at bathyal depths. The high present diversity of Squaliformes in the Indo-West Pacific supports the latter hypothesis (Table 2.6). Most taxa within the Lamniformes and Carcharhiniformes (with the notable exception of the scyliorhinids) are benthopelagic or pelagic with high vagility and wide dispersal. Even in these groups the vicariant closure of the Suez seaway in the Miocene and Panamanian seaway in the Pliocene were important in molding the present-day distribution of many taxa.

Overall, shark endemism and diversity in the tropics is highest in North Australia followed by the Indo-Malayan and Western Indian Ocean subregions (Table 2.7). The lowest diversity among tropical subregions is on the Pacific Plate followed by the eastern Pacific and the eastern Atlantic. This pattern is different from that for the tropical teleosts, which find their lowest diversity in the eastern Atlantic (Briggs, 1995). The apparent reason for the difference is that smaller benthic sharks such as the orectolobiforms that contribute substantially to shark diversity on coral reefs in most of the Indo-West

TABLE 2.6

Distribution of Squaliform Species by Zoogeographic Region and Subregions from Table 2.1

Regions and Subregions	Total	Total Endemics	Coastal	Coastal Endemics	Coastal/ Bathyal + Bathyal	Coastal Bathyal + Bathyal Endemics	Coastal/ Oceanic + Oceanic	Coastal/ Oceanic + Oceanic Endemics
A1a	26	2	1	0	23	2	2	0
A1b	23	1	0	0	22	1	1	0
A1c	12	1	1	0	8	1	3	0
A1d	31	12	0	0	27	12	4	0
A2	5	0	2	0	3	0	0	0
A3	23	2	1	0	21	2	1	0
A4	23	0	1	0	21	0	1	0
B1	41	9	2	0	35	9	4	0
B2	29	0	1	0	24	0	4	0
B3	24	0	1	0	18	0	4	0
B4	5	0	2	0	2	0	1	0
B5	11	2	1	0	7	0	3	2
B6	29	0	1	0	25	0	3	0
B7	14	0	1	0	9	0	4	0
B8	27	0	1	0	25	0	1	0
B9	20	0	0	0	19	0	1	0
C1	4	1	2	0	2	1	0	0
C2	26	0	2	0	23	0	2	0
C3	23	2	1	0	19	1	3	1
C4	16	1	1	0	13	1	2	0
C5	1	0	1	0	0	0	0	0
C6	7	0	1	0	6	0	0	0
C7	23	0	1	0	21	0	1	0
D	5	0	1	0	4	0	0	0
E	0	0	0	0	0	0	0	0

Note: Species numbers are approximate and may change as additional information accrues, particularly for bathyal sharks.

Pacific have not been able to reach the oceanic islands of the Pacific Plate because of low vagility. In addition, there is a lack of estuarine habitat there (Springer, 1982). The depauperate nature of the tropical eastern Pacific shark fauna is due to the lack of squaliforms discussed previously. Among warm temperate regions, the western North Pacific has the highest recorded diversity (Table 2.7). This may be a product of very active Japanese fisheries there and considerable scrutiny by Japanese ichthyologists (Nakabo, 2002). Other warm temperate areas with fairly high diversity include Australia, southern Africa, and both the western and eastern North Atlantic. Endemicity in the warm temperate zone is highest in the western North Pacific. Within the cold temperate zone (Table 2.7) Tasmania and the eastern Atlantic boreal regions have the highest diversity, and the eastern and western North Pacific the lowest. The apparent reason for these differences again are a high diversity of squaliforms in Tasmania and the eastern Atlantic and a dearth of squaliforms in the boreal Pacific discussed above.

Although patterns of regional diversity and endemism for coastal sharks are fairly well known, new species still are being discovered, and knowledge of the diversity and distribution of bathyal sharks is tentative at best. Further analyses of regional coastal shark faunas are possible, but beyond the scope of this chapter, which has grown well beyond that originally envisioned. Regional analyses of the bathyal shark faunas will require a much better understanding of the taxonomy of the squaliforms and scyliorhinids, and a more intensive collection effort in many areas of the world's oceans. It is a sobering thought that now at the dawn of the 21st century, science still remains ignorant of "what lives where" on the face of the Earth even for a moderate-sized charismatic vertebrate group like the Selachii. As deep-water fisheries proliferate and expand to new areas to exploit bathyal fish stocks, undoubtedly more bathyal shark material will become available for study (as it did in the last decade from Australia) (Last and

TABLE 2.7

Distribution of All Sharks by Zoogeographic Region and Subregions from Table 2.1

Regions and Subregions	Total	Total Endemics	Coastal	Coastal Endemics	Coastal/ Bathyal + Bathyal	Coastal/ Bathyal + Bathyal Endemics	Coastal/ Oceanic + Oceanic	Coastal/ Oceanic + Oceanic Endemics
A1a	110	12	36	3	48	9	26	0
A1b	124	15	46	4	51	11	27	0
A1c	49	1	6	0	18	1	25	0
A1d	140	47	44	15	67	32	29	0
A2	64	2	21	1	19	1	24	0
A3	93	9	23	2	47	7	23	0
A4	74	0	12	0	38	0	24	0
B1	120	37	28	13	67	24	25	0
B2	109	9	30	4	56	5	23	0
B3	88	7	18	3	43	4	27	0
B4	57	2	13	1	21	1	23	0
B5	54	2	13	0	20	1	21	1
B6	86	4	17	0	46	4	23	0
B7	60	2	17	1	19	1	26	0
B8	82	4	12	0	49	4	21	0
B9	64	0	12	0	35	0	17	0
C1	13	1	8	0	2	1	3	0
C2	80	0	19	0	48	0	13	0
C3	48	4	4	1	31	2	13	0
C4	40	1	8	0	24	1	8	0
C5	13	0	4	0	4	0	5	0
C6	39	0	9	0	15	0	15	0
C7	63	2	5	0	40	2	18	0
D	6	0	1	0	5	0	0	0
E	0	0	0	0	0	0	0	0

Note: Species numbers are approximate and may change as additional information accrues, particularly for bathyal sharks.

Stevens, 1994). Ironically, these same fisheries may be responsible for the demise of some shark species even before they are recognized by science (Musick et al., 2000).

Acknowledgments

Thanks are due to S.P. Applegate, G. Burgess, J. Bruner, E. Grogan, P. Last, R. Lund, J.D. McEachran, G. Naylor, J. Stevens, and many other colleagues who contributed immensely to this chapter through providing information or stimulating discussions. We are indebted to the Cambridge University Press, Tokai University Press, Hokkaido University Graduate School of Fisheries Science, and Elsevier Academic Press for providing permission to use previously published figures. This is VIMS Contribution 2557 and a contribution from the National Shark Research Consortium.

Appendix 2.1

Order	Family	Genus	Species	Body Size (cm): <100, 100–150, 150–300, >300	Benthic, Benthopelagic, Pelagic	Coastal, Oceanic, Bathyal	Number of FAO Regions
Hexanchiformes	Chlamydoselachidae	*Chlamydoselachus*	*anguineus*	150–300	Benthopelagic	Bathyal	10
	Hexanchidae	*Heptranchias*	*perlo*	100–150	Benthopelagic	Bathyal	12
		Hexanchus	*griseus*	>300	Benthopelagic	Bathyal	15
		Hexanchus	*nakamurai*	150–300	Benthopelagic	Bathyal	9
		Notorynchus	*cepedianus*	150–300	Benthopelagic	Coastal	11
Echinorhiniformes	Echinorhinidae	*Echinorhinus*	*brucus*	150–300	Benthopelagic	Bathyal	12
		Echinorhinus	*cookei*	150–300	Benthopelagic	Bathyal	5
Squaliformes	Squalidae	*Cirrhigaleus*	*asper*	100–150	Benthopelagic	Bathyal	5
		Cirrhigaleus	*barbifer*	100–150	Benthopelagic	Bathyal	3
		Squalus	*acanthias*	100–150	Benthopelagic	Coastal	15
		Squalus	*blainvillei*	<100	Benthopelagic	Bathyal	6
		Squalus	*brevirostris*	<100	Benthopelagic	Bathyal	1
		Squalus	*cubensis*	<100	Benthopelagic	Bathyal	2
		Squalus	*japonicus*	<100	Benthopelagic	Bathyal	2
		Squalus	*megalops*	<100	Benthopelagic	Bathyal	9
		Squalus	*melanurus*	<100	Benthopelagic	Bathyal	1
		Squalus	*mitsukurii*	<100	Benthopelagic	Bathyal	13
		Squalus	*rancureli*	<100	Benthopelagic	Bathyal	1
		Squalus	sp. A	<100	Benthopelagic	Bathyal	1
		Squalus	sp. B	<100	Benthopelagic	Bathyal	1
		Squalus	sp. C	<100	Benthopelagic	Bathyal	1
		Squalus	sp. D	<100	Benthopelagic	Bathyal	1
		Squalus	sp. E	<100	Benthopelagic	Bathyal	1
		Squalus	sp. F	<100	Benthopelagic	Bathyal	1
	Centrophoridae	*Centrophorus*	*acus*	150–300	Benthopelagic	Bathyal	3
		Centrophorus	*atromarginatus*	<100	Benthopelagic	Bathyal	3
		Centrophorus	*granulosus*	100–150	Benthopelagic	Bathyal	10
		Centrophorus	*harrissoni*	<100	Benthopelagic	Bathyal	5
		Centrophorus	*isodon*	<100	Benthopelagic	Bathyal	3
		Centrophorus	*lusitanicus*	<100	Benthopelagic	Bathyal	4
		Centrophorus	*moluccensis*	<100	Benthopelagic	Bathyal	5
		Centrophorus	*niaukang*	150–300	Benthopelagic	Bathyal	6
		Centrophorus	*squamosus*	150–300	Benthopelagic	Bathyal	10

Order	Family	Genus	Species	Body Size (cm): <100, 100–150, 150–300, >300	Benthic, Benthopelagic, Pelagic	Coastal, Oceanic, Bathyal	Number of FAO Regions
		Centrophorus	*tesselatus*	<100	Benthopelagic	Bathyal	5
		Centrophorus	*uyato*	<100	Benthopelagic	Bathyal	8
		Deania	*calcea*	100–150	Benthopelagic	Bathyal	8
		Deania	*hystricosum*	100–150	Benthopelagic	Bathyal	5
		Deania	*profundorum*	<100	Benthopelagic	Bathyal	6
		Deania	*quadrispinosum*	100–150	Benthopelagic	Bathyal	5
	Etmopteridae	*Aculeola*	*nigra*	<100	Benthopelagic	Coastal/bathyal	1
		Centroscyllium	*excelsum*	<100	Benthopelagic	Bathyal	1
		Centroscyllium	*fabricii*	<100	Benthopelagic	Bathyal	6
		Centroscyllium	*granulatum*	<100	Benthopelagic	Bathyal	2
		Centroscyllium	*kamoharai*	<100	Benthopelagic	Bathyal	2
		Centroscyllium	*nigrum*	<100	Benthopelagic	Bathyal	2
		Centroscyllium	*ornatum*	<100	Benthopelagic	Bathyal	2
		Centroscyllium	*ritteri*	<100	Benthopelagic	Bathyal	1
		Etmopterus	*baxteri*	<100	Benthopelagic	Bathyal	1
		Etmopterus	*bigelowi*	<100	Benthopelagic	Bathyal	8
		Etmopterus	*brachyurus*	<100	Benthopelagic	Bathyal	2
		Etmopterus	*bullisi*	<100	Benthopelagic	Bathyal	1
		Etmopterus	*carteri*	<100	Benthopelagic	Bathyal	1
		Etmopterus	*caudistigmus*	<100	Benthopelagic	Bathyal	2
		Etmopterus	*compagnoi*	<100	Benthopelagic	Bathyal	1
		Etmopterus	*decacuspidatus*	<100	Benthopelagic	Bathyal	1
		Etmopterus	*dianthus*	<100	Benthopelagic	Bathyal	1
		Etmopterus	*dislineatus*	<100	Benthopelagic	Bathyal	1
		Etmopterus	*evansi*	<100	Benthopelagic	Bathyal	1
		Etmopterus	*fusus*	<100	Benthopelagic	Bathyal	1
		Etmopterus	*gracilispinis*	<100	Benthopelagic	Bathyal	5
		Etmopterus	*granulosus*	<100	Benthopelagic	Bathyal	3
		Etmopterus	*hillianus*	<100	Benthopelagic	Bathyal	2
		Etmopterus	*litvinovi*	<100	Benthopelagic	Bathyal	1
		Etmopterus	*lucifer*	<100	Benthopelagic	Bathyal	7
		Etmopterus	*molleri*	<100	Benthopelagic	Bathyal	2
		Etmopterus	*perryi*	<100	Benthopelagic	Bathyal	1
		Etmopterus	*polli*	<100	Benthopelagic	Bathyal	2
		Etmopterus	*princeps*	<100	Benthopelagic	Bathyal	4
		Etmopterus	*pseudosqualiolus*	<100	Benthopelagic	Bathyal	2
		Etmopterus	*pusillus*	<100	Benthopelagic	Bathyal	7

Family	Genus	species	depth			n
	Etmopterus	*pycnolepis*	<100	Benthopelagic	Bathyal	1
	Etmopterus	*robinsi*	<100	Benthopelagic	Bathyal	1
	Etmopterus	*schultzi*	<100	Benthopelagic	Bathyal	1
	Etmopterus	*sentosus*	<100	Benthopelagic	Bathyal	1
	Etmopterus	*spinax*	<100	Benthopelagic	Bathyal	3
	Etmopterus	*splendidus*	<100	Benthopelagic	Bathyal	1
	Etmopterus	*unicolor*	<100	Benthopelagic	Bathyal	1
	Etmopterus	*villosus*	<100	Benthopelagic	Bathyal	1
	Etmopterus	*virens*	<100	Benthopelagic	Bathyal	1
	Miroscyllium	*sheikoi*	<100	Benthopelagic	Bathyal	2
	Trigonognathus	*kabeyai*	<100	Benthopelagic	Bathyal	8
Somniosidae	*Centroselachus*	*crepidater*	<100	Benthopelagic	Bathyal	10
	Centroscymnus	*coelolepis*	<100	Benthopelagic	Bathyal	7
	Centroscymnus	*owstoni*	<100	Benthopelagic	Bathyal	1
	Proscymnodon	*macracanthus*	<100	Benthopelagic	Bathyal	1
	Proscymnodon	*plunketi*	100–150	Benthopelagic	Bathyal	3
	Scymnodalatias	*albicauda*	100–150	Pelagic	Oceanic	1
	Scymnodalatias	*garricki*	<100	Pelagic	Oceanic	1
	Scymnodalatias	*oligodon*	<100	Pelagic	Oceanic	1
	Scymnodalatias	*sherwoodi*	<100	Pelagic	Oceanic	1
	Scymnodon	*ichiharai*	<100	Benthopelagic	Bathyal	2
	Scymnodon	*ringens*	<100	Benthopelagic	Bathyal	3
	Somniosus	*antarcticus*	>300	Benthopelagic	Bathyal	5
	Somniosus	*microcephalus*	>300	Benthopelagic	Coastal/bathyal	5
	Somniosus	*pacificus*	>300	Benthopelagic	Coastal/bathyal	3
	Somniosus	*rostratus*	100–150	Benthopelagic	Bathyal	4
	Zameus	*ichiharai*	<100	Benthopelagic	Bathyal	7
	Zameus	*squamulosus*	<100	Benthopelagic	Coastal/bathyal	2
	Oxynotus	*bruniensis*	<100	Benthic	Bathyal	1
Oxynotidae	*Oxynotus*	*caribbaeus*	<100	Benthic	Bathyal	5
	Oxynotus	*centrina*	100–150	Benthic	Coastal/bathyal	1
	Oxynotus	*japonicus*	<100	Benthic	Bathyal	2
	Oxynotus	*paradoxus*	100–150	Benthic	Bathyal	11
Dalatiidae	*Dalatias*	*licha*	150–300	Benthopelagic	Bathyal	2
	Euprotomicroides	*zantedeschia*	<100	Pelagic	Oceanic	9
	Euprotomicrus	*bispinatus*	<100	Pelagic	Oceanic	3
	Heteroscymnoides	*marleyi*	<100	Pelagic	Oceanic	10
	Isistius	*brasiliensis*	<100	Pelagic	Oceanic	3
	Isistius	*plutodus*	<100	Pelagic	Oceanic	1
	Mollisquama	*parini*	<100	Pelagic	Oceanic	1

Order	Family	Genus	Species	Body Size (cm): <100, 100–150, 150–300, >300	Benthic, Benthopelagic, Pelagic	Coastal, Oceanic, Bathyal	Number of FAO Regions
Pristiophoriformes	Pristiophoridae	Squaliolus	aliae	<100	Pelagic	Oceanic	1
		Squaliolus	laticaudus	<100	Pelagic	Oceanic	7
		Pliotrema	warreni	100–150	Benthic	Coastal/bathyal	2
		Pristiophorus	sp. A	100–150	Benthopelagic	Coastal	1
		Pristiophorus	sp. B	<100	Benthopelagic	Coastal	1
		Pristiophorus	cirratus	100–150	Benthopelagic	Coastal/bathyal	2
		Pristiophorus	japonicus	100–150	Benthopelagic	Coastal	1
		Pristiophorus	nudipinnis	100–150	Benthopelagic	Coastal	2
		Pristiophorus	schroederi	<100	Benthopelagic	Coastal/bathyal	1
Squatiniformes	Squatinidae	Squatina	sp. A	150–300	Benthic	Coastal/bathyal	2
		Squatina	sp. B	150–300	Benthic	Coastal/bathyal	2
		Squatina	aculeata	150–300	Benthic	Coastal/bathyal	3
		Squatina	africana	150–300	Benthic	Coastal/bathyal	1
		Squatina	argentina	150–300	Benthic	Coastal/bathyal	1
		Squatina	armata	100–150	Benthic	Coastal	2
		Squatina	australis	150–300	Benthic	Coastal/bathyal	2
		Squatina	californica	100–150	Benthic	Coastal	3
		Squatina	dumeril	100–150	Benthic	Coastal	2
		Squatina	formosa	<100	Benthic	Coastal/bathyal	2
		Squatina	guggenheim	<100	Benthic	Coastal	1
		Squatina	japonica	150–300	Benthic	Coastal	1
		Squatina	nebulosa	150–300	Benthic	Coastal	1
		Squatina	occulta	100–150	Benthic	Coastal	1
		Squatina	oculata	150–300	Benthic	Coastal/bathyal	4
		Squatina	squatina	150–300	Benthic	Coastal	3
		Squatina	tergocellata	<100	Benthic	Coastal/bathyal	2
		Squatina	tergocellatoides	<100	Benthic	Coastal/bathyal	2
Heterodontiformes	Heterodontidae	Heterodontus	francisci	100–150	Benthic	Coastal	2
		Heterodontus	galeatus	100–150	Benthic	Coastal	3
		Heterodontus	japonicus	100–150	Benthic	Coastal	1
		Heterodontus	mexicanus	<100	Benthic	Coastal	1
		Heterodontus	portujacksoni	100–150	Benthic	Coastal	3
		Heterodontus	quoyi	<100	Benthic	Coastal	1
		Heterodontus	ramalheira	<100	Benthic	Coastal	1
		Heterodontus	zebra	100–150	Benthic	Coastal	3
Orectolobiformes	Parascylliidae	Cirrhoscyllium	expolitum	150–300	Benthic	Coastal/bathyal	2
		Cirrhoscyllium	formosanum	<100	Benthic	Coastal/bathyal	1

Order	Family	Genus	species				
		Cirrhoscyllium	*japonicum*	<100	Benthic	Bathyal	1
		Parascyllium	sp. A	<100	Benthic	Coastal	1
		Parascyllium	*collare*	<100	Benthic	Coastal	3
		Parascyllium	*ferrugineum*	<100	Benthic	Coastal	2
		Parascyllium	*variolatum*	<100	Benthic	Coastal	2
	Brachaeluridae	*Brachaelurus*	*waddi*	<100	Benthic	coastal	3
		Heteroscyllium	*colcloughi*	<100	Benthic	Coastal	1
	Orectolobidae	*Eucrossorhinus*	*dasypogon*	150–300	Benthic	Coastal/bathyal	2
		Orectolobus	sp. A	150–300	Benthic	Coastal	1
		Orectolobus	*japonicus*	100–150	Benthic	Coastal	2
		Orectolobus	*maculatus*	150–300	Benthic	Coastal/bathyal	4
		Orectolobus	*ornatus*	150–300	Benthic	Coastal	4
		Orectolobus	*wardi*	<100	Benthic	Coastal	2
		Sutorectus	*tentaculatus*	<100	Benthic	Coastal	1
	Hemiscylliidae	*Chiloscyllium*	*arabicum*	<100	Benthic	Coastal	1
		Chiloscyllium	*burmensis*	<100	Benthic	Coastal	1
		Chiloscyllium	*griseum*	<100	Benthic	Coastal	4
		Chiloscyllium	*hasselti*	<100	Benthic	Coastal	3
		Chiloscyllium	*indicum*	<100	Benthic	Coastal	4
		Chiloscyllium	*plagiosum*	<100	Benthic	Coastal	4
		Chiloscyllium	*punctatum*	<100	Benthic	Coastal	4
		Hemiscyllium	*freycineti*	<100	Benthic	Coastal	1
		Hemiscyllium	*hallstromi*	<100	Benthic	Coastal	1
		Hemiscyllium	*ocellatum*	<100	Benthic	Coastal	3
		Hemiscyllium	*strahani*	<100	Benthic	Coastal	1
		Hemiscyllium	*trispeculare*	<100	Benthic	Coastal	2
	Ginglymostomatidae	*Ginglymostoma*	*cirratum*	150–300	Benthic	Coastal	8
		Nebrius	*ferrugineus*	150–300	Benthic	Coastal	5
		Pseudoginglymostoma	*brevicaudatum*	<100	Benthic	Coastal	1
	Stegostomatidae	*Stegostoma*	*fasciatum*	150–300	Benthic	Coastal	6
Lamniformes	Rhincodontidae	*Rhincodon*	*typus*	>300	Pelagic	Oceanic	13
	Mitsukurinidae	*Mitsukurina*	*owstoni*	>300	Benthopelagic	Bathyal	9
		Carcharias	*taurus*	>300	Benthopelagic	Coastal	12
	Odontaspididae	*Odontaspis*	*ferox*	>300	Benthopelagic	Bathyal	13
		Odontaspis	*noronhai*	>300	Benthopelagic	Bathyal	7
	Pseudocarchariidae	*Pseudocarcharias*	*kamoharai*	100–150	Pelagic	Oceanic	11
	Megachasmidae	*Megachasma*	*pelagios*	>300	Pelagic	Oceanic	6
	Alopiidae	*Alopias*	*pelagicus*	>300	Pelagic	Coastal/oceanic	8
		Alopias	*superciliosus*	>300	Pelagic	Coastal/oceanic	14
		Alopias	*vulpinus*	>300	Pelagic	Coastal/oceanic	16

Order	Family	Genus	Species	Body Size (cm): <100, 100–150, 150–300, >300	Benthic, Benthopelagic, Pelagic	Coastal, Oceanic, Bathyal	Number of FAO Regions
	Cetorhinidae	*Cetorhinus*	*maximus*	>300	Pelagic	Coastal/oceanic	15
	Lamnidae	*Carcharodon*	*carcharias*	>300	Benthopelagic	Coastal/oceanic	16
		Isurus	*oxyrinchus*	>300	Pelagic	Coastal/oceanic	15
		Isurus	*paucus*	>300	Pelagic	Coastal/oceanic	13
		Lamna	*ditropis*	150–300	Pelagic	Coastal/oceanic	4
		Lamna	*nasus*	>300	Pelagic	Coastal/oceanic	15
Carcharhiniformes	Scyliorhinidae	*Apristurus*	sp. A	<100	Benthic	Bathyal	2
		Apristurus	sp. B	<100	Benthic	Bathyal	2
		Apristurus	sp. C	<100	Benthic	Bathyal	2
		Apristurus	sp. D	<100	Benthic	Bathyal	2
		Apristurus	sp. E	<100	Benthic	Bathyal	2
		Apristurus	sp. F	<100	Benthic	Bathyal	1
		Apristurus	sp. G	<100	Benthic	Bathyal	2
		Apristurus	*acanutus*	<100	Benthic	Bathyal	1
		Apristurus	*aphyodes*	<100	Benthic	Bathyal	1
		Apristurus	*atlanticus*	<100	Benthic	Bathyal	1
		Apristurus	*brunneus*	<100	Benthic	Bathyal	3
		Apristurus	*canutus*	<100	Benthic	Bathyal	1
		Apristurus	*fedorvi*	<100	Benthic	Bathyal	1
		Apristurus	*gibbosus*	<100	Benthic	Bathyal	1
		Apristurus	*herklotsi*	<100	Benthic	Bathyal	1
		Apristurus	*indicus*	<100	Benthic	Bathyal	1
		Apristurus	*investigatoris*	<100	Benthic	Bathyal	1
		Apristurus	*japonicus*	<100	Benthic	Bathyal	1
		Apristurus	*kampae*	<100	Benthic	Bathyal	1
		Apristurus	*laurussoni*	<100	Benthic	Bathyal	4
		Apristurus	*longicephalus*	<100	Benthic	Bathyal	1
		Apristurus	*macrorhynchus*	<100	Benthic	Bathyal	1
		Apristurus	*macrostomus*	<100	Benthic	Bathyal	1
		Apristurus	*manis*	<100	Benthic	Bathyal	2
		Apristurus	*microps*	<100	Benthic	Bathyal	5
		Apristurus	*micropterygeus*	<100	Benthic	Bathyal	1
		Apristurus	*nasutus*	<100	Benthic	Bathyal	2
		Apristurus	*parvipinnis*	<100	Benthic	Bathyal	1
		Apristurus	*pinguis*	<100	Benthic	Bathyal	1
		Apristurus	*platyrhynchus*	<100	Benthic	Bathyal	1
		Apristurus	*profundorum*	<100	Benthic	Bathyal	2

Apristurus	*riveri*	<100	Benthic	Bathyal	1
Apristurus	*saldanha*	<100	Benthic	Bathyal	2
Apristurus	*sibogae*	<100	Benthic	Bathyal	1
Apristurus	*sinensis*	<100	Benthic	Bathyal	1
Apristurus	*spongiceps*	<100	Benthic	Bathyal	2
Apristurus	*stenseni*	<100	Benthic	Bathyal	1
Apristurus	*verweyi*	<100	Benthic	Bathyal	1
Apristurus	*sp. A*	<100	Benthic	Coastal/bathyal	1
Apristurus	*sp. B*	<100	Benthic	Bathyal	1
Apristurus	*sp. C*	<100	Benthic	Coastal/bathyal	1
Apristurus	*sp. D*	<100	Benthic	Coastal/bathyal	2
Apristurus	*sp. E*	<100	Benthic	Bathyal	1
Apristurus	*sp. F*	<100	Benthic	Coastal/bathyal	1
Asymbolus	*analis*	<100	Benthic	Coastal/bathyal	2
Asymbolus	*funebris*	<100	Benthic	Coastal	1
Asymbolus	*occiduus*	<100	Benthic	Coastal/bathyal	1
Asymbolus	*pallidus*	<100	Benthic	Bathyal	1
Asymbolus	*parvus*	<100	Benthic	Coastal/bathyal	1
Asymbolus	*rubiginosus*	<100	Benthic	Coastal/bathyal	3
Asymbolus	*submaculatus*	<100	Benthic	Coastal	1
Asymbolus	*vincenti*	<100	Benthic	Coastal/bathyal	2
Atelomycterus	*sp. A*	<100	Benthic	Coastal	2
Atelomycterus	*fasciatus*	<100	Benthic	Coastal	1
Atelomycterus	*macleayi*	<100	Benthic	Coastal	2
Atelomycterus	*marmoratus*	<100	Benthic	Coastal	4
Aulohalaelurus	*kanakorum*	<100	Benthic	Coastal	1
Aulohalaelurus	*labiosus*	<100	Benthic	Coastal	1
Bythaelurus	*alcocki*	<100	Benthic	Coastal/bathyal	1
Bythaelurus	*canescens*	<100	Benthic	Coastal/bathyal	1
Bythaelurus	*clevai*	<100	Benthic	Bathyal	1
Bythaelurus	*dawsoni*	<100	Benthic	Coastal	1
Bythaelurus	*hispidus*	<100	Benthic	Coastal/bathyal	2
Bythaelurus	*immaculatus*	<100	Benthic	Coastal/bathyal	1
Bythaelurus	*lutarius*	<100	Benthic	Coastal/bathyal	1
Cephaloscyllium	*sp. A*	<100	Benthic	Bathyal	2
Cephaloscyllium	*sp. B*	<100	Benthic	Bathyal	1
Cephaloscyllium	*sp. C*	<100	Benthic	Coastal/bathyal	1
Cephaloscyllium	*sp. D*	<100	Benthic	Bathyal	1
Cephaloscyllium	*sp. E*	<100	Benthic	Bathyal	2
Cephaloscyllium	*fasciatum*	<100	Benthic	Coastal/bathyal	3

Order	Family	Genus	Species	Body Size (cm): <100, 100–150, 150–300, >300	Benthic, Benthopelagic, Pelagic	Coastal, Oceanic, Bathyal	Number of FAO Regions
		Cephaloscyllium	*isabellum*	<100	Benthic	Coastal/bathyal	2
		Cephaloscyllium	*laticeps*	<100	Benthic	Coastal/bathyal	2
		Cephaloscyllium	*silasi*	<100	Benthic	Coastal/bathyal	1
		Cephaloscyllium	*sufflans*	<100	Benthic	Coastal/bathyal	1
		Cephaloscyllium	*umbratile*	100–150	Benthic	Coastal	1
		Cephaloscyllium	*ventriosum*	<100	Benthic	Coastal/bathyal	2
		Cephalurus	*cephalus*	<100	Benthic	Coastal/bathyal	1
		Galeus	sp. A	<100	Benthic	Coastal/bathyal	2
		Galeus	sp. B	<100	Benthic	Bathyal	1
		Galeus	*antillensis*	<100	Benthic	Bathyal	1
		Galeus	*arae*	<100	Benthic	Coastal/bathyal	1
		Galeus	*atlanticus*	<100	Benthic	Coastal/bathyal	1
		Galeus	*boardmani*	<100	Benthic	Coastal/bathyal	2
		Galeus	*cadenati*	<100	Benthic	Bathyal?	1
		Galeus	*eastmani*	<100	Benthic	Coastal/bathyal	2
		Galeus	*gracilis*	<100	Benthic	Coastal/bathyal	1
		Galeus	*longirostris*	<100	Benthic	Bathyal	1
		Galeus	*melastomus*	<100	Benthic	Coastal/bathyal	3
		Galeus	*murinus*	<100	Benthic	Coastal/bathyal	1
		Galeus	*nipponensis*	<100	Benthic	Coastal/bathyal	1
		Galeus	*piperatus*	<100	Benthic	Coastal/bathyal	1
		Galeus	*polli*	<100	Benthic	Coastal/bathyal	2
		Galeus	*sauteri*	<100	Benthic	Coastal/bathyal	2
		Galeus	*schultzi*	<100	Benthic	Coastal/bathyal	1
		Galeus	*spingeri*	<100	Benthic	Coastal/bathyal	1
		Halaelurus	sp. A	<100	Benthic	Bathyal	1
		Halaelurus	*boesemani*	<100	Benthic	Coastal	4
		Halaelurus	*buergeri*	<100	Benthic	Coastal	1
		Halaelurus	*lineatus*	<100	Benthic	Coastal/bathyal	1
		Halaelurus	*natalensis*	<100	Benthic	Coastal/bathyal	2
		Halaelurus	*quagga*	<100	Benthic	Coastal/bathyal	1
		Haploblepharus	*edwardsii*	<100	Benthic	Coastal/bathyal	2
		Haploblepharus	*fuscus*	<100	Benthic	Coastal	2
		Haploblepharus	*pictus*	<100	Benthic	Coastal	1
		Holohalaelurus	*punctatus*	<100	Benthic	Coastal/bathyal	2
		Holohalaelurus	*regani*	<100	Benthic	Coastal/bathyal	3
		Paramaturus	sp. A	<100	Benthic	Bathyal	1

Family	Genus	species	Depth			No.
	Paramaturus	campechiensis	<100	Benthic	Coastal/bathyal	1
	Paramaturus	macmillani	<100	Benthic	Bathyal	2
	Paramaturus	melanobranchius	<100	Benthic	Coastal/bathyal	1
	Paramaturus	pilosus	<100	Benthic	Coastal/bathyal	1
	Paramaturus	xaniurus	<100	Benthic	Coastal/bathyal	1
	Pentanchus	profundicolus	<100	Benthic	Coastal	2
	Poroderma	africanum	<100	Benthic	Coastal	2
	Poroderma	pantherinum	<100	Benthic	Coastal/bathyal	2
	Schroederichthys	bivius	<100	Benthic	Coastal/bathyal	2
	Schroederichthys	chilensis	<100	Benthic	Coastal/bathyal	1
	Schroederichthys	maculatus	<100	Benthic	Coastal/bathyal	1
	Schroederichthys	tenuis	<100	Benthic	Coastal/bathyal	2
	Scyliorhinus	besnardi	<100	Benthic	Coastal/bathyal	1
	Scyliorhinus	boa	<100	Benthic	Coastal/bathyal	1
	Scyliorhinus	canicula	<100	Benthic	Coastal/bathyal	3
	Scyliorhinus	capensis	<100	Benthic	Coastal/bathyal	2
	Scyliorhinus	cervigoni	<100	Benthic	Coastal/bathyal	2
	Scyliorhinus	comoroensis	<100	Benthic	Bathyal	1
	Scyliorhinus	garmani	<100	Benthic	Bathyal	2
	Scyliorhinus	haeckelii	<100	Benthic	Coastal/bathyal	2
	Scyliorhinus	hesperius	<100	Benthic	Coastal/bathyal	1
	Scyliorhinus	meadi	<100	Benthic	Coastal/bathyal	1
	Scyliorhinus	retifer	<100	Benthic	Coastal/bathyal	2
	Scyliorhinus	stellaris	100–150	Benthic	Coastal/bathyal	3
	Scyliorhinus	tokubee	<100	Benthic	Coastal	1
	Scyliorhinus	torazame	<100	Benthic	Coastal	2
	Scyliorhinus	torrei	<100	Benthic	Coastal/bathyal	1
Proscylliidae	Ctenacis	fehlmanni	<100	Benthic	Coastal/bathyal	1
	Eridacnis	barbouri	<100	Benthic	Coastal/bathyal	1
	Eridacnis	radcliffei	<100	Benthic	Coastal/bathyal	3
	Eridacnis	sinuans	<100	Benthic	Coastal/bathyal	1
	Proscyllium	habereri	<100	Benthic	Coastal	2
Pseudotriakidae	Gollum	attenuatus	<100	Benthic	Coastal/bathyal	2
	Pseudotriakis	microdon	150–300	Benthopelagic	Bathyal	8
Leptochariidae	Leptocharias	smithii	<100	Benthic	Coastal	2
Triakidae	Furgaleus	macki	100–150	Benthopelagic	Coastal	1
	Galeorhinus	galeus	150–300	Benthopelagic	Coastal/bathyal	11
	Gogolia	filewoodi	<100	Benthopelagic	Coastal	1
	Hemitriakis	sp. A	<100	Benthopelagic	Coastal/bathyal	1
	Hemitriakis	sp. B	<100	Benthopelagic	Bathyal	1

Order	Family	Genus	Species	Body Size (cm): <100, 100–150, 150–300, >300	Benthic, Benthopelagic, Pelagic	Coastal, Oceanic, Bathyal	Number of FAO Regions
		Hemitriakis	*abdita*	<100	Benthopelagic	Bathyal	1
		Hemitriakis	*falcata*	<100	Benthopelagic	Coastal	1
		Hemitriakis	*japonica*	100–150	Benthopelagic	Coastal/bathyal	1
		Hemitriakis	*leucoperiptera*	<100	Benthopelagic	Coastal	1
		Hypogaleus	*hyugaensis*	100–150	Benthopelagic	Coastal/bathyal	4
		Iago	*garricki*	100–150	Benthopelagic	Coastal/bathyal	1
		Iago	*omanensis*	<100	Benthopelagic	Coastal/bathyal	1
		Mustelus	sp. A	<100	Benthopelagic	Bathyal	2
		Mustelus	sp. B	100–150	Benthopelagic	Bathyal	2
		Mustelus	*antarcticus*	100–150	Benthopelagic	Coastal/bathyal	3
		Mustelus	*asterias*	100–150	Benthopelagic	Coastal/bathyal	3
		Mustelus	*californicus*	100–150	Benthopelagic	Coastal/bathyal	1
		Mustelus	*canis*	100–150	Benthopelagic	Coastal/bathyal	3
		Mustelus	*dorsalis*	<100	Benthopelagic	Coastal	2
		Mustelus	*fasciatus*	<100	Benthopelagic	Coastal	1
		Mustelus	*griseus*	<100	Benthopelagic	Coastal	2
		Mustelus	*henlei*	<100	Benthopelagic	Coastal/bathyal	2
		Mustelus	*higmani*	<100	Benthopelagic	Coastal/bathyal	2
		Mustelus	*lenticulatus*	100–150	Benthopelagic	Coastal/bathyal	1
		Mustelus	*lunulatus*	100–150	Benthopelagic	Coastal/bathyal	1
		Mustelus	*manazo*	100–150	Benthopelagic	Coastal	3
		Mustelus	*mento*	100–150	Benthopelagic	Coastal/bathyal	2
		Mustelus	*minicanis*	<100	Benthopelagic	Coastal	1
		Mustelus	*mosis*	100–150	Benthopelagic	Coastal/bathyal	2
		Mustelus	*mustelus*	100–150	Benthopelagic	Coastal/bathyal	5
		Mustelus	*norrisi*	<100	Benthopelagic	Coastal	2
		Mustelus	*palumbes*	100–150	Benthopelagic	Coastal/bathyal	2
		Mustelus	*punctulatus*	<100	Benthopelagic	Coastal	2
		Mustelus	*schmitti*	<100	Benthopelagic	Coastal	1
		Mustelus	*sinusmexicanus*	100–150	Benthopelagic	Coastal/bathyal	1
		Mustelus	*whitney*	<100	Benthopelagic	Coastal/bathyal	1
		Scylliogaleus	*quecketti*	<100	Benthopelagic	Coastal	2
		Triakis	*acutipinna*	<100	Benthopelagic	Coastal	1
		Triakis	*maculata*	150–300	Benthopelagic	Coastal	1
		Triakis	*megalopterus*	150–300	Benthopelagic	Coastal	2
		Triakis	*scyllium*	100–150	Benthopelagic	Coastal	2
		Triakis	*semifasciata*	150–300	Benthopelagic	Coastal	1

Family	Genus	species				
Hemigaleidae	Chaenogaleus	macrostoma	100–150	Benthopelagic	Coastal	4
	Hemigaleus	microstoma	100–150	Benthopelagic	Coastal	4
	Hemipristis	elongatus	150–300	Benthopelagic	Coastal	5
	Paragaleus	leucolomatus	<100	Benthopelagic	Coastal	1
	Paragaleus	pectoralis	100–150	Benthopelagic	Coastal	2
	Paragaleus	randalli	?	Benthopelagic	Coastal	2
	Paragaleus	tengi	<100	Benthopelagic	Coastal	2
Carcharhinidae	Carcharhinus	acronotus	100–150	Benthopelagic	Coastal	2
	Carcharhinus	albimarginatus	150–300	Benthopelagic	Coastal/oceanic	6
	Carcharhinus	altimus	150–300	Benthopelagic	Coastal/oceanic	8
	Carcharhinus	amblyrhychoides	150–300	Benthopelagic	Coastal/oceanic	4
	Carcharhinus	amblyrhynchos	150–300	Benthopelagic	Coastal/oceanic	6
	Carcharhinus	amboinensis	150–300	Benthopelagic	Coastal	5
	Carcharhinus	borneensis	<100	Benthopelagic	Coastal	3
	Carcharhinus	brachyurus	150–300	Benthopelagic	Coastal/oceanic	12
	Carcharhinus	brevipinna	150–300	Benthopelagic	Coastal	13
	Carcharhinus	cautus	100–150	Benthopelagic	Coastal	2
	Carcharhinus	dussumieri	100–150	Benthopelagic	Coastal	4
	Carcharhinus	falciformis	>300	Pelagic	Coastal/oceanic	13
	Carcharhinus	fitzroyensis	100–150	Benthopelagic	Coastal/oceanic	2
	Carcharhinus	galapagensis	>300	Benthopelagic	Coastal/oceanic	9
	Carcharhinus	hemiodon	100–150	Benthopelagic	Coastal/oceanic	4
	Carcharhinus	isodon	100–150	Benthopelagic	Coastal	3
	Carcharhinus	leiodon	<100	Benthopelagic	Coastal	1
	Carcharhinus	leucas	>300	Benthopelagic	Coastal	13
	Carcharhinus	limbatus	150–300	Benthopelagic	Coastal	13
	Carcharhinus	longimanus	>300	Benthopelagic	Oceanic	14
	Carcharhinus	macloti	<100	Pelagic	Coastal	4
	Carcharhinus	melanopterus	150–300	Benthopelagic	Coastal/oceanic	7
	Carcharhinus	obscurus	>300	Benthopelagic	Coastal/oceanic	14
	Carcharhinus	perezi	150–300	Benthopelagic	Coastal	2
	Carcharhinus	plumbeus	150–300	Benthopelagic	Coastal	14
	Carcharhinus	porosus	100–150	Benthopelagic	Coastal	4
	Carcharhinus	sealei	<100	Benthopelagic	Coastal	4
	Carcharhinus	signatus	150–300	Benthopelagic	Coastal/oceanic	6
	Carcharhinus	sorrah	150–300	Benthopelagic	Coastal	5
	Carcharhinus	tilsoni	150–300	Pelagic	Coastal	1
	Carcharhinus	wheeleri	150–300	Benthopelagic	Coastal	1
	Galeocerdo	cuvier	>300	Benthopelagic	Coastal/oceanic	13
	Glyphis	sp. A	150–300	Benthopelagic	Coastal	2

Order	Family	Genus	Species	Body Size (cm): <100, 100–150, 150–300, >300	Benthic, Benthopelagic, Pelagic	Coastal, Oceanic, Bathyal	Number of FAO Regions
		Glyphis	*gangeticus*	150–300	Benthopelagic	Coastal	3
		Glyphis	*glyphis*	150–300	Benthopelagic	Coastal	3
		Glyphis	*siamensis*	?	Benthopelagic	Coastal	1
		Isogomphodon	*oxyrhynchus*	100–150	Benthopelagic	Coastal	2
		Lamiopsis	*temmincki*	150–300	Benthopelagic	Coastal	4
		Loxodon	*macrorhinus*	<100	Benthopelagic	Coastal	4
		Nasolamia	*velox*	100–150	Benthopelagic	Coastal	2
		Negaprion	*acutidens*	150–300	Benthopelagic	Coastal	4
		Negaprion	*brevirostris*	150–300	Benthopelagic	Coastal	6
		Prionace	*glauca*	>300	Pelagic	Oceanic	15
		Rhizoprionodon	*acutus*	100–150	Benthopelagic	Coastal	7
		Rhizoprionodon	*lalandei*	<100	Benthopelagic	Coastal	2
		Rhizoprionodon	*longurio*	100–150	Benthopelagic	Coastal	2
		Rhizoprionodon	*oligolinx*	<100	Benthopelagic	Coastal	4
		Rhizoprionodon	*porosus*	<100	Benthopelagic	Coastal	2
		Rhizoprionodon	*taylori*	<100	Benthopelagic	Coastal	2
		Rhizoprionodon	*terraenovae*	<100	Benthopelagic	Coastal	2
		Scoliodon	*laticaudus*	<100	Benthopelagic	Coastal	4
		Triaenodon	*obesus*	100–150	Benthic	Coastal	6
	Sphyrnidae	*Eusphyra*	*blochii*	100–150	Benthopelagic	Coastal	4
		Sphyrna	*corona*	<100	Benthopelagic	Coastal	2
		Sphyrna	*lewini*	>300	Benthopelagic	Coastal/oceanic	14
		Sphyrna	*media*	100–150	Benthopelagic	Coastal	4
		Sphyrna	*mokarran*	>300	Benthopelagic	Coastal/oceanic	12
		Sphyrna	*tiburo*	100–150	Benthopelagic	Coastal	5
		Sphyrna	*tudes*	100–150	Benthopelagic	Coastal	2
		Sphyrna	*zygaena*	>300	Benthopelagic	Coastal/oceanic	16

References

Anderson, M.E., G.M. Cailliet, and B.S. Antrim. 1979. Notes on some uncommon deep-sea fishes from the Monterey Bay area. *Calif. Fish Game* 64(4):256–264.

Applegate, S.P. 1967. A survey of shark hard parts, in *Sharks, Skates and Rays*. P.W. Gilbert, R.F. Mathewson, and D.P. Ralls, Eds., Johns Hopkins University Press, Baltimore, 33–67.

Applegate, S.P. 1986. The El Cien Formation, Strata of Oligocene, and Early Miocene Age in Baja California Sur. *Univ. Nac. Auton. Mex. Inst. Geol. Rev.* 6:145–162.

Avise, J.C. 2000. *Phylogeography: The History and Formation of Species*. Harvard University Press, Cambridge, MA, 447 pp.

Backus, R.H. and J.E. Craddock. 1977. Pelagic faunal provinces and sound-scattering levels in the Atlantic Ocean, in *Ocean's Sound-Scattering Prediction*. N.R. Anderson and B.J. Zahuranes, Eds., Plenum Press, New York, 529–548.

Banford, H.M. 1998. Biogeography of Amphi-Atlantic and Amphi-American Fishes: The *Scomberomorus regalis* (Scombridae), *Strongylura marina* (Belonidae) and *Hyporhamphus unifasciatus* (Hemiramphidae) Species Groups. Ph.D. dissertation. School of Marine Science, College of William and Mary, Williamsburg, VA, 103 pp.

Bigelow, H.B. and W.C. Schroeder. 1948. Sharks, in *Fishes of the Western North Atlantic*. Sears Foundation for Marine Research. Yale University, New Haven, CT, 56–576.

Bigelow, H.B. and W.C. Schroeder. 1957. A study of the sharks of the suborder Squaloidea. *Bull. Mus. Comp. Zool. Harvard Univ.* 117(1):1–150 + 4 pl.

Bowen, B.W., A.L. Bass, L.A. Rocha, W.S. Grant, and D.R. Robertson. 2001. Phylogeography of the trumpetfishes (*Aulostomus*): ring species complex on a global scale. *Evolution* 55:1029–1039.

Bremer, K., 1992. Ancestral areas: a cladistic reinterpretation of the center of origin concept. *Syst. Biol.* 41:436–445.

Briggs, J.C., 1974. *Marine Zoogeography*. McGraw-Hill, New York, 475 pp.

Briggs, J.C., 1995. *Global Biogeography. Developments in Paleontology and Stratigraphy*. Elsevier, New York, 452 pp.

Brown, J.H. and M.V. Lomolino. 1998. *Biogeography*, 2nd ed. Sinauer Associates, Sunderland, MA, 691 pp.

Cappetta, H., 1987. *Chondrichthyes II, Mesozoic and Cenozoic Elasmobranchii*. Vol. 3B in *Handbook of Paleoichthyology*. H.P. Schultze, Ed., Verlag Dr. Friedrich Pfeil, Munich, Germany, 193 pp.

Cappetta, H., C. Duffin, and J. Zidela. 1993. Chondrichthyes, in *The Fossil Record*, Vol. 2. M.J. Benton, Ed., Chapman & Hall, London, 593–609.

Case, G.R. 1972. *A Pictorial Guide to Fossils*. Krieger, Melbourne, FL, 515 pp.

Castro, J.I. 1983. *The Sharks of North American Waters*. Texas A&M Press, College Station, 180 pp.

Castro, J.I., G.J.P. Naylor, and J.F. Marquez-Farias. 2003. A new species of nurse shark from the Pacific Ocean. *Proc. Annual Meeting of the American Elasmobranch Society*, Manaus, Brazil (abstr.).

Cherel, Y. and G. Duhamel. 2004. Antarctic jaws: cephalopod prey of sharks in Kerguelen waters. *Deep Sea Res.* 1(51):17–31.

Cioni, A.L. 1996. The extinct genus of *Notidanodon* (Neoselachii, Hexanchiformes), in *Mesozoic Fishes — Systematics and Paleoecology*. Arratia, G. and G. Viohl, Eds., Proceedings of the International Meeting Eichsttt, 1993. Verlag Dr. Friedrich Pfeil, Munich, Germany, 63–72.

Colvocoresses, J.A. and J.A. Musick. 1984. Species associations and community composition of Middle Atlantic Bight continental shelf demersal fishes. *Fish. Bull.* 82 (2):295–313.

Compagno, L.J.V. 1984a. *FAO Species Catalogue*. Vol. 4. *Sharks of the World. An Annotated and Illustrated Catalogue of Sharks Species Known to Date*. Part 1. *Hexanchiformes to Lamniformes*. FAO Fish. Synop., 125, FAO, Rome, 249 pp.

Compagno, L.J.V. 1984b. *FAO Species Catalogue*. Vol. 4. *Sharks of the World. An Annotated and Illustrated Catalogue of Sharks Species Known to Date*. Part 2. *Carcharhiniformes*. FAO Fish. Synop., 125, FAO, Rome, 251–655.

Compagno, L.J.V. 1988. *Sharks of the Order Carcharhiniformes*. Princeton University Press, Princeton, NJ.

Compagno, L.J.V. 1990. Alternate life history styles of cartilaginous fishes in time and space. *Environ. Biol. Fish.* 28(1–4):33–75.

Compagno, L.J.V. 1999. Systematics and body form, in *Sharks, Skates and Rays, The Biology of Elasmobranch Fishes*. W.C. Hamlett, Ed., Johns Hopkins University Press, Baltimore, 1–42.

Compagno, L.J.V. 2001. *Sharks of the World. An Annotated and Illustrated Catalogue of Shark Species Known to Date*. Vol. 2. *Bullhead, Mackerel and Carpet Sharks (Heterodontiformes, Lamniformes and Orectolobiformes)*. *FAO Species Catalogue for Fishery Purposes* 1, FAO, Rome, 269 pp.

Compagno, L.J.V. 2002. Review of the biodiversity of sharks and chimaeras in the South China Sea and adjacent areas, in *Elasmobranch Biodiversity, Conservation, and Management. Proceedings of the International Seminar and Workshop,* Sabah, Malaysia, July, 1997. S.L. Fowler, T.M. Reed, and A. Dopper, Eds., Occas. Paper IUCN Spec. Surv. Comm. No. 25, 52–63.

Compagno, L.J.V. In press. *Sharks of the World*. Vols. 1 and 3. *FAO Species Catalogue for Fishery Purposes*, 1, 2nd ed., FAO, Rome.

Cortés, E. 1999. Standardized diet composition and trophic levels of sharks. *ICES J. Mar. Sci.* 56:707–717.

Craw, R.C. 1988. Panbiogeography: method and synthesis in biogeography, in *Analytical Biogeography, an Integrated Approach to the Study of Animal and Plant Distributions*. A.A. Myers and P.S. Giller, Eds., Kluwer Academic Publishers, Dordrecht, 584 pp.

Craw, R.C., J.R. Grehan, and M.J. Heads. 1999. *Panbiogeography: Tracking the History of Life*. Oxford University Press, New York, 229 pp.

de Carvalho, M.R. 1996. Higher-level elasmobranch phylogeny, basal squaleans and paraphyly, in *Interrelationships of Fishes*. M.L.J. Stiassny, L.R. Parenti, and G.D. Johnson, Eds., Atlantic Press, New York, 35–62.

Douady, C.J., M. Dosay, M.S. Shivji, and M.J. Stanhope. 2003. Molecular phylogenetic evidence refuting the hypothesis of Batoidea (rays and skates) as derived sharks. *Mol. Phyl. Evol.* 26:215–221.

Duffin, C.J. 1988. The Upper Jurassic Selachian *Paleocarcharias* de Beaumont (1960). *Zool. J. Linn. Soc.* 94: 271–286.

Ebeling, A.W. 1967. Zoogeography of tropical deep-sea animals, in *Proceedings of the International Conference of Tropical Oceanography,* 1965. University of Miami, Institute of Marine Science, 593–613.

Eckert, S.A. and B.S. Stewart. 2001. Telemetry and satellite tracking of whale sharks, *Rhincodon typus* in the Sea of Cortez, Mexico and the North Pacific Ocean, in *The Behavior and Sensory Biology of Elasmobranch Fishes: An Anthology in Memory of Donald Richard Nelson*. T.C. Tricas and S.H. Gruber, Eds., Kluwer Academic, Dordrecht, 308 pp.

Ekman, S. 1953. *Zoogeography of the Sea*. Sidgwick & Jackson, London, 417 pp.

Fitch, J.E. 1964. The fish fauna of the Playa Del Rey locality, a Southern California marine Pleistocene deposit. *Los Angeles County Mus. Contrib. Sci.* 82:1–35.

Francis, M.P., J.D. Stevens, and P. Last. 1988. New records of *Somniosus* (Elasmobranchii: Squalidae) from Australia with comments on the taxonomy of the genus. *N.Z.. J. Mar. Freshwater Res.* 22:401–409.

Gage, J.D. and P.A. Tyler. 1991. *Deep Sea Biology*. Cambridge University Press, Cambridge, U.K.

Garrick, J.A.F. 1982. Sharks of the genus *Carcharhinus*. U.S. Department of Commerce, NOAA/NMFS Circ. 445, 194 pp.

Gibbs, R.H., R.H. Goodyear, M.J. Keene, and D.W. Brown. 1971. Biological Studies of the Bermuda Ocean Acre. II. Vertical Distribution and Ecology of the Lanternfishes (Family Myctophidae). Report to the U.S. Navy Underwater Systems Center, Smithsonian Institution, Washington, D.C.

Goto, M., T. Uyeno, and Y. Yabumoto. 1996. Summary of Mesozoic elasmobranch remains from Japan, in *Mesozoic Fishes — Systematics and Paleoecology*. G. Arratia and G. Viohl, Eds., Verlag Dr. Friedrich Pfeil, Munich, Germany, 73–83.

Goto, T. 2001. Comparative anatomy, phylogeny and cladistic classification of the order Orectolobiformes (Chondrichthyes, Elasmobranchii). *Mem. Grad. Sch. Fish. Sci. Hokkaido Univ.* 48(1):1–100.

Heemstra, P.C. 1973. A Revision of the Shark Genus *Mustelus* (Squaliformes, Carcharhinidae). Ph.D. dissertation. University of Miami, Miami, FL, 187 pp.

Heist, E., J.A. Musick, and J.E. Graves. 1996. Mitochondrial DNA diversity and divergence among sharpnose sharks *Rhizoprionodon terraenovae* from the Gulf of Mexico and mid-Atlantic Bight. *Fish. Bull.* 94:664–668.

Helfman, G.S., B.B. Collette, and D.E. Facey. 1997. *The Diversity of Fishes*. Blackwell Scientific, Oxford, 528 pp.

Herdendorf, C.E. and T.M. Berra. 1995. A Greenland shark from the wreck of the SS Central America at 2,200 meters. *Trans. Am. Fish. Soc.* 124(6):950–953.

Holland, K.N., B.M. Wetherbee, C.G. Lowe, and C.G. Meyer. 1999. Movements of tiger sharks (*Galeocerdo cuvieri*) in coastal Hawaiian Waters. *Mar. Biol.* 134:665–673.

Humphries, C.J. and L.R. Parenti. 1999. *Cladistic Biogeography,* 2nd ed. Clarendon Press, Oxford, 200 pp.

Kent, B.W. 1994. *Fossil Sharks of the Chesapeake Bay Region.* Egan Rees & Boyer, Columbia, MD, 146 pp.

Kohler, N.E., J.G. Casey, and P.E. Turner. 1998. NMFS Cooperative Shark Tagging Program, 1962–93: an atlas of shark tag and recapture data. *Mar. Fish. Rev.* 60(2): 1–87.

Last, P.R. and B. Séret. 1999. Comparative biogeography of the chondrichthyan faunas of the Tropical South-East Indian and South-West Pacific Oceans, in *Proc. 5th Indo-Pac Fish Conf.,* Paris, 293–306.

Last, P.R. and J.D. Stevens. 1994. *Sharks and Rays of Australia.* CSIRO, Collingwood, Victoria, Australia, 513 pp.

Last, P.R., G.H. Burgess, and B. Séret. 2002. Description of six new species of lantern sharks of the genus *Etmopterus* (Squaloidea: Etmopteridae) from the Australian region. *Cybium* 26(3):203–223.

Lieberman, B.S. 1999. *Paleobiogeography.* Kluwer Academic/Plenum Press, New York, 208 pp.

Longhurst, A. 1998. *Ecological Geography of the Sea.* Academic Press, San Diego, CA, 398 pp.

MacArthur, R.H. and E.O. Wilson. 1967. *The Theory of Island Biogeography.* Princeton University Press, Princeton, NJ, 203 pp.

Maisey, J.G., G.J.P. Naylor, and D.J. Ward. In press. Mesozoic Elasmobranchs, neoselachian phylogeny and the rise of modern elasmobranch diversity, in *Mesozoic Fishes,* Vol. 3, *Systematic and Fossil Record. Proceedings of the International Meeting,* Buckow. G. Arratia and H.-P. Schultze, Eds., Verlag Dr. Friedrich Pfeil, Munich, Germany.

Marshall, W.B. 1979. *Developments in Deep-Sea Biology.* Blanford Press, Dorset, U.K., 566 pp.

Martin, A.P. and G.J.P. Naylor. 1997. Independent origins of filter feeding in megamouth and basking sharks (Order Lamniformes) inferred from phylogenetic analysis of cytochrome *b* gene sequences, in *Biology of the Megamouth Shark.* K. Yano, J.F. Morrissey, Y. Yabumoto, and K. Nakaya, Eds., Tokai University Press, Tokyo, 39–50.

Martin, A.P., G.J.P. Naylor, and S.R. Palumbi. 1992. Rates of mitochondrial DNA evolution in sharks are slow compared to mammals. *Nature* 357:153–155.

Middlemiss, F.A. 1984. Distribution of Lower Cretaceous brachiopods and its relation to climate, in *Fossils and Climate.* P.J. Brenchley, Ed., John Wiley & Sons, New York, 165–170.

Mooi, R.D. and A.C. Gill. 2002. Historical biogeography of fishes, in *Handbook of Fish Biology and Fisheries,* Vol. I. P.S.B. Hart and J.D. Reynolds, Eds., Blackwell Science, Oxford, 43–68.

Morone, J.J. and J.V. Crisci. 1995. Historical biogeography: introduction and methods. *Annu. Rev. Ecol. Syst.,* 26:373–401.

Murdy, E.O., R.S. Birdsong, and J.A. Musick. 1997. *Fishes of Chesapeake Bay.* Smithsonian Institution Press, Washington, D.C., 324 pp.

Musick, J.A., J.A. Colvocoresses, and E.J. Foell. 1986. Seasonality and distribution, availability and composition of fish assemblages in Chesapeake Bight, in *Fish Community Ecology in Estuaries and Coastal Lagoons: Towards an Ecosystem Integration.* A. Yanez y Arancibia, Ed., University of Mexico Press, Mexico City, chap. 21, 451–474.

Musick, J.A., C.R. Tabit, and D. Evans. 1990. Body surface areas in galeoid sharks. *Copeia* 1990(4):1130–1133.

Musick, J.A., S. Branstetter, and J.A. Colvocoresses. 1993. Trends in Shark Abundance from 1974–1991 for the Chesapeake Bight of the U.S. Mid-Atlantic Coast. NOAA Tech. Rep. NMFS 115:1–18.

Musick, J.A., J.C. Defosse, S. Wilk, D. McMillan, and E. Grogan. 1996. Historical comparison of the structure of demersal fish communities near a deep-sea disposal site in the western North Atlantic. *J. Mar. Environ. Eng.* (3):149–171.

Musick, J.A., G. Burgess, G. Cailliet, M. Camhi, and S. Fordham. 2000. Management of sharks and their relatives (Elasmobranchii). *Fisheries* 25(3):9–13.

Nakabo, T., Ed. 2002. *Fishes of Japan.* Tokai University Press, Tokyo, 866 pp.

Naylor, G.J.P. 1989. The Phylogenetic Relationships of Carcharhiniform Sharks Inferred from Electrophoretic Data. Ph.D. dissertation. University of Maryland, College Park, MD.

Naylor, G.J.P. 1992. The phylogenetic relationships among requiem and hammerhead sharks: inferring phylogeny when thousands of equally most parsimonious trees result. *Cladistics* 8:295–318.

Naylor, G.J.P. and L. Marcus. 1994. Identifying isolated shark teeth of the genus *Carcharhinus* to species: reference for tracking phyletic change through the fossil record. *Am. Mus. Novit.* 3109:1–53.

Parenti, L.R. 1991. Ocean basins and the biogeography of freshwater fishes. *Aust. Syst. Bot.* 4:137–149.

Parr, A.E. 1933. A geographic ecological analysis of the seasonal changes in temperature conditions in shallow water along the Atlantic coast of the U.S. *Bull. Bingham Oceanogr. Coll. Yale Univ.* 4:1–90.

Pearcy, W.G., D.L. Stein, and R.S. Carney. 1982. The deep-sea benthic fish fauna of the northeastern Pacific Ocean on Cacadia and Tufts Abyssal Plains and adjoining continental slopes. *Biol. Oceanogr.* 1:342–375.

Platnick, N. and G. Nelson. 1978. A method of analysis in historical biogeography. *Syst. Zool.* 27:1–16.

Purdy, R.W., V.P. Schneider, S.P. Applegate, J.H. McLellan, R.L. Mayer, and B.H. Slaughter. 2001. The Neogene sharks, rays, and bony fishes from Lee Creek Mine, Aurora, North Carolina, in *Geology and Paleontology of the Lee Creek Mine, North Carolina, III.* C.E. Ray and D.J. Bohasku, Eds., *Smithsonian Contrib. Paleobiology* 90.

Randall, J.E. 1998. Zoogeography of shore fishes of the Indo-Pacific region. *Zool. Stud.* 37(4):227–268.

Reif, W.E. and C. Saure. 1987. Shark biogeography: vicariance is not even half the story. *N. Jahrb. Geol. Paleontol. Abh.* 175(1):1–17.

Rosen, D.E. 1975. A vicariance model of Caribbean biogeography. *Syst. Zool.* 24:431–464.

Rosenblatt, R.H. 1963. Some aspects of speciation in marine shore fishes, in *Speciation in the Sea, a Symposium.* J.P. Harding and N. Tebble, Eds., Systematics Assoc., London, 117–180.

Rosenblatt, R.H. and R.S. Waples. 1986. A genetic comparison of allopatric populations of shore fish species from the eastern and central Pacific Ocean, dispersal or vicariance? *Copeia* 1986:275–284.

Shirai, S. 1996. Phylogenetic interrelationships of Neoselachian (Chondrichthyes: Euselachii), in *Interrelationships of Fishes.* M.L.J. Stiaseny, L.R. Parenti and G.D. Johnson, Eds., Academic Press, New York, 9–32.

Simpson. G.G. 1952. Probabilities of dispersal in geological time. *Bull. Am. Mus. Nat. Hist.* 99:163–176.

Sims, D.W., E.J. Southall, and J.D. Metcalfe. 2002. Foraging and migratory behavior of basking sharks using seasonal scales as revealed by satellite archival telemetry. *Abst. Proc. 6th Annual European Elasmobranch Conference,* Cardiff, Wales.

Skomal, G. 2001. Shark tracking: DMF and collaborators break new ground. *DMF Newsl.,* Second Quarter, Mass. Div. of Marine Fish., Boston, 6–7.

Smith, A.G., D.G. Smith, and B.M. Funnell. 1994. *Atlas of Mesozoic and Cenozoic Coastlines.* Cambridge University Press, Cambridge, U.K., 99 pp.

Springer, S. and G.H. Burgess. 1985. Two new dwarf dogsharks (Etmopterus, Squalidae) found off the Caribbean coast of Colombia. *Copeia* 1985(3):584–591.

Springer, V.G. 1982. Pacific plate biogeography with special reference to shorefishes. *Smithsonian Contrib. Zool.* 465:1–82.

Stehmann, M., E.I. Kukuev, and I.I. Konovalenko. 1999. Three new records of the oceanic longnose pygmy shark, *Heteroscymnoides marleyi* (Squalidae, Chondrichthyes), from the southeastern Atlantic and southeastern Pacific. *Voprosy Ikhtiol.* 39(5):631–641.

Stuart, C.T., M.A. Rex, and R.J. Etter. 2003. Large scale spatial and temporal patterns of the deep-sea benthic species diversity, in *Ecosystems of the Deep Sea.* P.A. Tyler, Ed., Ecosystems of the Deep Oceans 28. Elsevier, Amsterdam, 295–312.

Sverdrup, H.U., M.U. Johnson, and R.H. Fleming. 1942. *The Oceans, Their Physics, Chemistry and General Biology.* Prentice-Hall, Englewood Cliffs, NJ, 1087 pp.

Thies, D. 1983. Jarazeitliche Neoselachian aus Deutschland und S. England. *Cour. Forsch. Inst. Senckenberg* 58:1–116.

Vakhrameev, V.A. and N.F. Hughes. 1991. *Jurassic and Cretaceous Floras and Climates of the Earth.* Cambridge University Press, Cambridge, U.K. 318 pp.

Welton, B.J. and R.F. Farrish. 1993. *The Collector's Guide to Fossil Sharks and Rays from the Cretaceous of Texas.* Before Time, Lewisville, TX, 204 pp.

West, G.J. and J.D. Stevens. 2001. Archival tagging of school shark, *Galeorhinus galeus,* in Australia: initial results, in *The Behavior and Sensory Biology of Elasmobranch Fishes: An Anthology in Memory of Donald Richard Nelson.* T.C. Tricas and S.H. Gruber, Eds., Kluwer Academic, Dordrecht, 283–298.

Wiley, E.O. 1981. *Phylogenetics: The Theory and Practice of Phylogenetic Systematics.* Wiley, New York, 439 pp.

Zezina, O.N. 1997. Biogeography of the bathyal zone, in *Advances in Marine Biology, The Biogeography of the Oceans.* J.H.S. Blaxter and A.J. Southward, Eds., Academic Press, San Diego, 389–426.

3

Phylogeny of Batoidea

John D. McEachran and Neil Aschliman

CONTENTS

3.1 Introduction

Batoidea (electric rays, sawfishes, guitarfishes, skates, and stingrays), have been variously classified within the neoselachians (modern cartilaginous fishes) (Séret, 1986). Müller and Henle (1841) considered batoids sister to the sharks, as did Günther (1870), Regan (1906), Garman (1913), White (1937), Bigelow and Schroeder (1953), Berg (1940), and Norman (1966). On the other hand, Goodrich (1909), Jordan (1923), Bertin (1939), Arambourg and Bertin (1958), Compagno (1973, 1977), Shirai (1996), and de Carvalho (1996) classified batoids as a subgroup rather than a taxon equivalent to the modern sharks. Although the above authors disagree on the interrelationships of the batoids and sharks, all authors agree that batoids constitute a monophyletic group.

Despite numerous classifications of the neoselachians, rigorous investigations of their interrelationships begin with Compagno (1973, 1977). Compagno provided a wealth of external and internal characters in his investigations that set the stage for Shirai's (1992a,b) well-supported classification of the taxon. Based on a large number of skeletal and muscle characters, Shirai considered batoids to be the sister group of the pristiophorids within the Squalea, one of two major taxa of neoselachians. Shirai's classification remained unchallenged until Douady et al. (2003), based on mt-DNA sequences, and Maisey et al. (in press), based on mt-DNA and nuclear DNA sequences and paleontological data, provided convincing evidence that batoids are sister to the modern sharks. Maisey et al. demonstrate that the character states employed by Shirai are related to a benthic mode of life and thus are very likely to be homoplasious.

Although monophyly of the batoids is widely accepted and well corroborated, the interrelationships within batoids remain controversial. Compagno (1973) divided batoids into four orders (Torpediniformes, Pristiformes, Rajiformes, and Myliobatiformes) and considered a clade within Rajiformes, Rhinobatoidei, to be the basal clade within the batoids. In 1977, Compagno considered Torpediniformes to be the basal clade based on primitive aspects of the gill arch structure. Heemstra and Smith (1980) considered Pristiformes sister to the Torpediniformes, Rajiformes, and Myliobatiformes. Maisey (1984) supported Compagno's claim that Torpediniformes were the basal clade, reasoning that the hyobranchial skeleton of Pristiformes was more specialized than that of generalized Jurassic batoids. Nishida (1990) considered pristiforms the basal clade of batoids, rhinobatoids to be polyphyletic, with *Rhynchobatus* and *Rhina* sister to the remaining batoids, and torpediniformes sister to the remainder of the rhinobatoids, the rajiforms, and the myliobatiforms. He divided the Myliobatiformes into basal clades (*Plesiobatis* and *Hexatrygon*) and four major clades (Urolophidae, Dasyatidae, Gymnuridae, and Myliobatidae). Lovejoy (1996) further elucidated the interrelationships within the Myliobatitformes. He proposed three major clades for Myliobatiformes: (1) *Urobatis* and *Urotrygon*; (2) *Taeniura*, amphi-American *Himantura*, *Paratrygon*, *Potamotrygon*, and *Plesiotrygon*; and (3) *Dasyatis*, Indo-West Pacific *Himantura*, *Gymnura*, *Myliobatis*, *Aetobatus,* and *Mobula*, with *Hexatrygon*, *Plesiobatis*, and *Urolophus* as basal clades. McEachran et al. (1996) investigated relationships within all of the batoids, and considered Torpediniformes sister to the remainder and pristiforms sister to rhinobatoids, rajoids, and myliobatiforms. The rhinobatoid genera *Rhinobatos, Zapteryx,* and *Trygonorrhina* and rajoids formed a polytomy with *Platyrhina* + *Platyrhinoidis* and *Zanobatus* + myliobatiforms. Within myliobatiforms, *Hexatrygon* was basal, *Pleisobatis* and *Urolophus* formed a polytomy with *Urobatis* + *Urotrygon* and the higher myliobatiforms, and the higher myliobatiforms formed two clades: (1) *Dasyatis* and *Gymnura* + the pelagic myliobatiforms and (2) *Taeniura* and amphi-American *Himantura* + Potamotrygonidae.

Rosenberger (2001b) examined the interrelationships within the dasyatid genus *Dasyatis* based on 14 of the 35 currently recognized species. Her study refutes monophyly of both *Dasyatis* and *Himantura*, and suggested that the amphi-American species of *Dasyatis* do not form a clade. Dunn et al. (2003) investigated the interrelationships of myliobatiforms based on mt-DNA sequences. This study supported the findings of Lovejoy (1996) and McEachran et al. (1996), in part, but placed *Taeniura* sister to *Dasyatis* and placed *Urobatis* sister to amphi-American *Himantura* + potamotrygonid clade. In Lovejoy and in McEachran et al. *Taeniura* were sister to amphi-American *Himantura* + potamotrygonid, and *Urobatis* were sister to *Urotrygon,* and this clade was sister to the two major clades of Myliobatiformes.

The purpose of the present study is to test and further resolve batoid interrelationships proposed in McEachran et al. (1996) by increasing the number of taxa surveyed, expanding the variety of morphological characters examined, and by selecting more appropriate outgroups. Outgroups are chosen from basal shark taxa and chimaeroids rather than from derived taxa of squaleans in accordance with the results of Douady et al. (2003) and Maisey et al. (in press). Many of the same ingroup taxa were used as in McEachran et al. (1996) but additional character complexes were analyzed and the character state polarities reached in the earlier study were reexamined with hopes of improving the resolution and accuracy of the revised study.

3.2 Analyses

Representatives of 32 of the 72 genera of batoids and four outgroup taxa were examined (Appendix 3.1). For species rich genera, e.g., *Torpedo, Narcine, Rhinobatos, Bathyraja, Raja, Urolophus, Urotrygon, Dasyatis, Himantura, Potamotrygon, Gymnura, Myliobatis, Rhinoptera*, and *Mobula,* several to many species were examined but only one to three were included in the data matrix because most of the character states employed were consistent for a majority or for all of the species in the genus. In cases where character states varied within a genus, additional species were included in the matrix to represent the character state variability. The outgroups, Chimaeridae, Heterodontidae, Chlamydoselachidae, and Hexanchidae, are either basal chondrichthians (Chimaeridae) or basal members of galeomorphs (Heterodontidae) or squaleomorphs (Chlamydoselachidae and Hexanchidae) (Didier, 1995; Douady et al., 2003; Maisey et al., in press). Galeomorphs and squaleomorphs are the two basal shark clades. The following character complexes were surveyed for characters that exhibited variation that was thought to reflect phylogenetic relationships: (1) external morphological structures; (2) squamation; (3) tooth root vascularization patterns; (4) lateral line patterns; (5) skeletal structures of neurocranium, branchial skeleton, vertebral column, scapulocoracoid, pelvic girdle, clasper, and fins; and (6) cephalic and branchial musculature. Specimens were cleared and stained, dissected or radiographed to observe internal structures. The anatomical terminology follows Miyake (1988), and data were also analyzed from Capapé and Desoutter (1979), Chu and Wen (1980), Compagno and Roberts (1982), Rosa (1985), Rosa et al. (1988), Miyake (1988, unpubl. data), Nishida (1990), Miyake and McEachran (1991), Miyake et al. (1992a,b), Shirai (1992a,b), Lovejoy (1996), McEachran et al. (1996), Herman et al. (1994, 1995, 1996, 1997, 1998, 1999), and Rosenberger (2001b). In nearly all cases, observations based on the literature were verified with independent observations. Most of the characters utilized are binary; those with multiple characters states were run unordered to reduce subjectivity.

The character matrix (Appendix 3.2) includes four outgroup taxa, 35 ingroup taxa, 82 characters, and 201 character states, and was analyzed using parsimony via PAUP version *4.0 Beta 10 (Swofford, unpubl.). Characters and character states are described in Appendix 3.3. The heuristic search option was used because of the size of the data matrix. The branch swapping algorithm was tree bisection-reconstruction (TBR). Strength of nodes was analyzed using Bremer decay indexes (Bremer, 1994). The data were subsequently reweighted by successive approximation (Farris, 1969; Carpenter, 1994) using the retention index values to select among equally parsimonious solutions.

3.3 Patterns

3.3.1 Claspers

The claspers of batoids, with the exception of the rajids, have not been broadly surveyed but, based on findings in this study, offer great potential in further resolving interrelationships. Externally claspers of batoids can be divided into two basic shapes: long to very long, slender, and depressed distally (*Rhynchobatus, Rhina, Rhinobatos, Zapteryx, Trygonorrhina, Platyrhina*, and rajids); and short, stout, and cylindrical to somewhat depressed (torpediniforms, *Platyrhinoidis, Zanobatus*, and myliobatiforms). The internal structure of batoid claspers is, with the exception of those of rajids, rather similar. The axial cartilage forms the axis of the clasper and it extends from the second intermediate cartilage (b2) to the tip of the glans. It is rodlike and calcified over most of its length but depressed, slightly expanded, and uncalcified near its tip. The dorsal and ventral marginal cartilages make up the proximal section of the clasper glans. They are moderately elongate to elongate, fused to the dorso-medial and lateral surfaces of the axial cartilage, slightly to moderately curved along their long axis to form a tube, and generally slightly to moderately expanded distally. The dorsal and ventral terminal cartilages make up the distal section of the clasper glans. They are generally relatively broad and short, and usually attached to the dorsomedial and lateral surfaces of the axial cartilage, respectively. Generally, a small, tapering, rodlike cartilage is loosely attached to the lateral margin of the ventral marginal cartilage near its junction with the ventral terminal cartilage. This rodlike cartilage forms the component claw in many taxa but in other taxa it is embedded in the integument and not visible externally. A large shieldlike cartilage (ventral covering piece) covers most or all of the ventral surface of the glans. It serves as site of insertion of the dilatator muscle that spreads the glans section of the clasper during copulation. The medial margin of the ventral covering piece forms at least part of the component pseudosiphon in many of the batoid taxa.

In Narcinidae and Narkidae the pseudosiphon is present, the dorsal marginal cartilage has a flange medial to its fusion with the axial cartilage, the cartilage forming the claw is embedded in the integument

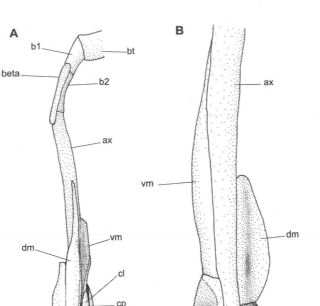

FIGURE 3.1 Clasper cartilages of *Narcine bancroftii* (TCWC 2923.01). (A) Dorsal view, (B) ventral view with ventral covering piece removed. ax = axial cartilage; b1 = first intermediate cartilage; b2 = second intermediate cartilage; beta = beta cartilage; bt = basipterygium; cl = cartilage forming claw; cp = ventral covering piece cartilage; dm = dorsal marginal cartilage; dt = dorsal terminal cartilage; vm = ventral marginal cartilage; vt = ventral terminal cartilage; bars = 1 cm.

and not visible externally, and the dorsal and ventral terminal cartilages are joined over their lengths to the axial cartilage (Figure 3.1 and Figure 3.2).

In *Rhinobatos, Zapteryx,* and *Platyrhinoidis* the pseudosiphon is present, dorsal marginal cartilage lacks a medial flange, the component claw is present and projects from the clasper groove, and the dorsal and ventral terminal cartilages are joined over their length to axial cartilage. Clasper of *Zanobatus* agrees with the above description except that the dorsal marginal cartilage has a medial flange; in addition, the clasper groove ends at the junction of the marginal and terminal cartilages and the terminal cartilages are bent ventrally to form an obtuse angle with the remainder of the clasper.

Claspers of *Urolophus, Urobatis,* and *Urotrygon* possess a pseudosiphon, a medial flange on dorsal marginal cartilage, a claw-forming cartilage embedded in integument and not visible externally, and a ventral terminal cartilage that is complexly formed, possesses a dorsally convex lateral flange, and is attached to the axial cartilage over its length (Figure 3.3 through Figure 3.5).

In *Pteroplatytrygon, Dasyatis americana, D. brevis, D. longa, D. sabina,* and *D. say* the pseudosiphon is absent, the medial flange on the dorsal marginal cartilage is absent, the spur-forming cartilage is absent, the dorsal terminal cartilage has a crenate lateral margin, and the ventral terminal cartilage is complexly formed, has a dorsally convex lateral flange, and is largely free of the axial cartilage (Figure 3.6).

Claspers of *Potamotrygon* possess a pseudosiphon, a medial flange on the dorsal marginal cartilage, a spur-forming cartilage embedded in the integument and not visible externally, and a ventral terminal cartilage that is complexly formed, has a dorsally convex lateral flange, and is joined to the axial oven length (Figure 3.7).

Claspers of *Gymnura, Aetobatus, Myliobatis,* and *Rhinoptera* possess a pseudosiphon, a medial flange on the dorsal marginal cartilage, the component claw projecting from clasper groove, and a ventral terminal cartilage that is complexly formed, has a dorsally convex lateral flange, and is joined to the axial over length (Figure 3.8).

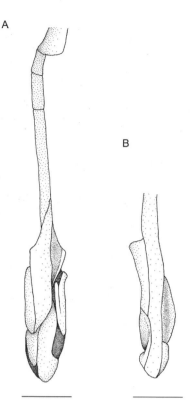

FIGURE 3.2 Clasper cartilages of *Discopyge tschudii* (FAKU 105040). (A) Dorsal view, (B) ventral view with ventral covering piece removed. Acronyms and bars as in Figure 3.1.

Claspers of rajids have been described by Ishiyama (1958, 1967), Hulley (1970, 1972a,b, 1973), Stehmann (1970, 1971a,b,c, 1976a,b, 1977, 1978, 1987), Menni (1971, 1972), McEachran (1977, 1982, 1983, 1984), McEachran and Stehmann (1977), McEachran and Martin (1978), McEachran and Compagno (1979, 1980, 1982), McEachran and Fechhelm (1982a,b), Ishihara and Ishiyama (1985, 1986), McEachran and Miyake (1986, 1988, 1990a), Ishihara (1987), McEachran et al. (1989), and McEachran and Last (1994). Unlike the remainder of the batoids, the covering piece of cartilage (called the dorsal terminal 1 cartilage) is on the dorsal aspect of the clasper glans and seldom covers the entire distal section of the glans. However, it does serve as the site of insertion of the dilatator muscle. In rajids the dorsal terminal cartilage is multiple and the individual cartilages are either arranged in series or in parallel. They are usually arranged in series in the subfamily Rajinae and in parallel in the subfamily Arhynchobatinae. The ventral terminal cartilage is also usually multiple and arranged in parallel. The ventral terminal cartilages are called accessory terminal cartilages in rajids. In some genera of Arhynchobatinae (*Bathyraja* and *Rhinoraja*) the lateral ventral terminal cartilage (called the accessory terminal 1 cartilage) is fused with the ventral marginal cartilage to form the component projection. The cartilage that forms the claw in rajids is located on the dorsal surface of the ventral marginal cartilage near its junction with the ventral terminal cartilages, rather than on the lateral margin of the ventral marginal cartilage as in the other batoid taxa. This cartilage is called the ventral terminal cartilage in rajids and forms the component shield.

3.3.2 Phylogenetic Analyses

The analysis of the data matrix of 39 taxa, including four outgroups, 82 characters, and 201 character states produced ten most parsimonious trees of 186 steps, with a consistency index (CI) of 0.6398, homoplasy index (HI) of 0.3602, and a retention index (RI) of 0.9016.

The strict consensus tree of the ten most parsimonious trees (Figure 3.9) is fairly well resolved but the nodes vary in both Bremer decay indices and unambiguous character state support. The phylogenetic

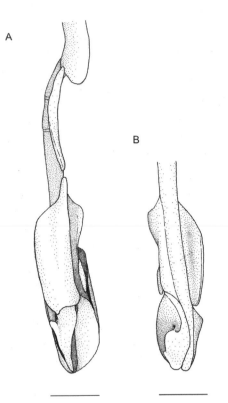

FIGURE 3.3 Clasper cartilages of *Urolophus bucculentus* (FSFRL EC 361). (A) Dorsal view, (B) ventral view with ventral covering piece removed. Acronyms and bars as in Figure 3.1.

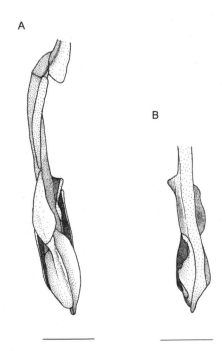

FIGURE 3.4 Clasper cartilages of *Urobatis halleri* (SIO uncat.) (A) Dorsal view, (B) ventral view with ventral covering piece removed. Acronyms and bars as in Figure 3.1.

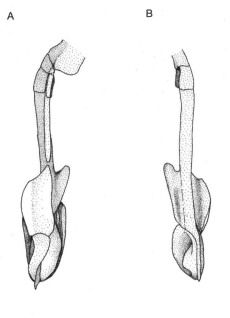

FIGURE 3.5 Clasper cartilages of *Urotrygon munda* (USNM 220612-4). (A) Dorsal view, (B) ventral view with ventral covering piece removed. Acronyms and bars as in Figure 3.1.

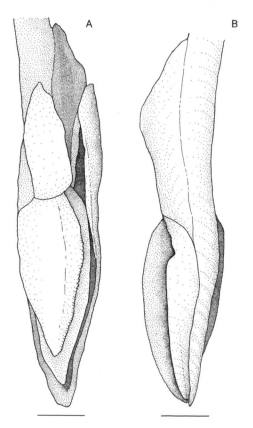

FIGURE 3.6 Clasper cartilages of *Dasyatis americana* (TCWC 2794.01). (A) Dorsal view, (B) ventral view with ventral covering piece removed. Acronyms and bars as in Figure 3.1.

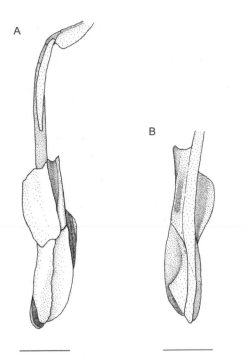

FIGURE 3.7 Clasper cartilages of *Potamotrygon magdalenae* (TCWC uncat.). (A) Dorsal view, (B) ventral view with ventral covering piece removed. Acronyms and bars as in Figure 3.1.

hypothesis supports batoid monophyly, and suggests that torpediniforms and *Pristis* form successive sister groups of the remainder batoids, that *Rhina* and *Rhynchobatus* are successive sister groups of Rhinobatidae (*Rhinobatos, Zapteryx, Trygonorrhina*) and Rajidae (*Bathyraja* and *Raja*), that *Platyrhina* + *Platyrhinoidis* and *Zanobatus* form successive sisters of myliobatiforms (terminal nodes above *Zanobatus*), that *Hexatrygon, Plesiobatis* + *Urolophus*, and *Urobatis* + *Urotrygon* form successive sister groups to the higher myliobatiforms, and that the higher myliobatiforms comprise a polytomy of partially to fully resolved clades. The ten most parsimonious trees varied in relationships between the rhinobatids and rajids and in the relationships of *Dasyatis kuhlii* with the higher myliobatiforms. In five of the trees *Trygonorrhina* are sister to rajids and *Rhinobatos* are sister to *Zapteryx*, and in the other five trees *Rhinobatos* are sister to rajids + (*Trygonorrhina* + *Zapteryx*) (not illustrated). In six of the trees *D. kuhlii* is sister to the three other clades of higher myliobatiforms: (1) *Taeniura* + [*Pteroplatytrygon* + (*Dasyatis brevis* + *D. longa*)]; (2) amphi-American *Himantura* + *Potamotrygon*; and (3) Indo-West Pacific *Himantura* + (*Gymnura* + the pelagic myliobatiforms) (not illustrated). In two of the trees *D. kuhlii* is sister to the *Taeniura* + [*Pteroplatytrygon* + (*Dasyatis brevis* + *D. longa*)] (not illustrated); and in two trees *D. kuhlii* is one of four unresolved clades of higher myliobatiforms, as in the strict consensus tree (Figure 3.9).

Successive approximation character weighting (Farris, 1969; Carpenter, 1988) resulted in five equally parsimonious trees solutions. The strict consensus tree of the five equally parsimonious solutions unites *Trygonorrhina* with rajids and *Rhinobatos* with *Zapteryx* but leaves the relationships of *D. kuhlii* unresolved (Figure 3.10).

3.4   Discussion

3.4.1   Phylogenetic Implications

The current analysis largely supports the relationships proposed in McEachran et al. (1996) with several exceptions. In the present study, *Rhina* and *Rhynchobatus* are successive sister groups of the rhinobatid

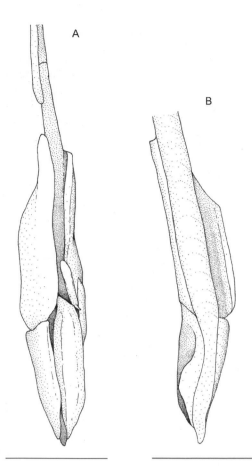

FIGURE 3.8 Clasper cartilages of *Gymnura micrura* (TCWC 642.08). (A) Dorsal view, (B) ventral view with ventral covering piece removed. Acronyms and bars as in Figure 3.1.

+ rajid clade; *Trygonorrhina* form a polychotomy with *Rhinobatos*, *Zapteryx*, and the rajids; *Plesiobatis* and *Urolophus* form a clade; *Dasyatis kuhlii* forms a polytomy with three other clades of myliobatiforms; and *Taeniura* forms a clade with *Pteroplatytrygon* and several *Dasyatis* species. In McEachran et al. (1996) *Rhina* were sister to *Rhynchobatus* and the remainder of the batoids above torpediniforms and *Pristis*. The *Rhina, Rhynchobatus*, rhinobatid, and rajid clade is supported by a Bremer decay index of 1, one unambiguous character (25,1: rostral appendix present), and an almost exclusive character state (66,1: elongated claspers). Rostral appendices do not occur in any other batoids, although they do occur in some Squalea (Shirai, 1992b, e.g., *Deania* and *Squalus*). Elongated claspers also occur in *Platyrhina*. In McEachran et al. (1996) *Rhinobatos, Zapteryx, Trygonorrhina*, and rajids form an unresolved node in the strict consensus tree of the successive approximation solutions, but this relationship was unsupported by unambiguous character states or by Bremer decay indices. In the present study the rhinobatid-rajid clade is supported by a Bremer decay index of 1, an unambiguous character (58,1: direct articulation of several pectoral radials with the scapulocoracoid posterior to the mesocondyle), and three character states (39,1: loss of labial cartilages; 53,1: mesocondyle of scapulocoracoid closer to procondyle than to metacondyle; and 54,1: antorbital cartilage directly articulating with propterygium and nasal capsule). Pectoral radials do not directly articulate with the scapulocoracoid posterior to mesopterygium in any other neoselachians. In *Zanobatus, Gymnura*, and the pelagic myliobatiforms some pectoral radials articulate directly with the scapulocoracoid but either the mesopterygium is absent or is displaced posteriorly and radials articulate with the scapulocoracoid anterior to the mesopterygium. No other batoids have scapulocoracoids with the mesocondyle located anterior to midlength. In the basal taxa and in *Platyrhina* and *Platyrhinoidis* the mesocondyle is located at midlength of the scapulocoracoid,

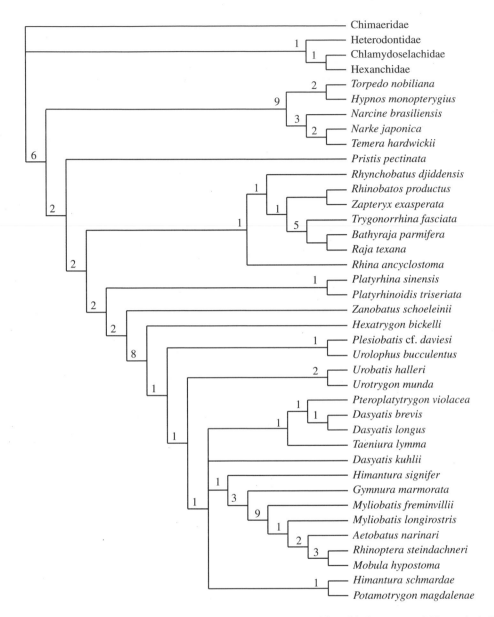

FIGURE 3.9 Strict consensus tree of ten most parsimonious trees generated from 82 characters and 39 taxa, including 4 outgroups. Bremer decay indices are given above nodes.

and in *Zanobatus* and the myliobatiforms the mesocondyle is either greatly elongated or located posterior to midlength of the scapulocoracoid. Labial cartilages are also absent in *Torpedo, Hypnos, Pristis*, and the myliobatiforms. This character state is unknown for *Trygonorrhina, Platyrhina*, and *Platyrhinoidis*. The antorbital cartilage also directly articulates with the propterygium of the pectoral girdle in the (*Platyrhina + Platyrhinoidis*) + (*Zanobatus* + the myliobatiforms) clade. In McEachran et al. (1996) *Plesiobatis* and *Urolophus* form a polytomy with *Urobatis + Urotrygon* and the remainder of the myliobatiforms in the consensus tree of the most parsimonious trees and form successive sister groups to *Urobatis + Urotrygon* and the remainder of the myliobatiforms in the consensus tree of the successive approximation solutions. In the present study, the *Plesiobatis + Urolophus* clade is supported by a Bremer decay index of 1 and one character state (33,1: fusion of the postorbital process and the posteriorly located triangular process of the supraorbital crest). This state also occurs in *Pteroplatytry-*

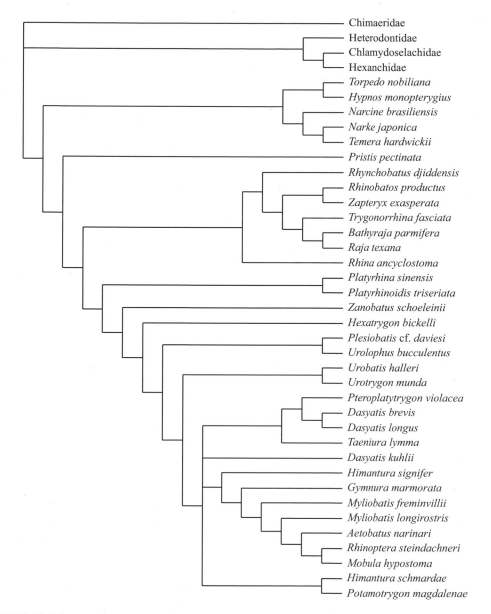

FIGURE 3.10 Strict consensus tree of five most parsimonious trees resulting from successive approximation character weighting.

gon, Aetobatus, Rhinoptera, and *Mobula.* In McEachran et al. (1996) *Taeniura* is sister to amphi-American *Himantura + Potamotrygon.* In the present study, the relationship of *Taeniura* with *Ptero-platytrygon* and *Dasyatis brevis + D. longa* is supported by a Bremer decay index of 1, one unambiguous character (69,1: dorsal terminal cartilage of clasper with a crenate margin), and five character states (16,3: tooth roots without a pulp cavity; 17,2: osteodentine widespread in tooth roots; 67,1: clasper component pseudosiphon absent; 68,0: dorsal marginal clasper cartilage without a medial flange; 72,1: ventral terminal cartilage largely free of axial cartilage). The pelagic myliobatiforms also lack a pulp cavity in tooth roots and have osteodentine widespread in tooth roots, although the states are unknown in *Mobula.* The clasper component pseudosiphon is also absent in *Bathyraja, Raja,* and *Mobula,* and the state is unknown in *Hypnos, Pristis, Rhina, Rhynchobatus, Hexatrygon, D. kuhlii,* and *Taeniura.* (A pseudosiphon is variably present in *Bathyraja* but this component is not considered homologous with the above state.) The dorsal marginal clasper cartilage also lacks a medial flange in the outgroups *Torpedo, Narcine, Rhinobatos, Zapteryx, Platyrhina, Platyrhinoidis,* and rajids, and the state is unknown in *Hypnos, Pristis, Rhina, Rhynchobatus, Trygonorrhina, Hexatrygon,* amphi-American *Himantura,* and *Taeniura.* The ventral terminal cartilage is also largely free of axial cartilage in rajids and the state is unknown in *Hypnos, Pristis, Rhina, Rhynchobatus, Trygonorrhina, Hexatrygon,* amphi-American *Himantura,* and *Taeniura.* However, lack of information on the clasper of *Taeniura* makes their association with this clade uncertain. In McEachran et al. (1996) *Taeniura* + (amphi-American *Himantura* + *Potamotrygon*) are sister to *Dasyatis* + (*Gymnura* + pelagic myliobatiforms). In the present study *Taeniura* and amphi-American *Himantura* + *Potamotrygon* are united in the same clade as in the earlier study but the relationships differ and are less resolved. The amphi-American *Himantura* + *Potamotrygon* clade comprise a polytomy with *D. kuhlii; Taeniura* + (*Pteroplatytrygon* + (*D. brevis* + *D. longa*)); and Indo-West Pacific *Himantura* + (*Gymnura* + the pelagic myliobatiforms). The lack of resolution in the present study is due in part to the inadequate sampling of the species-rich taxa (*Dasyatis* and *Himantura*) and lack of complete character sets for some of the taxa studied. Claspers were not available for *Taeniura,* amphi-American *Himantura,* and for the great majority of *Dasyatis* and Indo-West Pacific *Himantura* species. (Nishida, 1990, illustrates dorsal and ventral views of the left clasper of *D. kuhlii* but his illustrations are not adequate to evaluate all of the characters investigated in this study.) Given the lack of resolution and the inadequate sampling, the relationships within the polychotomy should be considered preliminary.

Despite the shortcomings in taxon sampling and availability of complete character sets, it is evident that both *Dasyatis* and *Himantura* (*sensu lato*) are nonmonophyletic. *Dasyatis kuhlii* is not grouped with any of the other three unresolved clades in the consensus tree (Figure 3.9). Unlike the *Taeniura, Pteroplatytrygon, D. brevis* + *D. longa* clade, *D. kuhlii* has tooth roots with broad pulp cavities (tooth roots lack pulp cavities in the latter clade), lacks osteodentine in tooth roots (osteodentine is widespread in tooth root in the later clade), lacks a crenate margin of the dorsal terminal cartilage (margin of dorsal terminal cartilage is crenate in latter clade), and has a ventral terminal cartilage that is attached to the axial cartilage (ventral terminal cartilage is free of axial cartilage in latter clade). Unlike the amphi-American *Himantura* + *Potamotrygon* clade, *D. kuhlii* lacks an angular cartilage in the maxillary-hyomandibular ligament. Unlike the Indo-West Pacific *Himantura,* and *Gymnura* + the pelagic myliobatiform clade, *D. kuhlii* lacks a subpleural loop of the hyomandibular canal that abruptly reverses directions, forming a deep hook or parallel course (Rosenberger, 2001b).

Indo-West Pacific *Himantura* is sister to *Gymnura* + the pelagic myliobatiforms whereas amphi-American *Himantura* is sister to *Potamotrygon* in the present study. The amphi-American *Himantura* possess an angular cartilage in the mandibular-hyomandibular ligament unlike the Indo-West Pacific *Himantura,* and the Indo-West Pacific *Himantura* unlike the amphi-American *Himantura* possess a subpleural loop of the hyomandibular canal that forms a lateral hook and the proximal segment of the pectoral fin propterygium is adjacent to or anterior to the anterior margin of the nasal capsule rather than adjacent nasal capsule. Further, the infraorbital loop of the suborbital and infraorbital canals forms a reticular pattern in the Indo-West Pacific *Himantura* but not in the amphi-American *Himantura.*

Recent studies of the interrelationships of batoids (Heemstra and Smith, 1980; Nishida, 1990; McEachran et al., 1996; and the present study) agree in some respects and disagree in others but illustrate that further research is needed to fully elucidate the phylogeny. Both Heemstra and Smith (1980) and Nishida (1990) consider pristids sister to the remainder of the batoids, while McEachran et al. (1996) and the present study consider torpediniforms sister to the remainder of the batoids. The discrepancies are due in part to the high degree of specializations and low degree of synapomorphies between each of these taxa and other batoid groups. Much of the skeletal structure in torpediniforms is affected by the massive development of the electric organs. Likewise, the relatively long bladelike tooth-bearing rostrum of pristids has apparently led to specializations in the posterior section of the cranium and the cervical vertebrae. In addition, both pristids and torpediniforms retain a number of primitive states that are unique to batoids. *Pristis* is the only batoid that has a basal cranial angle and aplesotic fins (ceratotrichia replacing radials at fin margins). In torpediniforms the three condyles of the scapulocoracoid (pro-, meso-, and metacondyles) are diagonally arranged rather than horizontally arranged as in the other batoids, and the arrangement in torpediniforms is considered primitive (McEachran et al., 1996).

The torpediniform family Narkidae, alone among the batoids, possesses a fully developed ceratohyal cartilage. In the remainder of the batoids the ceratohyal is partially or totally replaced by the pseudohyoid (Compagno, 1973). Torpediniforms and *Pristis* share several plesiomorphic character states with *Rhynchobatus* and *Rhina*: short cervicothoracic synarchials (anterior vertebrae fused into tube) resulting in the suprascapulae articulating with vertebrae rather than fused to synarchial; and propterygium of shoulder girdle failing to directly connect to the antorbial cartilage (Compagno, 1973).

Both Nishida (1990) and McEachran et al. (1996) considered rhinobatoids to be nonmonophyletic. Nishida placed *Rhynchobatus* and *Rhina* as sister taxa to all batoids other than pristids. McEachran et al. (1996) and the present study considered platyrhinids (*Platyrhina* and *Platyrhinoidis*, and *Zanobatus*) successive sisters of myliobatiforms, and placed rhinobatids (*Rhinobatos, Aptychotrema, Trygonorrhina*, and *Zapteryx*) in a clade with rajids. Additional study is needed to further resolve the interrelationships of the rhinobatids and rajids.

The recent studies of myliobatiforms by Nishida (1990), Lovejoy (1996), McEachran et al. (1996), Lovejoy et al. (1998), Rosenberger (2001b), Dunn et al. (2003), and the present study agree in a number of respects but suggest that further research is needed to fully resolve the relationships. There is consensus that amphi-American *Himantura* are sister to the Potamotrygonidae. Only *Potamotrygon* of the three genera of potamotrygonids was analyzed in the present study, but the family is a highly corroborated monophyletic group (Lovejoy, 1996; Lovejoy et al., 1998). There is a consensus that *Urolophus* is basal within the myliobatiforms, with the exception of *Hexatrygon* and *Plesiobatis*; that *Gymnura* is sister to the pelagic myliobatiforms; and that *Myliobatis* and *Aetobatus* are successive sister groups of *Rhinoptera* + *Mobula*. The relationships of *Urobatis, Dasyatis*, and *Taeniura* differ between the morphological and molecular studies and their status will remain equivocal until taxon sampling is increased and the taxa are subjected to extensive morphological and molecular analyses. The morphological studies of Lovejoy (1996), Rosenberger (2001b), and the present study suggest that both *Himantura* and *Dasyatis* are nonmonophyletic. Amphi-American *Himantura* are sister to *Potamotrygon* and should be classified in Potamotrygonidae, as suggested by McEachran et al. (1996). The type species of *Himantura* is the Indo-West Pacific species, *H. uarnak* Müller and Henle; thus, the amphi-American species *H. schmardae* and *H. pacifica* should be placed in another genus. Rosenberger (2001b) found that *D. kuhlii* is basal to a clade comprising *Pteroplatytrygon, Pastinachus sephen* (*Hypolophus sephen*), *Dasyatis*, Indo-Pacific *Himantura*, and *Gymnura*. In the present study *D. kuhlii* forms a polytomy with a clade including *Taeniura*, *Pteroplatytrygon*, and other species of *Dasyatis*, a clade containing amphi-American *Himantura* and *Potamotrygon*, and a clade comprising Indo-West Pacific *Himantura*, *Gymnura* + pelagic myliobatiforms. If these findings are corroborated in future studies, *D. kuhlii* will have to be placed in another genus. Rosenberger (2001b) classified *Pteroplatytrygon violacea* in *Dasyatis*, whereas Compagno et al. (1989) and McEachran and Fechhelm (1998) classified it in the monotypic genus *Pteroplatytrygon* on the basis of its disk shape and the fact that the eyes are not elevated above the disk as in other species of *Dasyatis*. Results of the present study support classification of *D. violacea* to *Pteroplatytrygon*. Confusion in the classification of Dasyatidae is not surprising given that it has been almost entirely based on the presence or absence of tail folds and ridges, a precedent that dates to Garman (1913), who considered it a matter of convenience and not necessarily reflective of phylogenetic relationships (Lovejoy, 1996). Lovejoy also stated that *Urobatis*, based on the structure of the synarchial and possibly the segmentation or lack of segmentation of the basihyal cartilage, might be paraphyletic. The western Atlantic species *U. jamaicensis* possesses the plesiomorphic condition for the lateral process of the synarchial and an unsegmented basihyal compared to the eastern Pacific *Urobatis* and *Urotrygon*. However, these observations need additional study.

Dunn et al. (2003) analyzed two molecular data sets to elucidate myliobatiform relationships: (1) 12S rRNA gene (1004 bp), tRNA genes valine, methionine, glycine, and isoleucine (290 bp), and portions of protein-coding genes NADH dehydrogenase 1 (ND1) and NADH dehydrogenase 2 (ND2) (234 bp); and (2) the first data set plus a portion of the cytochrome *b* gene (765 bp). The second data set, utilizing maximum likelihood analysis, produced a highly resolved cladogram similar to those of Lovejoy (1996), McEachran et al. (1996), and the present study. *Urolophus* were basal to the remainder of the myliobatiforms (*Hexatrygon* and *Plesiobatis* were not included in the study), *Gymnura* were sister to the pelagic myliobatiforms, and amphi-American *Himantura* were sister to *Potamotrygon*. Like the present study (in part), but unlike Lovejoy (1996) and McEachran et al. (1996), *Taeniura* were sister to *Dasyatis* and *Dasyatis* were part of a clade including *Potamotrygon* rather than sister to *Gymnura* + the pelagic myliobatiforms (Dunn et al., 2003). In the present study *D. brevis* and *D. longa* were in a clade separate from *Potamotrygon*. Unlike the three morphological studies, *Urobatis* were sister to amphi-America *Himantura* + *Potamotrygon* (Dunn et al., 2003).

Dunn et al. (2003) also analyzed the interrelationships of batoids based on their second molecular data set and hypothesized that *Pristis* + rhinobatids formed a trichotomy with rajids, and torpediniforms + the remainder of the batoids. These findings are closer to those of Heemstra and Smith (1980) and Nishida (1990), which placed pristids basal to the remainder of the batoids, than those of McEachran et al. (1996) and the present study. Conflicts among the morphological studies are in part due to the large number of autapomorphies and plesiomorphies of the torpediniforms and pristids, and the paucity of synapormorphies uniting either torpediniforms or pristids with other batoid taxa. Conflict between the molecular data and the present study may be resolved with more robust molecular data sets, in terms of taxa and genes sequenced.

3.4.2 Evolutionary Implications

The present study suggests that the depressed, disk-shaped morphology of batoids was achieved by two lineages (rajids and myliobatiforms) through separate ancestral taxa (rhinobatids and platyrhinids, respectively). The basal taxa retain a sharklike morphology with a thick and only moderately expanded disk formed by fusion of the pectoral fins with the head and trunk. Morphological differences between rhinobatids (*Rhinobatos*, *Zapteryx*, and *Trygonorrhina*) and platyrhinids (*Platyrhina* and *Platyrhinoidis*) appear to have constrained the manner in which rajids and myliobatiforms achieved their compressed disklike morphologies and may have likewise affected their respective modes of locomotion. Presumably, the trend for anteroposterior expansion of the scapulocoracoid that is evident in the rhinobatid–rajid clade and in the platyrhinid–*Zanobatus*–myliobatiform clade is related to undulatory–oscillatory modes of pectoral fin locomotion (Rosenberger, 2001a). The two clades, however, have achieved the expansion by alternative means. The rhinobatid–rajid clade has predominately expanded its scapulocoracoid posteriorly, between the mesocondyle and the metacondyle. The platyrhinid–*Zanobatus*–myliobatiform clade has predominately expanded its scapulocoracoid anteriorly, between the procondyle and the mesocondyle. The rhinobatid–rajid clade has some pectoral radials that articulate directly with the shoulder girdle posterior to the mesopterygium, and this condition may be the result of the posterior expansion of the scapulococraoid, without compensatory posterior expansion of the mesopterygium. In the platyrhinid–*Zanobatus*–myliobatiform clade the propterygium extends slightly behind the procondyle in *Platyrhina* and *Platyrhinoidis*, extends distinctly behind the procondyle in *Zanobatus*, and extends distinctly behind the procondyle and forms a synovial-like joint with the scapulocoracoid posterior to the procondyle in myliobatiforms (Howes, 1890; McEachran et al., 1996). It is a possibility that these trends established in the ancestral taxa of rajids and myliobatiforms constrained both their present-day anatomical structures and locomotor abilities. Expansion of the scapulocoracoid between the procondyle and the mesocondyle followed by posterior expansion of the posterior extension of the propterygium into a socketlike process, along with development of synovial joint between the scapular process and the synarchial, may have enabled the pelagic myliobatiforms to achieve an oscillatory mode of swimming. Rosenberger and Westneat (1999) and Rosenberger (2001a) found that dasyatids (*Taeniura*, *Dasyatis*, and *Pteroplatytrygon*) increase swimming velocity by increasing fin beat frequency and wave speed whereas rajids (*Raja*) increase velocity by decreasing wave number and increasing wave speed. Rosenberger (2001a) concludes that the difference between dasyatids and rajids may be due to the fact that the two taxa separately evolved pectoral fin locomotion. Both rajids and myliobatiforms have a full range of pectoral fin undulation but only the pelagic myliobatiforms achieved pectoral fin oscillation. Perhaps differences between the two clades in linear expansion of the scapulocoracoid (anterior to middle in myliobatiforms vs. middle to posterior in rajids) acted as an opportunity for a pelagic lifestyle in myliobatiforms but as a constraint for a pelagic lifestyle in rajids. It should be pointed out, however, that a pelagic lifestyle evolved twice in mylio-batiforms (separately in *Pteroplatytrygon violacea* and in myliobatids, rhinopterids, and mobulids) (Rosenberger, 2001a).

Rajids have a unique locomotor mode in which they employ the anterior lobe of their pelvic fins to push them off the bottom (walking according to Lucifora and Vassallo, 2002, and thrust and glide locomotion or punting according to Koester and Spirito, pers. comm.). Rajids have a unique pelvic fin that is partially or totally divided into an anterior lobe and a posterior lobe. The first or compound radial is triple jointed and the remainder of the radials in the anterior lobe articulate directly with the lateral aspect of the pectoral girdle (Holst and Bone, 1993; Lucifora and Vassallo, 2002; Koester and Spirito, pers. comm.). The anterior lobe thus acts as a leglike limb. The remainders of the batoids, with one exception (*Typhlonarke*), lack this specialization of the pelvic girdle and the thrust and glide mode of locomotion.

3.4.3 Phylogenetic Relationships within Rajidae

McEachran and Dunn (1998) investigated the interrelationships of rajids based on 31 taxa, including three outgroups, and 55 morphological characters. They divided rajids into two major taxa, Rajinae and Arhynchobatinae (Figure 3.11). The former taxon consisted of one fully resolved clade (Gurgesiellini) and two partially resolved clades (Rajini and Amblyrajini), and 15 genera and 2 unnamed generic level taxa, and 149 species. Three ambiguous characters defined Rajinae: scapulocoracoids without anterior bridges, distally expandable claspers, and the clasper component rhipidion. Arhynchobatinae were nearly fully resolved, consisted of Arhynchobatini and Riorajini, 11 genera and 79 species, and were defined by two unambiguous characters: basihyal cartilage with lateral extensions, and claspers with component projection. The strict consensus tree revealed considerable parallelisms in reduction of the rostral cartilage, concomitant forward extension of the pectoral fin radials and muscles, and enlarged nasal capsules. Given that the sister group of rajids are rhinobatids that are limited to shallow tropical to warm temperate marine habitats, it appears that the parallelisms are adaptations for deep-sea benthic habitats. Reduction of the rostral cartilage to a slender uncalcified rod and forward movement of the pectoral radials and muscles would provide deep-sea rajids with flexible, manipulatable snouts for grubbing in soft substrates. Enlarged nasal capsules could house large nasal rosettes, thus increasing chemosensitivity in regions with little light.

3.4.4 Classification of Batoids

A definitive classification of batoids is not possible at this time because interrelationships have not been completely resolved and some of the resolved nodes are not robustly supported. However, the following partially annotated classification is presented as a working hypothesis:

Class Chondrichthyes

Subclass Neoselachii

Cohort Batoidea

Order Torpediniformes

Family Torpedinidae Bonaparte, 1838

Subfamily Torpedininae Bonaparte, 1838: *Torpedo* Duméril, 1806

Subfamily Hypnidae Gill, 1862: *Hypnos* Duméril, 1852

> *Torpedo* and *Hypnos* are sister taxa with high bootstrap support suggesting that they form a monophyletic group. Because Torpedinidae and Hypnidae each consist of a single genus it makes sense to consider them subfamilies within the same family in agreement with Nelson (1994).

Family Narcinidae Gill, 1862

Subfamily *Narcininae* Gill, 1862: *Benthobatis* Alcock, 1898; *Diplobatis* Bigelow and Schroeder, 1948; *Narcine* Henle, 1834

Subfamily *Narkinae* Fowler, 1934: *Heteronarce* Regan, 1921; *Narke* Kaup, 1826; *Temera* Gray, 1831; *Typhlonarke* Waite, 1909

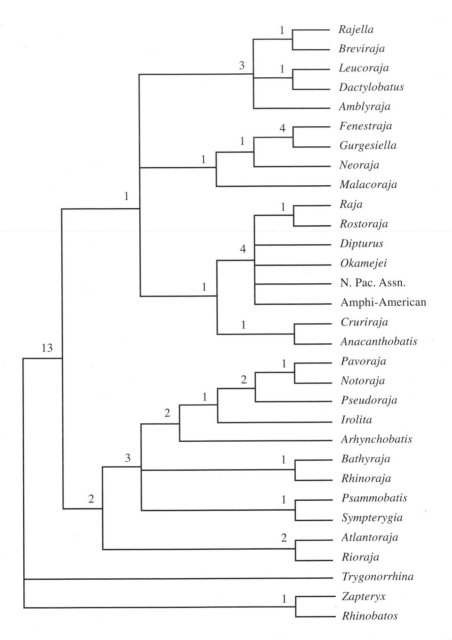

FIGURE 3.11 Strict consensus tree of 20 most parsimonious trees of 160 steps, based on 55 characters and 31 taxa of Rajidae and outgroups. (Adapted from McEachran and Dunn, 1998.) Bremer decay indices are given above nodes.

Narcine (representing Narcinidae) and *Narke* + *Temera* (representing Narkidae) form a clade with very high bootstrap support suggesting that they form a monophyletic group. Because Narcinidae and Narkidae comprise a total of 25 species in nine genera it makes sense to consider them subfamilies within the same family following Nelson (1994).

Order Pristiformes

Family Pristidae Bonaparte, 1838: *Anoxypristis* White and Moy-Thomas, 1941; *Pristis* Latham, 1794

Order Rajiformes

Incertae sedis Rhina Bloch and Schneider, 1801

Incertae sedis Rhynchobatus Müller and Henle, 1837

> These genera should be considered *incertae sedis* until they can be examined in greater detail. Skeletal structures were unavailable for this study.

Family Rhinobatidae Müller and Henle, 1837: *Aptychotrema* Norman, 1926; *Rhinobatos* Link, 1790; *Trygonorrhina* Müller and Henle, 1837; *Zapteryx* Jordan and Gilbert, 1880

> Placing *Rhinobatos, Zapteryx, Aptychotrema,* and *Trygonorrhina* in the same family is provisional because the relationships within the *Rhinobatos* (including *Aptychotrema*), *Zapteryx, Trygonorrhina*, rajid node are not fully resolved and the node is supported by a Bremer decay index of 1 and one unambiguous character. Study of the claspers of *Trygonorrhina*, which were not available, may further resolve the interrelationships.

Family Rajidae Bonaparte, 1831

Subfamily Rajinae: *Amblyraja* Malm, 1877; *Anacanthobatis* von Bonde and Swart, 1923; *Breviraja* Bigelow and Schroeder, 1948; *Cruriraja* Bigelow and Schroeder, 1948; *Dactylobatus* Bean and Weed, 1909; *Dipturus* Rafinesque, 1810; *Fenestraja* McEachran and Compagno, 1982; *Gurgesiella* de Buen, 1959; *Leucoraja* Malm, 1877; *Malacoraja* Stehmann, 1970; *Neoraja* McEachran and Compagno, 1982; *Okamejei* Ishiyama, 1959; *Raja* Linnaeus, 1758; *Rajella* Stehmann, 1970; *Rostroraja* Hulley, 1972; undescribed Genus *A* Assemblage of McEachran and Dunn (1998); undescribed Genus *B* Assemblage of McEachran and Dunn (1998)

Subfamily Arhynchobatinae: *Atlantoraja* Menni, 1972; *Arhynchobatis* Waite, 1909; *Bathyraja* Ishiyama, 1958; *Irolita* Whitley, 1931; *Notoraja* Ishiyama, 1958; *Pavoraja* Whitley, 1939; *Psammobatis* Günther, 1870; *Pseudoraja* Bigelow and Schroeder, 1954; *Rhinoraja* Ishiyama, 1952; *Rioraja* Whitley, 1939; *Symptergia* Müller and Henle, 1837

> Compagno (1999) divided Rajidae into three families: Rajidae, Arhynchobatidae, and Anacanthobatidae Hulley, 1972. Anacanthobatidae consists of *Anacanthobatis* and *Cruriraja*, genera that are nested in Rajidae according to McEachran and Dunn (1998). Elevating these genera to familial status would make Rajidae paraphyletic. Rajidae, as conceived herein, are a large taxon with about 250 species but the species are very similar in appearance and thus it does not seem practical to treat them in more than one family.

Order Myliobatiformes

This node is supported by a Bremer decay index of 2 and two unambiguous characters.

Suborder Platyrhinoidei

Family Platyrhinidae Jordan, 1923: *Platyrhina* Müller and Henle, 1838; *Platyrhinoidis* Garman, 1881

Suborder Zanobatoidei

Family Zanobatidae: *Zanobatus* Garman, 1913

Suborder Myliobatoidei

Superfamily Hexatrygonoidea Heemstra and Smith, 1980

Family Hexatrygonidae Heemstra and Smith, 1980: *Hexatrygon* Heemstra and Smith, 1980

Superfamily Urolophoidea

Family Urolophidae Müller and Henle, 1841: *Plesiobatis* Nishida, 1990; *Trygonoptera* Müller and Henle, 1841; *Urolophus* Müller and Henle, 1837

Superfamily Urotygonoidea McEachran et al., 1996

Family Urotrygonidae McEachran et al., 1996: *Urobatis* Garman, 1913; *Urotrygon* Gill, 1864

Superfamily Dasyatoidea

Incertae sedis Dasyatis kuhlii

Incertae sedis Pastinachus Rüppell, 1829

Incertae sedis Urogymnus Müller and Henle, 1837

Family Dasyatidae Jordan, 1888: *Dasyatis* Rafinesque, 1810; *Pteroplatytrygon* Fowler, 1910; *Taeniura* Müller and Henle, 1837

> Composition of this family is provisional because the node is supported by a Bremer decay index of 1 and one unambiguous character, and because a majority of the species in the genera have not been surveyed.

Family Potamotrygonidae Garman, 1913: amphi-American *Himantura* non-Müller and Henle, 1837; *Paratrygon* Duméril, 1865; *Plesiotrygon* Rosa, Castello, and Thorson, 1987; *Potamotrygon* Garman, 1877

Incertae sedis Indo-West Pacific Himantura Müller and Henle, 1837

> The node including Indo-West Pacific *Himantura* with the *Gymnura* + pelagic myliobatiform clade is weakly supported and few species of *Himantura* were surveyed in this study. Until further species are investigated the taxon is considered *incertae sedis*.

Family Gymnuridae Fowler, 1934: *Aetoplatea* Valenciennes, in Müller and Henle, 1841; *Gymnura* Kuhl in van Hasselt, 1823

Myliobatidae Bonaparte, 1816: *Aetobatus* Blainville, 1816; *Aetomylaeus* Garman, 1908; *Manta* Bancroft, 1828; *Mobula* Rafinesque, 1810; *Myliobatis* Cuvier, 1817; *Pteromylaeus* Garman, 1913; *Rhinoptera* Kuhl in Cuvier, 1829

3.4.5 Biogeography of Batoids

Batoids range from polar latitudes to tropical seas from the shoreline to depths of 3000 m. However, many clades within batoids are rather provincial in their distributions and analyses of their distributional patterns in conjunction with plate tectonic history may reveal evolutionary patterns. The major batoid taxa (torpediniforms, pristiforms, rajiforms, and myliobatiforms) date back to the late Jurassic to Paleocene in the fossil record (Cappetta, 1987). Thus, current distributional patterns should reflect the breakup of Gondwana during the latter part of the Mesozoic Era (Pitman et al., 1993).

The torpediniforms consist of two families, Torpedinidae, with 15 to 21 species in two genera (*Torpedo* with about 14 to 20 species and *Hypnos* with one species) and Narcinidae, with about 31 species in nine genera. *Torpedo* occur worldwide mostly in temperate to subtropical waters from near shore to about 1100 m. Several species are widely distributed across ocean basins while others are restricted to particular island groups or coastal areas (Michael, 1993; McEachran and de Carvalho, 2002). *Hypnos* are endemic to temperate and subtropical Australia between the surf zone to 220 m. Narcinidae are found worldwide mostly in warm temperate to tropical waters. *Narcine*, with about 15 species, are circumtropical, but the highest species diversity is in the Indo-West Pacific (de Carvalho, 1999). *Benthobatis* have two species, one in the western central Atlantic and the other in the central Indian Ocean, at depths between 274 and 1070 m. Most of the other genera are more restricted in their distributions. *Diplobatis* occurs in the tropical waters of the eastern Pacific and western Atlantic (Fechhelm and McEachran, 1984). *Discopyge* are limited to the austral waters of South America. *Crassinarke* occur in the western North Pacific off China, Japan, and Korea. *Heteronarce* occur in the western and central Indian Ocean (Compagno, 1999). *Narke* range from southern Africa to Japan in the Indo-West Pacific. *Temera* are limited to southeastern Asia. *Typhlonarke* are endemic to New Zealand.

The pristids occur in inshore tropical to warm temperate waters worldwide and freshwaters in tropical areas. There are about six species in two genera. Two or three species are wide ranging to worldwide. The highest diversity of species is in the Indo-West Pacific (Last and Stevens, 1994).

Rhynchobatus and *Rhina* occur in the tropical eastern Atlantic and the Indo-West Pacific in shallow tropical waters. There are about six species in two genera, with the highest diversity of species in the central western Pacific, from the Philippines, Indonesia, and northern Australia (Last and Stevens, 1994).

Rhinobatidae are found worldwide in tropical to warm temperate inshore waters but some occur as deep as 366 m. There are about 45 species in four genera. The genera *Aptychotrema* and *Trygonorrhina*, each with two species, are endemic to Australia (Last and Stevens, 1994). *Zapteryx* consist of three species, one in the warm temperate eastern Pacific from southern California to Baja California and the Gulf of California, a second from the tropical eastern Pacific from southern Mexico to Ecuador, and the third from southern Brazil. *Rhinobatos* occur worldwide in warm temperate to tropical seas in shallow water, and include 33 species. *Rhinobatos* species are classified in five putative subgenera (Compagno, 1999) that are partially limited to oceanic regions. *Rhinobatos* (*Acroteriobatus*) consist of seven species, six of which are limited to South Africa or the Indian Ocean. *Rhinobatos* (*Glaucostegus*) have 12 species and seven of these are amphi-American. The other species occur in the Indo-West Pacific. *Rhinobatos* (*Platypornax*) are monotypic and occur in the Indo-West Pacific. *Rhinobatos* (*Rhinobatos*) have ten species and seven of these occur in the Indo-West Pacific, including the Red Sea. The other species occur in the eastern Atlantic, in one case also the Mediterranean. *Rhinobatos* (*Scobatus*) include two species from the Indo-West Pacific. The remaining species, *R. prahli,* is unassigned to a subgenus but is likely related to *Glaucostegus* and occurs in the Caribbean Sea off Colombia. More than half of the species of *Rhinobatos* (19) are endemic to the Indo-West Pacific, five are endemic to the eastern Atlantic, and eight are amphi-American endemics.

Rajidae are found worldwide from polar latitudes to the equator between the shoreline and 3000 m, but in tropical waters they are limited to the outer portions of continental shelves (McEachran and Miyake, 1990b). Although widely distributed, various clades of rajids are restricted to specific geographical regions. Arhynchobatinae, one of the two major clades, is largely distributed in the Pacific Ocean, southwestern Atlantic Ocean, and southern Indo-West Pacific region. Within this subfamily Riorajini (*Atlantoraja* + *Rioraja*) are sister to the remainder of the clade and endemic to shallow warm temperate waters of southern Brazil, Uruguay, and northern Argentina. Arhynchobatini consist of an unresolved trichotomy ((*Psammobatis* + *Sympterygia*) + (*Bathyraja* + *Rhinoraja*) + (*Arhynchobatis* + (*Irolita* + (*Pseudoraja* + (*Notoraja* + *Pavoraja*)))))). *Psammobatis* and *Sympterygia*, with eight and four species, respectively, are endemic to the austral region of South America along the continental shelf. *Bathyraja* and *Rhinoraja*, with about 49 and 3 species, respectively, are found almost exclusively in the Pacific and western South Atlantic oceans. *Rhinoraja* are endemic to northern Japan. *Bathyraja* have three centers of species richness: the North Pacific and the Bering Sea; the austral region of South America; and the sub-Antarctic and Antarctic regions (Stehmann, 1986). Only six species of *Bathyraja* are known from areas outside of the three centers (one species from South Africa, one species from west coast of Africa, three species from the northern Atlantic, and two species from New Zealand) (Stehmann, 1986). One *Bathyraja* species described from New Zealand also occurs in the North Atlantic. Three additional undescribed species are known from western Australia, from Argentina, and from the North Pacific. The third clade of Arhynchobatinae, with one exception, is distributed in the Indo-West Pacific. *Arhynchobatis* are monotypic and endemic to New Zealand. *Irolita* consist of two species (one undescribed) and are endemic to western and southern Australia. *Notoraja*, with six described and a large number of undescribed species, are almost exclusively known from the southwestern Pacific and Indian Oceans. The sole exception occurs off southern Japan. The species of *Notoraja* from New Zealand and adjacent ridges and rises share a synapomorphy (paired thorns on the rostrum) and represent an undescribed genus that is sister to some or to all of the remainder of *Notoraja* (Last and McEachran, unpubl. ms.). Many of the undescribed species of *Notoraja* are found on isolated ridges and seamounts of the southern oceans and apparently are restricted to these topographic highs by surrounding abyssal depths. *Pavoraja* consist of two described and four undescribed species endemic to various regions around Australia. *Pseudoraja* are the biogeographic outlier of this Indo-West Pacific clade. The genus is monotypic and endemic to the Gulf of Mexico and Caribbean Sea. It is possible that *Pseudoraja* are misclassified in this clade. No mature males are known so clasper characters, which provided phylogenetic information in the analysis, were unavailable for study.

Rajinae, the other major clade of rajids, are largely restricted to the Atlantic, and to a lesser extent, the western North Pacific, western South Pacific, and eastern North Pacific. The subfamily contains an unresolved trichotomy (Rajini, Gugesiellini, and Amblyrajini). Rajini consist of two major clades: (*Anacanthobatis* + *Cruriraja*) and a largely unresolved clade ((*Raja* + *Rostroraja*) + *Dipturus* + *Okamejei*

+ North Pacific Assemblage + amphi-American Assemblage). The two assemblages are putative, unnamed generic-level taxa. *Cruriraja,* with eight species and *Anacanthobatis* with eight to ten species, have similar distributional patterns. They range from the Caribbean Sea and the Gulf of Mexico to South Africa to the eastern Indian Ocean and East China Sea. The remaining genera and generic-level taxa of Rajini have, with two exceptions, complementary distributions. *Raja,* with 13 species, and monotypic *Rostroraja* are endemic to the eastern Atlantic, Mediterranean Sea, and the southwestern Indian Ocean. Some of these species occur from Scandinavia to South Africa while others are restricted to relatively small ranges, e.g., Madeira, western Mediterranean, or the Mediterranean. *Dipturus,* with 30 species, occur worldwide, with the exception of the eastern North Pacific and the tropical western Pacific. They are most diverse in the Atlantic (13 species), southern African region (5 species), and Australia (9 species, 8 undescribed), and poorly represented in the Indian Ocean. *Okamejei,* with 13 species, are endemic to the tropical to warm temperate waters of the western Pacific (9 species) and central Indian Ocean (4 species). The North Pacific Assemblage, with 6 species, occurs in the temperate waters of the western North Pacific (1 species) and eastern North Pacific (5 species). The amphi-American Assemblage, with 7 species, occurs in the tropical to warm temperate waters of the eastern Pacific (2 species) and western Atlantic (5 species). The second major clade of Rajinae, Gurgesiellini, consists of (*Malacoraja* + (*Neoraja* + (*Gurgesiella* + *Fenestraja*))). *Malacoraja,* with 3 species, occur in the North Atlantic and eastern Atlantic. *Neoraja,* with 3 species, are endemic to the central western Atlantic and eastern Atlantic. *Gurgesiella,* with 3 species, occur in the tropical western Atlantic and eastern South Pacific. *Fenestraja,* with 8 species, are distributed in the western central Atlantic, western Indian Ocean, off Madagascar and India, and western Central Pacific off Celebes, and display a similar distributional pattern to those of *Anacanthobatis* and *Cruriraja.* Most of the species are endemic to the western Atlantic (5 species). Amblyrajini, third major clade of Rajinae, consist of a trichotomy (*Amblyraja* + (*Leucoraja* + *Dactylobatus*) + (*Breviraja* + *Rajella*)). *Amblyraja,* with 10 species, have an antitropical distribution. Most species occur in the North Atlantic (3 species) and western south Atlantic (3 species). One species is endemic to Antarctica and one may occur worldwide in temperate seas. *Leucoraja,* with 12 species, are primarily limited to the western North Atlantic (5 species), and the eastern Atlantic and southwestern Indian Ocean (7 species). One undescribed species of *Leucoraja* is endemic to western Australia (Last and Stevens, 1994). *Dactylobatus,* with 2 species, are endemic to the central western Atlantic. *Rajella,* with 14 species, occur in the western Atlantic, off South Africa, eastern South Pacific, eastern Indian Ocean, and southern Australia. Most species occur in the North Atlantic (7 species) and off South Africa (5 species). *Breviraja,* with 5 species, are endemic to the warm temperate to tropical western North Atlantic.

The myliobatiforms are circumglobal in tropical to warm temperate seas from the shoreline to the outer continental and insular shelves and slopes, and several taxa are distributed in freshwaters. For the most part, myliobatiforms are limited to continental and insular shelves.

Platyrhinidae occur in the Pacific; *Platyrhina,* with two species, occur in the western North Pacific; and *Platyrhinoidis,* with a single species, occur in the eastern North Pacific. *Zanobatus* is monotypic and is endemic to the tropical eastern Atlantic.

Hexatrygonidae consist of one to five species ranging from South Africa to Hawaii (Compagno, 1999). Urolophidae comprise about 22 species in three genera (*Urolophus,* with 17 species and *Trygonoptera,* with 4 species, and monotypic *Plesiobatis*). *Urolophus* occur in the western Pacific from southern Australia to Japan, with greatest species diversity in Australia. *Trygonoptera* are endemic to Australia. *Plesiobatis* range from South Africa to Hawaii. Urotrygonidae include 17 species in two genera (*Urobatis* and *Urotrygon*) endemic to tropical to warm temperate eastern Pacific and western Atlantic.

Dasyatoidea are circumglobal in tropical to warm temperate waters. *Dasyatis kuhlii* occurs throughout the Indo-West Pacific. Monotypic *Pastinachus* are widespread throughout the Indo-West Pacific. *Urogymnus,* with two or three species, occur in the tropical eastern Atlantic and the Indo-West Pacific. The Indo-West Pacific *Himantura* consist of 20 species in the Indo-West Pacific, and some species are restricted to freshwaters.

Dasyatidae, with 38 species in three genera, are circumglobal in tropical to warm temperate seas. *Dasyatis* have 34 species that occur worldwide in tropical to warm temperate waters, with some species exclusively occurring in tropical freshwaters (Berra, 2001). Greatest species richness is in the Indo-West

Pacific. *Pteroplatytrygon* is monotypic and has a circumtropical to warm temperate distribution. *Taeniura* have three species, two ranging throughout the Indo-West Pacific and one in the eastern Atlantic and Mediterranean Sea.

Potamotrygonidae, with 22 species in four genera (amphi-American *Himantura*, *Paratrygon*, *Plesiotrygon*, and *Potamotrygon*) are endemic to the New World. Amphi-American *Himantura* consist of two species, one in the eastern central Pacific and one in the western central Atlantic. The other three genera are found exclusively in the Atlantic drainages of South America, from the Rio Atrato to the Rio de la Plata (Berra, 2001).

Gymnuridae, with 11 species in two genera (*Aetoplatea* and *Gymnura*), are circumglobal in tropical to warm temperate seas. *Aetoplatea* have two species distributed in the Indian Ocean and western Pacific. *Gymnura* consist of nine species and are circumglobal in tropical to warm temperate seas.

Myliobatidae, with 33 species in seven genera (*Aetobatus*, with 2 species, *Aetomylaeus*, with 3 species, *Myliobatis*, with 11 species, *Pteromylaeus*, with 2 species, *Rhinoptera*, with 5 species, *Manta*, with 1 species, and *Mobula*, with 9 species), are circumglobal in tropical to warm temperate seas. *Aetomylaeus* occur throughout the Indo-West Pacific. *Pteromylaeus* occur in the eastern North Pacific and eastern Atlantic and Mediterranean. The other genera are circumtropical in tropical to warm temperate seas.

Several patterns are evident from the above distributions: (1) The Indo-West Pacific region has the highest taxon richness and thus appears to be the major center of radiation of batoids; (2) a slightly smaller number of genera are endemic to the New World, suggesting that the Americas served as a secondary center of radiation of the batoids; (3) some of the clades reflect a Tethyan Sea distribution, suggesting that the Tethys Sea was a major route into the proto-central Atlantic that formed in the Late Mesozoic and the eastern central Pacific; (4) several of the genera display distributional tracts from the western Central Atlantic to southern Africa and Madagascar to the eastern central Indian Ocean or the South China Sea; and (5) some of the major divisions within several the major clades have complementary distributions that may reflect ancient vicariant events.

About 22% (16 of 72) of the genera of batoids are endemic to the Indo-West Pacific, including 6 of the 10 genera of torpedinifoms, 6 of the 28 genera of rajids, and 9 of the 16 genera of myliobatiforms. The Indo-West Pacific also contains a greater percentage of species of batoids than any other region. The vast majority of basal taxa (torpediniforms, pristids, *Rhina*, *Rhynchobatus*, platyrhinids, *Hexatrygon*, and Urolophidae) occur in the Indo-West Pacific, including all but two of the genera of torpediniforms. *Zanobatus* is the only basal clade that does not occur in the Indo-West Pacific. This pattern suggests that the Indo-West Pacific was a major center of radiation of the batoids.

More than 19% (14 of 72) of the genera of batoids are endemic to the New World, including 2 of the 10 genera of torpediniforms, 5 of the 28 genera of rajids, and 3 of the 16 genera of myliobatiforms. However, none of the basal clades of batoids is endemic and only about 19% of the batoid species are endemic to the New World. These factors suggest that the New World radiations are more recent than the Indo-West Pacific radiations. Three patterns are evident in the distributions of amphi-American fauna. The first pattern is made up of genera that occur in the central or tropical eastern Pacific and western Atlantic (*Diplobatus*, *Zapteryx*, amphi-American rajid taxon, *Gurgesiella*, *Urobatis*, and *Urotrygon*). In all cases species occur either in the eastern Pacific and western Atlantic. Thus, emergence of the Isthmus of Panama in the Pliocene likely served as vicariant event for these genera. The second pattern includes genera that occur only in the central western Atlantic (*Pseudoraja*, *Dactylobatus*, and *Breviraja*). It is not known if these genera existed in the eastern Pacific prior to the emergence of the Isthmus of Panama and subsequently went extinct. However, there are a large number of batoid genera in the central western Atlantic that do not occur in the eastern Pacific (*Anacanthobatis*, *Cruriraja*, *Leucoraja*, *Fenestraja*, *Dactylobatus*, and *Breviraja*). The third pattern consists genera endemic to the austral region of South America (*Discopyge*, *Atlantoraja*, *Rioraja*, *Psammobatis*, and *Sympterygia*); however, *Atlantoraja* and *Rioraja* are known only from the southwestern Atlantic. Some of these genera may have had a more extensive, western Gondwana distribution in the early Cenozoic or vicariated after the separation of South America and Antarctic in the Eocene/Oligocene time.

Several genera and subgenera are distributed in the Indo-West Pacific and the eastern Atlantic: *Rhina*, *Rhynchobatus*, *Rhinobatos* (*Rhinobatos*), *Urogymnus*, and *Taeniura*. These distributions may represent historical Tethys Sea distributions. The Tethys Sea was a major tropical waterway that extended from the western Pacific to the proto-central Atlantic until the Miocene. It is possible that other genera with tropical distributions likewise had Tethyan distributions but subsequent dispersal events have erased the historical patterns.

Three genera of rajids display a distributional tract extending from the western central Atlantic to southern Africa and Madagascar to the eastern central Indian Ocean and South China Sea (*Anacanthobatis, Cruriraja,* and *Fenestraja*). This tract may reflect the historical distribution of these taxa and their route into the proto-Atlantic during the late Mesozoic to early Cenozoic. All three genera are limited to depths between 200 and 1000 m (McEachran and Miyake, 1990b). The southern route between Africa and Antarctica was the first deep-sea portal into the proto-Atlantic (Reyment, 1980; Pitman et al., 1993).

The major components of torpediniforms and rajids have complementary distributions that may reflect basal vicariant events in the two taxa. Torpedinidae have largely an antitropical distribution and Narcinidae have largely a circumtropical–warm temperate distribution. The rajid subfamily Arhynchobatinae, for the most part, is distributed in the Pacific, the western South Atlantic, and southern Indian and western South Pacific, and the rajid subfamily Rajinae, for the most part, is distributed in the western North Atlantic, the eastern Atlantic, and warm temperate regions to subtropical regions of the Indian Ocean and the western Pacific. It thus appears that much of the evolution within Arhynchobatinae occurred outside of the Atlantic and that much of the evolution of Rajinae occurred in the Atlantic.

Acknowledgments

Tsutomu Miyake prepared the illustrations of the claspers of *Narcine bancroftii, Discopyge tschudii, Urolophus bucculentus, Urobatus halleri, Urotrygon munda, Dasyatis americana, Potomotrygon magdalenae,* and *Gymnura micrura,* and provided illustrations of batoid species used but not illustrated in this chapter. Hera Konstantinou read an earlier draft of the manuscript and offered valuable suggestions. Hernán López-Fernández ran the PAUP version *4.0 Beta 10 analyses and prepared some of the cladogram illustrations. Heather Prestridge assisted with figure preparations. Curators of the following museums provided specimens used in this study: Bernice P. Bishop Museum (BPBM); California Academy of Sciences (CAS); JLB Smith Institute of Ichthyology (RUSI); Kyoto University, Department of Fisheries, Faculty of Agriculture (FAKU); Los Angeles County Museum (LACM); Museum of Comparative Zoology (MCZ); National Museum of Natural History, Smithsonian Institution (USNM); Scripps Institution of Oceanography (SIO); and the Texas Cooperative Wildlife Collection (TCWC). Acronyms are according to Leviton et al. (1985). Data used in the study were collected under grants from the National Science Foundation to J.D.M. (DEB82-04661 and BSR87-00292). Travel funds to visit the MCZ were provided by the Ernst Mayr Grant Fund to J.D.M. and Tsutomu Miyake.

Appendix 3.1: Specimens Examined

Chimaeridae: *Hydrolagus alberti* (TCWC 10940.01)

Heterodontidae: *Heterodontus francisci* (TCWC 3284.01); *H. mexicanus* (TCWC 7581.01)

Hexanchidae: *Heptranchias perlo* (TCWC 8534.01)

Torpedinidae: *Torpedo californica* (MCZ 43); *T. marmorata* (MCZ 42); *T. nobiliana* (TCWC uncataloged); *T. tremens* (TCWC 12124.01)

Hypnidae: *Hypnos monopterygius* (MCZ 38602)

Narkidae: *Narke japonica* (MCZ 1339); *Typhlonarke aysoni* (FAKU 46477, FAKU 47178)

Narcinidae: *Benthobatis marcida* (TCWC 442.01, TCWC 1903.01); *Diplobatis pictus* (TCWC 1900.01, TCWC 1909.01, TCWC 5291.01); *Discopyge tschudii* (FAKU 105040, FAKU 105043); *Narcine bancroftii* (TCWC 2923.01, TCWC 6808.01, TCWC 12125.01)

Pristidae: *Pristis pectinata* (MCZ 36960)

Rhinidae: *Rhina ancylostoma* (TCWC uncataloged); *Rhynchobatus djiddensis* (MCZ 806)

Rhinobatidae: *Rhinobatos lentiginosus* (TCWC 2191.02); *R. percellens* (MCZ 40025); *R. planiceps* (TCWC uncataloged); *R. productus* (CAS 65978); *Trygonorrhina fasciata* (MCZ 982S); *Zapteryx exasperata* (MCZ 833S, TCWC 7581.01); *Z. zyster* (TCWC 10846.01)

Rajidae: *Amblyraja hyperborea* (TCWC 3846.01); *A. radiata* (TCWC 2722.02); *Anacanthobatis americanus* (TCWC 2802.01); *Arhynchobatis asperrimus* (MCZ 40268); *Atlantoraja castelnaui* (TCWC uncataloged); *Bathyraja maculata* (TCWC 12040.07); *B. parmifera* (TCWC 6385.01); *Breviraja claramaculata* (TCWC 2728.02); *Cruriraja parcomaculata* (TCWC 3093.03); *C. rugosa* (UF 29861); *Dactylobatus clarkii* (TCWC 2703.01); *Dipturus batis* (TCWC 2819.05); *D. olseni* (TCWC 6839.29); *Fenestraja plutonia* (TCWC 6964.01); *Gurgesiella atlantica* (TCWC 3364.01); *Irolita waitei* (WAM P702); *Leucoraja circularis* (MNHN 1334); *L. erinacea* (TCWC 5260.01); *Malacoraja senta* (TCWC 4179.01); *Neoraja caerulea* (ISH 720/74); *Notoraja ochroderma* (CSIRO H248501); *Okamejei acutispina* (MCZ 40330); *Pavoraja alleni* (FSFRL EB-070); *Psammobatis extenta* (TCWC 3488.01); *Pseudoraja fischeri* (TCWC uncataloged); *Raja eglanteria* (TCWC 839.01); *R. miraletus* (TCWC 6454.01); *Rajella bigelowi* (TCWC 2811.01); *R. fuliginea* (TCWC 2701.01); *Rioraja agassizi* (FSFRL EM-101); *Rostroraja alba* (TCWC 3093.04); *Sympterygia brevicaudata* (TCWC 5445.01)

Platyrhinidae: *Platyrhina sinensis* (CAS 15919); *Platyrhinoidis triseriata* (CAS 31248)

Zanobatidae: *Zanobatus schoenleinii* (USNM 222120, TCWC uncataloged)

Plesiobatidae: *Plesiobatis daviesi* (RUSI 7861, BPBM 24578, RUSI 7861, TCWC uncataloged)

Urolophidae: *Urolophus bucculentus* (FSFRL EC-361)

Urotrygonidae: *Urobatis concentricus* (LACM 31771-2; TCWC 7563.07); *U. halleri* (TCWC 7586.05); *U. jamaicensis* (TCWC 815.01); *Urotrygon aspidura* (CAS 51834, CAS 51835-13); *U. asterias* (LACM 7013-4); *U. chilensis* (LACM 7013, USNM 29542); *U. microphthalmum* (USNM 222692); *U. munda* (USNM 220612-4); *U. rogersi* (LACM W50-51-12); *U. venezuelae* (USNM 121966, TCWC 7054.02)

Dasyatidae: *Dasyatis americana* (TCWC 2749.01, TCWC 5820.01); *D. brevis* (TCWC 12099.01); *D. kuhlii* (TCWC uncataloged); *D. longa* (TCWC 12102.01); *D. sabina* (TCWC 2790.01, TCWC 5824.01); *D. say* (TCWC 2791.01); *Himantura pacifica* (TCWC uncataloged); *Pteroplatytrygon violacea* (TCWC 10251.01); *Taeniura lymma* (TCWC 5278.01)

Potamotrygonidae: *Potamotrygon constellata* (MCA 2955); *P. magdalenae* (TCWC uncataloged)

Gymnuridae: *Gymnura marmorata* (TCWC uncataloged); *G. micrura* (TCWC 642.08)

Myliobatidae: *Aetobatus narinari* (MCZ 1400, TCWC 12107.01); *Myliobatis californicus* (MCZ 395, TCWC 12105.01); *M. freminvillii* (TCWC uncataloged); *M. goodei* (TCWC 3699.01); *M. longirostris* (TCWC 12106.01)

Rhinobatidae: *Rhinoptera bonasus* (TCWC 4423.01); *R. steindachneri* (TCWC uncataloged)

Mobulidae: *Mobula hypostoma* (MCZ 36406)

Appendix 3.2: Character Matrix

Chimaeridae	0000100000	01020??000	0000000?01	?000000000	0000000011	0000000000	000001?000	0000000000	00
Heterodontidae	0000000000	00000??000	0000000100	0000000000	0000000000	0000000000	000000?000	0000000000	00
Chlamydoselachidae	0000000000	00000??000	0000000000	0000000000	0000000000	0000000000	000000????	?000000000	00
Hexanchidae	0000000000	00000??000	0000000000	0000000000	0000000000	0000000000	000000????	?000000000	00
Torpedo nobiliana	1111111103	00020??100	0000001101	?1??000011	0002100001	0001000000	1001000001	0?00200110	01
Hypnos monopterygius	1111111103	00020??100	0000001101	?1??000011	0002100001	0?00100000	100100????	??00200110	01
Narcine brasiliensis	1111111103	00020??100	0000001101	?1??000000	0002100001	0000100000	1001000102	0000210111	01
Narke japonica	1111111103	00020??100	0000001101	?1??000000	0002000001	0000100000	1001000102	0?00210111	11
Temera hardwickii	1111111103	00020??100	0000001111	?1??000000	0002000001	0000100000	1001000102	0?00210111	11
Pristis pectinata	1111111100	0000020000	0000000000	?1??000000	0000100000	0100000000	000?00????	??01100000	02
Rhynchobatus djiddensis	1111111100	0000100100	0100100100	0000100000	0000110000	0100?00000	110?01????	??01100000	00
Rhina ancyclostoma	1111111100	0000100000	0000100100	0000100000	0000110000	0100?00000	120?01????	??01100000	00
Rhinobatos productus	1111111101	0000120000	0100100100	0000100010	0000110000	0111010000	1001010000	0001100000	00
Zapteryx exasperata	1111111101	0000120000	0200100100	0000100010	0000110000	0111010000	1001010000	0001100000	00
Trygonorrhina fasciata	1111111103	0000100000	0200100100	0000100010	0000110000	0111010100	0111010000	??01100000	00
Platyrhina sinensis	1111111101	0000120000	0201?00100	0000100?00	0000110000	0101010000	1001010?00	0001100000	03
Platyrhinoidis triseriata	1111111101	0000120000	0201?00100	0000100?00	0000110000	0101011000	1001010?00	0001100000	03
Zanobatus schoelleinii	1111111101	0000120000	0202000100	0000100?00	0000110000	0101011000	1101010000	2001100000	03
Bathyraja pamrifera	1111111103	0001101000	0000100100	0000100100	0000110000	0131021010	1101000100	1101100000	00
Raja texana	1111111103	0001101000	0100100100	0000100100	0000110000	0111000100	1101000100	1101100000	00
Hexatrygon bickelli	1111111002	00120100??	?0?2000100	0200000100	0000111110	0121021000	1101000?00	??01101?00	??
Plesiobatis cf. daviesi	1111111003	00110??010	0012001100	0210001100	1000111110	0121011000	1101000102	2001100200	23
Urolophus bucculentus	1111111003	0012110010	0012001100	0210000100	1010111110	0121011000	1101000102	2001100200	23
Urobatis halleri	1111111003	10121??010	1012001100	0200000010	1001111110	1121011000	1121000102	2001100200	23
Urotrygon munda	1111111003	1012110010	1012001100	0200000010	1002111110	1121011000	1121000102	2001100200	23
Pteroplatytrygon violacea	1111111003	0112132010	0012001100	0210000010	1001111110	1121031000	1101001011	2111100200	23
Dasyatis brevis	1111111003	0112132011	0012001100	0200000010	1001111110	1121031000	1101001011	2111100200	23
Dasyatis kuhlii	1111111003	0112110010	0012001100	0200000010	1001111110	1121021000	1101001011	2011100200	23
Dasyatis longa	1111111003	0112132021	0012001100	0200000010	1001111110	1121031000	1101001011	2111100200	23
Taeniura lymma	1111111003	0112132010	0012001100	0200000010	1001111110	1121021000	1101001011	??11100300	23
Himantura signifer	1111111003	0112110021	0012001100	0200000010	1001111110	1121031000	1101000101	2011100300	23
Himantura schmardae	1111111003	01121??010	0012001100	0200000010	1201111110	1121021000	1101000???	??11100300	23
Potamotrygon magdalenae	1111111003	0112110030	0012001100	0200000010	1201111110	0121021000	1301000101	2011100300	23
Gymnura marmorata	1111111003	0112010011	0012001100	0300000010	0000111110	0131031011	1101000102	2011100?00	23
Myliobatis freminvillii	1111111013	0112132012	0012001100	1301011110	1013111110	1131031011	1201000102	2011101000	24
Myliobatis longirostris	1111111013	0112132012	0012001100	1301011110	1012111110	1131031011	1201000102	2011101000	24
Aetobatus narinari	1111111023	0112032012	0012001100	1311011110	1013111110	1310?1011	1201000102	2011101?00	24
Rhinoptera steindachneri	1111111033	0112032012	0012011110	1311011110	1013111110	1310?1010	1201000102	2011101400	24
Mobula hypostoma	1111111043	01120??012	0012011110	1311010110	0013111110	1310?1010	1201001???	??11101?00	24

Appendix 3.3: Character Descriptions Supporting Batoid Monophyly

Character 1. Upper eyelid: 0 = present, 1 = absent. Character 2. Palatoquadrate: 0 = articulates with neurocranium, 1 = does not articulate with neurocranium. Character 3. Pseudohyal: 0 = absent, 1 = present. Character 4. Last ceratobranchial: 0 = free of scapulocoracoid, 1 = articulates with scapulocoracoid. Character 5. Synarchial: 0 = absent, 1 = present. Character 6. Suprascapula: 0 = free of vertebral column, 1 = articulates with vertebral column. Character 7. Antorbital cartilage: 0 = free of propterygium, 1 = articulates with propterygium and nasal capsule. Character 8. Levator and depressor rostri muscles: 0 = absent, 1 = present.

External Morphological Structures

Character 9. Cephalic lobes: 0 = absent, 1–4 = present. The pelagic stingrays (*Myliobatis, Aetobatus, Rhinoptera,* and *Mobula*) possess cephalic lobes anterior to the neurocranium supported by the pectoral girdle (McEachran et al., 1996). *Myliobatis* possess a single lobe that is continuous with the pectoral fin (9,1). *Aetobatus* possess a single lobe that is discontinuous with the pectoral fin (9,2). *Rhinoptera* and *Mobula* possess paired discontinuous lobes (9,3). The paired lobes in *Mobula* are extended to form cephalic fins (9,4). Character 10. Anterior nasal lobe: 0 = poorly developed, 1–3 = well developed. Lobe is moderately expanded medially to cover most of the medial half of the naris and extends medially onto the internarial space in *Zapteryx, Platyrhina, Platyrhinoidis,* and *Zanobatus* (10,1) (McEachran et al., 1996, Fig. 1). In *Hexatrygon* the anterior lobe extends medially to join its antimere and form a nasal curtain that falls short of the mouth (10,2). Torpediniforms, *Trygonorrhina, Plesiobatis,* rajids, and myliobatiforms possess nasal curtains that extend to or just anterior to the mouth (10,3) (McEachran et al., 1996, Fig. 1). In *Plesiobatis daviesi,* from South Africa, the nasal curtain falls short of the mouth according to Compagno et al. (1989) and Nishida (1990); thus, *P. cf. daviesi,* which is from Hawaii, may represent another taxon. Character 11. Spiracular tentacle: 0 = absent, 1 = present. *Urobatis* and *Urotrygon* are unique in possessing a tentacle on the inner margin of the spiracle during their later embryonic stages (11,1) (McEachran et al., 1996). Character 12. Radial cartilages in caudal fin. 0 = present, 1 = absent. *Pteroplatytrygon, Dasyatis, Taeniura, Himantura, Potamotrygon, Gymnura, Myliobatis, Aetobatus, Rhinoptera,* and *Mobula* lack caudal fins and caudal fin radials (12,1) (McEachran et al., 1996).

Squamation

Character 13. Serrated tail stings: 0 = absent, 1 = present. All but one myliobatiform genus and most species possess serrated stings (13,1) (McEachran et al., 1996). *Urogymnus* appears to have secondarily lost a serrated spine, as have several species within *Gymnura, Aetomylaeus,* and *Mobula.* Character 14. Placoid scales: 0 = uniformly present, 1–2 = limited to absent. Placoid scales are uniformly present in Galeomomorphi and Squaleomorphi, and very limited in holocephalids. Rajids, with very few exceptions, are sparsely to densely covered with placoid scales on the dorsal surface only (14,1). Some genera, *Atlantoraja, Rioraja, Irolita, Anacanthobatis, Dipturus, Okamejei, Raja, Rostroraja,* the North Pacific Assemblage, and the amphi-American Assemblage are largely free of denticles, but this state is considered derived within Rajidae (McEachran and Dunn, 1998) and only *Anacanthobatis* and *Irolita* are totally free of denticles. Torpediniforms, *Hexatrygon,* and the myliobatiforms are largely to totally free of denticles over their entire body surface (14,2). Character 15. Enlarged placoid scales: 0 = absent, 1 = present. According to Reif (1979) enlarged placloid scales are a derived character state of Rhinobatidae, Rajidae, and Dasyatidae. They also occur in *Rhynchobatus, Rhina, Platyrhina, Platyrhinoidis,* and *Zanobatus* (15,1).

Tooth Root Vascularization and Structure

Character 16. Pulp cavities in tooth roots: 0 = large, 1–3 = elongated to absent. *Rhynchobatus, Rhina, Trygonorrhina,* and rajids have tooth roots with large pulp cavities (16,0) (Herman et al., 1994, 1995,

1996, 1997). *Hexatrygon, Urolophus, Urotrygon, Himantura, Potamotrygon*, and *Gymnura* have broad and elongated pulp cavities in tooth roots (16,1) (Herman et al., 1997, 1998), *Pristis, Rhinobatos, Zapteryx, Platyrhina, Platyrhinoidis*, and *Zanobatus* have tooth roots with small pulp cavities (16,2), and *Dasyatis, Taeniura, Myliobatis, Aetobatus*, and *Rhinoptera* (Herman et al., 1998, 1999) have tooth roots that lack pulp cavities (16,3). Character 17. Osteodentine: 0 = absent, 1–2 = present to widespread. Osteodentine is present in the roots of large teeth only in rajids (Herman et al., 1994, 1995, 1996) and this state is though to be derived (17,1) and different from widespread occurrence of osteodentine in tooth roots (17,2) of *Dasyatis, Taeniura, Myliobatis, Aetobatus*, and *Rhinoptera* (Herman et al., 1998, 1999).

Lateral Line Canals

Character 18. Cephalic lateral line canal on ventral surface: 0 = present, 1 = absent. Cephalic lateral line is present on ventral side of body in all batoids and outgroups with the exception of the torpediniforms (18,1) (McEachran et al., 1996). Character 19. Infraorbital loop of suborbital and infraorbital canals: 0 = absent, 1–3 = present. Infraorbital loop is unique to myliobatiforms but the state in *Hexatrygon* is unknown (McEachran et al., 1996). In *Plesiobatis, Urolophus, Urobatis, Urotrygon, Pteroplatytrygon, Dasyatis brevis, D. kuhlii, Taeniura*, and amphi-America *Himantura* it forms a simple posterolaterally directed loop (19,1) (Lovejoy, 1996, Figs. 3a, b; Rosenberger, 2001b). In *D. longa* and Indo-West Pacific *Himantura* it forms a complex reticular pattern or a number of loops (19,2) (Lovejoy, 1996, Fig. 3c, d; Rosenberger, 2001b). In *Potamotrygon* the loop is directed forwardly (19,3) (Lovejoy, 1996, Fig. 4a). Character 20. Subpleural loop of the hyomandibular canal: 0 = broadly rounded, 1–2 = not broadly rounded. Loop forms lateral hook in *Dasyatis* (except in *D. kuhlii*), Indo-West Pacific *Himantura*, and *Gymnura* (20,1) (McEachran et al., 1996, Fig. 2a; Rosenberger, 2001b). In the pelagic stingrays (*Myliobatis, Aetobatus, Rhinoptera*, and *Mobula*) the lateral aspects of the subpleural loop are nearly parallel (20,2) (McEachran et al., 1996, Fig. 2b). Character 21. Lateral tubules of subpleural loop: 0 = unbranched, 1 = branched. In *Urobatis* and *Urotrygon* subpleural loop bear dichotomously branched lateral tubules (21,1) (McEachran et al., 1996, Fig. 4). Character 22. Abdominal canal on coracoid bar: 0 = absent, 1–2 = present. Cephalic lateral line forms abdominal canal on coracoid bar in *Rhynchobatus, Rhinobatos, Trygonorrhina, Zapteryx, Platyrhina, Platyrhinoidis, Zanobatus*, and *Raja*. In *Rhynchobatus, Rhinobatos*, and *Raja* the canal is in a groove (22,1) (McEachran et al., 1996, Fig. 3). In *Trygonorrhina, Zapteryx, Platyrhina, Platyrhinoidis*, and *Zanobatus* canals are represented by pores (22,2). Character 23. Scapular loops of scapular canals: 0 = absent, 1 = present. The trunk lateral line forms scapular loop dorsally over the shoulder girdle in myliobatiforms, but the condition is unknown for *Hexatrygon* (23,1) (McEachran et al., 1996, Fig. 4).

Skeletal Structures

Character 24. Rostral cartilage: 0 = complete, 1–2 = incomplete or absent. The rostral cartilage fails to reach the tip of the snout in *Platyrhina* and *Platyrhinoidis* (24,1) (McEachran et al., 1996, Fig. 5). In *Zanobatus* and myliobatiforms the rostral cartilage is either vestigial or completely lacking (24,2) (McEachran et al., 1996, Fig. 6). Character 25. Rostral appendices: 0 = absent, 1 = present. Rostral appendices are present in *Rhynchobatus, Rhina, Rhinobatos, Zapteryx, Trygonorrhina*, and Rajidae (25,1) (McEachran et al., 1996, Fig. 7). Character 26. Dorsolateral components of nasal capsule: 0 = absent, 1 = present. In *Rhinoptera* and *Mobula* the dorsolateral components of the nasal capsule form a pair of projections that support the cephalic lobes or cephalic fins (26,1) (McEachran et al., 1996). Character 27. Nasal capsules: 0 = laterally expanded, 1 = ventrolaterally expanded. Nasal capsules are ventrolaterally expanded in torpediniforms and myliobatiforms, except for *Hexatrygon* (27,1) (McEachran et al., 1996). Character 28. Basal angle of neurocranium: 0 = present, 1 = absent. Basal angle on the ventral surface of the neurocranium is absent in all batoids (28,1) except *Pristis* and outgroups Chlamydoselachidae and Hexanchidae but absent in Heterodontidae and unknown for Chimaeridae (Compagno, 1977; Shirai, 1992b). Character 29. Preorbital process: 0 = present, 1 = absent. Preorbital process is absent in *Temera, Rhinoptera*, and *Mobula* (29,1) (McEachran et al., 1996). Character 30. Supraorbital crest: 0 = present, 1 = absent. Supraorbital crest is absent in torpediniforms (30,1) (McEachran et al.,

1996). Chimaerids also lack the supraorbital crest (Didier, 1995), but this state is considered to be separately derived. Character 31. Anterior preorbital foramen: 0 = dorsally located, 1 = anteriorly located. Anterior preorbital foramen opens on the anterior aspect of the nasal capsule in pelagic myliobatiforms (31,1) (McEachran et al., 1996). The state in torpediniforms is unknown possibly because they lack a suparorbital crest (31?). Character 32. Postorbital process: 0 = narrow and in otic region, 1–3 = absent or broad. Postorbital process is absent in torpediniforms (32,1). In myliobatiforms the postorbital process is very broad and shelflike (32,2); furthermore, in *Gymnura* and pelagic myliobatiforms it is located in the orbital region (32,3) (Nishida, 1990, Figs. 10–17; McEachran et al., 1996). Character 33. Postorbital process: 0 = separated from triangular process, 1 = fused with triangular process. Postorbital process is distally fused with the triangular process of the supraorbital crest with the groove between the processes represented by a foramen in *Plesiobatis, Urolophus, Pteroplatytrygon, Aetobatus, Rhinoptera,* and *Mobula* (33,1) (Nishida, 1990, Fig. 17; McEachran et al., 1996; de Carvalho et al., in press). Character 34. Postorbital process: 0 = projects laterally, 1 = projects ventrolaterally. Lateral margin of postorbital process is prolonged and projects ventrolaterally to form a cylindrical protuberance in *Myliobatis, Aetobatus, Rhinoptera,* and *Mobula* (34,1) (McEachran et al., 1996). Character 35. Jugal arch: 0 = absent, 1 = present. Hyomandibular facet and posterior section of otic capsule are joined by an arch in *Pristis, Rhynchobatus, Rhina, Rhinobatos, Trygonorrhina, Zapteryx, Platyrhina, Platyrhinoidis, Zanobatus, Bathyraja,* and *Raja* (35,1) (McEachran et al., 1996, Fig. 7). Character 36. Antimeres of upper and lower jaws: 0 = separate, 1 = fused. Antimeres of upper and lower jaws are fused in *Aetobatus, Rhinoptera,* and *Mobula,* and in some species of *Myliobatis* (36,1) (Nishida, 1990, Fig. 21; McEachran et al., 1996). Character 37. Meckel's cartilage: 0 = not expanded medially, 1 = expanded medially. Meckel's cartilage is expanded and thickened near symphysis in *Myliobatis, Aetobatus,* and *Rhinoptera* (37,1) (Nishida, 1990, Figs. 20, 21; McEachran et al., 1996). Character 38. Winglike processes on Meckel's cartilage: 0 = absent, 1 = present. Meckel's cartilage has posteriorly expanded, winglike process in *Myliobatis, Aetobatus, Rhinoptera,* and *Mobula* (38,1) (Nishida, 1990, Figs. 20, 21; McEachran et al., 1996). Character 39. Labial cartilages: 0 = present, 1 = absent. Labial cartilages are present in the majority of squaleomorphs and galeomorphs, except in *Lamna, Pseudocarcharias,* and carcharhinoids (Shirai, 1992b) and in chimaerids (Didier, 1995). Labial cartilages are absent in *Torpedo, Hypnos, Pristis, Rhinobatos, Zapteryx,* rajids, and myliobatiforms (39,1) (Compagno, 1977; Nishida, 1990; Shirai, 1992b). Character 40. Medial section of hyomandibula: 0 = narrow, 1 = expanded. Medial section of the hyomandibula is longitudinally expanded and spans the entire length of the otico-occipital region of the neurocranium in *Torpedo* and *Hypnos* (40,1) (McEachran et al., 1996). Character 41. Hyomandibular–Meckelian ligament: 0 = absent, 1 = present. Distal tip of the hyomandibula and Meckel's cartilage are joined by a long ligament (hyomandibular–Meckelian ligament but called tendon in McEachran et al., 1996) in *Zanobatus, Plesiobatis, Urolophus, Urobatis, Urotrygon, Pteroplatytrygon, Dasyatis, Himantura, Potamotrygon, Taeniura, Myliobatis, Aetobatus,* and *Rhinoptera* (41,1) (Nishida, 1990, Figs. 20, 21; Lovejoy, 1996, Fig. 6; McEachran et al., 1996, Fig. 8). Character 42. Ligamentous cartilage(s): 0 = absent, 1–2 = present. A broad and triangular cartilage is embedded in the posterior section of the hyomandibular–Meckelian ligament in *Zanobatus* (42,1) (McEachran et al., 1996, Fig. 8b). Two small cartilages lie in parallel in ligament in *Himantura schmardae* (western Atlantic species) and *Potamotrygon* (42,2) (Nishida, 1990, Fig. 24; Lovejoy, 1996, Fig. 6; McEachran et al., 1996). Character 43. Small cartilage associated with hyomandibular–Meckelian ligament: 0 = absent, 1 = present. Small cartilage or cartilages, free of ligament, are located between the hyomandibula and Meckel's cartilage in *Urolophus, Myliobatis, Aetobatus, Rhinoptera,* and *Mobula* (43,1) (Garman, 1913, Pls. 73, 74, 75; Lovejoy, 1996, Fig. 6; McEachran et al., 1996). Character 44. Basihyal and first hypobranchial: 0 = both present and unsegmented, 1–3 = segmented or absent. The basihyal is located between the paired first hypobranchial cartilages in most neoselachians (Nelson, 1969; Shirai, 1993). In *Urobatis, Pteroplatytrygon, Dasyatis, Taeniura, Himantura,* and *Potamotrygon* the basihyal is segmented (44,1) (Nishida, 1990, Fig. 27; Miyake and McEachran, 1991, Fig. 8; Lovejoy, 1996, Fig. 7; McEachran et al., 1996). In torpediniforms and *Urotrygon* the basihyal is absent (44,2) (Nishida, 1990, Fig. 27d; Shirai, 1992, Plate 32b, Fig. 7; McEachran et al., 1996). In *Aetobatus, Rhinoptera,* and *Mobula* the basihyal and first hypobranchial cartilages are absent (44,3) (Nishida, 1990, Fig. 28; McEachran and Miyake, 1991, Fig. 8; McEachran et al., 1996). *Myliobatis* lack the basihyal and either have or lack the first hypobranchial cartilage (44,2 or 44,3, respectively). Character 45. Ceratohyal: 0 = fully developed, 1 = reduced or absent. Ceratohyal cartilage of the hyoid arch articulates with the basihyal and hyomandibula in most neoselachians (Nelson, 1969; Shirai, 1992). It is partially or totally replaced by the pseudohyal in all batoids (45,1) except for *Narke* (45,0) (Miyake and McEachran, 1991, Fig. 6; Lovejoy, 1996, Fig. 7; McEachran et al., 1996). Character 46. Suprascapula: 0 = articulates with vertebrae, 1 = fused with synarchial. Suprascapular cartilage of the shoulder girdle is fused with the synarchial in all batoids (46,1) except for torpediniforms and *Pristis* (46,0) (McEachran et al., 1996). Character 47. Ball and socket articulation between scapular process and synarchial: 0 = absent, 1 = present. Suprascapular process of the shoulder girdle forms a ball and socket articulation with the synarchial in the myliobatiforms (47,1) (McEachran et al., 1996). Character 48. Second synarchial: 0 = absent, 1 = present. Second synarchial, generally separated from the first synarchial by several free vertebral centra, is found in myliobatiforms (48,1) (Nishida, 1990, Fig. 36; McEachran et al., 1996). Character 49. Ribs: 0 = present, 1 = absent. Ribs are absent in myliobatiforms (49,1) (McEachran et al., 1996). Character 50. Scapular process: 0 = short, 1 = long. Scapular process of the shoulder girdle is long and posteriorly displaced in torpediniforms (50,1) (McEachran et al., 1996). Character 51. Scapular process: 0 = without fossa, 1 = with fossa. Scapular process of the shoulder girdle possesses a fossa or foramen in *Urobatis, Urotrygon, Taeniura, Himantura, Dasyatis, Myliobatis, Aetobatus, Rhinoptera,* and *Mobula* (51,1) (Nishida, 1990, Figs. 30, 31; Lovejoy, 1996, Fig. 9; McEachran et al., 1996). Character 52. Scapulocoracoid condyles: 0 = not horizontal, 1 = horizontal. Lateral aspect of the scapulocoracoid has three horizontally arranged condyles that articulate with the propterygium, mesopterygium, and the metapterygium, respectively, in all batoids (52,1) with exception of the torpediniforms; however, the character state is unknown for *Hypnos* (McEachran et al., 1996). Character 53. Mesocondyle: 0 = equidistant, 1–2 = closer to procondyle or to metacondyle. Scapulocoracoid is elongated between the mesocondyle and the metacondyle in *Rhino-batos, Zapteryx, Trygonorrhina, Bathyraja,* and *Raja* (53,1) (Nishida, 1990, Fig. 32; McEachran et al., 1996, Fig. 9). In myliobatiforms the scapulocoracoid is elongated between the procondyle and the mesocondyle (53,2) (Nishida, 1990, Figs. 30, 31, 32; Lovejoy, 1996, Fig. 9; McEachran et al., 1996, Fig. 9). Character 54. Antorbital cartilage: 0 = indirectly joins propterygium to nasal capsule, 1 = directly joins cartilages. Antorbital cartilage directly joins the propterygium of the shoulder girdle to the nasal capsule in all batoids (54,1) except for torpediniforms, *Pristis, Rhynchobatus,* and *Rhina* (McEachran et al., 1996). Character 55. Antorbital cartilage: 0 = not anteriorly expanded, 1 = anteriorly expanded. Antorbital cartilage is anteriorly expanded and fan- or antlerlike in torpediniforms (55,1) (Miyake et al., 1992b, Fig. 16; McEachran et al., 1996). Character 56. Segmentation of propterygium: 0 = posterior to mouth, 1–3 = anterior to mouth to anterior to nasal capsule. Proximal segment of propterygium of pectoral girdle is between mouth and antorbital cartilage in *Rhinobatos, Zapteryx, Trygonorrhina, Platyrhina, Platyrhinoidis, Plesiobatis, Urolophus, Urobatis,* and *Urotrygon* (56,1). In *Zanobatus, Hexa-trygon, Dasyatis kuhlii, Taeniura, Himantura schmardae,* and *Potamotrygon* the first segment is adjacent the nasal capsule (56,2) (Lovejoy, 1996, Fig. 10; Rosenberger, 2001b, Fig. 2). In the *Pteroplatytrygon,* remaining *Dasyatis, Himantura signifer, Gymnura,* and *Myliobatis* the first segment is adjacent to anterior margin of antorbital cartilage or anterior to margin of nasal capsule (56,3) (Lovejoy, 1996, Fig. 10; Rosenberger, 2001b, Fig. 2). Character 57. Proximal section of propterygium: 0 = does not extend posterior to procondyle, 1 = extends behind procondyle. Proximal section of propterygium of shoulder girdle extends behind procondyle and articulates with scapulocoracoid between pro- and mesocondyles in *Zanobatus, Platyrhina, Platyrhinoidis,* and myliobatiforms (57,1) (Nishida, 1990, Figs. 30, 31; Love-joy, 1996, Fig. 10; McEachran et al., 1996, Fig. 9; Rosenberger, 2001b, Fig. 2; de Carvalho et al., in press). Character 58. Pectoral fin radials: 0 = all articulate with pterygials or directly with scapulocoracoid between propterygium and mesopterygium, 1 = articulate with scapulocoracoid anterior to mesopterygium. Some pectoral fin radials articulate directly with scapulocoracoid posterior to mesopterygium in *Rhinobatos, Zapteryx, Trygonorrhina, Bathyraja,* and *Raja* (58,1) (Nishida, 1990, Fig. 32; McEachran et al., 1996, Fig. 9). Character 59. Mesopterygium: 0 = present and single, 1 = fragmented or absent. Mesopterygium is fragmented or absent in *Zanobatus, Gymnura, Myliobatis, Aetobatus, Rhinoptera,* and *Mobula* (59,1) (Nishida, 1990, Figs. 31, 32a,b; McEachran et al., 1996, Fig. 9). Character 60. Pectoral fin radials: 0 = not expanded distally, 1 = some pectoral fin radials expanded distally. Some of the fin radials supported by the propterygium are expanded distally and articulate with the surface of adjacent radials in *Gymnura, Myliobatis*, and *Aetobatus* (60,1) (Nishida, 1990, Fig. 34; McEachran et al., 1996). Character 61. Paired fin rays: 0 = aplesodic, 1 = plesodic. Pectoral and pelvic fins are plesodic, radials extend to margin of fins, and ceratotrichia are reduced or absent in all batoids but *Pristis* (61,1) (McEachran et al., 1996). Character 62. Puboischiadic bar: 0 = platelike, 1–3 = narrow and moderately to greatly arched. Puboischiadic bar of the pelvic girdle is narrow and moderately to strongly arched without distinct lateral prepelvic processes in *Rhynchobatus, Zanobatus, Hexatrygon, Plesiobatis, Urolophus, Urobatis, Urotrygon, Pteroplatytrygon, Dasyatis, Taeniura, Himantura*, and *Gymnura* (62,1) (Hulley, 1970, Fig. 1; Heemstra and Smith, 1980, Fig. 12; Nishida, 1990, Fig. 36; Lovejoy, 1996, Fig. 11; McEachran et al., 1996, Fig. 10; Rosenberger, 2001b, Fig. 6). In *Myliobatis, Aetobatus*, and *Mobula* the puboischiadic bar is narrow and strongly arched, with a triangular medial prepelvic process (62,2) (Hulley, 1972a, Fig. 1; Nishida,, 1990, Fig. 36; Lovejoy, 1996, Fig. 11; McEachran et al., 1996). In *Potamotrygon* the puboischiadic bar is narrow and moderately arched, with a barlike medial prepelvic process (62,3) (Nishida, 1990, Fig. 36; Lovejoy, 1996, Fig. 11; McEachran et al., 1996). Character 63. Puboischiadic bar: 0 = without triangular processes, 1 = with triangular processes. Puboischiadic bar of *Platyrhina* and *Platyrhinoidis* have two postpelvic processes (63,1) (Nishida, 1990, Fig. 36q; McEachran et al., 1996). Character 64. First pelvic radial: 0 = bandlike, 1–2 = variable. The first pelvic radial is thickened, variously shaped, and variably associated with distal radial segments, and variations in this segment are thought to have phylogenetic significance. In torpediniforms, *Rhinobatos, Zapteryx, Trygonorrhina, Platyrhina, Platyrhinoidis, Zanobatus*, and myliobatiforms the compound radial is bandlike and slightly expanded distally, and articulates with several radial segments in parallel fashion (64,1). In *Bathyraja* and *Raja* the compound radial is rodlike and articulates with single radial segments in serial fashion (64,2) (Holst and Bone, 1993, Fig. 1; Lucifora and Vassallo, 2002, Fig. 2). Character 65. Pelvic girdle condyles: 0 = close together, 1 = separated. Pelvic girdle condyles for the compound radial and the basipterygium are distinctly separated and several radials articulate directly with the pelvic girdle between the two condyles in *Bathyraja* and *Raja* (65,1). Character 66. Clasper length: 0 = short, 1 = long. Clasper is elongated and slender in Chimaeridae, *Rhynchobatus, Rhina, Rhinobatos, Zapteryx, Trygonorrhina, Platyrhina, Bathyraja*, and *Raja* (66,1) (Didier, 1995; Ishiyama, 1958; Last and Stevens, 1994). Claspers of the other outgroups and batoids are relatively short and usually rather stout (66,0) (Capapé and Desoutter, 1979, Fig. 1; Compagno and Roberts, 1982, Fig. 10; Fechhelm and McEachran, 1984, Fig. 14; Nishida, 1990, Figs. 59, 60). Character 67. Pseudosiphon: 0 = present, 1 = absent. Clasper component pseudosiphon, a blind cavity situated on the ventromedial aspect of the clasper and formed in part by the medial margin of the ventral covering piece cartilage, is absent in *Bathyraja, Raja*, at least some *Dasyatis* species, and *Mobula* (67,1) (Hulley, 1972a, Fig. 12; Nishida, 1990, Fig. 59). Character 68. Dorsal marginal clasper cartilage: 0 = lacks medial flange, 1 = possesses medial flange. The dorsal marginal clasper cartilage possesses a medial flange that extends most of the length of the cartilage in *Narcine, Narke, Temera, Zanobatus, Plesiobatis, Urolophus, Urobatis, Urotrygon, Dasyatis kuhlii, Himantura signifer, Potamotrygon, Gymnura, Myliobatis, Aetobatus*, and *Rhinoptera* (68,1) (Hulley, 1972a, Fig. 46; Compagno and Roberts, 1982, Fig. 10; Nishida, 1990, Fig. 60). Character 69. Dorsal terminal cartilage: 0 = smooth margin, 1 crenate margin. Dorsal terminal clasper cartilage of *Pteroplatytrygon, Dasyatis brevis, D. longa*, and *D. sabina* has a crenate lateral margin (69,1). Character 70. Cartilage forming component claw: 0 = present, 1–3 = absent, not visible externally, or forms component shield. Cartilage is absent in *Torpedo, Pteroplatytrygon, Dasyatis, Himantura signifer*, and *Potamotrygon* (70,1) (Capapé and Desoutter, 1979, Fig. 2; Rosa et al., 1988; Nishida, 1990, Fig. 60). In *Narcine, Narke, Temera, Urolophus, Urobatis, Urotrygon, Gymnura, Myliobatis, Aetobatus*, and *Rhinoptera* cartilage is embedded in integument and is not visible externally (70,2) (Nishida, 1990, Fig. 60, as small cartilage 1). In *Bathyraja* and *Raja* the ventral terminal clasper cartilage lines the inner ventral margin of the clasper glan and often forms the component shield (70,3) (Ishiyama, 1958; Stehmann, 1970; Hulley, 1972a; McEachran and Miyake, 1990). Character 71. Ventral terminal cartilage (accessory terminal 1 cartilage in rajids): 0 = simple, 1–2 = forming component sentinel or projection, or complex. Ventral terminal clasper cartilage is free distally and forms component sentinel or is fused with ventral marginal cartilage and forms component projection in *Bathyraja* and *Raja* (71,1) (Ishiyama, 1958; Stehmann, 1970; Hulley, 1972a; McEachran and Miyake, 1990). In *Zanobatus, Plesiobatis, Urolophus, Urobatis, Urotrygon, Pteroplatytrygon, Dasyatis, Himantura signifer, Potamotrygon, Gymnura, Myliobatis,*

Aetobatus, and *Rhinoptera* the ventral terminal cartilage is folded ventrally along its long axis to form a convex flange (71,2) (Hulley, 1972a, Fig 46, as accessory terminal 1 cartilage; Compagno and Roberts, 1982, Fig. 10, as ventral terminal cartilage). Character 72. Ventral terminal cartilage (accessory terminal 1 cartilage in rajids): 0 = attached over length to axial cartilage, 1 = free of axial. Ventral terminal clasper cartilage is free of axial cartilage in *Bathyraja, Raja, Pteroplatytrygon, Dasyatis brevis, D. longa,* and *D. sabina* (72,1) (Ishiyama, 1958; Stehmann, 1970; Hulley, 1972a; McEachran and Miyake, 1990). Character 73. Caudal vertebrae: 0 = diplospondylous, 1 = fused. Caudal vertebrae distal to serrated tail sting are fused into a tube in *Pteroplatytrygon, Dasyatis, Taeniura, Himantura, Potamotrygon, Myliobatis, Aetobatus, Rhinoptera,* and *Mobula* (73,1) (Lovejoy, 1996, Fig. 12; McEachran et al., 1996).

Cephalic and Branchial Musculature

Character 74. Ethmoideo-parethmoidalis: 0 = absent, 1 = present. Cranial muscle, ethmoideo-parethmoidalis, is present in all batoids (74,1) except for the torpediniforms (74,0) (Nishida, 1990, Figs. 43, 45, 46; McEachran et al., 1996). Character 75. Intermandibularis: 0 = present, 1–2 = absent or modified. The mandibular plate muscle, intermandibularis, is present in sharks but absent in batoids (75,1) except for torpediniforms (McEachran et al., 1996). In torpediniforms the intermandibularis muscle is a narrow band of muscle that originates on the hyomandibula and inserts on the posterior margin of Meckel's cartilage (75,2) (McEachran et al., 1996). Character 76. Ligamentous sling on Meckel's cartilage: 0 = absent, 1 = present. In *Narcine, Narke,* and *Temera* a ligamentous sling at the symphysis of Meckel's cartilage supports the intermandibularis, coracomandibularis, and depressor mandibularis muscles (76,1) (Miyake et al., 1992a; McEachran et al., 1996). Character 77. Depressor mandibularis: 0 = present, 1 = absent. The depressor mandibularis muscle is either absent or does not exist as an independent muscle in *Myliobatis, Aetobatus, Rhinoptera,* and *Mobula* (77,1) (McEachran et al., 1996). Character 78. Spiracularis: 0 = undivided, 1–4 = divided in various ways. Mandibular plate muscle, spiracularis, is divided and one bundle enters the dorsal oral membrane underlying the neurocranium (78,1) (Miyake et al., 1992a; McEachran et al., 1996). In *Plesiobatis, Urolophus, Urobatis, Urotrygon, Pteroplatytrygon,* and *Dasyatis* the spiracularis splits into lateral and medial bundles, and the medial bundle inserts onto the posterior surface of Meckel's cartilage and the lateral bundle inserts onto the dorsal edge of the hyomandibula (78,2) (Miyake et al., 1992a; McEachran et al., 1996). In *Taeniura, Himantura,* and *Potamotrygon* the muscle extends beyond the hyomandibula and Meckel's cartilage (78,3) (Miyake et al., 1992a; Lovejoy, 1996, Fig. 13b; McEachran et al., 1996). The spiracularis muscle is subdivided proximally and inserts separately onto the palatoquadrate and the hyomandibula in *Rhinoptera* (78,4) (McEachran et al., 1996). Character 79. Branchial electric organs: 0 = absent, 1 = present. Electric organs derived from branchial muscles are present in the torpediniforms (79,1). Character 80. Coracobranchialis muscle: 0 = consists of three to five components, 1 = single component. Coracobranchialis muscle of the branchial muscle plate consists of a single component in *Narcine, Narke,* and *Temera* (80,1) (Miyake et al., 1992a; McEachran et al., 1996). Character 81. Coracohymandibularis: 0 = single origin, 1–2 = separate origins. Coracohymandibularis of the hypobranchial muscle plate has separate origins on the facia supporting the insertion of the coracoarcualis and on the pericardial membrane in *Narke* and *Temera* (81,1) (McEachran et al., 1996). In the myliobatiforms the muscle has separate origins on the anterior portion of the ventral gill arch region and on the pericardial membrane (81,2) (McEachran et al., 1996). Character 82. Cocracohyoideus: 0 = parallel to body axis, 1–4 = absent, parallel to body axis or short, or diagonal to body axis. The cocracohyoideus of the hypobranchial muscle plate is absent in the torpediniforms (82,1) (Miyake et al., 1992a; McEachran et al., 1996). In *Pristis* the muscle runs parallel to the body axis and is very short (82,2) (Miyake et al., 1992a; McEachran et al., 1996). In *Platyrhina, Platyrhinoidis, Zanobatus,* and benthic myliobatiforms (*Plesiobatis, Urolophus, Urobatis, Urotrygon, Pteroplatytrygon, Dasyatis, Taeniura, Himantura, Potamotrygon,* and *Gymnura*) the muscle runs diagonally from the wall of the first two gill slits to the posteromedial aspect of the basihyal or first basibranch (82,3) (Nishida, 1990, Fig. 53a; McEachran et al., 1996, Fig. 11b). In the pelagic myliobatiforms each muscle fuses with its antimere by means of raphe near its insertion on the first hypobranch (82,4) (Nishida, 1990, Fig. 53b; McEachran et al., 1996).

References

Arambourg, C. and L. Bertin. 1958. Classe des Chondrichthyens, in *Traité de Zoologie*. P.P. Grassé, Ed., Masson, Paris, 2012–2056.

Berg, L. 1940. Classification of fishes, both recent and fossil. *Tr. Zool. Inst. Lenigrad* 5:87–517.

Berra, T.M. 2001. *Freshwater Fish Distribution.* Academic Press, New York.

Bertin, L. 1939. Essai de classification et de nomenclature des Poisson de la sous-classe des Sélaciens. *Bull. Inst. Oceanogr.* 775:1–23.

Bigelow, H.B. and W.C. Schroeder. 1953. Sawfishes, guitarfishes, skates, and rays; and chimaeroids, in *Fishes of the Western North Atlantic.* J. Tee-Van, C.M. Breeder, A.E. Parr, W.C. Schroeder, and L.P. Schultz, Eds., Sears Foundation of Marine Research, New Haven, CT.

Bremer, K. 1994. Branch support and tree stability. *Cladistics* 10:295–304.

Capapé, C. and M. Desoutter. 1979. Etude morphologique des pterygopodes de *Torpedo* (*Torpedo*) *marmorata* Risso, 1810 (Pisces, Torpedinidae). *Neth. J. Zool.* 29:443–449.

Cappetta, H. 1987. Chondrichthyes II: Mesozoic and Cenozoic Elasmobranchii, in *Handbook of Paleoichthyology,* Vol. 3B. H.-P. Schultze, Ed., Gustav Fischer Verlag, New York, 1–193.

Carpenter, J.M. 1988. Choosing among multiple equally parsimonious cladograms. *Cladistics* 4:291–296.

Carpenter, J.M. 1994. Successive weighting, reliability and evidence. *Cladistics* 10:215–220.

Chu, Y.T. and M.C. Wen. 1980. A study of the lateral-line canals system and that of Lorenzini ampullae and tubules of elasmobranchiate fishes of China. *Monograph of Fishes of China,* 2. Shanghai Science and Technology Press, Shanghai, People's Republic of China.

Compagno, L.J.V. 1973. Interrelationships of living elasmobranches, in *Interrelationships of Fishes.* P.A. Greenwood, R.S. Miles, and C. Patterson, Eds., Academic Press, New York, 15–61.

Compagno, L.J.V. 1977. Phyletic relationships of living sharks and rays. *Am. Zool.* 17:303–322.

Compagno, L.J.V. 1999. Checklist of living elasmobranchs, in *Sharks, Skates, and Rays.* W.C. Hamlett, Ed., Johns Hopkins University Press, Baltimore, 471–498.

Compagno, L.J.V. and T.R. Roberts. 1982. Freshwater stingrays (Dasyatidae) of southeast Asia, with a description of a new species of *Himantura* and reports of unidentified species. *Environ. Biol. Fish.* 7:321–339.

Compagno, L.J.V., D.A. Ebert, and M.J. Smale. 1989. *Guide to the Sharks and Rays of Southern Africa.* New Holland Publishers, London.

de Carvalho, M.R. 1996. Higher-level elasmobranch phylogeny, basal squaleans, and paraphyly, in *Interrelationships of Fishes.* M.L.J. Stiassny, L.R. Parenti, and G.D. Johnson, Eds., Academic Press, London, 35–62.

de Carvalho, M.R. 1999. A Systematic Revision of the Electric Ray Genus *Narcine* Henle, 1834 (Chondrichthyes: Torpediniformes: Narcinidae), and an Analysis of the Higher-Level Phylogenetic Relationships of the Orders of Elasmobranch Fishes (Chondrichthyes). Ph.D. dissertation, City University of New York, New York.

de Carvalho, M.R., J.G. Maisey, and L. Grande. In press. Freshwater stingrays of the Green River Formation of Wyoming (Early Eocene), with the descriptions of a new genus and species and an analysis of its phylogenetic relationships (Chondrichthyes: Myliobatiformes). *Bull. Amer. Mus. Nat. Hist.*

Didier, D.A. 1995. Phylogenetic systematics of extant chimaeroid fishes (Holocephali, Chimaeroidei). *Am. Mus. Novit.* 3119:1–86.

Douady, C.J., M. Dosay, M.S. Shivji, and M.J. Stanhope. 2003. Molecular phylogenetic evidence refuting the hypothesis of Batoidea (rays and skates) as derived sharks. *Mol. Phylog. Evol.* 26:215–221.

Dunn, K.A., J.D. McEachran, and R.L. Honeycutt. 2003. Molecular phylogenies of myliobatiform fishes (Chondrichthyes: Myliobatiformes), with comments on the effects of missing data on parsimony and likelihood. *Mol. Phylog. Evol.* 27:259–270.

Farris, J.S. 1969. A successive approximation approach to character weighting. *Syst. Zool.* 18:374–385.

Fechhelm, J.D. and J.D. McEachran. 1984. A revision of the electric ray genus *Diplobatus* with notes on the interrelationships of Narcinidae (Chondrichthyes, Torpediniformes). *Bull. Fla. State Mus. Biol. Sci.* 29:173–209.

Garman, S. 1913. The Plagiostomia (sharks, skates, and rays). *Mem. Mus. Comp. Zool. Harvard* 36:1–528.

Goodrich, E.S. 1909. Vertebrata Craniata, in *A Treatise on Zoology,* Vol. 9. R. Lankester, Ed., Black, London, 518 pp.

Günther, A. 1870. *Catalogue of the Fishes in the British Museum,* Vol. 8. British Museum (Natural History), London, 549 pp.

Heemstra, P.C. and M.M. Smith. 1980. Hexatrygonidae, a new family of stingrays (Myliobatiformes: Batoidea) from South Africa, with comments on the classification of batoid fishes. *Ichthyol. Bull. J.L.B. Smith Inst. Ichthyol.* 43:1–17.

Herman, J., M. Hovestadt-Euler, D.C. Hovestadt, and M. Stehmann. 1994. Contributions to the study of the comparative morphology of teeth and other relevant ichthyodorulites in living supra-specific taxa of Chondrichthyan fishes. Part B: Batomorphi No. 1a: Order Rajiformes-Suborder Rajoidei–Family: Rajidae Genera and Subgenera: *Anacanthobatis* (*Schroederobatis*), *Anacanthobatis* (*Springeria*), *Breviraja*, *Dactyobatus*, *Gurgesiella* (*Gurgesiella*), *Gurgesiella* (*Fenstraja*), *Malacoraja*, *Neoraja* and *Pavoraja*. *Bull. Inst. R. Sci. Nat. Belg.* 64:165–207.

Herman, J., M. Hovestadt-Euler, D.C. Hovestadt, and M. Stehmann. 1995. Contributions to the study of the comparative morphology of teeth and other relevant ichthyodorulites in living supra-specific taxa of Chondrichthyan fishes. Part B: Batomorphi No. 1b: Order Rajiformes-Suborder Rajoidei-Family: Rajidae-Genera and Subgenera: *Bathyraja* (with a deep-water, shallow water and transitional morphotypes), *Psammobatis*, *Raja* (*Amblyraja*), *Raja* (*Dipturus*), *Raja* (*Leucoraja*), *Raja* (*Raja*), *Raja* (*Rajella*) (with two morphotypes), *Raja* (*Rioraja*), *Raja* (*Rostroraja*), *Raja lintea*, and *Sympterygia*. *Bull. Inst. R. Sci. Nat. Belg.* 65:237–307.

Herman, J., M. Hovestadt-Euler, D.C. Hovestadt, and M. Stehmann. 1996. Contributions to the study of the comparative morphology of teeth and other relevant ichthyodorulites in living supra-specific taxa of Chondrichthyan fishes. Part B: Batomorphi No. 1c: Order Rajiformes-Suborder Rajoidei-Family: Rajidae-Genera and Subgenera: *Arhynchobatis*, *Bathyraja richardsoni*-type, *Cruriraja*, *Irolita*, *Notoraja*, *Pavoraja* (*Insentiraja*), *Pavoraja* (*Pavoraja*), *Pseudoraja*, *Raja* (*Atlantoraja*), *Raja* (*Okamejei*) and *Rhinoraja*. *Bull. Inst. R. Sci. Nat. Belg.* 66:179–236.

Herman, J., M. Hovestadt-Euler, D.C. Hovestadt, and M. Stehmann. 1997. Contributions to the study of the comparative morphology of teeth and other relevant ichthyodorulites in living supra-specific taxa of Chondrichthyan fishes. Part B: Batomorphi No. 2: Order Rajiformes-Suborder: Pristiodei-Family; Prisidae-Genera: *Anoxypristis* and *Pristis* No. 3: Suborder Rajoidei-Superfamily Rhinobatoidea-Families: Rhinidae-Genera *Rhina* and *Rhynchobatus* and Rhinobatidae-Genera: *Aptychotrema*, *Platyrhina*, *Platyrhinoidis*, *Rhinobatos*, *Trygonorrhyna*, *Zanobatus* and *Zapteryx*. *Bull. Inst. R. Sci. Nat. Belg.* 67:107–162.

Herman, J., M. Hovestadt-Euler, D.C. Hovestadt, and M. Stehmann. 1998. Contributions to the study of the comparative morphology of teeth and other relevant ichthyodorulites in living supra-specific taxa of Chondrichthyan fishes. Part B: Batomorphi No. 4a: Order Rajiformes-Suborder Mylioabtoidei-Superfamily Dasyatoidea-Family Dasyatidae-Subfamily Dasyatinae-Genera: *Amphotistius*, *Dasyatis*, *Himantura*, *Pastinachus*, *Pteroplatytrygon*, *Taeniura*, *Urogymnus* and *Urolophoides* (incl. Supraspecific taxa of uncertain status and validity), Superfamily Myliobatoidea-Family Gymnuridae-Genera: *Aetoplatea* and *Gymnura*, Superfamily Plesiobatoidea-Family Hexatrygonidae-Genus *Hexatrygon*. *Bull. Inst. R. Sci. Nat. Belg.* 68:145–197.

Herman, J., M. Hovestadt-Euler, D.C. Hovestadt, and M. Stehmann. 1999. Contributions to the study of the comparative morphology of teeth and other relevant ichthyodorulites in living supra-specific taxa of Chondrichthyan fishes. Part B: Batomorphi No. 4b. Order Rajiformes-Suborder Myliobatoidei-Superfamily Dastyatoidea-Family Dasyatidae-Subfamily Dasyatinae-Genera: *Taeniura*, *Urogymnus*, *Urolophoides*-Subfamily Potamotrygoninae-Genera: *Disceus*, *Plesiotrygon*, and *Potamotrygon* (incl. Supraspecific taxa of uncertain status and validity), Family Urolophidae-Genera: *Trygonoptera*, *Urolophus* and *Urotrygon*-Superfamily Myliobatidea-Family Gymnuridae-Genus: *Aetoplatea*. *Bull. Inst. R. Sci. Nat. Belg.* 69:161–200.

Holst, R.J. and Q. Bone, 1993. On bipedalism in skates and rays. *Philos. Trans. R. Soc. Lond.* 339:105–108.

Howes, G.B. 1890. Observations on the pectoral fin-structure of the living batoid fishes and of the extinct genus *Squaloraja*, with especial reference to the affinities of the same. *Proc. Zool. Soc. London*, pp. 675–688.

Hulley, P.A. 1970. An investigation of the Rajidae of the west and south coasts of southern Africa. *Ann. S. Afr. Mus.* 55:151–220.

Hulley, P.A. 1972a. The origin, interrelationships and distribution of southern African Rajidae (Chondrichthyes, Batoidei). *Ann. S. Afr. Mus.* 60:1–103.

Hulley, P.A. 1972b. A new species of southern African brevirajid skate (Chondrichthyes, Batoidei, Rajidae). *Ann. S. Afr. Mus.* 60:253–263.

Hulley, P.A. 1973. Interrelationships within Anacanthobatidae (Chondrichthyes, Batoidei), with a description of the lectotype of *Anacanthobatis marmoratus* Von Bonde and Swart, 1923. *Ann. S. Afr. Mus.* 62:131–158.

Ishihara, H. 1987. Revision of the western North Pacific skates of the genus *Raja*. *Jpn. J. Ichthyol.* 34:241–285.

Ishihara, H. and R. Ishiyama. 1985. Two new North Pacific skates (Rajidae) and a revised key to *Bathyraja* of the area. *Jpn. J. Ichthyol.* 32:143–179.

Ishihara, H. and R. Ishiyama. 1986. Systematics and distribution of the skates of the North Pacific (Chondrichthyes, Rajoidei), in *Indo-Pacific Fish Biology: Proc. Second Int. Conf. Indo-Pacific Fishes.* T. Uyeno, R. Arai, T. Taniuchi, and K. Matsuura, Eds., Ichthyological Society of Japan, Tokyo, 269–280.

Ishiyama, R. 1958. Studies on the rajid fishes (Rajidae) found in the waters around Japan. *J. Shimonoseki Coll. Fish.* 7:1–394.

Ishiyama, R. 1967. *Fauna Japonica, Rajidae (Pisces).* Tokyo Electrical Engineering College Press, Tokyo, 84 pp.

Jordan, D.S. 1923. A classification of fishes including families and genera as far as known. *Stanford Univ. Publ. (Biol.)* 3:77–243.

Last, P.R. and J.D. Stevens. 1994. *Sharks and Rays of Australia.* CSIRO, Collingwood, Victoria, Australia.

Leviton, A.E., R.H. Gibbs, Jr., E. Heal, and C.E. Dawson. 1985. Standards in herpetology and ichthyology: Part 1. Standard symbolic codes for institutional resource collections in herpetology and ichthyology. *Copeia* 1985:802–832.

Lovejoy, N.R. 1996. Systematics of myliobatoid elasmobranches: with emphasis on the phylogeny and historical biogeography of neotropical freshwater stingrays (Potamotrygonidae: Rajiformes). *Zool. J. Linn. Soc.* 117:207–257.

Lovejoy, N.R., E. Bermingham, and A.P. Andrew. 1998. Marine incursion into South America. *Nature* 396:421–422.

Lucifora, L.O. and A.I. Vassallo. 2002. Walking in skates (Chondrichthyes, Rajidae): anatomy, behavior and analogies to tetrapod locomotion. *Biol. J. Linn. Soc.* 77:35–41.

Maisey, J.C. 1984. Higher elasmobranch phylogeny and biostratigraphy. *Zool J. Linn. Soc.* 82:33–54.

Maisey, J.C., G.J.P. Naylor, and D.J. Ward. In press. Mesozoic elasmobranchs, neoselachian phylogeny and the rise of modern elasmobranch diversity, in *Mesozoic Fishes: Systematics and Paleoecology. Proceedings of the International Meeting,* Eichstät, 1993. Verlag Dr. Friedrich Pfeil, Munich, Germany.

McEachran, J.D. 1977. Variation in *Raja garmani* and the status of *Raja lentiginosa* (Pisces: Rajidae). *Bull. Mar. Sci.* 27:423–439.

McEachran, J.D. 1982. Revision of the South American skate genus *Symptergia* (Chondrichthyes, Rajiformes). *Copeia* 1982:867–890.

McEachran, J.D. 1983. Results of the research cruises of FRV "Walther Herwig" to South America. LXI: revision of the South American skate genus *Psammoabtis* Günther, 1879 (Elasmobranchii, Rajiformes, Rajidae). *Arch. Fischereiwiss.* 34:23–80.

McEachran, J.D. 1984. Anatomical investigation of the New Zealand skates, *Bathyraja asperula* and *B. spinifera*, with an evaluation of their classification within Rajoidei (Chondrichthyes, Rajiformes). *Copeia* 1984:45–58.

McEachran, J.D. and L.J.V. Compagno. 1979. A further description of *Gurgesiella furvescens*, with comments on the interrelationships of Gurgesiellidae and Pseudorajidae (Pisces, Rajoidei). *Bull. Mar. Sci.* 29:530–533.

McEachran, J.D. and L.J.V. Compagno. 1980. Results of the research cruises of FRV "Walther Herwig" to South America. LVI: A new species of skate from the southwestern Atlantic, *Gurgesiella dorsalifera* sp. nov. (Chondrichthyes, Rajoidei). *Arch. Fischereiwiss.* 31:1–14.

McEachran, J.D. and L.J.V. Compagno. 1982. Interrelationships of and within *Breviraja* based on anatomical structures (Pisces, Rajoidei). *Bull. Mar. Sci.* 32:399–425.

McEachran, J.D. and M.R. de Carvalho. 2002. Batoid fishes, in *Living Resources of the Western Central Atlantic,* Vol. 1: *Introduction, Molluscs, Crustaceans, Hagfishes, Sharks, Batoid Fishes, and Chimaeras.* K.E. Carpenter, Ed., FAO Species Identification Guide for Fishery Purposes and American Society of Ichthyologists and Herpetologists Species Publ. 5, FAO, Rome.

McEachran, J.D. and K.A. Dunn. 1998. Phylogenetic analysis of skates, a morphologically conservative clade of elasmobranches (Chondrichthyes: Rajidae). *Copeia* 1998:271–290.

McEachran, J.D. and J.D. Fechhelm. 1982a. A new species of skate from Western Australia, with comments on the status of *Pavoraja* Whitley, 1939 (Chondrichthyes, Rajiformes). *Proc. Biol. Soc. Wash.* 95:1012.

McEachran, J.D. and J.D. Fechhelm. 1982b. A new species of skate from the Indian Ocean, with comments on the status of *Raja* (*Okamejei*) (Elasmobranchii, Rajiformes). *Proc. Biol. Soc. Wash.* 95:440–450.

McEachran, J.D. and J.D. Fechhelm. 1998. *Fishes of the Gulf of Mexico*, Vol. 1. *Myxininformes to Gasterosteiformes*. University of Texas Press, Austin.

McEachran, J.D. and P.R. Last. 1994. New species of skate, *Notoraja ochroderma*, from off Queensland, Australia, with comments on the taxonomic limits of *Notoraja* (Chondrichthyes: Rajoidei). *Copeia* 1994:413–421.

McEachran, J.D. and C.O. Martin. 1978. Interrelationships and subgeneric classification of *Raja erinacea* and *R. ocellata* based on claspers, neurocrania and pelvic girdles (Pisces: Rajidae). *Copeia* 1978:593–601.

McEachran, J.D. and T. Miyake. 1986. Anatomical analysis of a putative monophyletic group of skates (Chondrichthyes, Rajoidei), in *Indo-Pacific Fish Biology: Proc. Second Int. Conf. Indo-Pacific Fishes.* T. Uyeno, R. Arai, T. Taniuchi, and K. Matsuura, Eds., Ichthyological Society of Japan, Tokyo, 281–290.

McEachran, J.D. and T. Miyake. 1988. A new species of skate from the Gulf of California (Chondrichthyes, Rajoidei). *Copeia* 1988:877–886.

McEachran, J.D. and T. Miyake. 1990a. Phylogenetic interrelationships of skates: a working hypothesis (Chondrichthyes, Rajoidei), in Elasmobranchs as Living Resources: Advances in the Biology, Ecology, Systematics, and the Status of the Fisheries. H.L. Pratt, Jr., S.H. Gruber, and T. Taniuchi, Eds., NOAA Tech. Rep. 90, 285–304.

McEachran, J.D. and T. Miyake. 1990b. Zoogeography and bathymetry of skates (Chondrichthyes, Rajoidei), in Elasmobranchs as Living Resources: Advances in the Biology, Ecology, Systematics, and the Status of the Fisheries. H.L. Pratt, Jr., S.H. Gruber, and T. Taniuchi, Eds., NOAA Tech. Rep. 90, 305–326.

McEachran, J.D. and M. Stehmann. 1977. Subgeneric placement of *Raja bathyphila* based on anatomical characters of the clasper, cranium and pelvic girdle. *Copeia* 1977:20–25.

McEachran, J.D., B. Séret, and T. Miyake. 1989. Morphological variation within *Raja miraletus* and the status of *R. ocellifera* (Chondrichthyes, Rajoidei). *Copeia* 1989:629–641.

McEachran, J.D., K.A. Dunn, and T. Miyake. 1996. Interrelationships of the batoid fishes (Chondrichthyes: Batoidea), in *Interrelationships of Fishes*. M.L.J. Stiassny, L.R. Parenti, and G.D Johnson, Eds., Academic Press, London, 63–82.

Menni, R.C. 1971. Anatomia del mixopterigio y posicion sistematics de *Raja flavirostris* Philippi, 1892. *Neotropica* (La Plata) 17:39–43.

Menni, R.C. 1972. Anatomia del mixopterigio y diferencias especificas en los generos *Psammobatis* y *Sympterygia* (Chondrichthyes). *Neotropica* (La Plata) 18:73–80.

Michael, S.W. 1993. *Reef Sharks & Rays of the World: A Guide to Their Identification, Behavior, and Ecology.* Sea Challengers, Petaluma, CA.

Miyake, T. 1988. The Systematics of the Stingray Genus *Urotrygon* with Comments of the Interrelationships within Urolophidae (Chondrichthyes: Myliobatiformes). Ph.D. dissertation, Texas A&M University, College Station.

Miyake, T. and J.D. McEachran. 1991. The morphology and evolution of the ventral gill arch skeleton in batoid fishes (Chondrichthyes: Batoidea). *Zool. J. Linn. Soc.* 102:75–100.

Miyake, T., J.D. McEachran, and B.K. Hall. 1992a. Edgeworth's legacy of cranial development with an analysis of muscles in the ventral gill arch region of batoid fishes (Chondrichthyes: Batoidea). *J. Morphol.* 212:213–256.

Miyake, T., J.D. McEachran, P.J. Walton, and B.K. Hall. 1992b. Development and morphology of rostral cartilages in batoid fishes (Chondrichthyes: Batoidea), with comments on homology within vertebrates. *Biol. J. Linn. Soc.* 46:259–298.

Müller, J. and J. Henle. 1841. *Systematische Beschreibung der Plagiostomen*. Veit, Berlin.

Nelson. G. J. 1969. Gill arches and the phylogeny of fishes, with notes on the classification of vertebrates. *Bull. Am. Mus. Nat. Hist.* 141:477–552.

Nelson, J.S. 1994. *Fishes of the World,* 3rd. ed. John Wiley & Sons, New York.

Nishida, K. 1990. Phylogeny of the suborder Myliobatidoidei. *Mem. Fac. Fish. Hokkaido Univ.* 37:1–108.

Norman, J.R. 1966. A Draft Synopsis of the Orders, Families, and Genera of Recent Fishes, and Fish-Like Vertebrates, British Museum (Natural History), London.

Pitman, W.C., III, S. Cande, J. LaBrecque, and J. Pindell. 1993. Fragmentation of Gondwana: the separation of Africa and South America, in *Biological Relationships between Africa and South America.* P. Goldblatt, Ed., Yale University Press, New Haven, CT, 15–34.

Regan, C.T. 1906. A classification of the selachian fishes. *Proc. Zool. Soc. London*, pp. 722–758.

Reif, W.-E. 1979. Morphogenesis and histology of large scales of batoids (Elasmobranchii). *Paläontol. Z.* 53:26–37.

Reyment, R.A. 1980. Paleo-oceanology and paleobiogeography of the Cretaceous South Atlantic Ocean. *Oceanol. Acta* 3:127–133.

Rosa, R.S. 1985. A Systematic Revision of the South American Freshwater Stingrays (Chondrichthyes: Potamotrygonidae). Ph.D. dissertation, College of William and Mary, Williamsburg. VA.

Rosa, R.S., H.P. Castello, and T.B. Thorson. 1988. *Plesiotrygon iwamae*, a new genus and species of neotropical freshwater stingray (Chondrichthyes: Potamotrygonidae). *Copeia* 1988:447–458.

Rosenberger, L.J. 2001a. Pectoral fin locomotion in batoid fishes: undulation versus oscillation. *J. Exp. Biol.* 204:379–394.

Rosenberger, L.J. 2001b. Phylogenetic relationships within the stingray genus *Dasyatis* (Chondrichthyes: Dasyatidae). *Copeia* 2001:615–627.

Rosenberger, L.J. and M.W. Westneat. 1999. Functional morphology of undulatory pectoral fin locomotion in the stingray *Taeniura lymma* (Chondrichthyes: Dasyatidae). *J. Exp. Biol.* 202:3523–3539.

Séret, B. 1986. Classification et phylogenèse des Chondrichthyens. *Oceanis* 12:161–180

Shirai, S. 1992a. Phylogenetic relationships of the angel sharks, with comments on elasmobranch phylogeny (Chondrichthyes, Squatinidae). *Copeia* 1992:505–518.

Shirai, S. 1992b. *Squalean Phylogeny: A New Framework of Squaloid Sharks and Related Taxa.* Hokkaido University Press, Sapporo.

Shirai, S. 1996. Phylogenetic interrelationships of neoselachians (Chondrichthyes: Euselachii), in *Interrelationships of Fishes.* M.L.J. Stiassny, L.R. Parenti, and G.D. Johnson, Eds., Academic Press, London, 9–34.

Stehmann, M. 1970. Vergleichende morphologische und anatomische Untersuchungen zur Neuordnung der Systematik der nordostatlantischen Rajidae (Chondrichthyes, Batoidei). *Arch. Wiss.* 21:73–164.

Stehmann, M. 1971a. *Raja (Leucoraja) leucosticta* spec. nov. (Pisces, Batoidei, Rajidae) eine neue Rochenart aus dem Seegebiet des tropischen Westafrika; gleichzeitig zur Frage des Vorkommens von Raja ackleyi Garman, 1881, im mittleren Ostatlantik. *Arch. Fischereiwiss.* 22:1–16.

Stehmann, M. 1971b. Ergebnisse der Forschungsreisen des FFS "Walther Herwig" nach Südamerika XVII: *Raja (Raja) herwigi* Krefft, 1965: ergänzende Untersuchungen zum subgenerischen Status der Art. *Arch. Fischereiwiss.* 22:85–97.

Stehmann, M. 1971c. Untersuchungen zur Validität von *Raja maderensis* Lowe, 1839, zur geographischen Variation von *Raja straeleni* Poll, 1951 und zum subgenerischen Status bei der Arten (Pisces, Batoidei, Rajidae). *Arch. Fischereiwiss.* 22:175–199.

Stehmann, M. 1976a. Revision der Rajoiden-Arten des nördlichen Indischen Ozean und Indopazifik (Elasmobranchii, Batoidea, Rajiformes). *Beaufortia* 315:133–175.

Stehmann, M. 1976b. *Breviraja caerulea* spec. nov. (Elasmobranchii, Batoirei); eine neue archibenthale Rochenart und zugleich ein Erstnachweis ihrer Gattung im Nordostatlantik. *Arch. Fischereiwiss.* 27:97–114.

Stehmann, M. 1977. Ein neuer archibenthaler Roche aus dem Nordostatlantik, *Raja kreffti*, spec. nov. (Elasmobranchii, Batoidei, Rajidae), die zweite Spezies im Subgenus *Malacoraja* Stehmann, 1970. *Arch. Fischereiwiss.* 28:77–93.

Stehmann, M. 1978. *Raja "bathyphila,"* eine Doppelart des subgenus *Rajella*: Wiederbeschreibung von *Raja bathyphila* Holt and Byrne, 1908 und *Raja bigelowi* spec. nov. (Pisces, Rajiformes, Rajidae). *Arch. Fischereiwiss.* 29:13–58.

Stehmann, M. 1986. Notes on the systematics of the rajid genus *Bathyraja* and its distribution in world oceans, in *Indo-Pacific Fish Biology: Proc. Second Int. Conf. Indo-Pacific Fishes.* T. Uyeno, R. Arai, T. Taniuchi, and K. Matsuura, Eds., Ichthyological Society of Japan, Tokyo, 261–268.

Stehmann, M. 1987. *Bathyraja meridionalis* sp. n. (Pisces, Elasmobranchii, Rajidae, a new deep-water skate from the eastern slope of subantarctic South Georgia Island. *Arch. Fischereiwiss.* 38:35–56.

White, E.G. 1937. Interrelationships of the elasmobranchs with a key to the order Galea. *Bull. Am. Mus. Nat. Hist.* 74:25–138.

4

Phylogeny and Classification of Extant Holocephali

Dominique A. Didier

CONTENTS

4.1 Overview of Living Holocephali

Members of Subclass Holocephali are distinguished from all other chondrichthyan fishes by a variety of morphological features, in particular the mode of fusion of the lower jaw to the cranium (holostyly) and the possession of non-replaceable, hypermineralized tooth plates among other characters (Maisey, 1986; Didier, 1995; Lund and Grogan, 1997). Fossil evidence indicates that holocephalans evolved from a common chondrichthyan ancestor, with the Holocephali arising from a vast radiation of Paleozoic Chondrichthyes (Lund and Grogan, 1997; Grogan and Lund, Chapter 1 of this volume). Today, there are only 33 described species of extant holocephalans (Didier, 1995; Eschmeyer, 1998; Table 4.1). Several new species are known to exist, but have yet to be described, and it is unlikely that the number of extant holocephalans exceeds 45 species. The living holocephalans and their closest fossil relatives belong to Order Chimaeriformes, which includes three families, Callorhinchidae, Rhinochimaeridae, and Chimaeridae, each of which is distinguished by a unique snout morphology (Figure 4.1). The living holocephalans as a group are commonly referred to as the chimaeroid fishes, or chimaeras; however; this general category for all living chimaeriform fishes is not to be confused with the chimaerid fishes, which refers specifically to members of Family Chimaeridae (also known as "ratfishes"). A current classification and a list of all valid species of chimaeroids are shown in Table 4.1.

The chimaeroid fishes have long intrigued taxonomists and evolutionary biologists because of their relationship to the better-known elasmobranchs, and their many seemingly primitive characters such as

TABLE 4.1

Classification and Geographic Range of Species of Extant Holocephali

Family Callorhinchidae
  Callorhinchus callorhynchus Linnaeus, 1758; South America; Argentina, Chile, Peru
  Callorhinchus capensis Duméril, 1865; South Africa
  Callorhinchus milii Bory de St. Vincent, 1823; Southern New Zealand and Australia
Family Rhinochimaeridae Garman, 1901
  Rhinochimaera pacifica (Mitsukurii, 1895); Pacific Ocean
  Rhinochimaera atlantica Holt and Byrne, 1909; Atlantic Ocean
  Rhinochimaera africana Compagno, Stehmann, and Ebert, 1990; Indian Ocean, Taiwan, Japan
  Harriotta raleighana Goode and Bean, 1895; Atlantic and Pacific Ocean/Circumglobal
  Harriotta haeckeli Karrer, 1972; Northeastern Atlantic, Southwestern Pacific (New Zealand)
  Neoharriotta pinnata (Schnakenbeck, 1929); Eastern Atlantic off Africa
  Neoharriotta pumila Didier and Stehmann, 1996; Northern Indian Ocean
  Neoharriotta carri Bullis and Carpenter, 1966; Caribbean Sea
Family Chimaeridae
  Chimaera monstrosa Linnaeus, 1758; Northern Atlantic, Mediterranean
  Chimaera cubana Howell-Rivero, 1936; Caribbean Sea
  Chimaera owstoni Tanaka, 1905; Japan
  Chimaera jordani Tanaka, 1905; Japan
  Chimaera lignaria Didier, 2002; Southern New Zealand, Tasmania
  Chimaera phantasma Jordan and Snyder, 1900; Western Pacific
  Chimaera panthera Didier, 1998; Southwestern Pacific (New Zealand)
  Hydrolagus affinis (Capello, 1867); North Atlantic
  Hydrolagus africanus (Gilchrist, 1922); Southeastern Atlantic (Africa)
  Hydrolagus alberti Bigelow and Schroeder, 1951; Gulf of Mexico, Caribbean
  Hydrolagus barbouri (Garman, 1908); Japan
  Hydrolagus bemisi Didier, 2002; New Zealand
  Hydrolagus colliei (Lay and Bennett, 1839); Northeastern Pacific; Gulf of California
  Hydrolagus lemures (Whitley, 1939); Australia
  Hydrolagus mirabilis (Collett, 1904); Northern Atlantic, Gulf of Mexico
  Hydrolagus mitsukurii (Dean, 1904b); Northwestern Pacific
  Hydrolagus macrophthalmus de Buen, 1959; Chile
  Hydrolagus novaezealandiae (Fowler, 1910); New Zealand
  Hydrolagus ogilbyi (Waite, 1898); Australia
  Hydrolagus purpurescens (Gilbert, 1905); Japan
  Hydrolagus pallidus Hardy and Stehmann, 1990; North Atlantic
  Hydrolagus trolli Didier and Séret, 2002; New Zealand, New Caledonia

the retention of a complete, unmodified hyoid arch. Most early work on the morphology and relationships of chimaeroid fishes focused primarily on studies of fossil forms (summarized in Didier, 1995). Most notable was the work of Dean who studied both fossil and living forms (Dean, 1903, 1904a,b,c, 1906, 1912). In terms of taxonomic work, most species of chimaeroids were described in the early part of the 20th century (Jordan and Snyder, 1900, 1904; Collett, 1904; Gilbert, 1905; Tanaka, 1905; Garman, 1908; Holt and Byrne, 1909; Fowler, 1910). Garman (1911) provided the first taxonomic review of the chimaeroid fishes that later served as the basis of Fowler's review published in 1941, and in 1953 Bigelow and Schroeder published a taxonomic summary of chimaeroid fishes of the northwestern Atlantic. Since the 1950s there has been a decline in active research on chimaeroid fishes. Since 1920, 12 new species have been described, 8 of which were described in the last 50 years (Gilchrist, 1922; Schnakenbeck, 1929; Whitley, 1939; Bigelow and Schroeder, 1951; de Buen, 1959; Bullis and Carpenter, 1966; Karrer, 1972; Compagno et al. 1990; Hardy and Stehmann, 1990; Didier and Stehmann, 1996; Didier, 1998). In the last decade there has been renewed interest in chimaeroid fishes and many new species have been discovered as a result of recent surveys and collecting efforts in deeper waters, often near or below 1000 m (Paulin et al., 1989; Compagno et al., 1990; Last and Stevens, 1994; Didier and Stehmann, 1996). Several new studies of chimaeroid fishes have been published including a phylogenetic study of chimaeroid fishes using a modern cladistic approach (Didier, 1995), embryological studies (Didier et al.,

FIGURE 4.1 Representative species from each of the three families of holocephalans. (A) *Callorhinchus milii* of Family Callorhinchidae; note the plow-shaped snout and heterocercal tail. Scale = 2 cm. (Courtesy of the American Museum of Natural History.) (B) *Rhinochimaera pacifica*, a long-snouted chimaera of Family Rhinochimaeridae (CBM-ZF6140). Scale = 10 cm. (Courtesy of the Ichthyological Society of Japan.) (C) *Chimaera panthera*, one of several new species in Family Chimaeridae (NSMT P32122). Scale = 5 cm. (Courtesy of the Ichthyological Society of Japan.)

1994, 1998; Grogan et al., 1999) and the description of several new species (Didier and Stehmann, 1996; Didier, 1998, 2002; Didier and Séret, 2002).

4.2 General Ecology and Behavior

The chimaeroids are marine fishes inhabiting all of the world's oceans with the exception of Arctic and Antarctic waters. Table 4.1 identifies the approximate geographic region where each species most commonly occurs. Most chimaeroids are deep-water dwellers of the shelf and slope off continental landmasses, oceanic islands, seamounts, and underwater ridges, generally occurring at depths of around 500 m and deeper. A few species inhabit shallower coastal waters, most notably *Hydrolagus colliei* off the west coast of the United States, and all three species in Family Callorhynchidae. Capture records indicate that chimaeroids tend to live on or near the bottom, apparently preferring bottom types of mud or ooze. Species in Family Callorhinchidae are restricted to the Southern Hemisphere and some other species are known from a relatively restricted range both vertically and horizontally (e.g., *H. barbouri, Chimaera panthera*); however, most species of chimaeroids seem to be widespread (e.g., *C. monstrosa, H. colliei*), sometimes occurring throughout an entire ocean basin (e.g., *Harriotta raleighana, Rhinochimaera pacifica*).

Very little information exists on the ecology and behavior of chimaeroid fishes due to the deep-water habitat of these fishes. Most information has been obtained through observations of the few commercially fished species that occur in near-shore waters (*Hydrolagus colliei, Callorhinchus milii, C. callorhynchus, and C. capensis*). These species appear to be locally migratory and exhibit seasonal inshore migration for breeding and spawning (Gorman, 1963). Studies of *H. colliei* provide evidence that this species aggregates by both sex and size. Males and females form separate groups, and juveniles tend to aggregate in deeper waters while larger fish exhibit seasonal migrations to shallower waters (Mathews, 1975; Quinn et al., 1980). Other species of chimaeroids may exhibit similar aggregation and migration patterns.

Chimaeroid diets consist primarily of benthic invertebrates and small fishes including conspecifics (Gorman, 1963; Johnson and Horton, 1972; Quinn et al., 1980). The tooth plates are used to crush hard-bodied prey such as crustaceans, mollusks, and echinoderms, including hydrothermal vent mussels (Graham, 1939; Macpherson and Roel, 1987; Marques and Porteiro, 2000); however, soft-bodied prey including salps, tunicates, and jellyfish are also consumed (Graham, 1956). Chimaeroid fishes appear to have few predators in the wild and it is likely that their shark relatives are their primary predators as evidenced by the consumption of adult *C. milii* by the New Zealand carpet shark, *Cephaloscyllium isabellum,* and school sharks, *Galeorhinus galeus* (Gorman, 1963; Didier, pers. obs.).

Methods for determining age and growth rates in chimaeroid fishes have been based on studies of *H. colliei* and *Callorhinchus milii*. Johnson and Horton (1972) tested a variety of morphological measures for *H. colliei* including: eye-lens weights, vertebral radii, spine sections, tooth plate ridges, and body-length frequencies, none of which proved effective. Because tooth plates are growing constantly, and are known to change morphology as the fish grows, it is unlikely that fish can be aged on the basis of tooth plate morphology (Garman, 1904; Bigelow and Schroeder, 1950; Didier et al., 1994). The most reliable method appears to be examination of the banding pattern in the dorsal fin spine. Based on the assumption of annual ring deposition, it was estimated that *C. milii* males mature at about 3 years and females at 4.5 years (Sullivan, 1978). In a more recent study of growth rates of *C. milli*, based on length-frequency and tag-recapture data, Francis (1997) found that growth rates differed among different populations. For example, males collected in the 1960s matured at 4+ years and males collected in the 1980s matured at 2+ to 3+ years. It is hypothesized that variation in growth rates may be related to biomass of the population. It is not known if these same growth trends are typical of other species of chimaeroids.

4.3 General Morphological Features

4.3.1 External Features

The chimaeroid fishes are characterized by long, tapering bodies and large heads. Adults range in size from small-bodied slender fishes averaging around 60 cm in total length (e.g., *Hydrolagus mirabilis*) to massive fishes exceeding 1 m in length with large bulky heads and bodies (e.g., *Chimaera lignaria*). The skin is completely scaleless in adults. Hatchlings and small juveniles have tiny denticles embedded in the skin along the dorsal surface of the trunk and head. These denticles are arranged in a horseshoe shape atop the head anterior to the first dorsal fin, and in two rows between the first and second dorsal fins and between the second dorsal and upper caudal fins. The skin in some species appears more fragile than in others and in preserved specimens is described as being deciduous or nondeciduous based on whether or not the skin easily flakes off in patches or remains intact.

The gill arches are concentrated underneath the neurocranium and covered by a fleshy operculum supported by cartilaginous rays. Only a single gill opening located anterior to the pectoral fin base is present on each side of the body. Adults lack a spiracle, although it is present in embryos (Didier et al., 1998). The ventrally positioned mouth is small, connected to the nostrils by deep grooves. The incisor-like anterior tooth plates and large nostrils give the appearance of a rabbitlike mouth and the common name "rabbitfish" is often applied to members of Family Chimaeridae. Most species possess large round eyes that appear a translucent green in color, although small eyes distinguish a few species.

4.3.1.1 Lateral Line Canals — The first comparative analysis of lateral line canals in chondrichthyans included descriptions of the morphology of the lateral line canals of several species of chimaeroids (Garman, 1888). Other early studies of the lateral line canals focused on innervation of the canals (Cole 1896a,b; Cole and Dakin, 1906) and histology (Reese, 1910). Reese (1910) was the first to note the morphological differences between the undilated and dilated canals, which he distinguished as "type 1" and "type 2" canals, respectively. More recently, several studies have focused on lateral line canal and neuromast morphology and evolution in primitive fishes including a variety of elasmobranchiomorphs, but not chimaeroids specifically (e.g., Northcutt, 1989; Webb and Northcutt, 1997; Maruska and Tricas, 1998; Peach and Rouse, 2000).

With the exception of the callorhynchids, which have enclosed, pored, lateral line canals that sit in the dermis, the living chimaeroid fishes have a unique lateral line system on the head and trunk that is composed of a series of open grooves supported by open, C-shaped, cartilaginous rings. Members of Family Chimaeridae are further distinguished by a modification of the grooves on the head in which the anterior portions that extend onto the snout are widened and have enlarged dilations between small series of cartilaginous rings. Aspects of lateral line canal morphology and distribution appear to have some taxonomic value at various levels. The morphology of the calcified rings is significant at higher taxonomic levels but the number of cartilaginous rings between dilations varies between three and seven for all species examined and does not seem to be a useful character for species identification. Preliminary evidence suggests that the number of lateral line canal dilations in chimaerids is correlated with size and may be significant for generic distinctions, but the number of canal dilations has not proved useful for species determination (Wilmot et al., 2001). The pattern of lateral line canals on the head, in particular the branching pattern of canals below the eye, is useful for distinguishing species (Didier, 1995, 1998; Didier and Stehmann, 1996; Didier and Nakaya, 1999; Didier and Séret, 2002). The terminology shown in Figure 4.2 follows a historical morphological approach to canal designation and is based on Garman (1888) with terminological modifications from Didier (1995, 1998) and Compagno et al. (1990). A different terminology was adopted by Fields et al. (1993) and was based on study of the nervous innervation of the canals in *Hydrolagus colliei*. Although terminology based on nervous innervation is more informative in terms of understanding homology, it cannot at present be adapted to represent the variation of canal branching patterns observed among all species of chimaeroids.

Adjacent to the lateral line canals of the head are clusters of ampullary pores that have most recently been described in detail for *H. colliei* (Fields et al., 1993). Morphological, behavioral, and neurophysiological studies of *H. colliei* confirm that these ampullar structures respond to electric fields and are homologous to the ampullae of Lorenzini in elasmobranchs (Fields and Lange, 1980). The location and number of ampullary pores have not proved to be of taxonomic significance.

4.3.1.2 Fins and Fin Spines

4.3.1.2 Fins and Fin Spines — All chimaeroids possess two dorsal fins, a caudal fin, and paired pectoral and pelvic fins, all with delicate fin webs supported by cartilaginous rays (ceratotrichia). The first dorsal fin is erectile, triangular in shape, and preceded by a long, stout spine. The dorsal fin spine is triangular in cross section with a keel-like ridge along the anterior surface and two rows of serrations present along the distal posterior edge of the upper half of spine. The lower half of the spine remains attached to the first dorsal fin. As individuals mature, the anterior keel becomes worn away so that the spine in large adults is often smooth or possesses only a very narrow keel, whereas juveniles and subadults will generally have a much more prominent keel. Likewise, the serrations along the distal posterior edge of the spine become worn with age; therefore, juveniles will have obvious serrations in two rows on the posterior surface of the spine while the serrations in adults are not as prominent and may be partly or mostly worn away. The fin spine of *Hydrolagus colliei* has been found to be mildly venomous (Halstead and Bunker, 1952). The fin spines of other species of chimaeroids, such as *Callorhinchus milii*, are known to inflict painful wounds that result in several days of swelling and redness (Didier, pers. obs.), and it is likely that the spines of most species of chimaeroids are venomous.

The second dorsal fin in most species is elongate and in some species may have a central indentation that nearly separates the second dorsal fin into two parts. All callorhinchids and *Neoharriottas* possess a prominent anal fin with an internal skeletal structure. The anal fin in chimaerids is small and lacks internal skeletal support. The caudal fin in most species is leptocercal with upper and lower caudal lobes of nearly equal size, although in some species of rhinochimaerids the caudal fin appears externally heterocercal. The callorhinchids are the exception in having a heterocercal tail. All species possess a distal caudal filament, which is usually in the form of an elongate whip, but may be quite short or absent in the adults of some species. The pectoral and pelvic fins are nearly uniform in shape for all species of chimaeroids. The large, broad triangular pectoral fins look and function like wings propelling the fish underwater by a flapping motion. The much smaller pelvic fins are usually squared or rounded along their distal edge. In a few species the shape of the pelvic fins is sexually dimorphic.

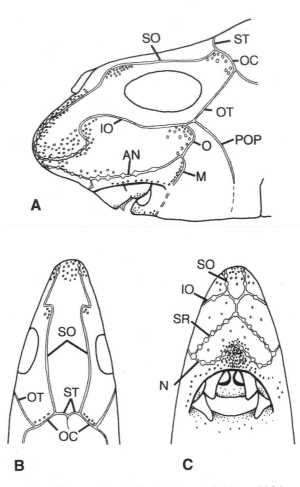

FIGURE 4.2 General pattern of lateral line canals on the head and snout of chimaeroid fishes as shown in a representative chimaerid fish: (A) lateral view, (B) dorsal view, (C) ventral view. Canals generally follow the same basic pattern in callorhinchids and rhinochimaerids although the canal positions will vary slightly due to elongation of snouts in members of these two families. Terminology of lateral line canals is as follows: AN = angular, IO = infraorbital, M = mandibular, N = nasal, O = oral, OC = occipital, OT = otic, POP = preopercular, SO = suborbital, SR = subrostral, ST = supratemporal. (Courtesy of the American Museum of Natural History.)

4.3.2 Skeleton

The skeleton is completely cartilaginous; however, like their elasmobranch relatives, chimaeroids possess calcified tissues in the dentition, denticles, and fin spines. Several unique skeletal features characterize the Holocephali; the most distinctive is the holostylic jaw in which the palatoquadrate is completely fused to the neurocranium. Holostyly in chimaeroids is derived from an ancestral autodiastylic state (Lund and Grogan, 1997; Grogan et al., 1999). Related to the holostylic jaw, and perhaps the most unusual skeletal feature, is the complete nonsuspensory hyoid arch. Articulating with the hyoid arch is the opercular cartilage and hyoid rays, which support the fleshy operculum. In addition to the hyoid arch are five regular gill arches, which are concentrated beneath the neurocranium.

Several unique features characterize the neurocranium of chimaeroids. Anterior to the orbits and dorsal to the nasal capsules is the ethmoid canal, which is an enclosed passage for nerves and blood vessels to the anterior-most region of the snout. The snout is supported by three rostral cartilages arranged as a single dorsal cartilage and paired ventral cartilages. An interorbital septum separates the orbits and is formed by a sheet of dense connective tissue rather than cartilage. Articulating with the occipital region of the neurocranium is the synarcual formed by the fusion of the first ten vertebral segments. The fin spine and first dorsal fin articulate with the synarcual. True vertebral centra are absent from the vertebral column of chimaeroids. Within the notochordal sheath are calcified rings, which are not segmentally organized, but appear to increase in number and density as the fish mature. Callorhynchids lack these notochordal rings. Other features of the skeletal anatomy of chimaeroids are summarized in Didier (1995).

4.3.2.1 Tooth Plates — Holocephalans are characterized by the possession of ever-growing, nonreplaceable hypermineralized tooth plates. All chimaeroids have six tooth plates in three pairs, a single pair in the lower jaw and two pairs in the upper jaw. The lower mandibular tooth plates are characterized by a large symphysial tritor, which is sculpted into a prominent point at the symphysial edge, and together the mandibular tooth plates form a distinct double-pointed beak at the symphysis. Incisor-like vomerine tooth plates are located at the anterior edge of the upper jaw and occlude with the mandibular tooth plates. Together the mandibular and vomerine tooth plates form a beaklike bite. Posterior to the vomerine tooth plates are the palatine tooth plates that lie flat on the roof of the mouth and occlude with the tongue and posterior edges of the mandibular tooth plates.

Because they fossilize well, the tooth plates are among the only fossil remains that are known to exist and have long been central to evolutionary studies of holocephalans (e.g., Dean, 1906, 1909; Moy-Thomas, 1939; Patterson, 1965; Ørvig, 1967, 1985; Bendix-Almgreen, 1968; Lund, 1977, 1986, 1988; Zangerl, 1981; see Stahl, 1999, for a recent review). Of particular interest are the mineralized tissues of the tooth plates. The bulk of the tooth plate comprises a matrix of trabecular dentine (Peyer, 1968; osteodentine of Ørvig, 1967) surrounding hypermineralized tritors that are composed of a tissue that has been identified by various workers as tubular dentine (Moy-Thomas, 1939), pleromin (Ørvig, 1967, 1985), and orthotrabeculine (Zangerl et al., 1993). On the oral surface of the tooth plate the tritors exhibit two distinct morphologies. Hypermineralized rods (Didier, 1995) are usually located at or near the edge of the tooth plate and appear as beads on a string ("pearlstrings" of Bargmann, 1933) while hypermineralized pads (Didier, 1995) are single large tritors located at or near the center of the tooth plate.

Details of the orientation, development, and growth of the tooth plates are important for understanding the evolution of the holocephalan dentition. Schauinsland (1903) observed that each tooth plate developed from a single primordium, and therefore it was interpreted that tooth plates probably did not evolve from separate tooth primordia like tooth families. A comparison of the development of the tooth plates of lungfishes and chimaeroids supported this hypothesis (Kemp, 1984). However, new embryological studies have shown that chimaeroid tooth plates exhibit a compound structure with individual tooth plates formed from multiple growth regions, suggesting that tooth plates may represent the fusion of members of a tooth family (Didier et al., 1994). Further support that tooth plates are derived from an ancestral chondrichthyan dentition is based on a reinterpretation of the growth and orientation of chimaeroid tooth plates and a new, more informative nomenclature for tooth plate surfaces that indicates that the chimaeroid dentition is lyodont (growing in a lingual to labial direction) and is similar to that of other chondrichthyans (Patterson, 1992).

4.3.3 Secondary Sexual Characteristics

Males and females are sexually dimorphic and males possess several secondary sexual structures including a frontal tenaculum, paired prepelvic tenacula, and paired pelvic claspers. Juvenile males lack the frontal tenaculum but have tiny developing pelvic claspers and small slitlike pouches on the ventral surface of the trunk. Development of the frontal tenaculum and growth of the prepelvic and pelvic claspers occur as sexual maturity is reached. Most early studies of secondary sexual characteristics focused on morphology and histology of the urogenital system, and these are summarized in the more recent comprehensive morphological work of Stanley (1963).

The frontal tenaculum is a small clublike structure with bulbous tip armed with numerous sharp denticles located on top of the head just anterior to the eyes. This structure, unique to chimaeroid fishes, has long been assumed to play a role in mating, and only recently has it been observed that males use the frontal tenaculum to grasp the pectoral fin of the female during copulation (D. Powell, Monterey Bay Aquarium, pers. comm.). The frontal tenaculum varies among species (Didier, 1995) and may be a useful character for species identification when considered in combination with other characters (e.g., Didier and Séret, 2002).

Paired prepelvic tenaculae are in the form of flat, spatulate blades with a row of prominent denticles along the medial edge. The prepelvic tenacualae articulate with the anterior edge of the pelvic girdle and are housed in pouches on the ventral side of the trunk. As they emerge from their pouches the tenaculae flex anteriorly and aid in anchoring the male to the ventral side of the female. The number of denticles on the medial edge of prepelvic tenaculae ranges from five to seven in every adult male specimen examined. Denticles are always located in a single line along the medial edge with the largest spines most proximal and distal spines the smallest. The only variation on this pattern occurs in *Hydrolagus africana,* and the presence of additional denticles on the prepelvic tenaculae appears to be a diagnostic character for this species.

The pelvic claspers extend from the medial edge of the pelvic fins and serve to transport sperm to the oviducts of the female. A comparative morphological study of the secondary sexual characteristics of elasmobranchs included descriptions of the morphology of pelvic claspers in several species of chimaeroids (Leigh-Sharpe, 1922, 1926). Pelvic claspers are phylogenetically useful characters (Didier, 1995); however, for species identification clasper characters are useful only when considered with other characters that are found in both males and females (e.g., Didier, 1998, 2002; Didier and Séret, 2002). Within Family Chimaeridae the pelvic claspers have been used as diagnostic characters at the genus level based on the distinction between a bipartite and tripartite condition (Garman, 1911; Fowler, 1941). However, based on the internal skeletal structure, all pelvic claspers of species in Family Chimaeridae should probably be interpreted as bifurcate because the central cartilaginous support of the pelvic clasper divides into only two branches, each with a fleshy denticulate lobe at its distal end (Figure 4.3). A separate fleshy lobe, continuous with the fleshy tissue of the lateral branch, encircles the base of the medial branch and this has been interpreted as the third branch in the "tripartite" claspers. This fleshy lobe usually lacks internal cartilaginous support, although in some species a thin strip of cartilage may support this third fleshy lobe; however, it does not originate from or articulate with the internal skeleton of the pelvic clasper and should not be interpreted as a third branch. In most individuals the fleshy lobe is closely associated with the medial branch, but in some it can be separated by the clasper groove, thus appearing as a third branch. The clasper groove is unrelated to the actual bifurcation of the clasper itself and runs the entire length of the clasper.

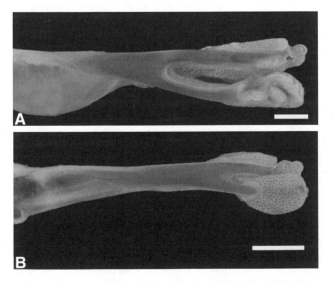

FIGURE 4.3 Cleared and stained pelvic claspers of representative specimens of *Chimaera* (A) and *Hydrolagus* (B) showing bifurcate internal skeletal morphology. Morphology of pelvic claspers, particularly the point at which the internal skeleton divides, is useful for species identification when used in combination with other morphological characters. Scale = 1 cm.

4.4 Reproduction and Development

Embryological development has been observed and described for only 2 of the 33 recognized species of chimaeroids. The earliest descriptive embryological studies were of *Callorhinchus milii* (Schauinsland, 1903) and *Hydrolagus colliei* (Dean, 1903, 1906). These studies were based on only a few embryos and lacked many critical early developmental stages. More recently Didier et al. (1998) described a complete, post-neurula, developmental series of *C. milii*. Details of reproduction and spawning are based primarily on studies of *C. milii* and *H. colliei* and it is assumed that reproductive biology for other species of chimaeroids is similar.

Chimaeroids, like their shark relatives, have internal fertilization in which males, equipped with pelvic claspers, transfer sperm directly into the female reproductive tract. All chimaeroids are oviparous. Females produce large, yolky eggs, pale yellow in color and similar in size to those of elasmobranchs. Fertilization occurs in the upper end of the reproductive tract and eggs pass through the nidamental gland where eggs are encased individually in a tough, leathery egg capsule within the oviduct (Dean, 1906; Didier, 1995; Didier et al., 1998). Egg capsules remain attached to the shell gland within the oviduct for several days (Dean, 1903, 1906; Sathyanesan, 1966) and are deposited directly on the seafloor. The shape of the egg capsules is characteristic for each family (Didier, 1995). Dean (1912) noted differences in the morphology of egg capsules among species of *Chimaera* and identified primitive and derived character states. He defined this gradual variation of egg capsule morphology as "orthogenesis," and these early studies indicate that egg capsule characters may be of phylogenetic significance (Dean, 1912). Egg capsules also may be taxonomically significant at the species level; however, it is difficult to reliably associate egg capsules with a species because the ranges of many species overlap, and the egg capsules are not usually collected in association with the females that laid them.

The egg is supported by a very fragile vitelline membrane and thick jellylike material that fills the inside of the egg capsule. As the embryo matures, the vitelline membrane will toughen and the jelly material breaks down. At the anterior, or blunt, end of the central spindle portion of the egg capsule is a raised seam that is tightly sealed when first laid, but as the embryo develops and the egg capsule wears with age the seal will gradually soften and open slightly. At the time of hatching the egg capsule will break open along this seam to release the fully developed embryo. Additional slits at the posterior end of the spindle will also gradually open to facilitate flow of water through the egg capsule for gas exchange and removal of waste products (Dean, 1903, 1904a, 1906; Didier, 1995).

Spawning generally occurs on flat, muddy or sandy substrates, but spawned egg capsules have also been observed on pebbly bottoms and in beds of seaweed. Females spawn two egg capsules simultaneously, one from each oviduct, which are deposited onto the ocean floor (Dean, 1906; Didier, 1995). Several pairs of eggs are laid each season, but the exact number is unknown. Captive *H. colliei* have been observed to lay a pair of eggs every 7 to 10 days (Sathyanesan, 1966; Didier et al., 1998; K. Wong, pers. obs.) and other species of chimaeroids probably spawn at a similar rate with females laying a pair of eggs every fortnight for several months (Gorman, 1963). It is likely that all females store sperm as evidenced by a recent study of *Callorhinchus milii* (Smith et al., 2001). A single collection of embryos in the field will contain embryos of all stages; therefore, it is likely that the spawning season lasts several months, perhaps up to 6 months, and the developmental period is suspected to take from 5 to 10 months before the fully developed embryo will hatch from its egg capsule (Dean, 1903, 1906; Gorman, 1963; Didier et al., 1998). Undoubtedly, temperature plays a role determining developmental rates and timing.

4.5 Taxonomic Review

4.5.1 Callorhinchidae (Figure 4.1A)

A prominent plow-shaped snout extending forward from the front of the head characterizes the callorhinchid fishes, also commonly known as the plow-nosed chimaeras or elephantfishes. A stiff cartilaginous rod supports the dorsal surface of the snout and at the distal end is a fleshy ovoid or leaf-shaped flap of tissue. There are three recognized species within this monogeneric family: *Callorhinchus milii* from New Zealand and Australia, *C. capensis* from southern Africa, and *C. callorynchus* from southern South America (Nelson, 1994; Didier, 1995; Eschmeyer, 1998). In addition to the unique snout morphology, callorhinchids differ from other chimaeroids in their more torpedo-like body shape, heterocercal tail, and a large skeletally supported anal fin. The callorhinchids are the most primitive living chimaeroids based on interpretation of a variety of characters of the tooth plates, skeleton, and musculature (Didier, 1995).

Morphologically the three species are nearly indistinguishable. All callorhinchids are silvery in color, black along the dorsal midline, with saddlelike bands on the dorsal side of the head and along the dorsal surface of the trunk, sometimes with dark blotches along the sides of the trunk as well. Unlike other chimaeroids, the callorhinchids have lateral line canals that are enclosed, visible on the body surface as narrow canals underneath the dermis, rather than open grooves. The eye is small. The second dorsal fin is not elongate, usually nearly equal to the length of the pectoral to pelvic space, very tall anteriorly, sloping posteriorly to a low, evenly tall fin with the height of the anterior portion about five times that of the posterior portion. Males possess simple scrolled pelvic claspers lacking fleshy lobes and denticulations (Figure 4.4A). The frontal tenaculum is flat, not deeply curved, with very short denticles on the distal bulb. There appear to be some distinctions among the frontal tenacula of males of the three species and this may prove a useful identifying feature, although useful only for distinguishing males of the species. Prepelvic tenacula are complex cartilaginous structures consisting of a flat cartilaginous blade, the fleshy portion adorned with flat multicuspid denticles, and a small cartilaginous tubelike structure lacking denticles. No large denticles are present along the medial edge of the blade. Rudimentary prepelvic pouches are present in females, visible on the ventral surface anterior to the pelvic girdle. Female callorhynchids produce large egg capsules, averaging 20 cm in length and 9 cm in width (Figure 4.5A). The single egg is encased in a central spindle-shaped cavity and extending around the lateral edge of the central spindle is a flexible ridged flange giving the egg capsule an overall ovoid shape. The dorsal side of the egg capsule is convex and the ventral side is concave.

In the absence of reliable morphological characters for distinguishing the three species the only means at present for species identification is by geographic location. Color pattern is highly variable and may be of limited usefulness for distinguishing species (note some suggested color variations that might be helpful for identification in the key below). Many authors have suggested that perhaps the traditional three species concept of Callorhinchidae is incorrect, suggesting they might all be one wide-ranging species (Norman, 1937; Bigelow and Schroeder, 1953; Krefft, 1990); however, differences in the shape of the egg capsules and some variation in the morphology of the frontal tenacula of males may support the validity of these species.

Not only is the identification of species difficult, but there has long been confusion regarding the taxonomy of Family Callorhinchidae. Callorhinchids were first described in various works by Gronovius (1756, 1763) and the first available description of a Callorhinchid was *Chimaera callorynchus* (Linnaeus, 1758). Lacépède (1798) recognized this species as separate from other chimaeras and he placed it in the genus *Callorhinchus*. Much later, Garman (1901) placed all chimaeroids distinguished by a flexible plow-shaped snout and tubular lateral line canals in Family Callorhynchidae. The family name Callorhynchidae was based on the type species, *Callorhinchus callorhynchus*, and is historically the most common family spelling used. The family name should correctly be Callorhinchidae (ICZN Code Article 32.3, 32.4; Fowler, 1941; Paxton et al., 1989; Didier, 1995). The first spelling of the genus was *Callorynchus* (Gronovius, 1754); however, that work was pre-Linnaean and confusion continues to this day regarding the correct spelling of the genus name. Recent research indicates that Lacépède (1798) is the first valid work containing the genus name and the currently accepted spelling is *Callorhinchus* (Didier, 1995; Eschmeyer, 1998).

4.5.2 Rhinochimaeridae (Figure 4.1B)

Family Rhinochimaeridae includes all chimaeroids possessing a long, tapering fleshy snout extending anterior to the head. Rhinochimaerids are commonly referred to as the long-nose chimaeras or spookfish. These fishes generally inhabit deep waters and are usually found at depths around 1000 m to more than 2000 m. This is a small family with only eight species in three genera. In general, rhinochimaerids are

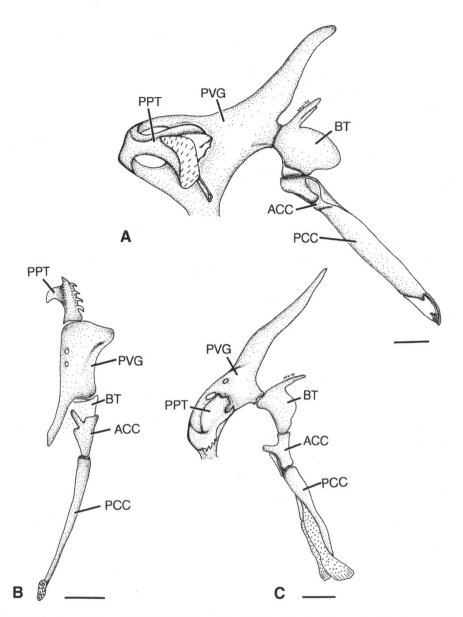

FIGURE 4.4 Skeletal morphology of the left side of the pelvic girdle showing the anatomy of pelvic claspers and prepelvic tenaculae in the three families of chimaeroids: (A) Callorhinchidae, (B) Rhinochimaeridae, (C) Chimaeridae. Scale = 2 cm. ACC = anterior clasper cartilage; BT = basipterygium; PCC = posterior clasper cartilage; PPT = prepelvic tenaculum; PVG = pelvic girdle. (Courtesy of the American Museum of Natural History.)

medium- to large-bodied fishes with an elongate spearlike snout and somewhat compressed, elongate bodies tapering to a narrow tail with an elongate distal filament. In some species the tail appears externally heterocercal with a dorsal caudal fin lobe that is very narrow and a much deeper ventral caudal fin web. In all species the color is usually grayish or brownish, often lighter or white ventrally, without any distinct color pattern. Hatchlings and very small juveniles tend to be much paler in color, dark in the region of the opercular flap, with very dark brown or black fins. The snout is also disproportionately long in juveniles when compared to adults of the same species. Adult males possess slender, rodlike pelvic claspers with small, fleshy denticulate tips (Figure 4.4B). Females produce an ovoid egg capsule, similar in shape to that of callorhynchids, in which a fanlike lateral web surrounds a hollow central spindle-shaped chamber; however, the lateral flange is usually much narrower than that of callorhinchid egg capsules and the central spindle is longer and somewhat indented at either or both ends (Figure 4.5B). As a result of their deep-water habitat, species of rhinochimaerids have been poorly studied and almost nothing is known of their biology and reproduction.

Morphologically there appear to be two distinct lineages of rhinochimaerids, the Rhinochimaerinae, which includes the genus *Rhinochimaera,* and the Harriottinae comprising the remaining two genera, *Harriotta* and *Neoharriotta* (Didier, 1995). Within the Harriottinae the genus *Neoharriotta* is distinguished from *Harriotta* by the possession of a distinct, separate anal fin (Bigelow and Schroeder, 1950). Gill (1893) first named the Harriottinae and Dean (1904c) supported the hypotheses that distinctions among rhinochimaerids warranted separation into two groups. Species of *Rhinochimaera* are distinguished from the harriottines by several morphological features including tooth plates in the form of smooth shearing blades rather than raised hypermineralized tritors on the surface, tubercles on the dorsal caudal fin, a dorsoventrally compressed neurocranium, and the presence of the retractor mesioventralis pectoralis muscle (Didier, 1995). The presence of tubercles on the dorsal caudal fin in *Neoharriotta pinnata* was noted by Bigelow and Schroeder (1950); however, this characteristic has not been observed in subsequent examination of this species (Didier, pers. obs.). The harriottine lineage is not supported by any synapomorphies and support for this separation within Rhinochimaeridae will require further investigation (Didier, 1995).

4.5.3   Chimaeridae (Figure 4.1C)

Commonly known as shortnose chimaeras, ratfishes, or ghost sharks, the chimaerid fishes are characterized by a conical fleshy snout that is bluntly pointed at the tip. Members of this family are distinguished from other chimaeroids by lateral line canals on the snout that are expanded with wide dilations. Species of chimaerids have somewhat compressed, elongate bodies tapering to a whiplike tail with an elongate filament. Most species are a uniform brown, gray, or black, but some species also exhibit color patterns with spots and stripes. In all species the eyes are large, usually a bright green in fresh specimens. Body size can be quite variable with some species remaining small and almost dwarflike at maturity (e.g., *Hydrolagus mirabilis*) while other species attain massive sizes with large bulky heads and bodies and maturing at over 1 m in length (e.g., *Chimaera lignaria*). All males have bifid pelvic claspers with fleshy denticulate tips (Figure 4.4C; discussed above in Section 3.4). Females produce egg capsules that are slender and spindle-shaped without broad lateral flanges (Figure 4.5C). Dean (1912) noted differences among the egg capsules of chimaerids and it is likely that egg capsules are species specific; however, egg capsules for most species have not been identified. The chimaerids are widespread geographically with species known from every ocean region with the exception of the far northern and southern Polar regions. They are known to occur at depths ranging from near-shore surface waters to deeper than 2000 m.

Family Chimaeridae is the most speciose with 22 recognized species in two separate genera. There are currently 7 recognized species of *Chimaera* and 15 species of *Hydrolagus* with at least 3 new species recognized in each genus. Morphologically the two genera are remarkably similar; the only difference is the presence of an anal fin separated from the ventral caudal fin by a notch in *Chimaera,* and a continuous ventral caudal fin without a notch separating the anal fin in *Hydrolagus.* Species of chimaerids are difficult to distinguish because of morphological similarity and are best identified by combinations of characters such as body color, lateral line canal pattern, fin shape, and the relative size or shape of the eyes, snout, or fin spine. It has been suggested that the anal fin may not be sufficient for separation of the two genera, and this single character does not hold up well when considered in combination with other characters (Hardy and Stehmann, 1990; Didier, pers. obs.). A revision of the family will be needed to fully resolve taxonomic difficulties among the chimaerids (see discussion in Appendix 4.1).

Acknowledgments

I am grateful to the many curators and collection managers who provided specimens and museum support for this research, especially T. Abe; N. Feinberg (AMNH); M. McGrouther and D. Hoese

FIGURE 4.5 Egg capsules from representative species of the three families of chimaeroid fishes. (A) *Callorhinchus milii*, shown in dorsal view (left) and ventral view (right). Scale = 3 cm. (B) *Rhinochimaera atlantica*, preserved specimen, shown in dorsal view (left) and ventral view (right). Scale = 3 cm. (C) *Hydrolagus colliei* shown in dorsolateral view (left) and ventral view (right). Scale = 1 cm.

(AMS), S. Schaefer, B. Saul (ANSP); N. Merrett (BMNH); D. Catania (CAS); P. Last, J. Stevens, A. Graham, and G. Yearsley (CSIRO, Department of Marine Research); Mary Ann Rodgers (FMNH); K. Nakaya and students (HUMZ); M. Stehmann (ISH); K. Hartel (MCZ); P. Pruvost, X. Gregorio, and B. Séret (MNHN); C. Roberts and A. Stewart (NMNZ); K. Matsuura and G. Shinohara (NSMT); L. Compagno and M. Van der Merwe (SAM); T. Pietsch (UW); G. Burgess (UF); L. Knapp, J. Finan, and S. Jewett (USNM). My thanks to students who have participated in aspects of chimaeroid research: L. Rosenberger, D. VanBuskirk, E. LeClair, B. Marquardt, and A. Wilmot. Funded by grants from the National Science Foundation (NSF DEB-9510735 and NSF DEB-0097541) and the National Geographic Society (5414-95). Additional support was received from the American Philosophical Society, JSPS, Jessup Fellowships (to E. LeClair and L. Rosenberger), a fellowship to the MNHN, and the New Zealand Foundation for Scientific Research and Technology, Biosystematics of NZ EEZ Fishes Project, contract MNZ603.

Appendix 4.1: Provisional Key to Species

This key is based on study of more than 1000 specimens, almost exclusively from museum collections. Live specimens may vary from preserved specimens, particularly in coloration and overall body shape, and subtle variations in color must be taken into consideration if using this key to identify fresh specimens. This is intended as a provisional key only and is designed for identification of adult or near-mature specimens of all valid species of chimaeroids. Known but as-yet-undescribed species are not included. Very small juveniles and hatchlings may not be identified using this key because they can vary considerably from adults in body proportions and coloration. It is for this reason that small juveniles and hatchings have not been positively identified for most species.

Species of *Hydrolagus* are particularly troublesome to identify and some species can only be positively identified on the basis of color. For example, couplet 22 distinguishes *H. ogilbyi* and *H. lemures* solely on the basis of subtle differences in body color, which may not be sufficient for accurate identification. *Hydrolagus lemures* was described from two small juvenile specimens and it is possible that this species is actually a color variant of *H. ogilbyi*. Likewise, *H. affinis* and *H. pallidus* are distinguished primarily by color and the distinction of light and dark colored species in couplet 24 is an important step in separating these two species. By far the greatest challenge is the separation of a complex of small-bodied species, all of which are pale brown with slender tapering bodies (step 23b). Separation of these species, particularly *H. alberti*, *H. bemisi*, and *H. mitsukurii* (step 29) may depend on geographic location. Although *H. mirabilis* can be distinguished from *H. alberti* by the indentation of the second dorsal fin (step 27), this character may not always be sufficient for identification of these very similar species, which also overlap in geographic range. Separation of these two species may require comparison of the head length to the distance from the junction of the common branch of the oral and preopercular canal to the junction of the trunk lateral line canal. The canal distance is usually longer in *H. mirabilis* and is <2.5 times the head length, and generally >2.5 times the head length in *H. alberti*.

1a. Plow-shaped snout extending forward from the head; body silvery with black saddlelike bands across the dorsal surface and dark blotches on the head and trunk; heterocercal tail; large anal fin precedes caudal; pelvic claspers in males unbranched, tubelike, lacking a fleshy denticulate tip 2

1b. Elongate, spear-shaped snout extending forward from the head; body color an even brown, without distinct markings; pelvic claspers unbranched, slender rods with denticulate bulbous tip 3

1c. Blunt fleshy snout, slightly pointed at the tip; body tapering to whiplike tail; lateral line canals on the snout expanded with large dilations; males with branched pelvic claspers bearing fleshy denticulate lobes at the tips 10

2a. Locality South America; dark spots on trunk along lateral line canal may be fused to form large blotches usually numbering less than 6 .... *Callorhinchus callorhynchus*

2b. Locality South Africa; trunk may be pale with few, usually about three, dark spots or blotches .... *Callorhinchus capensis*

2c. Locality New Zealand, Australia; spots on trunk along and above the lateral line canal often numbering 6 or greater; spots usually rounded, not fused into large blotches.......................... .... *Callorhinchus milii*

3a. Tooth plates with raised hypermineralized tritors on the surface..4

3b. Tooth plates smooth, lacking raised hypermineralized tritors on the surface...........................8

4a. Separate anal fin located anterior to the ventral lobe of the caudal fin 5

4b. Anal fin absent..7

5a. Pelvic fins rounded along distal margin; second dorsal fin uniform in height; oral and preopercular lateral line canals separated by a large space .... *Neoharriotta pinnata*

5b. Pelvic fins with straight posterior margin; second dorsal fin not uniform in height, sloping posteriorly; oral and preopercular lateral line canals separated by a narrow space 6

6a. Anal fin originates at, or anterior to, insertion of second dorsal fin; snout evenly slender along its length...*Neoharriotta carri*

6b. Anal fin originates posterior to insertion of second dorsal fin; snout wide at base, tapering to a slender distal tip.. *Neoharriotta pumila*

7a. Eye is large; dorsal fin spine equal to or longer than height of first dorsal fin.........................
....*Harriotta raleighana*

7b. Eye is small; dorsal fin spine significantly shorter than height of first dorsal fin.....................
....*Harriotta haeckeli*

8a. Body color an even dark brown, snout broad and paddle-shaped; eye is small; junction of supraorbital and infraorbital canals on ventral side of snout closer to the tip of the snout than to the nasal canal .... *Rhinochimaera africana*

8b. Body color a pale brownish gray with dark fins; snout narrow and conical shaped; junction of supraorbital and infraorbital canals on ventral side of snout nearly equidistant between the tip of the snout and the nasal canal... 9

9a. Locality Pacific Ocean; number of denticulation on upper lobe of caudal fin usually 41 to 68
.... *Rhinochimaera pacifica*

9b. Locality Atlantic Ocean; number of denticulations on upper lobe of caudal fin usually 19 to 33
.... *Rhinochimaera atlantica*

10a. Anal fin present, separated from the ventral caudal fin by a notch 11

10b. No anal fin present; ventral caudal fin a continuous ridge along base of tail 17

11a. Body color gray or brown with distinct mottled pattern 12

11b. Body evenly colored, pale silvery, tan, brown, black, gray-blue, or lavender........................ 14

12a. Body color gray or tan with chocolate brown reticulations and spots; posterior margin of first dorsal fin white; pelvic fins with rounded distal margin............................. *Chimaera panthera*

12b. Body color brown or red-brown with reticulations or pale mottling; pelvic fins with straight or squared distal margin... 13

13a. Pectoral fins broad, reaching to origin of pelvic fin base when depressed, rarely extending to posterior edge of pelvic fin base; males with long pelvic claspers, divided for the distal half of length, total length of claspers >20% of body length; preopercular and oral canals share a small common branch... *Chimaera monstrosa*

13b. Pectoral fins elongate and slender, reaching posterior edge of pelvic fin base or beyond when depressed; males with pelvic claspers divided for the distal one third of length, total length of claspers <20% of body length; preopercular and oral canals do not share a common branch
.... *Chimaera owstoni*

14a. Body color an even dark brown or black... *Chimaera jordani*

14b. Body color pale, silvery, tan, brown, gray-blue or lavender 15

15a. Preopercular and oral lateral line canals share a common branch from the infraorbital canal; body color gray-blue or lavender; adults massive in size ....*Chimaera lignaria*

15b. Preopercular and oral lateral line canals branch separately from the infraorbital canal 16

16a. Trunk lateral line canal with tight sinuous undulations along its entire length; faint dark longitudinal stripes along the lateral line canal and trunk .... *Chimaera phantasma*

16b. Trunk lateral line canal not undulated along its length; broad undulations, if present, only anterior to the level of the pelvic fin..*Chimaera cubana*

17a. Distinct pattern of white spots and or blotches on the head and trunk 18

17b. Body color even, lacking pattern of spots or stripes 20

18a. Oral and preopercular canals share a common branch from the infraorbital canal; body color dark brown or black with about nine large white spots; pelvic fins rounded............................
....*Hydrolagus barbouri*

18b. Oral and preopercular canals branch separately from the infraorbital canal......................... 19

19a. Body color brown or reddish brown with small white spots on head and trunk.......................
....*Hydrolagus colliei*

19b. Body color dark brown or gray, sometimes almost black, with white spots fusing to elongate blotches on the head and trunk .... *Hydrolagus novaezealandiae*

20a. Oral and preopercular canals branch separately from the infraorbital canal; body color an even purplish or purple-black; adults very large .... *Hydrolagus purpurescens*

20b. Oral and preopercular canals branch together or share a common branch from the infraorbital canal 21

21a. Trunk lateral line canal with regular, small sinuous undulations along its length 22

21b. Trunk lateral line canal lacking sinuous undulations 23

22a. Body color usually a uniform pale cream, brown, or tan, sometimes paler ventrally, fins darker with distal margins black; a pale indistinct brownish stripe may be visible along the trunk .... *Hydrolagus lemures*

22b. Body color pale, white, silvery, or tan, lighter ventrally, snout sometimes yellowish, fins dark, usually charcoal to black in color ....*Hydrolagus ogilbyi*

23a. Large-bodied fish, adults sometimes massive; body color dark black, purplish, blue, or gray.. 24

23b. Small-bodied, slender fish, some adults almost dwarf-like; body color pale, brown, tan, or silvery/gray... 27

24a. Body color dark, black, or purplish 25

24b. Body color pale, blue or gray... 26

25a. Massive body with large blunt head; body color a dark black or purplish-black ....
....*Hydrolagus affinis*

25b. Long slender body with pointed snout and large eye; body color an even dark black.............
.... *Hydrolagus macrophthalmus*

26a. Massive size with large blunt head; color pale gray or bluish..................*Hydrolagus pallidus*

26b. Large bodied with distinctly pointed snout; color pale blue or blue-gray with dark black margin around orbit and along trunk lateral line canal... *Hydrolagus trolli*

27a. Second dorsal fin deeply indented in the center, nearly dividing the fin into two parts; eye is large; body color a pale brown or gray-brown with dark fins; body shape with a tendency toward concentrated mass in the trunk, tapering rapidly to a long slender tail with long caudal filament.. *Hydrolagus mirabilis*

27b. Second dorsal fin straight along distal margin, or only slightly indented in the center......... 28

28a. Second dorsal fin only slightly indented in the center; long, curved fin spine usually equal to or sometimes exceeding height of the first dorsal fin; males with lateral patch of denticles on the prepelvic tenaculae; body color pale brown or tan .... *Hydrolagus africana*

28b. Second dorsal fin evenly tall along its length, not indented in center; prepelvic tenaculae possess only a single medial row of denticles 29

29a. Dorsal fin spine usually exceeds height of first dorsal fin and reaches beyond origin of second dorsal fin when depressed; color pale gray-brown with dark almost black fins; locality Japan ....*Hydrolagus mitsukurii*

29b. Dorsal fin spine usually equal to height of first dorsal fin, just reaches origin of second dorsal fin when depressed; color pale silvery/gray in life, pale brown or tan in fixative, pale white or cream ventrally; locality New Zealand .... *Hydrolagus bemisi*

29c. Dorsal fin spine usually nearly equal to height of first dorsal fin; body color an even brown with dark fins; locality Gulf of Mexico and Caribbean .... *Hydrolagus alberti*

References

Bargmann, W. 1933. Die Zahnplatten von *Chimaera monstrosa*. *Z. Zell. Mikr. Anat.* 19:537–561.

Bendix-Almgreen, S. A. 1968. The bradyodont elasmobranchs and their affinities; a discussion, in *Current Problems of Lower Vertebrate Phylogeny. Proc. of the 4th Nobel Symposium.* T. Ørvig, Ed., Amlqvist and Wiksell, Stockholm, 153–170.

Bigelow, H. B. and W. C. Schroeder. 1950. New and little known cartilaginous fishes from the Atlantic. *Bull. Mus. Comp. Zool.* 103:35–408.

Bigelow, H. B. and W. C. Schroeder. 1951. Three new skates and a new chimaeroid fish from the Gulf of Mexico. *J. Wash. Acad. Sci.* 41:383–392.

Bigelow, H. B. and W. C. Schroeder. 1953. Chimaeroids, in *Fishes of the Western North Atlantic*, Vol. 1, Pt. 2, Sears Foundation for Marine Research, Yale University, New Haven, CT, 539–541.

Bonaparte, C. L. 1831. *Saggio di una Distribuzione Metodica degli Animali Vettebrati*. Presso Antonio Boulzaler, Rome.

Bory de Saint-Vincent, J. B. 1823. *Dictionnaire Classique D'Histoire Naturelle*, Vol. 3, Paris.

Bullis, H. R., Jr. and J. C. Carpenter. 1966. *Neoharriotta carri* a new species of Rhinochimaeridae from the southern Caribbean Sea. *Copeia* 1966:443–450.

Capello, B. 1867. Descripção de dois peixes novos provendiéntes do mares de Portugal. *J. Sci. Math. Phys. Nat.* 1:314.

Cole, F. J. 1896a. On the cranial nerves of *Chimaera monstrosa* (Linn. 1754); with a discussion of the lateral line system, and of the morphology of the corda tympani. *Trans. R. Soc. Edinb.* 38:631–680.

Cole, F. J. 1896b. On the sensory and ampullary canals of *Chimaera*. *Anat. Anz.*, 12: 172–182.

Cole, F. J. and Dakin, W. J. 1906. Further observations on the cranial nerves of *Chimaera*. *Anat. Anz.* 28:595–599.

Collett, R. 1904. Diagnoses of four hitherto undescribed fishes from the depths south of the Faroe islands. Forhand. *Vidensk. Christiania* 9:5.

Compagno, L. J. V., M. Stehmann, and D. A. Ebert. 1990. *Rhinochimaera africana* a new longnose chimaera from southern Africa, with comments on the systematics and distribution of the genus *Rhinochimaera* Garman, 1901 (Chondrichthyes, Chimaeriformes, Rhinochimaeridae). *S. Afr. J. Mar. Sci.* 9:201–222.

Dean, B. 1903. An outline of the development of a chimaeroid. *Biol. Bull.* 4:270–286.

Dean, B. 1904a. The egg cases of chimaeroid fishes. *Am. Nat.* 38:486–487.

Dean, B. 1904b. Two Japanese species, *C. phantasma* Jordan and Snyder and *C. Mitsukurii* N.S., and their egg cases. *J. Coll. Sci. Imp. Univ. Tokyo* 19:5–9; pl. 1.

Dean, B. 1904c. Notes on the long-snouted chimaeroid of Japan, *Rhinochimaera (Harriotta) pacifica. J. Coll. Sci. Imp. Univ. Tokyo* 19:1–20.

Dean, B. 1906. *Chimaeroid Fishes and Their Development.* Carnegie Institute Publication 32, Washington, D.C., 156 pp.

Dean, B. 1909. Studies on fossil fishes (sharks, chimaeroids, and arthrodires). *Mem. Am. Mus. Nat. Hist.* 9:209–287.

Dean, B. 1912. Orthogenesis in the egg capsules of *Chimaera. Bull. Am. Mus. Nat. Hist.* 31:35–40.

de Buen, F. 1959. Notas preliminares sobre la fauna marina preabismal de Chile, con descripción de una familia de rayas, dos géneros y siete especies nuevos. *Bol. Mus. Nac. Hist. Nat.* 27:171–201.

Didier, D. A. 1995. Phylogenetic systematics of extant chimaeroid fishes (Holocephali, Chimaeroidei). *Am. Mus. Nat. Hist. Novit.* 3119:1–86.

Didier, D. A. 1998. The leopard *Chimaera*, a new species of chimaeroid fish from New Zealand (Holocephali, Chimaeriformes, Chimaeridae). *Ichthyol. Res.* 45:281–289.

Didier, D. A. 2002. Two new species of chimaeroid fishes from the southwestern Pacific Ocean (Holocephali, Chimaeridae). *Ichthyol. Res.* 49:299–306.

Didier, D. A. and K. Nakaya. 1999. Redescription of *Rhinochimaera pacifica* (Mitsukuri) and first record of *R. africana* Compagno, Stehmann & Ebert from Japan (Chimaeriformes: Rhinochimaeridae). *Ichthyol. Res.* 46:139–152.

Didier, D. A. and B. Séret. 2002. Chimaeroid fishes of New Caledonia with description of a new species of *Hydrolagus* (Chondrichthyes, Holocephali). *Cybium* 26:225–233.

Didier, D. A. and M. Stehmann. 1996. *Neoharriotta pumila*, a new species of longnose chimaera from the northwestern Indian Ocean (Pisces, Holocephali, Rhinochimaeridae). *Copeia* 1996: 955–965.

Didier, D. A., B. J. Stahl, and R. Zangerl. 1994. Compound tooth plates of chimaeroid fishes (Holocephali: Chimaeroidei). *J. Morphol.,* 222:73–89.

Didier, D. A., E. E. LeClair, and D. R. VanBuskirk. 1998. Embryonic staging and external features of development of the chimaeroid fish *Callorhinchus milii* (Holocephali, Callorhinchidae). *J. Morphol.* 236:25–47.

Duméril, A. 1865. *Histoire Naturelle des Poissons ou Ichthyologie Genéralé.* Vol. 1. *Elasmobranches, Plagiostomes et Holocéphales.* Paris, 689 pp.

Eschmeyer, B. 1998. *Catalog of Fishes*, Vols. 1–3. California Academy of Sciences, San Francisco.

Fields, R. D. and G. D. Lange. 1980. Electroreception in the ratfish (*Hydrolagus colliei*). *Science* 207:57–548.

Fields, R. D., T. H. Bullock, and G. D. Lange. 1993. Ampullary sense organs, peripheral, central and behavioral electroreception in chimeras (*Hydrolagus*, Holocephali, Chondrichthyes). *Brain Behav. Evol.* 41:269–289.

Fowler, H. W. 1908. A collection of fishes from Victoria Australia. *Proc. Acad. Nat. Sci.* 59:419–421.

Fowler, H. W. 1910. Notes on chimaeroid and ganoid fishes. *Proc. Acad. Nat. Sci.* 62:603–612.

Fowler, H. W. 1941. The fishes of the groups Elasmobranchii, Holocephali, Isospondyli and Ostarophysi obtained by the U.S. Bureau of Fisheries steamer "Albatross" in 1907 to 1910, chiefly in the Philippine Islands and adjacent seas. *Bull. U.S. Nat. Mus.* 100:486–510.

Francis, M. P. 1997. Spatial and temporal variation in the growth rate of elephantfish (*Callorhinchus milii*). *N.Z. J. Mar. Freshwater Res.* 31:9–23.

Francis, M., P. McMillan, R. Lasenby, and D. Didier. 1998. How to tell dark and pale ghost sharks apart. *Seafood N.Z.* 6:29–30.

Garman, S. 1888. On the lateral canal system of the Selachia and Holocephala. *Bull. Mus. Comp. Zool.*. 17:57–120.

Garman, S. 1901. Genera and families of the chimaeroids. *Proc. N. Engl. Zool. Club* 2:75–77.

Garman, S. 1904. The chismopnea especially *Rhinochimaera* and its allies. *Bull. Mus. Comp. Zool.* 41:245–272.

Garman, S. 1908. New plagiostomia and chismopnea. *Bull. Mus. Comp. Zool.* 51:251–256.

Garman, S. 1911. The chismopnea (chimaeroids). *Mem. Mus. Comp. Zool.* 15: 81–101.

Gilbert, C. H. 1905. Deep sea fishes of the Hawaiian Islands, in The Aquatic Resources of the Hawaiian Islands, D. S. Jordan and B. W. Evermann, Eds., *Bull. U.S. Fish Comm.* 23:575–713.

Gilchrist, J. D. F. 1922. Deep-sea fishes procured by the S.S. "Pickle" (Part 1). Union of South Africa, Fisheries and Marine Biological Survey, Report 2, pp. 41–79.

Gill, T. 1893. Families and subfamilies of fishes. *Nat. Acad. Sci. Mem.* 6:127–138.

Goode, G. B. and T. H. Bean. 1895. Oceanic Ichthyology. Special Bulletin of the United States National Museum. U.S. Government Printing Office, Washington, D.C., 553 pp.

Gorman, T. B. S. 1963. Biological and economic aspects of the elephant fish *Callorhynchus milii* Bory, in Pegasus Bay and the Canterbury Bight. Fisheries Technical Report, New Zealand Ministry of Agriculture and Fisheries 8: 1–54.

Graham, D. H. 1939. Food of the fishes of Otago Harbour and adjacent sea. *Trans. R. Soc. N.Z.* 68:421–436.

Graham, D. H. 1956. *A Treasury of New Zealand Fishes*. A. H. and A. W. Reed Publishers, Wellington, 424 pp.

Grogan, E. D., R. Lund, and D. Didier. 1999. A description of the chimaerid jaw and its phylogenetic origins. *J. Morphol.* 239:45–59.

Gronovius, L. T. 1756. Museum Ichthyologicum sistens piscium indegenorum quorumdam exoticorum, qui in museo Laurentii Theodori Gronovii, J. U. D. adservantur, descriptions ordine systematico… Historia Zoologica Lugduni Batavorum, sumptibus auctoris, prostat apud Theodorum Haak.

Gronovius, L. T. 1763. Zoophylacii Gronoviani Fasciculus primus exhibens animalia quadrupeda, amphibia atque pisces, quae in museo suo adservat, rite examinavit, systematice disposuit, descripsit atque iconibus illustravit laur. Theod. Gronovius, J. U. D. Lugduni Batavorum 1–136, 14 pls.

Halstead, B. W. and N. C. Bunker. 1952. The venom apparatus of the ratfish, *Hydrolagus colliei*. *Copeia* 1952:128–138.

Hardy, G. S. and M. Stehmann. 1990. A new deep-water ghost shark, *Hydrolagus pallidus* n.sp. (Holocephali, Chimaeridae), from the Eastern North Atlantic, and redescription of *Hydrolagus affinis* (Brito Capello, 1867). *Arch. Fischereiwiss.*, 40:229–248.

Holt, E. W. L. and L. W. Byrne. 1909. Preliminary note on some fishes from the Irish Atlantic Slope. *Ann. Mag. Nat. Hist.* 8:279–280.

Howell-Rivero, L. 1936. Some new, rare and little-known fishes from Cuba. *Proc. Boston Soc. Nat. Hist.* 41:41–76.

Johnson, A. G. and H. F. Horton. 1972. Length-weight relationship, food habits, parasites, and sex and age determination of the ratfish, *Hydrolagus colliei* (Lay and Bennett). *Fish. Bull. NOAA* 70:421–429.

Jordan, D. S. and J. O. Snyder. 1900. A list of fishes collected in Japan by Keinosuke Otaki, and by the United States Fish Commission steamer "Albatross," with descriptions of fourteen new species. *Proc. U.S. Nat. Mus.* 23:335–380.

Jordan, D. S. and J. O. Snyder. 1904. On the species of white chimaera from Japan. *Proc. U.S. Nat. Mus.* 27:223–226.

Karrer, C. 1972. Die Gattung *Harriotta* Goode and Bean, 1895 (Chondrichthyes, Chimaeriformes, Rhinochimaeridae). *Mitt. Zool. Mus.* (Berlin) 48:203–221.

Kemp, A. 1984. A comparison of the developing dentition of *Neoceratodus forsteri* and *Callorhynchus milii*. *Proc. Linn. Soc. N. S. W.* 107:245–262.

Krefft, G. 1990. Callorhynchidae, in *Check-list of the Fishes of the Eastern Tropical Atlantic*, Vol. 1. J. C. Quèro, J. C. Hureau, C. Karrer, A. Post, and L. Saldanha, Eds., UNESCO, Paris, 117 pp.

Lacépède, B. G. E. 1798. *Histoire Naturelle des Poissons*, Vol. 1. Paris, 400 pp.

Last, P. R. and J. D. Stevens. 1994. *Sharks and Rays of Australia*. CSIRO, Collingwood, Victoria, Australia, 513 pp.

Lay, G. T. and E. T. Bennett. 1839. Fishes, in *The Zoology of Captain Beechey's Voyage*. Henry G. Bohn, London, 71–75.

Leigh-Sharpe, W. H. 1922. The comparative morphology of the secondary sexual characters of elasmobranch fishes, memoirs IV and V. *J. Morphol.* 36:199–243.

Leigh-Sharpe, W. H. 1926. The comparative morphology of the secondary sexual characters of elasmobranch fishes, memoir X. *J. Morphol.* 42:335–348.

Linnaeus, C. 1758. Systema naturae sive regna tria naturae, systematice proposita per classes, ordines, genera, et species, cum characteribus, differentiis, synonymis, locis. Editio decima reformata, Vol. I, regnum animale, ii, 824 pp.

Lund, R. 1977. New information on the evolution of the bradyodont Chondrichthyes. *Fieldiana Geol.* 33:521–539.

Lund, R. 1986. The diversity and relationships of the Holocephali, in *Indo-Pacific Fish Biology. Proc. 2nd Int. Conf. Indo-Pacific Fish.* T. Uyeno, R. Arai, T. Taniuchi, and K. Matsuura, Eds., Ichthyological Society of Japan, Tokyo, 97–106.

Lund, R. 1988. New Mississippian Holocephali (Chondrichthyes) and the evolution of the Holocephali. *Mem. Mus. Nat. Hist. Nat.* (Paris) C 53:195–205.

Lund, R. and E. D. Grogan. 1997. Relationships of the Chimaeriformes and the basal radiation of the Chondrichthyes. *Rev. Fish Biol. Fish.* 7:65–123.

Macpherson, E. and B. A. Roel. 1987. Trophic relationships in the demersal fish community off Namibia. *S. Afr. J. Mar. Sci.* 5:585–596.

Maisey, J. G. 1986. Heads and tails: a chordate phylogeny. *Cladistics* 2:201–256.

Marques, A. and F. Porteiro. 2000. Hydrothermal vent mussel *Bathymodiolus* sp. (Mollusca: Mytilidae): diet item of *Hydrolagus affinis* (Pisces: Chimaeridae). *Copeia* 2000:806–807.

Maruska, K. F. and T. C. Tricas. 1998. Morphology of the mechanosensory lateral line system in the Atlantic stingray, *Dasyatis sabina*: The mechanotactile hypothesis. *J. Morphol.* 238:1–22.

Mathews, C. P. 1975. Note on the ecology of the ratfish, *Hydrolagus colliei*, in the Gulf of California. *Calif. Fish Game* 61:47–53.

Mitsukurii, K. 1895. On a new species of the chimaeroid group *Harriotta. Zool. Mag. Tokyo* 7:97–98.

Moy-Thomas, J. A. 1939. The early evolution and relationships of the elasmobranchs. *Biol. Rev.* 14:1–26.

Nelson, J. S. 1994. *Fishes of the World,* 2nd ed. John Wiley & Sons, New York, 43–47.

Norman, J. R. 1937. Coast fishes. II. The Patagonian region. *Discovery Rep.* 16:1–150.

Northcutt, G. R. 1989. The phylogenetic distribution and innervation of craniate mechanoreceptive lateral lines, in *The Mechanosensory Lateral Line — Neurobiology and Evolution.* S. Coombs, P. Gorner, and H. Munz, Eds., Springer, New York, 17–78.

Ørvig, T. 1967. Phylogeny of tooth tissues: evolution of some calcified tissues in early vertebrates, in *Structural and Chemical Organization of Teeth.* I. A. E. W. Miles, Ed., Academic Press, London, 45–110.

Ørvig, T. 1985. Histologic studies of ostracoderms, placoderms and fossil elasmobranchs 5. Ptyctodontid tooth plates and their bearing on holocephalans ancestry: the condition of chimaerids. *Zool. Sci.* 14:55–79.

Patterson, C. 1965. The phylogeny of the chimaeroids. *Philos. Trans. R. Soc. London B Biol. Sci.* 249:101–219.

Patterson, C. 1992. Interpretation of the toothplates of chimaeroid fishes. *Zool. J. Linn. Soc.* 106:33–61.

Paulin, C., A. Stewart, C. Roberts, and P. McMillan. 1989. *New Zealand Fish: A Complete Guide.* Miscellaneous Series 19, National Museum of New Zealand, Wellington.

Paxton, J. R., D. F. Hoese, G. R. Allen, and J. E. Hanley, Eds. 1989. *Zoological Catalogue of Australia,* Vol. 7, *Pisces.* Australian Government Publishing Service, Canberra, 664 pp.

Peach, M. B. and G. W. Rouse. 2000. The morphology of the pit organs and lateral line canal neuromasts of *Mustelus antarcticus* (Chondrichthyes: Triakidae). *J. Mar. Biol. Assoc. U.K.* 80:155–162.

Peyer, B. 1968. *Comparative Odontology.* University of Chicago Press, Chicago.

Quinn, T. P., B. S. Miller, and R. C. Wingert. 1980. Depth distribution and seasonal and diel movements of ratfish, *Hydrolagus colliei*, in Puget Sound. *Wash. Fish. Bull.* 78:816–821.

Reese, A. M. 1910. The lateral line system of *Chimaera colliei. J. Exp. Zool.* 9:349–371.

Sathyanesan, A. G. 1966. Egg-laying of the chimaeroid fish *Hydrolagus colliei. Copeia* 1966:132–134.

Schauinsland, H. 1903. Beiträge zur Entwicklungsgeschichte und Anatomie der Wirbeltiere. I. *Sphenodon, Callorhynchus, Chamaeleo. Zoologica* 16:5–32, 58–89.

Schnakenbeck, W. von. 1929. Über einige Meeresfische aus Südwestafrika. *Mitt. Zool. Staats. Zool. Mus. Hamburg* 44:38–45.

Smith, R. M., R. W. Day, T. I. Walker, and W. C. Hamlett. 2001. Microscopic organization and sperm storage in the oviducal gland of the elephant fish, *Callorhynchus milii*, at different stages of maturity. 6th Indo-Pacific Fish Conference (abstr.).

Stahl, B. J. 1999. Chondrichthyes III Holocephali, in *Handbook of Paleoichthyology,* Vol. 4. H.-P. Schultze, Ed., Verlag Dr. Friedrich Pfeil, Munich, 164 pp.

Stanley, H. P. 1963. Urogenital morphology in the chimaeroid fish *Hydrolagus colliei* (Lay and Bennett). *J. Morphol.* 112:97–127.

Sullivan, K. J. 1978. Age and growth of the elephant fish *Callorhinchus milii* (Elasmobranchii: Callorhynchidae). *N.Z. J. Mar. Freshwater Res.* 11:745–753.

Tanaka, S. 1905. On two new species of *Chimaera*. *J. Coll. Sci. Imp. Univ. Tokyo* 20:1–14.

Waite, E. R. 1898. Sea fisheries report upon trawling operations off the coast of New South Wales between the Manning River and Jervis Bay, carried on by H.M.C.S. "Thetis," Scientific Report on the Fishes, 41–42.

Webb, J. F. and G. R. Northcutt. 1997. Morphology and distribution of pit organs and canal neuromasts in non-teleost bony fishes. *Brain Behav. Evol.* 50:139–151.

Whitley, G. P. 1939. Taxonomic notes on sharks and rays. *Aust. Zool.* 9:227–262.

Wilmot, A., D. Didier, and J. Webb. 2001. Morphology of the lateral line canals of the head in chimaerid fishes (Family Chimaeridae). *Am. Zool.* 40:1261 (abstr.).

Zangerl, R. 1981. *Handbook of Paleoichthyology*, Vol. 3A, *Chondrichthyes I. Paleozoic Elasmobranchii*. Gustav Fischer Verlag, New York.

Zangerl, R., H. F. Winter and M. C. Hansen. 1993. Comparative microscopic dental anatomy in the Petalodontida (Chondrichthyes, Elasmobranchii). *Fieldiana Geol. Ser.* 26:1–43.

Part II

Form, Function,
and Physiological Processes

5

Biomechanics of Locomotion in Sharks, Rays, and Chimeras

Cheryl A.D. Wilga and George V. Lauder

CONTENTS

5.1 Introduction

The body form of sharks is notable for the distinctive heterocercal tail with external morphological asymmetry present in most taxa, and the ventrolateral winglike pectoral fins extending laterally from the body (Figure 5.1). These features are distinct from the variation in body form present in actinopterygian fishes (Lauder, 2000) and have long been of interest to researchers wishing to understand the functional design of sharks (Garman, 1913; Magnan, 1929; Grove and Newell, 1936; Harris, 1936; Aleev, 1969; Thomson, 1971).

5.1.1 Approaches to Studying Locomotion in Chondrichthyans

Historically, many attempts have been made to understand the function of the median and paired fins in sharks and rays, and these studies have included work with models (Harris, 1936; Affleck, 1950; Simons, 1970), experiments on fins removed from the body (Daniel, 1922; Harris, 1936; Alexander, 1965; Aleev, 1969), and quantification of body form and basic physical modeling (Thomson, 1976;

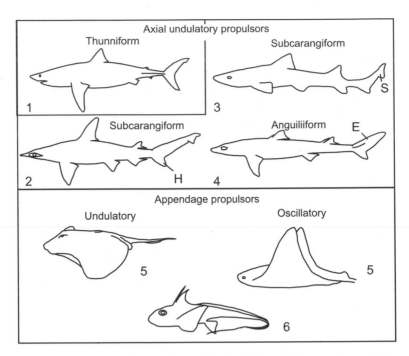

FIGURE 5.1 Propulsion mechanisms in chondrichthyans. Numbers indicate body groups (see text). E = epicaudal lobe; H = hypochordal lobe; S = subterminal lobe. (Based on Webb, 1984; Webb and Blake, 1985.)

Thomson and Simanek, 1977). More recently, direct quantification of fin movement using videography has allowed a better understanding of fin conformation and movement (Ferry and Lauder, 1996; Fish and Shannahan, 2000; Wilga and Lauder, 2000), although such studies have to date been limited to relatively few species. Obtaining high-resolution three-dimensional (3D) data on patterns of shark fin motion is a difficult task, and these studies have been confined to a highly controlled laboratory environment where sharks swim in a recirculating flow tank. Although locomotion of sharks and rays under these conditions does not allow the range of behaviors seen in the wild, the ability to obtain data from precisely controlled horizontal swimming as well as specific maneuvering behaviors has been vital to both testing classical hypotheses of fin function and to discovery of new aspects of locomotory mechanics. A key general lesson learned from recent experimental kinematic and hydrodynamic analyses of shark locomotion is the value of understanding the 3D pattern of fin movement, and the requirement for experimental laboratory studies that permit detailed analyses of fin kinematics and hydrodynamics.

Two new laboratory-based approaches in recent years have been particularly fruitful in clarifying the biomechanics of shark locomotion. Chief among these has been the use of two-camera high-speed video systems to quantify patterns of fin motion in 3D (e.g., Ferry and Lauder, 1996; Wilga and Lauder, 2000). Two-dimensional (2D) analyses are subject to very large errors when motion occurs in 3D, and the orientation of a planar surface element in 3D can be opposite to the angle appearing in a single 2D view; an example of this phenomenon relevant to the study of shark tails is given in Lauder (2000). The use of two simultaneous high-speed video cameras permits determination of the x, y, and z locations of individual tail points and hence the 3D orientation of fin and body surface elements. Using two separate synchronized cameras greatly increases the resolution in each view, as opposed to using a mirror to split a single camera image into two views, which produces low-resolution images of each view.

The second new approach to studying shark locomotor biomechanics has been the application of flow visualization techniques from the field of fluid mechanics. Briefly, the technique of digital particle image velocimetry (DPIV) (Willert and Gharib, 1991; Krothapalli and Lourenco, 1997) allows direct visualization of water flow around the fins of swimming sharks and quantification of the resulting body and fin wake (e.g., Lauder and Drucker, 2002; Wilga and Lauder, 2002; Lauder et al., 2003). We now have the ability to understand the hydrodynamic significance of different fin and body shapes, and to measure forces exerted on the water as a result of fin motion (Lauder and Drucker, 2002). This represents a real advance over more qualitative previous approaches such as injection of dye to gain an impression of how the fins of fishes function. Finally, more traditional experimental techniques such as electromyography to quantify the timing of muscle activation, in combination with newer techniques such as sonomicrometry (Donley and Shadwick, 2003), are revealing new aspects of shark muscle function during locomotion.

5.1.2 Diversity of Locomotory Modes in Chondrichthyans

Sharks, rays, and chimeras have had a long evolutionary history leading to the locomotor modes observed in extant forms (Carroll, 1988). Chondrichthyans have a remarkable diversity of body forms and locomotor modes for a group containing so few species (Figure 5.1). All sharks swim using continuous lateral undulations of the axial skeleton. However, angel sharks, which are dorsoventrally depressed, may supplement axial propulsion with undulations of their enlarged pectoral fins. Four modes of axial undulatory propulsion have been described, based on decreasing proportion of the body that is undulated during locomotion, which form a continuum from anguilliform to thunniform (Webb and Keyes, 1982; Webb and Blake, 1985; Donley and Shadwick, 2003). In anguilliform swimmers, the entire trunk and tail participate in lateral undulations where more than one wave is present. This mode is characteristic of many elongate sharks such as orectolobiforms, *Chlamydoselachus,* and more benthic carcharhiniform sharks like scyliorhinids. More pelagic sharks, such as squaliforms, most carcharhiniforms, and some lamniforms, are carangiform swimmers (Breder, 1926; Gray, 1968; Lindsey, 1978; Donley and Shadwick, 2003), where undulations are mostly confined to the posterior half of the body with less than one wave present. The amplitude of body motion increases markedly over the posterior half of the body (Webb and Keyes, 1982; Donley and Shadwick, 2003). Only the tail and caudal peduncle undulate in thunniform swimmers, which is a distinguishing feature of lamniform sharks, most of which are high-speed cruisers.

Most batoids (skates and rays) have short, stiff head and trunk regions with slender tails and therefore must swim by moving the pectoral fins. There are two modes of appendage propulsion exhibited by batoids: undulatory and oscillatory (Webb, 1984) (Figure 5.1). Similar to axial swimmers, undulatory appendage propulsors swim by passing undulatory waves down the pectoral fin from anterior to posterior (Daniel, 1922). Most batoids are undulatory appendage propulsors. However, some myliobatiforms, such as eagle and manta rays, swim by flapping their pectoral fins up and down in a mode known as oscillatory appendage propulsion. Holocephalans are appendage propulsors and utilize a combination of flapping and undulation of the pectoral fins for propulsion and maneuvering, much like many teleost fishes.

5.1.3 Body Form and Fin Shapes

Most species of sharks have a fusiform-shaped body that varies from elongate in species such as bamboo sharks to the more familiar torpedo shape of white sharks. However, angelsharks and wobbegong sharks are dorsoventrally depressed. There is great variability in the morphology of the paired and unpaired fins. Four general body forms have been described for sharks that encompass this variation (Thomson and Simanek, 1977), with two additional body forms that include batoids and holocephalans.

Sharks with body type 1 have a conical head, a large deep body, large pectoral fins, a narrow caudal peduncle with lateral keels, and a high aspect ratio tail (high heterocercal angle) that is externally symmetrical. These are typically fast-swimming pelagic sharks such as *Carcharodon, Isurus,* and *Lamna.* As is typical of most high-speed cruisers, these sharks have reduced pelvic, second dorsal, and anal fins, which act to increase streamlining and reduce drag. However, *Cetorhinus* and *Rhincodon,* which are slow-moving filter feeders, also fit into this category. In these sharks, the externally symmetrical tail presumably results in more efficient slow cruising speeds in large-bodied pelagic sharks and also aligns the mouth with the center of mass and the center of thrust from the tail and probably increases feeding efficiency.

Sharks with body type 2 have a more flattened ventral head and body surface, a less deep body, large pectoral fins, a lower heterocercal tail angle and lack keels. These are more generalized, continental swimmers such as *Alopias, Carcharias, Carcharhinus, Galeocerdo, Negaprion, Prionace, Sphyrna, Mustelus,* and *Triakis. Alopias* is similar to these sharks despite the elongate pectoral and caudal fins. Similarly, hammerheads, with the exception of the cephalofoil, also fit into this category. These sharks probably have the greatest range of swimming speeds. They also retain moderately sized pelvic, second dorsal, and anal fins and therefore remain highly maneuverable over their swimming range.

Sharks with body type 3 have relatively large heads, blunt snouts, more anterior pelvic fins, more posterior first dorsal fins, a low heterocercal tail angle with a small to absent hypochordal lobe and a large subterminal lobe. These sharks are slow-swimming epibenthic, benthic, and demersal sharks such as *Scyliorhinus, Ginglymostoma, Chiloscyllium, Galeus, Apristurus, Psudeotriakis,* and Hexanchiformes. Pristiophoriforms and pristiforms may fit best into this category. Although the body morphology of hexanchiform sharks is most similar to these, they have only one dorsal fin that is positioned more posterior on the body than the pelvic fins.

Body type 4 is united by only a few characteristics and encompasses a variety of body shapes. These sharks lack an anal fin and have a large epicaudal lobe. Only squalean or dogfish sharks are represented in this category. Most of these species are deep-sea sharks and have slightly higher pectoral fin insertions, i.e., *Squalus, Isistius, Centroscymus, Centroscyllium, Dalatius, Echinorhinus, Etmopterus,* and *Somniosus.* However, *Squalus* also frequents continental waters and have higher aspect tails similar to those in type 2.

A fifth body type can be described based on dorsoventral flattening of the body, enlarged pectoral fins, and a reduction in the caudal half of the body. This type would include batoids, except for pristiforms, pristiophoriforms, and angelsharks. These chondrichthyans are largely benthic, but also include the pelagic myliobatiform rays. Rajiforms and myliobatiforms locomote by undulating the pectoral fins, whereas torpediniforms undulate the tail and rhinobatiforms undulate both the pectoral fins and tail.

Holocephalans or chimeras represent the sixth body type. They resemble teleosts in that they are laterally compressed and undulate the pectoral fins rather than the axial body in steady horizontal swimming. Tail morphology ranges from a long and tapering (leptocercal) to distinctly heterocercal.

5.2   Locomotion in Sharks

5.2.1   Function of the Body during Steady Locomotion and Maneuvering

The anatomy of the various components of shark fin and body musculature and skeleton has recently been reviewed elsewhere (Bone, 1999; Compagno, 1999; Kemp, 1999; Liem and Summers, 1999), and is not covered again here, where our focus is the biomechanics of fin and body locomotion. However, it is worth noting that there are very few detailed studies of the musculature and connective tissue within fins, and knowledge of how myotomal musculature is modified at the caudal peduncle and how myotomal and skin connective tissue elements insert within the tail is poor at best (Reif and Weishampel, 1986; Wilga and Lauder, 2001). Such studies will be particularly valuable for understanding how muscular forces are transmitted to paired and median fins.

One of the most important factors in shark locomotion is the orientation of the body, as this is the primary means by which the overall force balance (considered in detail below) is achieved during swimming and maneuvering. When sharks are induced to swim horizontally so that the path of any point on the body is at all times parallel to the *x* (horizontal) axis with effectively no vertical (*y*) motion, the body is tilted up at a positive angle of attack to oncoming flow (Figure 5.2). This positive body angle occurs even though sharks are swimming steadily and not maneuvering, and are maintaining their vertical position in the water. This positive body angle ranges from 11° to 4° in *Triakis* and *Chiloscyllium,* respectively, at slow swimming speeds of 0.5 *l*/s. The angle of body attack varies with speed, decreasing to near zero at 2 *l*/s swimming speed (Figure 5.2). During vertical maneuvering in the water column, the angle of the body is altered as well (Figure 5.3). When leopard sharks rise so that all body points show increasing values along the *y*-axis, the body is tilted to a mean angle of 22° into the flow. During sinking in the water, the body is oriented at a negative angle of attack averaging −11° in *Triakis* (Figure

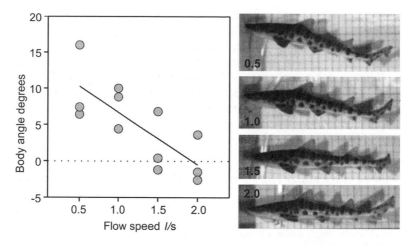

FIGURE 5.2 Plot of body angle vs. flow speed to show the decreasing angle of the body with increasing speed. Each symbol represents the mean of five body angle measurements (equally spaced in time) for five tail beats for four individuals. Images show body position at the corresponding flow speeds in *l*/s, where *l* is total body length (flow direction is left to right). At all speeds, sharks are holding both horizontal and vertical position in the flow, and not rising or sinking in the water column. Body angle was calculated using a line drawn along the ventral body surface from the pectoral fin base to the pelvic fin base and the horizontal (parallel to the flow). A linear regression ($y = 15.1 - 7.4x$, adjusted $r^2 = 0.43$; $P < 0.001$) was significant and gives the best fit to the data. (From Wilga, C.D. and G.V. Lauder. 2000. *J. Exp. Biol.* 203:2261–2278.)

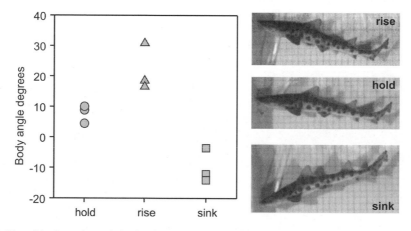

FIGURE 5.3 Plot of body angle vs. behavior during locomotion at 1.0 *l*/s. Circles indicate holding behavior, triangles show rising behavior and squares reflect sinking behavior. Body angle was calculated as in Figure 5.2. Each point represents the mean of five sequences for each of four individuals. To the right are representative images showing body position during rising, holding, and sinking behaviors. Body angle is significantly different among the three behaviors (ANOVA, $P = 0.0001$). (From Wilga, C.D. and G.V. Lauder. 2000. *J. Exp. Biol.* 203:2261–2278.)

5.3). These changes in body orientation undoubtedly reflect changes in lift forces necessary either to maintain body position given the negative buoyancy of most sharks or to effect vertical maneuvers.

Locomotor kinematics of the body in sharks at a variety of speeds has been studied by Webb and Keyes (1982). Recently, Donley and Shadwick (2003) have presented electromyographic recordings of body musculature to correlate activation patterns of red myotomal fibers with muscle strain patterns and body movement. Donley and Shadwick (2003) noted that red muscle fibers in the body myotomes of *Triakis* are activated to produce the body wave at a consistent relative time all along the length of the body. The onset of muscle activation always occurred as the red fibers were lengthening, and these fibers were deactivated consistently during muscle shortening. Donley and Shadwick (2003) concluded that the red muscle fibers along the entire length of the body produce positive power, and hence contribute to locomotor thrust generation, in contrast to some previous hypotheses, which suggested that locomotion in fishes is powered by anterior body muscles alone.

During propulsion and maneuvering in sharks, skates, and rays both median fins (caudal, dorsal, and anal) as well as paired fins (pectoral, pelvic) play an important role. In this chapter, however, we focus on the caudal fin and pectoral fins as virtually nothing quantitative is known about the function of dorsal, anal, and pelvic fins. Only Harris (1936) has conducted specific experiments designed to understand the function of multiple fins, and these studies were performed on model sharks placed in an unnatural body position in a wind tunnel. The role of the dorsal, anal, and pelvic fins during locomotion in elasmobranchs is a key area for future research on locomotor mechanics.

5.2.2 Function of the Caudal Fin during Steady Locomotion and Maneuvering

Motion of the tail is a key aspect of shark propulsion, and the heterocercal tail of sharks moves in a complex 3D manner during locomotion. Ferry and Lauder (1996) used two synchronized high-speed video cameras to quantify the motion of triangular segments of the leopard shark tail during steady horizontal locomotion. Sample video frames from that study are shown in Figure 5.4, which illustrates tail position at six times during a half tail stroke. One video camera viewed the tail laterally giving the x and y coordinates of identified locations on the tail, while a second camera aimed at a mirror downstream of the tail provided a posterior view giving z and y coordinates for those same locations. Tail marker locations were connected into triangular surface elements (Figure 5.5A and B) and their orientation was tracked through time. This approach is discussed in more detail by Lauder (2000). Analysis of surface element movement through time showed that for the majority of the tail beat cycle the caudal fin surface was inclined at an angle greater than 90° to the horizontal (Figure 5.5), suggesting that the downwash of water from the moving tail would be directed posteroventrally. These data provided kinematic corroboration of the classical model of shark heterocercal tail function, which hypothesized that the

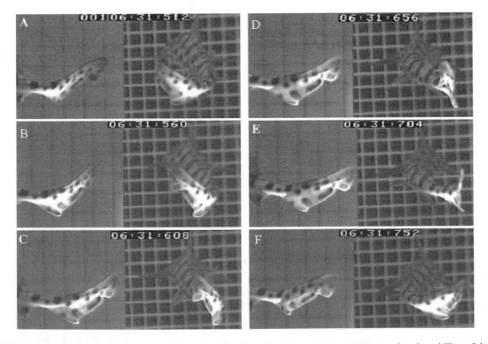

FIGURE 5.4 Composite video sequence of the tail beating from the leftmost extreme (A), crossing the midline of the beat (B, C, and D), and beating to the rightmost extreme or maximum lateral excursion (reached in E and F). In F, the tail has started its beat back to the left. Times for each image are shown at the top with the last three digits indicating elapsed time in milliseconds. Each panel contains images from two separate high-speed video cameras, composited into a split-screen view. (From Ferry, L.A. and G.V. Lauder. 1996. *J. Exp. Biol.* 199:2253–2268.)

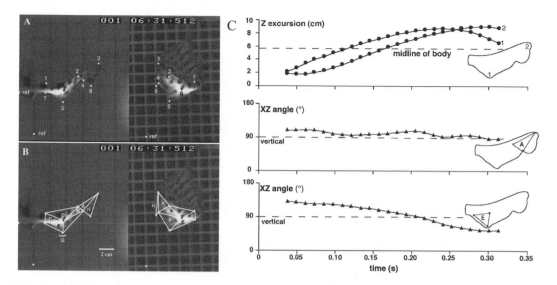

FIGURE 5.5 Images of the tail of a representative leopard shark, *Triakis semifasicata,* swimming in the flow tank. Landmarks (1–8) are shown in (A) with both lateral and posterior views, and (B) with the points joined to form the triangles (A–H) for analysis. Points marked "ref" were digitized as reference points. Both views were identically scaled using the grid in the lateral view (1 box = 2 cm); the smaller grid visible in posterior view is the upstream baffle reflected in the mirror, toward which the shark is swimming. (C) Heterocercal tail kinematics in a representative leopard shark swimming steadily at 1.2 *l*/s. Z-dimension excursions (upper panel) of two points on the tail and the 3D angles of two tail triangles with the *xz* plane. Note that for most of the tail beat, the orientation of these two triangular elements is greater than 90°, indicating that the tail is moving in accordance with the classical model of heterocercal tail function. (From Ferry, L.A. and G.V. Lauder. 1996. *J. Exp. Biol.* 199:2253–2268; and Lauder, G.V. 2000. *Am. Zool.* 40:101–122.)

shark caudal fin would generate both thrust and lift by moving water posteriorly and ventrally (Grove and Newell, 1936; Alexander, 1965; Lauder, 2000).

Although kinematic data provide strong evidence in support of the classical view of heterocercal tail function in sharks, they do not address what is in fact the primary direct prediction of that model: the direction of water movement. To determine if the heterocercal tail of sharks functions hydrodynamically as expected under the classical view, a new technique is needed that permits direct measurement of water flow. DPIV is such a technique and a schematic diagram of this approach as applied to shark locomotion is illustrated in Figure 5.6. Sharks swim in a recirculating flow tank, which has been seeded with small (12-μm mean diameter) reflective hollow glass beads. A 5 to 10 W laser is focused into a light sheet 1 to 2 mm thick and 10 to 15 cm wide and this beam is aimed into the flow tank using focusing lenses and mirrors (Figure 5.6). Sharks are induced to swim with the tail at the upstream edge of the light sheet so that the wake of the shark passes through the light sheet as this wake is carried downstream. A second synchronized high-speed video camera takes images of the shark body so that orientation and movements in the water column can be quantified.

Analysis of wake flow video images proceeds using standard DPIV processing techniques, and further details of DPIV as applied to problems in fish locomotion are provided in a number of recent papers (Drucker and Lauder, 1999; Wilga and Lauder, 1999, 2000, 2001, 2002; Lauder, 2000; Lauder and Drucker, 2002; Lauder et al., 2002, 2003). Briefly, cross-correlation of patterns of pixel intensity between homologous regions of images separated in time is used to generate a matrix of velocity vectors, which reflect the pattern of fluid flow through the light sheet. Sample DPIV data are presented in Figure 5.8. From these matrices of velocity vectors the orientation of fluid accelerated by the tail can be quantified, and any rotational movement measured as fluid vorticity. Recent research on fish caudal fin function has shown that the caudal fin of fishes sheds momentum in the form of vortex loops as the wake rolls up into discrete torus-shaped rings with a central high-velocity jet flow (Drucker and Lauder, 1999; Lauder and Drucker, 2002). By quantifying the morphology of these wake vortex rings, we can determine the direction of force application to the water by the heterocercal tail by measuring the direction of the

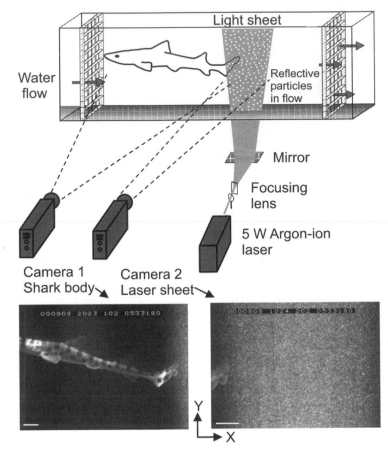

FIGURE 5.6 Schematic diagram of the working section of the flow tank illustrating the DPIV system. Sharks swam in the working section of the flow tank with the laser sheet oriented in a vertical (parasagittal, *xy*) plane. Lenses and mirrors are used to focus the laser beam into a thin light sheet that is directed vertically into the flow tank. The shark is shown with the tail cutting through the laser sheet. Two high-speed video cameras recorded synchronous images of the body (camera 1) and particles in the wake (camera 2) of the freely swimming sharks.

central vortex ring momentum jet. In addition, the absolute force exerted on the water by the tail can be calculated by measuring the strength and shape of the vortex rings (Dickinson, 1996; Drucker and Lauder, 1999; Lauder and Drucker, 2002).

Using the two-camera arrangement illustrated in Figure 5.6, Wilga and Lauder (2002) studied the hydrodynamics of the tail of leopard sharks during both steady horizontal locomotion and vertical maneuvering. They measured the orientation of the body relative to the horizontal, the path of motion of the body through the water, and the orientation and hydrodynamic characteristics of the vortex rings shed by the tail (Figure 5.7). Representative data from that study are shown in Figure 5.8, which illustrates the pattern of water velocity, and vortex ring orientation resulting from one tail beat in two species of sharks. Tail vortex rings are inclined significantly to the vertical and are tilted posterodorsally. The central high-velocity water jet through the center of each vortex ring is oriented posteroventrally at an angle between 40° and 45° below the horizontal. These data provide unequivocal support for the classical model of heterocercal tail function in sharks by demonstrating that the tail accelerates water posteroventrally and that there must necessarily be a corresponding reaction force with dorsal (lift) and anterior (thrust) components.

Analysis of the changing orientation of tail vortex rings as sharks maneuver vertically in the water demonstrates that the relationship between vortex ring angle and body angle remains constant as body angle changes during maneuvering (Figure 5.9). These data show that leopard sharks do not alter the direction of force application to the water by the tail during vertical maneuvering, in contrast to previous

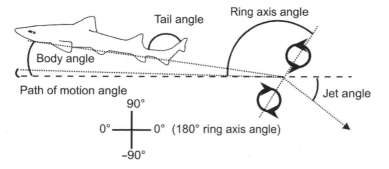

FIGURE 5.7 Schematic summary illustrating body and wake variables measured relative to the horizontal: body angle, from a line drawn along the ventral body surface; path of motion of the center of mass; tail angle between the caudal peduncle and dorsal tail lobe; ring axis angle from a line extending between the two centers of vorticity; and mean vortex jet angle. Angle measurements from the variables of interest (dotted lines) to the horizontal (dashed line) are indicated by the curved solid lines. Angles above the horizontal are considered positive and below the horizontal negative. Ring axis angle was measured from 0° to 180° (From Wilga, C.D. and G.V. Lauder. 2002. *J. Exp. Biol.* 205:2365–2374.)

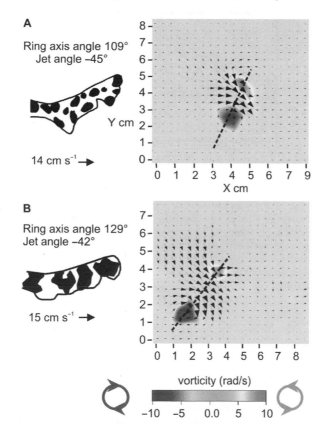

FIGURE 5.8 DPIV analysis of the wake of the tail of representative (A) *Triakis semifasciata* and (B) *Chiloscyllium punctatum* sharks during steady horizontal locomotion at 1.0 *l*/s. On the left is a tracing depicting the position of the tail relative to the shed vortex ring visible in this vertical section of the wake. The plot to the right shows fluid vorticity with the matrix of black velocity vectors representing the results of DPIV calculations based on particle displacements superimposed on top. A strong jet, indicated by the larger velocity vectors, passes between two counterrotating vortices representing a slice through the vortex ring shed from the tail at the end of each beat. The black dashed line represents the ring axis angle. *Note:* Light gray color indicates no fluid rotation, the dark gray color reflects clockwise fluid rotation, and medium gray color indicates counterclockwise fluid rotation. To assist in visualizing jet flow, a mean horizontal flow of *U* = 19 and *U* = 24 cm/s was subtracted from each vector for *T. semifasciata* and *C. punctatum,* respectively. (From Wilga, C.D. and G.V. Lauder. 2002. *J. Exp. Biol.* 205:2365–2374.)

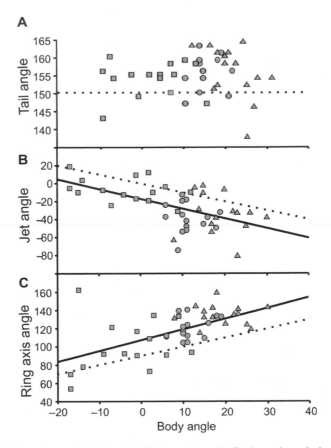

FIGURE 5.9 Plot of body angle vs. (A) tail angle, (B) jet angle, and (C) ring axis angle in leopard sharks, *Triakis semifasciata* while swimming at 1.0 *l*/s. Solid lines indicate a significant linear regression, and the dotted line represents the predicted relationship. The lack of significance of the tail vs. body angle regression ($P = 0.731$, $r^2 = 0.003$) indicates that the sharks are not altering tail angle as body angle changes, but instead are maintaining a constant angular relationship regardless of locomotor behavior. Jet angle decreases with increasing body angle ($P < 0.001$, $r^2 = 0.312$, $y = -17 - 1.087x$) at the same rate as the predicted parallel relationship, indicating that the vortex jet is generated at a constant angle to the body regardless of body position. Ring axis angle increases with body angle at the same rate as the predicted perpendicular relationship ($P < 0.001$, $r^2 = 0.401$, $y = 107 + 1.280x$). Circles, triangles, and squares represent holds, rises, and sinks, respectively. (From Wilga, C.D. and G.V. Lauder. 2002. *J. Exp. Biol.* 205:2365–2374.)

data from sturgeon that demonstrated the ability to actively alter tail vortex wake orientation as they maneuver (Liao and Lauder, 2000).

5.2.3   Function of the Pectoral Fins during Locomotion

5.2.3.1   Anatomy of the Pectoral Fins — There are two distinct types of pectoral fins in sharks based on skeletal morphology. In aplesodic fins, the cartilaginous radials are blunt and extend up to 50% into the fin with the distal web supported only by ceratotrichia. In contrast, plesodic fins have radials that extend more than 50% into the fin to stiffen it and supplement the support of the ceratotrichia (Compagno, 1988) (Figure 5.10). The last row of radials tapers to a point distally in plesodic fins. Plesodic fins appear in Lamniformes, hemigaleids, carcharhinids, sphyrnids, and batoids except for pristids; other groups have aplesodic fins (Shirai, 1996). The restricted distribution of plesodic pectoral fins in extant sharks, the different morphology in each group, and their occurrence in more derived members (by other characters) of each group strongly suggest that plesodic pectorals are derived and have evolved independently from aplesodic pectorals (Zangerl, 1973; Compagno, 1973, 1988; Bendix-Almgreen, 1975). The decreased skeletal support of aplesodic pectoral fins over plesodic fins allows

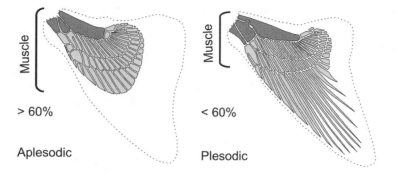

FIGURE 5.10 Skeletal structure of the pectoral fins in aplesodic sharks, such as leopard, bamboo, and dogfish (Wilga and Lauder, 2001) (left) and plesodic sharks, such as lemon, blacktip, and hammerhead (redrawn from Compagno, 1988) (right). The left pectoral fin for each species is shown in dorsal view. Dark gray elements are propterygium, mesopterygium, and metapterygium from anterior to posterior; light gray elements are radials; dotted line delimits extent of ceratotrichia into the fin web. Muscle insertion extends to the end of the third row of radials in aplesodic sharks and to the end of the second row or middle of the third row of radials in plesodic sharks.

greater freedom of motion in the distal web of the fin and may function to increase maneuverability. *Chiloscyllium* sp. (Orectolobiformes) frequently "walk" on the substrate using both the pectoral and pelvic fins (Pridmore, 1995) in a manner similar to that of salamanders. They can bend the pectoral fins such that an acute angle is formed ventrally when rising on the substrate and angles up to 165° are formed dorsally when station-holding on the substrate (pers. obs.). In contrast, the increased skeletal support of plesodic fins stiffens and streamlines the distal web, which reduces drag. Furthermore, the extent of muscle insertion into the pectoral fin appears to correlate with the extent of radial support into the fin and thus pectoral fin type. In sharks with aplesodic fins, the pectoral fin muscles insert as far as the third (and last) row of radial pterygiophores, well into the fin. In contrast, those sharks with plesodic fins have muscles that insert only as far as the second row (of three) of radials.

Streamlined rigid bodies are characteristic of fishes that are specialized for cruising and sprinting, whereas flexible bodies are characteristic of fishes that are specialized for accelerating or maneuvering (Webb, 1985, 1988). Applying this analogy to shark pectoral fins, it may be that plesodic fins are specialized for cruising (fast-swimming pelagic sharks) and aplesodic fins are specialized for accelerating or maneuvering (slow-cruising pelagic and benthic sharks).

5.2.3.2 Role of the Pectoral Fins during Steady Swimming — The function of the pectoral fins during steady horizontal swimming and vertical maneuvering (rising and sinking) has been tested experimentally in *Triakis semifasciata*, *Chiloscyllium plagiosum*, and *Squalus acanthias* (Wilga and Lauder, 2000, 2001, in prep.). Using 3D kinematics and fin marking (Figure 5.11), these studies have shown that the pectoral fins of these sharks are held in such a way that negligible lift is produced during

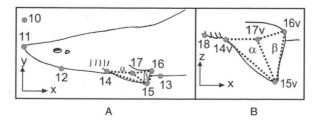

FIGURE 5.11 Schematic diagram of a shark illustrating the digitized points on the body and pectoral fin. (A) Lateral view of the head and pectoral fin, and (B) ventral view of pectoral fin region. Note that the reference axes differ for lateral (x, y) and ventral (x, z) views. Data from both views were recorded simultaneously. Points 14 to 16 are the same points in lateral and ventral views, while points 17 and 17v represent the same location on the dorsal and ventral fin surfaces. These 3D coordinate data were used to calculate a 3D planar angle between the anterior and posterior fin planes (α and β) as shown in B. (From Wilga, C.D. and G.V. Lauder. 2000. *J. Exp. Biol.* 203:2261–2278.)

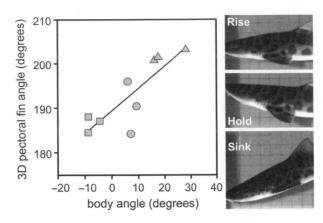

FIGURE 5.12 Graph of 3D pectoral fin angle vs. body angle for rising, holding, and sinking behaviors at 1.0 *l*/s in leopard sharks. Symbols are as in Figure 5.3. Body angle was calculated using the line connecting points 12 and 13 (see Figure 5.11) and the horizontal (parallel to the flow). Each point represents the mean of five sequences for each of four individuals. Images to the right show sample head and pectoral fin positions during each behavior. Pectoral fin angles equal to 180° indicate that the two fin triangles (see Figure 5.11) are coplanar; angles less than 180° show that the fin surface is concave dorsally; angles greater than 180° indicate that the fin surface is concave ventrally. The 3D internal pectoral fin angle is significantly different among the three behaviors (ANOVA, $P = 0.0001$). The least-squares regression line is significant (slope 0.41, $P < 0.001$; adjusted $r^2 = 0.39$). (From Wilga, C.D. and G.V. Lauder. 2000. *J. Exp. Biol.* 203:2261–2278.)

steady horizontal locomotion. The pectoral fins are cambered with an obtuse dorsal angle between the anterior and posterior regions of the fin (mean 190° to 191°) (Figure 5.12). Thus, the planar surface of the pectoral fin is held concave downward relative to the flow during steady swimming (Figure 5.13) as well as concave mediolaterally.

The posture of the pectoral fins relative to the flow during steady horizontal swimming in these sharks contrasts markedly to those of the wings in a cruising passenger aircraft. The anterior and posterior planes of the pectoral fins in these sharks during steady horizontal swimming are at negative and positive angles, respectively, to the direction of flow (Figure 5.13). When both planes are considered together, the chord angle is −4° to −5° to the flow. Conversely, the wings of most cruising passenger aircraft have a positive attack angle to the direction of oncoming air, which generates positive lift.

The planar surface of the pectoral fins of these sharks is held at a negative dihedral (fin angle relative to the horizontal) angle from −6° (*C. plagiosum*) to −23° (*T. semifasciata*) during steady horizontal swimming (Figure 5.14). The pectoral fins are destabilizing in this position (Smith, 1992; Simons, 1994; Wilga and Lauder, 2000) and promote rolling motions of the body, such as those made while maneuvering in the water column. For example, in a roll, the fin with the greatest angle to the horizontal meets the flow at a greater angle of attack, resulting in a greater force (F_x) directed into the roll, while the angle of attack of the more horizontally oriented fin is reduced by the same amount. The more horizontal fin therefore possesses a smaller force (F_x) opposing the roll while the more inclined fin has greater force directed into the roll, thereby contributing to the rolling motion. This is in direct contrast to previous studies suggesting that the pectoral fins of sharks are oriented to prevent rolling, as in the keel of a ship (Harris, 1936, 1953). Wings that are tilted at a positive angle with respect to the horizontal have a positive dihedral angle, as in passenger aircraft, and are self-stabilizing in that they resist rolling motions of the fuselage (Figure 5.14) (Smith, 1992; Simons, 1994). When a passenger aircraft rolls, the more horizontally oriented wing generates a greater lift force than the inclined wing (Smith, 1992; Simons, 1994). Thus, a corrective restoring moment arises from the more horizontal wing, which opposes the roll, and the aircraft is returned to the normal cruising position. Interestingly, the negative dihedral wings of fighter aircraft, which are manufactured for maneuverability, function similarly to that of shark pectoral fins.

The flow of water in the wake of the pectoral fins during locomotion in these three species was quantified using DPIV, to estimate fluid vorticity and the forces exerted by the fin on the fluid (see Drucker and Lauder, 1999; Wilga and Lauder, 2000). These results further corroborate the conclusion from the 3D kinematic data that the pectoral fins generate negligible lift during steady horizontal

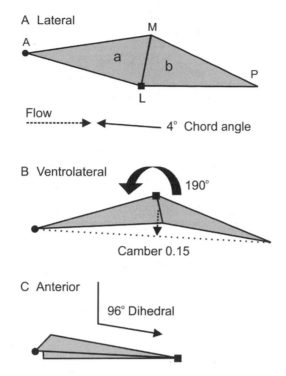

FIGURE 5.13 Orientation of the two pectoral fin planes (a and b) in 3D space during pelagic holding in bamboo sharks, *Chiloscyllium plagiosum* (leopard and dogfish sharks show similar conformations). Panels show (A) lateral, (B) ventrolateral, and (C) posterior views of the fin planes. Points defining the fin triangles correspond to the following digitized locations in Figure 5.11: A, anterior, point 14, black circle; L, point 15, black square; P, posterior, point 16; M, medial, point 17. Chord angle to the flow is given in the lateral view, camber and internal fin angle between planes a and b are given in the ventrolateral view, and the dihedral angle is shown in the posterior view (note that in the posterior view the angles are given as acute to the *xy* plane). (From Wilga, C.D. and G.V. Lauder. 2001. *J. Morphol.* 249:195–209.)

swimming. There was virtually no vorticity or downwash detected in the wake of the pectoral fins during steady horizontal swimming, which shows that little or no lift is being produced by the fins (Figure 5.15). According to Kelvin's law, vortices shed from the pectoral fin must be equivalent in magnitude but opposite in direction to the theoretical bound circulation around the fin (Kundu, 1990; Dickinson, 1996). Therefore, the circulation of the shed vortex can be used to estimate the force on the fin. Mean downstream vertical fluid impulse calculated in the wake of the pectoral fins during steady horizontal swimming was not significantly different from zero. This indicates that the sharks are holding their pectoral fins in such a way that the flow speed and pressure are equivalent on the dorsal and ventral surfaces of the fin. Furthermore, if the pectoral fins were generating lift to counteract moments generated by the heterocercal tail, there would necessarily be a downwash behind the wing to satisfy Kelvin's law. The lack of an observable and quantifiable downwash indicates clearly that, during holding behavior, pectoral fins generate negligible lift.

These results showing that the pectoral fins of these sharks do not generate lift during steady forward swimming stand in stark contrast to previous findings on sharks with bound or amputated fins (Daniel, 1922; Harris, 1936; Aleev, 1969). Although the results of such radical experiments are difficult to evaluate, it is likely that the lack of pectoral fin motion prevented the sharks from initiating changes in pitch and therefore limited their ability to achieve a horizontal position and adjust to perturbances in oncoming flow. Lift forces measured on the pectoral fins and body of a plaster model of *Mustelus canis* in a wind tunnel also suggested that the pectoral fins generated upward lift while the body generated no lift (Harris, 1936). However, the pectoral fins were modeled as rigid flat plates (2D) and tilted upward 8° to the flow while the longitudinal axis of the body was oriented at 0° to the flow. Although it is possible that *M. canis* locomotes with the body and pectoral fins in this position, the results of current studies on live

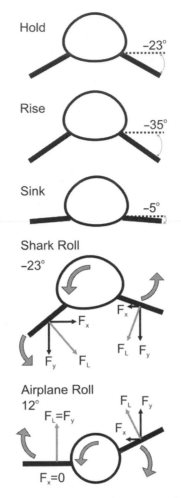

FIGURE 5.14 Schematic diagram of the dihedral orientation of the pectoral fins in a shark during holding, rising and sinking behaviors. Forces during a roll are illustrated below for the pectoral fins of a shark and the wings of an airplane. The body and fin are represented as a cross section at the level of plane α of the pectoral fin (see Figure 5.11). Thin gray double-headed arrows represent the dihedral angle between the plane α (dotted line) and pectoral fin. Thick arrows show the direction of movement of the body and fins or wing during a roll. Note that positive dihedrals (such as those used in aircraft design) are self-stabilizing, while fins oriented at a negative dihedral angle, as in sharks, are destabilizing in roll and tend to amplify roll forces. F_x, horizontal force; F_y, vertical force; F_L, resultant force. (From Wilga, C.D. and G.V. Lauder. 2000. *J. Exp. Biol.* 203:2261–2278.)

freely swimming and closely related *T. semifasciata*, as described herein, which has a very similar body shape, show a radically different orientation of the body and pectoral fins.

Three-dimensional kinematic analyses of swimming organisms are crucial to deriving accurate hypotheses about the function of the pectoral fins and body (Wilga and Lauder, 2000). The 2D angle of the anterior margin of the pectoral fin as a representation of the planar surface of the pectoral fin in sharks is extremely misleading. Although the pectoral fin appears to be oriented at a positive angle to the flow in lateral view, 3D kinematics reveals that the fin is actually concave downward with a negative dihedral. When viewed laterally, this negative-dihedral concave-downward orientation of the pectoral fin creates a perspective that suggests a positive angle of attack when the angle is, in fact, negative.

5.2.3.3 *Role of the Pectoral Fins during Vertical Maneuvering* — *Triakis semifasciata*, *Chiloscyllium plagiosum*, and *Squalus acanthias* actively adjust the angle of their pectoral fins to maneuver vertically in the water column (Wilga and Lauder, 2000, 2001, in prep.). Rising in the water

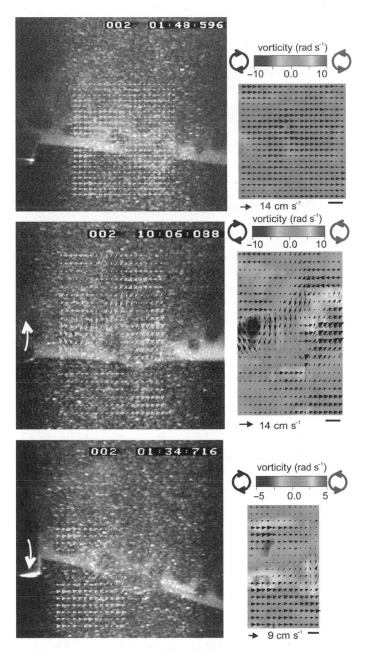

FIGURE 5.15 DPIV data from leopard shark pectoral fins during (top) holding vertical position, (middle) sinking, and (bottom) rising behaviors at 1.0 *l*/s (patterns for bamboo and dogfish sharks are similar). The video image (on the left) is a single image of a shark with the left pectoral fin located just anterior to the laser light sheet. Note that the ventral body margin is faintly visible through the light sheet. The plot on the right shows fluid vorticity with velocity vectors with conventions as in Figure 5.8. In the holding position, note that the fin is held in a horizontal position, and that the vorticity plot shows effectively no fluid rotation. Hence, the pectoral fins in this position do not generate lift forces. During sinking, note that there is a clockwise vortex (dark gray region of rotating fluid to the right) that resulted from the upward fin flip (curved white arrow) to initiate the sinking event. During rising, note that the fin has flipped ventrally (curved white arrow) to initiate the rising event, and that a counterclockwise vortex (medium gray region of rotating fluid to the right) has been shed from the fin. To assist in visualizing the flow pattern, a mean horizontal flow of $U = 33$ cm/s was subtracted from each vector. (Modified from Wilga and Lauder, 2001.)

column is initiated when the posterior plane of the fin is flipped downward to produce mean obtuse dorsal fin angles around 200°, while the leading edge of the fin is rotated upward relative to the flow. This downward flipping of the posterior plane of the fin increases the chord angle to +14, and as a result the shark rises in the water. In contrast, to sink in the water the posterior plane of the pectoral fin is flipped upward relative to the anterior plane, which produces a mean obtuse dorsal fin angle of 185°. At the same time, the leading edge of the fin is rotated downward relative to the flow such that the chord angle is decreased to −22°, and the shark sinks in the water.

The dihedral angle of shark pectoral fins changes significantly during vertical maneuvering in the water column (Figure 5.14). The dihedral angle increases to −35° during rising and decreases to −5° during sinking. This may be due to a need for greater stability during sinking behavior because the heterocercal tail generates a lift force that tends to drive the head ventrally. Holding the pectoral fins at a low dihedral angle results in greater stability during sinking compared to rising. The greater negative dihedral angle increases maneuverability and allows rapid changes in body orientation during rising.

These angular adjustments of the pectoral fins are used to maneuver vertically in the water column and generate negative and positive lift forces, which then initiate changes in the angle of the body relative to the flow. As the posterior plane of the pectoral fin is flipped down to ascend, a counterclockwise vortex, indicating upward lift force generation, is produced and shed from the trailing edge of the fin and pushes the head and anterior body upward (Figure 5.15). This vortex is readily visible in the wake as it rolls off the fin and is carried downstream. The opposite flow pattern occurs when sharks initiate a sinking maneuver in the water column. A clockwise vortex, indicating downward lift force generation, is visualized in the wake of the pectoral fin as a result of the dorsal fin flip and pulls the head and anterior body of the shark downward (Figure 5.15).

Lift forces produced by altering the planar surface of the pectoral fin to rise and sink appear to be a mechanism to reorient the position of the head and anterior body for maneuvering. Changing the orientation of the head will alter the force balance on the body as a result of interaction with the oncoming flow and will induce a change in vertical forces that will move the shark up or down in the water column. Forces generated by the pectoral fins are significantly greater in magnitude during sinking than during rising. This may be due to the necessity of reorienting the body through a greater angular change to sink from the positive body tilt adopted during steady swimming. A shark must reposition the body from a positive body tilt of 8° (mean holding angle) down through the horizontal to a negative body tilt of −11° (mean sinking angle), a change of 19°. In contrast, to rise a shark simply increases the positive tilt of the body by 14° (mean rise – hold difference), which should require less force given that the oncoming flow will assist the change from a slightly tilted steady horizontal swimming position to a more inclined rising body position.

5.2.3.4   *Function of the Pectoral Fins during Benthic Station-Holding* — *Chiloscyllium plagiosum* have a benthic lifestyle and spend much of their time resting on the substrate on and around coral reefs where current flows can be strong. To maintain position on the substrate during significant current flow, these sharks shift their body posture to reduce drag (Wilga and Lauder, 2001). The sharks reorient the longitudinal axis of the body to the flow with the head pointing upstream during current flow, but do not orient when current flow is negligible or absent. Body angle steadily decreases from 4° at 0 *l*/s to 0.6° at 1.0 *l*/s as they flatten their body against the substrate with increasing flow speed. This reduces drag in higher current flows thereby promoting station-holding. This behavior is advantageous in fusiform benthic fishes that experience a relatively high flow regime, such as streams where salmon parr are hatched (Arnold and Webb, 1989) and inshore coral reefs where bamboo sharks dwell (Compagno, 1984).

Chiloscyllium plagiosum also reorient the pectoral fins to generate negative lift, increase friction, and oppose downstream drag during station-holding in current flow (Wilga and Lauder, 2001). They hold the pectoral fins in a concave upward orientation, similar to that in sinking, which decreases from a mean planar angle of 174° at 0 *l*/s to a mean of 165° at 1.0 *l*/s. At the same time, the chord angle steadily decreases from a mean of 2.7° at 0 *l*/s to a mean of −3.9° at 1.0 *l*/s. Flattening the body against the substrate lowers the anterior edge of the fin, whereas elevating the posterior edge of the fin to decrease the planar angle significantly decreases the chord angle (Figure 5.16). In this orientation, water flow is deflected up and over the fin and produces a clockwise vortex that is shed from the fin tip. The clockwise vortex

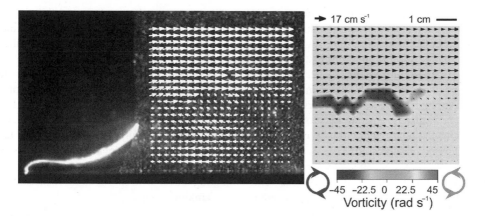

FIGURE 5.16 DPIV data from the pectoral fins of a representative bamboo shark *Chiloscyllium plagiosum* while holding station on the substrate. The video image on the left shows a shark with the left pectoral fin located in the anterior end of the laser light sheet; other conventions as in Figures 5.8 and 5.15. Note that the fin is held at a negative chord angle to the flow. A clockwise vortex (negative vorticity) was produced in the wake of the pectoral fins, which continued to rotate just behind the fin for several seconds until it was carried downstream by the flow (as seen here), after which a new vortex forms in the wake of the fin. (Modified from Wilga and Lauder, 2001.)

produces significant negative lift (mean –0.084 N) directed toward the substrate that is eight times greater than that generated during sinking. As the clockwise vortex shed from the fin rotates just behind the fin, flow recirculates upstream and pushes against the posterior surface of the fin, which opposes downstream drag. These movements generate negative lift that is directed toward the substrate and acts to increase total downward force and friction force, thereby promoting station-holding as predicted by previous studies (Arnold and Webb, 1991; Webb and Gerstner, 1996), as well as a novel mechanism leading to vortex shedding that opposes downstream drag to further aid benthic station-holding (Wilga and Lauder, 2001).

5.2.3.5 *Motor Activity in the Pectoral Fins* — Movement of the posterior plane of the pectoral fin during sinking and rising is actively controlled by *Triakis semifasciata*. At the beginning of a rise, the pectoral fin depressors (ventral fin muscles, adductors) are active to depress the posterior portion of the pectoral fin (Figure 5.17). Small bursts of activity in the lateral hypaxialis, protractor, and levator muscles are sometimes present during rising, probably to stabilize pectoral fin position. In contrast, the pectoral fin levators (dorsal fin muscles, adductors), as well as the cucullaris and ventral hypaxialis, are strongly active during elevation of the posterior portion of the fin at the beginning of sinking. Virtually

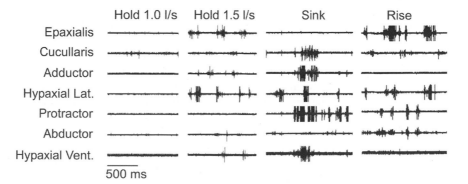

FIGURE 5.17 Electromyographic data from selected pectoral fin and body muscles during locomotion in *Triakis semifasciata* at 1.0 *l*/s for four behaviors: holding position at 1.0 and 1.5 *l*/s and sinking and rising at 1.0 *l*/s. Note the near absence of fin muscle activity while holding position at 1.0 *l*/s and recruitment of body and fin muscles at 1.5 *l*/s. The hypaxialis was implanted in both lateral (mid-lateral dorsal and posterior to pectoral fin base) and ventral (posterior to coracoid bar) positions. All panels are from the same individual. Scale bar represents 500 ms.

no motor activity is present in the pectoral fin muscles while holding position at 0.5 and 1.0 *l*/s, indicating that the pectoral fins are not actively held in any particular position during steady horizontal locomotion. However, at higher flow speeds (1.5 *l*/s), recruitment of epaxial and hypaxial muscles occurs with slight activity in the pectoral fin muscles that may function to maintain stability.

Epaxial or hypaxial muscles are recruited to elevate or depress the head and anterior body during rising or sinking, respectively. At the initiation of rising behavior, simultaneously with the head pitching upward, a strong burst of activity occurs in the cranial epaxialis, while it is virtually silent during holding and sinking. Similarly, a strong burst of activity occurs in the ventral hypaxialis during the initiation of sinking behavior, again with virtually no activity during holding and rising. This shows that the head is actively elevated or depressed to rise or sink, respectively, and that conformational changes in the anterior body assist the forces generated by the pectoral fins to accomplish vertical maneuvers. Finally, antagonistic pectoral fin muscles become active as rising or sinking slows or during braking (i.e., the levators are active as rising stops and the depressors are active as sinking stops).

5.2.4 Synthesis

The data presented above on pectoral and caudal fin function and body orientation in the shark species studied permit construction of a new model of the overall force balance during swimming (Figure 5.18). It is useful to discuss separately the vertical force balance and the rotational (torque) balance. During steady horizontal locomotion, when sharks are holding vertical position, body weight is balanced by lift forces generated by the heterocercal tail and ventral body surface. The ventral surface generates lift both anterior and posterior to the center of body mass by virtue of its positive angle of attack to the oncoming water. Sharks adjust their body angle to modulate the total lift force produced by the body, and can thus compensate for changes in body weight over both short and longer time frames.

Rotational balance is achieved by balancing the moments of forces around the center of mass. It has not been generally appreciated that the ventral body surface generates both positive and negative torques corresponding to the location of the ventral surface anterior and posterior to the center of mass. Water impacting the ventral body surface posterior to the center of mass will generate a counterclockwise torque of the same sign as that generated by the heterocercal tail. In contrast, water impacting the ventral body anterior to the center of mass will generate a clockwise torque, which is opposite in sign to that generated by the ventral body and tail posterior to the center of mass. As a result of experimental data demonstrating that shark pectoral fins do not generate lift during steady horizontal locomotion (Wilga and Lauder, 2000, 2001) as a result of their orientation relative to the flow, no role in generating either lift or torque is attributed to the pectoral fins during horizontal locomotion. This stands in contrast to the textbook depiction of shark locomotion in which the pectoral fins play a central role in controlling body position during horizontal locomotion. In our view, experimental kinematic and hydrodynamic data obtained over the last 5 years demonstrate that control of body orientation is the key to modulating lift and torques during horizontal swimming while the pectoral fins are not used for balancing forces during horizontal swimming.

However, during maneuvering the pectoral fins do play a key role in generating both positive and negative lift forces and hence torques about the center of mass (Figure 5.18). To rise in the water, sharks rapidly move the trailing pectoral fin edge ventrally, and a large vortex is shed, generating a corresponding lift force. This force has a clockwise rotational moment about the center of mass pitching the body up, increasing the angle of the body, and hence the overall lift force. As a result, sharks move vertically in the water even while maintaining horizontal position via increased thrust produced by the body and caudal fin.

To stop this vertical motion or to maneuver down (sink) in the water the trailing pectoral fin edge is rapidly elevated, shedding a large vortex, which produces a large negative lift force (Figure 5.18). This generates a counterclockwise torque about the center of mass, pitching the body down, exposing the dorsal surface to incident flow, and producing a net sinking motion. Pectoral fins thus modulate body pitch.

Overall, the force balance on swimming sharks is maintained and adjusted by small alterations in body angle and this in turn is achieved by elevation and depression of the pectoral fins. Pectoral fins thus play a critical role in shark locomotion by controlling body position and facilitating maneuvering, but they do not function to balance tail lift forces during steady horizontal locomotion.

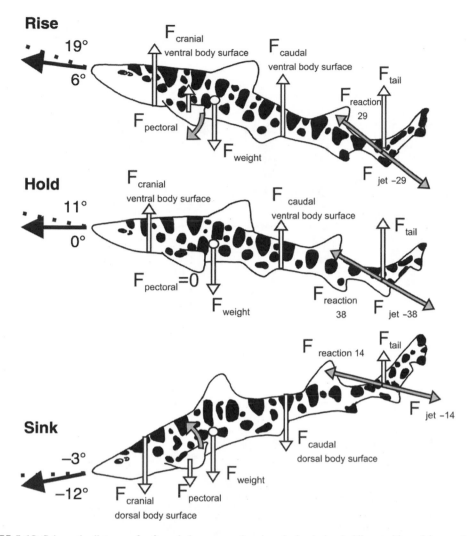

FIGURE 5.18 Schematic diagram of a force balance on swimming sharks during holding position, rising, and sinking behaviors (also representative of bamboo sharks, *Chiloscyllium punctatum*, and spiny dogfish, *Squalus acanthias*). The white circle represents the center of mass and vectors indicate forces *F* exerted by the fish on the fluid. Lift forces are generated by the ventral body surface, both anterior and posterior to the center of mass. The jet produced by the beating of the tail maintains a constant angle relative to body angle and path angle and results in an anterodorsally directed reaction force oriented dorsal to the center of mass during all three behaviors, supporting the classical model. Tail vortex jet angles are predicted means. (From Wilga, C.D. and G.V. Lauder. 2002. *J. Exp. Biol.* 205:2365–2374.)

5.3 Locomotion in Skates and Rays

As mentioned above, most batoids either undulate or oscillate the pectoral fins to move through the water (Figure 5.19). Basal batoids, such as guitarfishes, sawfishes, and electric rays, locomote by undulating their relatively thick tails similar to those of laterally undulating sharks (Rosenberger, 2001). Interestingly, *Rhinobatos lentiginosus*, which has a sharklike trunk and tail like all guitarfishes, also adopts a positive body angle to the flow during steady horizontal swimming (Rosenberger, 2001). Sawfishes and most electric rays are strict axial undulators and use only the tail for locomotion, whereas guitarfishes and some electric rays may supplement axial locomotion with undulations of the pectoral fin (Rosenberger, 2001). Most rays use strict pectoral fin locomotion (Rosenberger, 2001). However, some rays, such as *Rhinoptera* and *Gymnura*, fly through the water by oscillating the pectoral fins in

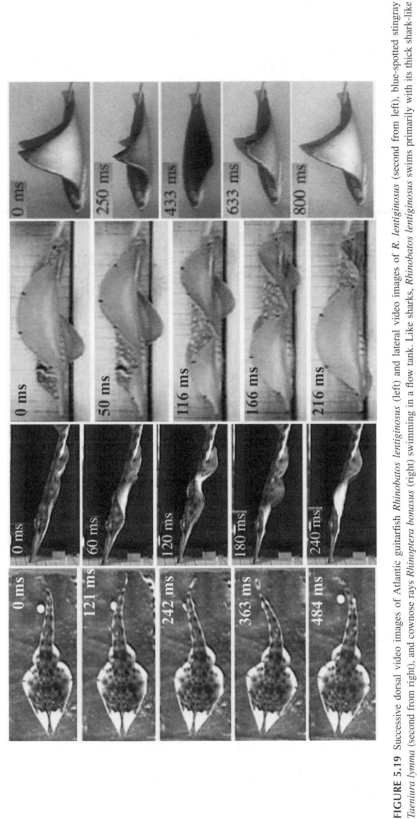

FIGURE 5.19 Successive dorsal video images of Atlantic guitarfish *Rhinobatos lentiginosus* (left) and lateral video images of *R. lentiginosus* (second from left), blue-spotted stingray *Taeniura lymma* (second from right), and cownose rays *Rhinoptera bonasus* (right) swimming in a flow tank. Like sharks, *Rhinobatos lentiginosus* swims primarily with its thick shark-like tail. (From Rosenberger, L.J. and M.W. Westneat. 1999. *J. Exp. Biol.* 202:3523–3539; Rosenberger, L.J. 2001. *J. Exp. Biol.* 204:379–394. With permission.)

broad up- and downstrokes in a manner that would provide vertical lift similar to that of aerial bird flight (Rosenberger, 2001). Although skates undulate the pectoral fins to swim when in the water column, they have enlarged muscular appendages on the pelvic fins that are modified for walking or "punting" off the substrate (Koester and Spirito, 1999) in a novel locomotor mechanism.

Some rays are able to vary the mechanics of the pectoral fins during locomotion (Rosenberger, 2001). There appears to be a trade-off between the amplitude of undulatory waves and fin beat frequency: those that have higher wave amplitudes have fewer waves, and vice versa (Rosenberger, 2001). This phenomenon appears to be correlated with lifestyle. Fully benthic rays and skates that are mostly sedentary, such as *Daysatis sabina* and *D. say*, have low-amplitude waves with high fin beat frequencies, permitting high maneuverability at low speeds, which is more suited for swimming slowly along the substrate to locate food items (Rosenberger, 2001). Fully pelagic rays are able to take advantage of the 3D environment of the water column and oscillate the pectoral fins using high-amplitude waves and low fin beat frequencies (Rosenberger, 2001). Rays and skates that have both benthic and pelagic lifestyles, such as *Raja* sp. and *D. violacea* and *D. americana*, are typically more active and have intermediate values of amplitude and frequency (Rosenberger, 2001).

However, oscillatory appendage propulsors that feed on benthic mollusks and crustaceans, such as cownose and butterfly rays, do not extend the fins below the ventral body axis during swimming, presumably so that they can use the lateral line canals to detect prey and also to avoid contact with the substrate (Rosenberger, 2001). In contrast, oscillatory appendage propulsors that feed in the water column, i.e., filter feeders such as manta and mobulid rays extend the pectoral fins equally above and below the body axis during swimming (Rosenberger, 2001). Some batoids are capable of modifying the swimming mechanism dependent on habitat; *Gymnura* undulates the pectoral fins when swimming along a substrate and oscillates them when swimming in the water column (Rosenberger, 2001). Undulatory mechanisms are efficient at slow speeds, have reduced body and fin drag, and are highly maneuverable (Blake, 1983a,b; Lighthill and Blake, 1990; Walker and Westneat, 2000; Rosenberger, 2001). In contrast, oscillatory mechanisms are efficient at fast cruising and generate greater lift, but are less well suited for maneuvering (Chopra, 1974; Blake, 1983b; Cheng and Zhuang, 1991; Rosenberger, 2001).

Different strategies are employed to increase swimming speed in various batoid species (Rosenberger, 2001). Most *Dasyatis* species increase fin beat frequency, wave speed, and stride length to increase swimming speed, while amplitude is held constant (Rosenberger and Westneat, 1999; Rosenberger, 2001). However, *Taeniura lymma* and *D. americana* increase fin beat frequency and wave speed but decrease wave number while holding amplitude constant to increase speed (Rosenberger and Westneat, 1999; Rosenberger, 2001). Similarly, *Raja elganteria* increases wave speed and decreases wave number to swim faster (Rosenberger and Westneat, 1999; Rosenberger, 2001). Oscillatory propulsors, *Rhinoptera* and *Gymnura*, increase wave speed in addition to fin-tip velocity to increase swimming speed (Rosenberger and Westneat, 1999; Rosenberger, 2001). Interestingly, *Gymnura* pauses between each fin beat at high flow speeds (Rosenberger and Westneat, 1999; Rosenberger, 2001), similar to the burst and glide flight mechanisms of aerial birds.

As expected, the dorsal and ventral fin muscles are alternately active during undulation of the pectoral fin from anterior to posterior (Figure 5.20) (Rosenberger and Westneat, 1999). The intensity of muscle contraction is increased to swim faster in *T. lymma* (Rosenberger and Westneat, 1999). The ventral muscles are also active longer than the respective dorsal muscles indicating that the downstroke is the major power-producing stroke (Rosenberger and Westneat, 1999). Chondrichthyans are negatively buoyant; thus lift must be generated to counter the weight of the fish as well as for locomotion. Interburst duration is decreased in *T. lymma* at higher swimming speeds with the fin muscles firing closer together (Rosenberger and Westneat, 1999).

5.4 Locomotion in Holocephalans

Chimeras have large flexible pectoral fins that have been described as both undulatory and oscillatory. The leading edge of the pectoral fin is flapped, which then passes an undulatory wave down the pectoral fin to the trailing edge (Combes and Daniel, 2001) (Figure 5.21). As expected, adult chimeras had a

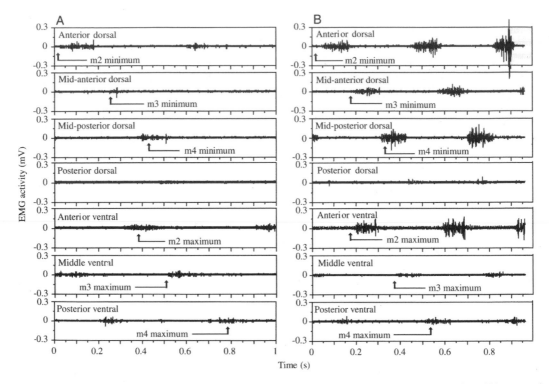

FIGURE 5.20 Electromyographic (EMG) data illustrating the muscle activity for the pectoral fin undulation of blue-spotted stingrays *Taeniura lymma* at a low speed of 1.2 disk length/s (A) and at a higher speed of 3.0 disk length/s (B). The electrode recordings are taken from the following muscles: anterior dorsal, mid-anterior dorsal, mid-posterior dorsal, posterior dorsal, anterior ventral, middle ventral, posterior ventral. The arrows below the EMG activity indicate the point during the fin-beat cycle at which the anterior, middle, and posterior fin markers are at their maximum (peak upstroke) and minimum (peak downstroke) excursion. (From Rosenberger, L.J. and M.W. Westneat. 1999. *J. Exp. Biol.* 202:3523–3539; Rosenberger, L.J. 2001. *J. Exp. Biol.* 204:379-394. With permission.)

larger-amplitude wave that was generated at a lower frequency than juvenile chimeras (Combes and Daniel, 2001). Interestingly, there is no net chordwise bend in the pectoral fin which averages a 0° angle of attack to the flow over a stroke cycle (Combes and Daniel, 2001). Potential flow models based on kinematic and morphological variables measured on the chimeras for realistic flexible fins and theoretical stiff fins emphasize the importance of considering flexion in models of animal locomotion (Combes and Daniel, 2001). Significantly higher values for thrust were calculated when the fin was assumed to be stiff rather than flexible as in reality (Combes and Daniel, 2001).

5.5 Future Directions

The diversity of shark species for which we have even basic functional data on locomotor mechanics is extremely limited. Most papers to date have focused on leopard (*Triakis*) and bamboo (*Chiloscyllium*) sharks swimming under controlled laboratory conditions. A high priority for future studies of locomotion in sharks, skates, and rays is to expand the diversity of taxa studied, especially for analyses of shark mechanics. The data obtained by Rosenberger (2001) on batoid locomotion are exemplary for their broadly comparative character, but studies like this are rare, perhaps necessarily so when detailed functional data must be obtained for a variety of behaviors.

Experimental studies of kinematics and hydrodynamics would benefit from increased spatial and temporal resolution so that a more detailed picture could be obtained of patterns of fin deformation and the resulting hydrodynamic wake, especially during unsteady maneuvering behaviors. New high-resolution digital video systems with greater than 1024 by 1024 pixels per frame operating at 500 frames/s

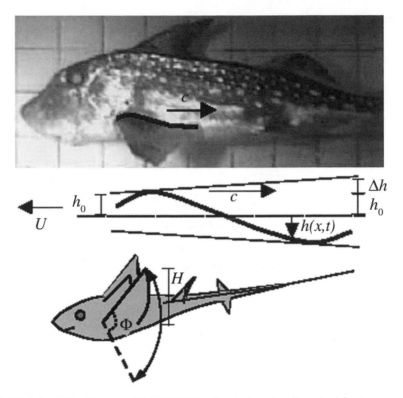

FIGURE 5.21 (Top) A ratfish with a wave (highlighted) traveling backward on its pectoral fin at wave speed c. (Middle) A 2D strip oscillating with amplitude h_0 and moving forward at velocity U while a wave passes rearward at velocity c. The amplitude changes from the leading to the trailing edge by a factor ε, the ratio of Δh to h_0. The instantaneous location of a point (x) on the strip is described by $h(x,t)$, where t is time. (Bottom) Diagram of a ratfish illustrating the angle (φ) subtended by a flapping fin and tip amplitude (H). (From Combes, S.A. and T.L. Daniel. 2001. *J. Exp. Biol.* 204:2073–2085. With permission.)

and faster will permit a new level of understanding of fin function and its impact on locomotor performance. Such increased resolution may also permit further observations of boundary layer flows in relation to surface denticle patterns to follow up on the observation by Anderson et al. (2001) that the boundary layer of *Mustelus* swimming at 0.5 *l*/s did not separate and remained attached along the length of the body.

There are effectively no data on the mechanical properties of elasmobranch connective tissue elements and of the role these play in transmitting forces to hydrodynamic fin control surfaces. This is a key area in which both *in vitro* studies of material properties and *in vivo* analyses of how elasmobranch connective tissues function can greatly enhance our understanding of elasmobranch locomotor mechanics.

Finally, to the extent that equipment and elasmobranch behavior permits, it would be extremely valuable to have quantitative 3D field data over the natural locomotor behavioral repertoire. For example, what are routine swimming speeds and what are typical vertical and lateral maneuvering velocities? What is the range of body angles observed during diverse locomotor behaviors? Such data, while difficult to obtain, would serve as a link between experimental laboratory studies of shark biomechanics and locomotor performance in nature.

Acknowledgments

Support for the research by the authors described here was provided by National Science Foundation Grants to C.D.W. (DBI 97-07846) and G.V.L. (IBN 98-07012) and from the University of Rhode Island to C.D.W. Preparation of the manuscript was supported by NSF IBN-0316675 to G.V.L.

References

Affleck, R. J. 1950. Some points in the function, development, and evolution of the tail in fishes. *Proc. Zool. Soc. Lond.* 120:349–368.

Aleev, Y. G. 1969. *Function and Gross Morphology in Fish*, trans. from the Russian by M. Raveh. Keter Press, Jerusalem.

Alexander, R. M. 1965. The lift produced by the heterocercal tails of Selachii. *J. Exp. Biol.* 43:131–138.

Anderson, E. J., W. McGillis, and M. A. Grosenbaugh. 2001. The boundary layer of swimming fish. *J. Exp. Biol.* 204:81–102.

Arnold, G. P. and P. W. Webb. 1991. The role of the pectoral fins in station-holding of Atlantic salmon parr (*Salmo salar* L.). *J. Exp. Biol.* 156:625–629.

Bendix-Almgreen, S. E. 1975. The paired fins and shoulder girdle in Cladoselache, their morphology and phyletic significance. *Colloq. Int. C. N. R. S.* (Paris) 218:111–123.

Blake, R. W. 1983a. *Fish Locomotion.* Cambridge University Press, Cambridge, U.K.

Blake, R. W. 1983b. Median and paired fin propulsion, in *Fish Biomechanics.* P. W. Webb and D. Weihs, Eds., Praeger, New York, 214–247.

Bone, Q. 1999. Muscular system: microscopical anatomy, physiology, and biochemistry of elasmobranch muscle fibers, in *Sharks, Skates, and Rays: The Biology of Elasmobranch Fishes.* W. C. Hamlett, Ed., Johns Hopkins University Press, Baltimore, 115–143.

Breder, C. M. 1926. The locomotion of fishes. *Zool. N.Y.* 4:159–256.

Carroll, R. L. 1988. *Vertebrate Paleontology and Evolution.* W. H. Freeman, New York.

Cheng, J. and L. Zhaung. 1991. Analysis of swimming three-dimensional waving plates. *J. Fluid Mech.* 232:341–355.

Chopra, M. G. 1974. Hydrodynamics of lunate-tail swimming propulsion. *J. Fluid Mech.* 64:375–391.

Combes, S. A., and T. L. Daniel. 2001. Shape, flapping and flexion: wing and fin design for forward flight. *J. Exp. Biol.* 204:2073–2085.

Compagno, L. J. V. 1973. Interrelationships of living elasmobranchs, in *Interrelationships of Fishes.* P. H. Greenwood, R. S. Miles, and C. Patterson, Eds., *Zool. J. Linn. Soc.,* Suppl. 1.

Compagno, L. J. V. 1984. *Sharks of the World.* United Nations Development Program, Rome.

Compagno, L. J. V. 1988. *Sharks of the Order Carcharhiniformes.* Princeton University Press, Princeton, NJ.

Compagno, L. J. V. 1999. Endoskeleton, in *Sharks, Skates, and Rays: The Biology of Elasmobranch Fishes.* W. C. Hamlett, Ed., Johns Hopkins University Press, Baltimore, 69–92.

Daniel, J. F. 1922. *The Elasmobranch Fishes.* University of California Press, Berkeley.

Dickinson, M. H. 1996. Unsteady mechanisms of force generation in aquatic and aerial locomotion. *Am. Zool.* 36:537–554.

Donley, J. and R. Shadwick. 2003. Steady swimming muscle dynamics in the leopard shark *Triakis semifasciata. J. Exp. Biol.* 206:1117–1126.

Drucker, E. G. and G. V. Lauder. 1999. Locomotor forces on a swimming fish: three-dimensional vortex wake dynamics quantified using digital particle image velocimetry. *J. Exp. Biol.* 202:2393–2412.

Ferry, L. A. and G. V. Lauder. 1996. Heterocercal tail function in leopard sharks: a three-dimensional kinematic analysis of two models. *J. Exp. Biol.* 199:2253–2268.

Fish, F. E. and L. D. Shannahan. 2000. The role of the pectoral fins in body trim of sharks. *J. Fish Biol.* 56:1062–1073.

Garman, S. 1913. The Plagiostoma (sharks, skates, and rays). *Mem. Mus. Comp. Zool. Harvard Coll.* 36.

Gray, J. 1968. *Animal Locomotion.* Weidenfeld and Nicolson, London.

Grove, A. J. and G. E. Newell. 1936. A mechanical investigation into the effectual action of the caudal fin of some aquatic chordates. *Ann. Mag. Nat. Hist.* 17:280–290.

Harris, J. E. 1936. The role of the fins in the equilibrium of the swimming fish. I. Wind tunnel tests on a model of *Mustelus canis* (Mitchell). *J. Exp. Biol.* 13:476–493.

Harris, J. E. 1953. Fin patterns and mode of life in fishes, in *Essays in Marine Biology.* S. M. Marshall and A. P. Orr, Eds., Oliver and Boyd, Edinburgh, 17–28.

Kemp, N. E. 1999. Integumentary system and teeth, in *Sharks, Skates, and Rays: The Biology of Elasmobranch Fishes.* W. C. Hamlett, Ed., Johns Hopkins University Press, Baltimore, 43–68.

Koester, D. M. and C. P. Spirito. 1999. Pelvic fin locomotion in the skate, *Leucoraja erinacea*. *Am. Zool.* 39:55A.

Krothapalli, A. and L. Lourenco. 1997. Visualization of velocity and vorticity fields, in *Atlas of Visualization*, Vol. 3. Y. Nakayama and Y. Tanida, Eds., CRC Press, Boca Raton, FL, 69–82.

Kundu, P. 1990. *Fluid Mechanics*. Academic Press, San Diego.

Lauder, G. V. 2000. Function of the caudal fin during locomotion in fishes: kinematics, flow visualization, and evolutionary patterns. *Am. Zool.* 40:101–122.

Lauder, G. V. and E. Drucker. 2002. Forces, fishes, and fluids: hydrodynamic mechanisms of aquatic locomotion. *News Physiol.. Sci.* 17:235–240.

Lauder, G. V., J. Nauen, and E. G. Drucker. 2002. Experimental hydrodynamics and evolution: function of median fins in ray-finned fishes. *Integ. Comp. Biol.* 42:1009–1017.

Lauder, G. V., E. G., Drucker, J. Nauen, and C. D. Wilga. 2003. Experimental hydrodynamics and evolution: caudal fin locomotion in fishes, in *Vertebrate Biomechanics and Evolution*. V. Bels, J.-P. Gasc, and A. Casinos, Eds., Bios Scientific Publishers, Oxford, 117–135.

Liao, J. and G. V. Lauder. 2000. Function of the heterocercal tail in white sturgeon: flow visualization during steady swimming and vertical maneuvering. *J. Exp. Biol.* 203:3585–3594.

Liem, K. F. and A. P. Summers. 1999. Muscular system, in *Sharks, Skates, and Rays: The Biology of Elasmobranch Fishes*. W. C. Hamlett, Ed., Johns Hopkins University Press, Baltimore, 93–114.

Lighthill, J. and R. Blake. 1990. Biofluid dynamics of balistiform and gymnotiform locomotion. Part 1. Biological background and analysis by elongated-body theory. *J. Fluid Mech.* 212:183–207.

Lindsey, C. C. 1978. Form, function, and locomotory habits in fish, in *Fish Physiology*, Vol. 7. *Locomotion*. W. S. Hoar and D. J. Randall, Eds., Academic Press, New York, 1–100.

Magnan, A. 1929. Les charactéristiques géométriqes et physiques des poissons. *Ann. Sci. Nat. Zool.* 10:1–132.

Pridmore, P. A. 1995. Submerged walking in the epaulette shark *Hemiscyllium ocellatum* (Hemiscyllidae) and its implications for locomotion in rhipidistian fishes and early tetrapods. *Zoology* 98:278–297.

Reif, W.-E. and D. B. Weishampel. 1986. Anatomy and mechanics of the lunate tail in lamnid sharks. *Zool. Jahrb. Anat.* 114:221–234.

Rosenberger, L. 2001. Pectoral fin locomotion in batoid fishes: undulation *versus* oscillation. *J. Exp. Biol.* 204:379–394.

Rosenberger, L. J. and M. W. Westneat. 1999. Functional morphology of undulatory pectoral fin locomotion in the stingray *Taeniura lymma* (Chondrichthyes: Dasyatidae). *J. Exp. Biol.* 202:3523–3539.

Shirai, S. 1996. Phylogenetic interrelationships of neoselachians (Chondrichthyes: Euselachii), in *Interrelationships of Fishes*. M. Stiassny, L. Parenti, and G. D. Johnson, Eds., Academic Press, San Diego, 9–34.

Simons, J. R. 1970. The direction of the thrust produced by the heterocercal tails of two dissimilar elasmobranchs: the Port Jackson shark, *Heterodontus portusjacksoni* (Meyer), and the piked dogfish, *Squalus megalops* (Macleay). *J. Exp. Biol.* 52:95–107.

Simons, M. 1994. *Model Aircraft Aerodynamics*. Argus Books, Herts, England.

Smith, H. C. 1992. *The Illustrated Guide to Aerodynamics*. TAB Books, New York.

Thomson, K. S. 1971. The adaptation and evolution of early fishes. *Q. Rev. Biol.* 46:139–166.

Thomson, K. S. 1976. On the heterocercal tail in sharks. *Paleobiology* 2:19–38.

Thomson, K. S. and D. E. Simanek. 1977. Body form and locomotion in sharks. *Am. Zool.* 17:343–354.

Walker, J. A. and M. W. Westneat. 2000. Mechanical performance of aquatic rowing and flying. *Proc. R. Soc. Lond. B* 267:1875–1881.

Webb, P. W. 1984. Form and function in fish swimming. *Sci. Am.* 251:72–82.

Webb, P. W. 1988. Simple physical principles and vertebrate aquatic locomotion. *Am. Zool.* 28:709–725.

Webb, P. W. and R. W. Blake. 1985. Swimming, in *Functional Vertebrate Morphology*. M. Hildebrand, D. M. Bramble, K. F. Liem, and D. B. Wake, Eds., Harvard University Press, Cambridge, MA, 110–128.

Webb, P. W. and C. L. Gerstner. 1996. Station-holding by the mottled sculpin, *Cottus bairdi* (Teleostei: Cottidae), and other fishes. *Copeia* 1996:488–493.

Webb, P. W. and R. S. Keyes. 1982. Swimming kinematics of sharks. *Fish. Bull.* 80:803–812.

Wilga, C. D. and G. V. Lauder. 1999. Locomotion in sturgeon: function of the pectoral fins. *J. Exp. Biol.* 202:2413–2432.

Wilga, C. D. and G. V. Lauder. 2000. Three-dimensional kinematics and wake structure of the pectoral fins during locomotion in leopard sharks *Triakis semifasciata*. *J. Exp. Biol.* 203:2261–2278.

Wilga, C. D. and G. V. Lauder. 2001. Functional morphology of the pectoral fins in bamboo sharks, *Chiloscyllium plagiosum*: benthic versus pelagic station holding. *J. Morphol.* 249:195–209.

Wilga, C. D. and G. V. Lauder. 2002. Function of the heterocercal tail in sharks: quantitative wake dynamics during steady horizontal swimming and vertical maneuvering. *J. Exp. Biol.* 205:2365–2374.

Willert, C. E. and M. Gharib. 1991. Digital particle image velocimetry. *Exp. Fluids* 10:181–193.

Zangerl, R. 1973. *Interrelationships of Early Chondrichthyans.* Academic Press, London.

6

Prey Capture Behavior and Feeding Mechanics of Elasmobranchs

Philip J. Motta

CONTENTS

6.1 Introduction

Perhaps the most remarkable thing about the elasmobranch feeding mechanism is its functional diversity despite its morphological simplicity. Compared to the teleost skull, which has approximately 63 bones (excluding the branchiostegal, circumorbital, and branchial bones), the feeding apparatus of a shark is composed of just 10 cartilaginous elements: the chondrocranium, paired palatoquadrate and Meckel's cartilages, hyomandibulae, ceratohyals, and a basihyal. Furthermore, the elasmobranchs lack pharyngeal jaws and the ability to further process food by this secondary set of decoupled jaws as do bony fishes. Despite this, sharks, skates, and rays display a diversity of feeding mechanisms and behaviors that, although they do not match those of the bony fishes, is truly remarkable, especially considering there are only approximately 1200 species of elasmobranchs compared to about 24,000 species of teleost fishes (Nelson, 1994; Compagno, 2001). The elasmobranchs capture prey by methods as diverse as ram, biting, suction, and filter feeding, and feed on prey ranging from plankton to marine mammals (Moss, 1972; Frazzetta, 1994; Motta and Wilga, 2001). Understanding the elasmobranch feeding mechanism will shed light on how this functional versatility is achieved and whether or not it parallels that of the bony fishes.

0-8493-1514-X/04/$0.00+$1.50
© 2004 by CRC Press LLC

Understanding the feeding mechanism of elasmobranchs is also important to biologists from an evolutionary perspective. The chondrichthyan fishes represent a basal group of jawed fishes that share a common ancestor with bony fishes (Schaeffer and Williams, 1977; Carroll, 1988; Long, 1995), and therefore they provide insight into the evolution of lower vertebrate feeding mechanisms. Studies on chondrichthyan fishes have provided an understanding of the evolution of the jaw depression mechanism in aquatic vertebrates (Wilga et al., 2000) and the evolution and function of jaw suspension systems in vertebrates (Grogan et al., 1999; Grogan and Lund, 2000; Wilga, 2002). Studies on elasmobranch teeth also provide insight into the evolution of dermal teeth and armor, and patterns of tooth replacement in vertebrates (Reif, 1978, 1980; Reif et al., 1978).

Despite a tremendous increase in the knowledge of bony fish feeding mechanisms in the last three decades (Liem, 1978; Lauder, 1985), there has been a relative paucity of studies on elasmobranchs, with even fewer on batoids (Marion, 1905; Bray and Hixon, 1978; Summers, 2000) than on sharks (Moss, 1972; Nobiling, 1977; Shirai and Nakaya, 1992). Numerous embryological and anatomical studies on the head of sharks in the previous century or early part of this century (reviewed in Motta and Wilga, 1995, 1999) were influential in our understanding of the evolution and development of the skull and branchial arches; however, following some earlier anatomical studies (Springer, 1961; Moss, 1972, 1977) there have been relatively few studies that incorporate cineradiography, high-speed photography, electromyography, and biomechanical modeling of the feeding apparatus (Wu, 1994; Motta et al., 1997; Ferry-Graham, 1998a,b; Wilga and Motta, 1998a,b, 2000).

The goal of this chapter is provide a review of the feeding behavior and mechanics of extant elasmobranchs with an emphasis on the structure and function of the feeding apparatus. To place prey capture and mechanics in a more meaningful framework it is necessary to outline how elasmobranchs approach their prey. Consequently, prey approach behavior is briefly discussed. Feeding behavior is considered to be those precapture behaviors (e.g., stalking, ambushing), whereas prey capture refers to the process beginning with opening of the mouth as the fish approaches the prey and usually ends with the prey grasped between the jaws. Because so little is known of the postcapture manipulation or processing it is only briefly covered. During manipulation the prey is reduced in size by cutting or crushing, often combined with head shaking, and then it is transported from the buccal cavity through the pharyngeal cavity into the esophagus. Similarly, as so little is known of batoid feeding mechanisms, sharks are emphasized more than skates or rays. In some instances *food* is used to refer to pieces of whole items offered to an animal under experimental conditions, whereas *prey* refers to dietary items captured during natural feeding. The review does not cover feeding ecology and diet (see Cortés, 1999; Wetherbee and Cortés, Chapter 8 of this volume), although diet is occasionally referred to when discussing feeding behaviors and mechanisms.

6.2 Ethology of Predation

6.2.1 Predatory Behaviors

Sharks, skates, and rays must first approach their prey before they can capture it. When the prey is within grasp of the predator the capture event is usually very rapid as compared to the approach, and at this point either the prey may be held within the grasp of the teeth or it may be transported directly through the mouth to the entrance of the esophagus. If the prey is grasped by the teeth one or a series of manipulation/processing bites can reduce the prey in size prior to the final transport event. In this manner we speak of capture bites, manipulation/processing bites, and hydraulic transport as the last invariably involves suction of the water with the entrained food (Motta and Wilga, 2001). The mechanics of swallowing, that is, getting the food into and through the esophagus, is still unresolved.

Because of the inherent difficulty of studying elasmobranchs in their natural environment, predatory behavior is generally poorly understood, especially as compared to that of bony fishes. Large or pelagic sharks are perhaps the least understood, but with the relatively recent advent of telemetry studies (Holland et al., 1999; Klimley et al., 2001) and attachment of small cameras or "CritterCams™" to free-swimming sharks (Heithaus et al., 2001, 2002a, Chapter 17 this volume) foraging patterns are being revealed. A great deal of what we know of predatory behavior is from anecdotal or one-of-a-kind observations (Pratt et al., 1982; Strong, 1990), telemetry studies (Klimley et al., 2001), behavioral studies of shallow-water benthic elasmobranchs (Strong, 1989; Fouts and Nelson, 1999), or is inferred from morphology (Compagno, 1990; Myrberg, 1991). Surprisingly, the more accessible batoids are vastly understudied as compared to sharks (Belbenoit and Bauer, 1972; Lowe et al., 1994).

How sharks and rays approach and hunt their prey is perhaps the least understood aspect of their feeding biology. Most elasmobranchs are probably very opportunistic in what they prey on (see Heithaus, Section 17.3.3, this volume) and how they acquire their prey (see Section 17.3 for a discussion of prey capture tactics).

When hunting by speculation, the fish searches an area it expects to have prey or it follows another organism expecting that animal to flush prey out by its presence (Curio, 1976). *Dasyatis* rays will position themselves at regions of higher tidal water movement such as near beach promontories waiting for prey organisms to be swept by. Large aggregations of rays may be found at these locations at periods of swift tidal movement (pers. obs.). Tiger sharks, *Galeocerdo cuvier,* in Shark Bay, Australia are most frequent during the season that dugongs and sea snakes are present, both of which are important prey items to the sharks, which show site fidelity to this area (Heithaus, 2001), and tiger sharks aggregate at the northwestern Hawaiian Islands during June and July coinciding with the summer fledging period of blackfooted and Laysan albatross birds, upon which they prey (Lowe et al., 2003). Each March and April whale sharks, *Rhincodon typus,* aggregate on the continental shelf of the central western Australian coast, particularly at Ningaloo Reef, in response to coral spawning events that occur each year (Gunn et al., 1999; Taylor and Pearce, 1999). White sharks, *Carcharodon carcharias,* spend a lot of time patrolling near seal colonies off the South Farallon Islands and Año Nuevo Island, California. Most of the shark's movement is back and forth parallel and near to the shoreline as it intercepts seals and sea lions departing and returning to the shore-based rookeries. In some cases the sharks pass within 2 m of the shore. Prey capture is, however, infrequent, compared to the time spent patrolling (Klimley et al., 2001).

Ambushing involves the predator trying to conceal or advertise (aggressive mimicry) its presence while lying in wait for the prey (Curio, 1976). By partially burying themselves in the soft substrate, Pacific angel sharks, *Squatina californica,* ambush demersal fishes. These sharks appear to actively select ambush sites within localized areas adjacent to reefs (Fouts, 1995; Fouts and Nelson, 1999). Pacific electric rays, *Torpedo californica,* either ambush their prey from the bottom or use a search-and-attack behavior from the water column. During the day the rays ambush their prey of mostly fishes by burying themselves in sand and jumping over the prey. Swimming over the prey, the rays cup their pectoral fins around the prey while electrically discharging. They then pivot over the stunned prey so as to swallow it head first. At night the rays are seen swimming or hovering in the water column 1 to 2 m above the substratum. The rays then lunge forward over the prey, cup their pectorals over the prey while discharging, then either pin the prey to the bottom or, using frontal somersaults and peristaltic-like movements of the disk, move the prey closer to the mouth for swallowing (Bray and Hixon, 1978; Lowe, 1991; Lowe et al., 1994). Similar stereotyped prey capture behavior has also been described for the electric rays, *T. marmorata, T. ocellata,* and *T. nobiliana* (Wilson, 1953; Belbenoit and Bauer, 1972; Michaelson et al., 1979; reviewed in Belbenoit, 1986).

Ambushing behavior of rays and sharks has been observed at the inshore spawning grounds of chokka squid (*Loligo vulgaris reynaudii*) off South Africa. Diamond rays, *Gymnura natalensis,* camouflage themselves in the substrate and then lunge out toward female squid as they try to spawn on the bottom. Large numbers of sharks and rays aggregate at these spawning grounds, and in addition to pajama catsharks, *Poroderma africanum,* and leopard catsharks, *P. pantherinum,* ambushing the spawning squid from the rocky reef substrate, the rays and sharks also chase down the squid to capture them, or simply bite off the attached egg masses from the substrate (Smale et al., 1995; 2001).

In contrast to ambushing, the stalking predator approaches the prey while concealed, then makes a sudden assault (Curio, 1976). White sharks, *C. carcharias,* will stalk prey downstream in oceanic or tidal currents (Pyle et al., 1996). Sevengill sharks, *Notorynchus cepedianus,* capture elusive prey by a stealthy underwater approach using very little body movement and only slight undulatory motions of the caudal fin. They move within striking distance, then make a quick dash at the prey, which can include fur seals (Ebert, 1991). Using tethered "CritterCams," Heithaus et al. (2002a) observed tiger sharks, *Galeocerdo cuvier,* stalking their benthic prey from above, in some cases getting as close as 2 m from large teleost fishes before the shark was detected.

Other elasmobranchs may lure prey to them. Luminescent tissue on the upper jaw of the megamouth shark, *Megachasma pelagios*, might attract euphausid shrimp and other prey into its mouth (Compagno, 1990). The white tips on the pectoral fins of oceanic whitetip sharks, *Carcharhinus longimanus*, might act as visual lures to aid in the capture of its rapid moving prey (Myrberg, 1991), and bioluminescence in the cookie-cutter shark, *Isistius brasiliensis*, might serve to lure pelagic predators from which it gouges chunks of flesh (Jones, 1971; Widder, 1998).

Most elasmobranchs will scavenge food when given the opportunity. Sevengill sharks, *N. cepedianus*, will feed on marine mammals including whale and dolphin carcasses, bait left on fishing hooks, and even human remains (Ebert, 1991). Tiger sharks, *G. cuvier*, are notorious opportunistic feeders, and in addition to their regular diet they will scavenge food ranging from dead dugongs to human refuse (Randall, 1992; Lowe et al., 1996; Smale and Cliff, 1998; Heithaus, 2001). Blue sharks, *Prionace glauca*, will similarly scavenge human refuse and dead or injured birds, although they have been observed to stalk resting birds (Stevens, 1973, cited in Henderson et al., 2001; Henderson et al., 2001). Large gray reef sharks, *C. amblyrhynchos*, at Enewetak Island, Marshall Islands, follow carangid jacks as both scavengers and predators (Au, 1991).

Although many species of sharks forage solitarily, in some cases aggregations of sharks will come together to feed. Blacktip reef sharks, *C. melanopterus*, and lemon sharks, *Negaprion brevirostris*, were observed to apparently herd schools of fish against the shoreline and then feed on them (Eibl-Eibesfeldt and Hass, 1959; Morrissey, 1991; McPherson et al., unpubl.), and oceanic whitetip sharks, *C. longimanus*, were observed to herd squid at night (Strasburg, 1958). Thresher sharks (*Alopias*) are reported to apparently work in groups to capture fish, using their long caudal fins to herd and stun fish (Coles, 1915; Budker, 1971; Compagno, 1984; Castro, 1996). Sevengill sharks, *Notorynchus cepedianus*, will circle a seal and prevent its escape. The circle is tightened and eventually one shark initiates the attack that stimulates the others to begin feeding (Ebert, 1991). Although some authors have considered these behaviors cooperative, they could simply reflect aggregations of animals at a prey, and not cooperative foraging (Motta and Wilga, 2001; see Heithaus, Section 17.3.3.8, this volume, for definition of cooperative foraging). So-called feeding frenzies of sharks appear to be nothing more than highly motivated feeding events involving generally many individuals. The sharks have been described as attacking prey or food items indiscriminately, moving at an accelerated speed, and disregarding any injuries they may receive in the attack. Injured or hooked sharks are often attacked and consumed by the other sharks. These feeding bouts, which can involve as few as six sharks to hundreds of sharks, can end abruptly as they begin (Gilbert, 1962; Vorenberg, 1962; Hobson, 1963; Springer, 1967; Nelson, 1969).

6.2.2   Feeding Location and Prey Capture

Sharks approach their prey on the surface, or in midwater, or on the bottom. One of the most popular misconceptions is that sharks must roll on their side to take prey in front of them because of their subterminal mouth (Budker, 1971). In fact, the mouth of modern sharks does not preclude them from feeding on prey in front of or above them, and sharks will approach surface or underwater food either with a direct head-on approach or roll on their side to bite at the food (Budker, 1971; Motta, pers. obs.). White sharks, *Carcharodon carcharias*, will approach in their normal orientation, roll on their side, or roll completely over so their ventral side is up when they feed on underwater bait or a floating whale carcass (Pratt et al., 1982; Tricas and McCosker, 1984). During surface feeding, *C. carcharias* often bite such prey as elephant seals, and then retreat until the prey lapses into shock or bleeds to death. The shark then returns to feed on the prey (Tricas and McCosker, 1984; McCosker, 1985). Tricas and McCosker referred to this as the "bite and spit" strategy. However, Klimley (1994) and Klimley et al. (1996) proposed that white sharks hold the pinniped prey tightly in their mouth and drag it below the surface, often removing a bite from the prey in the process. The prey may be released underwater, after which it floats or swims to the surface and dies by exsanguination. Meanwhile, the shark follows the prey to the surface to begin feeding after it dies. Blue sharks, *Prionace glauca*, approach schools of squid on the surface with an underwater approach or a surface charge. Small anchovies are captured from a normal swimming posture, but when capturing larger whole mackerel from behind, blue sharks

FIGURE 6.1 (A) Whale shark *R. typus* using pulsatile suction feeding on the surface, (B) view of head showing water flow in and around mouth. (Reproduced with permission from Colin McNulty.)

may roll on their side (Tricas, 1979). Large schools of oceanic whitetip sharks, *Carcharhinus longimanus*, have been observed swimming erratically in a sinuous course on the surface with their mouths wide open. These sharks made no attempt to snap up the small tuna through which they were swimming; rather, they appeared to simply wait for the fish to swim or leap into their mouths (Bullis, 1961). Surface-feeding blacknose *C. acronotus*, oceanic whitetip *C. longimanus*, white *Carcharodon carcharias*, and Caribbean reef *Carcharhinus perezi* sharks may raise the head just prior to prey capture (Bullis, 1961; Tricas and McCosker, 1984; Frazzetta and Prange, 1987; Motta and Wilga, 2001). This might place the open mouth in line with food as the shark approaches (Frazzetta and Prange, 1987). Whale sharks, *Rhincodon typus*, will make regular dives through the water column foraging for food. They will also swim slowly ($\sim$0.5 m s^{-1}) at or near the surface with their body at an angle and with the top of their head clear of the surface while feeding (Gunn et al., 1999) (Figure 6.1).

Rays and skates will also feed off the bottom. The ventral mouth of Pacific electric rays, *Torpedo californica*, does not preclude them from foraging in the water column in addition to sitting on the bottom. After stunning the prey, which can result in breaking of the vertebral column, they manipulate the prey toward the mouth with the pectoral fins or force the stunned prey to the substrate (Bray and Hixon, 1978; Lowe et al., 1994). The thorny skate, *Raja radiata*, is primarily a benthivorous feeder as juveniles and adolescents, but benthopelagic food items including fishes become important to larger individuals (Skjaeraasen and Bergstad, 2000). Dietary items indicate that *Dasyatis sayi* and *D. centroura* in Delaware Bay frequently feed off the bottom on free-swimming organisms (Hess, 1961).

Mobulid rays including *Manta birostris* and *Mobula tarapacana* filter-feed both at the surface and in midwater, extending their cephalic wings to funnel prey and water through the mouth. Upon encountering a patch of prey they will often swim in a circular formation or somersault while filter-feeding to stay within the patch (Notarbartolo-Di-Sciara and Hillyer, 1989; Motta, pers. obs.).

Some sharks will also take prey buried within the substrate or capture prey on the bottom. Leopard sharks, *Triakis semifasciata*, can apparently suck worms out of their burrows in addition to biting pieces off their benthic prey (Talent, 1976; Compagno, 1984). The epaulette shark, *Hemiscyllium ocellatum*, and the whitespotted bamboo shark, *Chiloscyllium plagiosum,* occasionally thrust their head into the sediment up to the level of the first gill slit, apparently using suction to capture their benthic prey of worms and crabs. They then winnow the prey from the sand in the buccopharyngeal cavity and eject the sand through the first gill slit (Heupel and Bennett, 1998; Wilga, pers. obs.). Skates and rays primarily feed in or on the bottom by biting pieces of sessile invertebrates or excavating buried prey, although they will feed in the water column (Hess, 1961; Babel, 1967; Holden and Tucker, 1974; Orth, 1975; VanBlaricom, 1976; Howard et al., 1977; Edwards, 1980; Ajayi, 1982; Sherman et al., 1983; Abd El-Aziz, 1986; Rudloe, 1989; Ebert et al., 1991; Thrush et al., 1991; Stokes and Holland, 1992; Gray et al., 1997; Hines et al., 1997; Goitein et al., 1998; Lucifora et al., 2000; Skjaeraasen and Bergstad, 2000; Muto et al., 2001; Valadez-Gonzalez et al., 2001). Rays dig up prey by pectoral "wing-flapping" and/or they hydraulically mine the prey by jetting water through the mouth (VanBlaricom, 1976; Howard et al., 1977; Gregory et al., 1979; Sasko, 2000; Muto et al., 2001; Sasko et al., unpubl.). The cownose ray, *Rhinoptera bonasus,* uses a combination of wing flapping and water jetting to expose prey in the wild (Schwartz, 1967, 1989; Sasko, 2000; Sasko et al., unpubl.); however, in the laboratory the rays rest on the substrate on the tips of their pectoral fins and use repeated jaw opening and closing movements at 2.4 to 2.9 cycles per second to generate water flow in and out of the buccal cavity. The ventrally directed jet of water resuspends the sand and bivalve food resulting in the effective separation of food and sand so the rays can capture the food. The large subrostral lobes are depressed, forming a chamber around the food item that it encloses laterally and partially anteriorly, and the two lobes have been observed to move independently and push food toward the mouth (Sasko, 2000; Sasko et al., unpubl. data). Large-scale destruction of eelgrass, *Zostera marina,* beds in the Chesapeake Bay has been attributed to the excavation behavior of *R. bonasus* (Orth, 1975). Excavation of benthic prey by rhythmic flapping of the rostrum and pectoral fins is common in other rays (Babel, 1967; VanBlaricom, 1976; Howard et al., 1977; Thrush et al., 1991; Hines et al., 1997). Southern stingrays, *D. americana,* excavate lancelets, *Branchiostoma floridae,* from the sandy substrate, and the presence in the gut of only medium- and large-sized prey led Stokes and Holland (1992) to speculate that the rays are winnowing out the sand and smaller lancelets while retaining the larger ones. Winnowing prey from ingested sediment is perhaps common in rays. The lesser electric ray, *Narcine brasiliensis*, which specializes in wormlike prey including polychaete worms and anguilliform fishes, uses suction to capture the prey along with some sediment, and ejects the latter out of the mouth, spiracle, or gill slits (Funicelli, 1975; Rudloe, 1989; Dean and Motta, unpubl.). Similarly, during food processing, *R. bonasus* can separate prey from sand, flushing the sand out of the mouth and gill slits. This ray can also strip unwanted parts of the food item, such as mussel shell, skin and vertebral column of fish, shell of shrimp, from the edible parts and eject the unwanted pieces. Larger pieces are ejected from the mouth and smaller particles such as sand exit through the gill slits (Sasko, 2000; Sasko et al., unpubl.).

Bottom-feeding horn sharks, *Heterodontus francisci,* use suction and biting to remove benthic invertebrates such as anemone tentacles, polychaetes, and urchins. They remove their prey with a "pecking-like" motion while they are often raised on their pectoral fins (Strong, 1989; Edmonds et al., 2001). Gray reef sharks, *Carcharhinus amblyrhynchos,* in Hawaii primarily feed near the bottom on reef associated teleosts and supplement their diet with invertebrates (Wetherbee et al., 1997). Rays are often taken by sharks, particularly hammerhead sharks (Gudger, 1907; Budker, 1971). Great hammerhead sharks, *Sphyrna mokarran,* have been observed to use their head to deliver powerful blows and to restrain rays on the substrate prior to biting pieces off the ray (Strong, 1990), as well as a "pin and pivot" behavior during which the shark forcibly presses the ray against the substrate with the ventral surface of the cephalofoil and then, with a twisting motion of the body, pivots its head while remaining atop the ray as it engulfs part or all of the ray (Chapman and Gruber, 2002). Small bonnethead sharks, *S. tiburo,* capture their food by depressing the mandible considerably as they swim over the food, catching the food either within the mouth or with the anterior mandibular teeth (Wilga, 1997; Wilga and Motta, 2000).

6.3 Feeding Mechanism

6.3.1 Mechanics of Prey Capture

When the shark, skate, or ray is within striking distance of the prey it begins the capture sequence. Prey capture is generally very rapid compared to the approach and typically lasts from about 100 to 400 ms. Capture begins when the mouth starts to open and lasts until the prey is grasped between the teeth or the jaws are closed on the prey (Motta et al., 2002). In some cases the mouth is briefly closed just prior to opening, and under those circumstances, this closing may be said to mark the initiation of capture. Capture may then be divided into three or four phases for heuristic purposes, although they are all continuous and rapid. If the slightly agape mouth is closed prior to mouth opening, this is termed the preparatory phase and is more common in suction-feeding bony fishes than elasmobranchs (Lauder, 1985). An expansive phase follows during which there might be cranial (head) elevation accompanied by depression of the lower jaw. The branchial apparatus may also be expanded and the paired labial cartilages that lie at the edges of the mouth extended during this phase. The compressive phase begins at peak gape, and as the lower jaw is elevated, the upper jaw (palatoquadrate cartilage) might be protruded toward the lower jaw. Cranial depression also occurs during this phase in many sharks, although surface feeding *Carcharodon carcharias* can keep the cranium elevated until the recovery phase. At the end of the compressive phase the prey is either grasped between the teeth or the food is already well within the buccal cavity. The recovery phase is marked by retraction of the upper jaw and the recovery of the other elements (hyomandibula, ceratohyal, basihyal, and branchial arches) back to their original resting positions (Figure 6.2) (Moss, 1972, 1977; Tricas and McCosker, 1984; Frazzetta and Prange, 1987; Frazzetta, 1994; Motta et al., 1997; Motta and Wilga, 2001).

Sharks and batoids capture their prey in a variety of ways. Ram feeding is perhaps the most common prey capture method in sharks, especially in carcharhinid and lamnid sharks. During ram capture, the shark swims over the relatively stationary prey, engulfs it whole or seizes it in its jaws. The food is then moved from the mouth through the pharyngeal cavity into the esophagus by hydraulic suction. Bonnet-head sharks, *Sphyrna tiburo*, ram-feed benthic food by depressing the mandible and scooping the food up as they swim over it (Wilga and Motta, 2000). White sharks, *Carcharodon carcharias*, primarily ram capture their food sometimes approaching the food at great speeds resulting in them leaving the water when feeding on surface-dwelling prey (Tricas and McCosker, 1984; Tricas, 1985; Klimley, 1994; Klimley et al., 1996).

Inertial suction feeding, or simply suction feeding, involves a decrease in the pressure of the buccopharyngeal chamber such that the prey or food is pulled into the mouth. There is a functional continuum from pure ram to pure inertial suction and fishes can, and often do, use a combination of both (Norton and Brainerd, 1993; Wilga and Motta, 1998a). Caribbean reef sharks, *Carcharhinus perezi*, taking pieces of food will primarily over-swim the food item by ram but also employ some suction as witnessed by the food being sucked into the mouth rapidly when it is very close to the approaching shark. Sixgill sharks, *Hexanchus griseus*, will also position themselves close to bait, sitting on the bottom and suck it into their mouth (Motta, pers. obs.).

Sharks specialized for suction prey capture such as the nurse shark, *Ginglymostoma cirratum*, exhibit a suite of kinematic and morphological characters including a relatively small mouth (generally less than one third head length) as compared to ram-feeding sharks, small teeth, a mouth laterally enclosed by large labial cartilages, hypertrophied abductor muscles, and rapid buccal expansion (Moss, 1965, 1977; Motta and Wilga, 1999; Motta et al., 2002). Suction feeding appears to be the predominant prey capture behavior in some clades including the orectolobiforms and batoids. Specialization for suction feeding apparently evolved independently in conjunction with a benthic lifestyle, and these suction specialists feed on both elusive and non-elusive prey that live in or on the substrate, are attached to it, or are associated with the bottom (Tanaka, 1973; Moss, 1977; Belbenoit, 1986; Wu, 1994; Fouts, 1995; Clark and Nelson, 1997; Wilga, 1997; Ferry-Graham, 1998b; Heupel and Bennett, 1998; Wilga and Motta 1998a,b; Fouts and Nelson, 1999; Edmonds et al., 2001; Motta et al., 2002; Robinson and Motta, 2002; Motta, pers. obs.). The prevalence of suction capture in batoids (Belbenoit and Bauer, 1972; Wilga

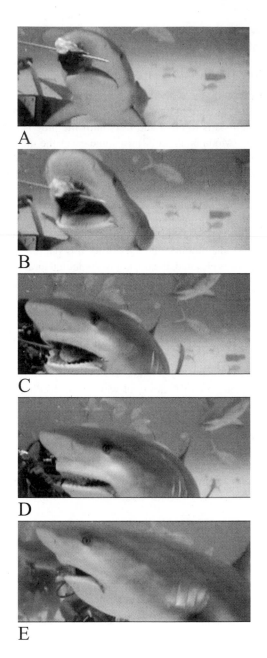

A

B

C

D

E

FIGURE 6.2 Food capture sequence of the Caribbean reef shark *Carcharhinus perezi*. (A) Start of prey capture by mandible depression; (B) the expansive phase characterized by mandible depression and head elevation; (C) during the compressive phase the mandible is elevated and the upper jaw protruded (note bulge of upper jaw); (D) the end of the compressive phase is marked by the food being grasped by the protruded upper jaw and elevated lower jaw; (E) during the recovery phase the upper jaw is retracted, while in this particular bite, the food is still held between the jaws.

and Motta. 1998b; Sasko et al., unpubl.) might be related to the fact that fish often comprise a significant portion of the diet in many rays and skates, particularly in larger individuals (Babel, 1967; Belbenoit and Bauer, 1972; Holden and Tucker, 1974; Funicelli, 1975; Bray and Hixon, 1978; Edwards, 1980; Ajayi, 1982; Abd El-Aziz, 1986; Ebert et al., 1991; Smale and Cowley, 1992; Lucifora et al., 2000; Skjaeraasen and Bergstad, 2000; Muto et al., 2001). Rapid suction combined with jaw protrusion might be an effective way to catch such elusive prey.

Biting, which may accompany ram feeding, may also occur when an elasmobranch approaches its prey or food, ceases swimming, and simply bites the prey or pieces off the prey. The cookie-cutter shark, *Isistius brasiliensis*, shows a unique biting behavior in which it employs its modified pharyngeal muscles, upper jaw, hyoid and branchial arches to suck onto its prey of pelagic fishes or marine mammals. Forming a seal with its fleshy lips, it then sinks its hooklike upper teeth and sawlike modified lower teeth into the prey, twists about its longitudinal axis to gouge out a plug of flesh, leaving a craterlike wound (Jones, 1971; Compagno, 1984; LeBoeuf et al., 1987; Shirai and Nakaya, 1992). The related kitefin shark, *Dalatias licha,* has dentition similar to that of the cookie-cutter shark and apparently feeds in the same manner (Clark and Kristof, 1990), as does the Greenland shark, *Somniosus microcephalus*. The latter apparently slowly stalks unsuspecting seals at breathing holes in the ice. Its slow movements and cryptic coloration may facilitate an element of surprise. Skomal and Benz (pers. obs.) observed Greenland sharks grasping seal carcasses in their jaws while oriented vertically in the water column. The sharks slowly rolled their bodies left and right allowing the band of closely opposed and elevated lower jaw teeth to carve out large hunks of flesh. In addition to ingesting whole sea turtles, Mediterranean white sharks, *Carcharodon carcharias,* and tiger sharks, *Galeocerdo cuvier,* in Shark Bay, Western Australia often bite off pieces of the turtle including limbs, often resulting in the turtle surviving (Fergusson et al., 2000; Heithaus et al., 2002b).

Continuous ram filter feeding, such as in the basking shark, *Cetorhinus maximus*, occurs when the shark continuously swims forward with the mouth open. In this manner, these sharks will actively seek and locate zooplankton patches on the surface. Basking sharks forage for longer periods in patches with high zooplankton density, and these high-density patches produce the most prolonged area-restricted searching during which the sharks follow convoluted swimming paths to stay within the plankton patches (Sims et al., 1997; Sims and Merrett, 1997; Sims and Quayle, 1998). The megamouth shark, *Megachasma pelagios*, and the whale shark, *Rhincodon typus*, can employ intermittent suction filter feeding, generating suction with aperiodic pulses (Taylor et al., 1983; Diamond, 1985; Compagno, 1990; Sanderson and Wassersug, 1993; Clark and Nelson, 1997; Martin and Naylor, 1997). Whale sharks can also use continuous ram filter feeding, or hang vertically in the water column. In the latter case they will suck prey into the mouth, or rise vertically out of the water and sink back under water, creating an inflow of water and prey into their open mouths (Gudger, 1941a,b; Springer, 1967; Budker, 1971; Colman, 1997).

6.3.2   Evolution of the Feeding Mechanism

The stem gnathostomes and early chondrichthyans had a jaw apparatus quite unlike modern sharks. In these the upper jaw was braced against the braincase at multiple locations. This type of jaw suspension termed *autodiastyly* was possibly the ancestral type for the Chondrichthyes. Autodiastyly is characterized by a nonsuspensory hyoid arch that articulated with the palatoquadrate, with the hyoid arch similar in morphology to the branchial arches. The palatoquadrate had ethmoidal and orbital articulations with the cranium (Figure 6.3) (Lund and Grogan, 1997; Grogan and Lund, 2000; reviewed in Wilga, 2002). The earliest sharks, the cladoselachians, had a large and almost terminal mouth with multicuspid teeth, relatively small labial cartilages, and a long palatoquadrate and Meckel's cartilage. The upper jaw of these sharks had an ethmoidal and a large postorbital articulation between the upper jaw and the cranium, and a hyomandibula that supposedly contributed little to jaw support. This type of jaw support is termed *amphistylic*. The body and caudal fin of these sharks were similar to modern fast-swimming pelagic sharks (Figure 6.4). Their teeth were suited for seizing and tearing prey rather than shearing or sawing, and it is speculated that they captured prey by overtaking and engulfing it, although suction may have played a role in prey capture (Schaeffer, 1967; Moy-Thomas and Miles, 1971; Carroll, 1988; Lund and Grogan, 1997). The xenacanthids that followed also had an amphistylic jaw suspension, a grasping dentition, long jaws, and a large gape suggesting a biting or ram feeding mechanism (Carroll, 1988; Wilga, 2002). The ctenacanthid sharks that followed likewise had an amphistylic jaw suspension, but gave rise to the neoselachians, which includes all modern sharks, skates, and rays (Figure 6.3). There was a general trend that involved shortening of the jaws and increased kinesis of the jaw suspension, including upper jaw protrusion. Modern sharks have a subterminal mouth, shorter jaws, more movable hyomandibula that suspends the jaws, more protrusible upper jaw with a smaller otic process, and a

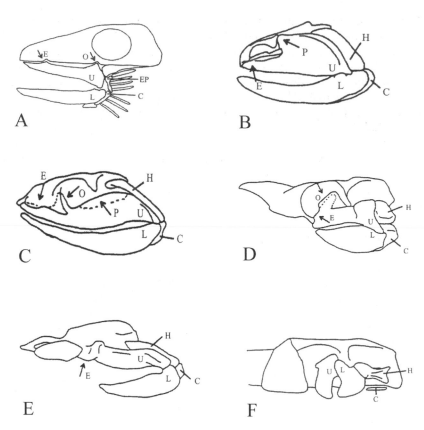

FIGURE 6.3 Left lateral views of select gnathostomes showing articulations involved with the jaw suspension. (A) Autodiastylic ancestor, (B) *Pleuracanthus*, Xenacanthida, (C) *Chlamydoselachus*, Chlamydoselachida, (D) *Squalus*, Squaliformes, (E) *Sphyrna*, Carcharhiniformes, (F) *Rhinobatos*, Batoidea. C = ceratohyal, E = ethmoidal articulation, EP = epihyal, H = hyomandibula, O = orbital articulation, L = lower jaw, P = postorbital articulation, U = upper jaw (Schaeffer, 1967; Lund and Grogan, 1997; Wilga and Motta, 1998a,b, 2000). (From Wilga, C.D. 2002. *Biol. J. Linn. Soc.* 75:483–502. With permission from Blackwell Publishing.)

dentition suited for sawing and shearing. In the modern galean sharks the ethmoidal articulation between the ethmoid process of the palatoquadrate and the ethmoid region of the cranium is the only anterior connection to the cranium, and is joined by an ethmopalatine ligament (Figure 6.5). This type of jaw suspension is termed *hyostylic* (Figure 6.3) (Schaeffer, 1967; Carroll, 1988; reviewed in Wilga, 2002). A few groups of sharks including the extinct *Chlamydoselachus* and the Squalea (e.g., dogfish) have an orbitostylic jaw suspension in which the orbital process articulates with the orbital wall (Maisey, 1980; Wilga, 2002). The batoids have a euhyostylic jaw suspension, which is perhaps the most kinetic jaw system. This type has no cranial-palatoquadrate articulation, the hyomandibula is the sole means of support for the jaws, and the hyoid arch is "broken up" with the hyomandibula losing its connection to the ceratohyal (Miyake and McEachran, 1991; Compagno, 1999; Wilga, 2002). The hyostylic and euhyostylic jaw suspension plays a key role in the functioning of the elasmobranch feeding mechanism.

6.3.3 Functional Morphology of the Feeding Mechanism

6.3.3.1   Sharks — Despite numerous studies on the anatomy of the head and cranium (e.g., Gadow, 1888; Luther, 1909; Goodey, 1910; Daniel, 1915, 1934; Allis, 1923; Edgeworth, 1935; Lightoller, 1939; Marinelli and Strenger, 1959; Gohar and Mazhar, 1964; Moss, 1972, 1977; Nobiling, 1977; Compagno, 1988; Waller and Baranes, 1991; Shirai and Okamura, 1992; Frazzetta, 1994; Wu, 1994; Motta and Wilga, 1995, 1999; Goto, 2001), the functional morphology of the feeding mechanism is only understood for some representative species, and is perhaps best understood for the carcharhiniform sharks.

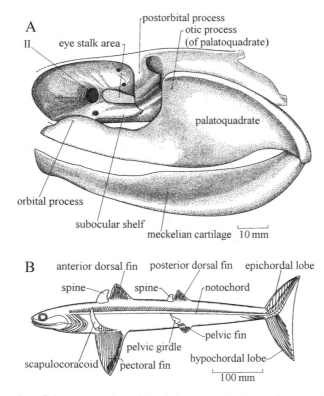

FIGURE 6.4 (A) Restoration of the neurocranium of *Cladodus*, (B) restoration of *Cladoselache fyleri*. The cladodont palatoquadrate in A had a large otic process that is not well represented in the whole animal reconstruction (B). The narrow suborbital ramus also extends anteriorly to the rostrum (Schaeffer, 1967). (From Moy-Thomas, J.A. and R.S. Miles. 1971. *Paleozoic Fishes*. Chapman & Hall, London. With permission from Kluwer Academic.)

The feeding apparatus is perhaps best known in the spiny dogfish, *Squalus acanthias*, and the lemon shark, *Negaprion brevirostris*. As previously discussed, squaloids have an orbitostylic jaw suspension in which the hyomandibula suspends the jaws from the cranium, and the palatoquadrate articulates with the orbital wall of the cranium by a relatively long orbital process (Marinelli and Strenger, 1959; Maisey, 1980; Wilga and Motta, 1998a) (Figure 6.6). The lemon shark has a hyostylic suspension in which the jaws are suspended from a more posteroventrally oriented hyomandibula, in contrast to the more laterally directed hyomandibula of the dogfish (Figure 6.5). The orbital process of the lemon shark is bound somewhat more loosely to the cranium by the elastic ethmopalatine ligament. The distal hyomandibula is braced against the mandibular knob of the mandible, and the ceratohyal is ligamentously bound to the distal hyomandibula and the mandible (Moss, 1965, 1972, 1977; Motta and Wilga, 1995). In both species, the hyomandibula is ligamentously bound to the ceratohyal and in turn to the ventral basihyal, which rests somewhat dorsal to the mandibular symphysis.

Electromyographic analyses reveal that during jaw opening a relatively conservative series of events occur in both species. Similar to the expansive phase described for teleost fishes (Liem, 1978; Lauder, 1985), the cranium is elevated by contraction of the epaxialis muscle, although cranial elevation need not occur (Motta et al., 1991, 1997; Wilga and Motta, 1998a) (Figure 6.7 and Figure 6.8). Almost simultaneously, the mandible is depressed primarily by the action of the coracomandibularis muscle, and the basihyal-ceratohyal apparatus begins to depress due to contraction of the coracoarcualis and coracohyoideus muscles. The branchial apparatus is depressed by action of the coracobranchiales muscles. In the dogfish in particular, the labial cartilages are extended as the mandible is depressed and laterally occlude the mouth (Motta et al., 1991, 1997; Wilga and Motta, 1998a). The compressive phase begins at peak gape as the mouth is maximally open, which is followed by the beginning of upper jaw protrusion and elevation of the mandible. Jaw adduction in both species is accomplished by contraction of the quadratomandibularis muscle. Various combinations of the preorbitalis and levator palatoquadrati

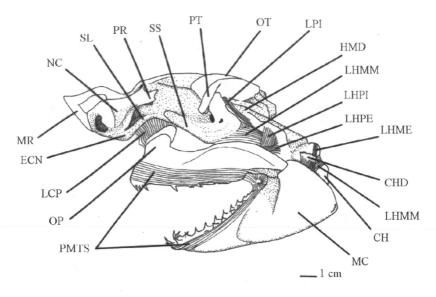

FIGURE 6.5 Left lateral view of the neurocranium, jaws, and hyoid arch of a 122 cm TL lemon shark, *Negaprion brevirostris*, with the skin and muscles removed. Tendons and ligaments are indicated. CH = ceratohyal, CHD = constrictor hyoideus dorsalis tendon, ECN = ectethmoid condyle, HMD = hyomandibula, LCP = ethmopalatine ligament, LHME = external hyoid-mandibular ligament, LHMM = medial hyoid-mandibular ligament, LHPE = external hyomandibula-palatoquadrate ligament, LHPI = internal hyomandibula-palatoquadrate ligament, LPI = postspicularis ligament, MC = Meckel's cartilage or lower jaw, MR = medial rostral cartilage, NC = nasal capsule, OP = orbital process of palatoquadrate, OT = otic capsule, PMTS = palatoquadrate-mandibular connective tissue sheath, PR = preorbital process, PT = postorbital process, SL = suborbital ledge, SS = suborbital shelf. (From Motta, P.J. and C.D. Wilga. 1995. *J. Morphol.* 226:309–329. Reprinted by permission of Wiley-Liss, Inc., a subsidiary of John Wiley & Sons, Inc.)

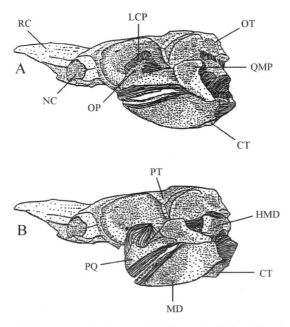

FIGURE 6.6 Left lateral view of the neurocranium, jaws, and hyoid arch of a 74.5 cm TL spiny dogfish, *Squalus acanthias*, with the skin and muscles removed. (A) At resting position, (B) at peak upper jaw protrusion. QMP = quadratomandibularis process of palatoquadrate, CT = ceratohyal, HMD = hyomandibula, LCP = ethmopalatine ligament, MD = mandible or lower jaw, NC = nasal capsule, OP = orbital process of palatoquadrate, OT = otic capsule of cranium, PQ = palatoquadrate cartilage or upper jaw, PT = postorbital process, RC = rostral cartilage. (From Wilga, C.D. and P.J. Motta. 1998. *J. Exp. Biol.* 201:1345–1358.)

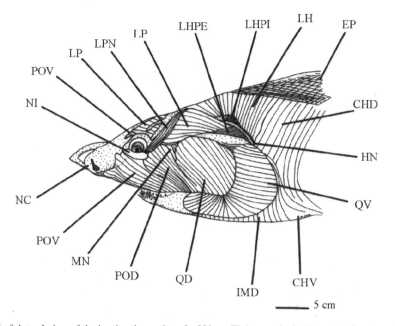

FIGURE 6.7 Left lateral view of the head and muscles of a 229 cm TL lemon shark, *Negaprion brevirostris*, with the skin removed and muscle fiber direction indicated. Myosepta of the epaxialis muscle (W-shape) are indicated in addition to the muscle fiber direction. The chondrocranial-palatoquadrate connective tissue sheath is removed. CHD = constrictor hyoideus dorsalis, CHV = constrictor hyoideus ventralis, EP = epaxialis, HN = hyomandibular nerve, IMD = intermandibularis, LH = levator hyomandibularis, LHPE = external hyomandibula-palatoquadrate ligament, LHPI = internal hyomandibula-palatoquadrate ligament, LP = levator palatoquadrati, LPN = levator palpebrae nictitantis, MN = mandibular branch of trigeminal nerve, NC = nasal capsule, NI = nictitating membrane, POD = dorsal preorbitalis, POV = ventral preorbitalis, QD = quadratomandibularis dorsal, QV = quadratomandibularis ventral. (From Motta, P.J. and C.D. Wilga. 1995. *J. Morphol.* 226:309–329. Reprinted by permission of Wiley-Liss, Inc., a subsidiary of John Wiley & Sons, Inc.)

muscles that are particular to each taxon protrude the upper jaw. In squaliform sharks such as in *S. acanthias*, the preorbitalis muscle (homologous to the ventral preorbitalis in carcharhiniform sharks; Moss, 1972; Compagno, 1988) produces an anteriorly directed force near the posterior region of the jaw (Figure 6.9). This forces the orbital process of the upper jaw to slide ventrally along the orbital wall and the ethmopalatine groove to protrude the upper jaw. As the upper jaw is protruding, the orbital process slides ventrally within the sleevelike ethmopalatine ligament until the ligament becomes taut, at which time upper jaw protrusion is complete. As the upper jaw protrudes, the entire jaw moves anteroventrally while the hyomandibula passively follows. The distal end of the hyomandibula is pulled ventrally and only slightly anteriorly. Because the action of an adductor muscle is to bring two elements closer together, contraction of the quadratomandibularis not only elevates the lower jaw, but may also pull the upper jaw away from the cranium toward the lower jaw. In this way, the quadratomandibularis may assist the preorbitalis in protruding the upper jaw (Wilga and Motta, 1998a).

The mechanism of upper jaw protrusion in carcharhiniform sharks differs slightly from that in squaliform sharks. The carcharhiniform mechanism has been proposed in several studies (Luther, 1909; Moss, 1972; Frazzetta and Prange, 1987; Frazzetta, 1994) and has largely been supported in functional studies of feeding in *N. brevirostris*, and the bonnethead shark, *Sphyrna tiburo* (Motta et al., 1997; Wilga, 1997; Wilga and Motta, 2000). Carcharhiniform sharks have a derived condition in which the levator palatoquadrati muscle is oriented more anteroposteriorly instead of dorsoventrally as in dogfish (Figure 6.7) (Moss, 1972; Nakaya, 1975; Compagno, 1988). In this orientation, the levator palatoquadrati muscle can assist the dorsal and ventral preorbitalis muscle (carcharhiniform sharks have two divisions of the preorbitalis muscle) in protruding the upper jaw (Figure 6.8). The dorsal division of the preorbitalis pulls the palatoquadrate ventrally as the ventral division of the preorbitalis and the levator palatoquadrati muscles pull it anterodorsally. Similar to the dogfish, the orbital process of the palatoquadrate is forced to glide on the ethmopalatine groove, and the resultant reaction force drives the upper jaw anteriorly and

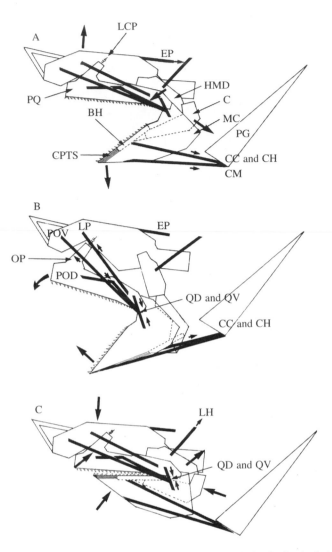

FIGURE 6.8 A model of chondrocranial, mandibular, and hyoid arch kinetics during feeding in the lemon shark, *Negaprion brevirostris*. (A) Expansive phase, characterized by depression of the mandible and elevation of the cranium; (B) compressive phase, characterized by elevation of the mandible, cranial depression, and palatoquadrate protrusion; (C) recovery phase, characterized by hyomandibular and palatoquadrate retraction. Only the major components of the chondrocranium, mandibular, and hyoid arch are represented; the branchial arches are not included. Thick dark lines indicate muscles, large arrows indicate the movement of specific elements, and small arrows indicate direction of muscle contraction. BH = basihyal, C = ceratohyal, CC = coracoarcualis, CH = coracohyoideus, CM = coracomandibularis, CPTS = chondrocranial-palatoquadrate connective tissue sheath, EP = epaxialis, HMD = hyomandibula, LCP = ethmopalatine ligament, LH = levator hyomandibularis, LP = levator palatoquadrati, MC = Meckel's cartilage or lower jaw, OP = orbital process of palatoquadrate, PG = pectoral girdle, POD = dorsal preorbitalis, POV = ventral preorbitalis, PQ = palatoquadrate cartilage or upper jaw, QD = quadratomandibularis dorsal, QV = quadratomandibularis ventral, RC = rostral cartilage. (From Motta, P.J. et al. 1997. *J. Exp. Biol.* 200:2765–2780.)

ventrally to protrude it. As the upper jaw is protruded, the ropelike ethmopalatine ligament unfolds (folded in the resting position) until it becomes taut, halting upper jaw protrusion. As the upper jaw protrudes, the jaws and the distal end of the hyomandibula also swing anteroventrally but to a greater extent than the spiny dogfish, and the distal ceratohyal and basihyal complex pivots posteroventrally. Contraction of the quadratomandibularis muscle might also assist upper jaw protrusion as described above (Moss, 1965). Peak hyoid depression occurs in the latter half of the compressive phase. In *Squalus acanthias*, *N. brevirostris,* and *Sphyrna tiburo* the mandible meets the maximally protruded upper jaw either with the

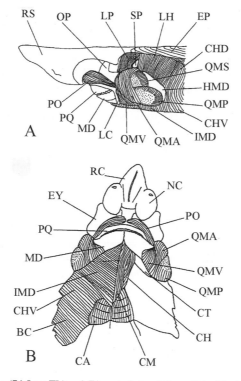

FIGURE 6.9 (A) Left lateral view (74.5 cm TL) and (B) ventral view (60 cm TL) of the head of the spiny dogfish, *Squalus acanthias*, with the skin and eye removed and muscle fiber direction indicated. Skin over the rostrum and cranium is left intact. Myosepta only of the epaxialis muscle are indicated. Raphe overlying quadratomandibularis is indicated by stippling. Anterior and posterior margins of the interhyoideus (deep to intermandibularis) are indicated by dotted lines. BC = branchial constrictors, CA = coracoarcualis, CH = coracohyoideus, CHD = constrictor hyoideus dorsalis, CHV = constrictor hyoideus ventralis, CM = coracomandibularis, CT = ceratohyal, EP = epaxialis, EY = eye, HMD = hyomandibula, IMD = interman-dibularis, LC = labial cartilages, LH = levator hyomandibularis, LP = levator palatoquadrati, MD = mandible or lower jaw, NC = nasal capsule, OP = orbital process of palatoquadrate, PO = preorbitalis, PQ = palatoquadrate or upper jaw, QMA = quadratomandibularis anterior, QMS = quadratomandibularis superficial, QMP = quadratomandibularis posterior, QMV = quadratomandibularis ventral, RC = rostral cartilage, RS = rostrum, SP = spiracularis. (From Wilga, C.D. and P.J. Motta. 1998. *J. Exp. Biol.* 201:1345–1358.)

food grasped between the teeth or after the food has been engulfed and passes through the buccal cavity. Finally, the recovery phase occurs as the palatoquadrate is retracted into its cranial seat. In the dogfish, the dorsoventrally oriented levator palatoquadrati assists in its retraction, whereas in the carcharhinids, the elastic ethmopalatine ligament assists. It is not known if the ethmopalatine ligament of squaloids is elastic. In both species, however, the levator hyomandibularis retracts the hyomandibula helping to elevate the entire jaw apparatus (Motta et al., 1997; Wilga and Motta, 1998a, 2000).

This kinematic sequence is similar to that reported for carcharhiniform sharks such as the blacknose (*Carcharhinus acronotus*), blacktip (*C. limbatus*), swell (*Cephaloscyllium ventriosum*), and Caribbean reef (*Carcharhinus perezi*) sharks, although cranial elevation and upper jaw protrusion may be lacking in some bites (Frazzetta and Prange, 1987; Ferry-Graham, 1997a, 1998a; Motta and Wilga, 2001). This differs somewhat for surface feeding in the lamnid white shark, *C. carcharias*, in that peak upper jaw protrusion occurs well before the lower jaw is completely elevated, and cranial depression does not occur until the recovery phase rather than during the compressive phase (Tricas and McCosker, 1984; Tricas, 1985). Prey capture, manipulation, and transport events in *N. brevirostris*, *Squalus acanthias*, and *Sphyrna tiburo* have a common kinematic and motor pattern sequence, but are distinguishable from each another by their duration and relative timing of individual kinematic events. Manipulation and transport events are typically shorter than capture events, although crushing manipulation events may be extensive in some species (Motta et al., 1997; Wilga, 1997; Wilga and Motta, 1998a,b, 2000; Motta and Wilga, 2001).

In contrast to these ram-feeding sharks, the mechanics of suction feeding is primarily understood from kinematic analyses, although electromyographic analysis is under way (Wu, 1994; Clark and Nelson, 1997; Ferry-Graham, 1997b, 1998b; Edmonds et al., 2001; Motta et al., 2002; Robinson and Motta, 2002). A variety of extant elasmobranchs use inertial suction to some degree as their primary feeding method — spiny dogfish, *Squalus acanthias* (Wilga and Motta, 1998a); leopard shark, *Triakis semifasciata* (Russo, 1975; Talent, 1976; Ferry-Graham, 1998b); wobbegong, *Orectolobus maculatus*, nurse shark, *Ginglymostoma cirratum*, whale shark, *Rhincodon typus*, zebra shark, *Stegostoma fasciatum* (Wu, 1994; Clark and Nelson, 1997; Motta et al., 2002; Robinson and Motta, 2002); horn shark, *Heterodontus francisci* (Strong, 1989; Edmonds et al., 2001); guitarfish, *Rhinobatos lentiginosus* (Wilga and Motta, 1998b); cownose ray, *Rhinoptera bonasus* (Sasko et al., unpubl.); lesser electric ray, *Narcine brasiliensis* (Dean and Motta, unpubl.); spotted torpedo ray, *Torpedo marmorata* (Wilson, 1953; Belbenoit and Bauer, 1972; Michaelson et al., 1979; Belbenoit, 1986); and perhaps the angel shark, *Squatina californica* (Fouts and Nelson, 1999). Inertial suction feeding elasmobranchs are found in at least eight families, often nested within clades that contain ram and compensatory suction feeders, indicating that specialization for inertial suction feeding has most likely evolved independently in several elasmobranch lineages (Motta and Wilga, 2001; Motta et al., 2002).

Ginglymostoma cirratum (Ginglymostomatidae), *T. semifasciata* (Triakidae), and *H. francisci* (Heterodontidae) appear to exhibit an abbreviated kinematic sequence in which cranial elevation is reduced or lacking during many capture bites. In contrast, carcharhiniform and lamniform sharks usually consume relatively large prey with their ventrally located mouth, and as such they elevate the cranium and depress the mandible to open the mouth as wide as possible and direct the gape more anteriorly toward the prey. However, *G. cirratum*, *T. semifasciata*, and *H. francisci* and perhaps most suction-feeding sharks primarily capture relatively small prey with a mouth that is almost terminal when maximally open (e.g., *G. cirratum*), or a mouth that is protruded anteroventrally to capture prey below them (e.g., *Squalus acanthias*). Consequently, lifting of the cranium during prey capture may not always be necessary (Motta et al., 2002). In these suction-feeding sharks, the labial cartilages protrude anteriorly as the lower jaw is depressed to effectively form a lateral enclosure of the mouth (Figure 6.10). This not only directs the suction anteriorly but may also prevent the food escaping from the sides of the mouth (Ferry-Graham, 1997b, 1998b; Wilga and Motta, 1998a; Motta and Wilga, 1999; Edmonds et al., 2001). Bite duration, from the beginning of mandible depression to retraction of the jaws to their resting position, is generally shorter for the suction-feeding sharks (*G. cirratum* 100 ms, *H. francisci* 113 to 148 ms, *T. semifasciata* 150 to 180 ms) than for ram-feeding sharks (*Sphyrna tiburo* 302 ms, *Negaprion brevirostris* 309 ms, *Carcharhinus perezi* 383 ms, *Cephaloscyllium ventriosum* 367 to 419 ms, *Carcharodon carcharias* 405 ms). Bite duration is 200 ms for suction and 280 ms for ram feeding sequences in the dogfish. Time to maximum gape from mouth opening is similarly much faster in suction feeding sharks (for example, *Orectolobus maculatus* 30 ms, *G. cirratum* 32 ms, *H. francisci* 47 to 64 ms) compared to the ram-feeding sharks (*N. brevirostris* 81 ms, *Carcharhinus perezi* 120 ms, and *S. tiburo* 162 ms; Tricas and McCosker, 1984; Tricas, 1985; Wu, 1994; Ferry-Graham, 1997a; Motta et al., 1997; Wilga and Motta, 1998a,b; Edmonds et al., 2001; Motta et al., 2002). This is expected, as suction pressure is directly related not only to the change in volume during the expansive phase, but also to the speed of buccal expansion (Lauder, 1980; Muller et al., 1982; Liem, 1993). *Ginglymostoma cirratum* can generate sub-ambient pressures as low as −98 kPA, and large nurse sharks can even dismember their food during suction (Tanaka, 1973; Motta and Wilga, 1999; Motta et al., 2002; Robinson and Motta, 2002; Motta et al., unpubl.).

Based on kinematic and cineradiograhic analysis and dissection, Wu (1994) proposed a mechanism for upper jaw protrusion in orectolobid sharks. First, the intermandibularis and interhyoideus muscles that span the inner margins of the mandible and ceratohyals, respectively, contract and medially compress the lower jaw and hyomandibulae. This results in a more acute symphyseal angle of the lower jaw such that the jaws move anteriorly similar to the change in height of a triangle when the base is shortened (Figure 6.11). As the lower jaw is depressed it pushes on the relatively large labial cartilages swinging them laterally and anteriorly, moving the oral aperture forward to form a round mouth opening. In addition, Wu proposes that the ceratohyals rotate around a process on the lower jaw, pushing the hyomandibulae anteroventrally, which in turn pushes the jaw articulation ventrally and anteriorly to protrude the jaws (Figure 6.11).

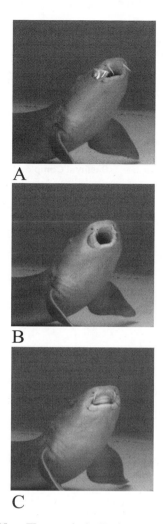

FIGURE 6.10 Suction food capture in a 85 cm TL nurse shark, *Ginglymostoma cirratum*. (A) Mandible depression, which averages 26 ms, occurring during the expansive phase; (B) peak gape, which occurs at 32 ms, is visible with the food entering the mouth (36 ms); (C) upper jaw protrusion visible as the white band inside the mouth during the compressive phase. Total bite time averages 92 ms.

The cookie-cutter shark, *Isistius brasiliensis,* employs a unique behavior and mechanism to gouge out pieces of its prey. It anchors itself to the prey with its hooklike upper teeth and sinks its large sawlike lower teeth into its prey as it apparently sucks onto its prey, forming a seal with its fleshy lips. Twisting about its longitudinal axis it gouges out a piece of flesh, leaving a craterlike wound (Jones, 1971; Compagno, 1984; LeBoeuf et al., 1987; Shirai and Nakaya, 1992). The upper jaw of this small shark is reduced in size and composed of two pieces, an anterior section that can pivot dorsally, and a posterior section. The lower jaw is relatively large and robust (Figure 6.12). Presumably the upper jaw pivots at this juncture when the shark has gripped its prey with its upper jaw, allowing the shark to pivot dorsally about this joint and sink its large lower jaw teeth into the prey. The adductor mandibulae and preorbitalis muscles are modified apparently to facilitate the gouging function of the lower jaw (Shirai and Nakaya, 1992).

The megamouth shark, *Megachasma pelagios* is apparently a slow, weak swimmer that filter-feeds on small deep-water prey such as euphausid shrimp. This shark has densely packed papillose gill rakers and relatively small gill openings, and the upper jaw is very protrusible. Anatomical investigation indicates that bioluminescent tissue in its mouth attracts prey. Protrusion of the upper jaw along with retraction of the mobile hyoid arch (ceratohyal and basihyal) creates suction that pulls the prey into the

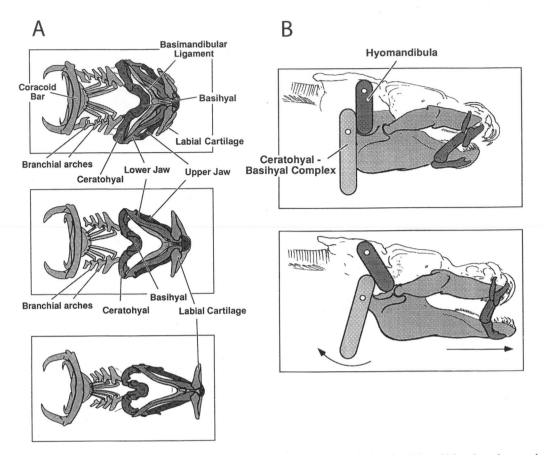

FIGURE 6.11 Feeding mechanism of *Orectolobus maculatus*. (A) Ventral view of the head, branchial arch, and pectoral girdle skeleton during jaw protrusion. In the top figure the shark is shown with its mouth closed. In the center figure the jaws are partly protruded, showing the retraction of the basihyal, the lateral compression of the jaw joints, and the anterolateral swing of the labial cartilages. In the bottom figure, the jaws are completely protruded showing the continued compression of the jaw joints and the branchial arches. The labial cartilages reach their maximum arc. (B) Schematic of the ceratohyal-hyomandibular mechanism of jaw protrusion. In the upper figure the ceratohyal and the hyomandibula are represented as two links of a kinematic chain. In the lower figure, as the ceratohyal rotates around the posterior process of the lower jaw, the dorsal end pushes against the hyomandibula. The hyomandibula rotates forward against the mandibular knob and pushes the lower jaw forward. (From Wu, E.H. 1994. *J. Morphol.* 222:175–190. Reprinted by permission of Wiley-Liss, Inc., a subsidiary of John Wiley & Sons, Inc.)

mouth. The shark would then close the mouth and protract the hyoid to push the water through the gills, filtering out the prey with the gill rakers. This mechanism is somewhat similar to that of the whale shark, *Rhinocodon typus*, which also employs a suction-filtering mechanism, but differs from that of the basking shark, *Carcharodon maximus*, which has slender jaws that are hardly protrusible. The jaws of *C. maximus* swing ventrally on the cranium and spread apart to form a circular hooplike mouth. Its gill raker denticles have hairlike crowns that do not greatly impede water flow through the gills and out the large gill openings, but catch microscopic crustaceans. The filtering apparatus of the basking shark is better suited for a higher rate of water flow than the megamouth shark, and the former is better suited for sustained, powerful swimming, which may average 0.85 m/s as it ram-filter-feeds (Gudger, 1941a,b; Taylor et al., 1983; Compagno, 1990; Clark and Nelson, 1997; Sims, 2000). Seasonal change in feeding morphology occurs in basking sharks, *C. maximus*. The gill raker sieve is apparently shed sporadically and nonsynchronously each year during winter, a period during which the sharks are believed not to feed. However, some basking sharks have been caught with gill rakers in autumn and winter, and it is now evident that basking sharks can continue to feed at plankton densities much lower than previously thought possible (Parker and Boeseman, 1954; Sims et al., 1997; Sims, 1999; Francis and Duffy, 2002).

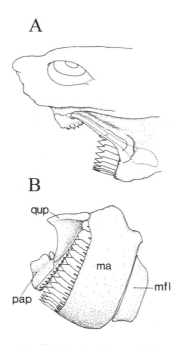

FIGURE 6.12 (A) Lateral view of the mouth of *Isistius brasiliensis* with the upper jaw protruded. Labial cartilages are indicated by broken line, (B) lateral view of the upper and lower jaw showing hinge on the upper jaw. Ma = mandibula or lower jaw; mfl = mandibular flap, a flexible, weakly chondrified plate at its posteroventral edge; pap = palatine process of palatoquadrate; qup = quadrate plate of palatoquadrate. (From Shirai, S. and K. Nakaya. 1992. *Zool. Sci.* 9:811–821. With permission.)

6.3.3.2 Batoids — The feeding mechanics of batoids differs from that of sharks in cranial anatomy and function. The hyoid arch of batoids is modified in that the hyomandibula is the only major support for the jaws, the euhyostylic jaw suspension, and the basihyal and ceratohyal are disconnected and separated ventrally from the hyomandibulae, becoming more or less degenerate or lost (Figure 6.13) (Heemstra and Smith, 1980; Miyake and McEachran, 1991; Compagno, 1999). The cranial muscles of batoids are basically similar to sharks, although some of the homologies are unclear (e.g., the "X" muscle of electric rays), the muscles are depressed in form (e.g., preorbitalis), some muscles may be lacking (e.g., intermandibularis), and some muscles may be unique to batoids (e.g., coracohyomandibularis) (Miyake et al., 1992).

There are very few studies on the feeding mechanism of batoids; two involve the guitarfish, *Rhinobatos lentiginosus,* and the lesser electric ray, *Negaprion brasiliensis.* The guitarfish captures its food by suction. The suction captures, manipulation bites, and suction transport of the food through the buccal cavity are all similar in the relative sequence of kinematic and motor activity, but differ in the absolute muscle activation time, the presence or absence of muscle activity, and in the duration of muscle activity (Figure 6.14). A preparatory phase, which is often present prior to food capture, is marked by activity of the levator palatoquadrati muscle as the upper jaw is being retracted. The expansive phase is characterized by mouth opening during which posteroventral depression of the lower jaw is initiated by the coracomandibularis. Midway through the expansive phase, the hyomandibula is depressed ventrally by the coracohyomandibularis and occasionally by the depressor hyomandibularis. This expands the orobranchial cavity. Movement of the food toward the mouth occurs during the activity of the hyomandibular depressors. The compressive phase begins with elevation of the lower jaw and the beginning of upper jaw protrusion. Maximum upper jaw protrusion is attained just prior to complete closure of the jaws. The compressive phase is represented by motor activity in the jaw adductors. Protrusion appears to be the coordinated effort of the quadratomandibularis and preorbitalis. The quadratomandibularis not only elevates the lower jaw but also protrudes the upper jaw by pulling the upper jaw ventrally toward the lower jaw. As the preorbitalis pulls the jaws anteroventrally, the upper jaw is protruded and the lower

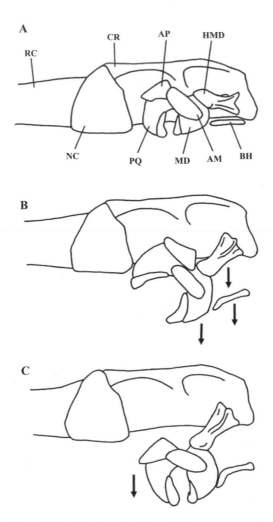

FIGURE 6.13 Left lateral view of the cranium, hyoid arch, and jaws of *Rhinobatos lentiginosus*. (A) Resting position with the jaw retracted; (B) depression of the lower jaw, hyomandibula, and basihyal opens the mouth; (C) maximally protruded position of the upper jaw. Arrows show movement of the cartilages. AM = adductor mandibulae process of the mandible, AP = adductor mandibulae process of the palatoquadrate, BH = basihyal, CR = cranium, HMD = hyomandibula, MD = mandible or lower jaw, NC = nasal capsule, PQ = palatoquadrate or upper jaw, RC = rostral cartilage. (From Wilga, C.D. and P.J. Motta. 1998. *J. Exp. Biol.* 201:3167–3184.)

jaw is elevated by the quadratomandibularis until the jaws are closed. In the final recovery phase, the head and jaws are returned to their resting position. The upper jaw is retracted by the levator palatoquadrati, and the hyomandibula retracted by the levator hyomandibularis. Hyomandibular elevation also elevates the jaws because the mandible is attached to the hyomandibula. The cranium is finally elevated to its resting position by the epaxialis and the levator rostri (Wilga and Motta, 1998).

The lesser electric ray, *N. brasiliensis,* has a remarkably protrusible and versatile mouth that it uses to probe beneath the substrate and suction feed on benthic invertebrates such as polychaete worms (Dean and Motta, unpubl.). Based on high-speed videographic analysis and anatomical dissection, Dean (unpubl.) proposed a novel mechanism for jaw protrusion that is similar to that proposed for *Orectolobus maculatus* by Wu (1994). During protrusion, the stout hyomandibulae are moved medioventrally, transmitting that motion to the attached mandible. This motion results in a medial compression of the entire jaw complex, shortening the distance between the right and left posterior corners of the jaws, forming a more acute symphyseal angle. As the angle between the mandibles is decreased, the jaws are forced anteroventrally in a manner similar to a scissors jack. The euhyostylic jaw suspension permits a degree

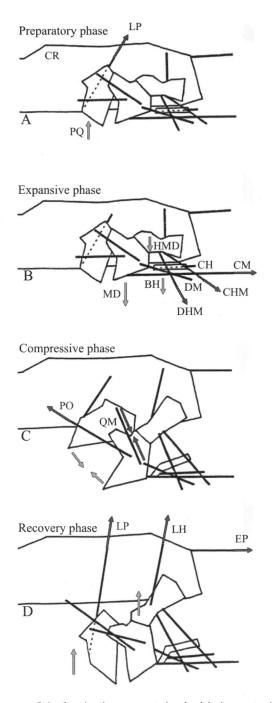

FIGURE 6.14 Schematic diagram of the functional components involved in jaw protrusion and jaw retraction during suction capture in *Rhinobatos lentiginosus*. (A) Upper jaw retraction during the preparatory phase, (B) lower jaw and hyomandibular depression during the expansive phase, (C) upper jaw protrusion and lower jaw elevation during the compressive phase, (D) hyomandibular, upper jaw, and lower jaw retraction during the recovery phase. Solid black lines represent muscles, with dark gray arrows indicating their direction of travel. Open elements represent skeletal elements, with their direction of movement indicated by light gray arrows. BH = basihyal, CH = coracohyoideus, CHM = coracohyomandibularis, CM = coracomandibularis, CR = cranium, DHM = depressor hyomandibularis, DM = depressor mandibularis, EP = epaxialis, HMD = hyomandibula, LH = levator hyomandibularis, LP = levator palatoquadrati, MD = mandible or lower jaw, PO = medial preorbitalis, PQ = palatoquadrate or upper jaw, QM = anterior quadratomandibularis. (From Wilga, C.D. and P.J. Motta. 1998. *J. Exp. Biol.* 201:3167–3184.)

of ventral protrusion that is impossible in the orectolobid sharks. The food item and sand are consequently sucked into the buccal region before maximum protrusion in reached. Food processing, when present, involves repeated, often asymmetrical protrusion of the jaws, while sand is expelled from the spiracles, gills, and mouth.

Cownose rays are pelagic rays that feed on benthic invertebrates such as mollusks and crustaceans (Orth, 1975; Smith, 1980; Schwartz, 1989; Nelson, 1994). The food is captured by suction in a conservative series of expansive, compressive, and recovery phases similar to that of other elasmobranchs, and then crushed between the platelike teeth. A 60-cm disk width cownose ray can generate bite forces ranging from 40 to 200 N (Sasko and Maschner, in Sasko, 2000; and Sasko et al., unpubl.). Coquinas (*Donax* sp.) of 20 to 30 mm crown length, a preferred prey of *Rhinoptera bonasus* in the Gulf of Mexico, fail at loads of 10 to 80 N (Maschner, 2000). An interesting question is how these elasmobranchs, with their cartilaginous jaws, and rays, which often have relatively loose mandibular and palatoquadrate symphyses, can crush such hard prey. Summers et al. (1998) and Summers (2000) noted that myliobatids have fused mandibular and palatoquadrate symphyses, flat, pavement-like tooth plates, multiple layers of calcified cartilage on the surface of the jaws, and calcified trabecular struts running through the jaws, making them well suited for their diet of hard prey. This trabecular cartilage is structurally and functionally convergent with trabecular bone and composed of mineralized struts, which are concentrated in the region where the tooth plates crush prey. Summers (2000) proposed a "nutcracker" model of jaw function whereby the food is positioned toward one side of the jaws. Because the lateral margins of the jaws are bound by strong ligaments and the mandibular and palatoquadrate symphyses are fused, the jaw can act as a nutcracker by asynchronous contraction of the jaw adductors (Summers, 1995), crushing the prey at the opposite end to that of the applied force. This system is calculated to amplify the closing force by two to four times.

6.4   Structure and Function of Elasmobranch Teeth

6.4.1   Arrangement and Terminology

Elasmobranch teeth are either arranged in rows on the palatoquadrate and Meckel's cartilage such as in most sharks and many rays, or they form large pavement-like tooth plates for crushing prey as in many batoids. Elasmobranch teeth are polyphyodont, which means they develop in rows similar to the teeth of bony fishes and are replaced at a regular interval. A tooth in the functional position at the edge of the jaw and its replacement teeth constitute a tooth row (file, family). The number of tooth rows/families varies from 1 per jaw in some rays to more than 300 in the whale shark. In most sharks there are 20 to 30 tooth rows. A series refers to a line of teeth along the jaws, which is parallel to the jaw axis and includes teeth from all rows (James, 1953; Reif, 1976, 1984; Compagno, 1984). The rate of replacement is species specific and affected by age of the animal, diet, seasonal changes, and water temperature. For most species only a few teeth are replaced at a time, although some sharks have different replacement rates for upper and lower jaws (Moss, 1967). The cookie-cutter shark, *Isistius brasiliensis*, differs in that its relatively large lower triangular teeth are shed together as a complete set (Strasburg, 1963). Replacement rates, as measured by the rate of movement of a tooth from the row lingual to the functional row to that of the functional row, varies from 9 to 12 days for the leopard shark, *Triakis semifasciata* (Reif et al., 1978), 9 to 28 days for the nurse shark, *Ginglymostoma cirratum* in summer, and 51-70 days in winter (Reif et al., 1978; Luer et al., 1990), 8 to 10 days for the lemon shark, *Negaprion brevirostris* (Moss, 1967), and about 4 weeks for *Heterodontus* (Reif, 1976). The teeth of *Myliobatis* rays are arranged quite differently as a central file of thick, flattened, usually hexagonal teeth that are fused together with three lateral files on each side of smaller teeth. Other myliobatid rays, for example, the spotted eagle ray, *Aetobatus narinari*, only have the central file of fused teeth (Figure 6.15). Together these teeth form a band on both the upper and lower tooth plate. Replacement teeth move toward the occlusal plane where they fuse and become functional. *Myliobatis* has three to ten rows of mature, unworn teeth behind the worn functional rows, and as they are replaced these teeth eventually pass aborally toward the mouth and are lost. *Aetobatus narinari* has an unusual condition in which the lower jaw teeth move anteriorly out of the crushing zone and remain attached to the tooth plate to form a

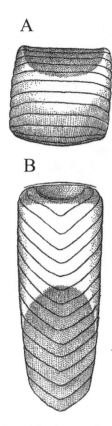

FIGURE 6.15 (A) Upper and (B) lower tooth plate of *Aetobatus narinari*. In the lower plate the front tooth is lowermost. (From Bigelow, H.B. and W.C. Schroeder. 1953. *Mem. Sears Found. Mar. Res.* 1(2):1–588. Courtesy of the Sears Foundation for Marine Research, Yale University.)

spadelike appendage, which is used to dig up prey items (Bigelow and Schroeder, 1953; Cappetta, 1986a,b; Summers, 2000; A. Barker, pers. obs.).

Within a jaw, homodont teeth are all the same shape and show no abrupt change in size. This is rare in recent and fossil sharks, but apparently exists in *Rhincodon* and *Cetorhinus*. Monognathic heterodonty refers to a significant change in size and shape of the teeth in different parts of the same jaw (upper or lower), and is common in recent and fossil sharks (Applegate, 1965; Compagno, 1988). Horn sharks (Heterodontidae) and bonnethead sharks (Sphyrnidae) both have anterior cuspidate teeth for grasping and posterior molariform crushing teeth (Figure 6.16) (Smith, 1942; Peyer, 1968; Budker, 1971; Taylor, 1972; Reif, 1976; Nobiling, 1977; Compagno, 1984). Carcharhinid sharks have dignathic heterodonty with more cuspidate lower jaw teeth lacking serrations, with more bladelike serrated upper teeth (Bigelow and Schroeder, 1948; Compagno, 1984, 1988). Sexual heterodonty occurs in many elasmobranchs, and in many cases, the teeth of adult males are different in shape from those of females and immature males. The dimorphism is often confined to the anterior teeth, and in the carcharhinid sharks, it is confined to smaller sharks that are less than 1 m in length. Sexual heterodonty in sharks and particularly rays appears to be related to courtship during which the males hold onto the females with their mouth, rather than to feeding (Springer, 1967; Compagno, 1970, 1988; Feduccia and Slaughter, 1974; McEachran, 1977; McCourt and Kerstitch, 1980; Cappetta, 1986b; Smale and Cowley, 1992; Nordell, 1994; Ellis and Shackley, 1995; Herman et al., 1995; Kajiura and Tricas, 1996).

6.4.2 Evolutionary and Functional Patterns

It is suggested that the earliest sharks for which there are no fossil teeth, just denticles (placoid scales), were microphagous filter feeders. Presumably, with selection for larger teeth, there was a concomitant

FIGURE 6.16 Dorsal view of the lower jaw teeth of the horn shark, *Heterodontus francisci*, showing the grasping teeth in the front of the jaw and the molariform or grinding teeth behind. Rostral tip of jaw at top of figure.

change to a macrophagous diet (Williams, 2001). Many of the early Paleozoic sharks including the cladodont, xenacanthid, hybodont, and ctenacanthid lineages had a dentition apparently suited for piercing, holding, and slashing. Most of the early Devonian and Carboniferous sharks have a tooth pattern often referred to as "cladodont" in form (Figure 6.17). These grasping teeth have a broad base with a single major cusp and smaller lateral cusps, and apparently slow replacement. In *Xenacanthus*, the lateral cusps are enlarged, and the central cusp is reduced. Hybodont and ctenacanthid sharks, in general, also had a tooth morphology that appears suited for piercing and holding prey; i.e., it was composed of two or more elongated cusps. Even within these early lineages, as in modern forms, there were repeated evolutionary forays into a benthic lifestyle and development of crushing, pavement-like teeth (Hotton, 1952; Schaeffer, 1967; Moy-Thomas and Miles, 1971; Zangerl, 1981; Cappetta, 1987; Carroll, 1988; Williams, 2001). Among extant sharks, the hexanchoids (*Hexanchus*, *Heptranchias*, *Notorhynchus*, *Chlamydoselachus*) and heterodontoids represent more primitive lineages. The teeth of

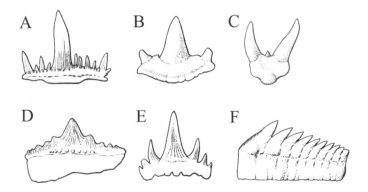

FIGURE 6.17 Ancestral shark tooth types: (A and B) acrodont teeth of *Cladodus* sp., (C) diplodus teeth of *Xenacanthus* sp., (D and E) hybodont type teeth, (F) tooth from extant *Hexanchus griseus*. (A to E, from Schaeffer, B., 1967, in *Sharks, Skates, and Rays*. P.W. Gilbert et al., Eds. Johns Hopkins University Press, Baltimore. With permission. F, from Bigelow, H.B. and W.C. Schroeder. 1948. *Mem. Sears Found. Mar. Res.* 1(1):1–576. Courtesy of the Sears Foundation for Marine Research, Yale University.)

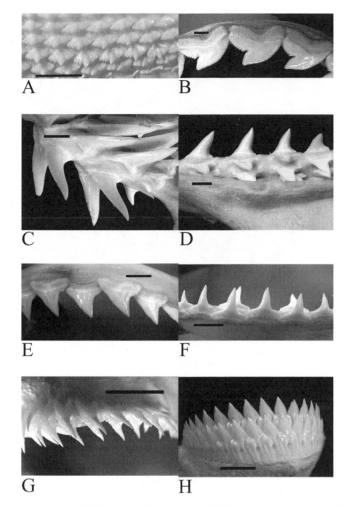

FIGURE 6.18 Modern tooth types: (A) lingual teeth of nurse shark *Ginglymostoma cirratum*, (B) upper lateral teeth of the tiger shark *Galeocerdo cuvier*, (C) upper anterior teeth of shortfin mako *Isurus oxyrinchus*, (D) lower lateral teeth of *I. oxyrinchus*, (E) upper anterior and lateral teeth of sandbar shark, *Carcharhinus plumbeus*, (F) lower anterior and lateral of *C. plumbeus*, (G) upper anterior and lateral teeth of kitefin shark, *Dalatias licha*, (H) lower teeth of *D. licha*. Scale bar is 1 cm in all cases.

hexanchoids are unlike older selachians and can be sawlike in *Hexanchus,* to three pronged and grasping-like in *Chlamydoselachus* (Figure 6.17) (Daniel, 1934; Pfeil, 1983; Cappetta, 1987; Carroll, 1988).

Modern extant sharks (and batoids) display a diversity of forms that are often ascribed functional roles (e.g., seizing/grasping, tearing, cutting, crushing, grinding), yet there are almost no quantitative functional studies of tooth use (Cappetta, 1986b, 1987). Teeth that apparently seize the prey prior to swallowing are generally small, with multiple rows of lateral cusplets. These may be found on benthic-associated sharks and rays such as in the Orectolobiformes (e.g., *Ginglymostoma cirratum*) and male dasyatid rays (Figure 6.18). Some teeth appear suited for seizing and tearing; i.e., they are long and pointed with narrow cusps (sand tiger, *Carcharias taurus*, shortfin mako, *Isurus oxyrincus*). *Isurus oxyrincus* has such teeth anteriorly and more triangular cutting teeth posteriorly. Sharks with bladelike cutting teeth tend to have one fully erect functional row forming an almost continuous blade. In the tiger shark *Galeocerdo cuvier*, the anterior and posterior margins have coarse serrations and are markedly asymmetrical with a distinct notch on the distal edge of the crown (Figure 6.19) (Bigelow and Schroeder, 1948; Cappetta, 1987; Williams, 2001). This more curved side of these teeth might serve to slice through tissue as the object is dragged across its surface, while the notch on the other side directs the object into the notch

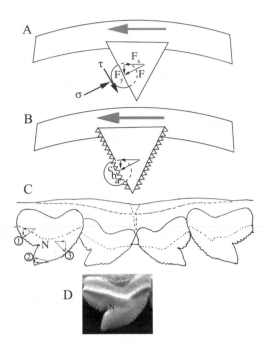

FIGURE 6.19 Proposed cutting mechanism for nonserrated, serrated, and notched shark teeth. (A) When tooth is drawn across an object as indicated by the large gray arrow, the object (denoted as a circular shape) exerts an impact on the leading edge of the beveled tooth (triangle). As the nonserrated tooth edge cuts into the object a force normal to the object (F) can be resolved into a force in the x plane (F_x) and a force in the y plane (F_y). These forces result in a stress normal to the tooth (σ) and a shear stress (τ) that result in the object being deflected toward the tip of the tooth as the tooth edge cuts into the object. The sharp leading edge of the tooth results in stress concentration that helps cut into the object (σ = Force/Area; the sharp leading edge has a very small area in contact with the object at any time). (B) As a serrated tooth is drawn across an object in the same way, the object is similarly deflected toward the tooth tip but the very small area at the tip of each serration further increases the stress, resulting in even greater penetration into the object. For example, as serration (a) encounters a region of the object it results in stress concentration, resulting in penetration of the tooth margin into the object, and similarly serrations (b) and (c) encounter additional uncut material as the tooth is driven toward (F_x), and across (F_y) the object; in this manner, serrations result in localized regions of high stress that facilitates cutting through the object. These serrations can be linearly arranged as they are on most fish teeth, and need not be laterally staggered as they are on a carpenter's wood saw. The latter serrations may reduce the entrapment of cut material from among the serrations. (C) Tiger shark, *Galeocerdo cuvier,* teeth are arranged and shaped in the indicated manner about the palatoquadrate symphysis. Different faces of these teeth may serve different functions. On the notched surface of these teeth, objects encountered at positions (1) or (2) are driven toward the notch (N), which is extremely narrow and thin, consequently increasing the stress in this region. This action serves to cut the material in a manner similar to a notched paper cutter or scissors. If the tooth is moving in the other direction the object (3) is driven toward the tooth tip and cut in the manner explained above. As the shark therefore swings its jaws from side to side while biting down on a prey item, the different faces of the teeth, which are arranged in a mirror image on the opposite jaw, cut through the prey by both of these methods. Tougher material such as ligaments, tendons, and bundles of collagen fibers may be cut more easily on the notched side of the tooth. (D) Fourth upper lateral tooth of *G. cuvier* with the notch (N) indicated.

while concentrating stress to sever more durable tissues such as collagen, cartilage, and bone (Figure 6.19). Witzell (1987) attributes the ability of *G. cuvier* to bite through whole large chelonid sea turtles to a suite of morphological and behavioral characters that include a single row of cusped and serrated teeth on the protruded jaw, a broad-based, heavily calcified jaw that is fused at the upper jaw symphysis (Moss, 1965, 1972), and head shaking that drags the teeth across the prey. Indigestible pieces of shell are regurgitated by stomach eversion, which has also been noted in other sharks and rays (Bell and Nichols, 1921; Budker, 1971; Witzell, 1987; Randall, 1992; Sims et al., 2000).

Many squaloid sharks, including *Etmopterus,* have a multicuspid grasping upper dentition and blade-like lower cutting teeth. A crushing-type dentition is found in *Mustelus.* Their teeth are low and have cutting edges with bluntly rounded apices (Bigelow and Schroeder, 1948; Cappetta, 1987). The crushing rear teeth of *Heterodontus* are closely opposed to each other such that the load or force on one or two teeth is distributed to adjacent teeth in the same tooth row (Nobiling, 1977). The lower jaw of *H. francisci* is also stiffer than the upper jaw, and the stiffness is greatest in the area of the molariform teeth. At the jaw rami where the cuspidate grasping teeth apparently experience less loading, the jaw is the least stiff (Summers et al., in press). Maschner (2000) also found that the teeth of *Rhinoptera bonasus* are interlocked so that a point load (force) on a tooth is effectively distributed to the jaw, resulting in less stress concentration at any one point. However, *Aetobatus narinari*, another hard prey specialist, shows a different pattern of jaw stiffness to that of the horn shark. The central part of this ray's jaw where the prey is crushed by the tooth plates is the stiffest (Summers et al., in press). In some sharks, the teeth are interlocked to form a continuous cutting edge (*Dalatias*, *Etmopterus*) (Figure 6.18), whereas others, including many lamnids (e.g., bigeye thresher, *Alopias superciliosus*, and *Carcharodon carcharias*), have edentulous spaces between teeth in the series (Shirai and Nakaya, 1990; Shimada, 2002a). Elasmobranchs, such as the ray *Aetobatus* that feed on hard benthic prey, have a grinding dentition of imbricated flattened teeth that form a dental plate (Cappetta, 1986a,b, 1987) (Figure 6.15).

The mechanics of piercing and cutting are poorly understood. Carcharhinid lower teeth may be used to grasp the prey and facilitate rapid penetration into the tissue (Figure 6.18). The mandible is then elevated and the prey grasped between the upper and lower jaw teeth. The serrated and triangular upper teeth saw through the prey, often facilitated by rapid head shaking (Springer, 1961; Moss, 1972, 1977; Frazzetta and Prange, 1987; Frazzetta, 1988, 1994; Smale et al., 1995; Motta et al., 1997). The mechanism by which serrated teeth cut compliant material such as skin and muscle has only been superficially investigated (Abler, 1992). Serrations may act to concentrate stress at their tips, hence piercing deeper at these regions, resulting in greater penetration as the tooth is dragged across the tissue (Figure 6.19). Squaloid sharks have bladelike teeth in both jaws with a large laterally pointed oblique smooth cusp that cuts through the prey during lateral head shaking (Compagno, 1984; Wilga and Motta, 1998a). Upper jaw protrusion in carcharhinid sharks might expose the serrated or bladelike upper teeth, facilitating their unobstructed lateral movement through the prey (Motta and Wilga, 2001). Frazzetta (1988, 1994) has proposed that the relatively loose fibrous connection of shark teeth to the jaw cartilage allows the teeth to conform to irregularities in soft tissue and guide around solid obstructions such as bone.

In contrast to many carcharhinid sharks, which have out-turned tooth tips, *C. carcharias* has the tips of the front teeth angled inward, perhaps making them more effective at gouging chunks of flesh, grasping prey items, or preventing prey escape from the mouth. During mouth closure, the tooth crown angle formed between the jaw and the center-most teeth increases by 8.7° as the jaw closes through an angle of 20° to 35°, and then decreases by 15.7° as the jaw is adducted through 35° or more. Although the mechanism is not clear, this is believed to facilitate a plucking action during feeding (Powlik, 1995). The upper anterior teeth of *Carcharias taurus* have a more pronounced inward inclination than the upper lateral and lower anterior teeth. Lucifora et al. (2001) speculates that the more outwardly inclined lower anterior teeth probably function in initial prey grasping while the upper anterior teeth puncture the prey. However, speculations such as these have not been experimentally tested.

Ontogenetic heterodonty refers to ontogenetic changes in dentition associated with ontogenetic changes in diet. The shape of the teeth and number of tooth cusps in horn sharks (Heterodontidae) changes with ontogeny. Rear replacement teeth gradually lose cusps, broaden at the base, and flatten along the crown. The more anterior recurved teeth have larger central cusps and fewer overall cusps with age. Juvenile Port Jackson shark, *Heterodontus portusjacksoni,* with more pointed teeth apparently take more soft-bodied prey than the adults (Smith, 1942; Peyer, 1968; McLaughlin and O'Gower, 1971; Taylor, 1972; Reif, 1976; Nobiling, 1977; Compagno, 1984; Shimada, 2002b). White sharks less than 1.5 m (TL, or total length) have relatively long and narrow teeth with lateral cusplets (Hubbell, 1996). Smaller white sharks feed primarily on fish, whereas larger animals with broader teeth prefer marine mammals (Tricas and McCosker, 1984). Lamniform sharks have an embryonic peglike dentition before parturition, and at about 30 to 60 cm TL their teeth transition into the adult lamnoid-type dentition just before or after birth. The early stage of the adult dentition often possesses bluntly pointed crowns without distinct cutting edges, serrations, and lateral cusplets of the adult teeth. This is perhaps to prevent the developing embryos, which are often consuming eggs and embryos *in utero*, from damaging the mother's uterus (Shimada, 2002b).

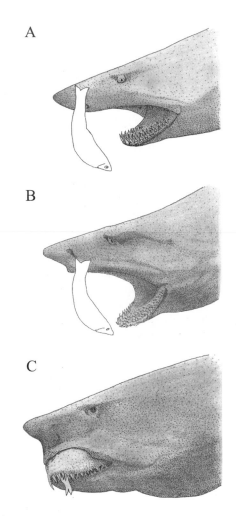

FIGURE 6.20 *Carcharias taurus* capturing food. (A) Shark approaches food with mouth partly open, (B) peak gape involving cranial elevation and mandible depression, (C) jaw closure showing extensive upper jaw protrusion. Illustrations reproduced from video. The position and size of the teeth are not illustrated with complete accuracy as they are not clear in the video image. (Video courtesy of D. Lowry, M. Matott, and D. Huber.)

6.5   The Enigma of Jaw Protrusion

Protrusion of the jaw during prey capture is an integral part of the feeding behavior in most elasmobranchs and likely serves numerous functions (Figure 6.20). Although mechanical models for upper jaw protrusion have been proposed for orectolobid sharks (Wu, 1994), *Squalus acanthias* (Wilga and Motta, 1998a), carcharhiniform sharks (Moss, 1977; Motta et al., 1997; Wilga and Motta, 2000; Wilga et al., 2001), and rays (Wilga and Motta, 1998b; Dean and Motta, unpubl.), we still do not understand its biological role. With little quantitative data, the following hypotheses for jaw protrusion in sharks have been proposed. Protrusion may shift the entire jaw apparatus away from the cranium and expose the teeth to allow more efficient bites and manipulation of prey; provide the shark with a versatile yet hydrodynamic subterminal mouth; facilitate the cutting action of the teeth, and allow deep gouging bites to be made into oversized prey; enable the shark to grasp items from the substrate with more precision; reorient the teeth of the upper jaw for increased grasping ability; further, nearly simultaneous protrusion of the upper jaw while the lower jaw is elevating may also provide the shark with a better grasp of struggling or elusive prey (Springer, 1961; Moss, 1972, 1977; Tricas and McCosker, 1984; Frazzetta and Prange, 1987; Frazzetta, 1994).

Because protrusion of the upper jaw occurs independently of cranial depression (Moss, 1972, 1977; Motta et al., 1997; Ferry-Graham, 1998b; Wilga and Motta, 1998a,b), protrusion in some sharks may assist in jaw closure as Frazzetta and Prange (1987) hypothesize. In *S. acanthias*, *Negaprion brevirostris* and *Rhinobatos lentiginosus,* protrusion of the upper jaw significantly decreases the jaw closing distance necessary for the lower jaw to travel before meeting the upper jaw by 27 to 64% (Motta et al., 1997; Wilga and Motta, 1998a,b).

Another function of protruding the upper jaw may be to disable prey. Capture of large prey often elicits vigorous lateral head shaking in many sharks including tiger (*Galeocerdo cuvier*), white (*Carcharodon carcharias*), blacknose (*Carcharhinus acronotus*), blacktip (*C. limbatus*), lemon (*Negaprion brevirostris*), bonnethead (*Sphyrna tiburo*), spiny dogfish (*Squalus acanthias*), and leopard (*Triakis semifasciata*) sharks (Springer, 1961; Moss, 1972; Frazzetta and Prange, 1987; Frazzetta, 1994; Motta et al., 1997; Wilga, 1997; Wilga and Motta, 1998a, 2000). This head shaking behavior is believed to cut or gouge smaller pieces from large prey. Cutting is facilitated by protrusion of the upper jaw into the prey (Springer, 1961; Hobson, 1963; Gilbert, 1970; Moss, 1972, 1977; Tricas and McCosker, 1984; Frazzetta and Prange, 1987; Frazzetta, 1988, 1994; Powlik, 1995). Rapid upper jaw protrusion (20 ms) in the horn shark, *Heterodontus francisci*, may be used like a striking chisel to remove attached prey, such as sea urchins, from the substrate (Edmonds et al., 2001). Furthermore, *Carcharodon carcharias* may prolong head elevation when feeding on whales to deliver multiple bites by repeated protrusion and retraction of the upper jaw until a consumable chunk is removed (Pratt et al., 1982). In rays, such as *Narcine brasiliensis,* which feed by inertial suction of benthic prey, protrusion of both the upper and lower jaw might serve to position the mouth closer to the prey to facilitate suction capture, as suction occurs only within a very short distance from the mouth in fishes (Wainwright et al., 2001; Dean, unpubl.).

6.6   Future Directions

Despite great advances in our understanding of the feeding biology of elasmobranchs, there still remain major gaps in our knowledge. Even though there have been consistent and excellent studies on the anatomy of the feeding apparatus and diet, the ethology of predation is particularly lacking, most likely because of the inherent difficulties of *in situ* studies of such large, mobile predators. Surprisingly, rays and skates would seem relatively easy to study, but are less understood and investigated than sharks. We are only beginning to understand how prey capture behavior differs within and among species, and the link between feeding behavior and morphology of the feeding apparatus. Feeding mechanisms, particularly that of jaw protrusion, are still investigated only in a handful of taxa representing a few families. Future investigations should go beyond inference studies based on moribund specimens and incorporate techniques such as electromyography, high-speed videography, pressure and displacement measurements, and the host of other techniques now available to functional morphologists. One promising line of investigation might involve the neural control of feeding behavior, particularly in elasmobranchs that appear to exhibit stereotyped capture patterns as compared to those that modulate the use of their jaws. Furthermore, the link between sensory input (for example, electrosensory vs. olfaction) and prey searching and capture could provide models for prey searching and sensory switching in aquatic vertebrates.

The mechanics of cutting and the function of teeth are poorly understood. Many questions remain: Are all cutting teeth alike? Exactly how do serrations benefit cutting and on which materials? Why are some teeth such as those of *Galeocerdo cuvier* asymmetrical? And finally, perhaps the most challenging task lies in our understanding the evolution of feeding types in the elasmobranchs, a task we can only accomplish with a thorough understanding of extant forms. As biologists, we are challenged with the wonderful task of unraveling the mysteries and truths of the feeding biology of elasmobranchs, dispelling the myths, and moving the field forward.

Acknowledgments

I gratefully acknowledge the editors for the invitation to contribute to this book. Many people and institutions contributed to the research conducted in my laboratory and their contributions are acknowledged. I thank the anonymous reviewers for providing their usual insightful feedback. The work would not be possible without the generous donations of specimens, facilities, and support by the University of South Florida and Mote Marine Laboratory. During the course of all experiments referred to here by the author, the animals were treated according to the University of South Florida and Mote Marine Laboratory Institutional Animal Care and Use Committee guidelines. Portions of the research reported were supported by grants from the National Science Foundation to the author and Robert E. Hueter (DEB 9117371 and IBN 9807863). This chapter is dedicated to my parents, Stanley and Elsie Motta, who serve as sources of support and inspiration.

References

Abd El-Aziz, S. H. 1986. Food and feeding habits of *Raja* species (Batoidei) in the Mediterranean waters of Alexandria. *Bull. Inst. Oceanogr. Fish.* (Arab. Repub. Egypt) 12:265–276.

Abler, W. L. 1992. The serrated teeth of tyrannosaurid dinosaurs, and biting structures in other animals. *Paleobiology* 18:161–183.

Ajayi, T. O. 1982. Food and feeding habits of *Raja* species (Batoidei) in Carmarthen Bay, Bristol Channel. *J. Mar. Biol. Assoc. U.K.* 62:215–223.

Allis, E. P. J. 1923. The cranial anatomy of *Chlamydoselachus anguineus*. *Acta Zool.* 4:123–221.

Applegate, S. P. 1965. Tooth terminology and variation in sharks with special reference to the sand shark, *Carcharias taurus* Rafinesque. *L.A. County Mus. Contrib. Sci.* 86:1–18.

Au, D. W. 1991. Polyspecific nature of tuna schools: shark, dolphin, and seabird associates. *U.S. Fish. Bull.* 89:343–354.

Babel, J. S. 1967. Reproduction, life history, and ecology of the round stingray, *Urolophus halleri* Cooper. *U.S. Fish. Bull.* 137:76–104.

Belbenoit, P. 1986. Fine analysis of predatory and defensive motor events in *Torpedo marmorata* (Pisces). *J. Exp. Biol.* 121:197–226.

Belbenoit, P. and R. Bauer. 1972. Video recordings of prey capture behaviour and associated electric organ discharge of *Torpedo marmorata* (Chondrichthyes). *Mar. Biol.* 17:93–99.

Bell, J. C. and J. T. Nichols. 1921. Notes on the food of Carolina sharks. *Copeia* 1921:17–20.

Bigelow, H. B. and W. C. Schroeder. 1948. Fishes of the Western North Atlantic. Lancelets, cyclostomes, sharks. *Mem. Sears Found. Mar. Res.* 1(1):1–576.

Bigelow, H. B. and W. C. Schroeder. 1953. Fishes of the Western North Atlantic. Sawfishes, guitarfishes, skates and rays. *Mem. Sears Found. Mar. Res.* 1(2):1–588.

Bray, R. N. and M. A. Hixon. 1978. Night-shocker: predatory behavior of the Pacific electric ray (*Torpedo californica*). *Science* 200:333–334.

Budker, P. 1971. *The Life of Sharks*. Columbia University Press, New York.

Bullis, H. R. 1961. Observations on the feeding behavior of white-tip sharks on schooling fishes. *Ecology* 42:194–195.

Cappetta, H. 1986a. Myliobatidae nouveaux (Neoselachii, Batomorphii) de l'Ypresien des Ouled Abdoun, Maroc. *Geol. Palaeontol.* 20:185–207.

Cappetta, H. 1986b. Types dentaires adaptatifs chez les selaciens actuels et post-paleozoiques. *Palaeovertebrata* 16:57–76.

Cappetta, H. 1987. *Chondrichthyes II. Handbook of Paleoichthyology*. Gustav Fischer Verlag, New York.

Carroll, R. L. 1988. *Vertebrate Paleontology and Evolution*. W. H. Freeman, New York.

Castro, J. I. 1996. *The Sharks of North American Waters*. Texas A&M University Press, College Station.

Chapman, D. D. and S. H. Gruber. 2002. A further observation of batoid prey handling by the Great hammerhead shark, *Sphyrna mokarran*, upon a Spotted eagle ray, *Aetobatus narinari*. *Bull. Mar. Sci.* 70:947–952.

Clark, E. and E. Kristof. 1990. Deep sea elasmobranchs observed from submersibles in Grand Cayman, Bermuda and Bahamas, in Elasmobranchs as Living Resources: Advances in the Biology, Ecology, Systematics, and the Status of the Fisheries. H. L. Pratt, Jr., S. H. Gruber, and T. Taniuchi, Eds., NOAA Tech. Rep. NMFS 90. U.S. Department of Commerce, Washington, D.C., 275–290.

Clark, E. and D. R. Nelson. 1997. Young whale sharks, *Rhincodon typus*, feeding on a copepod bloom near La Paz, Mexico. *Environ. Biol. Fish.* 50:63–73.

Coles, R. J. 1915. Notes on the sharks and rays of Cape Lookout. *N.C. Proc. Biol. Soc. Wash.* 28:89–94.

Colman, J. G. 1997. A review of the biology and ecology of the whale shark. *J. Fish Biol.* 51:1219–1234.

Compagno, L. J. V. 1970. Systematics of the genus *Hemitriakis* (Selachii: Carcharhinidae), and related genera. *Proc. Calif. Acad. Sci. Ser.* 4. 38:63–98.

Compagno, L. J. V. 1984. *Sharks of the World. An Annotated and Illustrated Catalogue of Shark Species Known to Date.* FAO Species Catalogue. Vol. 4. Part 1. *Hexanchiformes to Lamniformes.* U.N. Development Program, FAO, Rome.

Compagno, L. J. V. 1988. *Sharks of the Order Carcharhiniformes.* Princeton University Press, Princeton, NJ.

Compagno, L. J. V. 1990. Relationships of the megamouth shark, *Megachasma pelagios* (Lamniformes: Megachasmidae), with comments on its feeding habits, in Elasmobranchs as Living Resources: Advances in the Biology, Ecology, Systematics, and the Status of the Fisheries. H. L. Pratt Jr., S. H. Gruber, and T. Taniuchi, Eds., NOAA Tech. Rep. NMFS 90. U.S. Department of Commerce, Washington, D.C., 357–379.

Compagno, L. J. V. 1999. Endoskeleton, in *Sharks, Skates, and Rays: The Biology of Elasmobranch Fishes.* W. C. Hamlett, Ed., Johns Hopkins University Press, Baltimore, 69–92.

Compagno, L. J. V. 2001. *Sharks of the World. An Annotated and Illustrated Catalogue of Shark Species Known to Date. Bullhead, Mackerel and Carpet Sharks (Heterodontiformes, Lamniformes and Orectolobiformes).* FAO Species Catalogue for Fishery Purposes. No. 1, Vol. 2. FAO, Rome.

Cortés, E. 1999. Standardized diet compositions and trophic levels of sharks. *ICES J. Mar. Sci.* 56:707–717.

Curio, E. 1976. *The Ethology of Predation.* Springer-Verlag, Berlin.

Daniel, J. F. 1915. The anatomy of *Heterodontus francisci.* II. The endoskeleton. *J. Morphol.* 26:447–493.

Daniel, J. F. 1934. *The Elasmobranch Fishes.* University of California Press, Berkeley.

Diamond, J. M. 1985. Filter-feeding on a grand scale. *Nature* 316:679–680.

Ebert, D. A. 1991. Observations on the predatory behaviour of the sevengill shark *Notorynchus cepedianus.* *S. Afr. J. Mar. Sci.* 11:455–465.

Ebert, D. A., P. D. Cowley, and L. J. V. Compagno. 1991. A preliminary investigation of the feeding ecology of skates (Batoidea: Rajidae) off the West coast of Southern Africa. *S. Afr. J. Mar. Sci.* 10:71–81.

Edgeworth, F. H. 1935. *Cranial Muscles of Vertebrates.* Cambridge University Press, Cambridge, U.K.

Edmonds, M. A., P. J. Motta, and R. E. Hueter. 2001. Prey capture kinematics of the suction feeding horn shark, *Heterodontus francisci. Environ. Biol. Fish.* 62:415–427.

Edwards, R. R. C. 1980. Aspects of the population dynamics and ecology of the white spotted stingray, *Urolophus paucimaculatus* Dixon, in Port Phillip Bay, Victoria. *Aust. J. Mar. Freshwater Res.* 31:459–467.

Eibl-Eibesfeldt, I. and H. Hass. 1959. Erfahrungen mit Haien. *Z. Tierpsychol.* 16:733–746.

Ellis, J. R. and S. E. Shackley. 1995. Ontogenetic changes and sexual dimorphism in the head, mouth and teeth of the lesser spotted dogfish. *J. Fish Biol.* 47:155–164.

Feduccia, A. and B. H. Slaughter. 1974. Sexual dimorphism in skates (Rajidae) and its possible role in differential niche utilization. *Evolution* 28:164–168.

Fergusson, I. K., L. J. V. Compagno, and M. A. Marks. 2000. Predation by white sharks *Carcharodon carcharias* (Chondrichthyes: Lamnidae) upon chelonians, with new records from the Mediterranean Sea and a first record of the ocean sunfish *Mola mola* (Osteichthyes: Molidae) as stomach contents. *Environ. Biol. Fish.* 58:447–453.

Ferry-Graham, L. A. 1997a. Effects of prey size and elusivity on prey capture kinematics in leopard sharks, *Triakis semifasciata. Am. Zool.* 37:82A.

Ferry-Graham, L. A. 1997b. Feeding kinematics of juvenile swellsharks, *Cephaloscyllium ventriosum. J. Exp. Biol.* 200:1255–1269.

Ferry-Graham, L. A. 1998a. Feeding kinematics of hatchling swellsharks, *Cephaloscyllium ventriosum* (Scyliorhinidae): the importance of predator size. *Mar. Biol.* 131:703–718.

Ferry-Graham, L. A. 1998b. Effects of prey size and mobility on prey-capture kinematics in leopard sharks, *Triakis semifasciata. J. Exp. Biol.* 201:2433–2444.

Fouts, W. R. 1995. The Feeding Behavior and Associated Ambush Site Characteristics of the Pacific Angel
    Shark, *Squatina californica*, at Santa Catalina Island, California. Master's thesis, California State
    University, Long Beach.

Fouts, W. R. and D. R. Nelson. 1999. Prey capture by the Pacific angel shark, *Squatina californica*: visually
    mediated strikes and ambush-site characteristics. *Copeia* 1999:304–312.

Francis, M. P. and C. Duffy. 2002. Distribution, seasonal abundance and bycatch of basking sharks (*Cetorhinus
    maximus*) in New Zealand, with observations on their winter habitat. *Mar. Biol.* 140:831–842.

Frazzetta, T. H. 1988. The mechanics of cutting and the form of shark teeth (Chondrichthyes, Elasmobranchii).
    Zoomorphology 108:93–107.

Frazzetta, T. H. 1994. Feeding mechanisms in sharks and other elasmobranchs. *Adv. Comp. Environ. Physiol.*
    18:31–57.

Frazzetta, T. H. and C. D. Prange. 1987. Movements of cephalic components during feeding in some requiem
    sharks (Carcharhiniformes: Carcharhinidae). *Copeia* 1987:979–993.

Funicelli, N. A. 1975. Taxonomy, Feeding, Limiting Factors and Sex Ratios of *Dasyatis sabina*, *Dasyatis
    americana*, *Dasyatis sayi*, and *Narcine brasiliensis*. Ph.D. dissertation, University of Southern Missis-
    sippi, Hattiesburg.

Gadow, H. 1888. On the modifications of the first and second visceral arches, with special reference to the
    homologies of the auditory ossicles. *Philos. Trans. R. Soc. Lond.* 179B:451–485.

Gilbert, P. W. 1962. The behavior of sharks. *Sci. Am.* 207:60–68.

Gilbert, P. W. 1970. Studies on the anatomy, physiology and behavior of sharks. Final Report, Office of Naval
    Research, Contract No. 401(33): Project No. 104-471, 45 pp.

Gohar, H. A. F. and F. M. Mazhar. 1964. The internal anatomy of Selachii from the North Western Red Sea.
    Publ. Mar. Biol. Stat. (Al-Ghardaqa, Egypt) 13:145–240.

Goitein, R., F. S. Torres, and C. E. Signorini. 1998. Morphological aspects related to feeding of two marine
    skates *Narcine brasiliensis* and *Rhinobatos horkelli* Muller and Henle. *Acta Sci.* 20:165–169.

Goodey, T. 1910. A contribution to the skeletal anatomy of the frilled shark, *Chlamydoselachus anguineus*
    (Gar.). *Proc. Zool. Soc. Lond.* 2:540–571.

Goto, T. 2001. Comparative anatomy, phylogeny, and cladistic classification of the order Orectolobiformes
    (Chondrichthyes, Elasmobranchii). *Mem. Grad. Sch. Fish. Sci. Hokkaido Univ.* 48:1–100

Gray, A. E., T. J. Mulligan, and R. W. Hannah. 1997. Food habits, occurrence, and population structure of
    the bat ray, *Myliobatis californica*, in Humboldt Bay, California. *Environ. Biol. Fish.* 49:227–238.

Gregory, M. R., P. F. Ballance, G. W. Gibson, and A. M. Ayling. 1979. On how some rays (Elasmobranchia)
    excavate feeding depressions by jetting water. *J. Sed. Petrol.* 49:1125–1130.

Grogan, E. D. and R. Lund. 2000. *Debeerius ellefseni* (fam. nov., gen. nov., spec. nov.), an autodiastylic
    chondrichthyan from the Mississippian Bear Gulch Limestone of Montana (USA), the relationships of
    the Chondrichthyes, and comments on gnathostome evolution. *J. Morphol.* 243:219–245.

Grogan, E. D., R. Lund, and D. Didier. 1999. Description of the chimaerid jaw and its phylogenetic origins.
    J. Morphol. 239:45–59.

Gudger, E. W. 1907. A note on the hammerhead shark (*Sphyrna zygena*) and its food. *Science* 25:1005.

Gudger, E. W. 1941a. The food and feeding habits of the whale shark *Rhineodon typus*. *J. Elisha Mitchell
    Sci. Soc.* 57:57–72.

Gudger, E. W. 1941b. The feeding organs of the whale shark, *Rhineodon typus*. *J. Morphol.* 68:81–99.

Gunn, J. S., J. D. Stevens, T. L. O. Davis, and B. M. Norman. 1999. Observations on the short-term movements
    and behaviour of whale sharks (*Rhincodon typus*) at Ningaloo Reef, Western Australia. *Mar. Biol.*
    135:553–559.

Heemstra, P. C. and M. M. Smith. 1980. Hexatrygonidae, a new family of stingrays (Myliobatiformes:
    Batoidae) from South Africa, with comments on the classification of batoid fishes. *Ichthyol. Bull. J. L.
    B. Smith Inst. Ichthyol.* 43:1–17.

Heithaus, M. R. 2001. The biology of tiger sharks, *Galeocerdo cuvier*, in Shark Bay, Western Australia: sex
    ratio, size distribution, diet, and seasonal changes in catch rates. *Environ. Biol. Fish.* 61:25–36.

Heithaus, M. R., G. J. Marshall, B. Buhleier, and L. M. Dill. 2001. Employing CritterCam to study habitat
    use and behavior of large sharks. *Mar. Ecol. Prog. Ser.* 209:307–310.

Heithaus, M. R., L. M. Dill, G. J. Marshall, and B. Buhleier. 2002a. Habitat use and foraging behavior of
    tiger sharks (*Galeocerdo cuvier*) in a seagrass ecosystem. *Mar. Biol.* 140:237–248.

Heithaus, M. R., A. Frid, and L. M. Dill. 2002b. Shark-inflicted injury frequencies, escape ability, and habitat use of green and loggerhead turtles. *Mar. Biol.* 140:229–236.

Henderson, A. C., K. Flannery, and J. Dunne. 2001. Observations on the biology and ecology of the blue shark in the North-East Atlantic. *J. Fish Biol.* 58:1347–1358.

Herman, J., M. Hovestadt-Euler, D. C. Hovestadt, and M. Stehmann. 1995. Contributions to the study of the comparative morphology of teeth and other relevant ichthyodorulites in living supra-specific taxa of Chondrichthyan fishes. *Biologie* 65:237–307.

Hess, P. W. 1961. Food habits of two dasyatid rays in Delaware Bay. *Copeia* 1961:239–241.

Heupel, M. R. and M. B. Bennett. 1998. Observations on the diet and feeding habits of the epaulette shark, *Hemiscyllium ocellatum*, on Heron Island Reef, Great Barrier Reef. *Aust. Mar. Freshwater Res.* 49:753–756.

Hines, A. H., R. B. Whitlatch, S. F. Thrush, J. E. Hewitt, V. J. Cummings, P. K. Dayton, and P. Legendre. 1997. Nonlinear foraging response of a large marine predator to benthic prey: eagle ray pits and bivalves in a New Zealand sandflat. *J. Exp. Mar. Biol. Ecol.* 216:191–210.

Hobson, E. S. 1963. Feeding behavior in three species of sharks. *Pac. Sci.* 17:171–194.

Holden, M. J. and R. N. Tucker. 1974. The food of *Raja clavata* Linnaeus 1758, *Raja montagui* Fowler 1910, *Raja naevus* Muller and Henle 1841 and *Raja brachyura* Lafont 1873 in British waters. *Cons. Int. Explor. Mer.* 35:189–193.

Holland, K. N., B. M. Wetherbee, C. G. Lowe, and C. G. Meyer. 1999. Movements of tiger sharks (*Galeocerdo cuvier*) in coastal Hawaiian waters. *Mar. Biol.* 134:665–673.

Hotton, N., III. 1952. Jaws and teeth of American xenacanth sharks. *J. Paleontol.* 26:489–500.

Howard, J. D., T. V. Mayou, and R. W. Heard. 1977. Biogenic sedimentary structures formed by rays. *J. Sed. Petrol.* 47:339–346.

Hubbell, G. 1996. Using tooth structure to determine the evolutionary history of the white shark, in *Great White Sharks, the Biology of Carcharodon carcharias.* A. P. Klimley and D. G. Ainley, Eds., Academic Press, New York, 9–18.

James, W. W. 1953. The succession of teeth in elasmobranchs. *Proc. Zool. Soc. Lond.* 123:419–475.

Jones, E. C. 1971. *Isitius brasiliensis*, a squaloid shark, the probable of crater wounds on fishes and cetaceans. *U.S. Fish. Bull.* 69:791–798.

Kajiura, S. M. and T. C. Tricas. 1996. Seasonal dynamics of dental sexual dimorphism in the Atlantic stingray, *Dasyatis sabina. J. Exp. Biol.* 199:2297–2306.

Klimley, P. A. 1994. The predatory behavior of the white shark. *Am. Sci.* 82:122–133.

Klimley, P. A., P. Pyle, and S. D. Anderson. 1996. The behavior of white sharks and their pinniped prey during predatory attacks, in *Great White Sharks, the Biology of Carcharodon carcharias.* A. P. Klimley and D. G. Ainley, Eds., Academic Press, New York, 175–191.

Klimley, P. A., B. J. Leboeuf, K. M. Cantara, J. E. Richert, S. F. Davis, S. Van Sommeran, and J. T. Kelly. 2001. The hunting strategy of white sharks (*Carcharodon carcharias*) near a seal colony. *Mar. Biol.* 138:617–636.

Lauder, G. V. 1980. The suction feeding mechanism in sunfishes (*Lepomis*): an experimental analysis. *J. Exp. Biol.* 88:49–72.

Lauder, G. V. 1985. Aquatic feeding in lower vertebrates, in *Functional Vertebrate Morphology.* M. Hildebrand, D. M. Bramble, K. F. Liem, and D. B. Wake, Eds., Belknap Press, Cambridge, 210–229.

LeBoeuf, B. J., J. E. McCosker and J. Hewitt. 1987. Crater wounds on northern elephant seals: the cookiecutter shark strikes again. *U.S. Fish. Bull.* 85:387–392.

Liem, K. F. 1978. Modulatory multiplicity in the functional repertoire of the feeding mechanisms in cichlid fishes. *J. Morphol.* 158:323–360.

Liem, K. F. 1993. Ecomorphology of the teleostean skull, in *The Skull. Functional and Evolutionary Mechanisms,* Vol. 3. J. Hanken and B. K. Hall, Eds., University of Chicago Press, Chicago, 422–452.

Lightoller, G. H. S. 1939. Probable homologues. A study of the comparative anatomy of the mandibular and hyoid arches and their musculature. Part I: Comparative morphology. *Trans. Zool. Soc. Lond.* 24:349–444.

Long, J. A. 1995. *The Rise of Fishes.* Johns Hopkins University Press, Baltimore.

Lowe, C. G. 1991. The *in Situ* Feeding Behavior and Associated Electric Organ Discharge of the Pacific Electric Ray, *Torpedo californica.* Master's thesis, California State University, Long Beach.

Lowe, C. G., R. N. Bray, and D. R. Nelson. 1994. Feeding and associated electrical behavior of the Pacific electric ray *Torpedo californica* in the field. *Mar. Biol.* 120:161–169.

Lowe, C. G., B. M. Wetherbee, G. L. Crow, and A. L. Tester. 1996. Ontogenetic dietary shifts and feeding behavior of the tiger shark, *Galeocerdo cuvier*, in Hawaiian waters. *Environ. Biol. Fish.* 47:203–211.

Lowe, C. G., B. M. Wetherbee, K.N. Holland, and C.G. Meyer. 2003. Movement patterns of tiger and Galapagos sharks around French Frigate Shoals, Hawaii. Abstr. American Society of Ichthyologists and Herpetologists joint meeting, June 26 – July 1, Manaus, Brazil.

Lucifora, L. O., J. L. Valero, C. S. Bremec, and M. L. Lasta. 2000. Feeding habits and prey selection by the skate *Dipterus chilensis* (Elasmobranchii: Rijidae) from the South-Western Atlantic. *J. Mar. Biol. Assoc. U.K.* 80:953–954.

Lucifora, L. O., R. C. Menni, and A. H. Escalante. 2001. Analysis of dental insertion angles in the sand tiger shark, *Carcharias taurus* (Chondrichthyes: Lamniformes). *Cybium* 25:23–31.

Luer, C. A., P. C. Blum, and P. W. Gilbert. 1990. Rate of tooth replacement in the nurse shark, *Ginglymostoma cirratum*. *Copeia* 1990:182–191.

Lund, R. and E. D. Grogan. 1997. Relationships of the Chimaeriformes and the basal radiation of the Chondrichthyes. *Rev. Fish Biol. Fish.* 7:65–123.

Luther, A. 1909. Untersuchungen über die vom n. trigeminus innervierte Muskulatur der Selachier (Haie und Rochen) unter Berücksichtigung ihrer Beziehungen zu benachbarten Organen. *Acta Soc. Sci. Fenn.* 36:1–176.

Maisey, J. G. 1980. An evaluation of jaw suspension in sharks. *Am. Mus. Novit.* 2706:1–17.

Marinelli, W. and A. Strenger. 1959. *Vergleichende Anatomie und Morphologie der Wirbeltiere*. III. *Lieferung (Squalus acanthias)*. Franz Deuticke, Vienna.

Marion, G. E. 1905. Mandibular and pharyngeal muscles of acanthias and raia. *Tufts Coll. Stud.* 2:1–34.

Martin, A. P. and G. J. P. Naylor. 1997. Independent origins of filter-feeding in megamouth and basking sharks (order Lamniformes) inferred from phylogenetic analysis of cytochrome *b* gene sequences, in *Biology of Megamouth Shark*. K. Yano, J. F. Morrissey, Y. Yabumoto, and K. Nakaya, Eds., Tokai University Press, Tokyo, 39–50.

Maschner, R. P., Jr. 2000. Studies of the Tooth Strength of the Atlantic Cow-nose Ray, *Rhinoptera bonasus*. Master's thesis, California State Polytechnic University, Pomona.

McCosker, J. E. 1985. White shark attack behavior: observations of and speculations about predator and prey strategies. *Mem. South. Calif. Acad. Sci.* 9:123–135.

McCourt, R. M. and A. N. Kerstitch. 1980. Mating behavior and sexual dimorphism in dentition in the stingray *Urolophus concentricus* from the Gulf of California. *Copeia* 1980:900–901.

McEachran, J. D. 1977. Reply to "sexual dimorphism in skates (Rajidae)." *Evolution* 31:218–220.

McLaughlin, R. H. and A. K. O'Gower. 1971. Life history and underwater studies of a heterodont shark. *Ecol. Monogr.* 41:271–289.

Michaelson, D. M., D. Sternberg, and L. Fishelson. 1979. Observations on feeding, growth and electric discharge of newborn *Torpedo ocellata* (Chondrichthyes, Batoidei). *J. Fish Biol.* 15:159–163.

Miyake, T. and J. D. McEachran. 1991. The morphology and evolution of the ventral gill arch skeleton in batoid fishes (Chondrichthyes: Batoidea). *Zool. J. Linn. Soc.* 102:75–100.

Miyake, T., J. D. McEachran, and B. K. Hall. 1992. Edgeworth's legacy of cranial muscle development with an analysis of muscles in the ventral gill arch region of batoid fishes (Chondrichthyes: Batoidea). *J. Morphol.* 212:213–256.

Morrissey, J. F. 1991. Home range of juvenile lemon sharks, in *Discovering Sharks*. S. H. Gruber, Ed., American Littoral Society, Highlands, NJ, 85–86.

Moss, S. A. 1965. The Feeding Mechanisms of Three Sharks: *Galeocerdo cuvieri* (Peron & LeSueur), *Negaprion brevirostris* (Poey), and *Ginglymostoma cirratum* (Bonnaterre). Ph.D. dissertation, Cornell University, Ithaca, NY.

Moss, S. A. 1967. Tooth replacement in the lemon shark, *Negaprion brevirostris*, in *Sharks, Skates, and Rays*. P. W. Gilbert, R. F. Mathewson, and D. P. Rall, Eds., Johns Hopkins University Press, Baltimore, 319–329.

Moss, S. A. 1972. The feeding mechanism of sharks of the family Carcharhinidae. *J. Zool. Lond.* 167:423–436.

Moss, S. A. 1977. Feeding mechanisms in sharks. *Am. Zool.* 17:355–364.

Motta, P. J. and C. D. Wilga. 1995. Anatomy of the feeding apparatus of the lemon shark, *Negaprion brevirostris*. *J. Morphol.* 226:309–329.

Motta, P. J. and C. D. Wilga. 1999. Anatomy of the feeding apparatus of the nurse shark, *Ginglymostoma cirratum. J. Morphol.* 241:1–29.

Motta, P. J. and C. D. Wilga. 2001. Advances in the study of feeding behaviors, mechanisms, and mechanics of sharks. *Environ. Biol. Fish.* 60:131–156.

Motta, P. J., R. E. Hueter, and T. C. Tricas. 1991. An electromyographic analysis of the biting mechanism of the lemon shark, *Negaprion brevirostris*: functional and evolutionary implications. *J. Morphol.* 201:55–69.

Motta, P. J., R. E. Hueter, T. C. Tricas, and A. P. Summers. 1997. Feeding mechanism and functional morphology of the jaws of the lemon shark, *Negaprion brevirostris* (Chondrichthyes, Carcharhinidae). *J. Exp. Biol.* 200:2765–2780.

Motta, P. J., R. E. Hueter, T. C. Tricas, and A. P. Summers. 2002. Kinematic analysis of suction feeding in the nurse shark, *Ginglymostoma cirratum* (Orectolobiformes, Ginglymostomatidae). *Copeia* 2002:24–38.

Moy-Thomas, J. A. and R. S. Miles. 1971. *Paleozoic Fishes*. Chapman & Hall, London.

Muller, M., J. W. M. Osse, and J. H. G. Verhagen. 1982. A quantitative hydrodynamic model of suction feeding in fish. *J. Theor. Biol.* 95:49–79.

Muto, E. Y., L. S. H. Soares, and R. Goitein. 2001. Food resource utilization of the skates *Rioraja agassizii* (Muller and Henle, 1841) and *Psammobatis extenta* (Garman, 1913) on the continental shelf off Ubatuba, South-Eastern Brazil. *Rev. Brasil. Biol.* 61:217–238.

Myrberg, A. A., Jr. 1991. Distinctive markings of sharks: ethological considerations of visual function. *J. Exp. Zool.* 5:156–166.

Nakaya, K. 1975. Taxonomy, comparative anatomy and phylogeny of Japanese catsharks, Scyliorhinidae. *Mem. Fac. Fish. Hokkaido Univ.* 23:1–94.

Nelson, D. R. 1969. The silent savages. *Oceans* 1:8–22.

Nelson, J. S. 1994. *Fishes of the World*. John Wiley & Sons, New York.

Nobiling, G. 1977. Die Biomechanik des Kieferapparates beim Stierkopfhai (*Heterodontus portusjacksoni = Heterodontus philippi*). *Adv. Anat. Embr. Cell Biol.* 52:1–52.

Nordell, S. E. 1994. Observations of the mating behavior and dentition of the round stingray, *Urolophus halleri. Environ. Biol. Fish.* 39:219–229.

Norton, S. F. and E. L. Brainerd. 1993. Convergence in the feeding mechanics of ecomorphologically similar species in the Centrarchidae and Cichlidae. *J. Exp. Biol.* 176:11–29.

Notarbartolo-Di-Sciara, G. and E. V. Hillyer. 1989. Mobulid rays off Eastern Venezuela. *Copeia* 1989:607–614.

Orth, R. J. 1975. Destruction of eelgrass, *Zostera marina*, by the cownose ray, *Rhinoptera bonasus*, in the Chesapeake Bay. *Chesapeake Sci.* 16:205–208.

Parker, H. W. and M. Boeseman. 1954. The basking shark (*Cetorhinus maximus*) in winter. *Proc. Zool. Soc. Lond.* 124:185–194.

Peyer, B. 1968. *Comparative Odontology*. University of Chicago Press, Chicago.

Pfeil, F. H. 1983. Zahmorphologische Untersuchungen an rezenten und fossilen Haien der Ordnungen Chlamy-doselachiformes und Echinorhiniformes. *Palaeoichthyologica* 1:1–135.

Powlik, J. J. 1995. On the geometry and mechanics of tooth position in the white shark, *Carcharodon carcharias. J. Morphol.* 226:277–288.

Pratt, H. L., Jr., J. G. Casey, and R. B. Conklin. 1982. Observations on large white shark, *Carcharodon, carcharias*, off Long Island, New York. *U.S. Fish. Bull.* 80:153–156.

Pretlow-Edmonds, M. A. 1999. Prey Capture Kinematics of the Horn Shark, *Heterodontus francisci*. Master's thesis, University of South Florida, Tampa.

Pyle, P., A. P. Klimley, S. D. Anderson, and R. P. Henderson. 1996. Environmental factors affecting the occurrence and behavior of white sharks at the Farrallon Islands, California, in *Great White Sharks, the Biology of Carcharodon carcharias*. A. P. Klimley and D. G. Ainley, Eds., Academic Press, New York, 281–291.

Randall, J. E. 1992. Review of the biology of the tiger shark (*Galeocerdo cuvier*). *Aust. J. Mar. Freshwater Res.* 43:21–31.

Reif, W. E. 1976. Morphogenesis, pattern formation, and function of the dentition of *Heterodontus* (Selachii). *Zoomorphology* 83:1–46.

Reif, W. E. 1978. Shark dentitions: morphogenetic processes and evolution. *Geol. Paleontol. Abh.* 157:107–115.

Reif, W. E. 1980. Development of dentition and dermal skeleton in embryonic *Scyliorhinus canicula*. *J. Morphol.* 166:275–288.

Reif, W. E. 1982. Evolution of dermal skeleton and dentition in vertebrates: the odontode regulation theory. *Evol. Biol.* 15:287–368.

Reif, W. E. 1984. Pattern regulation in shark dentitions, in *Pattern Formation. A Primer in Developmental Biology.* G. M. Malacinski, Ed., Macmillan, New York.

Reif, W. E., D. McGill, and P. Motta. 1978. Tooth replacement rates of the sharks *Triakis semifasciata* and *Ginglymostoma cirratum*. *Zoll. Jahrb. Anat. Bd.* 99:151–156.

Robinson, M. P. and P. J. Motta. 2002. Patterns of growth and the effects of scale on the feeding kinematics of the nurse shark (*Ginglymostoma cirratum*). *J. Zool.* 256:449–462.

Rudloe, A. 1989. Captive maintenance of the lesser electric ray, with observations of feeding behavior. *Prog. Fish Cult.* 51:37–41.

Russo, R. A. 1975. Observations on the food habits of leopard sharks (*Triakis semifasciata*) and brown smooth-hounds (*Mustelus henlei*). *Calif. Fish Game* 61:95–103.

Sanderson, S. L. and R. Wassersug. 1993. Convergent and alternative designs for vertebrate suspension feeding, in *The Skull.*, Vol. 3. J. Hanken and B. K. Hall, Eds., University of Chicago Press, Chicago, 37–112.

Sasko, D. E. 2000. The Prey Capture Behavior of the Atlantic Cownose Ray, *Rhinoptera bonasus*. Master's thesis, University of South Florida, Tampa.

Schaeffer, B. 1967. Comments on elasmobranch evolution, in *Sharks, Skates, and Rays*. P. W. Gilbert, R. F. Matthewson, and D. P. Rall, Eds., Johns Hopkins University Press, Baltimore, 3–35.

Schaeffer, B. and M. Williams. 1977. Relationship of fossil and living elasmobranchs. *Am. Zool.* 17:293–302.

Schwartz, F. J. 1967. Embryology and feeding behavior of the Atlantic cownose ray *Rhinoptera bonasus*, *Assoc. Isl. Mar. Lab. Car.* 1967:15.

Schwartz, F. J. 1989. Feeding behavior of the cownose ray, *Rhinoptera bonasus* (family Myliobatidae). *Assoc. Southeast. Biol. Bull.* 36:66.

Sherman, K. M., J. A. Reidenauer, D. Thistle, and D. Meeter. 1983. Role of a natural disturbance in an assemblage of marine free-living nematodes. *Mar. Ecol. Prog. Ser.* 11:23–30.

Shimada, K. 2002a. Dental homologies in Lamniform sharks (Chondrichthyes: Elasmobranchii). *J. Morphol.* 251:38–72.

Shimada, K. 2002b. Teeth of embryos in lamniform sharks (Chondrichthyes: Elasmobranchii). *Environ. Biol. Fish.* 63:309–319.

Shirai, S. and K. Nakaya. 1990. Interrelationships of the Etmopterinae (Chondrichthyes, Squaliformes), in Elasmobranchs as Living Resources: Advances in the Biology, Ecology, Systematics, and the Status of the Fisheries. H. L. Pratt, Jr., S. H. Gruber, and T. Taniuchi, Eds., NOAA Tech. Rep. 90, pp. 347–356.

Shirai, S. and K. Nakaya. 1992. Functional morphology of feeding apparatus of the cookie-cutter shark, *Isistius brasiliensis* (Elasmobranchii, Dalatiinae). *Zool. Sci.* 9:811–821.

Shirai, S. and O. Okamura. 1992. Anatomy of *Trigonognathus kabeyai*, with comments on feeding mechanism and phylogenetic relationships (Elasmobranchii, Squalidae). *Jpn. J. Icthyol.* 39:139–150.

Sims, D. W. 1999. Threshold foraging behaviour of basking sharks on zooplankton: life on an energetic knife-edge? *Proc. R. Soc. Lond.* 266:1437–1443.

Sims, D. W. 2000. Filter-feeding and cruising swimming speeds of basking sharks compared with optimal models: they filter-feed slower than predicted for their size. *J. Exp. Mar. Biol. Ecol.* 249:65–76

Sims, D. W. and D. A. Merrett. 1997. Determination of zooplankton characteristics in the presence of surface feeding basking sharks *Cetorhinus maximus*. *Mar. Ecol. Prog. Ser.* 158:297–302.

Sims, D. W. and V. A. Quayle. 1998. Selective foraging behaviour of basking sharks on zooplankton in a small-scale front. *Nature* 393:460–464.

Sims, D. W., A. M. Fox, and D. A. Merrett. 1997. Basking shark occurrence off South-West England in relation to zooplankton abundance. *J. Fish Biol.* 51:436–440.

Sims, D. W., P. L. R. Andrews, and J. Z. Young. 2000. Stomach rinsing in rays. *Nature* 404:566.

Skjaeraasen, J. E. and O. A. Bergstad. 2000. Distribution and feeding ecology of *Raja radiata* in the north-eastern North Sea and Skagerrak (Norwegian Deep). *ICES J. Mar. Sci.* 57:1249–1260.

Smale, M. J. and G. Cliff. 1998. Cephalopods in the diets of four shark species (*Galeocerdo cuvier, Sphyrna lewini, S. zygaena* and *S. mokarran*) from KwaZulu-Natal, South Africa. *S. Afr. J. Mar. Sci.* 20:241–253.

Smale, M. J. and P. D. Cowley. 1992. The feeding ecology of skates (Batoidea: Rajidae) off the Cape South coast, South Africa. *S. Afr. J. Mar. Sci.* 12:823–834.

Smale, M. J., W. H. H. Sauer, and R. T. Hanlon. 1995. Attempted ambush predation on spawning squids *Loligo vulgaris reynaudii* by benthic pyjama sharks, *Poroderma africanum*, off South Africa. *J. Mar. Biol. Assoc. U.K.* 75:739–742.

Smale, M. J., W. H. H. Sauer, and M. J. Roberts. 2001. Behavioural interactions of predators and spawning chokka squid off South Africa: towards quantification. *Mar. Biol.* 139:1095–1105.

Smith, H. M. 1942. The heterodontid sharks: their natural history and the external development of *Heterodontus* (*Cestracion*) *japonicus* based on notes and drawings by Bashford Dean, in *The Bashford Dean Memorial Volume — Archaic Fishes*, Art. 8, American Museum of Natural History, New York.

Smith, J. W. 1980. The Life History of the Cownose Ray, *Rhinoptera bonasus* (Mitchell 1815), in lower Chesapeake Bay, with Notes on the Management of the Species. Master's thesis, College of William and Mary, Williamsburg, VA.

Springer, S. 1961. Dynamics of the feeding mechanism of large galeoid sharks. *Am. Zool.* 1:183–185.

Springer, S. 1967. Social organization of shark populations, in *Sharks, Skates, and Rays*. P. W. Gilbert, R. F. Mathewson, and D. P. Rall, Eds., Johns Hopkins University Press, Baltimore.

Stokes, M. D. and N. D. Holland. 1992. Southern stingray (*Dasyatis americana*) feeding on lancelets (*Branchiostoma floridae*). *J. Fish Biol.* 41:1043–1044.

Strasburg, D. W. 1958. Distribution, abundance, and habits of pelagic sharks in the central Pacific Ocean. *U.S. Fish. Bull.* 58:335–361.

Strasburg, D. W. 1963. The diet and dentition of *Isistius brasiliensis*, with remarks on tooth replacement in other sharks. *Copeia* 1963:33–40.

Strong, W. R., Jr. 1989. Behavioral Ecology of Horn Sharks, *Heterodontus francisci*, at Santa Catalina Island, California, with Emphasis on Patterns of Space Utilization. Master's thesis, California State University, Long Beach.

Strong, W. R., Jr. 1990. Hammerhead shark predation on stingrays: an observation of prey handling by *Sphyrna mokarran*. *Copeia* 1990:836–840.

Summers, A. P. 1995. Is there really asymmetry in the muscle activation patterns of skates? Abstract, American Elasmobranch Society Annual Meeting, Edmondton, Alberta, Canada.

Summers, A. P. 2000. Stiffening the stingray skeleton — an investigation of durophagy in myliobatid stingrays (Chondrichthyes, Batoidea, Myliobatidae). *J. Morphol.* 243:113–126.

Summers, A. P., T. J. Koob, and E. L. Brainerd. 1998. Stingray jaws strut their stuff. *Nature* 395.

Summers, A. P., R. A. Ketcham, and T. Rowe. In press. Structure and function of the horn shark (*Heterodontus francisci*) cranium through ontogeny — the development of a hard prey specialist. *J. Morphol.*

Talent, L. G. 1976. Food habits of the leopard shark, *Triakis semifasciata*, in Elkhorn Slough, Monterey Bay, California. *Calif. Fish Game* 62:286–298.

Tanaka, S. K. 1973. Suction feeding by the nurse shark. *Copeia* 1973:606–608.

Taylor, J. G. and A. F. Pearce. 1999. Ningaloo reef currents: implications for coral spawn dispersal, zooplankton and whale shark abundance. *J. Roy. Soc. West. Aust.* 82:57–65.

Taylor, L. R. 1972. A Revision of the Sharks of the Family Heterodontidae (Heterodontiformes, Selachii). Ph.D. dissertation, University of California at San Diego, San Diego.

Taylor, L. R., L. J. V. Compagno, and P. J. Struhsaker. 1983. Megamouth — a new species, genus, and family of lamnoid shark (*Megachasma pelagios*, family Megachasmidae) from the Hawaiian Islands. *Proc. Calif. Acad. Sci.* 43:87–110.

Thrush, S. F., R. D. Pridmore, J. E. Hewitt, and V. J. Cummings. 1991. Impact of ray feeding disturbances on sandflat macrobenthos: do communities dominated by polychaetes or shellfish respond differently? *Mar. Ecol. Prog. Ser.* 69:245–252.

Tricas, T. C. 1979. Relationships of the blue shark, *Prionace glauca*, and its prey species near Santa Catalina Island, California. *U.S. Fish. Bull.* 77:175–182.

Tricas, T. C. 1985. Feeding ethology of the white shark, *Carcharodon carcharias*. *Mem. South. Calif. Acad. Sci.* 9:81–91.

Tricas, T. C. and J. E. McCosker. 1984. Predatory behavior of the white shark (*Carcharodon carcharias*) with notes on its biology. *Proc. Calif. Acad. Sci.* 43:221–238.

Valadez-Gonzalez, C., B. Anguilar-Palomino, and S. Hernandez-Vazquez. 2001. Feeding habits of the round stingray *Urobatis halleri* (Cooper, 1863) (Chonrichthyes: Urolophidae) from the continental shelf of Jalisco and Colima, Mexico. *Cien. Mar.* 27:91–104.

VanBlaricom, G. R. 1976. Preliminary observations on interactions between two bottom-feeding rays and a community of potential prey in a sublittoral sand habitat in southern California, in *1st Pacific Northwestern Technical Workshop*. Washington Sea Grant Division of Marine Resources, Astoria, OR, 153–162.

Vorenberg, M. M. 1962. Cannibalistic tendencies of lemon and bull sharks. *Copeia* 1962:455–456.

Wainwright, P. C., L. A. Ferry-Graham, T. B. Waltzek, A. M. Carroll, C. D. Hulsey, and J. R. Grubich. 2001. Evaluating the use of ram and suction during prey capture by cichlid fishes. *J. Exp. Biol.* 204:3039–3051.

Waller, G. N. H. and A. Baranes. 1991. Chondrocranium morphology of Northern Red Sea triakid sharks and relationships to feeding habits. *J. Fish Biol.* 38:715–730.

Wetherbee, B. M., G. L. Crow, and C. G. Lowe. 1997. Distribution, reproduction, and diet of the gray reef shark *Carcharhinus amblyrhynchos* in Hawaii. *Mar. Ecol. Prog. Ser.* 151:181–189.

Widder, E. A. 1998. A predatory use of counterillumination by the squaloid shark, *Isistius brasiliensis*. *Environ. Biol. Fish.* 53:267–273.

Wilga, C. D. 1997. Evolution of Feeding Mechanisms in Elasmobranchs: A Functional Morphological Approach. Ph.D. dissertation, University of South Florida, Tampa.

Wilga, C. D. 2002. A functional analysis of jaw suspension in elasmobranchs. *Biol. J. Linn. Soc.* 75:483–502.

Wilga, C. D. and P. J. Motta. 1998a. Conservation and variation in the feeding mechanism of the spiny dogfish, *Squalus acanthias*. *J. Exp. Biol.* 201:1345–1358.

Wilga, C. D. and P. J. Motta. 1998b. Feeding mechanism of the Atlantic guitarfish, *Rhinobatos lentiginosus*: modulation of kinematic and motor activity. *J. Exp. Biol.* 201:3167–3184.

Wilga, C. D. and P. J. Motta. 2000. Durophagy in sharks: feeding mechanics of the hammerhead *Sphyrna tiburo*. *J. Exp. Biol.* 203:2781–2796.

Wilga, C. D., P. C. Wainwright, and P. J. Motta. 2000. Evolution of jaw depression mechanics in aquatic vertebrates: insights from Chondrichthyes. *Biol. J. Linn. Soc.* 71:165–185.

Williams, M. 2001. Tooth retention in cladodont sharks: with a comparison between primitive grasping and swallowing, and modern cutting and gouging feeding mechanisms. *J. Vertebr. Paleontol.* 21:214–226.

Wilson, D. P. 1953. Notes from the Plymouth Aquarium II. *J. Mar. Biol. Assoc. U.K.* 32:199–208.

Witzell, W. N. 1987. Selective predation on large cheloniid sea turtles by tiger sharks (*Galeocerdo cuvier*). *Jpn. J. Herpetol.* 12:22–29.

Wu, E. H. 1994. Kinematic analysis of jaw protrusion in orectolobiform sharks: a new mechanism for jaw protrusion in elasmobranchs. *J. Morphol.* 222:175–190.

Zangerl, R. 1981. *Chondrichthyes I. Paleozoic Elasmobranchii*. Gustav Fischer Verlag, New York.

7

Metabolism, Energetic Demand, and Endothermy

John K. Carlson, Kenneth J. Goldman, and Christopher G. Lowe

CONTENTS

7.1 Introduction

Despite the ecological significance of elasmobranchs as top-level predators in most marine ecosystems (Cortés, 1999), information on their energetics and metabolism is meager. Metabolism is an important component of an organism's daily energy budget and may account for its greatest, yet most variable proportion (Lowe, 2001). It was hypothesized that sharks had lower metabolic rates than comparable teleosts because most of the original work on the metabolic rate of sharks focused on relatively inactive, cooler-water sharks such as spotted dogfish, *Scyliorhinus canicula* (Piiper and Schumann, 1967; Metcalf and Butler, 1984) and spiny dogfish, *Squalus acanthias* (Brett and Blackburn, 1978). Over time, better techniques have evolved that allow study of more active elasmobranch species that were typically considered difficult to work with in captivity. These advances in technology have expanded our knowledge of ecology, activity level, morphology, cellular physiology, and kinematics of elasmobranchs that exhibit a wide range of lifestyles, indicating that elasmobranchs have metabolic rates comparable to teleost fishes of similar size and lifestyle.

Elasmobranchs vary in their ability to pump water over their gills through buccal pumping. Variation in this ability is directly linked to variability in metabolism and lifestyle. For example, elasmobranchs in Orders Heterodontiformes and Rajiiformes are relatively less active and demersal, and oxygenate their gills via buccal pumping. However, more active pelagic species such as those found in Orders Myliobatiformes and Carcharhiniformes (Families Carcharhinidae and Sphyrnidae) utilize ram ventilation, which allows the organism to ventilate its gills by holding the mouth open while swimming (Brown and Muir, 1970). A shift to this mode occurs when swimming velocity reaches a rate at which flow volume is adequate to supply respiratory needs. Among other species of elasmobranchs particularly lamnid, carcharhinid, and sphyrnid sharks, branchiostegal systems are reduced and thus inadequate to force water over the gills when forward movement has slowed or movement has ceased. These sharks are termed obligate ram ventilators because they must maintain constant forward movement for respiration (Roberts, 1978). Like tunas and mackerels, these sharks possess morphological, behavioral, and physiological adaptations for continuous activity (Parsons, 1990). Active swimming not only furnishes adequate gill ventilation, but also generates lift, needed because these species lack a means of buoyancy regulation (Weihs, 1981). However, the requirement for continuous activity results in an increased metabolic cost. While many carcharhiniform sharks have specializations for continuous swimming, lamniform sharks also swim continuously, and several members of this group possess additional characteristics linked to their evolution of endothermy, which may further increase energetic requirements.

The goal of this chapter is to provide an overview of current knowledge on metabolism and energetic requirements of elasmobranchs. In this chapter, we (1) discuss methods used to estimate metabolic rate in elasmobranchs; (2) compare and contrast energetic requirements for elasmobranchs within and among taxa, and document factors that affect these requirements; and (3) discuss potential techniques to stimulate future research and to further our understanding of elasmobranch energetics.

7.2 Methods of Metabolic Rate Estimation

7.2.1 Respirometry

Because oxygen is needed for maximal aerobic conversion of foodstuffs to energy, measuring oxygen consumption rate (mg O_2 kg^{-1} h^{-1}), also known as indirect calorimetry, has become the standard in determination of aerobic metabolism in postabsorptive (i.e., metabolic rate excluding energy devoted to digestion and assimilation) elasmobranchs. Oxygen consumption (VO$_2$) is typically measured using an oxygen electrode to quantify reduction in dissolved oxygen in water as the animal respires. The amount of oxygen consumed over time can be used to calculate the metabolic rate. Several types of respirometers have been used to measure VO$_2$ of elasmobranchs. Closed respirometers are common and are simple to use: they require a single O_2 electrode to measure the decrease in O_2 as water is continuously recirculated in a sealed chamber. Open respirometers are a bit more sophisticated and require the use of two O_2 probes to measure the difference in O_2 concentrations before water enters a fish-holding chamber and after water leaves the chamber. A further review of respirometers and their advantages and disadvantages can be found in Cech (1990). With elasmobranchs, both design and complexity in respirometers have varied depending on the study and the component of metabolism of interest.

7.2.1.1 Annular/Circular Respirometers — Because of their relatively simple construction and low costs, many of the estimates of metabolism for elasmobranchs have been obtained using open (Du Preez et al., 1988; Bushnell et al., 1989; Howe, 1990) or closed annular and circular respirometers

(Parsons, 1990; Sims et al., 1993; Carlson et al., 1999). These types of respirometers permit elasmobranchs to swim freely in a circular pattern or to rest on the bottom, and both types allow for estimation of routine (RMR; the metabolic rate of a postabsorptive fish under volitional activity) or standard (SMR; the metabolic rate of a postabsorptive fish completely at rest) metabolic rate. Although annular respirometers are easy to build and simple to operate, there are trade-offs in making them large enough so elasmobranchs can swim freely, but sufficiently small in volume to provide adequate O_2 measurement resolution. Bosclair and Tang (1993) indicated that there is an associated energetic cost of turning and accelerating with swimming in a circular respirometer. However, Boggs (1984) tested a theoretical model devised by Weihs (1981) to correct for fish swimming in a circular path and concluded that there was no substantial bias in determination of metabolic rate made in circular tanks, at least for skipjack tuna, *Katsuwonus pelamis*. Because tunas and sharks (excluding lamnids) differ in their swimming kinematics, it has not been fully resolved whether Boggs' (1984) study is applicable to sharks.

Some problems can arise in closed, static systems as a result of lack of water mixing. This problem can be overcome if a species studied swims continuously, which causes water mixing (Parsons, 1990; Carlson et al., 1999). Because an elasmobranch is permitted to swim voluntarily, direct continuous observation or motion sensors are required to determine when the fish is active or inactive in order to calculate SMR. Annular respirometers are acceptable for determining SMR or RMR, but they may not be sufficient for quantifying costs of swimming because, in most cases, the elasmobranch will not maintain a steady swimming speed over a long enough period of time.

7.2.1.2 Swim Tunnel Respirometers —

A number of studies have used closed swim tunnel respirometers (Figure 7.1) to obtain more accurate measures of metabolic rate. Swim tunnels are analogous to treadmills, wherein water is moved through the holding chamber, and the fish or elasmobranch swims in place against the on-flowing current (Brett, 1964). Because swimming velocity is controlled over a range of water speeds, oxygen consumption rates can be more precisely measured for a given level of activity and are typically used to measure active metabolic rate (total cost of standard metabolic rate and activity). In the late 1980s, Graham and colleagues at Scripps Oceanographic Institution in San Diego, CA developed a large "Brett-type" seagoing swim tunnel respirometer that could accommodate larger sharks (Graham et al., 1990). As part of their work, metabolic rates and swimming performance studies have been determined for leopard, *Triakis semifasciata* (Scharold et al., 1989), lemon, *Negaprion brevirostris*, and shortfin mako sharks, *Isurus oxyrinchus* (Graham et al., 1990).

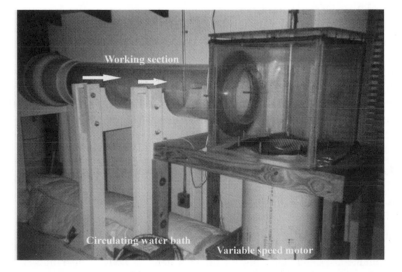

FIGURE 7.1 A "Brett"-type recirculating swim tunnel respirometer. The working section of the water tunnel houses the shark during experimentation and arrows indicate flow direction. Flow filters (not seen) within the tunnel promote rectilinear flow. A heating/cooling circulating water bath pumps heated or cooled water to regulate swim tunnel temperature. The water tunnel is currently housed at the Department of Biology, University of Mississippi, Oxford.

Recently, a smaller version of the "Brett-type" swim tunnel was constructed and used for estimation of swimming performance, kinematics, and metabolism of juvenile scalloped hammerhead sharks, *Sphyrna lewini* (Lowe, 1996, 2001).

Although swim tunnel respirometers may be better for some species (e.g., ram ventilators), their use requires the ability to induce the fish to swim, and the associated stress of being confined can result in increased metabolic expenditures. Brett and Blackburn (1978) attempted to measure swimming performance and metabolism of spiny dogfish in a swim tunnel originally developed for work on sockeye salmon, *Oncorhynchus nerka*. However, spiny dogfish placed in the swim tunnel would not swim continuously, so the authors estimated metabolism using a closed annular respirometer where the shark was permitted to swim freely. Lowe (1996) demonstrated that scalloped hammerhead sharks swimming at similar velocities in a pond beat their tails up to 21% slower than those in a swimming tunnel, which suggests sharks expend more energy while swimming in the tunnel (Lowe, 1996). However, Lowe (2001) developed an adjusted oxygen consumption rate for sharks swimming in a respirometer using a power–performance relationship of tailbeat frequency and relative swimming speed.

The use of respirometry to determine metabolic rates in elasmobranchs has not been without complications. As with any fish, confinement in a respirometer may stress the animal and affect estimates of metabolism. It is difficult to design respirometers that can accommodate the entire size range of a species or allow the animal to move about as it would in the wild because of their size and the associated scaling effect of mass on metabolic rate (Schmidt-Nielsen, 1984). The process of capturing, holding, and transporting sharks to the laboratory for experimentation can also prove to be difficult. Nevertheless, respirometry techniques offer the best means of quantifying metabolic expenditure of ectothermic fishes.

7.2.2 Biotelemetry

Much of what is known about the physiological ecology of many elasmobranchs has come from laboratory studies, because of logistical difficulties in studying marine fishes in their natural environment (e.g., Bushnell et al., 1989; Scharold et al., 1989; Carlson, 1998; Lowe, 1998). Controlled laboratory studies have shown how some elasmobranchs respond to changes in their environment, although there can be problems in extrapolating results from laboratory studies to free-swimming animals in the field or to other unstudied species (Lowe et al., 1998; Lowe and Goldman, 2001). Conversely, large size and high mobility of many elasmobranchs make controlled laboratory studies extremely difficult, and these animals can only be studied in the field. Thus, comparative laboratory and field studies are essential.

The ongoing evolution of acoustic telemetry techniques continues to enhance our ability to gather physiological data from captive and free-swimming elasmobranchs (see Lowe and Goldman, 2001, for a thorough review). A variety of sensors have been used to telemeter data on physiological parameters that are linked to metabolic rate, such as muscle temperature (Carey et al., 1982), heart rate (Scharold et al., 1989; Scharold and Gruber, 1991), swimming speed (Sundström and Gruber, 1998; Parsons and Carlson, 1998), and tailbeat frequency (Lowe et al., 1998; Lowe, 2002). As discussed below, several of these studies have used biotelemetry in combination with respirometry to gauge whether a particular physiological parameter could serve as an accurate estimator (or indicator) of metabolic rate for elasmobranchs in the field.

7.2.2.1 Muscle Temperature Telemetry — Telemetering fish muscle temperature involves placing a rigid thermistor deep into the internal epaxial red muscle then measuring changes in muscle temperature as the transmitter pulse rate changes (Carey and Lawson, 1973; Carey and Robison, 1981). Carey and colleagues (1982) designed a multitransmitter package consisting of an epaxial muscle thermistor, an ambient water thermistor, and depth-sensing transmitters that could be harpooned into the dorsal musculature of a shark. Each transmitter operated at a different frequency so data could be telemetered simultaneously, thus allowing for direct water and body temperature comparisons as the shark swam at different depths.

Carey et al. (1982) found that a large white shark, *Carcharodon carcharias*, tracked in the northwest Atlantic exhibited a 3 to 5°C elevation in muscle temperature over ambient water temperature. This

TABLE 7.1

Summary of Standard Metabolic Rates (VO_2) for a Variety of Elasmobranch Species

Species	Temp. (°C)	Mass (kg)	N	Methods	Metabolic rate (mg O_2 kg^{-1} h^{-1})	Ref.
Isurus oxyrinchus	16–20	3.9	1	Swimming closed	240*	Graham et al. (1990)
Carcharhinus acronotus	28	0.5–0.8	10	Circular closed	239*	Carlson et al. (1999)
Sphyrna lewini	21–28	0.5–0.9	17	Swimming closed	189*	Lowe (2001)
Sphyrna lewini	22–29	0.6–1.2	5	Biotelemetry	170	Lowe (2002)
Sphyrna tiburo	28	1.0	8	Open flow-through	168	Carlson and Parsons (2003)
Sphyrna tiburo	25	0.8–1.4	12	Circular closed	156*	Carlson (1998)
Negaprion brevirostris	25	1.6	7	Annular closed	153*	Scharold and Gruber (1991)
Ginglymostoma cirratum	23	1.3–4.0	5	Flow-through	106	Fournier (1996)
Negaprion brevirostris	22	0.8–1.3	13	Annular closed	95	Bushnell et al. (1989)
Scyliorhinus stellaris	25	2.5	12	Circular flow-through	92	Piiper et al. (1977)
Triakis semifasciata	14–18	2.2–5.8	5	Swimming closed	91.7*	Scharold et al. (1989)
Carcharodon carcharias	15	~943	1	Biotelemetry	60	Carey et al. (1982)
Cetorhinus maximus	—	~1000	—	Modeling	62–91	Sims (2000)
Scyliorhinus canicula	15	1.0	33	Circular closed	38.2	Sims (1996)
Squalus acanthias	10	2.0	6	Circular closed	32.4	Brett and Blackburn (1978)
Squalus suckleyi	10	2.2–4.3	9	Flow-through	31.0	Hanson and Johansen (1970)
Dasyatis americana	20	0.3	6	Flow-through	164	Fournier (1996)
Rhinobatus annulatus	15	1.0	10	Circular flow-through	61	Du Preez et al (1988)
Myliobatus californica	14	5.0	6	Circular flow-through	50	Hopkins and Cech (1994)
Myliobatus aquila	10	1.1–2.1	5	Flow-through	44.4	Du Preez et al (1988)
Dasayatis violacea	20	10.7	9	Circular flow-through	39.1	Ezcurra (2001)
Raja erinacea	10	0.5	6	Circular flow-through	20	Hove and Moss (1997)

Note: Standard metabolic rates estimated through extrapolation to zero velocity are indicated by an asterisk. Methods indicates the type of respirometer used to measure metabolic rate except *Carcharodon carcharias* and *Sphyrna lewini* (Lowe, 2002) VO_2 estimates obtained from biotelemetry field experiments and *Cetorhinus maximus* VO_2 estimate obtained from models.

shark showed a distinct preference for swimming in the thermocline, which is not uncommon behavior for endothermic fishes (Carey et al., 1971, 1981; Carey and Robison, 1981; Holland et al., 1990; Holts and Bedford, 1993; Block et al., 1998; Brill et al., 1999). Because of what Carey et al. (1982) termed, "the shark's fortuitous movements from cold to warm water," they were able to estimate its rate of metabolism from the rate of change in its muscle temperature. Their estimated metabolic rate for an approximately 943-kg white shark was 60 mg O_2 kg^{-1} h^{-1} (Table 7.1). Carey et al. (1982) lacked no savvy when putting their calculation into appropriate terms stating, "Our metabolic rate for the white shark is about three times higher than that estimated for a one ton spiny dogfish at 20°C after the latter had been adjusted for temperature and scaled for size." While they indicated their procedure overestimated the shark's metabolic rate, Carey et al. (1982) did not "pretend to great accuracy" in their calculation but thought it better than extrapolating from smaller specimens of other species, which is almost certainly true. Although this method of determining metabolism is applicable only to endothermic fishes (Carey et al., 1982), it still may not yield results. For example, Tricas and McCosker (1984) tracked a white shark off South Australia that also exhibited a similar 3 to 5°C elevation in muscle temperature over ambient temperature. However, in this study no thermocline was present and water temperature did not vary over the time of the track; hence metabolic rate could not be examined.

7.2.2.2 Heart Rate Telemetry

— The first applications of heart rate telemetry for estimating metabolic rates of elasmobranchs were tested on leopard and lemon sharks (Scharold et al., 1989; Scharold and Gruber, 1991). In these studies, sharks instrumented with electrocardiogram (EKG) acoustic transmitters were observed in respirometers to determine relationships between heart rate and VO_2. Scharold et al. (1989) exercised instrumented leopard sharks in a swim tunnel respirometer over a range of aerobic swimming speeds, and the authors found that heart rate increased at a significantly linear rate

with increased swimming speed; however, heart rate varied considerably with increases in VO_2 ($r^2 = 0.38$). As a result, heart rate was found to account for only 32% of the rise in VO_2. Similar results were obtained from juvenile lemon sharks observed swimming voluntarily in an annular respirometer (Scharold and Gruber, 1991). Heart rates of lemon sharks increased at a significant linear rate with increases in swimming speed, but also varied considerably with increases in VO_2 ($r^2 = 0.35$). Heart rate only accounted for an 18% rise in VO_2 for the lemon shark.

Both Scharold et al. (1989) and Scharold and Gruber (1991) concluded that heart rate makes a relatively small percentage contribution to changes in VO_2 and that cardiac output is likely facilitated by increases in stroke volume and/or arteriovenous oxygen differences with increased activity. As a result, heart rate was not considered to be an adequate indicator of metabolic rate for these two species, and no field experiments were conducted. These findings were supported by those of Lai et al. (1989), who measured changes in heart rate and stroke volume of resting and swimming leopard sharks in a swim tunnel and found that these sharks modulated stroke volume more than heart rates. Based on similarities in heart structure, it has been suggested that other ectothermic elasmobranch species may also exhibit this cardiac response (Emery, 1985; Farrell, 1991; Tota and Gattusa, 1996). However, this may not be true for endothermic elasmobranchs. Recent studies of cardiac physiology in shortfin mako shark indicate that their hearts resemble those of birds and mammals and these sharks may modulate heart rate more than stroke volume (Lai et al., 1997). Heart rate alone thus may provide an adequate field indicator of metabolic rate for some taxa.

7.2.2.3 Swimming Speed Telemetry

— A number of studies have used speed-sensing transmitters to measure swimming speeds and energy consumption of elasmobranchs in the field (Gruber et al., 1988; Carey and Scharold, 1990; Parsons and Carlson, 1998; Sundström and Gruber, 1998). In the early 1980s Gruber and colleagues were able to systematically quantify all components of the energy budget of lemon sharks in the laboratory, albeit focusing on smaller individuals (Gruber, 1984). As part of this effort, Bushnell et al. (1989) determined the relationship between swimming speed and oxygen consumption rate for juvenile lemon sharks in an annular respirometer. Gruber et al. (1988) attached speed-sensing acoustic transmitters to two mature lemon sharks (178 and 210 cm TL, or total length) and made direct measurements of swimming speeds. Later, Sundström and Gruber (1998) tracked three immature lemon sharks (154 to 188 cm TL) using similar speed-sensing transmitters. Using VO_2-swimming speed data from Bushnell et al. (1989), they were able to estimate the metabolic rates of the tracked lemon sharks based on measured swimming speeds. These data allowed Sundström and Gruber (1998) to construct the first field-derived energy budget for an elasmobranch and to bridge a key research gap between laboratory-based data and field measurements of free-swimming sharks.

Although these studies on lemon sharks have developed the most detailed description of a shark energy budget to date, these data are only from larger sharks tracked in the field. Smaller sharks can be studied in respirometers but cannot be tracked using the speed-sensing transmitters because of the current large size of the transmitter package. Conversely, large sharks can be tracked in the field but not studied in respirometers due to restrictions on respirometer size (Graham et al., 1990). Extrapolating metabolic rate data from juvenile sharks to adult sharks remains problematic, requiring mass-specific corrections to account for differences in size between sharks studied in the laboratory and those tracked in the field.

Parsons and Carlson (1998) used speed-sensing (propeller style) acoustic transmitters to quantify swimming speeds of bonnethead sharks, *Sphyrna tiburo*, under normoxic and hypoxic conditions in an artificial lagoon. The authors also compared VO_2 of sharks at different swimming speeds and under different oxygen concentrations in a circular respirometer. Bonnethead sharks swam significantly faster and increased their mouth gape in hypoxic conditions compared with normoxic conditions. As a result, sharks experienced higher VO_2 in hypoxic conditions, which the authors attributed to the increased swimming speeds. Parsons and Carlson (1998) concluded that speed-sensing transmitters provided a more accurate measure of activity than measuring the distance it took for a shark to swim between two points. These authors also noted that differences in swimming speeds between sharks *in situ* and those measured in the respirometer may be attributable to the added stress of handling and confinement. Speed-sensing transmitters undoubtedly increase the accuracy of measuring swimming speeds of elasmobranchs in the field (Parsons and Carlson, 1998; Sundström and Gruber, 1998). Much like using heart rate, use of swimming speed to quantify metabolism of free-swimming elasmobranchs in the field requires laboratory calibration and thereby limits the use of this technique.

7.2.2.4 *Tailbeat Frequency Telemetry* — Tailbeat frequency (TBF) has also been used as a correlate of energy consumption. Laboratory studies have shown that most fishes increase TBF in proportion to increases in swimming speed, while some species modulate their tailbeat amplitude or propulsive wavelength in addition to TBF to increase forward thrust (e.g., Bainbridge, 1958; Hunter and Zweifel, 1971; Dewar and Graham, 1994; Lowe, 1996). TBF, therefore, provides a reliable indicator of activity and exertion, although detailed laboratory calibrations are required to determine these relationships as well as energy expenditures of fishes in the field (Stasko and Horrall, 1976; Briggs and Post, 1997; Lowe et al., 1998).

A variety of sensors have been developed to measure TBF in fishes. The most common method uses electromyogram electrodes placed in the epaxial swimming muscle to monitor rhythmic body flexing. However, another type of tailbeat sensor developed uses a simple magnetized pivoting vane, which passes over a reed switch on the caudal peduncle with every lateral sweep of the tail (Lowe et al., 1998).

The first study to use acoustic tailbeat telemetry to quantify energy expenditure of an elasmobranch was conducted by Lowe (2002). Five scalloped hammerhead shark pups instrumented with tailbeat transmitters were tracked for periods up to 50 h continuously, while TBF was recorded from every successive tailbeat and averaged over 15-min periods. These data and previously determined laboratory relationships of VO_2 were used to determine swimming speeds and VO_2 rates over the course of the track. Tracks indicated that these sharks have higher metabolic requirements than those estimated for other species of tropical elasmobranchs and that they swim relatively faster than other species studied. Because of the direct coupling of laboratory and field experiments on the same size sharks, Lowe (1998) likely represents the most accurate estimates of field-based energy consumption for an elasmobranch to date.

Several physiological correlates (e.g., heart rate) do not initially appear to be good indicators for accessing actual metabolic rate; however, their relationships to metabolic rate require further investigation, and while these correlates may or may not provide good indicators of metabolism, they may provide good physiological correlates to other environmental stressors. Additionally, these types of studies have permitted investigation of increased drag and O_2 consumption on animals carrying transmitter packages (Scharold and Gruber, 1991; Lowe, 1998, 2002). Such studies will greatly assist researchers in examining energetic effects of external transmitters on elasmobranch fishes, and estimates of metabolic rates for animals carrying transmitters in the field.

Although each of the bioenergetics studies to date has provided valuable information on physiological correlates of elasmobranch metabolism, it is still very difficult to compare metabolic rates among species because of differences in experimental technique, size of animals used, and water temperature. Although the use of telemetry has certain logistic difficulties and limitations, bioenergetics of near-shore elasmobranchs could be determined by using these methods in direct comparisons between laboratory and field. Use of the latest technologies, such as acoustic transponders, underwater listening stations, and satellite telemetry, along with improvements in captive animal husbandry, will eventually allow study of bioenergetics in more active and pelagic species.

7.3 Estimates and Comparisons of Metabolic Rate

7.3.1 Standard Metabolic Rate

SMR is the metabolic rate of a postabsorptive fish completely at rest (Fry, 1957). SMR can be measured directly for animals that rest or estimated indirectly for species that are obligate ram ventilators. Because obligate ram ventilators such as lamnid and sphyrnid sharks swim continuously, standard metabolism has been estimated by extrapolation to zero velocity based on the oxygen consumption–swimming speed relationship. However, estimating standard VO_2 by extrapolating to zero activity is potentially problematic. This method may bias estimates due to extrapolating beyond the measured swimming speed range and could be overestimated if the swimming speed and VO_2 functions were elevated or the regression slope was reduced as a result of inefficient swimming at low swimming speeds (Brett, 1964). The use of paralyzed fish has proved to be useful to validate extrapolated standard metabolic rates. Standard metabolic rates determined by extrapolation for yellowfin, *Thunnus albacares,* and skipjack tunas by Dewar and Graham (1994) were similar to that reported by Brill (1987) for paralyzed tunas. Carlson and Parsons (2003) measured a standard VO_2 of 168 mg O_2 kg^{-1} h^{-1} for paralyzed bonnethead sharks, which was close to extrapolation values of 156 mg O_2 kg^{-1} h^{-1}. The few SMRs available for sharks show a wide variation, ranging from 31.0 mg O_2 kg^{-1} h^{-1} for 2.2 to 4.3 kg Pacific dogfish sharks, *Scyliorhinus suckleyi,* at 10°C to 240 mg O_2 kg^{-1} h^{-1} for a 3.9 kg shortfin mako shark at 16 to 20°C (Table 7.1).

Ectothermic tropical and subtropical sharks appear to have standard metabolic rates similar to active ectothermic teleosts of comparable lifestyles. Metabolic rates for sharks 0.5 to 1.5 kg in body mass at temperatures from 22 to 28°C range from 95 mg O_2 kg^{-1} h^{-1} for lemon sharks (Bushnell et al., 1989) to 189 mg O_2 kg^{-1} h^{-1} for scalloped hammerhead sharks at 26°C (Lowe, 2001). In general, species that are obligate ram ventilators and swim continuously have the highest measure of metabolism. Lower estimates of VO_2 are generally found for cooler-water (10 to 20°C), less-active species such as leopard shark (91.7 mg O_2 kg^{-1} h^{-1}; Scharold et al., 1989), spotted dogfish (38.2 mg O_2 kg^{-1} h^{-1}; Sims, 1996), and spiny dogfish (32.4 mg O_2 kg^{-1} h^{-1}; Brett and Blackburn, 1978). The highest SMR determined for an ectothermic obligate ram-ventilating shark was for 0.5 kg blacknose sharks, *Carcharhinus acronotus,* at 28°C (239 mg O_2 kg^{-1} h^{-1}; Carlson et al., 1999), although this measure may be slightly higher than expected due to few data points at slow swimming speeds. Comparably sized moderately active teleosts have SMR ranging from 125 mg O_2 kg^{-1} h^{-1} for 0.9 kg largemouth bass (*Micropterus salmoides*; Beamish, 1970) to 158 mg O_2 kg^{-1} h^{-1} for 1.6 kg smallmouth buffalofish (*Ictiobus bubalus*; Adams and Parsons, 1998) at 25 to 27°C. The high SMR found for more-active fishes was hypothesized to reflect increased gill surface areas and associated osmoregulatory costs (Brill, 1996). However, recent measures of Na$^+$-K$^+$-ATPase activity in the gills of skipjack and yellowfin tuna (species that possess some of the highest estimates of SMR; Brill, 1996), estimated the costs of osmoregulation to be at most 9 to 13% of standard metabolic rate (Brill et al., 2001). Thus, the reasons for elevated SMR in tunas remain unexplained. Because osmoregulatory costs of more-active elasmobranchs have not been estimated, it cannot be determined if the higher SMR found for more-active sharks is due to osmoregulation or other metabolic processes.

Standard metabolic rates of skates and rays are similar to those determined for similar-sized, cooler-water, less-active sharks. At temperatures from 10 to 15°C, metabolic rates ranged from 20 to 61 mg O_2 kg^{-1} h^{-1} (Table 7.1). Although metabolic rates are in general low, bat rays, *Myliobatus californica,* have a high temperature sensitivity (Q_{10} or the increase in a rate caused by a 10°C increase in temperature; Schmidt-Nielsen, 1983), which is reflected in their VO_2 (see Section 7.6.1). Standard metabolic rate in bat rays increased from about 50 to 170 mg O_2 kg^{-1} h^{-1} over a temperature range of 14 to 20°C (Hopkins and Cech, 1994). Thus, bat rays exposed to thermally heterogeneous environments have marked changes in energetic requirements. To accommodate this, it is thought that bat rays behaviorally thermoregulate by moving to cooler water to reduce energetic demands and moving to warmer water to exploit increases in metabolism for feeding (Matern et al., 2000).

Standard metabolic rates have been measured for embryonic sharks and skates. Diez and Davenport (1987) measured the VO_2 of unencapsulated 3.0 g embryonic spotted dogfish, *Scyliorhinus canicula,* which consumed 0.087 ml O_2 g^{-1} h^{-1}. These rates were slightly higher than those measured from encapsulated 5 to 7 g little skates, *Raja erinacea,* which exhibited average SMR of 0.032 ml O_2 g^{-1} h^{-1} (Leonard et al., 1999).

7.3.2   Maximum Metabolic Rate

Sharks that are more active have higher maximum metabolic rates (MMR) when contrasted to sharks that are more sedentary. Even when standardizing for temperature effects ($Q_{10} = 2.0$), MMR were about 1.5 to 2.3 times greater for active species. At 25°C, a 2.0 kg spiny dogfish consumed a maximum of 250 mg O_2 kg^{-1} h^{-1} (Brett and Blackburn, 1978) compared to 620 mg O_2 kg^{-1} h^{-1} for a 1.6 kg lemon shark (Graham et al., 1990). Scalloped hammerhead sharks swimming at 1.0 body length per second (bl s^{-1}) consumed up to 500 mg O_2 kg^{-1} h^{-1} at 26°C (Lowe, 2001), while Scharold et al. (1989) measured metabolic rates to 384 mg O_2 kg^{-1} h^{-1} for less-active leopard sharks swimming at 0.9 bl s^{-1}.

7.3.3 Specific Dynamic Action

Specific dynamic action (SDA) refers to the energetic costs associated with digestion and assimilation (Jobling, 1981). Among teleosts, specific dynamic action can account for 15 to 20% of ingested energy and is generally measured by the increase in metabolic rate following feeding (Brett and Groves, 1979).

Although few estimates have been made in elasmobranchs, results suggest that costs of digestion are similar to those found for teleosts. Du Preez et al. (1988) reported energy losses with feeding of 17.3% for guitarfish, *Rhinobatus annulatus*, and 12.9% for bullray, *Myliobatus aquila*, although variables such as prefeeding levels, period of starvation, and activity levels were not controlled or similar between species. Based on controlled feeding studies of lesser spotted dogfish, specific dynamic action was estimated at 6.0 to 12.5% for juvenile and adult dogfish, respectively (Sims and Davies, 1994). The results suggest that juvenile sharks have reduced energy costs in terms of digestion and assimilation despite higher levels of food consumption. Sims and Davies (1994) hypothesized this relationship was due more to efficient conservation of metabolic energy (which could then be used for growth) rather than a reduced rate of biosynthesis.

7.3.4 Anaerobic Metabolism

Anaerobic metabolism is powered by white muscle, which comprises the majority of muscle in ecto-thermic elasmobranchs. White muscle is the primary muscle used during burst swimming, and some sharks appear to have high burst swimming capacities. Telemetry data on blue sharks, *Prionace glauca*, indicate that they are capable of short duration bursts up to 2 m s^{-1} (Carey and Scharold, 1990). There are descriptions of blacktip, *Carcharhinus limbatus*, and spinner sharks, *C. brevipinna*, leaping and spinning on their body axes above the water surface, which requires considerable exertion to propel themselves out of the water (Castro, 1996; Carlson, pers. obs.).

In general, elasmobranchs and teleosts with similar activity levels have comparable levels of anaerobic metabolism. By using biochemical indices, low levels of citrate synthase (an index of aerobic capacity) and lactate dehydrogenase (an index of anaerobic capacity) in white mytomal muscle were reported for benthic skates and rays, similar to those of demersal teleosts (Dickson et al., 1993). Intermediate levels of citrate synthase and lactate dehydrogenase were similar among moderately active teleosts and elas-mobranchs. The greatest capacity for anaerobic metabolism was observed for shortfin mako shark, which, along with tunas, have significantly greater white muscle citrate synthase and lactate dehydrogenase levels and buffering capacities than ectothermic fishes (Dickson et al., 1993). Shortfin mako sharks also have higher white muscle activities of creatine phosphokinase (an index of ATP production rate during burst swimming) than active ectothermic sharks and teleosts, which allows for redox balance to be retained during anaerobiosis (Dickson, 1996; Bernal et al., 2001a).

7.4 Energetic Costs of Swimming

7.4.1 Swimming Efficiency

The relationship between relative swimming speed (bl s^{-1}) and metabolic rate (log transformed mg O$_2$ kg^{-1} h^{-1}) is similar among comparable size ectothermic sharks (Figure 7.2). The slope of the relationships ranges from 0.27 for blacknose shark (Carlson et al., 1999) to 0.36 for lemon shark (Bushnell et al., 1989). These relationships indicate that the energy required to move a given amount of mass per measure of distance is the same. Although kinematic variables such as TBF and amplitude, and propulsive wavelength may vary among sharks and over swimming speeds, the similarity in rate of change in metabolic rate with swimming speed may be attributable to morphological adaptations for drag reduction that all of these species share. For example, leopard, lemon, and juvenile scalloped hammerhead sharks have similar TBF at a given speed; however, yellowfin tuna have significantly lower TBF at a given speed (Figure 7.3A). In general, TBF increases at about the same rate for each species. While it may appear the tuna is more efficient due to lower TBF at speed, these sharks have a lower cost per tailbeat (Figure 7.3B). This higher cost of propulsion for the tuna is likely attributable to higher SMR. Lemon,

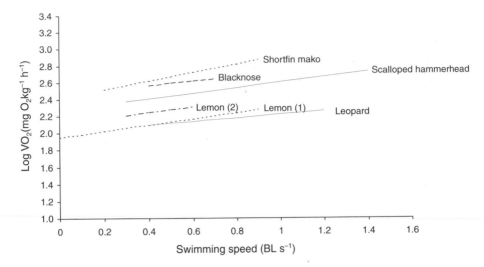

FIGURE 7.2 The relationship of swimming speed and log-transformed metabolic rate of shortfin mako shark (Graham et al., 1990), blacknose shark (Carlson et al., 1999), scalloped hammerhead shark (Lowe, 2001), lemon shark, and leopard shark (Scharold et al., 1989). Relationships are shown over speeds at which data were collected. Lemon (1) refers to data collected by Bushnell et al. (1989) and Lemon (2) for data collected by Scharold and Gruber (1991). Estimates for scalloped hammerhead, blacknose, and lemon shark (2) were collected at 25 to 28°C while shortfin mako, leopard, and lemon (1) were collected between 16 and 22°C.

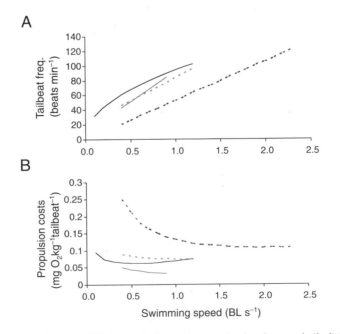

FIGURE 7.3 (A) Relationship between TBF (beats min⁻¹) over a range of swimming speeds (l s⁻¹) for scalloped hammerhead sharks (dashed gray line) (0.5 kg at 26°C; Lowe, 1996), lemon sharks (solid black line) (1.2 kg at 22°C; Graham et al., 1990), leopard sharks (solid gray line) (2 to 5 kg at 16°C; Scharold et al., 1989), and yellowfin tuna (dashed black line) (2 kg at 24°C; Dewar and Graham, 1994). (B) Relationship between propulsion cost (mg O₂ kg⁻¹ tailbeat⁻¹) over a range of swimming speeds (l s⁻¹) for the same species.

blacknose, and scalloped hammerhead sharks are characterized by a high aspect ratio caudal fin, high values of dorsal thrust angle, and a moderate heterocercal angle in the tail. The body is fusiform and moderately deep with very large pectoral fins. Sharks with these characters are considered a Group 2 body form (Thompson and Simanek, 1977). Although most swimming speed and metabolic rate relationships have been determined for this body form of shark, it is likely that sharks with less fusiform body, low tail, and dorsal fin insertion more posterior (Groups 3 and 4; Thompson and Simanek, 1977) will have higher energetic costs with increasing swimming speed.

Interestingly, the relationship of relative swimming speed to metabolic rate for shortfin mako shark demonstrates a greater rate of increase of metabolic rate with swimming speed (Graham et al., 1990), suggesting less efficient swimming. This contrasts to the hypothesis that shortfin mako shark energetic capacities and swimming performance approach those of tuna. However, this relationship is based on one individual over limited speeds. Based on other physiological and morphological evidence (Bernal et al., 2001a; Section 7.5), it is likely that shortfin mako sharks (and lamnid sharks overall) possess swimming efficiencies close to tunas.

7.4.2 Critical Swimming Speed and Sustainable Swimming

Critical swimming speed, an index of aerobically sustainable swimming capacity, has been determined only for leopard, lemon (Graham et al., 1990), and scalloped hammerhead sharks (Lowe, 1996). Critical swimming speeds were found to be comparable (~0.9 to 1.7 bl s^{-1}) for sharks of similar lengths (50 to 70 cm total length). The similarity in critical swimming speed among these three species is surprising given leopard sharks have a body design more adapted for a sedentary, demersal life, whereas lemon and scalloped hammerhead sharks have body designs adapted for cruising (Thompson and Simanek, 1977). For scalloped hammerhead sharks, one possible explanation for low critical swimming speeds is the effects of the swimming tunnel (see Section 7.2.1.2) on estimates of critical swimming speed (Lowe, 1996). However, Dickson et al. (1993) found no significant differences in red muscle citrate synthase levels between blue and leopard shark, suggesting similar aerobic capacities despite differences in body design and ecology.

Although ectothermic elasmobranchs lack large quantities of red muscle for sustained swimming, evidence suggests that white muscle contributes to intermediate-speed sustained swimming. Moreover, the division of labor among red and white muscle may be more interchangeable for elasmobranchs. A study on endurance training in leopard sharks found increases in citrate synthase, lactate dehydrogenase, and muscle fiber diameter, suggesting that white muscle can be used for sustained swimming (Gruber and Dickson, 1997).

7.4.3 Cost of Transport

The overall impact of swimming and energy costs (maintenance, SDA, and locomotion) is expressed as the total cost of transport (cal g^{-1} km^{-1}; Schmidt-Nielsen, 1972). Total cost of transport examines use of all energy available. Within a species, larger sharks have a lower cost of transport than smaller sharks (Figure 7.4). For example, Parsons (1990) determined the total cost of transport for a 0.9 kg bonnethead was 1.21 cal g^{-1} km^{-1}, whereas an 8.0 kg bonnethead would expend only 0.4 cal g^{-1} km^{-1}, when swimming at their theoretical optimal velocities (Weihs, 1977). Total cost of transport generally decreases with mass by an exponent of about 0.3 (Schmidt-Nielsen, 1984). It is worth noting that, despite differences in body shape and swimming mode, the total energetic cost of transport appears to be similar across a variety of teleost species (Schmidt-Nielsen, 1984). This relationship indicates that it is more efficient for a large shark to transport 1 kg of body mass over a given distance than it is for a small shark. Among endotherms, this is likely because the mass-specific metabolic rate of a large animal is lower than that of a smaller animal due to lower surface-to-volume ratios for larger animals (Schmidt-Neilsen, 1984). The relationship in sharks and bony fishes could be because larger sharks have higher optimal swimming speeds, presumably attainable due to their increased stride length (Videler and Nolet, 1990).

Total cost of transport demonstrates a U-shaped relationship when plotted against swimming speed. Total costs of transport are initially high, because swimming speed (U) is too slow to overcome inertial

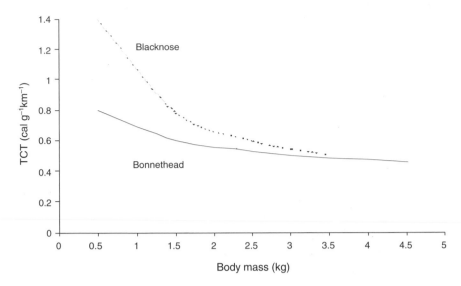

FIGURE 7.4 The effect of total cost of transport and body mass for bonnethead and blacknose shark. (Data are from Parsons, 1990, and Carlson et al., 1999.)

drag. As U increases, a fish overcomes inertial drag and minimizes friction drag. However, as U exceeds this threshold, friction drag will substantially increase (at the rate of $U^{2.5-2.8}$) and result in an increased swimming cost (Videler and Nolet, 1990). This trend has been shown for a variety of species including blacknose shark (Carlson et al., 1999), sockeye salmon (Brett, 1963), white crappie, *Pomoxis annularis* (Parsons and Sylvester, 1992), and yellowfin tuna (Dewar and Graham, 1994). At similar swimming speeds, total cost of transport is generally higher for more active sharks, likely due to the influence of higher standard metabolic rates.

When comparing only costs of swimming, the net cost of transport (the difference between standard and active metabolic rates) is the preferred variable. Some sharks have been shown to demonstrate higher energetic costs at slower swimming speeds. Lowe (2001) proposed the higher net cost of transport for hammerhead sharks at slower speeds could be due to increased drag created by the wing-shaped head of the hammerhead shark, which forces the sharks to swim at suboptimal velocities; whereas at intermediate speeds the shape of the head could increase hydrostatic lift, thereby decreasing energetic costs. Gruber and Dickson (1997) found higher energetic costs when forcing leopard sharks to swim at slower speeds but could not discern whether these costs were due to the respirometer (see Section 7.2.1.2) or to inefficient swimming at slower speeds. Lowe (2002) found that laboratory measures of optimal swimming speed (speed at the lowest net cost of transport; 0.75 bl s^{-1}) were similar to typical swimming speeds of free-ranging sharks (0.81 bl s^{-1}) in their natural environment. At optimal swimming speed, juvenile sharks swimming at 0.8 bl s^{-1} would only be operating at 25% of their metabolic scope. This scope was very similar to that estimated for lemon sharks even though they exhibited an optimal swimming speed of 0.4 bl s^{-1} (Bushnell et al., 1989).

7.5 Endothermy

7.5.1 Background

The steady-state body temperature of most fishes is similar to ambient water temperature as a result of the linkage between aerobic heat production and heat loss via the gills and body surface (Brill et al., 1994). However, lamnid sharks have the capacity to conserve metabolic heat via vascular countercurrent heat exchangers (*retia mirabilia*), thereby maintaining a steady-state body temperature that is elevated over ambient water temperature (Carey et al., 1981, 1985; Goldman, 1997; Bernal et al., 2001a; Goldman et al., in press). *Retia* in lamnid sharks are located in the cranium near the eyes (orbital *retia*), in locomotor musculature (lateral cutaneous *retia*), and the viscera (suprahepatic *rete*). *Lamna* spp. possess an additional visceral *rete* (kidney *rete*). Lamnids have developed a distinct anterior and medial red muscle position, in conjunction with evolution of *retia* and elevated body temperatures, with the red muscle internalized and lying close to the spine instead of near the body wall as in ectothermic fishes (Carey et al., 1985; Bernal et al., 2001a).

The average body core temperature of lamnid sharks ranges between 22 and 26.6°C, depending on species (Lowe and Goldman, 2001). The maximum reported elevation of body temperature over ambient water temperature is 8.0°C for shortfin mako sharks (Carey et al., 1981), 14.3°C for white sharks (Goldman, 1997), and 21.2°C for salmon sharks, *L. ditropis* (Goldman, 2002; Goldman et al., in press). Lamnid sharks not only possess elevated body temperatures, but also regulate body temperature via physiological means (i.e., regulate heat balance by altering their whole-body thermal rate coefficient, *k*), at least in mako and salmon sharks (Bernal et al., 2001b; Goldman, 2002; Goldman et al., in press).

Several other elasmobranch species have been shown to possess *retia*. Alopiid sharks (threshers) not only possess *retia*, but also have anterior-medial red muscle placement similar to lamnids, a lateral circulation pattern to the red muscle, and exhibit evidence of endothermy (Carey et al., 1971; Bone and Chubb, 1983; Bernal, pers. comm.; Goldman, unpubl. data). Three species of myliobatoid rays possess *retia* (Alexander, 1995, 1996); however, no temperature measurements have been obtained from these species, so their body temperatures and thermoregulatory abilities (if any) are still unknown.

7.5.2 Indirect Calorimetry: Endotherms vs. Ectotherms

Because lamnid sharks are endotherms, they should have higher SMRs than ectothermic sharks, as endothermy increases the total aerobic capacity of an organism. To date, the only lamnid metabolic data to support this hypothesis comes from a single 3.9 kg mako shark with an SMR of 240 mg O_2 kg^{-1} h^{-1} at 16 to 20°C (Graham et al., 1990). However, this value was extrapolated to zero swimming speed, which may not be the most representative metabolic rate estimate for an obligate ram ventilator, whose swimming speed is never zero. Even with the difficulties of controlling for water temperature, animal weight, respirometer type, and swimming speed, comparisons of routine metabolic rates (RMR) would be more meaningful when examining obligate ram ventilators. Additionally, comparing obligate ram ventilators to other obligate ram ventilators would be more meaningful than extrapolating their RMR to zero for comparison to non-obligate ram ventilating species (e.g., lemon sharks). Aside from the shortfin mako shark, the only other obligate ram ventilating sharks for which VO_2 data exist are for two ectothermic species; bonnethead (Parsons, 1990) and blacknose sharks (Carlson et al., 1999). Both studies tested sharks of similar sizes (up to 4.7 kg for bonnethead and 3.5 kg for blacknose sharks) to the 3.9 kg shortfin mako shark tested by Graham et al. (1990).

Routine metabolic rate of the endothermic shortfin mako shark swimming at 24.6 cm s^{-1} was 262 mg O_2 kg^{-1} h^{-1} in 16°C water and 507 mg O_2 kg^{-1} h^{-1} in 20°C water (Graham et al., 1990). Mean RMR over the course of the 36 h experiment (mean swimming speed = 24.6 cm s^{-1}) was 369 mg O_2 kg^{-1} h^{-1}. In contrast, mean RMR for a 3.5 kg ectothermic blacknose was lower (278.5 mg O_2 kg^{-1} h^{-1}) than for the shortfin mako even though the blacknose swam in considerably warmer water (28°C) and at a faster mean swimming speed (31.4 cm s^{-1}; Carlson et al., 1999). Parsons (1990) studied RMR of bonnethead sharks ranging in size from 0.095 to 4.7 kg. Using his equation to estimate VO_2, a 3.9 kg bonnethead, at 25°C, would have a VO_2 of 195.5 mg O_2 kg^{-1} h^{-1}. As with the blacknose shark, the estimated RMR of this ectothermic obligate ram ventilator was less than that of the endothermic shortfin mako. Had water temperatures been the same in all studies, the mako's RMR would be even higher at 28°C or, oppositely, bonnethead RMR would be lower at 16 to 20°C. Although no direct comparisons for weight, swimming speed, temperature, and respirometer type can be made between endothermic and ectothermic elasmobranchs, it appears that endothermic sharks possess higher metabolic rates than ectothermic sharks under similar conditions.

7.5.3 Indirect Evidence of Higher Metabolic Rates in Endothermic Sharks

Along with the evolution of *retia* and endothermy, lamnid sharks possess several characteristics indicating they have high aerobic and anaerobic capacities, and higher metabolic rates than ectothermic sharks. These features show a remarkable evolutionary convergence with endothermic tunas and reflect specializations related to efficient, high-performance swimming and an active lifestyle (see Bernal et al., 2001a, for thorough review).

In addition to red muscle that is internalized with anterior-medial placement, lamnid sharks also show a partial separation (shear) between adjacent red and white muscle, making red muscle free to contract relative to white muscle during slow-speed swimming (Carey et al., 1985; Bernal et al., 2001a). Ectothermic elasmobranchs do not possess this feature; their white muscle appears to contribute to intermediate-speed sustainable swimming, as it is connected to externalized red muscle (and skin). It has been predicted that this "muscle shear" characteristic in lamnids, along with the distinct red muscle position in the body cavity, may decrease energy output requirements and enhance swimming performance by allowing the red muscle to transfer power directly to the caudal peduncle and caudal fin (Bernal et al., 2001a).

High-performance swimming adaptations in lamnid sharks include features that enhance uptake (large gill-surface area) and delivery of a large amount of O_2 to the red muscle, including large heart, and blood hemoglobin and hematocrit levels similar to those of birds and mammals (Emery, 1985, 1986; Emery and Szczepanski, 1986; Oikawa and Kanda, 1997; Tota et al., 1983; Tota, 1999; Bernal, pers. comm.). Elevated red and white muscle temperatures speed the contraction–relaxation cycle and increase muscle power output, which may result in faster cruising speeds (Johnston and Brill, 1984; Dickson et al., 1993; Altringham and Block, 1997; Bernal, pers. comm.).

Lamnid sharks also have been shown to possess modified biochemical characteristics in white myotomal muscle and heart ventricle that enhance greater aerobic and anaerobic metabolic capacities (Dickson et al., 1993; Bernal et al., 2001a; Bernal, pers. comm.). Compared with other active sharks, shortfin mako possesses higher white muscle activities of citrate synthase, lactate dehydrogenase, and creatine phosphokinase (Dickson et al., 1993; Bernal et al., 2001a; Bernal, pers. comm.). Thus, lamnid sharks appear to have high aerobic and anaerobic scopes and a high capacity for anaerobic ATP production during burst swimming. Although no modifications of biochemical characteristics have been found in red muscle, at *in vivo* temperatures both the shortfin mako shark and salmon shark have been estimated to increase red muscle enzyme activities by 48 and 123%, respectively (Bernal et al., 2003).

Elasmobranchs and teleosts deal with similar acid loads in the blood during and after periods of high exertion, but elasmobranchs have a lower capacity to buffer acid at the site of production and differ from teleosts in their tolerance to blood acidification (Dickson et al., 1993). Wells and Davies (1985) found that hemoglobin-O_2 binding and blood O_2-carrying capacity in the shortfin mako shark were not significantly affected by a large drop in blood pH that occurred after periods of high activity. The higher hemoglobin content of blood and the ability to uptake and deliver more oxygen to muscle may reduce oxygen debt, or decrease the amount of time necessary to offset the oxygen debt after periods of sustained burst swimming (Dickson et al., 1993; Bernal et al., 2001a).

Stable tissue temperatures would conserve metabolic function during ambient temperature changes. The thermal buffer created by metabolic heat retention may reduce the impact of ambient temperature fluctuations on metabolic rates of young (small) lamnid sharks, and the buffer may eliminate any effects on larger individuals because they possess greater thermal inertia (Bernal et al., 2001a; Goldman et al., in press). This thermal buffer likely permits lamnids to exploit cool and boreal waters.

7.6 Environmental Effects on Metabolism

7.6.1 Temperature

Ambient temperature is a key variable and plays a major role in controlling metabolic rates of ectothermic elasmobranchs, whereas fluctuations in ambient temperature may have a reduced or no impact on the metabolic rates of endothermic sharks. Metabolic rate typically increases by a Q_{10} of 2 to 3 for every 10°C rise in temperature, although this rate varies among species (Brett and Groves, 1979; Schmidt-Nielsen, 1983). Du Preez et al. (1988) reported a Q_{10} response of 2.27 between 15 and 25°C for guitarfish whereas the Q_{10} response for bullray was 1.87 between 10 and 25°C. Hopkins and Cech (1994) determined a Q_{10} of 6.8 for bat rays over a temperature range of 14 to 20°C. Among sharks, bonnetheads were found to have a Q_{10} of 2.34 at 20 to 30°C (Carlson and Parsons, 1999), while scalloped hammerhead sharks have a Q_{10} of 1.34 at 21 to 29°C (Lowe, 2001). These trends suggest interspecific differences in Q_{10} among elasmobranchs, but length of acclimation (e.g., acute or seasonal) at each experimental temperature may result in an increased or decreased sensitivity to metabolic rate. For example, Hopkins and Cech (1994) determined a Q_{10} for bat rays exposed to acute changes in temperature, while Carlson and Parsons (1999) and Lowe (2001) measured changes in oxygen consumption rate sensitivity to seasonally acclimatized sharks.

As pointed out by Brett (1971), most ectothermic fishes are thermal conformers and generally inhabit an optimal temperature range between upper and lower lethal temperatures. The optimal temperature range is thought to be where physiological rates (e.g., metabolism, growth, digestion) would be optimized to enhance fitness. Recent studies using telemetry suggest that elasmobranchs found in thermally heterogeneous environments will feed in warmer waters and rest in cooler waters. Matern et al. (2000) proposed that bat rays took advantage of their elevated metabolism by feeding in warmer waters then moving to cooler waters to lower metabolism (i.e., energetic demands) and possibly gastric evacuation rate while maintaining assimilation efficiency. Although this behavioral thermoregulation hypothesis has only been proposed for bat rays, the possibility exists that other species exhibiting diel movements may be taking advantage of thermally heterogeneous environments. For example, blue sharks displayed daily vertical dives from the surface to depths of 250 m and experienced water temperature changes of 7 to 9°C (Carey and Scharold, 1990).

7.6.2 Salinity

Most species of elasmobranchs are found in marine environments and likely would not encounter radical changes in salinity. However, bull shark, *Carcharhinus leucas* (Snelson et al., 1984), and several species of rays (Schwartz, 1995; Meloni et al., 2002) are found in brackish waters. Despite the hypothesis that elasmobranchs are osmoconformers (using solutes to maintain osmolarity) studies on Atlantic stingray, *Dasyatis sabina*, suggest osmoregulatory energy costs associated with decreases in salinity (Janech and Piermarini, 1997; Janech et al., 1998; Piermarini and Evans, 2000).

Increasing osmoregulatory costs could raise SMR (Brett and Groves, 1979). Evidence for this was provided by Meloni et al. (2002) for 0.4 to 1.7 kg bat rays exposed to various levels of salinity: SMR increased from 12.6 mg O_2 h^{-1} at 33 and 36‰ to 24.1 mg O_2 h^{-1} at 15 and 25‰.

7.6.3 Dissolved Oxygen

Oxygen levels throughout marine environments vary in relation to depth, productivity, time of day, and other factors. Sharks have been captured in areas with decreased dissolved oxygen levels suggesting sharks encounter and deal with areas of low dissolved oxygen (Grace and Henwood, 1998; Carlson and Parsons, 2001). Metabolic responses of elasmobranchs to oxygen depletion differ among species depending on behavior and physiology. Metabolic rate and activity were found to decrease in response to low dissolved oxygen in spotted dogfish and Florida smoothhound, *Mustelus norrisi*, species that increase buccal pumping rate to augment the flow of water over the gills (Metcalf and Butler, 1984; Carlson and Parsons, 2001). Reduction in activity during hypoxic exposure is thought to reduce energy expenditure, as a considerable amount of energy may be used for swimming. Energy saved may then be dedicated to additional respiratory needs such as increased buccal pumping rate.

The behavioral response to hypoxia for obligate ram-ventilating sharks is to increase swimming speed and metabolism (Parsons and Carlson, 1998; Carlson and Parsons, 2001). The increase in metabolism has not been determined to be independent of or dependent on hypoxia, but increased swimming speed as a mechanism for regulating respiration would appear to be metabolically costly and would seem to increase the problem of obtaining sufficient oxygen to meet increased swimming speed. Increased swimming speed could be a flight response as determined for tunas (Bushnell and Brill, 1991). However, Parsons and Carlson (1998) suggested that because sharks have reduced metabolic demands with respect to tunas, increased swimming speed and gape may be energetically similar to other mechanisms for oxygen regulation such as increased buccal pumping rates found in non-obligate ram-ventilator species.

7.6.4   Time of Day

Elasmobranchs exhibit changes in diurnal activity patterns. Higher activity levels at night have been reported for horn, *Heterodontus francisci,* and swell sharks, *Cephaloscylium ventriosum* (Nelson and Johnson, 1970) and for lemon shark (Gruber et al., 1988) *in situ,* suggesting these animals are nocturnal. Lesser spotted dogfish (Sims et al., 1993), bonnethead (Parsons and Carlson, 1998), and little skate (Hove and Moss, 1997) increased swimming at night or under dark conditions. Lowe (2002) found that juvenile scalloped hammerhead sharks significantly increased their swimming speed at night and thus incurred a higher metabolic cost. Nixon and Gruber (1988), Sims et al. (1993), and Hove and Moss (1997) measured increases in metabolic rate coinciding with increased activity. Although activity was not measured, Du Preez et al. (1988) also found nocturnal peaks in routine oxygen consumption rates for guitarfish and bull ray. In these studies, elasmobranchs were exposed to various cycles of light and dark that suggest activity is controlled by an exogenous circadian rhythm influenced by light. Further, experiments conducted on blacknose shark, bonnethead shark, and Florida smoothhound under constant light found no predictable changes in swimming speed and oxygen consumption rate with time of day (Carlson and Parsons, 2001).

Increases in activity and metabolism are likely influenced by stimulation of the pineal organ, which causes the secretion of melatonin and, in fishes, melatonin influences almost all body processes including locomotion, skin color, and reproductive cycle (Bonga, 1993). Sharks possess a pigment-free patch of skin over the epiphysis in the chondrocranium, which could allow for light transmission. Gruber et al. (1975) noted an area of reduced opacity in the top of the chondrocraniums in lemon and bull sharks and in smooth dogfish.

7.7   Conclusions and Future Directions

Most studies on elasmobranch metabolism have been concerned with the juvenile stage and with species confined to coastal areas. Little or no research on adult or pelagic species has been performed despite evidence that metabolism varies by species, size, and life stage. Estimates of metabolism for a variety of ecologically diverse species are becoming increasingly important because bioenergetics have applications to population and ecosystem modeling (Kitchell et al., 2002; Lowe, 2002; Schindler et al., 2002).

Despite the obvious problems of the large size of sharks and construction of a respirometer large enough to accommodate these highly active species, new techniques need to be developed to obtain estimates of metabolism. In lieu of constructing very large water tunnels, mathematical models show promise. Using independent estimates of lower threshold prey densities, Sims (2000) developed a threshold foraging behavior model for estimation of metabolic rate in basking sharks, *Cetorhinus maximus.* The "best estimates" of RMR from Sims' threshold model were 62 to 91 mg O_2 kg^{-1} h^{-1} for a 1000 kg basking shark, which agreed fairly well with his shark and fish VO_2–mass scaling relationships. However, the large number of estimators used in Sims' (2000) threshold model also creates a large uncertainty in his estimates. Nevertheless, the continuing development of models may lead to a viable way of estimating metabolic rates of large sharks in the field and to subsequent comparisons of large endothermic and ectothermic species.

Because elasmobranchs are considered to be iso-osmotic to seawater and the resulting water flux rate would be predicted to be negligible, Parsons and Carlson (unpubl. data) proposed to use the doubly labeled water method (Nagy, 1987) to estimate field metabolic rates in sharks. Unfortunately, bonnetheads exposed to two levels of salinity (30 and 25‰) experienced high levels of water flux. Sharks injected with tritiated water had no detectable levels of the isotope within the blood after 30 min from injection. These results suggest that this method may be appropriate only when salinity remains high (≥35‰).

The increasing sophistication and technology associated with biotelemetry likely hold the most promise for determining metabolism of these larger species (Lowe and Goldman, 2001). Biotelemetry may be particularly useful for species that are large and difficult to maintain in captivity. In addition, there is a great need for bridging the gap between laboratory and field studies. Development of new physiological sensors and transponding systems may greatly facilitate collection of energetics data for elasmobranch species. Elasmobranch models may also provide the best insight into our understanding of free-ranging fish physiology, because these animals are large enough to carry integrated transmitter packages capable of recording environmental data simultaneously.

Many of the comparisons that we have made among and within taxa are limited. Variability in experimental temperature, mass effects, and experimental design and apparatus make comparisons difficult. To provide a better picture of the energetics and metabolic capacities of sharks, comparisons among taxa must be performed using similarly sized animals that control for temperature under identical experimental protocols. Graham et al. (1990) examined relationships in three species of sharks varying in body form, activity level, and physiology using the same protocols and revealed many of the trends reported herein.

Much remains to be learned about the energetics of elasmobranchs. It is evident that species adapted for continuous activity possess higher energetic capacities, but the details of the swimming performance and its relation to aerobic and anaerobic capacities remain to be quantified. Obtaining large sample sizes will always prove difficult with these animals. Improvements in experimentation techniques, capture, and husbandry of elasmobranchs will aid in elucidating energetic relationships.

References

Adams, S.R. and G.R. Parsons. 1998. Laboratory based measurements of swimming performance and related metabolic rates of field sampled smallmouth buffalo (*Ictiobus bubulus*): a study of seasonal changes. *Phys. Zool.* 71:350–358.

Alexander, R.L. 1995. Evidence of a counter-current heat exchanger in the ray *Mobula tarapacana* (Chondrichthyes: Elasmobranchii: Batiodea: Myliobatiformes). *J. Zool. Lond.* 237:377–384.

Alexander, R.L. 1996. Evidence of brain-warming in the mobulid rays, *Mobula tarapacana* and *Manta birostris* (Chondrichthyes: Elasmobranchii: Batiodea: Myliobatiformes). *J. Linn. Soc.* 188:151–164.

Altringham, J.D. and B.A. Block. 1997. Why do tuna maintain elevated slow muscle temperatures? Power output of muscle isolated from endothermic and ectothermic fish. *J. Exp. Biol.* 200:2617–2627.

Bainbridge, R. 1958. The speed of swimming of fish as related to size and to the frequency and amplitude of the tail beat. *J. Exp. Biol.* 35:1183–1226.

Beamish, F.W.H. 1970. Oxygen consumption of largemouth bass, *Micropterus salmoides*, in relation to swimming speed and temperature. *Can. J. Zool.* 48:1221–1228.

Bernal, D., K.A. Dickson, R.E. Shadwick, and J.B. Graham. 2001a. Review: analysis of the evolutionary convergence for high performance swimming in lamnid sharks and tunas. *Comp. Biochem. Physiol.* 129A: 695–726.

Bernal, D., C. Sepulveda, and J.B. Graham. 2001b. Water tunnel studies of heat balance in swimming mako sharks. *J. Exp. Biol.* 204:4043–4054.

Bernal, D., D. Smith, G. Lopez, D. Weitz, T. Grimminger, K. Dickson, and J.B. Graham. 2003. Comparative studies of high performance swimming in sharks. II. Metabolic biochemistry of locomotor and myocardial muscle in endothermic and ectothermic sharks. *J. Exp. Biol.* 206:2845–2857.

Block, B.A., H. Dewar, T. Williams, E.D. Prince, C. Farwell, and D. Fudge. 1998. Archival tagging of Atlantic bluefin tuna, *Thunnus thynnus. Mar. Tech. Soc. J.* 32:37–46.

Boggs, C.H. 1984. Tuna Bioenergetics and Hydrodynamics. Ph.D. dissertation. University of Wisconsin–Madison, 115 pp.

Bone, Q. and A.D. Chubb. 1983. The retial system of the locomotor muscle in the thresher shark. *J. Mar. Biol. Assoc. U.K.* 63:239–241.

Bonga, S.E.W. 1993. Endocrinology, in *Fish Physiology.* D.H. Evans, Ed., CRC Press, Boca Raton, FL, 469–502.

Bosclair, D. and M. Tang. 1993. Empirical analysis of the influence of swimming pattern on the net energetic cost of swimming in fishes. *J. Fish. Biol.* 42:169–183.

Brett, J.R. 1963. The energy required for swimming by young sockeye salmon with a comparison of the drag force on a dead fish. *Trans. R. Soc. Can.* 27:1637–1652.

Brett, J.R. 1964. The respiratory metabolism and swimming performance of young sockeye salmon. *J. Fish. Res. Board Can.* 21:1183–1225.

Brett, J.R. 1971. Energetic responses of salmon to temperature. A study of some thermal relations in the physiology and freshwater ecology of sockeye salmon (*Oncorhynchus nerka*). *Am. Zool.* 11:99–113.

Brett, J.R. and J.M. Blackburn. 1978. Metabolic rate and energy expenditure of the spinydogfish, *Squalus acanthias. J. Fish. Res. Board Can.* 35:816–821.

Brett, J.R. and T.D.D. Groves. 1979. Physiological energetics, in *Fish Physiology,* Vol. 8. W.S. Hoar and D.J. Randall, Eds., Academic Press, New York, 279–352.

Briggs, C.T. and J.R. Post. 1997. *In situ* activity metabolism of rainbow trout (*Oncorhynchus mykiss*): estimates obtained from telemetry of axial muscle electromyograms. *Can. J. Fish. Aquat. Sci.* 54:859–866.

Brill, R.W. 1987. On the standard metabolic rate of tropical tunas, including the effect of body size and acute temperature change. *Fish. Bull. U.S.* 85:25–35.

Brill, R.W. 1996. Selective advantage conferred by the high performance physiology of tunas, billfishes, and dolphin fish. *Comp. Biochem. Physiol.* 113A:2–15.

Brill, R.W., H. Dewar, and J.B. Graham. 1994. Basic concepts relevant to heat transfer in fishes, and their use in measuring the physiological thermoregulatory abilities of tunas. *Environ. Biol. Fish.* 40:109–124.

Brill, R.W., B.A. Block, C.H. Boggs, K.A. Bigelow, E.V. Freund, and D.J. Marcinek. 1999. Horizontal movements and depth distribution of large adult yellowfin tuna, *Thunnus albacores*, near the Hawaiian Islands, recorded using ultrasonic telemetry: implications for the physiological ecology of pelagic fishes. *Mar. Biol.* 133:395–408.

Brill, R.W., Y. Swimmer, C. Taxboel, K. Cousins, and T. Lowe. 2001. Gill and intestinal N^+-K^+ ATPase activity, and estimated maximal osmoregulatory costs on three high-energy-demand teleosts: yellowfin tuna (*Thunnus albacares*), skipjack tuna (*Katsuwonus pelamis*), and dolphin fish (*Coryphaena hippurus*). *Mar. Biol.* 138:935–944.

Brown, C.E. and B.S. Muir. 1970. Analysis of ram ventilation of fish gills with application to skipjack tuna (*Katsuwonus pelamis*). *J. Fish. Res. Board. Can.* 27:1637–1652.

Bushnell, P.G. and R.W. Brill. 1991. Responses of swimming skipjack *Katsuwonus pelamis* and yellowfin *Thunnus albacares* tunas to acute hypoxia, and a model of their cardiovascular function. *Physiol. Zool.* 64:787–811.

Bushnell, P.G., P.L. Lutz, and S.H. Gruber. 1989. The metabolic rate of an active, tropical elasmobranch, the lemon shark (*Negaprion brevirostris*). *Exp. Biol.* 48:279–283.

Carey, F.G. and K.D. Lawson. 1973. Temperature regulation in free-swimming bluefin tuna. *Comp. Biochem. Physiol.* 44A:375–392.

Carey, F.G. and B.H. Robison. 1981. Daily patterns in the activities of swordfish, *Xiphias gladius*, observed by acoustic telemetry. *Fish. Bull. U.S.* 79:277–292.

Carey, F.G. and J.V. Scharold. 1990. Movements of blue sharks (*Prionace glauca*) in depth and course. *Mar. Biol.* 106:329–342.

Carey, F.G., J.M. Teal, J.W. Kanwisher, K.D. Lawson, and J.S. Beckett. 1971. Warm-bodied fish. *Am. Zool.* 11:137–145.

Carey, F.G., J.M. Teal, and J.W. Kanwisher. 1981. The visceral temperatures of mackerel sharks (Lamnidae). *Physiol. Zool.* 54:334–344.

Carey, F.G., J.W. Kanwisher, O. Brazier, G. Gabrielson, J.G. Casey, and H.L. Pratt, Jr. 1982. Temperature and activities of a white shark, *Carcharodon carcharias. Copeia* 1982:254–260.

Carey, F.G., J.G. Casey, H.L. Pratt, D. Urquhart, and J.E. McCosker. 1985. Temperature, heat production, and heat exchange in lamnid sharks. *South. Calif. Acad. Sci. Mem.* 9:92–108.

Carlson, J.K. 1998. The Physiological Ecology of the Bonnethead Shark, *Sphyrna tiburo*, Blacknose Shark, *Carcharhinus acronotus*, and Florida Smoothhound Shark, *Mustelus norrisi*: Effects of Dissolved Oxygen and Temperature. Ph.D. dissertation, University of Mississippi, Oxford, 106 pp.

Carlson, J.K. and G.R. Parsons. 1999. Seasonal differences in routine oxygen consumption rates of the bonnethead shark. *J. Fish Biol.* 55:876–879.

Carlson, J.K. and G.R. Parsons. 2001. The effects of hypoxia on three sympatric shark species: physiological and behavioral responses. *Environ. Biol. Fish.* 61:427–433.

Carlson, J.K. and G.R. Parsons. 2003. Respiratory and hematological responses of the bonnethead shark, *Sphyrna tiburo*, to acute changes in dissolved oxygen. *J. Exp. Mar. Biol. Ecol.* 294:15–26.

Carlson, J.K., C.P. Palmer, and G.R. Parsons. 1999. Oxygen consumption rate and swimming efficiency of the blacknose shark, *Carcharhinus acronotus*. *Copeia* 1999:34–39.

Castro, J.I. 1996. Biology of the blacktip shark, *Carcharhinus limbatus*, off the southeastern United States. *Bull. Mar. Sci.* 59:508–522.

Cech, J.J. 1990. Respirometry, in *Methods of Fish Biology*. C.B. Schreck and P.B. Moyle, Eds., American Fisheries Society, Bethesda, MD, 335–362.

Cortés, E. 1999. Standardized diet compositions and trophic levels of sharks. *ICES J. Mar. Sci.* 56:707–717.

Dewar, H. and J.B. Graham. 1994. Studies of tropical tuna swimming performance in a largewater tunnel. I. Energetics. *J. Exp. Biol.* 192:13–31.

Dickson, K.A. 1996. Locomotor muscle of high performance fishes: what do comparisons of tunas with other ectothermic taxa reveal? *Comp. Biochem. Physiol.* 113A:39–49.

Dickson, K.A., M.O. Gregorio, S.J. Gruber, K.L. Loefler, M. Tran, and C. Terrel. 1993. Biochemical indices of aerobic and anaerobic capacity in muscle tissues of California elasmobranch fishes differing in typical activity level. *Mar. Biol.* 117:185–193.

Diez, J.M. and J. Davenport. 1987. Embryonic respiration in the dogfish (*Scyliorhinus canicula* L.). *J. Mar. Biol. Assoc. U.K.* 67:249–261.

Du Preez, H.H., A. McLachlan, and J.F.K. Marias. 1988. Oxygen consumption of two nearshore elasmobranchs, *Rhinobatus annulatus* (Muller & Henle, 1841) and *Myliobatis aquila* (Linnaeus, 1758). *Comp. Biochem. Physiol.* 89A:283–294.

Emery, S.H. 1985. Hematology and cardiac morphology in the great white shark, *Carcharodon carcharias*. *Mem. South. Calif. Acad. Sci.* 9:73–80.

Emery, S.H. 1986. Hematological comparisons of endothermic vs. ectothermic elasmobranch fishes. *Copeia* 1986:700–705.

Emery, S.H. and A. Szczepanski. 1986. Gill dimensions in pelagic elasmobranch fishes. *Biol. Bull.* 171:441–449.

Ezcurra, J.M. 2001. The Mass-Specific Routine Metabolic Rate of Captive Pelagic Stingrays, *Dasyatis violacea*, with Comments on Energetics. M.S. thesis, Moss Marine Laboratory, California State University, Stanislaus, 64 pp.

Farrell, A.P. 1991. From hagfish to tuna — a perspective on cardiac function. *Physiol. Zool.* 64:1137–1164.

Fournier, R.W. 1996. The Metabolic Rates of Two Species of Benthic Elasmobranchs, Nurse Sharks and Southern Stingrays. M.S. thesis, Hofstra University, Hempstead, NY, 29 pp.

Fry, F.E.J. 1957. The aquatic respiration of fish, in *The Physiology of Fishes*, Vol. 1. M.E. Brown, Ed., Academic Press, New York, 1–63.

Goldman, K.J. 1997. Regulation of body temperature in the white shark, *Carcharodon carcharias*. *J. Comp. Phys. B* 167:423–429.

Goldman, K.J. 2002. Aspects of Age, Growth, Demographics and Thermal Biology of Two Lamniform Shark Species. Ph.D. dissertation, College of William and Mary, School of Marine Science, Virginia Institute of Marine Science, Williamsburg, 220 pp.

Goldman, K.J., S.D. Anderson, R.J. Latour, and J.A. Musick. In press. Homeothermy in adult salmon sharks, *Lamna ditropis*. *Environ. Biol. Fish.*

Grace, M. and T. Henwood. 1998. Assessment of the distribution and abundance of coastal sharks in the U.S. Gulf of Mexico and eastern seaboard, 1995 and 1996. *Mar. Fish. Rev.* 59:23–32.

Graham, J.B., H. Dewar, N.C. Lai, W.R. Lowell, and S.M. Arce. 1990. Aspects of shark swimming performance determined using a large water tunnel. *J. Exp. Biol.* 151:175–192.

Gruber, S.H. 1984. Bioenergetics of captive and free-ranging lemon sharks. *AAZPA Ann. Conf. Proc.* 340–373.

Gruber, S.H., D.I. Hamasaki, and B.L. Davis. 1975. Window to the epiphysis in sharks. *Copeia* 1975:375–380.

Gruber, S.H., D.R. Nelson, and J.F. Morrissey. 1988. Patterns of activity and space utilization of lemon sharks, *Negaprion brevirostris*, in a shallow Bahamian lagoon. *Bull. Mar. Sci.* 43:61–77.

Gruber, S.J. and K.A. Dickson. 1997. Effects of endurance training in the leopard shark, *Triakis semifasciata*. *Physiol. Zool.* 70:481–492.

Hanson, D. and K. Johansen. 1970. Relationships of gill ventilation and perfusion in Pacific dogfish, *Squalus suckleyi*. *J. Fish. Res. Bd. Can.* 27:551–564.

Holland, K.N., R.W. Brill, and R.K.C. Chang. 1990. Horizontal and vertical movements of yellowfin and bigeye tuna associated with fish aggregating devices. *Fish. Bull. U.S.* 88:493–507.

Holts, D.B. and D.W. Bedford. 1993. Horizontal and vertical movements of the shortfin mako shark, *Isurus oxyrinchus*, in the southern California bight. *Aust. J. Mar. Freshwater Res.* 44:901–909

Hopkins, T.E. and J.J. Cech. 1994. Effect of temperature on oxygen consumption of the bat ray, *Myliobatis californica* (Chondrichthyes, Myliobatidae). *Copeia* 1994:529–532.

Hove, J.R. and S.A. Moss. 1997. Effect of MS-222 on response to light and rate of metabolism of the little skate *Raja erinacea*. *Mar. Biol.* 128:579–583.

Howe, J.C. 1990. Oxygen consumption rate in juvenile scalloped hammerhead sharks [*Sphyrna lewini* (Griffith and Smith)]: a preliminary study. *J. Aquaricult. Aquat. Sci.* 5:28–31.

Hunter, J.R. and Z.R. Zweifel. 1971. Swimming speed, tail beat frequency, tail beat amplitude, and size in jack mackerel, *Trachurus symmetricus*, and other fishes. *Fish. Bull. U.S.* 69:253–256.

Janech, M.G. and P.M. Piermarini. 1997. Urine flow rate and urine composition of freshwater Atlantic stingrays, *Dasyatis sabina*, from the St. Johns River, Florida. *Am. Zool.* 37:147A.

Janech, M.G., W.R. Fitzgibbon, D.H. Miller, E.R. Lacy, and D.W. Ploth. 1998. Effect of dilution of renal excretory function of the Atlantic stingray, *Dasyatis sabina*. *FASEB J.* 12:A423.

Jobling, M. 1981. The influences of feeding on the metabolic rate of fishes: a short review. *J. Fish. Biol.* 18:385–400.

Johnston, I.A. and R.W. Brill. 1984. Thermal dependence of contractile properties of single skinned muscle fibers from Antarctic and various warm water marine fishes including skipjack tuna (*Katsuwonus pelamis*) and kawakawa (*Euthynnus affinis*). *J. Comp. Physiol.* 155B:63–70.

Kitchell, J.F., T.E. Essington, C.H. Boggs, D.E. Schindler, and C.J. Walters. 2002. The role of sharks and longline fisheries in a pelagic ecosystem of the central Pacific. *Ecosystems* 5:202–216.

Lai, N.C., J.B. Graham, W.R. Lowell, and R. Shabetai. 1989. Elevated pericardial pressure and cardiac output in the leopard shark *Triakis semifasciata* during exercise: the role of the pericardioperitoneal canal. *J. Exp. Biol.* 147:263–277.

Lai, N.C., K.E. Korsmeyer, S. Katz, D.B. Holts, L.M. Laughlin, and J.B. Graham. 1997. Hemodynamics and blood properties of shortfin mako (*Isurus oxyrinchus*). *Copeia* 1997:424–428.

Leonard, J.B.K., A.P. Summers, and T.J. Koob. 1999. Metabolic rate of embryonic little skate, *Raja erinacea* (Chondrichthyes: Batiodea): the cost of active pumping. *J. Exp. Zool.* 283:13–18.

Lowe, C.G. 1996. Kinematics and critical swimming speeds of juvenile scalloped hammerhead sharks. *J. Exp. Biol.* 199:2605–2610.

Lowe, C.G. 1998. Swimming Efficiency and Bioenergetics of Juvenile Scalloped Hammerhead Sharks in Kaneohe Bay, Hawaii. Ph.D. dissertation. University of Hawaii, Honolulu, 130 pp.

Lowe, C.G. 2001. Metabolic rates of juvenile scalloped hammerhead sharks (*Sphyrna lewini*). *Mar. Biol.* 139:447–453.

Lowe, C.G. 2002. Bioenergetics of free-ranging scalloped hammerhead sharks (*Sphyrna lewini*) in Kaneohe Bay, Oahu, HI. *J. Exp. Mar. Biol. Ecol.* 278:141–156.

Lowe, C.G. and K.J. Goldman. 2001. Physiological telemetry of elasmobranchs: bridging the gap. *Environ. Biol. Fish.* 60:251–256.

Lowe, C.G., K.N. Holland, and T.G. Wolcott. 1998. A new acoustic tailbeat transmitter for fishes. *Fish. Res.* 36:275–283.

Matern, S.A., J.J. Cech, and T.E. Hopkins. 2000. Diel movements of bat rays, *Myliobatus californica*, in Tomales Bay, California: evidence for behavioral thermoregulation. *Environ. Biol. Fish.* 58:173–182.

Meloni, C.J., J.J. Cech, and S.M. Katzman. 2002. Effects of brackish salinities on oxygen consumption of bat rays (*Myliobatus californica*). *Copeia* 2002:462–465.

Metcalf, J.D. and P.J. Butler. 1984. Changes in activity and ventilation response to hypoxia in unrestrained, unoperated dogfish, *Scyliorhinus canicula*. *J. Exp. Biol.* 108:411–418.

Nagy, K.A. 1987. Field metabolic rate and food requirement scaling in mammals and birds. *Ecol. Monogr.* 57:111–128.

Nelson, D.R. and R.H. Johnson. 1970. Diel activity rhythms in the nocturnal, bottom-dwelling sharks, *Heterodontus francisci* and *Cephaloscylium ventriosum*. *Copeia* 1970:732–739.

Nixon, A.J. and S.H. Gruber. 1988. Diel metabolic and activity patterns of the lemon shark (*Negaprion brevirostris*). *J. Exp. Zool.* 248:1–6.

Oikawa, S. and T. Kanda. 1997. Some features of the gills of a megamouth and a shortfin mako with reference to metabolic activity, in *Biology of the Megamouth Shark*. K. Yano, J.F. Morrissey, Y. Yambumoto, and K. Nakaya, Eds., Tokai University Press, Tokyo, 93–104.

Parsons, G.R. 1990. Metabolism and swimming efficiency of the bonnethead shark, *Sphyrna tiburo*. *Mar. Biol.* 104:363–367.

Parsons, G.R. and J.K. Carlson. 1998. Physiological and behavioral responses to hypoxia in the bonnethead shark, *Sphyrna tiburo*: routine swimming and respiratory regulation. *Fish Physiol. Biochem.* 19:189–196.

Parsons, G.R. and J.L. Sylvester. 1992. Swimming efficiency of the white crappie, *Pomoxisannularis*. *Copeia* 1992:1033–1038.

Piermarini, P.M. and D.H. Evans. 2000. Effects of environmental salinity on Na+/K+-ATPase in the gills of a euryhaline elasmobranch (*Dasyatis sabina*). *J. Exp. Biol.* 203:2957–2966.

Piiper, J. and D. Schumann. 1967. Efficiency of oxygen exchange in the gills of the dogfish, *Scyliorhinus stellaris*. *Respir. Physiol.* 2:135–148.

Piiper, J., M. Meyer, H. Worth, and H. Willmer. 1977. Respiration and circulation during swimming activity in the dogfish, *Scyliorhinus stellaris*. *Respir. Physiol.* 30:221–239.

Roberts, J.L. 1978. Ram gill ventilation in fishes, in *The Physiological Ecology of Tunas*. G.D. Sharp and A.E. Dizon, Eds., Academic Press, New York, 83–88.

Scharold, J. and S.H. Gruber. 1991. Telemetered heart rate as a measure of metabolic rate in the lemon shark, *Negaprion brevirostris*. *Copeia* 1991:942–953.

Scharold, J., N.C. Lai, W.R. Lowell, and J.B. Graham. 1989. Metabolic rate, heart rate, and tailbeat frequency during sustained swimming in the leopard shark *Triakis semifasciata*. *Exp. Biol.* 48:223–230.

Schindler, D.E., T.E. Essington, J.F. Kitchell, C. Boggs, and R. Hilborn. 2002. Sharks and tunas: fisheries impacts on predators with contrasting life histories. *Ecol. Appl.* 12:735–748.

Schmidt–Nielsen, K. 1972. Locomotion: Energy cost of swimming, flying, and running. *Science* 177:222–228.

Schmidt–Nielsen, K. 1983. *Animal Physiology: Adaptation and Environment*. Cambridge University Press, New York, 619 pp.

Schmidt–Nielsen, K. 1984. *Scaling: Why Is Animal Size So Important?* Cambridge University Press, New York, 241 pp.

Schwartz, F.J. 1995. The biology of freshwater elasmobranchs. *J. Aquaricult. Aquat. Sci.* 7:45–51.

Sims, D.W. 1996. The effect of body size on the metabolic rate of the lesser spotted dogfish. *J. Fish. Biol.* 48:542–544.

Sims, D.W. 2000. Can threshold foraging of basking shark be used to estimate their metabolic rate? *Mar. Ecol. Prog. Ser.* 200:289–296.

Sims, D.W. and S.J. Davies. 1994. Does specific dynamic action (SDA) regulate return of appetite in the lesser spotted dogfish, *Scyliorhinus canicula*? *J. Fish. Biol.* 45:341–348.

Sims, D.W., S.J. Davies, and Q. Bone. 1993. On the diel rhythms in metabolism and activity of post-hatchling lesser spotted dogfish, *Scyliorhinus canicula*. *J. Fish. Biol.* 43:749–754.

Snelson, F.F., T.J. Mulligan, and S.E. Williams. 1984. Food habits, occurrence, and population structure of bull shark, *Carcharhinus leucas*, in Florida coastal lagoons. *Bull. Mar. Sci.* 34:71–80.

Stasko, A.B. and R.M. Horrall. 1976. Method of counting tailbeats of free swimming fish by ultrasonic telemetry techniques. *J. Fish. Res. Board Can.* 33:2596–2598.

Sundström, L.F. and S.H. Gruber. 1998. Using speed-sensing transmitters to construct a bioenergetics model for subadult lemon sharks, *Negaprion brevirostris* (Poey), in the field. *Hydrobiologia* 371/372:241–247.

Thompson, K.S. and D.E. Simanek. 1977. Body form and locomotion in sharks. *Am. Zool.* 17:343–354.

Tota, B. 1999. Heart, in *Sharks, Skates and Rays. The Biology of Elasmobranch Fishes*. C.W. Hamlett, Ed., John Hopkins University Press, Baltimore, MD, 238–272.

Tota, B. and A. Gattuso. 1996. Heart ventricle pumps in teleost and elasmobranchs: a morphometric approach. *J. Exp. Zool.* 275:162–171.

Tota, B., V. Cimini, G. Salvatore, and G. Zummo. 1983. Comparative study of the arterial lacunary systems of the ventricular myocardium of elasmobranch and teleosts fishes. *Am. J. Anat.* 167:15–32.

Tricas, T.C. and J.E. McCosker. 1984. Predatory behavior of the white shark (*Carcharodon carcharias*), with notes on its biology. *Proc. Calif. Acad. Sci.* 43:221–238.

Videler, J.J. and N.R. Nolet. 1990. Cost of swimming measured at optimum speed: scale effects, differences between swimming styles, taxonomic groups, and submerged and surface swimming. *Comp. Biochem. Physiol.* 97A:91–99.

Weihs, D. 1977. Effects of size on sustained swimming speeds of aquatic organisms, in *Scale Effects in Animal Locomotion.* T.J. Pedley, Ed., Academic Press, New York, 333–339.

Weihs, D. 1981. Voluntary swimming speeds of two species of large carcharhinid sharks. *Copeia* 1981:222–224.

Wells, R.M.G. and P.S. Davies. 1985. Oxygen binding by the blood and hematological effects of capture stress in two big gamefish: mako shark and striped marlin. *Comp. Biochem. Physiol.* 81A:643–646.

8

Food Consumption and Feeding Habits

Bradley M. Wetherbee and Enric Cortés

CONTENTS

8.1 Introduction

Although it is widely recognized that sharks and other elasmobranchs often play a role in the transfer of energy between upper trophic levels within marine ecosystems, our understanding of the dynamics of prey consumption and processing of food in elasmobranchs remains rudimentary. To fully comprehend energy flow through elasmobranchs in marine communities it is necessary not only to know what they eat, but also to characterize the rates at which they ingest, digest, and process energy and nutrients contained in prey that is consumed. As with other areas of elasmobranch biology, investigations on dynamics of feeding and processing food lag behind such studies on other marine fishes and vertebrates. By far the most common elasmobranch feeding studies simply describe stomach contents of a particular species in a particular location. Rate of consumption, feeding patterns, and the fate of food once ingested have been examined for very few species of elasmobranchs.

The spiral valve-type intestine present in elasmobranchs has been referred to as a primitive design and there has been speculation that food is processed differently as it passes through the digestive systems of elasmobranchs than for most teleost fishes. The different digestive morphology present in elasmo-branchs might be expected to influence time for passage of food through the alimentary canal, the efficiency of energy and nutrient absorption, the rate of consumption, and ultimately the amount of energy available for growth and other needs.

In this chapter we review information on patterns of food consumption and processing of food in the digestive tracts of elasmobranchs, with special emphasis on sharks. In general terms we examine food consumption from several perspectives: what is eaten, feeding patterns, and how much is eaten. Our discussion includes dietary overlap and dietary breadth among species of elasmobranchs as well as presumptions that have been made about food partitioning in these species. Second, we review the current state of knowledge concerning processing of food once ingested by elasmobranchs, including rates of digestion and evacuation of food from the stomachs and entire intestinal tracts of elasmobranchs. Absorption, assimilation, and conversion of ingested food into new tissue are also discussed. For most topics, we include methodological considerations relevant for experimental design and interpretation of results for past or future elasmobranch feeding studies. We conclude by offering some recommendations for future work.

8.2 Diet

The feeding biology of elasmobranchs has been investigated to understand the natural history of a particular species, the role of elasmobranchs in marine ecosystems, the impact of elasmobranch predation on economically valuable or endangered prey, and various other reasons. For these reasons researchers have attempted to describe the diets of elasmobranchs, ranging from the stomach contents of a single shark, to detailed examination of the quantity of each prey item, feeding periodicity, and frequency.

8.2.1 Quantification of Diet

Many early descriptions of the diets of different elasmobranch species were simply lists of prey items recovered from their stomachs (Coles, 1919; Breeder, 1921; Gudger, 1949; Clark and von Schmidt, 1965; Randall, 1967; Dahlberg and Heard, 1969). Other studies have quantified prey types found in stomachs using counts: the number of stomachs with a specific prey (frequency of occurrence, O), the total number of a specific prey found in stomachs (N), or using total weight (W) or volume (V) of a specific prey item (Stevens, 1973; Matallanas, 1982; Stillwell and Kohler, 1982; Snelson et al., 1984; to cite a few). Each of these terms has shortcomings for accurately expressing the amount of various prey that constitute the diet of a consumer (Bowen, 1996; Mumtaz Tirasin and Jorgensen, 1999; Liao et al., 2001). For example, expression of stomach contents with counts may give the impression that a specific prey item that occurs very frequently in stomachs represents one of the most important prey items. However, if these prey are small, they may represent only a small proportion of the total food consumed. Similarly, if diet is expressed in terms of weight or volume, consumption of a single large prey item would imply that this prey is a major component of the diet, when in fact very few individuals may have consumed it. To overcome such limitations, diet has often been reported in terms of a combination of several indices, such as the index of relative importance (Cortés, 1997, 1999):

$$(IRI) = \%F(\%W + \%N) \tag{8.1}$$

Compound expressions of diet provide less biased estimates of the contribution of various prey in the diet of a consumer, but their use remains controversial (Cortés, 1998; Hansson, 1998). Nonetheless Cortés (1997) suggested that presentation of stomach contents of sharks in terms of %IRI would both provide estimates of the diet that were intuitive and that would allow more direct comparison among studies.

Reliance on stomach contents to quantify diet of an animal also has limits. For example, rate of digestion of prey items in the stomach may vary with size and type of prey, and therefore items that are digested slowly may be overrepresented in stomachs examined. Capture technique may also influence contents in stomachs. Stomach contents of sharks captured at depth may be regurgitated, or differentially regurgitated, as the sharks are brought to the surface. Similar presumptions have been made in a number of studies where sharks were captured using gillnets.

Ecological energetics are a common framework for consideration of the fate of food consumed by animals, relating consumption to life activities through a common unit of measure, the calorie or joule (Kleiber, 1975; Brafield and Llewellyn, 1982). Diet in energetic terms would refer to the amount of energy that each item ingested contributes toward the total amount of energy consumed by an animal. The first law of thermodynamics (conservation of energy) necessitates that all energy consumed by an animal be balanced by energy used (for growth, metabolism, or reproduction) and energy lost (in feces and urine) (Kleiber, 1975). Therefore, quantification of diet in energetic terms (the amount of energy contributed by each prey type) might provide a method for expressing diet in standardized and biologically meaningful terms. Difficulties of such an approach include determination of initial size of each prey item consumed and energy content of each prey type (Scharf et al., 1998). An additional consideration far beyond simply quantifying stomach contents would be the inclusion of the energetic costs of capturing various types of prey. Although such analyses would be extremely challenging given current technology available, a general understanding of the amount of energy expenditure required to capture specific prey would provide insight into net energy gains resulting from capture and consumption of particular prey types.

8.2.2 Broad Dietary Groups

As carnivores, elasmobranchs consume a limited array of prey in comparison to teleosts, which also include omnivores and herbivores. However, there is a wide range of prey consumed by elasmobranchs, ranging from very small plankton to whales. Plankton or small crustaceans are consumed by large, filter-feeding elasmobranch species, including manta rays (*Manta birostris*) and basking (*Cetorhinus maximus*), whale (*Rhincodon typus*), and megamouth sharks (*Megachasma pelagios*) (Gudger, 1941; Hallacher, 1977; Compagno, 1990; Sims and Merrett, 1997; Sims and Quayle, 1998). The diet of most species of sharks includes teleosts, and for many species the percentage of stomachs containing teleosts exceeds 90%, particularly for sharks in the genus *Carcharhinus* (Bass et al., 1973; Stevens and Wiley, 1986; Stevens and McLoughlin, 1991; Cliff and Dudley, 1992; Salini et al., 1992; Castro, 1993; Dudley and Cliff, 1993), closely related sharpnose (*Rhizoprionodon*) and hammerhead (*Sphyrna*) species (Stevens and Lyle, 1989; Stevens and McLoughlin, 1991; Simpfendorfer and Milward, 1993) as well as mackerel sharks (Lamnidae) (Stillwell and Kohler, 1982; Gauld, 1989). Elasmobranchs are common prey of many sharks and may form a large portion of the diet of some large carcharhinids (Cliff and Dudley, 1991a; Dudley and Cliff, 1993; Wetherbee et al., 1996; Gelsleichter et al., 1999), hammerheads (Stevens and Lyle, 1989; Cliff, 1995), sixgill (*Hexanchus griseus*) and sevengill (*Notorynchus cepedianus*) sharks (Ebert, 1991, 1994), and white (*Carcharodon carcharias*) and tiger (*Galeocerdo cuvier*) sharks (Gudger, 1932; Cliff et al., 1989; Lowe et al., 1996).

Cephalopods are also common prey items. Many pelagic sharks feed on squid (Backus et al., 1956; Stillwell and Casey, 1976; Kohler, 1987; Smale, 1991), and demersal sharks often feed on octopus (Relini Orsi and Wurtz, 1977; Mauchline and Gordon, 1983; Baba et al., 1987; Castro et al., 1988; Kubota et al., 1991; Stevens and McLoughlin, 1991; Carrassón et al., 1992; Ebert et al., 1992; Ebert, 1994; Waller and Baranes, 1994). Small, benthic catsharks (Scyliorhinidae), smoothhounds (Triakidae), and hornsharks (Heterodontidae) frequently prey upon mollusks (Talent, 1976; Lyle, 1983; Menni, 1985; Segura-Zarzosa et al., 1997; Gelsleichter et al., 1999), and crustaceans form a large portion of the diet of a number of bottom-feeding carcharhinid species (Medved et al., 1985; Lyle, 1987; Stevens and McLoughlin, 1991; Salini et al., 1992, 1994; Simpfendorfer and Milward, 1993), hammerheads (Castro, 1989; Cortés et al., 1996; Bush, 2002), sharpnose (Gómez Fermin and Bashirulah, 1984; Devadoss, 1989; Gelsleichter et al., 1999), smoothhounds (Talent, 1982; Taniuchi et al., 1983; King and Clark, 1984; Vianna and Amorim, 1995; Rountree and Able, 1996; Smale and Compagno, 1997), catsharks (Ford, 1921; Macpherson, 1980; Lyle, 1983; Cross, 1988; Ebert et al., 1996; Heupel and Bennett, 1998), and batoids (Ajayi, 1982; Smith and Merriner, 1985; Ebert et al., 1991; Smale and Cowley, 1992; Barry et al., 1996; Ellis et al., 1996; Schwartz, 1996).

Large sharks occasionally consume vertebrates other than fish. Birds have been found in the stomach of bull sharks (*Carcharhinus leucas*; Tuma, 1976) and tiger sharks (Saunders and Clark, 1962; Dodrill and Gilmore, 1978; Heithaus, 2001a; Carlson et al., 2002) and may compose a large part of the diet of tiger (Bass et al., 1973; Stevens, 1984; Simpfendorfer, 1992; Lowe et al., 1996) and white sharks (Randall et al., 1988). Reptiles (turtles and snakes) are occasionally eaten by carcharhinid sharks (Heatwole et al., 1974; Tuma, 1976; Lyle, 1987; Lyle and Timms, 1987; Cliff and Dudley, 1991a) and white sharks

(Long, 1996; Fergusson et al., 2000) and are common in the stomachs of tiger sharks (Witzell, 1987; Stevens and McLaughlin, 1991; Simpfendorfer, 1992; Lowe et al., 1996; Heithaus, 2001a). Marine mammals are frequently preyed upon by large sharks such as white and tiger sharks (Bell and Nichols, 1921; LeBoeuf et al., 1982; Stevens, 1984; Corkeron et al., 1987; Cliff et al., 1989; Lowe et al., 1996; Dudley et al., 2000; Heithaus, 2001a) and have been found in stomachs of carcharhinid sharks (Bass et al, 1973; Cliff and Dudley, 1991a; Wetherbee et al., 1996) and of sleeper sharks (*Somniosus*) (Scofield, 1920), sixgill and sevengill sharks (Hexanchidae) (Ebert, 1991, 1994). The unusual tooth and jaw morphology of cookie-cutter sharks (*Isistius brasiliensis* and presumably *I. plutodon*) enables these sharks to maintain an essentially parasitic lifestyle by removing plugs of flesh from large vertebrates (tunas, billfish, dolphins, and whales) and from squid (Strasburg, 1963; Jones, 1971; Jahn and Haedrich, 1988; Muñoz-Chapuli et al., 1988; Shirai and Nakaya, 1992). Readers are referred to Cortés (1999) for a summary of standardized diet compositions of 149 shark species.

8.2.3 Diet Shifts

Adequate representation of the diet of a species of elasmobranch is complicated by differences in diet that occur within species among individuals of different sizes, geographical locations, and during different seasons. Ontogenetic change in feeding habits is an almost universal phenomenon in fishes and thus its occurrence in elasmobranchs is not surprising considering that, as many species of sharks and rays increase in size, there also are changes in habitat occupied, movement patterns, swimming speed, size of jaws, teeth and stomachs, energy requirements, experience with prey, vulnerability to predation, and other factors that result in variable exposure to prey or improved ability of larger sharks to capture different prey items (Graeber, 1974; Weihs et al., 1981; Stillwell and Kohler, 1982; Lowe et al., 1996).

Although diet shifts are more often reported qualitatively rather than based on rigorous statistical analysis, there are many reports of a shift from a diet of invertebrates to a diet that is more varied or that includes more teleosts (Olsen, 1954; Capapé, 1974, 1975; Capapé and Zaouali, 1976; Talent, 1976; Jones and Geen, 1977; Mauchline and Gordon, 1983; Smale and Cowley, 1992; Stillwell and Kohler, 1993; García de la Rosa and Sánchez, 1997; Platell et al., 1998; Smale and Goosen, 1999; Kao, 2000; Jakobsdóttir, 2001). There are also multiple studies that document increased consumption of elasmobranchs (Matallanas, 1982; Cortés and Gruber, 1990; Cliff and Dudley, 1991a; Smale, 1991; Lowe et al., 1996; Simpfendorfer et al., 2001a,b) and marine mammals (Tricas and McCosker, 1984; Ebert, 1994) with increasing size of shark. A number of studies, however, found no ontogenetic dietary changes (Kohler, 1987; Cliff and Dudley, 1991b; Matallanas et al., 1993; Clarke et al., 1996; Cortés et al., 1996; Segura-Zarzosa et al., 1997; Avsar, 2001; Jakobsdóttir, 2001).

There are also examples of geographical differences in the diets of several wide-ranging species of sharks. For example, the diets of spiny dogfish (*Squalus acanthias*), blue (*Prionace glauca*), sandbar (*Carcharhinus plumbeus*), blacktip (*C. limbatus*), and bull sharks all differed among locations in the Atlantic, Pacific, and Indian Oceans (Gudger, 1948, 1949; Holden, 1966; Rae, 1967; Wass, 1971; Stevens, 1973; Gubanov and Grigoryev, 1975; Tuma, 1976; Jones and Geen, 1977; Tricas, 1979; Kondyurin and Myagkov, 1982; Stevens et al., 1982; Sarangadhar, 1983; Snelson et al., 1984; Medved, 1985; Cliff et al., 1988; Harvey, 1989; Cliff and Dudley, 1991a; Dudley and Cliff, 1993; Lowe et al., 1996). Variation of diet among locations is exemplified by the tiger shark, which has a diet that differed substantially among areas sampled worldwide (DeCrosta et al., 1984; Simpfendorfer, 1992; Lowe et al., 1996; Simpfendorfer et al., 2001a). Diet may differ within a species even between locations that are relatively close, as has been found for sandbar (Lawler, 1976; Medved et al., 1985; Stillwell and Kohler, 1993) and lemon sharks (*Negaprion brevirostris*; Springer, 1950; Schmidt, 1986; Cortés and Gruber, 1990), and the starspotted smoothhound (*Mustelus manazo*; Yamaguchi and Taniuchi, 2000). Habitat type and water depth have also been found to influence diet composition (Stillwell and Kohler, 1982, 1993; Kohler, 1987; Cortés et al., 1996; Smale and Compagno, 1997; Webber and Cech, 1998). Several authors have reported differences in the diet between sexes of sharks (Bonham, 1954; Matallanas, 1982; Hanchet, 1991; Stillwell and Kohler, 1993; Simpfendorfer et al., 2001a), which may be due to sexual segregation within species and different sizes attained by males and females. In all, findings of geographical differences in diet of sharks are not surprising considering the diversity of prey in different regions and the apparent plasticity of feeding behaviors among sharks (see Heithaus, Chapter 17 of this volume, for a more complete discussion).

Variation in feeding of sharks is further demonstrated by seasonal differences in diet that have been reported within species (Capapé, 1974; Talent, 1976; Jones and Geen, 1977; Tricas, 1979; Lyle, 1983; Olsen, 1984; Kohler, 1987; Dudley and Cliff, 1993; Waller and Baranes, 1994; Cortés et al., 1996; Nagasawa, 1998; Platell et al., 1998; Allen and Cliff, 2000; Horie and Tanaka, 2000). Seasonal differences in diet presumably reflect seasonal migration of sharks or of their prey. For example, Matallanas (1982) reported seasonal shifts in the most important teleosts in the diet of kitefin sharks (*Dalatias licha*) and Stillwell and Kohler (1982) described seasonal shifts between consumption of fish and cephalopods by the mako shark (*Isurus oxyrinchus*). There is also evidence of a diet shift in leopard sharks (*Triakis semifasciata*) sampled at a single location during two periods 25 years apart, which may be indicative of community changes (Kao, 2000).

8.2.4   Feeding Relationships

There have been relatively few investigations comparing diets of sympatric species of elasmobranchs. In several studies, standard ecological indices of similarity were used to calculate dietary overlap among elasmobranch species, among elasmobranchs and teleosts caught in the same location, or among different size classes of a single species. Such comparisons represent initial attempts to characterize food partitioning and competition among elasmobranchs and co-occurring teleosts. Ecological indices of dietary breadth or diversity have also been calculated for several species of elasmobranchs to examine the degree of feeding specialization.

The available evidence indicates that both food partitioning and competition for food resources are likely to occur in marine communities where elasmobranchs occur. Dietary overlap among sympatric species of elasmobranchs has been characterized — qualitatively or using quantitative indices — as low (Macpherson, 1981; Baba et al., 1987; Carrassón et al., 1992; Orlov, 1998), moderate (Relini Orsi and Wurtz, 1977; Smale and Compagno, 1997; Orlov, 1998) to substantial (Macpherson, 1980; Ellis et al., 1996), high (Salini et al., 1990; Platell et al., 1998), and variable depending on the species compared (Macpherson, 1981; Euzen, 1987). Varying degrees of diet overlap have also been described for co-occurring elasmobranchs and teleosts (Blaber and Bulman, 1987; Ali et al., 1993; Clarke et al., 1996), or for elasmobranchs and marine mammals (Clarke et al., 1996; Heithaus, 2001b). At the intraspecific or intrapopulation level, increased dietary overlap is most often encountered between pairs of consecutive size classes (Cortés et al., 1996; Wetherbee et al., 1996, 1997; García de la Rosa and Sánchez, 1997; Platell et al., 1998; Kao, 2000; Simpfendorfer et al., 2001a; Koen Alonso et al., 2002), or between similar size classes of elasmobranchs and teleosts (Platell et al., 1998). Food overlap also tends to be high between adjacent geographic locations (Yamaguchi and Taniuchi, 2000; Simpfendorfer et al., 2001a).

Diets of elasmobranchs vary from highly specialized to very generalized. Specialized diets include those of elasmobranchs that consume zooplankton, crustaceans, and cephalopods as discussed in an earlier section. In contrast, a number of top predators, such as bull and tiger sharks, have very generalized diets. Varying degrees of specialization have been reported in studies that calculated true measures of diversity (Macpherson, 1981; Blaber and Bulman, 1987; Clarke et al., 1989; Carrassón et al., 1992; Ali et al., 1993; Cortés et al., 1996; Ellis et al., 1996; Simpfendorfer et al., 2001a) or that reported only the total number of different prey types or contained qualitative statements about dietary diversity (Chatwin and Forrester, 1953; Capapé and Zaouali, 1976; Segura-Zarzosa et al., 1997; Smale and Compagno, 1997; Gelsleichter et al., 1999). Dietary breadth tends to increase with size or age in some cases (Talent, 1976; Cortés and Gruber, 1990; Lowe et al., 1996; Wetherbee et al., 1996, 1997) and decrease in others (Smale and Compagno, 1997; Platell et al., 1998; Yamaguchi and Taniuchi, 2000; Simpfendorfer et al., 2001a).

Because of the widespread occurrence of ontogenetic, geographical, and seasonal changes in feeding habits discussed above, very few studies on the diet of sharks have been extensive enough to provide a comprehensive description of the diet for a species. Additionally, the diversity of prey found in stomachs generally increases with the number of stomachs sampled. The issue of sample sufficiency can be addressed by using cumulative prey curves to determine whether a sufficient number of stomachs have been examined to describe precisely the diet of the species in question (see Ferry and Cailliet, 1996;

Cortés, 1997, and references therein). Clearly, there is ample opportunity for improving our understanding of aspects of the feeding ecology of elasmobranchs at the organism, population, community, and ecosystem level through additional and more focused research.

8.2.5   Feeding Patterns

Understanding a consumer's feeding patterns requires more than knowledge of the prey items that make up its diet. The dynamics of the feeding process must be accounted for, and thus to understand the ecological interaction between predator and prey we must have knowledge of the amount of food ingested and the feeding frequency of the predator. Analysis of stomach contents allows inference of feeding patterns through reconstruction of meal sizes, ingestion times, feeding duration, and feeding frequency. The frequency of occurrence of empty stomachs, the number, weight, and stage of digestion of food items, in combination with knowledge on the gastric evacuation dynamics of each food item, all give insight into the feeding pattern of a predator.

The occurrence of high proportions of empty stomachs in shark diet studies and in commercial fisheries operations is common (Wetherbee et al., 1990). Use of longlines to capture sharks may attract more animals with empty stomachs, but this is unlikely when using passive gear such as gillnets or active gear such as trawls. Frequent occurrence of empty stomachs, combined with the observation that there are often few food items — many of them in advanced stages of digestion — in shark stomachs — e.g., in the juvenile sandbar shark (Medved et al., 1985) and the juvenile lemon shark (Cortés and Gruber, 1990) — lends support to the notion that many sharks are intermittent rather than continuous feeders, because otherwise one would expect to regularly find multiple food items at different stages of digestion and few empty stomachs. Demersal carnivores that feed on invertebrate prey, such as many skates and rays (Bradley, 1996), and filter feeding zooplanktivorous sharks are obvious exceptions to this pattern (Baduini, 1995; Sims and Quayle, 1998), as they feed more continuously.

Feeding frequency can be estimated from the total time required to complete gastric evacuation and the proportion of empty stomachs in a sample (Diana, 1979). Based on this method, Jones and Geen (1977) estimated that mature spiny dogfish would feed only every 10 to 16 days after completely filling their stomachs, whereas Medved et al. (1985) and Cortés and Gruber (1990) estimated a feeding frequency of 95 h and 33 to 47 h for juvenile sandbar and lemon sharks, respectively.

Gastric evacuation experiments (Section 8.3.2) allow development of qualitative scales describing the various stages of digestion of food items. These qualitative scales can then be used to calculate the difference between the least and most advanced stages of digestion of food items found in stomachs of field-sampled animals, and infer feeding duration. Medved et al. (1985), Cortés and Gruber (1990), and Bush and Holland (2002) used this approach to obtain estimates of feeding duration for juvenile sandbar (7 to 9 h), lemon (11 h), and scalloped hammerhead sharks (*Sphyrna lewini*; 9 to 10 h). The occurrence of food items in different stages of digestion in stomachs of juvenile lemon and sandbar sharks caught at the same time also indicated that feeding in these two species was asynchronous; i.e., there was no preferred feeding time for all individuals of a population, a pattern believed to be prevalent in most shark species. Conversely, Kao (2000) reported some evidence for feeding synchronicity in the leopard shark off the central California coast. Results from Medved et al. (1985) and Cortés and Gruber (1990) for juvenile sandbar and lemon sharks, respectively, did not reveal increased food consumption at night or during a particular tidal phase. However, these studies did not estimate meal ingestion times, as we explain in the next paragraph.

Cortés (1997) reviewed the numerous methodological issues that can affect the interpretation of diel feeding chronology in fishes and elasmobranchs. In addition to the effect of passive vs. active sampling gear, experimental design, and statistical analysis of results, he cautioned against using the weight of stomach contents alone to assess diel feeding (dis)continuity and to interpret diel feeding chronology. To estimate preferred feeding times it is also necessary to reconstruct meal ingestion times using qualitative stage-of-digestion scales. In captivity, Longval et al. (1982) found a cyclical feeding pattern in juvenile lemon sharks, with peak consumption followed by several days of reduced food intake. The evidence for sharks, as exemplified by work on juvenile lemon sharks, supports the concept of a cyclical pattern of feeding motivation observed in many vertebrates, whereby relatively short feeding bouts would be followed by longer periods of reduced predatory activity until the return of appetite, which in the lesser spotted dogfish (*Scyliorhinus canicula*) was found to be inversely correlated with gastric evacuation rate (Sims et al., 1996).

8.2.6 Trophic Levels

It is commonly accepted that sharks are top predators in many marine communities. However, until recently, virtually no quantitative estimates of trophic levels existed for sharks. Cortés (1999) calculated standardized diet compositions and estimated trophic levels for 149 shark species belonging to 23 families using published trophic levels of prey categories, largely based on the Ecopath II model (Christensen and Pauly, 1992). He concluded that sharks as a group are tertiary consumers (trophic level > 4) that occupy trophic positions similar to those of marine mammals and higher than those of seabirds. Measurement of stable isotopes of nitrogen and carbon in tissues of marine consumers is an alternative approach to estimating trophic level based on stomach contents. To date, only two studies on sharks have used stable isotope analysis to estimate trophic level; in the basking shark (Ostrom et al., 1993) and Greenland shark (*Somniosus microcephalus*; Fisk et al., 2002). Fisk et al. (2002) also used concentrations of organochlorine contaminants to estimate the trophic level of Greenland sharks, concluding that results from stable isotope analysis and this technique did not agree. They attributed the lower trophic level obtained through stable isotope ($\delta^{15}N$) analysis compared to that from contaminant analysis to urea retention in elasmobranch tissues for osmoregulation, which could result in lower levels of $\delta^{15}N$ and thus underestimate trophic level. Further investigation of the effect of urea retention on $\delta^{15}N$ levels is thus required (Fisk et al., 2002) along with comparisons of stable isotope and dietary-based estimation of trophic levels.

8.3 Food Consumption

Feeding ecology is an important aspect of the life-history strategy of a species that can be adequately expressed through determination of food consumption rates. Daily rates of food consumption are in turn dependent on gastric evacuation rates. Measurement of daily rates of food consumption and digestion rates require regular collection of stomach contents of fish caught in the wild and fish held in captivity in the laboratory or field. This poses a particularly difficult problem for those studying elasmobranchs and sharks in particular, because of the difficulty of keeping them in captivity and the logistical requirements of extended field sampling. Additionally, rates of consumption in teleost fishes may vary depending upon a myriad of intrinsic (e.g., age, feeding history, reproductive status) and extrinsic factors (e.g., geographical location, habitat type, water temperature, prey availability). The scarcity of information on food consumption rates of elasmobranchs is thus hardly surprising.

8.3.1 Daily Ration

Daily ration is the mean amount of food consumed on a daily basis by individuals of a population, generally expressed as a proportion of mean body weight. Although an individual does not ingest the same amount of food everyday and may not even feed daily, daily ration is a good measure for comparative studies (Wetherbee et al., 1990). There are two basic approaches for estimating daily ration: (1) *in situ* (field-derived) methods, which require knowledge of the amount of food found in stomachs of fish sampled in the wild and of the gastric evacuation dynamics of the ingested foodstuffs, and (2) bioenergetic models, which estimate food consumption based on the other components of the bioenergetic equation (growth, metabolism, excretion, and egestion).

With field-based methods, daily ration cannot be estimated by simply examining stomach contents because the amount of food found in stomachs is a function of both ingestion and digestion rates (Wetherbee et al., 1990). Cortés (1997) reported that there has been very little investigation of the applicability to elasmobranchs of the most common models used to estimate daily ration in teleosts. *In situ* methods of estimation that have been used for elasmobranchs include those by Elliott and Persson

(1978), Diana (1979), Eggers (1979), Pennington (1985), and Olson and Mullen (1986). Cortés (1997) concluded that the Diana and Olson–Mullen methods applied better to intermittent feeders, such as most sharks, and that these models were also based on less restrictive assumptions and required comparatively less demanding sampling regimens. Given the absence of error analyses of the estimates of daily ration in elasmobranch studies, Cortés (1997) advocated the use of resampling techniques, such as bootstrapping, or Monte Carlo simulation to enable statistical testing of differences between estimates obtained through different models and generally to provide a picture of the variability associated with those point estimates.

Laboratory approaches to estimating daily ration are based on a bioenergetic or energy budget equation (Winberg, 1960), which relates consumption (C) to growth (G), metabolism (M), excretion (urine, U), and egestion (feces, F):

$$C = G + M + U + F \qquad (8.2)$$

The daily energy required for growth (J day^{-1}) can be derived from laboratory or field estimates of growth (g day^{-1}) multiplied by the energy equivalent of shark tissue (J g^{-1}), which to date has only been determined for juvenile lemon sharks (5.41 kJ g^{-1} [wet weight]; Cortés and Gruber, 1994) and scalloped hammerhead pups (6.07 kJ g^{-1}; Lowe, 2002). The daily energy required for total metabolic expenditures (J day^{-1}) can be obtained from average daily metabolic rate (for example), expressed as mg O$_2$ kg shark^{-1} day^{-1}, multiplied by a standard oxycalorific value of 3.25 cal ml O$_2$$^{-1}$ (Elliott and Davidson, 1975) or 13.59 J ml O$_2$$^{-1}$, and adjusting for shark mass (kg). The energy lost as non-assimilated food (urine and feces) has only been measured in the lemon shark (Wetherbee and Gruber, 1993), where it represented approximately 27% of the total ingested energy. This proportion of energy corresponding to $F + U$ can be substituted into the bioenergetic equation by multiplying $G + M$ by a factor of 1.37 (to account for energy losses). The final step is to use the energy value of food consumed (J g^{-1}), divide it into 1.37($G + M$), and express the result as a percentage of body weight. Cortés and Gruber (1990) used a variation of this bioenergetic approach to estimate daily ration for juvenile lemon sharks; i.e., they used a laboratory-derived feeding rate–growth rate curve (also known as G–R curve) to estimate daily ration in the wild as the food intake level that corresponded to field-observed growth.

Table 8.1 summarizes studies of food consumption rates in elasmobranchs, including the shape of the model that best described the rate of gastric evacuation, total gastric evacuation time, estimates of daily ration, and gross conversion efficiency. Feeding rates of elasmobranchs — at least on a body weight basis — are considerably lower than those of many teleosts (Brett and Groves, 1979), even with the inclusion of sharks fed to satiation in captivity, and rarely surpass 3% BW day^{-1} (Table 8.1). In addition, consumption rates of adults may decrease by an order of magnitude with respect to those of pups, as found for captive sevengill sharks (*Notorynchus cepedianus*) fed to satiation (Van Dykhuizen and Mollet, 1992; Table 8.1) and in bioenergetic estimates for the bonnethead (*Sphyrna tiburo*; E. Cortés, unpubl.).

8.3.2   Gastric Evacuation

Estimation of daily ration through *in situ* methods requires knowledge of gastric evacuation rates. As in many areas of elasmobranch research, our ability to conduct controlled field or laboratory experiments is severely impaired by the difficulty of maintaining large individuals, which has resulted in experiments conducted on small species or juvenile stages of larger species (Cortés, 1997).

Cortés (1997) pointed out that there is still considerable debate about the adequacy of the most common mathematical models (linear, exponential, square root, surface area) used to describe gastric evacuation in fishes, and that no single model can be used to represent the dynamics of different species consuming different prey under different environmental conditions in all cases. The physiological rationale for the various models of gastric evacuation and the statistical adequacy of the criteria used to select the best model of evacuation have been extensively reviewed elsewhere (see references in Cortés, 1997). Cortés (1997) advocated the use of multiple measures of statistical fit along with formal residual analysis and an examination of residual plots before selecting a model, but pointed out that even with thorough analyses results may still be inconclusive. A sensible approach for estimating daily ration through *in situ* methods is therefore to evaluate the effects of various evacuation models.

TABLE 8.1

Summary of Gastric Evacuation, Daily Ration, and Food Conversion Efficiency Estimates for Elasmobranchs

Species	Stage	GE Curve	TGET (h)	Daily Ration ($\%BW\ day^{-1}$)	K_1 (%)	Ref.
Carcharhinus acronotus	Juvenile, adult	—	—	0.87–1.56[a] (28)	—	Carlson and Parsons (1998)
Carcharhinus dussumieri	Juvenile (10)	—	—	2.91[b] (26–30)	—	Salini et al. (1999)
Carcharhinus leucas	Pup (6)	—	—	0.50[c] (24)	—	Schmid et al. (1990)
Carcharhinus leucas	Pup (5)	—	—		5–12[c] (23–25)	Schmid and Murru (1994)
Carcharhinus melanopterus	Juvenile (20)	—	—	0.3–0.8[c] (22–28)	20[e] (22–28)	Taylor and Wisner (1989)
Carcharhinus plumbeus	Juvenile	Gompertz	81–104 (22–26; 17)	0.9–1.3[d] (25; 414)	14.1 (25)	Medved (1985); Medved et al. (1988)
	Nr (3)		>48 (nr)			Wass (1973)
	Pup			1.43[a] (18.5)		Stillwell and Kohler (1993)
	Juvenile, adult			0.86[a] (18.5)		Stillwell and Kohler (1993)
	Adult (6)			0.47[c] (24)		Schmid et al. (1990)
Carcharhinus tilstoni	Juvenile (4)			3.44[b] (26–30)		Salini et al. (1999)
Negaprion acutidens	Juvenile (4)			3.35[b] (26–30)		Salini et al. (1999)
Negaprion brevirostris	Juvenile	Linear	28–41 (20–29; 48)	1.5–2.1[d] (23–32; 86)	9.4–13.1 (32) [–64–25][e] (25; 80)	Cortés and Gruber (1990, 1992, 1994)
	Juvenile	Exponential	24 (25; 20)	2.7[b] (25; 6)	[22.4][e] (25; 3)	Schurdak and Gruber (1989)
	Juvenile					Gruber (1984); Longval et al. (1982)
	Juvenile (1), adult (1)			0.5–1.4[b] (21–29)		Clark (1963)
Prionace glauca	Nr		>24 (14–16; 3)	—	—	Tricas (1979)
	Adult	Exponential[f]	164 (19; 2)	0.40–0.65[d] (17; 54)	17.1 (17)	Kohler (1987)
Sphyrna lewini	Juvenile	Multiple[g]	>5–29 (21–29; 64)	2.12–3.54 (22–28; 451)	—	Bush and Holland (2002)
Sphyrna lewini	Juvenile				2.9–9.4[a] (26)	Lowe (2002)
Sphyrna tiburo	All	Logistic	>50 (20–30; 46)	2.16–4.34[d] (20–30; 53)	—	Tyminski et al. (1999)
Triakis semifasciata	All	Linear	28–32 (13–18; 30)	0.85–2.20 (nr; 138)	—	Kao (2000)
Schroederichthys chilensis	Nr	Exponential	74 (16; 18)			Aedo and Arancibia (2001)
Scyliorhinus canicula	All	Surface area[h]	50–>70 (14; 237)			Macpherson et al. (1989)
	Adult	Exponential	>200 (15; 20)			Sims et al. (1996)
Isurus oxyrinchus	Adult		36–48[i]	2.2–3.0[a] (19)		Stillwell and Kohler (1982)
Carcharias taurus	Adult (13)			0.27[c] (24)		Schmid et al. (1990)
Ginglymostoma cirratum	Adult (6)			0.31[c] (24)		Schmid et al. (1990)
Notorynchus cepedianus	Pup			2[c] (12–14)	25–40	Van Dykhuizen and Mollet (1992)
	Juvenile			0.6[c] (12–14)	10–15	
	Adult			0.2[c] (12–14)	—	

TABLE 8.1 (Continued)

Summary of Gastric Evacuation, Daily Ration, and Food Conversion Efficiency Estimates for Elasmobranchs

Species	Stage	GE Curve	TGET (h)	Daily Ration ($\%BW\ day^{-1}$)	K_1 (%)	Ref.
Squalus acanthias	Juvenile, adult	—	124 (10; 75)	1.3[b] (10; 5)	6.1–10.7[j]	Jones and Geen (1977)
	—	—	—	0.4[k] (10)	—	Holden (1966)
	Adult	—	—	1.5–2.0[a] (10)	—	Brett and Blackburn (1978)
	Adult	—	>48 (15)	—	—	Van Slyke and White (1911)
	All	—	—	2.60[l] (nr; 3396)	—	Tanasichuk et al. (1991)
Dasyatis sabina	All	Exponential[f]	—	—	—	Bradley (1996)
Gymnura altavela	Adult (2)	—	—	2.52 (27–33; 48)	10.8[c] (23)	Henningsen (1996)
Raja erinacea	All	Multiple[g]	12–52[m] (10 and 16; 28)	—	—	Nelson and Ross (1995)
Callorhynchus callorhynchus[n]	All	—	>24 (13, 113)	1.36 (11.5–13, 181)	—	Di Giácomo et al. (1994)

Note: Abbreviations: GE curve is the mathematical model that best describes gastric evacuation; TGET is total gastric evacuation time; K_1 is gross conversion efficiency (annual production divided by annual consumption estimates); Nr is not reported; single values in parentheses denote temperature range in degrees Celsius, except for the Stage column, where they indicate sample size; a second value indicates sample size.

[a] Bioenergetic estimate(s) only.

[b] Captive sharks fed experimental meal to satiation.

[c] Food consumed by captive sharks in display aquarium.

[d] Includes both *in situ* and bioenergetic estimates.

[e] Derived in laboratory or aquarium experiments where sharks were fed at varying ration sizes and growth recorded.

[f] Assumed functional relationship.

[g] Different models provided the best fit depending on temperature, food type, or meal size.

[h] Gastric evacuation of small prey items was adequately described by exponential model.

[i] Assumed values.

[j] 6.1% is for age 1 dogfish, 10.7% is for age 0 dogfish.

[k] "Working" bioenergetic estimate.

[l] Estimated from mean stomach fullness indices.

[m] Depending on temperature and food type.

[n] A holocephalan.

In addition to the well-known accelerating effect of temperature (Brett and Groves, 1979), meal size and food type also seem to affect the gastric evacuation dynamics of elasmobranchs. Larger meal sizes generally take longer to digest and evacuate (Sims et al., 1996; Bush and Holland, 2002). In general, it appears that small, more friable, and easily digestible items are evacuated more quickly than larger items with lower surface-to-volume ratios (Medved, 1985; Schurdak and Gruber, 1989; Cortés and Gruber, 1992; Nelson and Ross, 1995). Surface area models provided the best fit to gastric evacuation data for the lesser spotted dogfish, especially when the meal included more than one prey item (Macpherson et al., 1989). Most species of elasmobranchs consume different types of prey, which in turn may be evacuated from the stomach at different rates, and thus greatly influence estimates of daily ration based on gastric evacuation rate. For example, Medved (1985) found that time required for evacuation of crab and teleost prey from the stomachs of sandbar sharks could differ by as much as 20 h. In general, the effects of food type, number and digestibility of prey, and meal size on gastric evacuation dynamics of elasmobranchs would clearly improve the accuracy of estimates of daily ration and overall rates of consumption.

The sequence of digestion and gastric evacuation of foodstuffs in elasmobranchs has not been fully elucidated. An initial lag phase before the start of gastric evacuation into the intestine, attributed to the time required for gastric juices and enzymatic reactions to take effect, was reported for the sandbar shark (Medved, 1985); however, this delay in the onset of digestion may have resulted from handling and force feeding of experimental animals (Wetherbee et al., 1990). In fishes, initial chemical digestion is generally attributed to pepsin, an acid protease (Holmgren and Nilsson, 1999). Plots of the change in energy content of the ingested meal with time suggested that tissues with higher energy, such as muscle, were evacuated before lower-energy tissues, such as exoskeleton, during the earlier stages of gastric evacuation in gray smoothhound sharks (*Mustelus californicus*; San Filippo, 1995). In contrast, Schurdak and Gruber (1989) reported that carbohydrates were evacuated from stomachs of lemon sharks prior to evacuation of proteins. For a detailed description of the anatomy and physiology of the digestive system of elasmobranchs readers are referred to Holmgren and Nilsson (1999).

Although research for skates and rays is extremely scarce, emptying of food from the stomachs of elasmobranchs takes considerably longer than in teleosts. With very few exceptions, it takes a minimum of one to — often — several days to completely evacuate a meal from the stomach of elasmobranchs (Table 8.1). Presumably, lamnid sharks, such as the white shark, and other species capable of elevating stomach temperature above ambient water temperature through countercurrent mechanisms (McCosker, 1987; see Carlson et al., Section 7.5 of this volume) could have rapid rates of digestion, but no gastric evacuation measurements have been made to date on such heterothermic species.

8.4 Excretion and Egestion

A portion of food that is consumed by elasmobranchs is not absorbed by the digestive tract and is egested as feces. Additionally, a portion of the food that is absorbed by intestinal cells is not available for the energetic demands of the animal and is excreted as nitrogenous waste in urine and gill effluent.

8.4.1 Excretion

Energetic losses in gill effluent and urine have not been measured in elasmobranchs, but have been presumed to be similar in scale to losses (about 7% of the energy budget) estimated for teleost fishes (Brett and Groves, 1979). Quantification of energy losses through the gills and kidneys of elasmobranchs is problematic due to the large quantity of water involved in housing elasmobranchs, as well as retention of nitrogenous wastes in the form of urea and trimethylamine oxide in blood and tissues for osmoregulatory purposes (Perlman and Goldstein, 1988; Wood, 1993; Evans et al., Chapter 9 of this volume).

8.4.2 Egestion

Elasmobranchs have a spiral valve intestine, which functions to increase surface area for digestion and absorption of food, but which also conserves space in the body cavity for a large liver and development of large embryos (Moss, 1984). The digestive capability of the spiral valve intestine has been investigated in only one species of elasmobranch, the lemon shark (Wetherbee and Gruber, 1993). These authors used an indirect method of measurement incorporating an inert, naturally occurring marker (acid-insoluble ash) into food. In this study, lemon sharks were capable of absorbing energy and nutrients in food with an average efficiency close to 80%, which is similar to many carnivorous teleosts. However, the time required for a meal to be completely eliminated from the digestive tract of lemon sharks was prolonged (70 to 100 h) in comparison to most teleosts (Wetherbee et al., 1987; Wetherbee and Gruber, 1990). Other studies have reported that food remains in the digestive tract of elasmobranchs for long periods of time (up to 18 days) in comparison to most teleosts (Wetherbee et al., 1990; Sims et al., 1996). The protracted periods of time required for complete food passage, in addition to difficulties involved with maintaining sharks in captivity and the labor-intensive methods required for fecal collection, present major obstacles for studies on digestive efficiency of sharks (Wetherbee and Gruber, 1993).

Prolonged passage of food through digestive tracts of elasmobranchs may be required for spiral valve intestines to accomplish digestion and absorption of food at levels comparable with those of teleosts. There have been several studies on enzymatic digestion in the stomachs of elasmobranchs, but few studies on pancreatic and brush border enzymes that function to break down macromolecules to smaller subunits for absorption across the intestinal epithelium (Sullivan, 1907; Van Slyke and White, 1911; Fänge and Grove, 1979; Caira and Jolitz, 1989; Papastamatiou, 2003). Although the relationship between prolonged food passage time and limitation of enzymatic digestion in elasmobranchs is unknown, it is apparent that prolonged food passage is related to a low rate of consumption in sharks, which in turn limits growth and reproductive rates. Although low rates of food consumption may provide evolutionary advantages for elasmobranch populations, the associated low growth and reproductive rates are life history characteristics that contribute to the vulnerability of the majority of elasmobranch populations to overfishing.

8.5 Production

Production, or growth in body mass, can be measured through laboratory experiments, field mark–recapture methods, or indirectly through size at age relationships. Relative rates of production (expressed as percent body weight) of most teleost species are considerably higher than those of elasmobranchs (Wetherbee el al., 1990), with many teleosts doubling their body weight in less than a week after birth (Brett and Groves, 1979). Relative growth rates in length and mass are much higher for immature than mature individuals in most elasmobranch species (see Cailliet and Goldman, Chapter 14 of this volume), especially during the first year of life. Branstetter (1990) estimated first-year growth in body length for several shark species, with values ranging from 16 to 100% per year. Wetherbee et al. (1990) reported values of first-year growth in mass of 33, 79, and 138% for the spiny dogfish, sandbar shark, and lemon shark, respectively. In relative terms, small coastal and pelagic species tend to grow at a faster rate than their large coastal counterparts, probably reflecting differences in the risk of predation faced by juveniles. As very few estimates of food consumption are available, it is unclear whether differences in production are a result of different levels of food consumption or differences in energy partitioning.

Growth efficiency measures have very seldom been calculated in elasmobranchs. The efficiency of food conversion to somatic growth, or gross conversion efficiency (K_1), is important ecologically because it measures the proportion of ingested food that will be available to the next trophic level (Warren and Davis, 1967). K_1 values reported for elasmobranchs range from about 3 to 40% (Table 8.1). Van Dykhuizen and Mollet (1992) reported that K_1 values (which they referred to as cumulative total efficiency) decreased with increasing age, from 25 to 40% at age 1 to 3 years to 10 to 15% at age 5 to 6 years in aquarium-fed sevengill sharks. Most K_1 values for elasmobranchs (Table 8.1) are comparable to values reported for teleosts (10 to 25%; Brett and Groves, 1979), indicating that elasmobranchs are generally capable of converting energy to growth as efficiently as teleosts.

The rate of production and K_1 are functions of the rate of food consumption. Only one study has examined this relationship in elasmobranchs. Cortés and Gruber (1994) found that the relationship between production rate and feeding rate in juvenile lemon sharks was best described by a von Bertalanffy growth-like equation of the form:

$$G_r = G_{\max}(1 - e^{-k(R-R_m)})$$ (8.3)

where G_r is growth rate, $G_{\max}$ is maximum growth rate, k is the rate of change in growth rate with feeding rate, R is feeding rate, and R_m is the maintenance ration (no growth). They reported very similar values of $R_m = 1.06\%$ wet BW day^{-1} and G_s (loss in weight due to starvation) = 1.11% BW day^{-1}. Cortés (1991) also estimated a value for R_{opt}, the optimal ration (Pandian, 1982), of 2.15 BW day^{-1} for a 2-kg lemon shark in its first year of life, by drawing a tangent from the origin of coordinates in the G–R curve to the point in the curve with the steepest slope. Cortés and Gruber (1994) found values of K_1 ranging from –64% to 25%, and that K_1 slowed, but continued to increase, at ration levels above maintenance. This finding did not support those from several studies with teleosts where a dome-shaped curve was found (Paloheimo and Dickie, 1966), and K_1 rapidly decreased after reaching a peak at an optimum feeding rate. The efficiency of conversion of absorbed food to growth, or net conversion efficiency (K_2), has not been measured for any elasmobranchs, except for an estimate of 33% provided by Gruber (1984) for juvenile lemon sharks.

8.6   Conclusions

The major prey item consumed by elasmobranchs is teleost fishes; however, there are numerous exceptions to this generalization. Accurate descriptions of the diets of elasmobranchs are complicated by the plasticity of their feeding habits, which regularly result in ontogenetic and spatiotemporal shifts. Based on determinations for a limited number of species, sharks appear to exhibit short feeding bouts followed by longer periods of digestion. The food consumption dynamics of elasmobranchs may ultimately be governed by a morphological peculiarity of this group of predators, a spiral valve intestine. This digestive morphology likely dictates slower rates of gastrointestinal emptying, lower food consumption rates, lower production rates, and generally slower food dynamics for elasmobranchs compared to teleosts. From our limited knowledge, however, it appears that elasmobranchs are capable of absorbing food and converting it to growth with efficiencies comparable to those of teleosts.

Another peculiarity, the physiological adaptation of elasmobranchs for retention of high levels of urea in their blood and tissues, may complicate estimation of trophic levels through stable isotope analysis and quantification of energy losses in gill effluent and urine for bioenergetic studies. Clearly, much remains to be learned about food consumption and feeding habits of elasmobranchs. Because of the difficulty of conducting controlled experiments with large, adult individuals of many elasmobranch species, we advocate a pragmatic approach to advance our knowledge of the feeding ecology of this group.

References

Aedo, G. and H. Arancibia. 2001. Gastric evacuation of the redspotted catshark under laboratory conditions. *J. Fish Biol.* 58:1454–1457.

Ajayi, T. O. 1982. Food and feeding habits of *Raja* species (Batoidei) in Carmarthen Bay, Bristol Channel. *J. Mar. Biol. Assoc. U.K.* 62:215–223.

Ali, T. S., A. R. M. Mohamed, and N. A. Hussain. 1993. Trophic interrelationships of the demersal fish assemblage in the Northwest Arabian Gulf, Iraq. *Asian Fish. Sci.* 6:255–264.

Allen, B. R. and G. Cliff. 2000. Sharks caught in the protective gill nets of Kwazulu-Natal, South Africa. 9. The spinner shark (*Carcharhinus brevipinna*) (Müller and Henle). *S. Afr. J. Mar. Sci.* 22:199–215.

Avsar, D. 2001. Age, growth, reproduction and feeding of the spurdog (*Squalus acanthias* Linnaeus, 1758) in the South-eastern Black Sea. *Estuarine Coastal Shelf Sci.* 52:269–278.

Baba, O., T. Taniuchi, and Y. Nose. 1987. Depth distribution and food habits of three species of small squaloid sharks off Choshi. *Nippon Suisan Gakkaishi* 53:417–424.

Backus, R. H., S. Springer, and E. L. Arnold. 1956. A contribution to the natural history of the white-tip shark, *Pterolamiops longimanus* (Poey). *Deep Sea Res.* 3:178–188.

Baduini, C. L. 1995. Feeding Ecology of the Basking Shark (*Cetorhinus maximus*) Relative to Distribution and Abundance of Prey. M.S. thesis, Moss Landing Marine Laboratories, San Jose State University, San Jose, CA.

Barry, J. P., M. M. Yoklavich, G. M. Cailliet, D. A. Ambrose, and B. S. Antrim. 1996. Trophic ecology of the dominant fishes in Elkhorn Slough, California, 1974–1980. *Estuaries* 19:115–138.

Bass, A. J., J. D. D'Aubrey, and N. Kistnasamy. 1973. *Sharks of the East Coast of Southern Africa. 1. The Genus* Carcharhinus *(Carcharhinidae).* Invest. Rep. Oceanogr. Res. Inst. Durban 33, 168 pp.

Bell, J. C. and J. T. Nichols. 1921. Notes on the food of Carolina sharks. *Copeia* 1921:17–20.

Blaber, S. J. M. and C. M. Bulman. 1987. Diets of fishes of the upper continental slope of eastern Tasmania: content, calorific values, dietary overlap and trophic relationships. *Mar. Biol.* 95:345–356.

Bonham, K. 1954. Food of the spiny dogfish *Squalus acanthias*. *Fish. Res. Pap.* 1:25–36. Washington Dept. Fish.

Bowen, S. 1996. Quantitative description of the diet, in *Fisheries Techniques*. B. R. Murphy and D. W. Willis, Eds., American Fisheries Society, Bethesda, MD, 513–532.

Bradley, J. L. 1996. Prey Energy Content and Selection, Habitat Use and Daily Ration of the Atlantic Stingray, *Dasyatis sabina*. M.S. thesis. Florida Institute of Technology, Melbourne.

Brafield, A. E. and M. J. Llewellyn. 1982. *Animal Energetics*. Blackie and Sons, Glasgow.

Branstetter, S. 1990. Early life-history implications of selected carcharhinoid and lamnoid sharks of the northwest Atlantic, in Elasmobranchs as Living Resources: Advances in the Biology, Ecology, Systematics, and the Status of the Fisheries. H. L. Pratt, Jr., S. H. Gruber, and T. Taniuchi, Eds., NOAA Tech. Rep. NMFS 90, U.S. Department of Commerce, Washington, D.C.

Breeder, C. M., Jr. 1921. The food of *Mustelus canis* (Mitchell) in mid-summer. *Copeia* 101:85–86.

Brett, J. R. and J. M. Blackburn. 1978. Metabolic rate and energy expenditure of the spiny dogfish, *Squalus acanthias*. *J. Fish. Res. Board Can.* 35:816–821.

Brett, J. R. and T. D.D. Groves. 1979. Physiological energetics, in *Fish Physiology,* Vol. 8. W. S. Hoar, D. J. Randall, and J. R. Brett, Eds., Academic Press, New York, 279–352.

Bush, A. C. 2002. The Feeding Ecology of Juvenile Scalloped Hammerhead Sharks (*Sphyrna lewini*) in Kāne'ohe Bay, Ō'ahu, Hawai'i. Ph.D. dissertation, University of Hawaii, Honolulu.

Bush, A. C. and K. N. Holland. 2002. Food limitation in a nursery area: estimates of daily ration in juvenile scalloped hammerheads, *Sphyrna lewini* (Griffith and Smith, 1834) in Kāne'ohe Bay, Ō'ahu, Hawai'i. *J. Exp. Mar. Biol. Ecol.* 278:157–178.

Caira, J. N. and E. C. Jolitz. 1989. Gut pH in the nurse shark, *Ginglymostoma cirratum* (Bonnaterre). *Copeia* 1989:192–194.

Capapé, C. 1974. Contribution à la biologie des Scyliorhinidae des côtes tunisiennes. II. *Scyliorhinus canicula* Linné, 1758: régime alimentaire. *Ann. Inst. Michel Pacha* 7:13–29.

Capapé, C. 1975. Contribution à la biologie des Scyliorhinidae des côtes tunisiennes. IV. *Scyliorhinus stellaris* (Linné, 1758): régime alimentaire. *Arch. Inst. Pasteur Tunis* 52:383–394.

Capapé, C. and J. Zaouali. 1976. Contribution à la biologie des Scyliorhinidae des côtes tunisiennes. V. *Galeus melastomus* Rafinesque, 1810: régime alimentaire. *Arch. Inst. Pasteur Tunis* 53:281–292.

Carlson, J. K. and G. R. Parsons. 1998. Estimates of daily ration and a bioenergetic model for the blacknose shark, *Carcharhinus acronotus*. American Society of Ichthyologists and Herpetologists. 78th Annual Meeting, University of Guelph, Guelph, Ontario, Canada, July 16–22 (abstr.).

Carlson, J. K., M. A. Grace, and P. K. Lago. 2002. An observation of juvenile tiger sharks feeding on clapper rails off the southeastern Coast of the United States. *Southeast. Nat.* 1:307–310.

Carrassón, M., C. Stefanescu, and J. E. Cartes. 1992. Diets and bathymetric distributions of two bathyal sharks of the Catalan deep sea (western Mediterranean). *Mar. Ecol. Prog. Ser.* 82:21–30.

Castro, J. I. 1989. The biology of the golden hammerhead, *Sphyrna tudes*, off Trinidad. *Environ. Biol. Fish.* 24:3–11.

Castro, J. I. 1993. The biology of the finetooth shark, *Carcharhinus isodon*. *Environ. Biol. Fish.* 36:219–239.

Castro, J. I., P. M. Bubucis, and N. A. Overstrom. 1988. The reproductive biology of the chain dogfish, *Scyliorhinus retifer*. *Copeia* 1988:740–746.

Chatwin, B. M. and C. F. Forrester. 1953. Feeding habits of dogfish *Squalus suckleyi* (Girard). *Fish. Res. Bd. Can. Prog. Rep. Pac. Coast Sta.* 95:35–38.

Christensen, V. and D. Pauly. 1992. The Ecopath II — a software for balancing steady-state models and calculating network characteristics. *Ecol. Model.* 61:169–185.

Clark, E. 1963. The maintenance of sharks in captivity, with a report on their instrumental conditioning, in *Sharks and Survival*. P.W. Gilbert, Ed., D.C. Heath, Boston, 115–149.

Clark, E. and K. von Schmidt. 1965. Sharks of the central Gulf coast of Florida. *Bull. Mar. Sci.* 15:13–83.

Clarke, M. R., K. J. King, and P. J. McMillan. 1989. The food and feeding relationships of black oreo, *Allocyttus niger*, smooth oreo, *Pseudocyttus maculatus*, and eight other fish species from the continental slope of the southwest Chatham Rise, New Zealand. *J. Fish. Biol.* 35:465–484.

Clarke, M. R., D. C. Clarke, H. R. Martins, and H. M. da Silva. 1996. The diet of the blue shark (*Prionace glauca* L.) in Azorean waters. Arquipélago. *Life Mar. Sci.* (Ponta Delgada) 14A:41–56.

Cliff, G. 1995. Sharks caught in the protective gill nets off Kwazulu-Natal, South Africa. 8. The great hammerhead shark (*Sphyrna mokarran*) (Ruppell). *S. Afr. J. Mar. Sci.* 15:105–114.

Cliff, G. and S. F. J. Dudley. 1991a. Sharks caught in the protective gill nets of Natal, South Africa. 4. The bull shark (*Carcharhinus leucas*) (Valenciennes). *S. Afr. J. Mar. Sci.* 10:253–270.

Cliff, G. and S. F. J. Dudley. 1991b. Sharks caught in the protective gill nets of Natal, South Africa. 5. The Java shark (*Carcharhinus amboinensis*) (Muller and Henle). *S. Afr. J. Mar. Sci.* 11:443–453.

Cliff, G. and S. F. J. Dudley. 1992. Sharks caught in the protective gill nets of Natal, South Africa. 6. The copper shark (*Carcharhinus brachyurus*) (Gunther). *S. Afr. J. Mar. Sci.* 12:663–674.

Cliff, G., S. F. J. Dudley, and B. Davis. 1988. Sharks caught in the protective gill nets of Natal, South Africa. 1. The sandbar shark (*Carcharhinus plumbeus*) (Nardo). *S. Afr. J. Mar. Sci.* 7:255–265.

Cliff, G., S. F. J. Dudley, and B. Davis. 1989. Sharks caught in the protective gill nets of Natal, South Africa. 2. The great white shark (*Carcharodon carcharias*) (Linnaeus). *S. Afr. J. Mar. Sci.* 8:131–144.

Coles, R. J. 1919. The large sharks off Cape Lookout, North Carolina. The white sharks or maneater, tiger shark and hammerhead. *Copeia* 69:34–43.

Compagno, L. J. V. 1990. Relationships of the megamouth shark, *Megachasma pelagios* (Lamniformes: Megachasmidae), with comments on its feeding habits, in Elasmobranchs as Living Resources: Advances in the Biology, Ecology, Systematics, and the Status of the Fisheries. H. L. Pratt, Jr., S. H. Gruber, and T. Taniuchi, Eds., NOAA Tech. Rep. NMFS 90, U.S. Department of Commerce, Seattle, WA, 357–379.

Corkeron, P. J., R. J. Morris, and M. M. Bryden. 1987. Interactions between bottlenose dolphins and sharks in Moreton Bay, Queensland. *Aquat. Mammals* 13:109–113.

Cortés, E. 1991. Alimentación en el Tiburón Galano, *Negaprion brevirostris* (Poey): Dieta, Hábitos Alimentarios, Digestión, Consumo y Crecimiento. Ph.D. dissertation, University of Barcelona, Spain.

Cortés, E. 1997. A critical review of methods of studying fish feeding based on analysis of stomach contents: application to elasmobranch fishes. *Can. J. Fish. Aquat. Sci.* 54:726–738.

Cortés, E. 1998. Methods of studying fish feeding: reply. *Can. J. Fish. Aquat. Sci.* 55:2708.

Cortés, E. 1999. Standardized diet compositions and trophic levels of sharks. *ICES J. Mar. Sci.* 56:707–717.

Cortés, E. and S. H. Gruber. 1990. Diet, feeding habits, and estimates of daily ration of young lemon sharks, *Negaprion brevirostris* (Poey). *Copeia* 1990:204–218.

Cortés, E. and S. H. Gruber. 1992. Gastric evacuation in the young lemon shark, *Negaprion brevirostris*, under field conditions. *Environ. Biol. Fish.* 35:205–212.

Cortés, E. and S. H. Gruber. 1994. Effect of ration size on growth and gross conversion efficiency of young lemon sharks, *Negaprion brevirostris*. *J. Fish. Biol.* 44:331–341.

Cortés, E., C. A. Manire, and R. E. Hueter. 1996. Diet, feeding habits, and diel feeding chronology of the bonnethead shark, *Sphyrna tiburo*, in southwest Florida. *Bull. Mar. Sci.* 58:353–367.

Cross, J. N. 1988. Aspects of the biology of two scyliorhinid sharks, *Apristurus brunneus* and *Parmaturus xaniurus*, from the upper continental slope off southern California. *Fish. Bull.* 86:691–702.

Dahlberg, M. D. and R.W. Heard. 1969. Observations on elasmobranchs from Georgia. *Q. J. Fla. Acad. Sci.* 32:21–25.

DeCrosta, M. A., L. R. Taylor, Jr., and J. D. Parrish. 1984. Age determination, growth, and energetics of three species of carcharhinid sharks in Hawaii, in *Proceedings of the Second Symposium on Resource Investigations in the Northwest Hawaiian Islands*, Vol. 2, May 25–27, 1983. University of Hawaii Sea Grant MR-84-01, pp. 75–95.

Devadoss, P. 1989. Observations on the length-weight relationship and food and feeding habits of spade nose shark, *Scoliodon laticaudus* Muller and Henle. *Indian J. Fish.* 36:169–174.

Diana, J. S. 1979. The feeding pattern and daily ration of a top carnivore, the northern pike, *Esox lucius. Can. J. Zool.* 57:977–991.

Di Giácomo, E., A. M. Parma, and J. M. Orensanz. 1994. Food consumption by the cock fish, *Callorhynchus callorhynchus* (Holocephali: Callorhynchidae), from Patagonia (Argentina). *Environ. Biol. Fish.* 40:199–211.

Dodrill, J. W. and G. R. Gilmore. 1978. Land birds in the stomachs of tiger sharks *Galeocerdo cuvieri* (Peron and Leseur). *Auk* 95:585–586.

Dudley, S. F. J. and G. Cliff. 1993. Sharks caught in the protective gill nets of Natal, South Africa. 7. The blacktip shark (*Carcharhinus limbatus*) (Valenciennes). *S. Afr. J. Mar. Sci.* 13:237–254.

Dudley, S. F. J., M. D. Anderson-Reade, G. S. Thompson, and P. B. McMullen. 2000. Concurrent scavenging off a whale carcass by great white sharks, *Carcharodon carcharias*, and tiger sharks, *Galeocerdo cuvier. Fish. Bull.* 98:646–649.

Ebert, D. A. 1991. Diet of the sevengill shark *Notorynchus cepedianus* in the temperate coastal waters of southern Africa. *S. Afr. J. Mar. Sci.* 11:565–572.

Ebert, D. A. 1994. Diet of the sixgill shark *Hexanchus griseus* off southern Africa. *S. Afr. J. Mar. Sci.* 14:213–218.

Ebert, D. A., P. D. Cowley, and L. J. V. Compagno. 1991. A preliminary investigation of the feeding ecology of skates (Batoidea: Rajidae) off the west coast of Southern Africa. *S. Afr. J. Mar. Sci.* 10:71–81.

Ebert, D. A., L. J. V. Compagno, and P. D. Cowley. 1992. A preliminary investigation of the feeding ecology of squaloid sharks off the west coast of southern Africa. *S. Afr. J. Mar. Sci.* 12:601–609.

Ebert, D. A., P. D. Cowley, and L. J. V. Compagno.1996. A preliminary investigation of the feeding ecology of catsharks (Scyliorhinidae) off the west coast of southern Africa. *S. Afr. J. Mar. Sci.* 17:233–240.

Eggers, D. M. 1979. Comments on some recent methods for estimating food consumption by fish. *J. Fish. Res. Board Can.* 36:1018–1019.

Elliott, J. M. and W. Davidson. 1975. Energy equivalents of oxygen consumption in animal energetics. *Oecologia* 19:195–201.

Elliott, J. M. and L. Persson. 1978. The estimation of daily rates of food consumption for fish. *J. Anim. Ecol.* 47:977–991.

Ellis, J. R., M. G. Pawson, and S. E. Shackley. 1996. The comparative feeding ecology of six species of shark and four species of ray (Elasmobranchii) in the north-east Atlantic. *J. Mar. Biol. Assoc. U.K.* 76:89–106.

Euzen, O. 1987. Food habits and diet comparison of some fish of Kuwait. *Kuwait Bull. Mar. Sci.* 1987:65–85.

Fänge, R. and D. Grove. 1979. Digestion, in *Fish Physiology*, Vol. 8. W.S. Hoar, D.J. Randall, and J.R. Brett, Eds., Academic Press, New York, 161–260.

Fergusson, I. K., L. J. V. Compagno, and M. A. Marks. 2000. Predation by white sharks *Carcharodon carcharias* (Chondrichthyes: Lamnidae) upon chelonians, with new records from the Mediterranean Sea and a first record of the ocean sunfish *Mola mola* (Osteichthyes: Molidae) as stomach contents. *Environ. Biol. Fish.* 58:447–453.

Ferry, L. A. and G. M. Cailliet. 1996. Sample size and data analysis: are we characterizing and comparing diet properly? in *Feeding Ecology and Nutrition in Fish, International Congress of the Biology of Fishes.* D. MacKinlay and K. Shearer, Eds., American Fisheries Society, Bethesda, MD, 71–80.

Fisk, A.T., S. A. Tittlemier, J. L. Pranschke, and R. J. Norstrom. 2002. Using anthropogenic contaminants and stable isotopes to assess the feeding ecology of Greenland sharks. *Ecology* 83:2162–2172.

Ford, E. 1921. A contribution to our knowledge of the life histories of the dogfishes landed at Plymouth. *J. Mar. Biol. Assoc. U.K.* 12:468–505.

García de la Rosa, S. B. and F. Sánchez. 1997. Alimentación de *Squalus acanthias* y predación sobre *Merluccius hubbsi* en el Mar Argentino entre 34°47′– 47°S. *Rev. Invest. Des. Pesq.* 11:119–133.

Gauld, J. A. 1989. Records of porbeagles landed in Scotland, with observations on the biology, distribution and exploitation of the species. DAFS Scottish Fish. Res. Rep. 45, 15 pp.

Gelsleichter, J., J. A. Musick, and S. Nichols. 1999. Food habits of the smooth dogfish, *Mustelus canis*, dusky shark, *Carcharhinus obscurus*, Atlantic sharpnose shark, *Rhizoprionodon terraenovae*, and the sand tiger, *Carcharias taurus*, from the northwest Atlantic Ocean. *Environ. Biol. Fish.* 54:205–217.

Gómez Fermin, E. and A. K. M. Bashirulah. 1984. Relación longitud-peso y hábitos alimenticios de *Rhizoprionodon porosus* Poey 1861 (Fam. Carcharhinidae) en el oriente de Venezuela. *Bol. Inst. Oceanogr. Venezuela Univ. Oriente* 23:49–54.

Graeber, R. C. 1974. Food intake patterns in captive juvenile lemon sharks, *Negaprion brevirostris. Copeia* 1974:554–556.

Gruber, S. H. 1984. Bioenergetics of the captive and free-ranging lemon shark (*Negaprion brevirostris*). *Proc. Am. Assoc. Parks Aquar.* 1984:341–373.

Gubanov, Y. P. and V. N. Grigoryev. 1975. Observations on the distribution and biology of the blue shark *Prionace glauca* (Carcharhinidae) of the Indian Ocean. *J. Ichthyol.* 15:37–43.

Gudger, E. W. 1932. Cannibalism among the sharks and rays. *Sci. Mon.* 34:403–419.

Gudger, E. W. 1941. The food and feeding habits of the whale shark, *Rhineodon typus*. *J. Elisha Mitchell Soc.* July:57–72.

Gudger, E. W. 1948. The tiger shark, *Galeocerdo tigrinus*, on the North Carolina coast and its food and feeding habits there. *J. Elisha Mitchell Sci. Soc.* 64:221–233.

Gudger, E. W. 1949. Natural history notes on tiger sharks *Galeocerdo tigrinus*, caught at Key West, Florida, with emphasis on food and feeding habits. *Copeia* 1949:39–47.

Hallacher, L. E. 1977. On the feeding behavior of the basking shark, *Cetorhinus maximus*. *Environ. Biol. Fish.* 2:297–298.

Hanchet, S. 1991. Diet of spiny dogfish, *Squalus acanthias* Linnaeus, on the east coast, South Island, New Zealand. *J. Fish Biol.* 39:313–323.

Hansson, S. 1998. Methods of studying fish feeding: a comment. *Can. J. Fish. Aquat. Sci.* 55:2706–2707.

Harvey, J. T. 1989. Food habits, seasonal abundance, size, and sex of the blue shark, *Prionace glauca*, in Monterey Bay, California. *Calif. Fish Game* 75:33–44.

Heatwole, J., E. Heatwole, and C. R. Johnson. 1974. Shark predation on sea snakes. *Copeia* 1974:780–781.

Heithaus, M. R. 2001a. The biology of tiger sharks, *Galeocerdo cuvier*, in Shark Bay, Western Australia: sex ratio, size distribution, diet, and seasonal changes in catch rates. *Environ. Biol. Fish.* 61:25–36.

Heithaus, M. R. 2001b. Predator–prey and competitive interactions between sharks (order Selachii) and dolphins (suborder Odontoceti): a review. *J. Zool. Lond.* 253:53–68.

Henningsen, A. D. 1996. Captive husbandry and bioenergetics of the spiny butterfly ray, *Gymnura altavela* (Linnaeus). *Zoo Biol.* 15:135–142.

Heupel, M. R. and M. B. Bennett. 1998. Observations on the diet and feeding habits of the epaulette shark, *Hemiscyllium ocellatum* (Bonnaterre), on Heron Island Reef, Great Barrier Reef, Australia. *Mar. Freshwater Res.* 49:753–756.

Holden, M. J. 1966. The food of the spurdog, *Squalus acanthias* (L.). *J. Cons. Perm. Int. Explor. Mer* 30:255–266.

Holmgren, S. and S. Nilsson. 1999. Digestive system, in *Sharks, Skates, and Rays. The Biology of Elasmobranch Fishes.* W. C. Hamlett, Ed., Johns Hopkins University Press, Baltimore, 144–173.

Horie, T. and S. Tanaka. 2000. Reproduction and food habits of two species of sawtail sharks, *Galeus eastmani* and *G. nipponenis*, in Suruga Bay, Japan. *Fish. Sci.* 66:812–825.

Jahn, A. E. and R. L. Haedrich. 1988. Notes on the pelagic squaloid shark *Isistius brasiliensis*. *Biol. Oceanogr.* 5:297–309.

Jakobsdóttir, K. B. 2001. Biological aspects of two deep-water squalid sharks: *Centroscyllium fabricii* (Reinhardt, 1825) and *Etmopterus princeps* (Collett, 1904) in Icelandic waters. *Fish. Res.* 51:247–265.

Jones, B. C. and G. H. Geen. 1977. Food and feeding of spiny dogfish (*Squalus acanthias*) in British Columbia waters. *J. Fish. Res. Board Can.* 34:2067–2078.

Jones, E. C. 1971. *Isistius brasiliensis*, a squaloid shark, the probable cause of crater wounds on fishes and cetaceans. *Fish. Bull.* 69:791–798.

Kao, J. S. 2000. Diet, Daily Ration and Gastric Evacuation of the Leopard Shark (*Triakis semifasciata*). M.S. thesis, California State University, Hayward.

King, K. J. and M. R. Clark. 1984. The food of rig (*Mustelus lenticulatus*) and the relationship of feeding to reproduction and condition in Golden Bay. *N. Z. J. Freshwater Res.* 18:29–42.

Kleiber, M. 1975. *The Fire of Life: An Introduction to Animal Energetics*. Krieger, Huntington, NY.

Koen Alonso, M., E. A. Crespo, N. A. García, S. N. Pedraza, P. A. Mariotti, and N. J. Mora. 2002. Fishery and ontogenetic driven changes in the diet of the spiny dogfish, *Squalus acanthias*, in Patagonian waters, Argentina. *Environ. Biol. Fish.* 63:193–202.

Kohler, N. E. 1987. Aspects of the Feeding Ecology of the Blue Shark, *Prionace glauca* in the Western North Atlantic. Ph.D. dissertation, University of Rhode Island, Kingston.

Kondyurin, V. V. and N. A. Myagkov. 1982. Morphological characteristics of two species of spiny dogfish, *Squalus acanthias* and *Squalus fernandinus* (Squalidae, Elasmobranchii), from the southeastern Atlantic. *J. Ichthyol.* 22:41–51.

Kubota, T., Y. Shiobara, and T. Kubodera. 1991. Food habits of the frilled shark, *Chlamydoselachus anguineus* collected from Suruga Bay, Central Japan. *Nippon Suisan Gakkaishi* 57:15–20.

Lawler, E. F. 1976. The Biology of the Sandbar Shark *Carcharhinus plumbeus* (Nardo, 1827) in the Lower Chesapeake Bay and Adjacent Waters. M.S. thesis, College of William and Mary, Williamsburg, VA.

LeBoeuf, B. L., M. Riedman, and R. S. Keyes. 1982. White shark predation on pinnipeds in California coastal waters. *Fish. Bull.* 80:891–895.

Liao, H, C. L. Pierce, and J. G. Larscheid. 2001. Empirical assessment of indices of prey importance in the diets of predacious fish. *Trans. Am. Fish. Soc.* 130:583–591.

Long, D. J. 1996. Records of white shark-bitten leatherback sea turtles along the California coast, in *Great White Sharks. The Biology of Carcharodon carcharias.* A. P. Klimley and D. J. Ainley, Eds., Academic Press, New York, 317–319.

Longval, M. J., R. M. Warner, and S. H. Gruber. 1982. Cyclical patterns of food intake in the lemon shark *Negaprion brevirostris* under controlled conditions. *Fla. Sci.* 45:25–33.

Lowe, C. G. 2002. Bioenergetics of free-ranging juvenile scalloped hammerhead sharks (*Sphyrna lewini*) in Kāne'ohe Bay, Ō'ahu, HI. *J. Exp. Mar. Biol. Ecol.* 278:141–156.

Lowe, C. G., B. M. Wetherbee, G. L. Crow, and A. L. Tester. 1996. Ontogenetic dietary shifts and feeding behavior of the tiger shark, *Galeocerdo cuvier*, in Hawaiian waters. *Environ. Biol. Fish.* 47:203–211.

Lyle, J. M. 1983. Food and feeding habits of the lesser spotted dogfish, *Scyliorhinus canicula* (L.), in Isle of Man waters. *J. Fish. Biol.* 23:725–737.

Lyle, J. M. 1987. Observations on the biology of *Carcharhinus cautus* (Whitley), *C. melanopterus* (Quoy and Gaimard) and *C. fitzroyensis* (Whitley) from Northern Australia. *Aust. J. Mar. Freshwater Res.* 38:701–710.

Lyle, J. M. and G. J. Timms. 1987. Predation on aquatic snakes by sharks from northern Australia. *Copeia* 1987:802–803.

Macpherson, E. 1980. Régime alimentaire de *Galeus melastomus* Rafinesque, 1810, *Etmopterus spinax* (L., 1758) et *Scymnorhinus licha* (Bonnaterre, 1788) en Méditerranée occidentale. *Vie Milieu* 30:139–148.

Macpherson, E. 1981. Resource partitioning in a Mediterranean demersal fish community. *Mar. Ecol. Prog. Ser.* 4:183–193.

Macpherson, E., J. Lleonart, and P. Sánchez. 1989. Gastric emptying in *Scyliorhinus canicula* (L.): a comparison of surface-dependent and non-surface dependent models. *J. Fish Biol.* 35:37–48.

Matallanas, J. 1982. Feeding habits of *Scymnorhinus licha* in Catalan waters. *J. Fish Biol.* 20:155–163.

Matallanas, J., M. Carrasón, and M. Casadevall. 1993. Observations on the feeding habits of the narrow mouthed cat shark *Schroederichthys bivius* (Chondrichthyes, Scyliorhinidae) in the Beagle Channel. *Cybium* 17:55–61.

Mauchline, J. and J. D. M. Gordon. 1983. Diets of the sharks and chimaeroids of the Rockall Trough, northeastern Atlantic Ocean. *Mar. Biol.* 75:269–278.

McCosker, J. E. 1987. The white shark, *Carcharodon carcharias*, has a warm stomach. *Copeia* 1987:195–197.

Medved, R. J. 1985. Gastric evacuation in the sandbar shark, *Carcharhinus plumbeus. J. Fish Biol.* 26:239–253.

Medved, R. J., C. E. Stillwell, and J. G. Casey. 1985. Stomach contents of young sandbar sharks, *Carcharhinus plumbeus*, in Chincoteague Bay, Virginia. *Fish. Bull.* 83:395–402.

Medved, R. J., C. E. Stillwell, and J. G. Casey. 1988. The rate of food consumption of young sandbar sharks (*Carcharhinus plumbeus*) in Chincoteague Bay, Virginia. *Copeia* 1988:956–963.

Menni, R. C. 1985. Distribución y biología de *Squalus acanthias, Mustelus schmitti* y *Galeorhinus vitaminicus* en el mar Argentino en agosto-septiembre de 1978 (Chondrichthyes). *Rev. Mus. Plata (Nueva Serie) Sec. Zool.* 13:151–182.

Moss, S. A. 1984. *Sharks — An Introduction for the Amateur Naturalist.* Prentice-Hall, Englewood Cliffs, NJ.

Mumtaz Tirasin, E. and T. Jorgensen. 1999. An evaluation of the precision of diet description. *Mar. Ecol. Prog. Ser.* 182:243–252.

Muñoz-Chapuli, R., J. C. R. Salgado, and J. M. de La Serna. 1988. Biogeography of *Isistius brasiliensis* in the north-eastern Atlantic, inferred from crater wounds on swordfish (*Xiphas gladius*). *J. Mar. Biol. Assoc. U.K.* 68:315–321.

Nagasawa, K. 1998. Predation by salmon sharks (*Lamna ditropis*) on Pacific salmon (*Oncorhynchus* spp.) in the North Pacific Ocean. *N. Pac. Anadr. Fish Comm. Bull.* 1:419–433.

Nelson, G. A. and M. R. Ross. 1995. Gastric evacuation in little skate. *J. Fish Biol.* 46:977–986.

Olsen, A. M. 1954. The biology, migration and growth rate of the school shark, *Galeorhinus australis* (Mcleay) (Carcharhinidae) in south-eastern Australian waters. *Aust. J. Mar. Freshwater Res.* 5:353–410.

Olsen, A. M. 1984. Synopsis of biological data on the school shark *Galeorhinus australis*. FAO Fisheries Synopsis 139. FAO Fisheries, Rome.

Olson, R. J. and A. J. Mullen. 1986. Recent developments for making gastric evacuation and daily ration determinations. *Environ. Biol. Fish.* 16:183–191.

Orlov, A. M. 1998. On feeding of mass species of deep-sea skates (*Bathyraja* spp., Rajidae) from the Pacific waters of the northern Kurils and southeastern Kamchatka. *J. Ichthyol.* 38:635–644.

Ostrom, P. H., J. Lien, and S. A. Macko. 1993. Evaluation of the diet of Sowerby's beaked whale, *Mesoplodon bidens*, based on isotopic comparisons among northwest Atlantic cetaceans. *Can. J. Zool.* 71:858–861.

Paloheimo, J. and L. M. Dickie. 1966. Food and growth of fishes. III. Relations among food, body size and growth efficiency. *J. Fish. Res. Board Can.* 23:1209–1248.

Pandian, T. J. 1982. Contributions to the bioenergetics of a tropical fish, in *Gutshop'81 Fish Food Habits Studies*. G. M. Cailliet and C. A. Simenstad, Eds., Washington Sea Grant Program, Seattle, WA, 124–131.

Papastamatiou, Y. 2003. Gastric pH Changes Associated with Feeding in Leopard Sharks (*Triakis semifasciata*): Can pH Be Used to Study the Foraging Ecology of Sharks? M.S. thesis, California State University, Long Beach.

Pennington, M. 1985. Estimating the average food consumption by fish in the field from stomach contents data. *Dana* 5:81–86.

Perlman, D. F. and L. Goldstein. 1988. Nitrogen metabolism, in *Physiology of Elasmobranch Fishes*. T. J. Shuttleworth, Ed., Springer-Verlag, Berlin, 253–276.

Platell, M. E., I. C. Potter, and K. R. Clarke. 1998. Resource partitioning by four species of elasmobranchs (Batoidea: Urolophidae) in coastal waters of temperate Australia. *Mar. Biol.* 131:719–734.

Rae, B. B. 1967. The food of the dogfish, *Squalus acanthias* L. Department of Agriculture and Fisheries for Scotland. Her Majesty's Stationery Office, Edinburgh.

Randall, B. M., R. M. Randall, and L. J. V. Compagno. 1988. Injuries to jackass penguins (*Spheniscus demersus*): evidence for shark involvement. *J. Zool.* 214:589–600.

Randall, J. E. 1967. Food habits of reef fishes of the West Indies. *Stud. Trop. Oceanogr.* 5:665–847.

Relini Orsi, L. and M. Wurtz. 1977. Patterns and overlap in the feeding of two selachians of bathyal fishing grounds in the Ligurian Sea. *Rapp. Comm. Int. Mer Médit.* 24:89–94.

Rountree, R. A. and K. W. Able. 1996. Seasonal abundance, growth, and foraging habits of juvenile smooth dogfish, *Mustelus canis*, in a New Jersey estuary. *Fish. Bull.* 94:522–534.

Salini, J. P., S. J. M. Blaber, and D. T. Brewer. 1990. Diets of piscivorous fishes in a tropical Australian estuary, with special reference to predation on penaeid prawn. *Mar. Biol.* 105:363–374.

Salini, J. P., S. J. M. Blaber, and D. T. Brewer. 1992. Diets of sharks from estuaries and adjacent waters of the north-eastern Gulf of Carpentaria, Australia. *Aust. J. Mar. Freshwater Res.* 43:87–96.

Salini, J. P., S. J. M. Blaber, and D. T. Brewer. 1994. Diets of trawled predatory fish of the Gulf of Carpentaria, Australia, with particular reference to predation on prawns. *Aust. J. Mar. Freshwater Res.* 45:397–411.

Salini, J. P., M. Tonks, S. J. M. Blaber, and J. Ross. 1999. Feeding of captive, tropical carcharhinid sharks from the Embley River estuary, northern Australia. *Mar. Ecol. Prog. Ser.* 184:309–314.

San Filippo, R. A. 1995. Diet, Gastric Evacuation and Estimates of Daily Ration of the Gray Smoothhound, *Mustelus californicus*. M.S. thesis, San Jose State University, San Jose, CA.

Sarangadhar, P. N. 1983. Tiger shark — *Galeocerdo tigrinus* — Muller and Henle. Feeding and breeding habits. *J. Bombay Nat. Hist. Soc.* 44:101–110.

Saunders, G. B. and E. Clark. 1962. Yellow-billed cuckoo in stomach of tiger shark. *Auk* 79:118.

Scharf, F. S., R. M. Yetter, A. P. Summers, and F. Juanes. 1998. Enhancing diet analysis of piscivorous fishes in the Northwest Atlantic through identification and reconstruction of original prey sizes from ingested remains. *Fish. Bull.* 96:575–588.

Schmid, T. H. and F. L. Murru. 1994. Bioenergetics of the bull shark, *Carcharhinus leucas*, maintained in captivity. *Zoo Biol.* 13:177–185.

Schmid, T. H., F. L. Murru, and F. McDonald. 1990. Feeding habits and growth rates of bull (*Carcharhinus leucas* (Valenciennes)), sandbar (*Carcharhinus plumbeus* (Nardo)), sandtiger (*Eugomphodus taurus* (Rafinesque)) and nurse (*Ginglymostoma cirratum* (Bonnaterre)) sharks maintained in captivity. *J. Aquar. Aquat. Sci.* 5:100–105.

Schmidt, T. W. 1986. Food of young juvenile lemon sharks, *Negaprion brevirostris* (Poey), near Sandy Key, western Florida Bay. *Fla. Sci.* 49:7–10.

Schurdak, M. E. and S. H. Gruber. 1989. Gastric evacuation of the lemon shark *Negaprion brevirostris* (Poey) under controlled conditions. *Exp. Biol.* 48:77–82.

Schwartz, F. J. 1996. Biology of the clearnose skate, *Raja eglanteria*, from North Carolina. *Fla. Sci.* 59:82–95.

Scofield, N. B. 1920. Sleeper shark captured. *Calif. Fish Game* 6:80.

Segura-Zarzosa, J. C., L. A. Abitia-Cárdenas, and F. Galván-Magaña. 1997. Observations on the feeding habits of the shark *Heterodontus francisci* Girard 1854 (Chondrichthyes: Heterodontidae), in San Ignacio Lagoon, Baja California Sur, México. *Cien. Mar.* 23:111–128.

Shirai, S. and K. Nakaya. 1992. Functional morphology of feeding apparatus of the cookie-cutter shark, *Isistius brasiliensis* (Elasmobranchii, Dalatiinae). *Zool. Sci.* 9:811–821.

Simpfendorfer, C. A. 1992. Biology of tiger sharks (*Galeocerdo cuvier*) caught by the Queensland shark meshing program off Townsville, Australia. *Aust. J. Mar. Freshwater Res.* 43:33–43.

Simpfendorfer, C. A. and N. E. Milward. 1993. Utilisation of a tropical bay as a nursery area by sharks of the families Carcharhinidae and Sphyrnidae. *Environ. Biol. Fish.* 37:337–345.

Simpfendorfer, C. A., A. B. Goodreid, and R. B. McAuley. 2001a. Size, sex and geographic variation in the diet of the tiger shark, *Galeocerdo cuvier*, from Western Australian waters. *Environ. Biol. Fish.* 61:37–46.

Simpfendorfer, C. A., A. B. Goodreid, and R. B. McAuley. 2001b. Diet of three commercially important shark species from Western Australian waters. *Mar. Freshwater Res.* 52:975–985.

Sims, D. W. and D. A. Merrett. 1997. Determination of zooplankton characteristics in the presence of surface feeding basking sharks *Cetorhinus maximus*. *Mar. Ecol. Prog. Ser.* 158:297–302.

Sims, D. W. and V. A. Quayle. 1998. Selective foraging behaviour of basking sharks on zooplankton in a small-scale front. *Nature* 393:460–464.

Sims, D. W., S. J. Davies, and Q. Bone. 1996. Gastric emptying rate and return of appetite in lesser spotted dogfish, *Scyliorhinus canicula* (Chondrichthyes: Elasmobranchii). *J. Mar. Biol. Assoc. U.K.* 76:479–491.

Smale, M. J. 1991. Occurrence and feeding of three shark species, *Carcharhinus brachyurus*, *C. obscurus* and *Sphyrna zygaena*, on the Eastern Cape coast of South Africa. *S. Afr. J. Mar. Sci.* 11:31–42.

Smale, M. J. and L. J. V. Compagno. 1997. Life history and diet of two southern African smoothhound sharks, *Mustelus mustelus* (Linnaeus, 1758) and *Mustelus palumbes* Smith, 1957 (Pisces: Triakidae). *S. Afr. J. Mar. Sci.* 18:229–248.

Smale, M. J. and P. D. Cowley. 1992. The feeding ecology of skates (Batoidea: Rajidae) off the Cape south coast, South Africa. *S. Afr. J. Mar. Sci.* 12:823–834.

Smale, M. J. and A. J. J. Goosen. 1999. Reproduction and feeding of spotted gully shark, *Triakis megalopterus* off the Eastern Cape, South Africa. *Fish. Bull.* 97:987–998.

Smith, J. W. and J. V. Merriner. 1985. Food habits and feeding behavior of the cownose ray, *Rhinoptera bonasus*, in lower Chesapeake Bay. *Estuaries* 8:305–310.

Snelson, F. F., Jr., T. J. Mulligan, and S. E. Williams. 1984. Food habits, occurrence, and population structure of the bull shark, *Carcharhinus leucas* in Florida coastal lagoons. *Bull. Mar. Sci.* 34:71–80.

Springer, S. 1950. Natural history notes on the lemon shark *Negaprion brevirostris*. *Tex. J. Sci.* 2:349–359.

Stevens, J. D. 1973. Stomach contents of the blue shark (*Prionace glauca* L.) off southwest England. *J. Mar. Biol. Assoc. U.K.* 53:357–361.

Stevens, J. D. 1984. Biological observations on sharks caught by sport fishermen off New South Wales. *Aust. J. Mar. Freshwater Res.* 35:573–590.

Stevens, J. D. and J. M. Lyle. 1989. Biology of three hammerhead sharks (*Eusphyra blochii, Sphyrna mokarran*, and *S. lewini*), from northern Australia. *Aust. J. Mar. Freshwater Res.* 40:129–146.

Stevens, J. D. and K. J. McLoughlin. 1991. Distribution, size, and sex composition, reproductive biology and diet of sharks from northern Australia. *Aust. J. Mar. Freshwater Res.* 40:129–146.

Stevens, J. D. and P. D. Wiley. 1986. Biology of two commercially important carcharhinid sharks from northern Australia. *Aust. J. Mar. Freshwater Res.* 37:671–688.

Stevens, J. D., T. L. O. Davis, and A. G. Church. 1982. NT shark gillnetting survey shows potential for Australian fishermen. *Aust. Fish.* 41:39–43.

Stillwell, C. E. and J. G. Casey. 1976. Observations on the bigeye thresher shark, *Alopias superciliosus* in the western North Atlantic. *Fish. Bull.* 74:221–225.

Stillwell, C. E. and N. E. Kohler. 1982. Food, feeding habits, and estimates of daily ration of the shortfin mako (*Isurus oxyrinchus*) in the Northwest Atlantic. *Can. J. Fish. Aquat. Sci.* 39:407–414.

Stillwell, C. E. and N. E. Kohler. 1993. Food habits of the sandbar shark *Carcharhinus plumbeus* off the U.S. northeast coast, with estimates of daily ration. *Fish. Bull.* 91:138–150.

Strasburg, D. W. 1963. The diet and dentition of *Isistius brasiliensis*, with remarks on tooth replacement in other sharks. *Copeia* 1963:33–40.

Sullivan, M. S. 1907. The physiology of the digestive tract of elasmobranchs. *U.S. Fish Wildl. Serv. Fish. Bull.* 27:3–27.

Talent, L. G. 1976. Food habits of the leopard shark, *Triakis semifasciata*, in Elkhorn Slough, Monterey Bay, California. *Calif. Fish Game* 62:286–298.

Talent, L. G. 1982. Food habits of the gray smoothhound, *Mustelus californicus*, the brown-smoothhound, *Mustelus henlei*, the shovelnose guitarfish, *Rhinobatos productus*, and the bat ray, *Myliobatus californica*, in Elkhorn Slough, California. *Calif. Fish Game* 68:224–234.

Tanasichuk, R. W., D. M. Ware, W. Shaw, and G. A. McFarlane. 1991. Variations in diet, daily ration, and feeding periodicity of Pacific hake (*Merluccius productus*) and spiny dogfish (*Squalus acanthias*) off the lower west coast of Vancouver Island. *Can. J. Fish. Aquat. Sci.* 48:2118–2128.

Taniuchi, T., N. Kuroda, and Y. Nose. 1983. Age, growth, reproduction, and food habits of the star-spotted dogfish, *Mustelus manazo* collected from Choshi. *Bull. Jpn. Soc. Sci. Fish.* 49:1325–1334.

Taylor, L. and M. Wisner. 1989. Growth rates of captive blacktip reef sharks (*Carcharhinus melanopterus*). *Bull. Inst. Océanogr. Monaco* 5:211–217.

Tricas, T. C. 1979. Relationships of the blue shark, *Prionace glauca*, and its prey species near Santa Catalina Island, California. *Fish. Bull.* 77:175–182.

Tricas, T. C. and J. E. McCosker. 1984. Predatory behavior of the white shark (*Carcharodon carcharias*), with notes on its biology. *Proc. Calif. Acad. Sci.* 43:221–238.

Tuma, R. E. 1976. An investigation of the feeding habits of the bull shark, *Carcharhinus leucas*, in the lake Nicaragua-Rio San Juan system, in *Investigations of the Ichthyofauna of Nicaraguan Lakes*. T.B. Thorson, Ed., School of Life Sciences, University of Nebraska, Lincoln, 533–538.

Tyminski, J. P., E. Cortés, C. A. Manire, and R. E. Hueter. 1999. Gastric evacuation and estimates of daily ration in the bonnethead shark *Sphyrna tiburo*. American Society of Ichthyologists and Herpetologists, 79th Annual Meeting, Pennsylvania State University, University Park, June 24–30 (abstr.).

Van Dykhuizen, G. and H. F. Mollet. 1992. Growth, age estimation and feeding of captive sevengill sharks, *Notorynchus cepedianus*, at the Monterey Bay aquarium. *Aust. J. Mar. Freshwater Res.* 43:297–318.

Van Slyke, D. D. and G. F. White. 1911. Digestion of protein in the stomach and in intestine of the dogfish. *J. Biol. Chem.* 9:209–217.

Vianna, M. and A. F. de Amorim. 1995. Feeding habits of the shark *Mustelus canis* (Mitchill, 1815), caught in southern Brazil. VII Reunião do grupo de trabalho sobre pesca e pesquisa de tubarões e raias no Brasil. Fundação Universidade do Rio Grande, Rio Grande, Brasil, 20–24 November 1995 (abstr.).

Waller, G. N. H. and A. Baranes. 1994. Food of *Iago omanensis*, a deep water shark from the northern Red Sea. *J. Fish Biol.* 45:37–45.

Warren, C. E. and G. E. Davis. 1967. Laboratory studies on the feeding, bioenergetics, and growth of fish, in *The Biological Basis of Freshwater Fish Production*. S.D. Gerking, Ed., Blackwell Scientific, London, 175–214.

Wass, R. C. 1971. A Comparative Study of the Life History, Distribution and Ecology of the Sandbar Shark and the Gray Reef Shark in Hawaii. Ph.D. dissertation, University of Hawaii, Honolulu.

Wass, R. C. 1973. Size, growth, and reproduction of the sandbar shark, *Carcharhinus milberti*, in Hawaii. *Pac. Sci.* 27:305–318.

Webber, J. D. and J. J. Cech, Jr. 1998. Nondestructive diet analysis of the leopard shark from two sites in Tomales Bay, California. *Calif. Fish Game* 84:18–24.

Weihs, D., R. S. Keyes, and D. M. Stalls. 1981. Voluntary swimming speeds of two species of large carcharhinid sharks. *Copeia* 1981:219–222.

Wetherbee, B. M. and S. H. Gruber. 1990. The effect of ration level on food retention time in juvenile lemon sharks, *Negaprion brevirostris*. *Environ. Biol. Fish.* 29:59–65.

Wetherbee, B. M. and S. H. Gruber. 1993. Absorption efficiency of the lemon shark *Negaprion brevirostris* at varying rates of energy intake. *Copeia* 1993:416–425.

Wetherbee, B. M., S. H. Gruber, and A. L. Ramsey. 1987. X-radiographic observations of food passage through digestive tracts of lemon sharks. *Trans. Am. Fish. Soc.* 116:763–767.

Wetherbee, B. M., S. H. Gruber, and E. Cortés. 1990. Diet, feeding habits, digestion, and consumption in sharks, with special reference to the lemon shark, *Negaprion brevirostris*, in Elasmobranchs as Living Resources: Advances in the Biology, Ecology, Systematics, and the Status of the Fisheries. H. L. Pratt, Jr., S. H. Gruber, and T. Taniuchi, Eds., NOAA Tech. Rep. NMFS 90, U.S. Department of Commerce, Seattle, WA, 29–47.

Wetherbee, B. M., C. G. Lowe, and G. L. Crow. 1996. Biology of the Galapagos shark, *Carcharhinus galapagensis*, in Hawai'i. *Environ. Biol. Fish.* 45:299–310.

Wetherbee, B. M., C. G. Crow, and C. G. Lowe. 1997. Distribution, reproduction, and diet of the gray reef shark *Carcharhinus amblyrhychos* in Hawaii. *Mar. Ecol. Prog. Ser.* 151:181–189.

Winberg, G. G. 1960. Rate of metabolism and food requirements of fishes. *Fish. Res. Board. Can. Trans. Ser.* 194:1–202.

Witzell, W. N. 1987. Selective predation on large cheloniid sea turtles by tiger sharks (*Galeocerdo cuvier*). *Jpn. J. Herpetol.* 12:22–29.

Wood, C. M. 1993. Ammonia and urea metabolism and excretion, in *The Physiology of Fishes*. D. H. Evans, Ed., CRC Press, Boca Raton, FL, 379–425.

Yamaguchi, A. and T. Taniuchi. 2000. Food variations and ontogenetic dietary shift of the starspotted-dogfish *Mustelus manazo* at five locations in Japan and Taiwan. *Fish. Sci.* 66:1039–1048.

9

Homeostasis: Osmoregulation, pH Regulation, and Nitrogen Excretion

David H. Evans, Peter M. Piermarini, and Keith P. Choe

CONTENTS

9.1 Introduction

The osmotic composition of the body fluids of all sharks, skates, and rays is distinctly different from that of their environment (Table 9.1). Of special note is the extremely high concentration of urea that characterizes marine elasmobranch plasma (and cytoplasm). Enzyme function is maintained in the face of such high concentrations of urea by protein structural modifications to offset the denaturing effects of urea and/or by the presence of a stabilizing solute, trimethylamine-N-oxide (TMAO). Elasmobranch blood TMAO levels are usually approximately 75 mmol, but intracellular concentrations approach 200 mmol or 50% of urea levels, a ratio that has been shown to be of general occurrence (e.g., Hochachka and Somero, 2002). Because of the salt and water gradients across their permeable gill epithelium, these fishes must utilize various organs to maintain blood ionic and osmotic consistency (osmoregulation). Specifically, the osmotic gain of water across the gills of marine elasmobranchs must be balanced by renal excretion, and the diffusional gain of NaCl must be balanced by salt excretory mechanisms in the rectal gland, kidney, or gills. The few species of elasmobranchs that enter or reside in more dilute salinities (including freshwater) must balance the large osmotic gain of water by increased urinary water excretion. In addition, the diffusional loss of NaCl across the gills must be balanced by gill salt uptake mechanisms, and the kidney tubules must reabsorb salt to minimize urinary salt loss.

TABLE 9.1

Blood Osmolarity and Major Solutes of Some Representative Elasmobranchs

Species	Osmolarity (mOsm.l⁻¹)	Na (mmol.l⁻¹)	Cl (mmol.l⁻¹)	Urea (mmol.l⁻¹)	Ref.
Squalus acanthias (SW)	1018	286	246	351	Burger and Hess, 1960
Dasyatis sabina (SW)	1034	310	300	394	Piermarini and Evans, 1998
Seawater	930	440	495	0	Burger and Hess, 1960
Dasyatis sabina (FW)	621	212	208	196	Piermarini and Evans, 1998
Potamotrygon sp. (FW)	282	164	152	1.1	Griffith et al., 1973
Potamotrygon sp. (FW)	320	178	146	1.2	Wood et al., 2002
Freshwater	38	3.0	3.7	ND	Piermarini and Evans, 1998

The pH of the body fluids of elasmobranchs is regulated tightly despite the internal synthesis of acid–base metabolites (e.g., ammonia, bicarbonate, urea, and fixed acids) and changes in the acid–base chemistry of environmental water (e.g., hypercapnia, temperature change, and low pH). This is accomplished by rapid pH buffering, followed by net excretion of acid and/or base into environmental water by the same organs that maintain osmotic balance (reviewed by Claiborne, 1998). Nitrogenous metabolites, produced by the catabolism of amino acids and nucleic acids, are also excreted by these organs (reviewed by Wood, 1993). Most of this nitrogen is excreted as urea, as would be expected from its high concentrations in the blood (Table 9.1). However, urea excretion is counterproductive to osmoregulation in seawater; therefore, the gills and kidneys are optimized for retaining this needed osmolyte, and these mechanisms are discussed with osmoregulation. Ammonia is produced in lower quantities but is more toxic than urea. It is secreted by mechanisms that are associated with acid and/or base secretion, and therefore are discussed with pH regulation.

Earlier, general discussions of these physiological problems and solutions can be found in a variety of reviews (e.g., Shuttleworth, 1988a; Evans, 1993, 1998; Hamlett, 1999; McNab, 2002); this chapter focuses on more recent research advances, especially those using cellular and molecular approaches.

9.2   Anatomy of the Rectal Gland, Kidney, and Gill

9.2.1   Rectal Gland

The digitiform rectal gland is located near the caudal region of the intestinal tract, where a single artery (posterior mesenteric artery) supplies the gland with blood from the dorsal aorta and a single vein (rectal gland vein) drains the gland of blood into the posterior intestinal vein (Figure 9.1). The caudal end of the gland contains a short duct (rectal duct) that connects the gland to the digestive tract, into which the gland's secretions are expelled. This simple arrangement of blood vessels and secretory duct led to the development of the *in vitro* perfused gland preparation (reviewed by Silva et al., 1990), which has served as a model to study the mechanisms of secretory fluid production and secondary active NaCl transport (see Section 9.3.1).

Externally, the rectal gland is surrounded by a capsule that is composed of an outer connective tissue layer with small blood vessels and nerve fibers, and an inner, circumferential smooth muscle layer (e.g., Bulger, 1963; Evans and Piermarini, 2001). Internally, the rectal gland is composed of several thousand secretory tubules that all empty into a central lumen that is continuous with the rectal duct and digestive tract (Figure 9.1). In selachians, these tubules are often arranged radially, and directly or indirectly branch into the central lumen, while in batoids the tubules are arranged in discrete lobules, which are formed by several thin extensions of the capsule that radiate toward the central lumen (Bonting, 1966). The arterial blood that feeds the tubules originates from the posterior mesenteric artery and runs along the gland's longitudinal axis via the rectal gland artery. Several circumferential arteries branch off this vessel to supply blood to numerous arterioles that drain into sinusoidal capillary beds, which surround the tubules (Figure 9.1). Venous return from the sinusoid capillaries is via venules that drain into larger

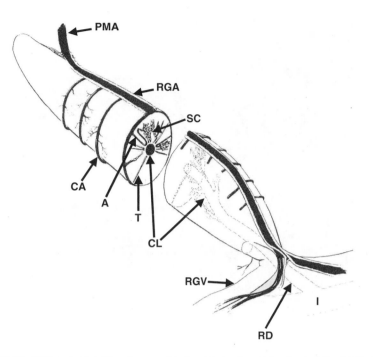

FIGURE 9.1 Schematic of spiny dogfish (*Squalus acanthias*) rectal gland anatomy. See text for details. PMA = posterior mesenteric artery, RGA = rectal gland artery, CA = circumferential artery, A = arteriole, SC = sinusoidal capillary, T = tubule, CL = central lumen, RGV = rectal gland vein, I = intensine. (Modified from Kent, B. and K.R. Olson. 1982. *Am. J. Physiol.* 243:R296–303. With permission.)

veins, which eventually collect into a central vein near the caudal portion of the gland that flows into the posterior intenstinal vein (Kent and Olson, 1982; Olson, 1999).

The secretory tubules are composed of a columnar epithelium, which in cross section consists of several mitochondrion-rich cells arranged circumferentially (Bulger, 1963; Eveloff et al., 1979; Ernst et al., 1981) (Figure 9.2). These polarized cells have extensive basolateral plasma membrane infoldings that provide a large surface area for insertion of ion transport proteins involved with NaCl secretion (see

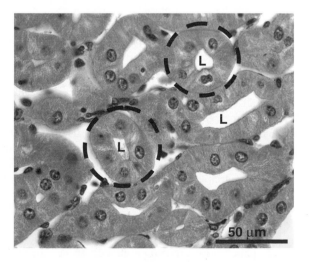

FIGURE 9.2 Light photomicrograph of secretory tubules from *S. acanthias* rectal gland. Note lumens (L) of secretory tubules. Tubules cut in cross section are circled. Bar = 50 μm.

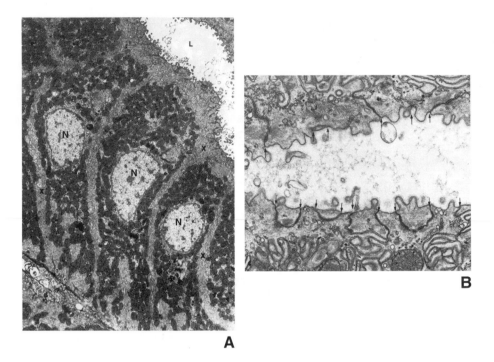

FIGURE 9.3 (A) Electron micrograph (4800×, original magnification) of secretory tubule cells from *S. acanthias* rectal gland. Note the nuclei (N) surrounded by clusters of mitochondria and extensive basolateral membrane infoldings (X). L = tubule lumen. (B) Electron micrograph (20,300×, original magnification) of apical regions of secretory tubule cells. Arrows indicate the shallow tight junctions between tubule cells. (Modified from Ernst, S.A. et al. 1981. *J. Membr. Biol.* 58:101–114. With permission.)

Section 9.3.1 and Figure 9.3A). The apical plasma membrane is less intricate and contains subtle microvilli and/or microridges (mircoplicae) (Figure 9.3B). However, the subapical cytoplasm contains numerous membranous vesicles that likely contain a chloride channel (cystic fibrosis transmembrane regultor; CFTR) that plays a critical role in active NaCl secretion (see Section 9.3.1). Another key anatomical feature of these cells is the tight junctions between the apical regions of neighboring secretory tubule cells. Ultrastructural studies have determined that these junctions are not extensive and are considered shallow (Ernst et al., 1981; Forrest et al., 1982) (Figure 9.3B). Importantly, these shallow tight junctions are "leaky" to Na^+, and are assumed to provide a route of paracellular Na^+ transport (see Section 9.3.1).

In freshwater elasmobranchs (e.g., *Potamotrygon* spp., *Carcharhinus lecuas*, *Dasyatis sabina*), the mass of the rectal gland is reduced compared to marine species (Thorson et al., 1978, 1983; Piermarini and Evans, 1998), but the overall organization and anatomy of the gland is identical to their marine counterparts (Oguri, 1964; Thorson et al., 1978). The smaller mass of the gland can be attributed to fewer secretory tubules compared to marine species (Oguri, 1964; Thorson et al., 1978). Ultrastructural comparisons of rectal gland tubule cells between freshwater and marine elasmobranchs have not been made, but it would be interesting to determine if any morphological features of the tubule cells associated with NaCl transport (e.g., mitochondria abundance, basolateral membrane infoldings, apical vesicles, tight junctions) differ between the two groups.

9.2.2 Kidney

The elasmobranch kidneys are paired organs (one on each side of the vertebral column) found on the dorsal wall of the abdominal cavity. In selachians, the kidneys are long and narrow; they first appear as fine threadlike structures at the cranial portion of the body cavity and gradually become more robust toward the caudal end of the body cavity. In batoids, the kidneys are relatively short and wide and occur

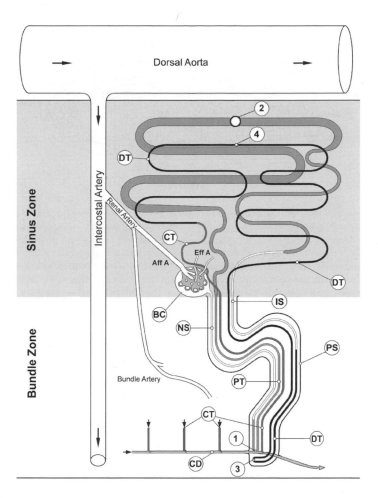

FIGURE 9.4 Schematic of elasmobranch nephron and renal arterial blood flow. The neck segment (NS) arises from the distal end of Bowman's capsule and extends into the bundle zone, where it becomes the proximal tubule (PT). The PT continues into the bundle zone, but takes a sharp turn (loop 1) that changes the direction of the tubule toward the sinus zone. Deep in the sinus zone, the PT turns back toward the bundle zone (loop 2), and before reaching the bundle zone, the nephron transforms into the intermediate segment (IS). The IS extends into the bundle zone, where it transitions to the distal tubule (DT). The DT continues through the bundle zone, but takes a sharp turn (loop 3) and changes direction toward the sinus zone. The DT progresses deep into the sinus zone, where it turns (loop 4) back toward the bundle zone. Before reaching the bundle zone, the DT transforms into the collecting tubule (CT), which continues through the bundle zone and empties into a collecting duct (CD). The CD will eventually empty into a ventral ureter, which will carry urine to the cloaca for excretion. Aff A = afferent arteriole, PS = peritubular sheath. Arrows indicate direction of blood or urine flow. See text for details on blood flow. (Courtesy of Dr. Hartmut Hentschel; modified by P.M. Piermarini.)

primarily in the caudal portion of the body cavity. In cross section, elasmobranch kidneys appear to have distinct "bundle zones" that are found in the dorsal and lateral part of the kidney and "sinus zones" that are found in the ventral and medial regions of the kidney (Lacy et al., 1985; Lacy and Reale, 1999). This zonation is due to the elaborate arrangement of nephrons and vasculature, which is discussed below.

Arterial blood is supplied to the kidneys by segmental, intercostal arteries that branch off the dorsal aorta (Figure 9.4). These arteries further subdivide into renal arteries that become afferent arterioles of the glomeruli and bundle arteries that provide blood to interstitial capillaries of bundle zones (Hentschel, 1988) (Figure 9.4 and Figure 9.5). Short efferent arterioles drain blood from the glomeruli to sinus zones, which are blood sinuses that bathe certain nephron segments (Figure 9.4 and Figure 9.5). Sinus zones also receive efferent blood from the bundle interstitial capillaries and the renal portal veins that drain blood from caudal parts of the animal into the afferent renal and afferent intra-renal veins (Figure 9.5).

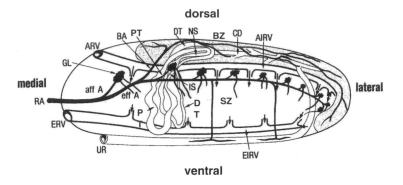

FIGURE 9.5 Schematic of a cross section through a skate (*Raja erinacea*) kidney showing general arrangement of blood vessels and a single nephron. Arterial circulation (including capillaries and glomeruli) is solid black. Venous circulation is thick-lined structures. Kidney tubules (including nephron, collecting duct, and ureter) are thin-lined structures. RA = renal artery, aff A = afferent arteriole, GL = glomerulus, eff A = efferent arteriole, BA = bundle artery, ARV = afferent renal vein, AIRV = afferent intrarenal vein, EIRV = efferent intrarenal vein, ERV = efferent renal vein, NS = neck segment, PT = proximal tubule, IS = intermediate segment, DT = distal tubule, CD = collecting duct, UR = ureter, BZ = bundle zone, SZ = sinus zone. (Modified from Hentschel, H. 1988. *Am. J. Anat.* 183:130–147. With permission.)

Efferent intrarenal veins collect blood from sinus zones and return it to the systemic venous circulation via an efferent renal vein (Figure 9.5).

In addition to this intricate renal vascular anatomy, the elasmobranch nephron displays a complexity that rivals that of mammalian nephrons. Although the specific nomenclature of certain nephron segment subregions varies between researchers, there is a general consensus that the elasmobranch nephron is composed of five distinct tubule segments (reviewed by Hentschel and Elger, 1989; Lacy and Reale, 1999). However, where these segments begin and end is still a matter of debate. The nephron segments are defined by their general morphological appearance (e.g., size and shape) and the ultrastructure of the epithelial cells that compose the segments. From proximal to distal nephron, the segments are termed: neck, proximal, intermediate, distal, and collecting (Figure 9.4). Ultrastructural features of these cells have been thoroughly reviewed by Hentschel and Elger (1987, 1989) and Lacy and Reale (1999) and are beyond the scope of this chapter.

All nephrons begin at renal corpuscles that lie near the interface between the bundle and sinus zones (Figure 9.4). The renal corpuscle is responsible for urine formation via ultrafiltration and is composed of a glomerulus surrounded by an epithelial Bowman's capsule. At the distal end of Bowman's capsule, the first nephron segment (neck) arises and enters the bundle zone, where the tubule begins the first of four loops through the bundle and sinus zones (Figure 9.4). The bundle zone contains the first and third loops of the nephron, and the sinus zone contains the second and fourth loops (Figure 9.4). In the bundle zone, the tubules run countercurrent to one another, are highly organized, and are tightly packed together by a cellular, peritubular sheath that surrounds the tubules and isolates the tubules from blood of the sinus zone (Lacy et al., 1985; Lacy and Reale, 1986). This arrangement of tubules in the bundle zone has been hypothesized to function as a countercurrent multiplier for urea and to play a role in urea reabsorption (Lacy et al., 1985; Hentschel and Elger, 1987). In the sinus zone, the tubules are relatively loosely organized and not packed together tightly. This arrangement has been hypothesized to maximize the surface area for diffusion between the urine in the tubules and blood in the sinus, which may allow for osmotic equilibration between the two fluids (Friedman and Hebert, 1990; Lacy and Reale, 1995).

Morphological studies of kidneys from stenohaline freshwater stingrays (*Potmatotrygon* sp.) have provided indirect evidence that the bundle zone is involved with urea reabsorption. These stingrays have abandoned a ureosmotic osmoregulatory strategy (Table 9.1) and have a much simpler nephron anatomy than other elasmobranchs in that the tubules only make two loops through a "simple" and "complex" zone (Lacy and Reale, 1995, 1999). These zones are not physically separated by a peritubular sheath and, therefore, the tubules are bathed by the same blood sinus throughout their length. Moreover, the tubules do not have a discrete bundle zone with the putative countercurrent multiplier for urea (Hentschel

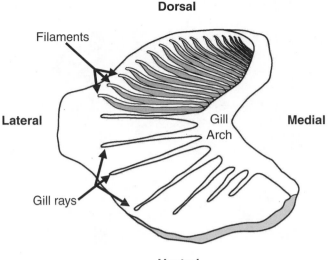

FIGURE 9.6 Schematic of a gill arch with (top) and without gill filaments. Gill filaments lay upon the interbrachial septum, which is composed of cartilaginous gill rays with connective tissue between them (see text for details).

and Elger, 1987; Lacy and Reale, 1995, 1999), which suggests the bundle zone is critical for the efficient renal reabsorption of urea seen in most elasmobranch fishes.

9.2.3 Gill

In most elasmobranchs, the gills are composed of ten cartilaginous gill arches, five on each side of the pharynx. Radiating laterally from each arch, in a fanlike manner, are several supportive, cartilaginous rods (gill rays) with sheets of connective tissue between them that form the interbranchial septum. Parallel to the gill rays, on both cranial and caudal sides of the septum, there is a row of several gill filaments (hemibranch), except for the first gill arch, which has only a hemibranch on the caudal side of the septum (Figure 9.6).

Gill filaments are the functional unit of gills, and each filament is supplied with blood through an afferent filamental artery that branches off an arch's afferent branchial artery, which receives blood from the ventral aorta. Each afferent filamental artery runs the length of a filament and is intersected perpendicularly several times along its length by afferent lamellar arterioles. These arterioles perfuse individual lamellae, which are thin protrusions of the gill filament that run perpendicular to the filament's longitudinal axis (Figure 9.7). In simplest terms, a lamella is two flat epithelial sheets held apart by a discontinuous group of cells (pillar cells); the space between the epithelial sheets and pillar cells is perfused with blood (Figure 9.8). Lamellae provide a short diffusion distance for gas exchange between the fish's blood and its environment, and dramatically increase the surface area of the filamental epithelium. Efferent lamellar arterioles drain blood that perfuses the lamellae into an efferent filamental artery, which runs parallel to and counter to its afferent counterpart (Figure 9.7). Each efferent filamental artery empties into an efferent branchial artery that connects to the dorsal aorta (Olson, 2002a,b).

The epithelium of the gill filaments and lamellae of elasmobranchs is composed of several cell types (reviewed by Laurent and Dunel, 1980; Laurent, 1984; Wilson and Laurent, 2002), but pavement cells and mitochondrion-rich cells are the two most relevant to this chapter. Most of the gill epithelium surface area (~90%) is composed of pavement cells that are typically squamous in shape, studded with microvilli and/or microplicae on their apical membrane, sparsely populated with mitochondria, and have a simple basolateral membrane (no infoldings). The tight junctions that form between the apical regions of pavement cells, and any adjacent cells are considered "deep" and form a barrier that is impermeable to ions (Wilson and Laurent, 2002). Overall, the thin shape and simple ultrastructure of pavement cells

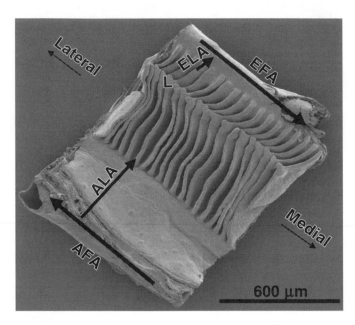

FIGURE 9.7 Scanning electron micrograph (ventral view) of a piece of a gill filament from the Atlantic stingray (*D. sabina*). Thick arrows indicate direction of blood flow through filament and lamellae (L). Thin arrows indicate orientation of filament. See text for details. AFA = afferent filamental artery, ALA = afferent lamellar arteriole, ELA = efferent lamellar arteriole, EFA = efferent filamental artery.

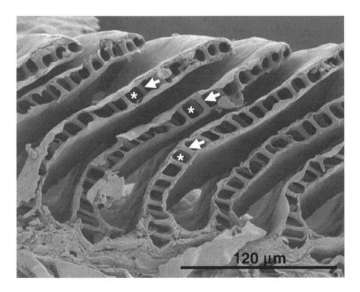

FIGURE 9.8 Scanning electron micrograph of a cross section through *D. sabina* gill lamellae. Arrows indicate pillar cells and asterisks indicate blood space. See text for details.

suggest they are primarily involved with gas exchange. In teleost fishes, evidence suggests pavement cells can play an active role in acid–base balance as well as Na^+ transport (Goss et al., 2001; Galvez et al., 2002), but data are lacking for such a role of pavement cells in elasmobranch fishes.

Mitochondrion-rich cells (MRCs) of elasmobranchs are typically large, ovoid cells that have high densities of mitochondria, complex apical membrane morphologies, numerous subapical vesicles, and a convoluted basolateral membrane (Wilson and Laurent, 2002; Wilson et al., 2002). MRCs are usually found between lamellae on the afferent side of the filament, interspersed among pavement cells. However, certain environmental conditions, such as low salinity, are associated with the appearance of MCRs on gill lamellae as well (Piermarini and Evans, 2000, 2001). The presence of these relatively large, meta-bolically active cells on lamellae may compromise the effectiveness of lamellae as gas exchange surfaces (Bindon et al., 1994; Gilmour et al., 1995; Perry, 1997). The numerous mitochondria of these cells suggest they may be sites of active ion transport relevant to NaCl and/or acid–base regulation. However, unlike the NaCl-secreting MRCs of marine teleost gills and the elasmobranch rectal gland, gill MRCs exist singly in the epithelium and form "deep" tight junctions with adjacent pavement cells (Wilson and Laurent, 2002; Wilson et al., 2002).

Morphological studies have identified two different MRC types in the elasmobranch gill epithelium based on apical membrane morphology (Crespo, 1982; Laurent, 1984; Wilson and Laurent, 2002). We have recently corroborated these findings with immunohistochemical studies in the gills of *Dasyatis sabina* (Piermarini and Evans, 2001) and have hypothesized that these MRC types may be responsible for distinct ion regulatory functions in the elasmobranch gill (see Section 9.3.3).

9.3 Role of the Rectal Gland, Kidney, and Gill in Osmoregulation

9.3.1 Rectal Gland

The rectal gland is the major, but not sole, site of needed salt secretion by marine elasmobranchs. There are surprisingly few, and no recent, determinations of the ionic concentration of rectal gland fluids from intact animals (Table 9.2), but it is clear that the secretory fluid contains high concentrations of NaCl, even higher than levels in the surrounding seawater. However, it is now known that the gland responds to a variety of hormones (see Silva et al., 1996), so these *in vivo* values may have been biased by stress. *In vivo*, rates of rectal gland secretion are of the order of 50 µl 100 g^{-1} h^{-1}, so the net secretion of Na^+ and Cl^- is approximately 20 µmol 100 g^{-1} h^{-1} (e.g., Shuttleworth, 1988b). Isolated, unstimulated perfused glands secrete at approximately 10% of these values, because of a much lower flow rate rather than a lower fluid salt concentration. Stimulation with cyclic AMP increases the flow rate tenfold in glands isolated from either *Squalus acanthias* or *Scyliorhinus canicula* (e.g., Shuttleworth, 1988b), the only two species for which recent data are available.

Because the rectal gland, at least in *Squalus acanthias*, is relatively easy to isolate and perfuse, this preparation has been utilized to study the mechanisms of fluid production (see Riordan et al., 1994; Silva, 1996; Silva et al., 1997; Olson, 1999 for recent reviews). In fact, this preparation has become a model for epithelial salt transport in general, because the transport proteins expressed are common to a variety of mammalian tissues, including the intestine and kidney. The production of a fluid containing NaCl at higher concentrations than the perfusing blood is mediated by basolateral uptake of Na^+, Cl^-, and K^+ by a cotransport protein (termed NKCC because it carries 2 Cl^-, 1 K^+, and 1 Na^+) that is driven electrochemically by the adjacent, basolateral Na-K-activated ATPase (termed NKA). This protein produces the Na^+ gradient to drive the system by recycling Na^+ back into the blood. K^+ is recycled back into the blood via a basolateral K^+ channel, and Cl^- is secreted into the rectal gland lumen (down its electrochemical gradient) via an apical Cl^- channel (CFTR). The net secretion of Cl^- produces an electrical gradient that draws Na^+ into the fluid through the paracellular pathways between adjacent cells (Figure 9.9). This original model was proposed 25 years ago (Silva et al., 1977), largely based on inhibitor

TABLE 9.2

Rectal Gland Fluid Solute Concentrations in Representative Elasmobranchs

Species	Na (mmol.l^{-1})	Cl (mmol.l^{-1})	Ref.
Squalus acanthias	540	533	Burger and Hess, 1960
Scyliorhinus canicula	554	ND	Payan and Maetz, 1970
Dasyatis sabina	ND	583	Beitz, 1977
Raja ocellata	490	499–505	Holt and Idler, 1975
Seawater	440	495	Burger and Hess, 1960

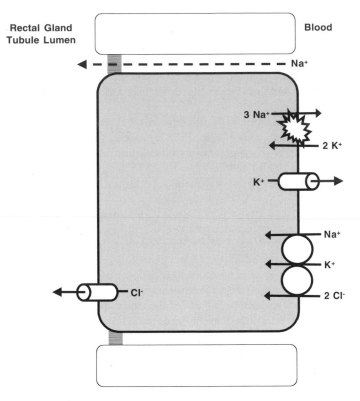

FIGURE 9.9 Current model of NaCl transport into the lumen of the elasmobranch rectal gland. See text for details.

studies with perfused glands, but modern electrophysiological, biochemical, and molecular techniques have confirmed the basic pathways. In addition, in the past 12 years, isolated rectal gland tubules and epithelial sheets produced by rectal gland tissue culture have been utilized to confirm the model and study intracellular events, as well as hormonal control of the rectal gland (e.g., Greger et al., 1986; Karnaky et al., 1991; Riordan et al., 1994; Vallentich et al., 1995; Forrest, 1996). The genes for shark rectal gland NKCC (Xu et al., 1994) and the apical CFTR (Marshall et al., 1991) have been cloned; however, there is currently only a partial gene sequence for rectal gland NKA (MacKenzie et al., 2002). Using immunohistochemistry, both the NKCC and NKA have been localized to the basolateral membrane of rectal gland tubule cells (Lytle et al., 1992), and CFTR has been localized to the apical membrane (Marshall et al., 1991).

Secretion from the rectal gland is stimulated by a variety of circulating and local signaling agents, including adenosine, vasoactive intestinal peptide, and natriuretic peptides (e.g., Shuttleworth, 1988b; Silva et al., 1996; Olson, 1999). In addition to direct affects on epithelial transport proteins, secretion may also be controlled by alterations in perfusion produced by release of catecholamines or other vasoactive agents (e.g., Shuttleworth, 1988b; Evans, 2001; Fellner and Parker, 2002). Moreover, the gland itself may respond to circulating agents that affect smooth muscle, because a circumferential band of smooth muscle cells is found within the rectal gland (Bulger, 1963; Evans and Piermarini, 2001) and cross-sectional rings of gland can respond to a variety of agents, such as acetylcholine, endothelin, nitric oxide, and natriuretic peptides (Evans and Piermarini, 2001). The relative role or roles, *in vivo*, played by direct vs. indirect effects of secretagogues on the gland, its perfusion, or epithelial transport are unknown at present.

Despite its ability to secrete a fluid that is hypertonic to the blood, the rectal gland apparently is not the only site of net salt secretion from elasmobranchs. Burger demonstrated nearly 40 years ago that extirpation of the rectal gland from *S. acanthias* was followed by little or no change in plasma ion levels, and the procedure did not kill the animal (Burger, 1965); this has been confirmed by subsequent studies (e.g., Evans et al., 1982; Shuttleworth, 1988b; Wilson, 2002). Thus, other pathways for net salt secretion

TABLE 9.3

Renal Function in Representative Elasmobranchs

Species	Osmolarity (mOsmol.l⁻¹)	Volume (ml.kg⁻¹.h⁻¹)	Na⁺ Conc. (mmol.l⁻¹)	Urea Conc. (mmol.l⁻¹)	Ref.
Squalus acanthias	800	0.5	240	100	Burger, 1967
Scyliorhinus canicula	960	0.19	238	124	Henderson et al., 1988
Hemiscyllium plagiosum	797	0.36	249	248	Wong and Chan, 1977
Raja erinacea (SW)	967	ND	180	ND	Stolte et al., 1977
Dasyatis sabina (FW)	53	10	8	20	Janech and Piermarini, 2002
Pristin microdon (FW)	55	10	ND	14	Smith, 1931

by marine elasmobranchs must exist. Intuitively, these must be the kidney and/or the gills, but evidence for compensatory NaCl secretion by these organs does not exist (see Sections 9.3.2 and 9.3.3).

It is assumed that the smaller-sized rectal gland of freshwater elasmobranchs is associated with a reduced function of the gland, because there is no osmoregulatory need to excrete NaCl in a freshwater environment. Although physiological measurements on rectal fluid secretion rates or composition are lacking, studies have shown that overall ATPase activity (Gerzeli et al., 1976) and NKA activity and immunoreactivity (Piermarini and Evans, 2000) are reduced compared to their marine counterparts, which suggests the gland has a decreased biochemical potential for NaCl secretion.

9.3.2 Kidney

Unfortunately, renal function in elasmobranchs has not been a focus of recent research, so most data sets are decades old and often consist of determination of a few functional parameters. However, a recent description of an *in situ* perfused shark kidney suggests that advances may be made in the near future (Wells et al., 2002). The glomerular filtration rate (GFR) of the elasmobranch kidney is high (~1 to 3 ml kg⁻¹ h⁻¹), of the same order as that described for teleosts in freshwater, and significantly higher than that described for marine teleosts (~0.5 ml kg⁻¹ h⁻¹; e.g., Evans, 1979; Wells et al., 2002). Presumably, this high GFR in elasmobranchs has evolved because of the inwardly directed osmotic gradient across the elasmobranch gill and its relatively high water permeability (e.g., Evans, 1979). After filtration, the urine is processed in the kidney tubules to produce urine with the volume and solute characteristics listed in Table 9.3. Of note is the fact that the total osmotic concentration of the urine is below that of the plasma (compare with Table 9.1), suggesting that the elasmobranch kidney is unable to produce a net secretion of salt and is, therefore, apparently not the organ that can compensate for the rectal gland after extirpation (see above; Section 9.3.1). Burger found that urinary salt loss did increase in *S. acanthias* after rectal gland removal, but the urine salt concentration was still below that of the plasma (Burger, 1965) and, therefore, did not provide for net salt elimination.

The urine flow rate is below the GFR, so fluid reabsorption must take place in the renal tubules; urea and TMAO also must be absorbed because their urine concentrations are far below plasma levels (Table 9.1 and Table 9.3). The dominant monovalent ions, Na⁺ and Cl⁻, are also reabsorbed, but to a lesser extent. On the other hand, like other marine animals (teleosts, invertebrates), elasmobranchs secrete divalent ions (Mg^{2+}, SO_4^{2-}) into the urine (see Henderson et al., 1988). As one might expect, the urine flow rates in the freshwater adapted *D. sabina*, and *Pristis microdon* in freshwater, are far above those recorded in marine elasmobranchs (Table 9.3). However, as GFRs have not been determined for euryhaline species, it is not known if the increased urine flow is secondary to an increased GFR or decreased tubular water reabsorption. Moreover, it is unclear if an increase in GFR would be secondary to increased single nephron filtration rate or glomerular recruitment, as both have been described when the GFR is increased in marine elasmobranch kidneys (e.g., Henderson et al., 1988).

The complexity of the elasmobranch nephron (see Section 9.2.2) has made difficult many of the experimental approaches (perfused tubules, micropuncture, tissue culture, etc.) that have allowed the physiological dissection of kidney function in other vertebrates. Interestingly, like other fish species, it appears that elasmobranchs can secrete NaCl into the lumen of the proximal tubule, driving fluid secretion, which aids the GFR (Beyenbach and Fromter, 1985; Sawyer and Beyenbach, 1985). Current evidence suggests that subsequent, distal Na$^+$ and urea reabsorption are functionally linked, with Na$^+$ uptake from the urine in the early tubule (near loop 3) via an NKCC cotransporter, which is homologous to the NKCC cotransporter expressed in the mammalian thick ascending limb of the loop of Henle (Friedman and Hebert, 1990). The shark NKCC isoform has been cloned from the rectal gland of *S. acanthias* (see Section 9.3.1), and RNA transcripts have been identified in the kidney by Northern blotting (Xu et al., 1994). More recently, transcripts have been localized in the apical membranes of the distal segments (Biemesderfer et al., 1996), corroborating the electrophysiological data (Hebert and Friedman, 1990). Boylan proposed more than 30 years ago that the countercurrent arrangement of the elasmobranch nephron segments in the bundle zone could produce a multiplication effect, driving the reabsorption of urea as it does in the mammalian nephron (Boylan, 1972). More recent anatomical studies have provided morphological support for such a proposition (e.g., Lacy et al., 1985; Hentschel and Elger, 1987; Hentschel et al., 1993; Hentschel et al., 1998; Lacy and Reale, 1999). Friedman and Hebert (1990) have proposed that urea uptake takes place in the collecting tubule within the peritubular sheath, driven by the osmotic and urea gradients produced by the NKCC transporter in the early distal tubule. An elasmobranch urea transporter has been cloned recently from *S. acanthias*, and transcripts are expressed in kidney extracts (Smith and Wright, 1999); however, tubular localization has not been published. Further electrophysiological and/or molecular studies are warranted, but it is apparent that this coupled Na$^+$ and urea transport system may account for the substantial renal reabsorption (at least 95%) of urea that is commonly seen in marine elasmobranchs (e.g., Forster, 1967). Indeed, the gills, not kidney, are the site of the vast majority of urea excretion (e.g., Wood et al., 1995).

Because urinary TMAO concentrations are below that of the blood, it is apparent that this organic solute is also reabsorbed after filtration at the glomerulus, but there are few data to suggest specific mechanisms. Cohen et al (1959) demonstrated that structural analogues of TMAO, such as trimethylamine and dimethylamine, inhibited reabsorption of TMAO, but methylamine had no effect, which suggested a specific transporter. A TMAO-specific transporter has recently been described in a bacterium that is able to absorb TMAO from the medium (Raymond and Plopper, 2002), and the transport is unaffected by methylamine, but similar data are lacking for the putative TMAO transporter in elasmobranch renal tubules. However, renal reabsorption of TMAO may be passive (via the countercurrent system that reabsorbs urea) because recent studies with *S. acanthias* erythrocytes have demonstrated volume-activated TMAO transport via the organic osmolyte channel (Koomoa et al., 2001). It is clear that molecular characterization and localization studies are necessary before the mechanisms of TMAO reabsorption in the elasmobranch kidney are known.

9.3.3 Gills

The large size of elasmobranchs has precluded the physiological protocols that allowed initial characterization of the transport mechanisms expressed in the teleost gill epithelium (reviewed by Evans, 1979), and no opercular epithelium is present, like that which has allowed the biophysical characterization of teleost gill transport (reviewed by Karnaky, 1998). Therefore, most evidence for salt transport by the elasmobranch gill epithelium is anatomical, biochemical, and molecular (rather than physiological), and recent studies indicate the elasmobranch gill is involved with salt uptake and acid–base regulation, rather than salt secretion.

In contrast to the marine teleost gill, the elasmobranch gill epithelium is not considered to play a major role in NaCl extrusion; in fact, early isotopic flux experiments suggested elasmobranch gills are sites of net NaCl uptake, not excretion (Bentley et al., 1976). Although MRCs are present in the marine elasmobranch gill, the "shallow" tight junctions that are crucial to NaCl secretion by the teleost gill and elasmobranch rectal gland (see Section 9.2.1) are not present in the elasmobranch gill (see Section 9.2.3). Nevertheless, MRCs with elevated activity and immunoreactivity for basolateral NKA (Na,K-ATPase-rich cells) have been detected in the elasmobranch gill (Conley and Mallatt, 1988; Piermarini and Evans, 2000; Wilson et al., 2002), but evidence for other important proteins of the NaCl secretory pathway (e.g., NKCC and CFTR) in Na,K-ATPase-rich cells is lacking. Moreover, in *S. acanthias*, removal of the rectal gland does not increase gill MRC cell number or NKA activity (Wilson et al., 2002). Therefore, if the marine elasmobranch gill is involved with NaCl extrusion, it probably uses mechanisms different from those found in the marine teleost gill and elasmobranch rectal gland.

The marine elasmobranch gill also contains MRCs with measurable immunoreactivity for vacuolar proton-ATPase (V-H$^+$-ATPase-rich cells) (Wilson et al., 1997; Piermarini and Evans, 2001) that have no detectable NKA immunostaining (Piermarini and Evans, 2001). It has been shown that the V-H$^+$-ATPase immunostaining in *D. sabina* gill is basolateral (Piermarini and Evans, 2001), and the V-H$^+$-ATPase-rich cells contain apical immunoreactivity for an anion exchanger (pendrin) (Piermarini et al., 2002) that is involved with Cl$^-$/HCO$_3^-$ exchange in the mammalian kidney (Royaux et al., 2001). This suggests the V-H$^+$-ATPase-rich cells are involved with chloride uptake and bicarbonate excretion, pathways important for osmoregulation in freshwater and acid–base regulation, but not marine osmoregulation.

Studies on euryhaline and freshwater elasmobranchs have provided further evidence that the elasmobranch gill is involved with ion uptake and acid–base regulation (e.g., Piermarini et al., 2002; Wood et al., 2002). For example, NKA activity and immunoreactivity (immunohistochemistry and Western blotting) are enhanced in gills of freshwater *D. sabina* relative to marine *D. sabina*, and these parameters decrease when freshwater animals are acclimated to seawater (Piermarini and Evans, 2000). Likewise, V-H$^+$-ATPase and pendrin immunoreactivity follows a similar pattern to NKA, albeit in separate cells (Piermarini and Evans, 2001; Piermarini et al., 2002). The results from these studies suggest that the elasmobranch gill epithelium has a greater potential for active transport in freshwater individuals, which is consistent with the greater need for NaCl uptake in freshwater environments. Specifically, we have proposed that the V-H$^+$-ATPase-rich cells are the site of pendrin-mediated Cl$^-$ uptake in exchange for HCO$_3^-$ excretion. This would be driven by basolateral V-H$^+$-ATPase that establishes a gradient for HCO$_3^-$ secretion. We also proposed that the NKA-rich cells are a potential site of Na$^+$-uptake in exchange for H$^+$ excretion, via a Na$^+$/H$^+$ exchange mechanism that could be driven by NKA (Figure 9.10). Although direct physiological measurements of NaCl transport are lacking in the *D. sabina*, Wood and colleagues (2002) have recently provided measurements on ion transport kinetics in stenohaline, freshwater *Potamotrygon* sp. that are consistent with our proposed model.

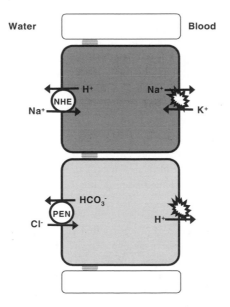

FIGURE 9.10 Hypothetical model of NaCl and acid–base transport in the elasmobranch gill. We propose that V-H-ATPase-rich cells are a site of chloride uptake and bicarbonate secretion via an apical pendrin-like anion exchanger (PEN). Basolateral V-H-ATPase would actively pump protons across the basolateral membrane, which would create an intracellular bicarbonate concentration that is favorable for apical bicarbonate secretion. We hypothesize that Na,K-ATPase-rich cells are a site of proton secretion and sodium uptake through a sodium/hydrogen exchange mechanism (NHE). Basolateral Na,K-ATPase would actively pump sodium ions across the basolateral membrane to maintain low intracellular sodium concentrations that are favorable for apical sodium uptake.

9.4 Role of the Rectal Gland, Kidney, and Gill in pH Regulation and Ammonia Excretion

9.4.1 Rectal Gland

Unfortunately, net acid excretion in rectal gland fluid has never been quantified, so the contribution of the rectal gland to systemic pH regulation is not known. However, some characteristics of rectal gland fluid suggest that it probably does not have a large role, if any. For example, it has close to a neutral pH (~6.8) (Burger, 1965) and is produced at rates that are too low (<1 ml kg^{-1} h^{-1}) (Burger, 1972) for it to be a major quantitative contributor to systemic acid or base excretion. For comparison, even urine, which is more acidic (~5.8; see Section 9.4.2) and is produced at higher rates (Section 9.3.1 vs. Section 9.3.2), usually has a small quantitative role in systemic pH regulation (see Section 9.4.2) relative to the gills (see Section 9.4.3). Reports of ammonia excretion rates by rectal glands are also lacking. However, as is the case for acid and base excretion, the rectal gland's low flow rates probably make any ammonia secretion inconsequential relative to the gills (see Section 9.4.3).

9.4.2 Kidney

The kidneys of elasmobranchs are generally not considered to play a large role in systemic pH regulation, because urinary acid excretion accounts for less than 1% of whole-animal net acid excretion during respiratory and metabolic acidosis (Heisler et al., 1976; Evans et al., 1979a; Holeton and Heisler, 1983) and because urinary net base excretion is negligible during metabolic alkalosis (Wood et al., 1995). Similarly, urinary urea and ammonia excretion account for less than 1% of whole-animal nitrogen excretion (Wood et al., 1995; and reviewed by Wood, 1993). As already discussed in Sections 9.1 and 9.3.2, urea serves as an osmolyte in marine elasmobranchs and is reabsorbed by the kidneys to limit its excretion. Mechanisms of ammonia secretion by elasmobranch kidneys have not been studied, and therefore are not reviewed here.

The small contribution of the kidney to pH regulation is thought to reflect two characteristics of marine elasmobranchs: (1) urine flow rates (~1 ml kg^{-1} h^{-1}) that are inconsequential relative to high gill perfusion rates (Cross et al., 1969; Holeton and Heisler, 1983; Swenson and Maren, 1986), and (2) a urinary pH that is fixed at below 6.0 to prevent precipitation of divalent ion salts (Smith, 1939) that are concentrated in the urine. A recent study suggests that even the kidneys of a euryhaline elasmobranch in freshwater (*D. sabina*) have little or no response to acid–base disturbances, even though they have high urine flow rates and no need for a fixed urinary pH (Choe and Evans, 2003). This lack of a renal response to acidosis by *D. sabina* may be due to a short history of freshwater existence (<100,000 years) and regulatory mechanisms that can take advantage of high urine flow rates have not evolved (Cook, 1939). In contrast, teleost fishes have occurred in freshwater for probably more than 100 million years and do have renal contributions to systemic pH regulation (Cameron and Kormanik, 1982; Hyde et al., 1987; Perry et al., 1987).

Although the kidneys do not appear to play a significant role in compensating for acid–base disturbances, they do possess acid–base transport processes that are vital for reabsorbing filtered bicarbonate, and maintaining an acidic urine to prevent precipitation of divalent ion salts. Bicarbonate, the major extracellular buffer (reviewed by Claiborne, 1998), enters nephrons at the glomeruli and must be reabsorbed to prevent base loss in the urine. As in mammals, this is thought to occur by acid secretion from the apical membrane of nephron cells, which converts filtered bicarbonate to CO_2. CO_2 then diffuses into cells, across the apical membrane, where it is converted back into bicarbonate, which is transported across the basolateral membrane, into the surrounding blood (Maren, 1987). Acid secretion from renal nephron cells maintains a urinary pH that is almost two units below that of the blood. Unfortunately, the cellular and biochemical mechanisms of acid secretion in the kidneys of elasmobranchs have not received much attention. Early work focused on the enzyme carbonic anhydrase (Hodler et al., 1955), which was later studied in detail by Maren and colleagues who showed that, in contrast to mammals, urinary pH, titratable acid excretion, and bicarbonate reabsorption are not affected by carbonic anhydrase inhibition in *S. acanthias* (reviewed by Maren, 1987). These results suggest that acid–base transport in elasmobranch kidneys is accomplished by mechanisms that are independent of carbonic anhydrase, which is in contrast to renal transport mechanisms in mammals (Dobyan and Bulger, 1982).

As mentioned in Section 9.3.2, identification of transport processes in elasmobranch kidneys by conventional methods (perfused tubules, micropuncture, tissue culture, etc.) has been complicated by their complex anatomy (see Section 9.2.2). However, because the identities of many specific acid-transporting proteins in mammalian kidneys are known, it is now possible to use biochemical, immunological, and molecular techniques on elasmobranch kidneys. Mammalian nephron cells are known to express at least three major types of acid-secreting transporters along their luminal membranes: Na^+/H^+ exchangers, V-H^+-ATPases, and H^+/K^+-ATPases (Gluck and Nelson, 1992; Wingo and Smolka, 1995; Paillard, 1997).

Na^+/H^+ exchangers (NHEs) are responsible for most of acid secretion and/or bicarbonate reabsorption by mammalian proximal tubule cells (Weinstein, 1994; Paillard, 1997). In elasmobranchs, Na^+/H^+ exchange was detected with biochemical assays of membranes isolated from spiny dogfish kidneys (Shetlar et al., 1987; Bevan et al., 1989). The authors speculated that the membranes were from proximal tubules, suggesting that elasmobranchs use NHEs for acid secretion and/or bicarbonate reabsorption similar to mammals. However, immunological and/or nucleic acid hybridization techniques are needed to positively identify and locate NHE(s) to specific elasmobranch nephron segments.

Immunohistochemical studies in our laboratory suggest that vacuolar type H^+-ATPase (V-H^+-ATPase), a protein responsible for acid excretion and/or bicarbonate reabsorption in proximal tubule and distal tubule cells of mammalian kidneys (Gluck et al., 1996), is expressed in several segments of Atlantic stingray kidneys (Choe and Evans, unpubl. results). The strongest staining appears throughout proximal tubules of the sinus zone with weaker staining in some distal tubules of the sinus and bundle zones. These results suggests that V-H^+-ATPase participates in acid secretion and/or bicarbonate reabsorption in elasmobranch kidneys, as it does in mammal kidneys (Gluck et al., 1996).

In the last decade, it has become clear that H^+/K^+-ATPases are expressed in mammal kidneys where they function in acid–base and ion transport (Wingo and Cain, 1993; Wingo and Smolka, 1995; Silver and Soleimani, 1999). This is a diverse group of transporters that includes at least three different homologous forms (Caviston et al., 1999). The best studied is $HK\alpha1$, which was first identified as the proton pump of mammalian stomachs. Two studies have provided evidence that this transporter is expressed in elasmobranch kidneys by showing that antibodies for mammalian $HK\alpha1$ label the apical membranes of nephron cells (Hentschel et al., 1993; Swenson et al., 1994). Swenson et al. (1994) also showed that almost all of titratable acid excretion from kidneys was inhibited by intra-arterial injections of an inhibitor of $HK\alpha1$ (SCH-28080). These studies provide strong evidence that $HK\alpha1$ secretes acid in many segments of elasmobranch kidneys, but further studies are needed to determine the molecular identity of these putative H^+/K^+-ATPases.

9.4.3 Gill

Several studies on fishes have shown that compensation for pH disturbances caused by either internal synthesis of acid–base metabolites or stressful environmental water chemistry is accomplished by net excretion of acid and/or base into environmental water (reviewed by Heisler, 1986a, 1988, 1993; Claiborne, 1998). Gills are responsible for more than 90% of this transport in elasmobranchs, with secretion from the kidneys, skin, rectal gland, and abdominal pores contributing less that 5% combined (reviewed by Choe and Evans, 2003; Heisler, 1986b; 1988). Like the gills of marine teleosts, the gills of marine elasmobranchs respond to pH disturbances rapidly and efficiently, suggesting well-developed mechanisms of acid–base transport that are under tight regulation. For example, rates of net acid excretion as high as 1000 mmol kg^{-1} h^{-1} are attained within a few hours of acute respiratory or metabolic acidosis in *S. acanthias* and *Scyliorhinus stellaris*, with blood pH recovering to control levels within 8 to 10 h (Heisler et al., 1976; Claiborne and Evans, 1992; Wood et al., 1995). Similarly, rates of net base excretion above 1000 mmol kg^{-1} h^{-1} are attained within a few hours of metabolic alkalosis (Wood et al., 1995).

Like net acid and base excretion, almost all ammonia is excreted by extrarenal routes, generally thought to be dominated by the gills (Heisler, 1988; Wood, 1993; Wood et al., 1995). Ammonia is a weak base (pK ~9 to 10), and therefore occurs as both a gas, ammonia (NH_3), and an ion, ammonium (NH_4^+), in aqueous solutions; the sum of both forms is total ammonia (T_{amm}). In elasmobranchs, there is evidence that both forms of ammonia can cross the branchial epithelium, with their relative proportions determined by environmental salinity and the relative pHs and T_{amm} concentrations of blood and water at the gills. For marine elasmobranchs under control conditions, most T_{amm} is thought to be excreted by diffusion of NH_3, with the rest by diffusion of NH_4^+ (Heisler, 1988). However, some studies have measured increases in T_{amm} excretion during increased net acid excretion (Evans, 1982; Claiborne and Evans, 1992). This suggests that T_{amm} and acid secretion can be linked in elasmobranch gills, either by transporters like NHEs, which can substitute NH_4^+ for H^+ (Paillard, 1997) or by NH_3 reacting with H^+ to form NH_4^+ at the gill's boundary (Wood, 1993). Note that in the latter case, ammonia and acid secretion are complementary; acid secretion helps to maintain a diffusion gradient for NH_3 by converting it to NH_4^+, and NH_3 buffers H^+ secretion minimizing the pH gradient that acid transporters must work against. Therefore, acid–base conditions should always be considered when evaluating ammonia excretion, and ammonia should always be considered when evaluating acid transport.

Mechanisms of acid–base transport have been studied extensively in teleosts gills, but have only recently received much attention in elasmobranchs gills. As discussed above, the gills of elasmobranchs in freshwater are thought to function in ion (specifically, Na^+ and Cl^-) absorption by mechanisms that link Na^+ absorption with H^+ (and/or NH_4^+) secretion and Cl^- absorption with HCO_3^- secretion. Interestingly, these transport mechanisms probably evolved for acid–base regulation in seawater, and later assumed functions in ion regulation when the ancestors of euryhaline elasmobranchs entered freshwater (Evans, 1984). Evans and colleagaues (1979b; Evans, 1982) provided the first evidence that acid secretion is Na^+ dependent in elasmobranchs, by showing that titratable acid excretion from *Squalus acanthias* and *Raja erinacea* was completely abolished when they were placed in Na^+-free water, and that amiloride (a Na^+ transport blocker) inhibited half of titratable acid excretion from little skates (Evans et al., 1979b; Evans, 1982).

More recent studies have focused on identifying specific branchial acid transporters with immunological and molecular biology techniques. Two models have been proposed to explain the dependence of acid excretion on external Na^+: (1) apical $Na^+/H^+(NH_4^+)$-exchangers, and (2) apical V-H^+-ATPases that are electrically balanced by absorption of Na^+ through apical channels. Evidence supporting the role of fish gill Na^+/H^+ exchangers in acid–base regulation was reviewed recently by Claiborne et al. (2002), and, therefore, is only briefly summarized here. Claiborne et al. have cloned and sequenced mRNA transcripts from *S. acanthias* and *R. erinacea* gills that are homologous to the NHE-2 isoform of mammals (pers. comm.). A few studies have used antibodies made for mammalian NHE proteins to detect and/or locate NHE-like protein expression in marine elasmobranch gills (Wall et al., 2001; Edwards et al., 2002; Choe et al., 2002). The most comprehensive of these was by Edwards et al. (2002), which showed that NHE immunoreactivity occurs in Na^+/K^+-ATPase-rich cells in four species of marine elasmobranchs. Therefore, it is clear that NHEs are expressed in marine elasmobranch gills, and further studies are now needed to determine their physiological function.

The first direct evidence for expression of V-H^+-ATPase comes from an immunological study (Wilson et al., 1997) that localized staining for the A subunit of V-H^+-ATPase to interlamellar cells of *S. acanthias* gills. However, it was not clear if the staining was in the apical or basolateral membrane. Recent immunological studies in our laboratory, using confocal microscopy, suggest that most of V-H^+-ATPase is located in the basolateral membranes of gills of *S. acanthias* (Figure 9.11B), matching the location of this transporter in the euryhaline stingray, *D. sabina* (Piermarini and Evans, 2001). In addition, Pendrin, a Cl^-/HCO_3^- exchanger, appears to stain the apical membranes of the same cells (Figure 9.11B), matching the model for Cl^- uptake that was proposed for *D. sabina* (Piermarini et al., 2002). Therefore, most of V-H^+-ATPase appears to be in the basolateral membranes of the gills of both the marine *S. acanthias* and the euryhaline *D. sabina*, where it may contribute to base excretion, in exchange for Cl^- absorption, instead of acid excretion.

Preliminary studies in our laboratory are the first to suggest that a protein homologous to an H^+/K^+-ATPase is also expressed in the gills of marine elasmobranchs. RT-PCR was used with mRNA from *S. acanthias* to obtain a partial sequence (1050 base pairs) that is greater than 83% identical to HKα1 sequences at the amino acid level (Choe, unpubl. results). In addition, two antibodies made for different

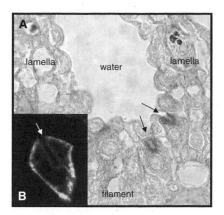

FIGURE 9.11 Immunohistochemical staining for pendrin (A) and vacuolar H+-ATPase (B) on gills from spiny dogfish (*S. acanthias*). Pendrin staining occurs in canicula-like apical regions (arrows) of cells in the filament (arrows). Vacuolar H+-ATPase staining occurs in the basolateral membrane of cells with the same morphology.

parts of mammalian HKα1 stain the gills of marine *D. sabina* (Choe and Evans, unpubl. results). However, the subcellular location of this staining is not clear, and on Western blots bands of the gills and stomach were slightly different in size. This makes imperative further studies to positively identify this putative elasmobranch gill H+/K+-ATPase and determine its function.

9.5 Summary and Perspectives

Osmotic, ionic, pH, and nitrogen homeostasis is maintained by an integrated suite of transport processes in the rectal gland, kidney, and gill of elasmobranchs. Many of the basic physiological parameters have been measured in only a few species, so the data need to be updated and expanded to other species. However, the morphology of the relevant structures is well studied, and new molecular techniques are being applied to determine the genes and proteins that are involved in these processes. Current data have demonstrated that the cellular and molecular processes that are involved in elasmobranch physiology are homologous to those present in mammals and humans. This provides an avenue for future genomic and proteomic approaches that will delineate the evolution of important transport processes, and possibly provide information important in biomedical research.

Acknowledgments

Research in our laboratory has been supported by various grants from the National Science Foundation, most recently IBN-0089943. We are especially indebted to Dr. Hartmut Hentschel, Max-Planck-Institut für Molekulare Physiologie, Dortmund, Germany who kindly allowed us to modify his unpublished figure to produce Figure 9.4.

References

Beitz, B. E. 1977. Secretion of rectal gland fluid in the Atlantic stingray, *Dasyatis sabina*. *Copeia* 1977:585–587.

Bentley, P. J., J. Maetz, and P. Payan. 1976. A study of the unidirectional fluxes of Na and Cl across the gills of the dogfish *Scyliorhinus canicula* (Chondrichthyes). *J. Exp. Biol.* 64:629–637.

Bevan, C., R. K. H. Kinne, R. E. Shetlar, and E. Kinne-Saffran. 1989. Presence of a Na/H exchanger in brush border membranes isolated from the kidney of the spiny dogfish, *Squalus acanthias. J. Comp. Physiol.* 159:339–348.

Beyenbach, K. W. and E. Fromter. 1985. Electrophysiological evidence for Cl secretion in shark renal proximal tubules. *Am. J. Physiol.* 248:F282–95.

Biemesderfer, D., J. A. Payne, C. Y. Lytle, and B. Forbush, III. 1996. Immunocytochemical studies of the Na-K-Cl cotransporter of shark kidney. *Am. J. Physiol.* 270:F927–936.

Bindon, S., K. Gilmour, J. Fenwick, and S. Perry. 1994. The effects of branchial chloride cell proliferation on respiratory function in the rainbow trout *Oncorhynchus mykiss. J. Exp. Biol.* 197:47–63.

Bonting, S. L. 1966. Studies on sodium-potassium-activated adenosinetriphosphatase. XV. The rectal gland of the elasmobranchs. *Comp. Biochem. Physiol.* 17:953–966.

Boylan, J. W. 1972. A model for passive urea reabsorption in the elasmobranch kidney. *Comp. Biochem. Physiol. A* 42:27–30.

Bulger, R. E. 1963. Fine structure of the rectal (salt-secreting) gland of the spiny dogfish, *Squalus acanthias. Anat. Rec.* 147:95–127.

Burger, J. W. 1965. Roles of the rectal gland and kidneys in salt and water excretion in the spiny dogfish. *Physiol. Zool.* 38:191–196.

Burger, J. W. 1967. Problems in the electrolyte economy of the spiny dogfish, *Squalus acanthias,* in *Sharks, Skates, and Rays.* P. W. Gilbert, R. F. Mathewson, and D. P. Rall, Eds., Johns Hopkins University Press, Baltimore, 177–185.

Burger, J. W. 1972. Rectal gland secretion in the stingray, *Dasyatis sabina. Comp. Biochem. Physiol. A* 42:31–32.

Burger, J. W. and W. N. Hess. 1960. Function of the rectal gland of the spiny dogfish. *Science* 131:670–671.

Cameron, J. N. and G. A. Kormanik. 1982. The acid base status responses of gills and kidneys to infused acid and base loads in the channel catfish *Ictalurus punctatus. J. Exp. Biol.* 99:143–160.

Caviston, T. L., W. G. Campbell, C. S. Wingo, and B. D. Cain. 1999. Molecular identification of the renal H+,K+-ATPases. *Semin. Nephrol.* 19:431–437.

Choe, K. P. and D. Evans. 2003. Compensation for hypercapnia by a euryhaline elasmobranch: effect of salinity and roles of gills and kidneys in fresh water. *J. Exp. Zool.* 297A:52–63.

Choe, K. P., A. I. Morrison-Shetlar, B. P. Wall, and J. B. Claiborne. 2002. Immunological detection of Na+/H+ exchangers in the gills of a hagfish, *Myxine glutinosa,* an elasmobranch, *Raja erinacea,* and a teleost, *Fundulus heteroclitus. Comp. Biochem. Physiol.* 131A:375–385.

Claiborne, J. B. 1998. Acid-base regulation, in *The Physiology of Fishes.* D. H. Evans, Ed., CRC Press, Boca Raton, FL, 177–198.

Claiborne, J. B. and D. H. Evans. 1992. Acid–base balance and ion transfers in the spiny dogfish (*Squalus acanthias*) during hypercapnia: a role for ammonia excretion. *J. Exp. Zool.* 261:9–17.

Claiborne, J. B., S. L. Edwards, and A. I. Morrison-Shetlar. 2002. Acid-base regulation in fishes: cellular and molecular mechanisms. *J. Exp. Zool.* 293:302–319.

Cohen, J. J., M. A. Krupp, C. A. Chidsey, and C. L. Blitz. 1959. Effect of TMA and its homologues on renal conservation of TMA-oxide in the spiny dogfish *Squalus acanthias. Am. J. Physiol.* 194:229–235.

Conley, D. M. and J. Mallatt. 1988. Histochemical localization of Na+K-ATPase and carbonic anhydrase activity in the gills of 17 fish species. *Can. J. Zool.* 66:2398–2405.

Cook, C. W. 1939. Scenery of Florida interpreted by a geologist. *Fla. Geol. Surv. Geol. Bull.* 17:1–118.

Crespo, S. 1982. Surface morphology of dogfish (*Scyliorhinus canicula*) gill epithelium and surface morphological changes following treatment with zinc sulfate: a scanning electron microscope study. *Mar. Biol.* 67:159–166.

Cross, C. E., B. S. Packer, J. M. Linta, H. V. Murdaugh, Jr., and E. D. Robin. 1969. H+ buffering and excretion in response to acute hypercapnia in the dogfish *Squalus acanthias. Am. J. Physiol.* 216:440–452.

Dobyan, D. C. and R. E. Bulger. 1982. Renal carbonic anhydrase. *Am. J. Physiol.* 243:F311–324.

Edwards, S. L., J. A. Donald, T. Toop, M. Donowitz, and C.-M. Tse. 2002. Immunolocalization of sodium/proton exchanger-like proteins in the gills of elasmobranchs. *Comp. Biochem. Physiol.* 131A:257–265.

Ernst, S. A., S. R. Hootman, J. H. Schreiber, and C. V. Riddle. 1981. Freeze-fracture and morphometric analysis of occluding junctions in rectal glands of elasmobranch fish. *J. Membr. Biol.* 58:101–114.

Evans, D. H. 1979. Fish, in *Comparative Physiology of Osmoregulation in Animals,* Vol. 1. G. M. O. Maloiy, Ed., Academic Press, Orlando, FL, 305–390.

Evans, D. H. 1982. Mechanisms of acid extrusion by two marine fishes: the teleost, *Opsanus beta*, and the elasmobranch, *Squalus acanthias. J. Exp. Biol.* 97:289–299.

Evans, D. H. 1984. Gill Na/H and Cl/HCO₃ exchange systems evolved before the vertebrates entered fresh water. *J. Exp. Biol.* 113:464–470.

Evans, D. H. 1993. *The Physiology of Fishes*. CRC Press, Boca Raton, FL.

Evans, D. H. 1998. *The Physiology of Fishes*, 2nd ed. CRC Press, Boca Raton, FL.

Evans, D. H. 2001. Vasoactive receptors in abdominal blood vessels of the dogfish shark, *Squalus acanthias. Physiol. Biochem. Zool.* 74:120–126.

Evans, D. H. and P. M. Piermarini. 2001. Contractile properties of the elasmobranch rectal gland. *J. Exp. Biol.* 204:59–67.

Evans, D., G. Kormanik, and E. J. Krasny, Jr. 1979a. Mechanisms of ammonia and acid extrusion by the little skate, *Raja erinacea. J. Exp. Zool.* 208:431–437.

Evans, D. H., G. A. Kormanik, and E. Krasny, Jr. 1979b. Mechanisms of ammonia and acid extrusion by the little skate *Raja erinacea. J. Exp. Zool.* 208:431–437.

Evans, D. H., A. Oikari, G. A. Kormanik, and L. Mansberger. 1982. Osmoregulation by the prenatal spiny dogfish, *Squalus acanthias. J. Exp. Biol.* 101:295–305.

Eveloff, J., K. J. Karnaky, Jr., P. Silva, F. H. Epstein, and W. B. Kinter. 1979. Elasmobranch rectal gland cell: autoradiographic localization of ³Houabain-sensitive Na, K-ATPase in rectal gland of dogfish, *Squalus acanthias. J. Cell. Biol.*. 83:16–32.

Fellner, S. K. and L. Parker. 2002. A Ca$^{(2+)}$-sensing receptor modulates shark rectal gland function. *J. Exp. Biol.* 205:1889–1897.

Forrest, J. N., Jr. 1996. Cellular and molecular biology of chloride secretion in the shark rectal gland: regulation by adenosine receptors. *Kidney Int.* 49:1557–1562.

Forrest, J. N., Jr., J. L. Boyer, T. A. Ardito, H. V. Murdaugh, Jr., and J. B. Wade. 1982. Structure of tight junctions during Cl secretion in the perfused rectal gland of the dogfish shark. *Am. J. Physiol.* 242:C388–392.

Forster, R. P. 1967. Osmoregulatory role of the kidney in cartilaginous fishes (Chondrichthyes), in *Sharks, Skates and Rays*. P. W. Gilbert, R. F. Mathewson, and D. P. Rall, Eds., Johns Hopkins University Press, Baltimore, 187–195.

Friedman, P. A. and S. C. Hebert. 1990. Diluting segment in kidney of dogfish shark. I. Localization and characterization of chloride absorption. *Am. J. Physiol.* 258:R398–408.

Galvez, F., S. D. Reid, G. Hawkings, and G. G. Goss. 2002. Isolation and characterization of mitochondria-rich cell types from the gill of freshwater rainbow trout. *Am. J. Physiol.* 282:R658–668.

Gerzeli, G., G. F. De Stefano, L. Bolognani, K. W. Koenig, M. V. Gervaso, and M. F. Omodeo-Salé. 1976. The rectal gland in relation to the osmoregulatory mechanisms of marine and freshwater elasmobranchs, in *Investigations of the Ichthyofauna of Nicaraguan Lakes*. T. B. Thorson, Ed., School of Life Sciences, University of Nebraska, Lincoln, 619–627.

Gilmour, A., J. Fenwick, and S. Perry. 1995. The effects of softwater acclimation on respiratory gas transfer in the rainbow trout *Oncorhynchus mykiss. J. Exp. Biol.* 198:2557–2567.

Gluck, S. and R. Nelson. 1992. The role of the V-ATPase in renal epithelial H⁺-transport. *J. Exp. Biol.* 172:205–218.

Gluck, S., D. Underhill, M. Iyori, S. Holliday, T. Kostrominova, and B. Lee. 1996. Physiology and biochemistry of kidney vacuolar H⁺-ATPase. *Annu. Rev. Physiol.* 58:427–445.

Goss, G. G., S. Adamia, and F. Galvez. 2001. Peanut lectin binds to a subpopulation of mitochondria-rich cells in the rainbow trout gill epithelium. *Am. J. Physiol.* 281:R1718–1725.

Greger, R., E. Schlatter, and H. Gogelein. 1986. Sodium chloride secretion in rectal gland of dogfish, *Squalus acanthias. NIPS.* 1:134–136.

Griffith, R. W., P. K. T. Pang, A. K. Srivastava, and G. E. Pickford. 1973. Serum composition of freshwater stingrays (Potamotrygonidae) adapted to fresh and dilute sea water. *Biol. Bull.* 144:304–320.

Hamlett, W. C. 1999. *Sharks, Skates, and Rays*. Johns Hopkins University Press, Baltimore.

Hebert, S. C. and P. A. Friedman. 1990. Diluting segment in kidney of dogfish shark. II. Electrophysiology of apical membranes and cellular resistances. *Am. J. Physiol.* 258:R409–417.

Heisler, N. 1986a. Acid-base regulation in fishes, in *Acid–Base Regulation in Animals,* Vol. 1. N. Heisler, Ed., Elsevier Science, Amsterdam, 309–356.

Heisler, N. 1986b. Comparative aspects of acid–base regulation, in *Acid–Base Regulation in Animals*. N. Heisler, Ed., Elsevier, Science, Amsterdam, 397–450.

Heisler, N. 1988. Acid–base regulation, in *Physiology of Elasmobranch Fishes*. T. J. Shuttleworth, Ed., Springer-Verlag, Berlin, 215–252.

Heisler, N. 1993. Acid-base regulation, in *The Physiology of Fishes*. D. H. Evans, Ed., CRC Press, Boca Raton, FL, 343–378.

Heisler, N., H. Weitz, and A. M. Weitz. 1976. Hypercapnia and resultant bicarbonate transfer processes in an elasmobranch fish (*Scyliorhinus stellaris*). *Bull. Eur. Physiolpathol. Respir.* 12:77–85.

Henderson, I. W., L. B. O'Toole, and N. Hazon. 1988. Kidney function, in *Physiology of Elasmobranch Fishes*. T. J. Shuttleworth, Ed., Springer-Verlag, Berlin, 201–214.

Hentschel, H. 1988. Renal blood vascular system in the elasmobranch, *Raja erinacea* Mitchill, in relation to kidney zones. *Am. J. Anat.* 183:130–147.

Hentschel, H. and M. Elger. 1987. The distal nephron in the kidney of fishes. *Adv. Anat. Embryol. Cell Biol.* 108:1–151.

Hentschel, H. and M. Elger. 1989. Morphology of glomlerular and aglomerular kidneys, in *Structure and Function of the Kidney*, Vol. 1. R. K. H. Kinne, Ed., Karger, Basel, 1–72.

Hentschel, H., S. Mahler, P. Herter, and M. Elger. 1993. Renal tubule of dogfish, *Scyliorhinus canicula*: a comprehensive study of structure with emphasis on intramembrane particles and immunoreactivity for $H^{(+)}$-$K^{(+)}$-adenosine triphosphatase. *Anat. Rec.* 235:511–532.

Hentschel, H., U. Storb, L. Teckhaus, and M. Elger. 1998. The central vessel of the renal countercurrent bundles of two marine elasmobranchs — dogfish (*Scyliorhinus canicula*) and skate (*Raja erinacea*) — as revealed by light and electron microscopy with computer-assisted reconstruction. *Anat. Embryol.* (Berlin) 198:73–89.

Hochachka, P. W. and G. N. Somero. 2002. *Biochemical Adaptation. Mechanism and Process in Physiological Evolution*. Oxford University Press, Oxford.

Hodler, J., O. Heinemann, A. P. Fishman, and H. W. Smith. 1955. Urine pH and carbonic anhydrase activity in the marine dogfish. *Am. J. Physiol.* 183:155–162.

Holeton, G. and N. Heisler. 1983. Contribution of net ion transfer mechanisms to acid–base regulation after exhausting activity in the larger spotted dogfish (*Scyliorhinus stellaris*). *J. Exp. Biol.* 103:31–46.

Holt, W. F. and D. R. Idler. 1975. Influence of the interrenal gland on the rectal gland of a skate. *Comp. Biochem. Physiol.* 50C:111–119.

Hyde, D. A., T. W. Moon, and S. F. Perry. 1987. Physiological consequences of prolonged aerial exposure in the American eel, *Anguilla rostrata*: blood respiratory and acid–base status. *J. Comp. Physiol.* 157:635–642.

Janech, M. G. and P. M. Piermarini. 2002. Renal water and solute excretion in the Atlantic stingray in fresh water. *J. Fish. Biol.* 60:1–5.

Karnaky, K. J., Jr. 1998. Osmotic and ionic regulation, in *The Physiology of Fishes*. D. H. Evans, Ed., CRC Press, Boca Raton, FL, 157–176.

Karnaky, K. J., Jr., J. D. Valentich, M. G. Currie, W. F. Oehlenschlager, and M. P. Kennedy. 1991. Atriopeptin stimulates chloride secretion in cultured shark rectal gland cells. *Am. J. Physiol.* 260:C1125–1130.

Kent, B. and K. R. Olson. 1982. Blood flow in the rectal gland of *Squalus acanthias*. *Am. J. Physiol.* 243:R296–303.

Koomoa, D. L., M. W. Musch, A. V. MacLean, and L. Goldstein. 2001. Volume-activated trimethylamine oxide efflux in red blood cells of spiny dogfish (*Squalus acanthias*). *Am. J. Physiol.* 281:R803–810.

Lacy, E. R. and E. Reale. 1986. The elasmobranch kidney. III. Fine structure of the peritubular sheath. *Anat. Embryol.* 173:299–305.

Lacy, E. R. and E. Reale. 1995. Functional morphology of the elasmobranch nephron and retention of urea, in *Cellular and Molecular Approaches to Fish Ionic Regulation*. Vol. 14. C. M. Wood and T. J. Shuttleworth, Eds., Academic Press, San Diego, 107–146.

Lacy, E. R. and E. Reale. 1999. Urinary system, in *Sharks, Skates, and Rays*. W. C. Hamlett, Ed., Johns Hopkins University Press, Baltimore, 353–397.

Lacy, E. R., E. Reak, D. S. Schlusselberg, W. K. Smith, and D. J. Woodward. 1985. A renal countercurrent system in marine elasmobranch fish: a computer-assisted reconstruction. *Science* 227:1351–1354.

Laurent, P. 1984. Gill internal morphology, in *Fish Physiology*. Vol. 10A. W. S. Hoar and D. J. Randall, Eds., Academic Press, Orlando, 73–183.

Laurent, P. and S. Dunel. 1980. Morphology of gill epithelia in fish. *Am. J. Physiol.* 238:R147–159.

Lytle, C., J. C. Xu, D. Biemesderfer, M. Haas, and B. Forbush III. 1992. The Na-K-Cl cotransport protein of shark rectal gland. I. Development of monoclonal antibodies, immunoaffinity purification, and partial biochemical characterization. *J. Biol. Chem.* 267:25428–25437.

MacKenzie, S., C. P. Cutler, N. Hazon, and G. Cramb. 2002. The effects of dietary sodium loading on the activity and expression of Na, K-ATPase in the rectal gland of the European dogfish (*Scyliorhinus canicula*). *Comp. Biochem. Physiol.* B 131:185–200.

Maren, T. 1987. Renal acidification in marine fish fifty years after Homer Smith, in *Symposium on Renal, Fluid, and Electrolyte Physiology.* P. Cala and L. Goldstein, Eds., *Bull. Mt. Desert Isl. Biol. Lab.,* Suppl. 1:28–35.

Marshall, J., K. A. Martin, M. Picciotto, S. Hockfield, A. C. Nairn, and L. K. Kaczmarek. 1991. Identification and localization of a dogfish homolog of human cystic fibrosis transmembrane conductance regulator. *J. Biol. Chem.* 266:22749–22754.

McNab, B. K. 2002. *The Physiological Ecology of Vertebrates. A View from Energetics.* Cornell University Press, Ithaca, NY.

Oguri, M. 1964. Rectal glands of marine and fresh-water sharks: comparative histology. *Science* 144:1151–1152.

Olson, K. R. 1999. Rectal gland and volume homeostasis, in *Sharks, Skates, and Rays.* W. C. Hamlett, Ed., Johns Hopkins University Press, Baltimore, 329–352.

Olson, K. R. 2002a. Gill circulation: regulation of perfusion distribution and metabolism of regulatory molecules. *J. Exp. Zool.* 293:320–335.

Olson, K. R. 2002b. Vascular anatomy of the fish gill. *J. Exp. Zool.* 293:214–231.

Paillard, M. 1997. Na+/H+ exchanger subtypes in the renal tubule: function and regulation in physiology and disease. *Exp. Nephrol.* 5:277–284.

Payan, P. and J. Maetz. 1970. Balance hydrique et minerale chez les elasmobranches: arguments en faveur d'un controle endocrinien. *Bull. Inf. Sci. Tech. CEA.* 146:77–96.

Perry, S. F. 1997. The chloride cell: structure and function in the gills of freshwater fishes. *Annu. Rev. Physiol.* 59:325–347.

Perry, S., S. Malone, and D. Ewing. 1987. Hypercapnic acidosis in the rainbow trout. II. Renal ionic fluxes. *Can. J. Zool.* 65:896–902.

Piermarini, P. M. and D. H. Evans. 1998. Osmoregulation of the Atlantic stingray (*Dasyatis sabina*) from the freshwater Lake Jesup of the St. Johns River, Florida. *Physiol. Zool.* 71:553–560.

Piermarini, P. M. and D. H. Evans. 2000. Effects of environmental salinity on Na(+)/K(+)-ATPase in the gills and rectal gland of a euryhaline elasmobranch (*Dasyatis sabina*). *J. Exp. Biol.* 203(19):2957–2966.

Piermarini, P. M. and D. H. Evans. 2001. Immunochemical analysis of the vacuolar proton-ATPase B-subunit in the gills of a euryhaline stingray (*Dasyatis sabina*): effects of salinity and relation to Na+/K+-ATPase. *J. Exp. Biol.* 204:3251–3259.

Piermarini, P. M., J. W. Verlander, I. E. Royaux, and D. H. Evans. 2002. Pendrin immunoreactivity in the gill epithelium of a euryhaline elasmobranch. *Am. J. Physiol.* 283: R983–992.

Raymond, J. and G. Plopper. 2002. A bacterial TMAO transporter. *Comp. Biochem. Physiol.* B 133:29.

Riordan, J. R., B. Forbush III, and J. W. Hanrahan. 1994. The molecular basis of chloride transport in shark rectal gland. *J. Exp. Biol.* 196:405–418.

Royaux, I. E., S. M. Wall, L. P. Karniski, L. A. Everett, K. Suzuki, M. A. Knepper, and E. D. Green. 2001. Pendrin, encoded by the Pendred syndrome gene, resides in the apical region of renal intercalated cells and mediates bicarbonate secretion. *Proc. Natl. Acad. Sci. U.S.A.* 98:4221–4226.

Sawyer, D. B. and K. W. Beyenbach. 1985. Mechanism of fluid secretion in isolated shark renal proximal tubules. *Am. J. Physiol.* 249:F884–890.

Shetlar, R., C. Bevan, T. Maren, and E. Kinne-Saffran. 1987. Presence of a Na+/H+ exchanger in brush border membranes isolated from dogfish (*Squalus acanthias*) kidney. *Bull. Mt. Desert Isl. Biol. Lab.* 27:50.

Shuttleworth, T. J. 1988a. *Physiology of Elasmobranch Fishes.* Springer-Verlag, Berlin, 324.

Shuttleworth, T. J. 1988b. Salt and water balance — extrarenal mechanisms, in *Physiology of Elasmobranch Fishes.* T. J. Shuttleworth, Ed., Springer-Verlag, Berlin, 171–199.

Silva, P., J. Stoff, M. Field, L. Fine, J. N. Forrest, and F. H. Epstein. 1977. Mechanism of active chloride secretion by shark rectal gland: role of Na-K-ATPase in chloride transport. *Am. J. Physiol.* 233:F298–306.

Silva, P., R. J. Solomon, and F. H. Epstein. 1990. Shark rectal gland. *Meth. Enzymol.* 192:754–766.

Silva, P., R. J. Solomon, and F. H. Epstein. 1996. The rectal gland of *Squalus acanthias*: a model for the transport of chloride. *Kidney Int.* 49:1552–1556.

Silva, P., R. J. Solomon, and F. H. Epstein. 1997. Transport mechanisms that mediate the secretion of chloride by the rectal gland of *Squalus acanthias*. *J. Exp. Zool.* 279:504–508.

Silver, R. and M. Soleimani. 1999. H⁺-K⁺-ATPases: regulation and role in pathophysiological states. *Am. J. Physiol.* 276:F799–811.

Smith, C. P. and P. A. Wright. 1999. Molecular characterization of an elasmobranch urea transporter. *Am. J. Physiol.* 276:R622–626.

Smith, H. W. 1931. The absorption and secretion of water and salts by the elasmobranch fishes. II. Marine elasmobranchs. *Am. J. Physiol.* 98:296–310.

Smith, H. W. 1939. The excretion of phosphate in the dogfish *Squalus acanthias*. *J. Cell Comp. Physiol.* 14:95–102.

Stolte, H., R. G. Galaske, G. M. Eisenbach, C. Lechene, B. Schmidt-Nielsen, and J. W. Boylan. 1977. Renal tubule ion transport and collecting duct function in the elasmobranch little skate *Raja erinacea*. *J. Exp. Zool.* 199:403–410.

Swenson, E. and T. Maren. 1986. Dissociation of CO₂ hydration and renal acid secretion in the dogfish, *Squalus acanthias*. *Am. J. Physiol.* 250:F288–F293.

Swenson, E. R., A. D. Fine, T. H. Maren, E. Reale, E. R. Lacy, and A. J. Smolka. 1994. Physiological and immunocytochemical evidence for a putative H-K-ATPase in elasmobranch renal acid secretion. *Am. J. Physiol.* 267:F639–645.

Thorson, T. B., R. M. Wotton, and T. A. Georgi. 1978. Rectal gland of freshwater stingrays, *Potamotrygon* spp. (Chondrichthyes: Potamotrygonidae). *Biol. Bull.* 154:508–516.

Thorson, T. B., D. R. Brooks, and M. A. Mayes. 1983. The evolution of freshwater adaptation in stingrays. *Natl. Geogr. Res. Rep.* 15:663–694.

Vallentich, J. D., K. J. Karnaky, and W. M. Moran. 1995. Phenotypic expression and natriuretic peptide-activated chloride secretion in cultured shark (*Squalus acanthias*) rectal gland epithelial cells, in *Cellular and Molecular Approaches to Fish Ionic Regulation*, Vol. 14. C. Wood and T. J. Shuttleworth, Eds., Academic Press, San Diego, 173–205.

Wall, B., A. Morrison-Shetlar, and J. B. Claiborne. 2001. Expression of NHE-like proteins in the gills of the little skate (*Raja erinacea*); effect of hypercapnia. *Bull. Mt. Desert Isl. Biol. Lab.* 40:60–61.

Weinstein, A. M. 1994. Ammonia transport in a mathematical model of rat proximal tubule. *Am. J. Physiol.* 267:F237–248.

Wells, A., W. G. Anderson, and N. Hazon. 2002. Development of an *in situ* perfused kidney preparation for elasmobranch fish: action of arginine vasotocin. *Am. J. Physiol.* 282:R1636–642.

Wilson, J., D. J. Randall, A. W. Vogl, and G. K. Iwama. 1997. Immunolocalization of proton-ATPase in the gills of the elasmobranch, *Squalus acanthias*. *J. Exp. Zool.* 278:78–86.

Wilson, J. M. and P. Laurent. 2002. Fish gill morphology: inside out. *J. Exp. Zool.* 293:192–213.

Wilson, J. M., J. D. Morgan, A. W. Vogl, and D. J. Randall. 2002. Branchial mitochondria-rich cells in the dogfish *Squalus acanthias*. *Comp. Biochem. Physiol.* A. 132:365–374.

Wingo, C. S. and B. D. Cain. 1993. The renal H-K-ATPases: physiological significance and role in potassium homeostasis. *Annu. Rev. Physiol.* 55:323–347.

Wingo, C. S. and A. J. Smolka. 1995. Function and structure of H-K-ATPase in the kidney. *Am. J. Physiol.* 269:F1–16.

Wong, T. M. and D. K. O. Chan. 1977. Physiological adjustments to dilution of the external medium in the lip-shark *Hemiscyllium plagiosum* (Bennett). II. Branchial, renal, and rectal gland function. *J. Exp. Zool.* 200:85–96.

Wood, C. 1993. Ammonia and urea metabolism and excretion, in *The Physiology of Fishes*, Vol. 1. D. Evans, Ed., CRC Press, Boca Raton, FL, 379–425.

Wood, C., P. Pärt, and P. Wright. 1995. Ammonia and urea metabolism in relation to gill function and acid–base balance in a marine elasmobranch, the spiny dogfish (*Squalus acanthias*). *J. Exp. Biol.* 198:1545–1558.

Wood, C. M., A. Y. Matsuo, R. J. Gonzalez, R. W. Wilson, M. L. Patrick, and A. L. Val. 2002. Mechanisms of ion transport in *Potamotrygon*, a stenohaline freshwater elasmobranch native to the ion-poor blackwaters of the Rio Negro. *J. Exp. Biol.* 205:3039–3054.

Xu, J. C., C. Lytle, T. T. Zhu, J. A. Payne, E. Benz, Jr., and B. Forbush III. 1994. Molecular cloning and functional expression of the bumetanide-sensitive Na-K-Cl cotransporter. *Proc. Natl. Acad. Sci. U.S.A.* 91:2201–2205.

10

Reproductive Biology of Elasmobranchs

Jeffrey C. Carrier, Harold L. Pratt, Jr., and José I. Castro

CONTENTS

10.1 Introduction

The elasmobranchs have had an incredibly long evolutionary history: more than 400 million years. During this extensive period elasmobranchs separately evolved many adaptations such as exquisite senses and complex reproductive modes that rival those of the most advanced tetrapods. In this chapter we review the reproductive adaptations of the elasmobranchs and show how these adaptations have contributed to their evolutionary success and genetic continuity. It is not intended here to produce a complete review of elasmobranch reproduction, as there are several excellent reviews already: Budker (1958), Wourms (1977), and Dodd (1983). Our goal is to produce a brief overview of elasmobranch reproduction that will lead the reader to the more specialized literature. We have used examples of anatomy and modes of reproduction for the few elasmobranchs that are well known. However, we must add the caveat that, although elasmobranch reproduction has proceeded along only a few paths, there is great diversity among congeners. This diversity is often expressed as different brood sizes, ovarian cycles, gestation periods, mating systems, use of different nurseries, etc. We must also state that the reproductive processes for most sharks remain unknown. Unraveling the many secrets of elasmobranch reproduction will remain a challenge for future researchers.

The primitive mode of reproduction in fishes is the production of very large numbers of eggs and sperm, which are shed into the water, where fertilization occurs. This process is known as oviparity and is typical of bony fishes. The embryos of oviparous fishes are provided with only a small amount of yolk and consequently hatch in an undeveloped or larval condition and require weeks or months to complete development. Both eggs and young are highly vulnerable to predators and environmental factors for prolonged periods and suffer heavy mortality.

The evolutionary success of sharks is partly due to the efficiency of their reproductive adaptations that depart from simple oviparity. The most significant of these adaptations are internal fertilization and the production of small numbers of large young, which hatch or are born as active, fully developed miniature sharks. The embryos spend their developmental stages within their mother's body and so receive protection during their most vulnerable stages. The young are born at relatively large size, reducing the number of potential predators and competitors while increasing the number of potential prey, thus enhancing their chances of survival (Castro, 1983).

10.2 Internal Fertilization

All elasmobranchs have internal fertilization. Internal fertilization ensures that energy-expensive eggs are not consumed by other animals, and that the energy allocated to reproduction is passed to the embryos and not wasted. In addition internal fertilization improves the likelihood and efficiency of fertilization and avoids sperm wastage. Female elasmobranchs retain the fertilized eggs or embryos for varying periods of time, thus protecting the embryos during their most vulnerable stages. Depending on how long females retain the fertilized eggs, elasmobranchs can be divided into two groups: oviparous (egg-laying) forms, and viviparous (live-bearing) forms. Oviparous elasmobranchs retain their eggs for short periods of time and then deposit them on the substrate or attach them to bottom structures. Viviparous forms retain their embryos until the embryos have completed development, and then give birth to live young. The young of both oviparous and viviparous forms hatch, or are born, fully developed, as miniature copies of the adult. By bypassing a larval stage, elasmobranchs reduce losses to predation, and by hatching or being born as an active, relatively large fish, they also have a greater number of potential prey. These distinctions between oviparous and viviparous forms were first made by Aristotle around 343 B.C. (Aristotle, 343 B.C., Thompson translation, 1949). Aristotle also noted copulation in elasmobranchs, as well as their claspers, the egg cases of oviparous forms, and the yolk sac placenta of viviparous elasmobranchs, as distinct from the mammalian placenta.

10.2.1 Oviparous Forms

Oviparous elasmobranchs enclose their eggs in tough horny egg cases and then lay them on the substrate or attach them to bottom structures. Once deposited, the egg cases receive no further care from the parent. Parental care is unknown in elasmobranchs. Embryos are nourished solely by yolk stored in the yolk sac (Figure 10.1). The incubation periods may last from a few months to a year or longer. After a few weeks of development, a small slit opens on each side of the egg case, to allow for ventilation and oxygenation of the egg case. The embryo aids in ventilation by constantly fanning its tail, promoting water flow and exchange with the surrounding water. The tough egg case forms the only protective barrier against predators that the embryo has. Egg cases are preyed upon by numerous predators, from gastropods to other sharks, and may suffer heavy mortality. Oviparous elasmobranchs are benthic, primarily littoral or bathyal in habit, and rarely of large size (Tortonese, 1950). Oviparous forms produce small young because the amount of nutrients available to the developing embryo is limited by what is stored in the yolk sac. Oviparity is found only in three families of sharks (the Heterondotidae, the Scyliorhinidae, and the Orectolobidae) and in the skates (Rajiformes). Oviparity is probably the primitive or ancestral condition in sharks. It is from this condition that viviparity arose by the prolongation of the time that the embryos are retained by the female.

FIGURE 10.1 (Color figure follows p. 304.) Finetooth shark (*Carcharhinus porosus*); 15 mm embryo attached to yolk sac. (Photo copyright José I. Castro. Used with permission.)

10.2.2 Viviparity

Viviparous forms retain their embryos in the uterus during the entire period of development. Thus, the embryos are born fully developed, as miniature copies of the adults. Viviparous forms can be divided into aplacental and placental forms, depending on whether a placental connection is developed between mother and offspring (Budker, 1958). However, it must be realized, as Budker (1958) and Wourms (1977) have pointed out, that the elasmobranchs are a diverse group with a continuum of reproductive adaptations, where oviparity and viviparity are the ends of the continuum.

10.2.2.1 Aplacental Viviparity — Aplacental forms do not form a placental connection between mother and offspring. In the recent past aplacental viviparity was often called ovoviviparity, a term borrowed from the vertebrate literature. Aplacental viviparous elasmobranchs vary widely in their modes of nourishing their embryos. Wourms divided aplacental forms into three functional groups depending on whether the embryos (1) depend solely on yolk reserves, (2) are oophagous (egg-eating), (3) are nourished through placental analogues.

10.2.2.2 Yolk Dependency — In this form, represented by many major groups including the Squaliformes, the Hexanchiformes, Squatinaformes, some Orectolobiformes, some Carcharhiniformes, and many primitive batoid groups, the embryos depend solely on the yolk deposited in the egg at the time of ovulation. The embryos are retained in the uterus simply to protect them, but they do not receive any supplemental nourishment from the mother during gestation. Consequently, like the young of oviparous forms, the young of these elasmobranchs are relatively small at birth, given the finite amount of nutrients available in the yolk sac.

10.2.2.3 Oophagy — In the oophagous elasmobranchs, the ovary grows to a tremendous size, often weighing over 5 kg (Gilmore et al., 1983; Gilmore, 1993). The eggs of oophagous sharks are small; usually 5 to 7 mm in diameter and most exist to nourish the developing young. The embryo is dependent on yolk for a very short time, perhaps just a couple of weeks, because the minute egg holds very little yolk. When the embryos are only about 5 cm long, they begin to ingest other eggs in the uterus. These small embryos have a temporary precocious dentition that allows them to rupture the eggs and ingest their contents. Usually only a few fertilized eggs are produced at the beginning of gestation.

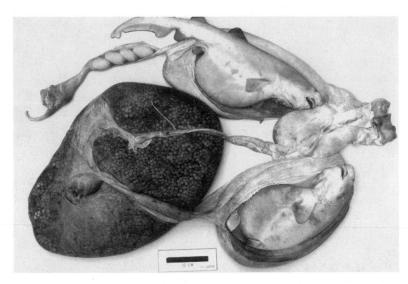

FIGURE 10.2 (Color figure follows p. 304.) Reproductive tract of a sandtiger shark (*Carcharias taurus*) carrying one oophagous embryo in each uterus. The ovary can be seen on the left and several egg cases are being shunted to the embryos. (Photo copyright José I. Castro. Used with permission.)

In the case of the smalltooth thresher, *Alopias pelagicus*, only one fertilized egg is released into each oviduct, a single egg inside an egg case. In the mako (*Isurus oxyrinchus*) and white shark (*Carcharodon carcharias*), six to ten fertilized eggs are released into each oviduct. In the unique case of the sandtiger shark, *Carcharias taurus*, as many as a dozen embryos may be produced per oviduct. After a number of egg cases containing a single fertilized egg have been produced, the female shifts to producing egg cases containing several unfertilized eggs. These egg cases are known as "feeding egg cases" as their function is to convey yolk to the embryos. When the largest embryo reaches 10 to 12 cm in length, it seeks out other embryos and kills them by biting them. As the surviving embryo grows larger, it ingests the long-dead embryos along with the contents of the feeding egg cases. This intrauterine consumption of siblings has been called intrauterine cannibalism. The embryo continues to ingest eggs in large quantities, acquiring a large, protruding yolk stomach (Figure 10.2). Sandtiger embryos quit feeding in the last month or two of gestation and consume the yolk stored in their yolk stomachs. Because oophagous embryos have a very large supply of yolk (energy) available to them, they often attain relatively large size at birth. Sandtiger young attain over 1 m at birth, or about a third of the size of the mother. Oophagy appeared very early in the evolution of the cartilaginous fishes, and by any standards, it is a very effective method of nourishing embryos to a large size.

10.2.2.4   Placental Analogues — Placental analogues are regions of the uterine epithelium that secrete a nutritive "uterine milk" or embryotrophe (histotroph) that is secreted by long villi called trophonemata (Alcock, 1892) on the uterine lining and ingested by the embryos (Amoroso, 1960). This mode of nutrition is common in the rays (Myliobatiformes).

10.2.2.5   Placental Viviparity — The embryos of placental viviparous forms are nourished by yolk stored in the yolk sac during their first few weeks (Figure 10.3 and Figure 10.4). As the nourishing yolk is exhausted, the yolk sac elongates and its distal surface becomes highly vascularized. Where the vascularized surface of the yolk sac touches the uterine wall, the tissues of mother and offspring grow into intimate contact, forming the yolk sac placenta. Once the placenta is formed, nutrients can be shunted to the developing embryo directly from the bloodstream of the mother. The embryos of placental elasmobranchs have practically an unlimited supply of energy. Nutrients are limited only by the health of the mother and her food supply.

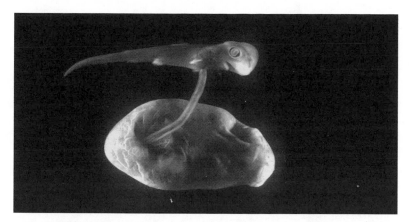

FIGURE 10.3 (Color figure follows p. 304.) Atlantic sharpnose shark (*Rhizoprionodon terranovae*); 43 mm embryo. Note yolk sac becoming flaccid as yolk is being consumed. (Photo copyright José I. Castro. Used with permission.)

FIGURE 10.4 (Color figure follows p. 304.) Atlantic sharpnose shark (*Rhizoprionodon terranovae*); 250 mm embryo at midterm. (Photo copyright José I. Castro. Used with permission.)

The selachian placenta has been known for a long time. Aristotle, around 343 B.C., was the first to note the umbilical connection and the placenta of elasmobranchs. Later, Rondelet (1558) illustrated a *Mustelus* attached to its fetus by an umbilical cord. As early as 1673 Steno schematically illustrated the placental relationship and the connection of the umbilical cord into the intestine of a *Mustelus* (Maar, 1910). For an excellent review of the history of elasmobranch reproduction see Wourms (1977).

10.3 Male Reproductive System

The male reproductive system consists of the testes, the genital ducts (efferent ducts, epididymides, and the storage organ, the ampullae epididymides), the urogenital papilla, the siphon sacs, and the claspers (intromittent secondary sex organs). The reproductive system of male elasmobranchs has been described in detail by many authors (Borcea, 1906; Dean, 1906; Daniel, 1928; Matthews, 1950). In the elasmobranchs, the testes are paired, symmetrical structures situated at the anterior end of the coelom, dorsal to the liver. Usually, each testis is suspended from the middorsal body wall by a mesorchium. In some species, the testes are embedded at the anterior end of the epigonal organ. In immature animals the testes

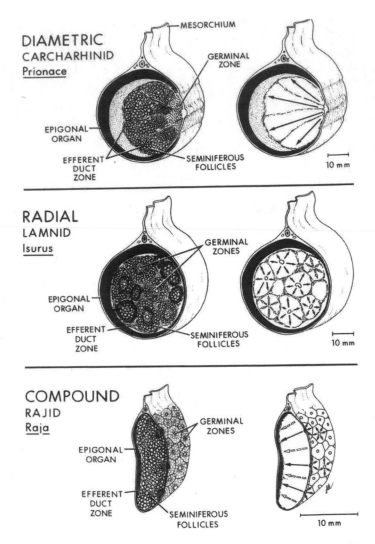

FIGURE 10.5 Schematic representations of three different forms of elasmobranch testes. (From Pratt, H.L. 1988. *Copeia* 1988:719–729. With kind permission from the American Society of Ichthyologists and Herpetologists.)

may appear as an inconspicuous mass of whitish tissue or a light streak on the surface of the epigonal organ. In adult animals the testes usually vary greatly in size during the year, enlarging and swelling during the breeding season, and regressing at other times.

The testes also vary greatly among species in morphology and functional arrangement. In the basking shark, *Cetorhinus maximus*, the testes are enclosed in the epigonal organ and they are made up of numerous separate lobules separated by connective tissue (Matthews, 1950). Pratt (1988) has referred to this type of testis as radial, because the germinative zone is at the center of the lobule and the development of the seminiferous follicles proceeds radially from the center of the lobule toward the circumference, where efferent ductules collect the spermatozoa as they mature (Figure 10.5). The lamniform sharks share this type of testis. In the requiem sharks (Carcharhinidae) and the hammerhead sharks (Sphyrnidae), the testes protrude from the surface of the epigonal organ. In this type of testis, called diametric by Pratt (1988), the development of the seminiferous follicles proceeds from one wall across the diameter of the testis to the opposite wall, where efferent ductules collect the spermatozoa. A third type of testis shows elements of both diametric and radial morphology and is found in many batoids. Pratt (1988) describes this form as a compound testis where lobular divisions are absent, the

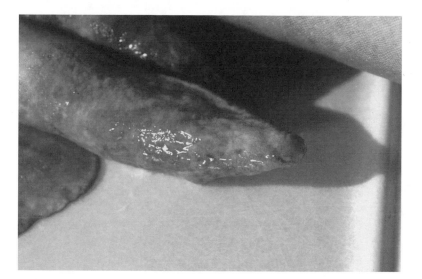

FIGURE 10.6 Clasper of a mature nurse shark (*Ginglymostoma cirratum*). (Photo copyright Jeffrey C. Carrier. Used with permission.)

germinal tissue is located on the surface in regions referred to as "islands," and efferent ductules are also located close to the surface to receive maturing spermatozoa.

Spermatogenesis occurs in the testis, in spherical follicles termed ampullae (Stanley, 1966), situated at the end of a highly branched collecting tubule system. Stanley (1966) described the functional unit of the elasmobranch testis as the spermatocyst, a spherical structure that comprises many spermatoblasts, which consist of Sertoli cells and their associated germ cells (Parsons and Grier, 1992). The developmental process progresses through formation of the spermatocyst and ends when the spermatocyst bursts open and the Sertoli cells fragment. At this point, the spermatozoa are released into an efferent ductule that has formed a patent connection to the spermatocyst during the last stages of spermatogenesis. Sperm are conveyed through minute ductules (ductus efferens) into a highly convoluted epididymis, which can be seen along the vertebral column on either side of the dorsal aorta. Posteriorly the epididymis passes into the ductus deferens (or ampulla epididymis; Jones and Jones, 1982), which is a sperm-storage organ. In immature males the epididymis is a straight tube on the ventral surface of the kidneys; in sexually mature males the anterior portion is highly coiled, while the posterior becomes a septated thick-walled straight tube that enlarges to form the ampulla. The posterior ends of the seminal vesicles share common lumina with the paired sperm sacs and the ureters. The size and location of the sperm sac is highly variable depending on species. The two sperm sacs end posteriorly in the urogenital sinus into which also protrude the conical ends of the ampullae. The small cavity is enclosed by and exits through a conical projection, the urogenital papillae, which protrudes into the cloaca.

In male elasmobranchs, the median edges of the pelvic fins form paired, tubelike copulatory organs known as claspers (Figure 10.6). In immature elasmobranchs the claspers are small and flexible. Upon reaching maturity, the claspers calcify, harden, and form articulations with the pelvic fin base. The calcification and rigidity of the clasper, and the ability of the rhipidion to splay open and erect the spur, are the best standards for determining maturity in elasmobranchs (Clark and von Schmidt, 1965). During copulation one of these intromittent organs is inserted into the female to transfer sperm (Figure 10.7). The particular clasper used is determined by the nature of the male's grip on the female. A grip on the right pectoral fin that positions a female on the male's left side (Figure 10.8) would result in the use of the male's right clasper (Carrier et al., 1994; Pratt and Carrier, 2001). Although most studies reveal the use of a single clasper during copulation, accounts indicate that some species may insert both claspers, including *Scyliorhinus canicula* (Leigh-Sharpe, 1920). Opening of the terminal rhipidion serves to anchor the clasper within the reproductive tract of the female (Figure 10.9), often with a sharp hook or spur (Gilbert and Heath, 1972).

FIGURE 10.7 Clasper insertion during copulation in nurse sharks (*Ginglymostoma cirratum*). (Photo copyright Jeffrey C. Carrier. Used with permission.)

FIGURE 10.8 (Color figure follows p. 304.) Male nurse shark (*Ginglymostoma cirratum*) (left) gripping a female's pectoral fin as a prelude to copulation. (Photo copyright Jeffrey C. Carrier. Used with permission.)

During mating the claspers are rotated forward and one is inserted into the female. Sperm is forcibly ejected with the aid of contractile saclike organs known as siphon sacs. The siphon sacs are paired muscular bladders that lie just beneath the skin of the ventral side of the body and extend anteriorly almost to the level of the pectoral fins, where they end blindly (Gilbert and Heath, 1972). Posteriorly, the siphon sacs connect to an opening at the proximal end of the clasper called the apopyle. The clasper tube itself runs posteriorly from the apopyle to its distal opening called the hypopyle. The apopyle has no connection to the urogenital papilla, which is the site of sperm release just inside the cloacal opening. Although the siphon sacs have no direct connection to the urogenital papilla, most male sharks appear to rotate one clasper forward during mating, forming a connection between the clasper apopyle and the urogenital papilla. This was first hypothesized by Leigh-Sharpe (1920) and has been supported by most

FIGURE 10.9 Claspers of a mature nurse shark (*Ginglymostoma cirratum*) spread open to reveal the rigid structures and spur, evident in upper left. (Photo copyright Jeffrey C. Carrier. Used with permission.)

observations of mating elasmobranchs (reviewed by Pratt and Carrier, 2001). The function of the siphon sacs seems to be to force sperm from the cloaca through the claspers and out into the female by means of a seawater current produced upon contraction. There is also speculation that they may serve to flush the female reproductive tract of semen from previous matings (Leigh-Sharpe, 1920; Gilbert and Heath, 1972; Eberhard, 1985). Recent analysis of detailed videotaped mating in *Triaenodon obesus* shows siphon sac filling when males grasp a female, immediately prior to clasper insertion. The sacs gradually deflated during copulation (Whitney et al., unpubl.). Such gradual deflation tends to support a sperm propulsive role for siphon sacs.

In many elasmobranchs sperm are aggregated and packed in rounded, ovoid, or tubular small matrices each containing a very large number of sperm. Following the definitions of Nielson et al. (1968), Pratt and Tanaka (1994) clarified the identity of the elasmobranch sperm packets as spermatophores, which are sperm encapsulated in a matrix, and spermozeugma, sperm that are embedded but unencapsulated. In shark spermozeugma, the long sperm heads are embedded and tails beat freely around and inside the mass. Spermatophores can be large, and range from 10 mm as in *Carcharodon carcharias* to 300 mm long in *Carcharias taurus* (Pratt and Tanaka, 1994). It was noted by Matthews (1950) that the function of the spermatophore is also to protect the sperm and prevent loss of sperm by leakage into the water during copulation. In *Carcharhinus isodon*, Castro (1993) found evidence that the sperm are transferred in a dissociated state and embedded in a matrix he referred to as a spermozeugma (Castro, 1993). These aggregations may serve to protect spermatozoa from breakdown or loss on exposure to seawater during copulation, or other physical or chemical conditions that might compromise sperm viability. Matthews (1950) believes that the function of the spermatophore is also to protect the sperm and to prevent loss of sperm by leakage into the water during copulation. There is much variation in these temporary structures and little is known about their function or nomenclature.

Associated with the epididymis and the ductus deferens is the accessory Leydig gland, described by Jones and Jones (1982) as a branched tubular gland that connects with the initial segment of the epididymis. Dissection reveals some cells that appear to be specialized for protein production, and may account for the elevated levels of protein present in seminal fluids (Jones and Lin, 1993). Additionally, the alkaline (Marshall's) gland, located near the seminal vesicle, has also been shown in some skates to produce a highly alkaline secretion that may assist with the production of sperm or have a role in the formation of copulatory plugs (Hamlett, 1999).

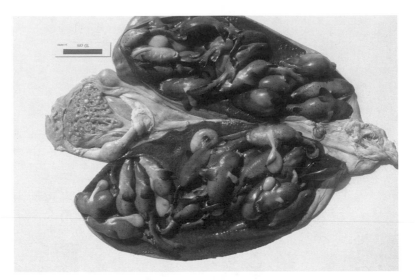

FIGURE 10.10 (Color figure follows p. 304.) Entire reproductive system of pregnant nurse shark (*Ginglymostoma cirratum*) showing young in both uteri. (Photo copyright José I. Castro. Used with permission.)

10.4   Female Reproductive System

10.4.1   Anatomy

The female reproductive system (Figure 10.10) consists of the ovaries and the oviducts. The ovaries are paired or single structures located at the anterior end of the body cavity dorsal to the liver. Usually, both ovaries are functional in the more primitive forms, while in the galeoid forms, such as the genera *Scyliorhinus, Carcharhinus, Mustelus,* and *Sphyrna*, only the right ovary is functional and the left ovary is vestigial or absent. The appearance of the ovaries varies with the sexual condition of the specimen. In immature females they are small and are only visible as a thin strip of granulated tissue (minute oocytes); in mature females the ovaries are large, often bearing bright yellow, vitellogenic oocytes on the surface. In some of the advanced forms, the single ovary is embedded in the anterior end of a long epigonal organ. Ovaries can be of two types (Figure 10.11). Most are of the compact (Nelsen, 1953) or naked type (gymnovarium). This type of ovary produces a few eggs of very large size, usually from 20 to 60 mm in diameter. A second type, found in lamnoid sharks, produces countless very small ova, 3 to 5 mm in diameter, that are fed to their oophagous embryos (Matthews, 1950; Pratt, 1988).

The oviducts are paired, long tubular structures that run the length of the body cavity on both sides of the vertebral column. The oviducts unite at their anterior ends at the ostium, which forms a wide funnel that apposes the ovary or ovaries. From the ostium, the oviducts curve caudally into thin tubes that lead to the shell or oviducal gland.* The shell gland appears as a slight swelling on the anterior part of the oviduct in immature animals, and as a well-differentiated gland several times the diameter of the oviducts in mature animals. The shell gland secretes an egg membrane or a shell that encloses the eggs as they pass through the oviduct. The shell gland may also act as the site of sperm storage and fertilization. The shell gland enlarges to about twice its resting size immediately after fertilization and during egg passage. The posterior part of each oviduct is enlarged to form the uterus, where the embryonic shark develops. In most cases, the embryos are loose in the uterus and in close contact with each other. In some of the carcharhinid sharks and in the hammerhead sharks, the embryos are in separate compartments or crypts in the uterus. The two uteri unite posteriorly to form a chamber, the vagina, which

---

* This gland has been called shell gland, oviducal gland, and nidamental gland. The authors prefer to call it shell gland because its function is to produce the eggshell. Oviducal simply means that it is a gland in the oviduct or that it conveys the egg. Nidamental means "nest building," a function that the gland does not perform. Thus, shell gland seems more appropriate, and it is a shorter term.

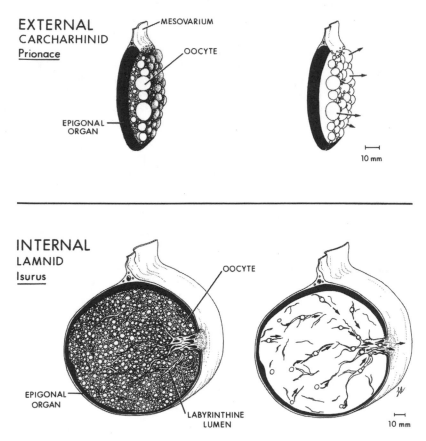

FIGURE 10.11 Schematic representations of two different forms of elasmobranch ovaries. (From Pratt, H.L. 1988. *Copeia* 1988:719–729. With kind permission from the American Society of Ichthyologists and Herpetologists.)

opens into the cloaca by a large aperture. In juvenile sharks the vagina is sealed by a membrane or hymen that thins and eventually breaks or is broken as the female reaches maturity.

10.4.2   Elasmobranch Ova

Elasmobranch eggs are generally large and contain an enormous amount of yolk; consequently, they are produced in relatively small numbers. The oophagous lamnoid sharks produce very large numbers of small eggs. When eggs are released from the ovary, they pass through the ostium into the oviducts. Generally, two eggs are ovulated at one time, one going into each oviduct. Eggs acquire the egg membrane or egg case after passage through the shell gland. In oviparous forms, the egg case is tough and provides protection for the embryo during its long developmental period, usually lasting from several months to a year. In viviparous forms where the embryos are retained through development, the egg case is thin and diaphanous as it no longer serves to protect the embryos from predation. The egg case is incorporated into the placenta. Because of its large size, the elasmobranch egg is strongly telolecithal, and only a small disk at the animal pole takes part in the cleavage process. Cleavage is generally of the meroblastic type. The part of the egg not taking part in the cleavage process becomes the yolk sac.

10.4.3   Sperm Storage

Fertilization occurs prior to encapsulation and is generally considered in lamnoid sharks to occur in or before the shell gland in the upper oviduct (Gilmore et al., 1983). Pratt (1993) reviewed sperm storage in several elasmobranch species and localized sperm histologically within the shell gland in dusky sharks (*Carcharhinus obscurus*), blue sharks (*Prionace glauca*), Atlantic sharpnose sharks (*Rhizoprionodon terrenovae*), and scalloped hammerhead sharks (*Sphyrna lewini*). Lamniform sharks showed no evidence of long-term storage.

Storage times range from 4 weeks in *C. cautus* (White et al., 2002), to more than 5 months in *Galeorhinus galeus* (Peres and Vooren, 1991), 6 months in *Furgaleus macki* (Simpfendorfer and Unsworth, 1998), and 12 months in *P. glauca* (Pratt, 1993). *Scyliorhinus retifer* may store sperm well beyond a year (Castro et al., 1988). In this study, egg-laying females were isolated from males and produced fertile eggs that hatched normally after 401 days. Some normal and abnormal embryos were produced in eggs that hatched 469 days later, and normal embryos hatched at intervals as long as 843 days following isolation.

The fine network of tubules within the shell gland seems to provide ample space for sperm to be stored, although the presence of sperm in the lumen of the shell gland may not by itself imply storage. Additional studies encompassing different species are necessary to identify the full extent of regions where sperm are stored within the gland's tubular network. Hamlett et al. (2002) found evidence of a uterine epithelial-sperm interaction, more specifically, a sperm attachment, in the female uterus of *Mustelus canis* in addition to the sperm storage in the terminal zone of the shell gland.

10.5 Reproductive Cycles

The reproductive cycles of elasmobranchs are complex and poorly understood. The reproductive cycle encompasses the ovarian cycle and the gestation period. The ovarian cycle is how often a female develops a batch of vitellogenic oocytes and ovulates a batch of eggs. The gestation period is the length of time between fertilization and parturition. These two processes may run concurrently or consecutively. For example, in the spiny dogfish, a new batch of vitellogenic oocytes is developing in the ovary at the same time that a brood of embryos is growing in the uteri. Thus, the ovarian cycle and gestation proceed concurrently. Both the ovarian cycle and the gestation period last almost 2 years. Thus, a female reproduces every 2 years, and the reproductive cycle is said to be biennial. In many carcharhinid sharks, such as the blacktip and finetooth sharks, the ovarian cycle and the gestation period run consecutively (Castro, 1996). For example, a female may become pregnant in May, continue gestation for about a year, and give birth the following May. After parturition, the female enters a "resting stage" where she begins to store lipids in her liver. In late winter, the next batch of oocytes begins to grow rapidly. By late spring, the oocytes have grown to their maximum size, and the female mates again in May and ovulates, starting the cycle again. Thus, from ovulation to ovulation, or from parturition to parturition, the reproductive cycle takes 2 years (biennial). Some species, such as the Atlantic sharpnose shark (*Rhizoprionodon terraenovae*), the smooth dogfish (*M. canis*) (Conrath and Musick, 2002), and the scalloped hammerhead (*Sphyrna lewini*), have annual cycles. In these sharks, the ovarian cycle and the gestation period run concurrently, each lasting about a year. Females carry developing oocytes and embryos at the same time. These females give birth and mate again shortly afterward. Thus, these sharks reproduce annually. Longer reproductive cycles, up to 3.5 years, have been postulated.

10.6 Mating and Reproductive Behaviors

Little is known of mating behavior in elasmobranchs. While the mechanism of mating has been known at least since the time of Aristotle (Agassiz, 1871), actual observations of mating are few, mostly in captive animals. Pratt and Carrier (Pratt and Carrier, 2001) summarized these and other studies in a review of reproductive behavior. Table 10.1 summarizes the behaviors that were observed in these early investigations. Because fertilization is internal, precopulatory activities involve actions that result in the male ultimately grasping the female in a manner that provides an appropriate alignment for insertion of a clasper into the female. Many sharks, skates, and rays orally grasp fins, but some of the more sinuous species may bite and hold the body and flanks. Some courtship bites are preliminary and may serve to stop the female or signal male intent. These bites are less tenacious than feeding bites and usually do not employ full force or full closure of the jaw.

TABLE 10.1

Summary of Observed Courtship and Mating Behaviors in Elasmobranch Fishes

General Behavior and Species	Descriptions/Notes	Ref.
Sharks		
Precopulatory and courtship		
Following		
Carcharhinus melanopterus	"Close follow" near female's vent possibly olfactory-mediated	Johnson and Nelson, 1978
Ginglymostoma cirratum	Male and female swim parallel and synchronously side by side	Klimley, 1980
Negaprion brevirostris	Swimming with body axes in parallel	Clark, 1963
Female avoidance		
Carcharias taurus	Female "shielding" with pelvics close to substrate	
Ginglymostoma cirratum	"Lying on back" the female rests motionless and rigid	Gordon, 1993
	Female "pivots and rolls" on her back when a male bites her pectoral fin	Klimley, 1980
Female acceptance		
Carcharias taurus	"Submissive" body, "cupping" and "flaring" of pelvic fins	Gordon, 1993
Ginglymostoma cirratum	Female arches body toward male, "cups" pelvic fins	Carrier et al., 1994
Biting		
Heterodontus francisci	Male bites and wraps female pectoral fin body, tail, gills	Dempster and Herald, 1961
Scyliorhinus retifer	Male bites and wraps female pectoral fin body, tail, gills	Castro et al., 1988
Scyliorhinus torazame	Male bites and wraps female pectoral fin body, tail, gills	Uchida et al., 1990
Ginglymostoma cirratum	Male bites and holds female's pectoral fin	Klimley, 1980; Carrier et al., 1994
Carcharhinus sp.	Male bites and holds female's pectoral fin	Clark, 1975
Triaenodon obesus	Male bites and holds female's pectoral fin	Uchida et al., 1990; Tricas and Lefeuvre, 1985
Positioning and alignment		
Ginglymostoma cirratum	"Nudging" female into position with head	Klimley, 1980
Ginglymostoma cirratum	After "pectoral bite" male rolls female, then aligns for insertion	Carrier et al., 1994
Ginglymostoma cirratum		Carrier et al., 1994
Sphyrna lewini	"Torso thrust" with "clasper flexion" possibly filling siphon sacs	Klimley, 1985
Carcharias taurus	"Crossing" or "splaying" claspers as position requires	Gordon, 1993
Group		
Ginglymostoma cirratum	Multiple males compete or cooperate for a mate; a cooperative behavior or a single male "blocking" a mating pair.	Carrier et al., 1994
Insertion and copulation	Insertion of one or more claspers into the cloaca leading to ejaculation	Carrier et al., 1994
Copulatory		
Male bites female while at rest		
Heterodontus francisci	Male wraps around female's body	Dempster and Herald, 1961
Scyliorhinus retifer	Male wraps around female's body	Castro et al., 1988
Smaller shark sp.	Male wraps around female's body	Dempster and Herald, 1961; Castro et al., 1988; Gilbert and Heath, 1972; Dral, 1980
Triaenodon obesus	Heads to substrate, sharks undulate to keep tails elevated	Tricas and Lefeuvre, 1985

TABLE 10.1 (continued)
Summary of Observed Courtship and Mating Behaviors in Elasmobranch Fishes

General Behavior and Species	Descriptions/Notes	Ref.
Ginglymostoma cirratum	"Lying parallel on substrate" less than two pectoral widths apart during bouts of "parallel swimming"	Klimley, 1980
Ginglymostoma cirratum	Heads to substrate, tails elevated or lying parallel	Carrier et al., 1994
	"Copulation" sometimes in groups of many males	
Heterodontus francisci	Male crosses female's body, rhythmic motion for up to 35 m	Dempster and Herald, 1961
Parallel swimming "in copula"		
Negaprion brevirostris	Coordinated pair swimming while copulating	Clark, 1963
Carcharodon carcharias	Possible coordinated pair swimming while copulating	Francis, 1996
Polygyny		
Ginglymostoma cirratum	Males will mate with many females over several weeks	Pratt and Carrier, 2001
Polyandry		
Ginglymostoma cirratum	Females will mate with many males over several weeks	Pratt and Carrier, 2001
Postcopulatory	Pair remains together or departs rapidly	Carrier et al., 1994
Stalking		
Carcharias taurus	Male aggression toward other species in a captive environment	Gordon, 1993
Batoids		
Precopulatory and courtship		
Following		
Aetobatus narinari	Rapid "chase," close to tail of female	Uchida et al., 1990
Manta birostris	Rapid "chase," close to tail of female	Yano et al., 1999
Myliobatis californica	Male ventral to female with wingbeats synchronized	Tricas, 1980
Myliobatis californica	Males "follow" females	Feder et al., 1974
Female avoidance		
Urolophus halleri	Females bury in sand to "avoid" males	Tricas et al., 1995
Aetobatus narinari	Females raise back out of water and slap wings on surface in response to male "nipping"	Tricas, 1980
Urolophus halleri	Females spine males with caudal spine	Michaels, 1993
Female acceptance		
Raja eglanteria	"Back arching," "pectoral fin undulations" to attract males	Luer and Gilbert, 1985
Biting		
Aetobatus narinari	"Gouging," "nibbling" bites on female dorsal surface	Uchida et al., 1990
Rhinoptera bonasus	"Gouging," "nibbling" bites on female dorsal surface	Tricas, 1980
Rhinoptera javanica	"Gouging," "nibbling" bites on female dorsal surface	Uchida et al., 1990
Manta birostris	Male grasps pectoral fin tips (nipping)	Yano et al., 1999
Group		
Dasyatid and Myliobatid rays	Common for multiple males to "follow" single females	Uchida et al., 1990; Tricas, 1980; Feder et al., 1974
Rhinoptera javanica	Many captive males overwhelmed a female for multiple matings	Uchida et al., 1990
	Mortality sometimes resulted from wounds and exhaustion	

TABLE 10.1 (continued)

Summary of Observed Courtship and Mating Behaviors in Elasmobranch Fishes

General Behavior and Species	Descriptions/Notes	Ref.
Other behaviors		
Aetobatus narinari	Males "bob" and "sway" while "following" "avoiding" females	Tricas, 1980
Copulatory		
While reposed on bottom		
Raja eglanteria	Copulate for 1–4 h while at rest on bottom; male holds trailing edge of female's pectoral fin, swings tail beneath hers and inserts one clasper	Luer and Gilbert, 1985
While swimming		
Manta birostris	Copulation near the surface, abdomen to abdomen	Yano et al., 1999
Aetobatus narinari	Mating abdomen to abdomen in the mid-depths of the tank; insertion time was 0.5 to 1.5 min	Uchida et al., 1990
Rhinoptera javanica	Starts at the surface or mid-depth, abdomen to abdomen, continues on the bottom	Uchida et al., 1990
Rhinoptera bonasus	Starts at the surface or mid-depth, abdomen to abdomen, continues on the bottom	Uchida et al., 1990
Polyandry		
Aetobatus narinari	A captive female mated many times in succession with three to four males in 1 h	Uchida et al., 1990
Rhinoptera javanica	Multiple matings common	Uchida et al., 1990
Postcopulatory		
Manta birostris	Male remains attached to pectoral fin tip briefly	Yano et al., 1999

Source: Pratt, H.L., Jr. and J. C. Carrier. 2001. *Environ. Biol. Fish.* 60:157–188. With permission.

The precopulatory activities may often be prolonged and involve mate recognition and rejection of some males by females. It is clear from studies of *Ginglymostoma cirratum* that females often employ selective behavior strategies such as refuging, avoidance, arching, and shielding that can prevent copulation. If interested in mating, females may acquiesce and exhibit acceptance behaviors such as flaring or cupping of the pelvic fins that facilitate an adequate grip for copulation (Pratt and Carrier, 2001). Once the male has an adequate grip on the female, a clasper is inserted and copulation ensues (Figure 10.7). Males also exhibit cooperative behaviors between males, such as blocking, a helping behavior that reveals a degree of social interaction not previously described in elasmobranchs (Carrier et al., 1994).

Nurse shark females mate with multiple males and genetic analysis of broods reveal multiple paternity (Ohta et al., 2000; Saville et al., 2002; Carrier et al., 2003). Similar patterns of paternity have also been shown in *Negaprion brevirostris* (Feldheim et al., 2001) and may well represent a general strategy to maintain a level of genetic diversity in animals that produce only a few broods in a lifetime (see Heist, Chapter 16 of this volume).

10.7 Conclusion

The general trend in elasmobranch evolution is a progression from a modified oviparity to viviparity, with the production of small numbers of fully developed young. This evolution has proceeded along relatively few pathways, but there is great diversity in adaptations to nourish the young, reproductive cycles, small brood size, etc. It is likely that future research will reveal many other novel adaptations within the known pathways.

The reproductive adaptations that have made elasmobranchs evolutionarily successful for eons, delayed maturity, long reproductive cycles, and small broods, now threaten their survival. Today the ocean environment has changed with humans assuming the role as the apex predator, and elasmobranchs are being fished in quantities that may exceed their capacity to reproduce. Caught in diverse directed fisheries as well as bycatch in many other fisheries, many elasmobranchs are declining at a very rapid pace (Baum et al., 2003). Given their evolutionary pathways, it is highly unlikely that elasmobranchs will be able to compensate for the current fishing mortality that they are experiencing.

Understanding the many aspects of elasmobranch reproductive systems and processes will be an important factor in shaping conservation and management initiatives (Pratt and Casey, 1990). Much work needs to be done to learn the specific details of reproductive parameters of different species because management and conservation models will require accurate specific data and not generalized data from related species. We also need to know more about the specific habitat requirements of coastal species and whether these specific habitats can continue to contribute to successful reproduction with human-kind's impact on the environment. Ultimately, as human populations continue to grow, only enlightenment and restraint will ensure the survival of these interesting and important fishes.

References

Agassiz, L. 1871. On the method of copulation among selachians. *Proc. Boston Soc. Nat. Hist.* 14:339–341.

Alcock, A. 1892. On the utero-gestation in *Trygon bleekeri*. *Ann. Mag. Nat. Hist. 6th Ser.* 9:417–427.

Amoroso, E. C. 1960. Viviparity in fishes, in *Hormones in Fish: Proceedings of a Symposium Held in the Zoological Society of London*. I. C. Jones, Ed., Zoological Society of London, London, 153–180.

Aristotle. 343 B.C. (Thompson translation 1949). Historia Animalium, in *The Works of Aristotle,* Vol. 4. Translated by D. W. Thompson. Clarendon Press, Oxford, 485–633.

Baum, J. K., R. A. Myers, D. G. Kehler, B. Worm, S. J. Harley, and P. A. Doherty. 2003. Collapse and conservation of shark populations in the northwest Atlantic. *Science* 299:389–392.

Borcea, I. 1906. Recherches sur le systèm uro-génital des Elasmobranches. *Arch. Zool. Exp. Gen.* 4:199–485.

Budker, P. 1958. La viviparité chez les sélachiens, in *Traité de Zoologie: Anatomie, Systematique, Biologie.* P. P. Grassé, Ed., Maison et Cie Editeurs Libraires de l'Académie de Médecine, Paris, 1755–1790.

Carrier, J. C., H. L. Pratt, Jr., and L. K. Martin. 1994. Group reproductive behaviors in free-living nurse sharks, *Ginglymostoma cirratum*. *Copeia* 1994:646–656.

Carrier, J. C., F. L. Murru, M. T. Walsh, and H. L. Pratt, Jr. 2003. Assessing reproductive potential and monitoring gestation in nurse sharks (Ginglymostoma cirratum): bridging the gap between field research and captive studies. *Zoo Biol.* 62:142–156.

Castro, J. I. 1983. *The Sharks of North American Waters.* Texas A&M University Press, College Station.

Castro, J. I. 1993. The biology of the finetooth shark, *Carcharhinus isodon. Environ. Biol. Fish..* 36:219–232.

Castro, J. I. 1996. Biology of the blacktip shark, *Carcharhinus limbatus*, off the southeastern United States. *Bull. Mar. Sci.* 59:508–522.

Castro, J. I., P. M. Bubucis, and N. A. Overstrom. 1988. The reproductive biology of the chain dogfish, *Scyliorhinus retifer. Copeia.* 1988:740–746.

Clark, E. 1963. The maintenance of sharks in captivity, with a report on their instrumental conditioning, in *Sharks and Survival*. P. W. Gilbert, Ed., D.C. Heath, Boston, 115–149.

Clark, E. 1975. The strangest sea. *Natl. Geogr. Mag.* 148:338–343.

Clark, E. and K. von Schmidt. 1965. Sharks of the central Gulf Coast of Florida. *Bull. Mar. Sci.* 15:13–83.

Conrath, C. L. and J. A. Musick. 2002. Reproductive biology of the smooth dogfish, *Mustelus canis*, in the northwest Atlantic Ocean. *Environ. Biol. Fish.* 64:367–377.

Daniel, J. F. 1928. *The Elasmobranch Fishes.* University of California Press, Berkeley.

Dean, B. 1906. *Chimaeroid fishes and Their Development*. Carnegie Institution, Washington, D.C., 171 pp.

Dempster, R. P. and E. S. Herald. 1961. Notes on the hornshark, *Heterodontus francisci*, with observations on mating activities. *Occ. Pap. Calif. Acad. Sci.* 33:1–7.

Dodd, J. M. 1983. Reproduction in cartilaginous fishes (Chondrichthyes), in *Fish Physiology*, Vol. 9A. W. S. Hoar, D. J. Randall, and E. M. Donaldson, Eds., Academic Press, New York.

Dral, A. J. 1980. Reproduction en aquarium du requin de fond tropical, *Chiloscyllium griseum* Mull. et Henle (Orectolobides). *Rev. Fr. Aquariol.* 7:99–104.

Eberhard, W. G. 1985. *Sexual Selection and Male Genitalia.* Harvard University Press, Cambridge, MA.

Feder, H. M., C. H. Turner, and C. Limbaugh. 1974. Observations on fishes associated with kelp beds in Southern California. *Calif. Fish Biol.* 1160:1–144.

Feldheim, K. A., S. H. Gruber, and M. V. Ashley. 2001. Multiple paternity of a lemon shark litter (Chondrichthyes : Carcharhinidae). *Copeia.* 2001:781–786.

Francis, M. P. 1996. Observations on a pregnant white shark with a review of reproductive biology, in *Great White Sharks: The Biology of Carcharodon carcharias.* A. P. Klimley and D. G. Ainley, Eds., Academic Press, San Diego, 157–172.

Gilbert, P. W. and G. W. Heath. 1972. The clasper-siphon sac mechanism in *Squalus acanthias* and *Mustelus canis. Comp. Biochem. Physiol.* 42A:97–119.

Gilmore, R. G. 1993. Reproductive biology of lamnoid sharks. *Environ. Biol. Fish.* 38:95–114.

Gilmore, R. G., J. W. Dodrill, and P. A. Linley. 1983. Reproduction and embryonic development of the sand tiger shark, *Odontaspis taurus* (Rafinesque). *Fish. Bull.* 81:201–225.

Gordon, I. 1993. Pre-copulatory behavior of captive sand tiger sharks, *Carcharius taurus. Environ. Biol. Fish.* 38:159–164.

Hamlett, W. C. 1999. Male reproductive system, in *Sharks, Skates, and Rays: The Biology of Elasmobranch Fishes.* W. C. Hamlett, Ed., Johns Hopkins University Press, Baltimore, 444–471.

Hamlett, W. C., J. A. Musick, C. K. Hysell, and D. M. Sever. 2002. Uterine epithelial-sperm interaction, endometrial cycle and sperm storage in the terminal zone of the oviducal gland in the placental smoothhound, *Mustelus canis. J. Exp. Zool.* 292:129–144.

Johnson, R. H. and D. R. Nelson. 1978. Copulation and possible olfaction-mediated pair formation in two species of carcharhinid sharks. *Copeia* 1978:539–542.

Jones, R. C. and M. J. Lin. 1993. Structure and functions of the genital ducts of the male Port Jackson shark, *Heterodontus portusjacksoni. Environ. Biol. Fish.* 38:127–138.

Jones, S. N. and R. C. Jones. 1982. The structure of the male genital system of the Port Jackson shark, *Heterodontus potusjacksoni* with particular reference to genital ducts. *Aust. J. Zool.* 30:523–541.

Klimley, A. P. 1980. Observations of courtship and copulation in the nurse shark, *Ginglymostoma cirratum. Copeia* 1980:878–882.

Klimley, A. P. 1985. Schooling in the large predator, *Sphryna lewini*, a species with low risk of predation: a non-egalitarian state. *Z. Tierpsychol.* 70:297–319.

Leigh-Sharpe, W. H. 1920. The comparative morphology of the secondary sexual characteristics of elasmobranch fishes. *Mem. J. Morphol.* 34:245–265.

Luer, C. A. and P. W. Gilbert. 1985. Mating behavior, egg deposition, incubation period and hatching in the clearnosed skate, *Raja eglanteria. Environ. Biol. Fish.* 13:161–171.

Maar, V. 1910. *Nicolai Stenonis Opera Philosophica*, Vol. I. Vilhelm Tryde, Copenhagen, 264 pp.

Matthews, L. H. 1950. Reproduction in the basking shark, *Cetorhinus maximus* (Gunner). *Philos. Trans. R. Soc. Lond.* 234:247–316.

Michaels, S. 1993. *Reef Sharks and Rays of the World.* Sea Challengers, Monterey, 107 pp.

Nelsen, O. E. 1953. *Comparative Endocrinology of the Vertebrates.* Blackiston, New York.

Nielson, J. G., A. Jespersen, and O. Munk. 1968. Spermatophores in Ophidioidea (Pisces, Percomorphi), in *Galathea Report.* T. Wolff, Ed., Danish Science Press, Copenhagen, 239–254.

Ohta, Y., K. Okamura, E. C. McKinney, S. Bartl, K. Hashimoto, and M. F. Flajnik. 2000. Primitive synteny of vertebrate major histocompatibility complex class I and class II genes. *Proc. Natl. Acad. Sci. U.S.A.* 97:4712–4717.

Parsons, G. R. and H. J. Grier. 1992. Seasonal changes in shark testicular structure and spermatogenesis. *J. Exp. Zool.* 261:173–184.

Peres, M. B. and C. M. Vooren. 1991. Sexual development, reproductive cycle, and fecundity of the school shark *Galeorhinus galeus* off southern Brazil. *Fish. Bull.* 89:655–667.

Pratt, H. L. 1988. Elasmobranch gonad structure — a description and survey. *Copeia* 1988:719–729.

Pratt, H. L., Jr. 1993. The storage of spermatozoa in the oviducal glands of Western North Atlantic sharks. *Environ. Biol. Fish.* 38:139–149.

Pratt, H. L., Jr. and J. C. Carrier. 2001. A review of elasmobranch reproductive behavior with a case study on the nurse shark, *Ginglymostoma cirratum. Environ. Biol. Fish.* 60:157–188.

Pratt, H. L., Jr. and J. Casey. 1990. Shark reproductive strategies as a limiting factor in directed fisheries, with a review of Holden's method of estimating growth parameters, in NOAA Tech. Rep. NMFS 90: Elasmobranchs as Living Resources: Advances in the Biology, Ecology, Systematics, and the Status of the Fisheries. H. L. Pratt, Jr., S. H. Gruber, and T. Taniuchi, Eds., U.S. Department of Commerce, Washington, D.C., 97–111.

Pratt, H. L. and S. Tanaka. 1994. Sperm storage in male elasmobranchs — a description and survey. *J. Morphol.* 219:297–308.

Rondelet, G. 1558. L'histoire entiére de poissons. Facsimile 2002, in *Éditions de CTHS*. Part One: 1–418; Part Two: 1–181, La Comité des Travaux Historiques et Scientifiques, Tours.

Saville, K. J., A. M. Lindley, E. G. Maries, J. C. Carrier, and H. L. Pratt, Jr. 2002. Multiple paternity in the nurse shark, *Ginglymostoma cirratum*. *Environ. Biol. Fish.* 63:347–351.

Simpfendorfer, C. A. and P. Unsworth. 1998. Reproductive biology of the whiskery shark, *Furgaleus macki*, off south-western Australia. *Mar. Freshwater Res.* 49:687–693.

Stanley, H. P. 1966. The structure and development of the seminiferous follicle in *Scyliorhinus caniculus* and *Torpedo marmorata* (Elasmobranchii). *Z. Zellforsch.* 75:453–468.

Tortonese, E. 1950. Studi dui plagiostomi. III. Un fondamentale carattere biologico degli Squali. *Arch. Ital. Zool.* 1950:101–155.

Tricas, T. C. 1980. Courtship and mating-related behaviors in myliobatid rays. *Copeia* 1980:553–556.

Tricas, T. C. and E. M. Lefeuvre. 1985. Mating in the reef white-tip shark *Triaenodon obesus*. *Mar. Biol.* 84:233–237.

Tricas, T. C., S. W. Michael, and J. A. Sisneros. 1995. Electrosensory optimization to conspecific phasic signals for mating. *Neurosci. Lett.* 202:129–132.

Uchida, S., M. Toda, and Y. Kamei. 1990. Reproduction of elasmobranchs in captivity, in Elasmobranchs as Living Resources: Advances in the Biology, Ecology, Systematics, and Status of the Fisheries. H. L. Pratt, Jr., S. H. Gruber, and T. Taniuchi, Eds., NOAA Tech. Rep. NMFS 90, 211–237.

White, T., G. Hall, and C. Potter. 2002. Size and age compositions and reproductive biology of the nervous shark *Carcharhinus cautus* in a large subtropical embayment, including an analysis of growth during pre- and postnatal life. *Mar. Biol.* 141:1153–1166.

Wourms, J. P. 1977. Reproduction and development in chondrichthyan fishes. *Am. Zool.* 17:379–410.

Yano, K., F. Sato, and T. Takahashi. 1999. Observation of the mating behavior of the manta ray, *Manta birostris*, at the Ogasawara Islands. *Ichthyol. Res.* 46:289–296.

11

Hormonal Regulation of Elasmobranch Physiology

James Gelsleichter

CONTENTS

11.1   Introduction

The field of "elasmobranch endocrinology" began at the same time as the field of vertebrate endocrinology itself, when Bayliss and Starling (1903) used extracts from the intestines of sharks and skates to demonstrate the actions of secretin, the first described vertebrate hormone. Although perhaps coincidental, a pivotal role for sharks and their relatives in the birth of this field is prophetic to some extent, given that many vertebrate hormones appear to have first appeared in the cartilaginous fishes. Because sharks and their relatives occupy such a critical position in the evolution of the vertebrate endocrine system, studies on endocrinology of these fishes contribute to a better understanding of the roles that hormones exert in all higher vertebrates. Furthermore, because hormones regulate virtually all aspects of elasmo-branch physiology, knowledge concerning the function of the elasmobranch endocrine system is essential for developing a full comprehension of how these fishes develop, grow, reproduce, and survive.

   Because the structure and comparative aspects of the elasmobranch endocrine system are generally well addressed in most comparative endocrinology texts (e.g., Norris, 1997), this chapter focuses on the manner in which hormones participate in the regulation of processes vital for survival of sharks and their relatives. Although a "functional approach" has been used in this chapter, an extensive list of references regarding the structure and comparative homologies of elasmobranch hormones in addition to their known or putative actions has been provided for the reader more concerned with these topics.

11.2   Digestion and Energy Metabolism

11.2.1   Overview

The survival of an individual elasmobranch depends on its ability to convert food items into usable nutrients through actions of the digestive system. With the exception of certain specialized adaptations such as the spiral valve intestine, both the structure and function of the elasmobranch gastrointestinal tract are generally similar to that in other vertebrate groups (see review by Holmgren and Nilsson, 1999). Following its capture and maceration by the oral cavity, food is transferred through the esophagus to a two-chambered stomach, where it is stored and partially disrupted through the actions of acid-secreting and proteolytic enzyme-secreting cells. Afterward, the acidic slurry of incompletely digested food (generally referred to as chyme) enters the duodenum, where it is broken down further by intestinal and pancreatic enzymes, the latter of which are transferred to the duodenum via the pancreatic duct. Bile produced by the liver and stored by the gallbladder also contributes to food digestion, particularly the hydrolysis of fat, following its transport to the duodenum via the bile duct. Bile salts emulsify dietary fat globules, a process that causes them to be dispersed as smaller droplets more prone to digestion by pancreatic enzymes. The duodenum also receives bicarbonate-rich pancreatic secretions, which neutralize chyme prior to its movement to more delicate sites of nutrient absorption in the spiral valve intestine. Once this passage occurs, nutrients are assimilated presumably through both passive and active forms of uptake. Nondigested material is transported through the rectum and discharged to the environment via the cloaca.

11.2.2   Digestive Hormones

Based on the presence and distribution of major vertebrate gut hormones in the elasmobranch gastrointes-tinal tract, endocrine regulation of digestion in these fishes is likely to be similar to that occurring in higher vertebrates (Figure 11.1). The secretion of digestive acids in the foregut may be hormonally regulated by gastrin, which is capable of stimulating this process in spiny dogfish (Vigna, 1983) and has been localized in endocrine cells of the stomach, intestine, and/or pancreas of this and other shark species (Holmgren and Nilsson, 1983, El-Salhy, 1984; Aldman et al., 1989; Jonsson, 1995; Johnsen et al., 1997). Once chyme enters the duodenum, localized declines in pH likely trigger intestinal release of the hormone secretin, which stimulates secretion of bicarbonate-rich pancreatic juices in higher

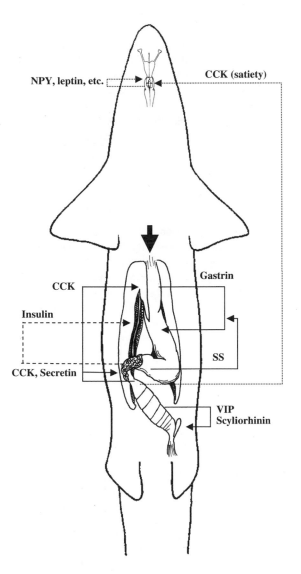

FIGURE 11.1 Proposed mechanism for the hormonal regulation of digestion and energy metabolism in elasmobranchs. Following ingestion of food (large arrow), gastric acid secretion is likely stimulated by gastrin, but may be inhibited by SS at some point in the digestive process. Transport of chyme into the intestine causes release of secretin and CCK, which increases production and supply of pancreatic enzymes/bicarbonate secretions and hepatic bile. Release of CCK may also influence satiation and/or the production of other putative satiety hormones at the level of the hypothalamus (dotted lines). Ingestion of prey with high salt content may influence salt release by the rectal gland via the actions of VIP or scyliorhinins, but the physiological significance of these hormones remains unclear. Following absorption of energy substrates, the pancreatic hormone insulin appears to promote energy storage in the liver and other tissues (dashed line). Other hormones are believed to influence gut motility and/or circulation (see text), but are not included in the present figure.

vertebrates and has been identified in the intestine of skates and sharks (Bayliss and Starling, 1903). The arrival of chyme in the elasmobranch midgut also is believed to stimulate intestinal release of cholecystokinin (CCK), the hormone primarily responsible for regulating the supply of bile and pancreatic enzymes to the duodenum in mammals. A similar role for CCK in sharks and their relatives is supported by detection of CCK-like substances in the elasmobranch intestine and pancreas (Hansen, 1975; Vigna, 1979; Holmgren and Nilsson, 1983; El-Salhy, 1984; Aldman et al., 1989; Jonsson et al.,

1995, Johnsen et al., 1997), as well as evidence for CCK-binding activity (Oliver and Vigna, 1996) and CCK-like actions (Andrews and Young, 1988) in the elasmobranch gallbladder. Last, inhibition of the digestive process may be regulated by somatostatin (SS), which is known to suppress production of gastric acid via inhibition of gastrin release. Cells containing SS have been localized in several components of the elasmobranch gut, including the gastric mucosa (King and Millar, 1979; Holmgren and Nilsson, 1983; El-Salhy, 1984; Conlon et al., 1985; Tagliafierro et al., 1985, 1989). Virtually all of these compounds also have been detected in nerves of the elasmobranch gut and some (e.g., CCK, SS) have been shown to be capable of influencing gut motility in dogfish and skates (Lundin et al., 1984; Andrews and Young, 1988; Aldman et al., 1989). Therefore, they may additionally function as neurotransmitters and play important roles in regulating the passage of food through the gastrointestinal tract (Holmgren and Nilsson, 1999).

11.2.3 Hormones Involved in Gut Motility and/or Blood Flow

Although numerous other hormones have been detected in the gastrointestinal system of elasmobranchs, the digestive functions of few have been investigated. Four of these compounds, peptide YY (PYY), the chemically similar neuropeptide Y (NPY), bombesin/gastrin-releasing peptide (GRP), vasoactive intestinal polypeptide (VIP), and the family of peptides known as the tachykinins are of notable interest because they are believed to exert significant actions on gut motility and/or circulation in higher vertebrates and have been consistently localized in the elasmobranch digestive tract (Holmgren and Nilsson, 1983; El-Salhy, 1984; Cimini et al., 1985, 1989, 1992; Dimaline et al., 1986, 1987; Conlon et al., 1987, 1992; Bjenning and Holmgren, 1988; Bjenning et al., 1990, 1991, 1993; Shaw et al., 1990; Pan et al., 1992, 1994; Chiba et al., 1995; Chiba, 1998). In general, bombesin/GRP, VIP, and the tachykinins appear to promote vertebrate digestion by increasing blood flow to the gut in addition to exerting varied effects on acid or enzyme secretion and/or gut motility. In contrast, both NPY and PYY are believed to suppress vertebrate digestion by reducing gastrointestinal blood flow and inhibiting gastric acid secretion, pancreatic enzyme release, and gallbladder contraction (see reviews by Sheikh, 1991; Berglund et al., 2003). Studies on elasmobranchs have observed increased blood flow to the gut in response to treatment with bombesin/GRP, NPY, and two tachykinins, scyliorhinin I and II (Bjenning et al., 1990; Holmgren et al., 1992b; Kågstrom et al., 1996). In contrast, VIP has been shown to reduce both the motility and perfusion of the dogfish gut (Lundin et al., 1984; Holmgren et al., 1992a), the opposite of that observed in mammals. Although the physiological significance of these findings remains unclear, the present data support potentially important roles for these peptides in elasmobranch digestion.

11.2.4 Interactions between Feeding and Ion Homeostasis

Given that the ingestion of marine invertebrates and fish represents a route for significant salt intake, it is interesting to note that several gastrointestinal hormones also appear to influence salt secretion by the elasmobranch rectal gland *in vitro*. In particular, VIP has been shown to be a potent stimulant of this process in spiny dogfish by causing vasodilation of rectal gland vasculature as well as increases in cellular cAMP, one of the enzymes responsible for regulating secretory activity in this organ (Epstein et al., 1981; Chipkin et al., 1988; Ecay and Valentich, 1991; Lehrich et al., 1998). Although VIP is unable to elicit this response in the common dogfish, *Scyliorhinus canicula* (Thorndyke and Shuttleworth, 1985), rectal gland secretion in this species is similarly increased by treatment with scyliorhinin II (Anderson et al., 1995). In contrast, SS, bombesin, and NPY inhibit rectal gland secretion, although the effects of SS and bombesin appear to be mediated via inhibition of VIP-stimulated responses (Silva et al., 1985, 1990, 1993). Based on the localization of VIP, SS, and bombesin in the elasmobranch rectal gland (Holmgren and Nilsson, 1983), the effects of these compounds in intact animals may be a consequence of local rather than gastrointestinal sources. However, because rectal gland secretion increases significantly following feeding activity (MacKenzie et al., 2002), hormones originating from the gut likely exert at least some physiological role in regulating salt release by this organ.

11.2.5   Hormones Involved in Energy Metabolism

Following their absorption by the gastrointestinal system, the molecular products of food digestion — i.e., monosaccharides (e.g., glucose), amino acids, fatty acids, glycerol — are directly utilized for production of energy or taken up into cells and transformed into compounds that contribute to growth and/or energy storage. As in other vertebrates, the uptake, conversion, and storage of energy substrates in elasmobranchs appears to be promoted by the hormone insulin, which has been detected in and/or isolated from the pancreas of several shark and ray species (Sekine and Yui, 1981; Bajaj et al., 1983; El-Salhy, 1984; Conlon and Thim, 1986; Anderson et al., 2002). The release of insulin from the elasmobranch pancreas appears to be at least partially regulated by circulating nutrient levels based on the rise in plasma insulin concentrations during periods of increased feeding in *S. canicula* (Gutiérrez et al., 1988). Treatment of elasmobranchs with insulin has been shown to decrease plasma glucose levels and increase muscle and liver glycogen stores (Leibson and Plisetskaia, 1972; deRoos and deRoos, 1979; deRoos et al., 1985; Anderson et al., 2002), effects consistent with the active deposition of metabolic substrates made available following a feeding event. However, insulin-provoked hypoglycemia and the cellular uptake of injected glucose generally occur more slowly in elasmobranchs compared with mammals (Patent, 1970; deRoos and deRoos, 1979; deRoos et al., 1985; Anderson et al., 2002). These phenomena may be due to a lack of insulin-dependent glucose transporters in elasmobranch tissues, the factors responsible for the rapid clearance of circulating glucose in mammals following insulin treatment (Anderson et al., 2002). Although the absence of these transporters in elasmobranch tissues remains unconfirmed, this argument is persuasive based on the relative lack and limited importance of direct sources of glucose (i.e., carbohydrates) in the protein- and fat-rich diet of sharks and their relatives. Furthermore, because insulin also promotes a reduction in circulating amino acid levels in elasmobranchs (deRoos et al., 1985), it likely plays a more important role in stimulating the cellular uptake, utilization, and/or transformation of these compounds compared with its actions on dietary glucose.

Despite the importance of hepatic lipid storage in elasmobranchs, little is known regarding the role that insulin may play in regulating this process. Nonetheless, insulin may be involved in stimulating postprandial uptake of lipids in the elasmobranch liver based on the association between increases in feeding activity, total plasma lipids, circulating insulin concentrations, and hepatosomatic index in *S. canicula* (Gutiérrez et al., 1988). Although insulin further promotes lipid storage in mammals by inhibiting the mobilization of stored fats, this does not appear to be the case for sharks and their relatives. Treatment of elasmobranchs with insulin has no effect on circulating levels of ketone bodies, the primary end products of hepatic lipid metabolism in these fishes. The maintenance of high endogenous levels of ketone bodies in even recently fed elasmobranchs appears to be due to their use as key fuels for aerobic metabolism, a practice that is unusual in nonstarved vertebrates (deRoos et al., 1985; Gutiérrez et al., 1988; Watson and Dickson, 2001; Anderson et al., 2002). As suggested by several authors, the use of ketones as an energy source in cartilaginous fish is likely due to their limited capacity for the transport and utilization of non-esterified fatty acids compared with that in other vertebrate groups (deRoos et al., 1985; Watson and Dickson, 2001; Anderson et al., 2002).

Although it is produced by the elasmobranch pancreas (Sekine and Yui, 1981; El-Salhy, 1984; Conlon and Thim, 1985; Gutiérrez et al., 1986; Faraldi et al., 1988; Tagliafierro et al., 1989; Conlon et al., 1994), the insulin-antagonist glucagon does not appear to stimulate a rise in circulating glucose levels in sharks (Patent, 1970). However, because unfed elasmobranchs appear to derive energy primarily from ketone bodies (deRoos et al., 1985; Anderson et al., 2002), these observations may reflect the greater reliance of these animals on lipid stores rather than glycogen reserves during periods of undernourishment. Although the metabolism of stored lipids in fasting mammals and birds is regulated by glucagon, no studies have directly investigated if this hormone has similar actions in sharks and their relatives. Clearly, this topic should be addressed in future studies, especially due to the often sporadic feeding habits of large migratory sharks.

In addition to the pancreas, a number of other hormonal systems also appear to influence energy metabolism in sharks and their relatives. For example, thyroid hormones have been shown to alter levels of enzymes involved in amino acid and lipid metabolism in *Squalus acanthias* (Battersby et al., 1996), an action similar to that observed in higher vertebrates. Hormones involved in regulating growth and the stress response in elasmobranchs also contribute to the regulation of energy metabolism in these fishes, and are discussed in later sections of this chapter. In contrast, the peptides (i.e., urotensin I and II) of the caudal neurosecretory system, the urophysis, do not appear to influence carbohydrate or lipid metabolism in elasmobranchs, as they have been shown to do in certain teleosts (Conlon et al., 1994a).

11.2.6   Possible Mechanisms of Satiation

Considering the general fascination with elasmobranch feeding behavior, it is interesting to note that most of the hormones believed to regulate appetite in mammals and some nonmammalian vertebrates (see review by Jensen, 2001) also are present in sharks and their relatives. This includes CCK and bombesin, both of which appear to suppress the intake of food in mammals, birds, and teleosts following their release from the gastrointestinal system. Given that these compounds may regulate hunger by actions on the hypothalamus, it is noteworthy to mention that CCK-like binding activity has been detected in the brain of the mako shark, *Isurus oxyrinchus* (Oliver and Vigna, 1996). Hypothalamic factors that stimulate food intake in higher vertebrates, particularly NPY and galanin, also have been detected in the elasmobranch brain in several studies (Vallarino et al., 1988a, 1991; Chiba and Honma, 1992; Conlon et al., 1992; McVey et al., 1996; Chiba et al., 2002). Last, evidence for encephalic expression of leptin, a hormone considered to be a major factor regulating satiety in birds and mammals, has been observed in the bonnethead shark, *Sphyrna tiburo,* and the smooth dogfish, *Mustelus canis* (Londraville, pers. comm.). As no studies to date have investigated the effects of these or other potential "satiety hormones" on elasmobranch feeding behavior, this is a topic in need of considerable attention.

11.3   Growth

The factors that regulate elasmobranch growth are of interest to several parties, particularly fishery scientists, who use estimates of growth rate to determine the resilience of shark and ray populations to exploitation. Unfortunately, very little is known regarding the hormonal control of growth in cartilaginous fishes. However, virtually all major hormones involved in the endocrine growth axis of higher vertebrates also are present in elasmobranchs, so the regulation of growth in sharks and their relatives is probably similar to that in mammals (see review by Le Roith et al., 2001). If this is the case, the primary factor controlling elasmobranch growth is likely to be growth hormone (GH), which has been isolated from the pituitary gland of two elasmobranchs, the blue shark, *Prionace glauca* (Hayashida and Lewis, 1978; Yamaguchi et al., 1989) and *Squalus acanthias* (Kawauchi et al., unpubl. data). Secretion of GH is generally regulated by stimulatory (growth hormone-releasing hormone, or GHRH) and inhibitory (SS) factors originating from the hypothalamus, both of which have been detected in the elasmobranch brain (Conlon et al., 1985; Plesch et al., 2000). In mammals, GH promotes somatic and skeletal growth by stimulating cell proliferation and differentiation in skeletal muscle and cartilage. In addition, GH increases production of insulin-like growth factors (IGFs), highly conserved compounds (Bautista et al., 1990) that stimulate cell hypertrophy and/or extracellular matrix production in skeletal muscle, fat, and cartilage through autocrine and paracrine mechanisms. Evidence for anabolic actions of these growth factors in elasmobranchs has been provided by Gelsleichter and Musick (1999), who observed increased growth of skate vertebral cartilage in response to treatment with IGF-I. However, until relationships among production of GHRH, SS, GH, and IGF-I in elasmobranchs have been characterized, the regulatory scheme proposed in Figure 11.2 is largely speculative.

11.4   Stress

11.4.1   Chromaffin Tissue and Catecholamines

As the primary factors responsible for maintaining vertebrate homeostasis, hormones play key roles in the response to physiological imbalances caused by exposure to stressful stimuli (Figure 11.3). Like that

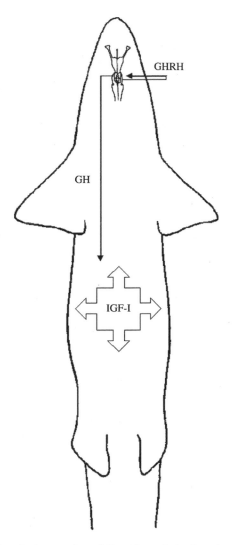

FIGURE 11.2 Proposed mechanism for hormonal regulation of growth in elasmobranchs. Release of GHRH from the hypothalamus is presumed to stimulate pituitary release of GH into general circulation. As recently proposed in mammals, the peripheral effects of GH are likely to cause local release of IGF-I, which stimulates tissue growth through autocrine and/or paracrine mechanisms.

in other vertebrates, the response to acute forms of stress in elasmobranchs appears to be partially regulated by the chromaffin tissue, small masses of neurosecretory cells distributed along the dorsal surface of the kidney. In response to neural signals resulting from exposure of elasmobranchs to diverse physiological stressors (e.g., hypoxia, hemorrhage, capture, handling, and exercise), chromaffin cells secrete epinephrine and norepinephrine, the neurohormones known collectively as catecholamines (Opdyke et al., 1982, 1983; Carroll et al., 1984; Metcalfe and Butler, 1984; Butler et al., 1986). Catecholamines promote the mobilization of energy reserves in sharks and their relatives, as demonstrated by the reduction in hepatic lipid stores and/or increase in circulating nutrient levels in elasmobranchs following treatment with these compounds (Grant et al., 1969; Patent, 1970; Lipshaw et al., 1972; deRoos and deRoos, 1978) or during stressful events (Torres et al., 1986; Tort et al., 1994; Hoffmayer and Parsons, 2001). Catecholamines also increase blood pressure in elasmobranchs (Opdyke et al., 1982), an action that promotes the transport of metabolic substrates to muscles and organs such as the brain and heart. In contrast, blood flow to the elasmobranch gut is reduced in response to catecholamine treatment (Holmgren et al., 1992a), a logical outcome considering that digestion is an unnecessary

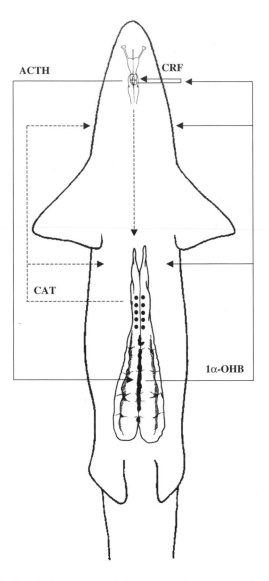

FIGURE 11.3 Proposed mechanism for hormonal regulation of stress in elasmobranchs. Perception of stressful stimuli causes neurally mediated release of catecholamines (CAT) from chromaffin tissue, which is distributed as small, isolated pockets of neurosecretory cells along the dorsal surface of the kidney (dashed line, only the anterior portion of chromaffin cell masses are demonstrated [small black dots]). Stress is also believed to cause release of corticotropin-releasing factor (CRF) from the hypothalamus, which promotes secretion of corticotropin (ACTH) from the pituitary gland. Release of ACTH stimulates production of the unique corticosteroid 1α-OHB from the interrenal gland, which is situated along the dorsomedial surface of the kidney. Both catecholamines and 1α-OHB are believed to have effects on branchial function and cardiovascular pressure, as well as the utilization of energy substrates.

process during stressful periods. Catecholamines also have been shown to increase perfusion and ventilation frequency of elasmobranch gills, effects that stimulate the uptake of oxygen and its delivery to tissues (Metcalfe and Butler, 1984; Butler et al., 1986).

11.4.2 Hypothalamo–Pituitary–Interrenal Axis

In most vertebrates, exposure to stress (both acute and chronic) also stimulates production of corticosteroids by the adrenal gland or its nonmammalian homologue, the interrenal body. The release of these compounds is regulated by increased expression of the pituitary factor adrenocorticotropic hormone (ACTH), which in turn is stimulated by the hypothalamic compound corticotrophin-releasing factor (CRF). Like catecholamines, corticosteroids promote the mobilization of nutrient provisions in stressed vertebrates and also inhibit growth and energy storage. Although pituitary CRF has yet to be characterized in elasmobranchs, they appear to possess a functional hypothalamic–pituitary–interrenal (HPI) axis based on presence of ACTH (Lowry et al., 1974; Shimamura et al., 1978; Okamoto et al., 1979; Denning-Kendall et al., 1982; Vallarino and Ottonello, 1987; Amemiya et al., 2000) and its ability to regulate interrenal production of 1α-hydroxycorticosterone (1α-OHB), the unique corticosteroid produced only in these fishes (Idler and Truscott, 1966). Nonetheless, few studies have investigated the response of the elasmobranch HPI axis to stress.

The lack of studies on the role of the HPI axis in regulating the stress response in elasmobranchs may in part result from the difficulties in measuring circulating levels of 1α-OHB via immunoassays due to lack of specific antibodies for this compound. Because of this, Manire and Rasmussen (unpubl. data) measured serum concentrations of corticosterone (B), the precursor to 1α-OHB, to assess changes in the HPI of bonnethead sharks subjected to the stresses of capture, transport, and captive maintenance. These researchers observed a positive correlation between the duration of animal restraint and serum B concentrations in *Sphyrna tiburo*, suggesting at least some degree of relationship between acute stress and the activity of the elasmobranch interrenal gland. If corticosteroids are involved in the stress response in elasmobranchs, they are likely to influence energy metabolism in these fishes as they do in other vertebrates. This premise is based on the ability of ACTH to induce hyperglycemia in sharks, perhaps by stimulating increased production of substrates (e.g., amino acids, lactate) for glucose synthesis (deRoos and deRoos, 1973, 1992). Corticosteroids also may suppress elasmobranch growth given that B is capable of inhibiting extracellular matrix production in skate vertebral cartilage *in vitro* (Gelsleichter and Musick, 1999).

11.5 Osmoregulation

11.5.1 Overview

Marine and euryhaline elasmobranchs adapt to changes in environmental salinity primarily by regulating endogenous concentrations of the organic salt urea (see Evans et al., Chapter 9 of this volume). In marine elasmobranchs, urea retention stems the potential osmotic loss of water, whereas salt secretion by the rectal gland counteracts the influx of sodium and chloride across the gills and gut. In elasmobranchs capable of surviving in freshwater systems (e.g., bull sharks), reductions in the retention of urea lower blood osmolarity and diminish water gain to some extent. However, because these animals remain hyperosmotic to the environment, some uptake of water does occur and is compensated for by an increase in urine output. The maintenance of solute concentrations in freshwater-adapted elasmobranchs appears to be regulated by an increase in ion uptake in the gills, as well as a decline in the secretory activity of the rectal gland.

Although the endocrine control of elasmobranch osmoregulation remains poorly understood, at least four hormonal systems appear to play major roles in regulating water and ion balance in these fishes (Figure 11.4). These factors include the HPI axis, the renin–angiotensin system (RAS), VIP, and C-type natriuretic peptide (CNP). A number of other endocrine factors including thyroid hormones, catecholamines, and peptides of the gut, urophysis, and neurohypophysis also may influence osmoregulation and ionic regulation in sharks and their relatives, but their possible roles in these processes have not been extensively studied.

11.5.2 Interrenal Corticosteroids

In tetrapods, mineralocorticosteroids (e.g., aldosterone, corticosterone) produced by the adrenal or interrenal gland influence ion homeostasis by stimulating the retention of sodium through actions on the kidney, gut, urinary bladder, and/or accessory organs. A similar role for the HPI axis in regulating ion levels in

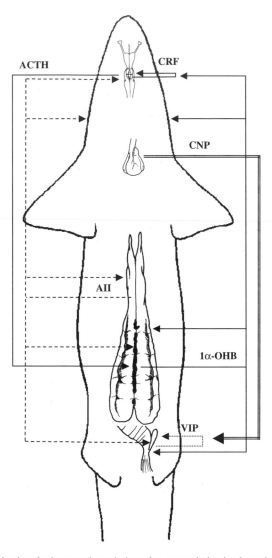

FIGURE 11.4 Proposed mechanism for hormonal regulation of osmoregulation in elasmobranchs. Water and ion balance in sharks and their relatives are modulated through four major systems: the HPI axis (solid line), the RAS (dashed line), VIP (dotted line), and CNP (double line). Although their specific functions remain unclear, both the HPI axis and the RAS are believed to influence osmotic balance through effects on the gill, kidney, and rectal gland that promote retention of salt. The RAS also stimulates production of 1α-OHB and drinking behavior. Production of VIP in the rectal gland occurs in response to changes in blood volume, and leads to increased secretion of sodium and chloride from this organ. The release of CNP from the heart also occurs in response to changes in cardiovascular pressure, and increases both the production of VIP and the release of salt from the rectal gland.

elasmobranchs is supported by the presence of 1α-OHB-binding activity in the gills, kidney, and rectal gland of these fishes (Idler and Kane, 1980; Burton and Idler, 1986). Changes in secretory activity of the rectal gland have been observed in interrenalectomized skates (Holt and Idler, 1975; Idler and Kane, 1976). As in higher vertebrates, corticosteroids appear to function in the retention of sodium in cartilaginous fishes, a premise supported by the increase in circulating 1α-OHB concentrations in dogfish adapted to 50% seawater (Armour et al., 1993a). Although this action would clearly reduce sodium loss in hypoosmotically challenged (i.e., freshwater-adapted) elasmobranchs, it also may contribute to osmoregulation in marine sharks or rays, especially in cases when urea homeostasis is compromised. For example, Armour et al. (1993a) observed a rise in plasma 1α-OHB concentrations in *Scyliorhinus canicula* maintained in 130% seawater and fed a low-protein diet. Because this dietary restriction limits urea biosynthesis, increased levels of 1α-OHB are believed to have been involved in stimulating the retention of sodium and chloride, the alternative osmoregulatory strategy employed by these fishes.

11.5.3 Renin–Angiotensin System

The RAS includes a series of biochemical steps that begin with the conversion of the hepatic glycoprotein angiotensinogen to angiotensin I (ANG I) via actions of renin, an enzyme secreted by the juxtaglomerular (JG) cells of the kidney of all vertebrates, except agnathans. The biologically active form of this peptide, angiotensin II (ANG II), is formed through subsequent cleavage of ANG I by the compound, angiotensin-converting enzyme (ACE). In terrestrial vertebrates and bony fish, ANG II participates in the regulation of ion and water balance by stimulating corticosteroid production, drinking behavior, and changes in kidney function, actions that generally result in the uptake and/or retention of sodium as well as an increase in cardiovascular pressure. Although it was initially thought to be absent in elasmobranchs (Bean, 1942; Nishimura et al., 1970), both the presence of the RAS (Henderson et al., 1981; Uva et al., 1992; Takei et al., 1993) and its actions on ion and water homeostasis in these fishes have been confirmed. Specific receptors for ANG II have been conclusively or at least tentatively detected in a number of elasmobranch tissues including interrenal gland, gill, rectal gland, and intestine (Tierney et al., 1997). The most significant action that ANG II appears to have in elasmobranchs is to modulate the HPI axis, as demonstrated by the superlative number of ANG II receptors in the interrenal gland (Tierney et al., 1997) and its ability to stimulate 1α-OHB secretion *in vitro* (O'Toole et al., 1990; Armour et al., 1993b; Anderson et al., 2001a) and *in vivo* (Hazon and Henderson, 1984, 1985). In addition to promoting sodium retention via effects on interrenal corticosteroidogenesis, ANG II also appears to influence electrolyte balance in elasmobranchs by reducing rates of glomerular filtration (GFR) and urine flow (UFR) in the kidney (Anderson et al., 2001a) and inhibiting salt release by rectal gland (Anderson et al., 1995, 2001a). As recently demonstrated, both responses are likely to result from ANG II–stimulated reductions in blood flow through these organs (Anderson et al., 2001a). Last, ANG II also has been shown to be capable of increasing drinking rate in elasmobranchs despite the earlier belief that this process was both unnecessary and not practiced with regard to these fishes (Hazon et al., 1989; Anderson et al., 2001b). Much like the retention of salt stimulated by 1α-OHB, ingestion of seawater may enable elasmobranchs to adapt to hyperosmotic environments, a premise supported by the positive correlation between salinity and drinking rate in *S. canicula* (Hazon et al., 1997).

11.5.4 Vasoactive Intestinal Polypeptide

As previously discussed in the section regarding gut hormones, VIP influences ion homeostasis in at least some elasmobranchs by stimulating salt secretion by epithelial cells of the rectal gland (Epstein et al., 1981, 1983; Chipkin et al., 1988; Ecay and Valentich, 1991; Lehrich et al., 1998). Although this process may partially result from gastrointestinal sources of this peptide, it is primarily regulated by the release of VIP from nerves surrounding this organ. The effect of VIP on epithelial cell salt secretion is mediated by increased production of intracellular cAMP, which activates the efflux of chloride to the rectal gland lumen via Cl⁻ channels (Olson, 1999). The transport of sodium from the epithelial cells to the neighboring blood supply and, afterward, to the rectal gland lumen is largely a consequence of decreased intracellular chloride levels and the negative potential in the lumen, both of which are established by chloride secretion. VIP also appears to regulate salt secretion by analogous organs in other vertebrates, such as the salt glands of certain reptiles (Belfry and Cowan, 1995; Franklin et al., 1996; Reina and Cooper, 2000) and birds (Gerstberger, 1988; Gerstberger et al., 1988; Martin and Shuttleworth, 1994). Therefore, much like its structure, the physiological functions of this compound appear to be highly conserved throughout vertebrate evolution.

11.5.5 C-Type Natriuretic Peptide

Increased production of VIP and a subsequent rise in salt secretion by the elasmobranch rectal gland are stimulated by CNP, the only natriuretic peptide characterized to date in sharks and their relatives

(Takei, 2000). Although CNP is generally present only in the brain in most vertebrates, it is also expressed by the heart in elasmobranchs and is secreted in response to increased cardiac pressure associated with an elevation in blood volume. In addition to its indirect, VIP-mediated effect on rectal gland activity, CNP also appears to stimulate salt secretion by this organ through direct actions on epithelial cells. Binding of CNP to natriuretic peptide type-B receptors (NPR-B) in the rectal gland epithelium causes an increase in intracellular levels of cyclic granosine monophosphate (cGMP) and protein kinase C (PKC), two signaling factors (i.e., secondary messengers) that produce a synergistic effect on chloride transport via cellular Cl⁻ channels (Silva et al., 1999). Last, CNP also is capable of causing dilation of the rectal gland, a process that results in a rise in salt release through increased perfusion of this organ (Evans and Piermarini, 2001). Similar processes that promote salt extrusion in higher vertebrates also appear to be regulated by natriuretic peptides. However, these actions are a result of compounds other than CNP, which generally acts as paracrine factor in the brain in teleosts, amphibians, birds, and mammals (Takei, 2000).

11.5.6   Neurohypophysial Hormones

Because it is a major factor influencing ion and water balance in most vertebrates, the neurohypophyseal hormone arginine vasotocin (AVT, also known as arginine vasopressin and antidiuretic hormone, or AVP, ADH, in mammals) also may play a significant role in regulating these phenomena in cartilaginous fishes. This premise is supported by the high degree of homology between AVT/AVP from elasmobranchs and other vertebrate groups, whereas the diversity of compounds from the other major neurohypophyseal hormone (i.e., oxytocin) lineage may reflect a lack of conserved function (Acher et al., 1999). In terrestrial vertebrates, AVT/AVP reduces urinary water loss (i.e., diuresis) and GFR in response to rises in osmotic pressure (e.g., dehydration). Similar actions also have been reported in the teleost kidney (Amer and Brown, 1995), along with possible effects of this hormone on branchial ion transfer (Guibbolini et al., 1988). Recently, Wells et al. (2002) have demonstrated that AVT is capable of reducing diuresis in isolated kidney preparations from *S. canicula* previously adapted to reduced salinities. Therefore, it may be involved in regulating osmoregulation in sharks and rays by modulating urine production. Other studies have suggested that AVT also may influence urea reabsorption in elasmobranchs as it does in other vertebrates (Acher et al., 1999), but this has yet to be directly investigated.

11.6   Physiological Color Change

11.6.1   Overview

As first described by Schaeffer (1921), many elasmobranchs are capable of dramatically altering their skin color in response to the color or shade of their immediate environment. This process, which is generally termed physiological color change, occurs through the migration of pigment-containing organelles within specialized dermal cells known as chromatophores. The most abundant type of chromatophore in elasmobranch skin is the melanophore, a dermal cell that contains the brown-black pigment melanin within organelles called melanosomes. Elasmobranch melanophores are normally "punctuate" in appearance; that is, melanosomes are concentrated in the center of the cell, an arrangement that is associated with pallor or "lightening" of the skin. When certain sharks, skates, or rays are situated above a dark background, melanosomes become dispersed throughout the cytoplasm of the melanophore, giving the skin a darker appearance. The ability to chromatically adapt to their background benefits elasmobranchs by reducing the risk of predation, as well as enhancing opportunities for prey capture.

11.6.2   Melanocyte-Stimulating Hormone

Physiological color change in elasmobranchs is primarily regulated by α-melanocyte-stimulating hormone (α-MSH), a 13-amino acid peptide produced in the neurointermediate lobe (NIL) of the pituitary gland. The presence of this regulatory system was established by early studies that demonstrated that removal

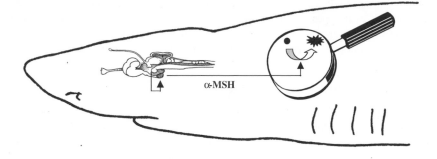

FIGURE 11.5 Proposed mechanism for hormonal regulation of physiological color change in elasmobranchs. Pituitary release of the hormone α-MSH causes expansion of dermal melanophores and darkening of the skin. Under conditions that favor skin pallor, the release of α-MSH is suppressed by hypothalamic signals, resulting in the contraction of melanophores and skin lightening.

of the NIL resulted in skin lightening in sharks and skates, which could be reversed by treatment with NIL extracts and/or blood from dark-adapted animals (Lundstrom and Bard, 1932; Parker, 1936; Waring, 1936; Chevins and Dodd, 1970). Subsequent experiments that examined the effects of the multiple types of MSH produced by the NIL (i.e., α-, β-, and γ-MSH only; δ-MSH was not yet described) on dogfish skin coloration confirmed that the alpha form is the principal factor influencing elasmobranch color change (Wilson and Dodd, 1973a; Sumpter et al., 1984). Although the elasmobranch pituitary produces both acetylated and deacetylated forms of α-MSH (Lowry and Chadwick, 1970; Bennett et al., 1974; Love and Pickering, 1974; Eberle et al., 1978; Denning-Kendall et al., 1982), the acetylated form appears to be more active in regulating skin coloration based on its greater effect in both *in vitro* and *in vivo* melanophore bioassays (Sumpter et al., 1984). α-MSH also contributes to the regulation of morphological color change, a gradual, long-term adjustment in elasmobranch skin color associated with changes in the total amount of melanin present in an animal's epidermis (Wilson and Dodd, 1973b). Furthermore, α-MSH also may function as a neurotransmitter and/or neuromodulator in elasmobranchs since its presence has been detected in the brains of several shark species (Vallarino et al., 1988b; Chiba, 2001).

The manner by which α-MSH regulates physiological color change in elasmobranchs (Figure 11.5) is believed to be similar to that first proposed for amphibians (Hogben, 1942). Under environmental conditions that favor skin pallor, the release of α-MSH appears to be suppressed by neural signals originating from the rostral lobe of the hypothalamus. This premise has been supported by experimental studies, which have observed irreversible skin darkening in sharks and skates following removal of the rostral lobe or damage to the connections between it and the NIL (Chevins and Dodd, 1970). When an elasmobranch is repositioned above a dark surface, the visual system likely receives new stimuli resulting from a reduction in the amount of light that is reflected from the background to the upper portion of the retina. Neural information associated with differential stimulation of the upper and lower retina is subsequently conveyed to the hypothalamus, which relaxes the normal inhibition of α-MSH release and/or promotes the production of this hormone through stimulatory factors. Following its release, circulating α-MSH binds to hormone receptors (i.e., MC1 receptors; Mountjoy et al., 1992) in melanophores and promotes skin darkening by eliciting melanosome dispersion, presumably via actions of cytoskeletal organelles such as microtubules and/or microfilaments.

The importance of the visual system in triggering physiological color change in elasmobranchs has been validated by studies on blinded dogfish, which lack the ability to undergo this process in response to changes in background coloration (Wilson and Dodd, 1973c). However, as these animals do exhibit limited pallor when maintained in complete darkness, the presence of nonvisual factors that influence melanophore function is likely. In particular, the pineal gland is believed to regulate changes in skin coloration resulting from nonvisual perception of light levels based on the lack of such responses in pinealecotomized dogfish. These findings suggest that melatonin, the hormone primarily secreted by the pineal gland during the dark cycle, may be responsible for inducing skin pallor in elasmobranchs during nocturnal periods as it appears to do in other vertebrates. If so, the effect of melatonin on elasmobranch skin coloration may be mediated through changes in α-MSH release given that it is unable to influence melanosome dispersion in freshwater ray skin *in vitro* (Visconti and Castrucci, 1993).

11.6.3   Other Factors Potentially Influencing Color Change

Recently, Visconti et al. (1999) determined that prolactin (PRL), a 190- to 200-amino acid peptide produced in the par distalis of the pituitary gland, is as potent as α-MSH in stimulating melanosome dispersion in freshwater ray, *Potamotrygon reticulatus,* skin *in vitro*. Based on these observations, they suggested that circulating PRL also may function in regulating physiological color change in elasmo-branchs, a role previously proposed for this hormone in amphibians (Camargo et al., 1999). The same researchers did not observe significant melanosome translocation in *P. reticulatus* skin in response to treatment with endothelins, catecholamines, or purines, compounds, which have been shown to influence color change in other vertebrates (Visconti and Castrucci, 1993; Visconti et al., 1999). Treatment with melanin-concentrating hormone (MCH), a 17-amino acid peptide localized in the brain and pars distalis of elasmobranchs (Vallarino et al., 1989), also had no effect on *P. reticulatus* skin color *in vitro* despite its well-described ability to cause melanosome aggregation and skin lightening in teleosts. However, physiological color change in teleosts differs greatly from that in elasmobranchs in that it is regulated by neural as well as hormonal signals via direct innervation of dermal melanophores. MCH may influence skin pigmentation in elasmobranchs indirectly through effects on α-MSH release, a regulatory process that would be overlooked in *in vitro* studies.

11.7   Reproduction

11.7.1   Overview

The diversity of breeding strategies in sharks and their relatives (see Carrier et al., Chapter 10 of this volume) makes it imprudent to generalize concerning the hormonal control of elasmobranch reproduc-tion. However, it is valid to assume that the brain–pituitary–gonadal (BPG) axis is the primary endocrine system involved in regulating procreation in most, if not all, cartilaginous fishes (Figure 11.6). Environ-mental signals likely initiate this endocrine cascade, which begins with the secretion of gonadotropin-releasing hormone (GnRH) from neurons in the hypothalamus and other portions of the elasmobranch brain. Release of GnRH stimulates the production of gonadotropins (GTHs) from the elasmobranch pituitary gland, which in turn promote gametogenesis and the secretion of reproductive steroids (i.e., androgens, estrogens, and/or progestins) in the gonads. In addition to regulating gamete production via autocrine and/or paracrine mechanisms, gonadal steroids are presumably involved in modulating repro-ductive behavior, as well as the development and function of secondary sex organs. Furthermore, these compounds are likely to influence the production of GnRH and GTHs via feedback mechanisms, based on the presence of steroid binding sites in the elasmobranch hypothalamus (Jenkins et al., 1980). Last, gonadal steroids and perhaps other aspects of the BPG axis also have the potential to alter production of other hormones such as relaxin, calcitonin, and thyroid hormones, which appear to play accessory roles in regulating reproduction in certain elasmobranchs.

11.7.2   Gonadotropin-Releasing Hormone

Multiple forms (i.e., as many as seven) of GnRH have been detected in the brain of several chondrichthyan species (King and Millar, 1980; Sherwood and Sower, 1985; Powell et al., 1986; Wright and Demski, 1991; King et al., 1992; Lovejoy et al., 1991, 1992a,b; Calvin et al., 1993; Sherwood and Lovejoy, 1993;

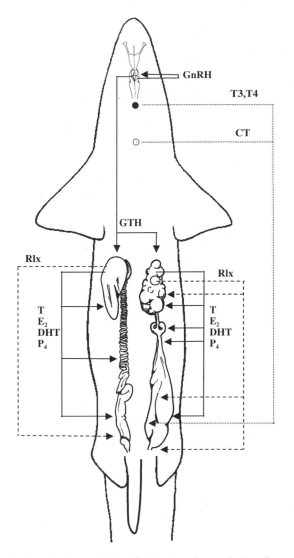

FIGURE 11.6 Proposed mechanism for the regulation of elasmobranch reproduction. Perception of environmental cues that signal seasonality are believed to trigger release of GnRH from the hypothalamus. The release of GnRH stimulates the production of GTH from the ventral lobe of the pituitary gland, which in turn promote production of reproductive steroids (e.g., T, E_2, DHT, and P_4) from the testis and ovary of male (left) and female (right) elasmobranchs. Reproductive steroids regulate gonadal steroidogenesis and gametogenesis, as well as various aspects of reproductive tract development. Both the testis and ovary also produce relaxin (Rlx), which appears to modulate certain aspects of reproduction in both sexes (dashed line). Last, both the ultimobranchial gland (open circle) and thyroid gland (closed circle) are believed to function in regulating reproduction in pregnant, viviparous elasmobranchs through production of calcitonin (CT) and the thyroid hormones T_3 and T_4, respectively.

D'Antonio et al., 1995; Forlano et al., 2000). As recently demonstrated by Forlano et al. (2000), variations in the neuroanatomical distribution of certain GnRH subtypes in the elasmobranch brain suggest that these compounds may function in regulating discrete aspects of reproduction. Neurons present in the hypothalamus and regions of the forebrain that primarily express the dogfish form of GnRH (dfGnRH) are generally considered to be the principal elements responsible for regulating GTH production in the ventral lobe of the pituitary gland, the primary site of gonadotropic activity. This premise is particularly well supported for male Atlantic stingrays, *Dasyatis sabina*, in which changes in dfGnRH expression in certain regions of the forebrain appear to be associated with the seasonal reproductive cycle (Forlano et al., 2000). Because elasmobranchs lack a neural or vascular conduit between the hypothalamus and the ventral lobe, transport of GnRH to pituitary gonadotrophs presumably occurs via the general circulation. This suggestion appears feasible based on the presence of both GnRH and GnRH-binding proteins (GnRHBP) in the blood of certain elasmobranchs (King and Millar, 1980; Powell et al., 1986; King et al., 1992; Pierantoni et al., 1993; Sherwood and Lovejoy, 1993; D'Antonio et al., 1995). The transport of GnRH in systemic circulation also provides a route for its direct actions on the gonads, which are likely to occur because removal of the pituitary gland is capable of only partially impairing steroidogenesis and gametogenesis in these fishes (Dobson and Dodd, 1977a,b; Sumpter et al., 1978b). More specific evidence for a direct relationship between GnRH bioactivity and gonadal function in elasmobranchs has been provided by Jenkins et al. (1980), Fasano et al. (1989), and Callard et al. (1993), all of which observed changes in gonadal (i.e., ovarian and testicular) steroidogenesis following administration of GnRH-like substances.

The terminal nerve (TN, also known as the nervus terminalis, or NT), a cranial nerve that connects the brain and peripheral olfactory structures, represents an additional major site of GnRH production in the elasmobranch forebrain (Nozaki et al., 1984; Stell, 1984; Demski et al., 1987; Lovejoy et al., 1992b; Wright and Demski, 1993; White and Meredith, 1995; Chiba et al., 1996; Chiba, 2000; Forlano et al., 2000). As a result of its direct association with chemoreceptive structures, GnRH-producing cells in the TN have been implicated in the regulation of reproductive processes and/or behaviors resulting from perception of olfactory cues linked with breeding (i.e., pheromones) (Demski and Northcutt, 1983). Because GnRH-positive fibers in this nerve project to sites in the forebrain generally believed to regulate ventral lobe function (Forlano et al., 2000), the putative effects of the TN on GTH and/or gonadal steroid production may be indirectly mediated through increased GnRH secretion in these regions. Additionally, GnRH originating from the TN may have direct effects on ventral lobe and/or gonadal function via transport in systemic circulation and/or cerebrospinal fluid (CSF), the latter of which contains increased GnRH levels following electrical stimulation of the TN (Moeller and Meredith, 1998). However, while the scenarios proposed above appear both logical and plausible, concrete evidence for a change in TN activity in response to olfactory stimuli has yet to be demonstrated in sharks (Bullock and Northcutt, 1984; White and Meredith, 1995). Nonetheless, since the elasmobranch TN does appear to respond to at least some peripheral signals (White and Meredith, 1995), a function for this nerve in regulating sensory-mediated reproductive events remains a possibility. A reciprocal role for the TN in modulating the responsiveness of elasmobranch sensory organs to chemosensory and visual (Demski et al., 1987) cues through efferent pathways also has been proposed, but similarly requires confirmation.

In addition to locations in the forebrain, sizable populations of GnRH-containing neurons also have been detected in the midbrain and hindbrain of certain elasmobranchs (Wright and Demski, 1991, 1993; Forlano et al., 2000). Because fibers from these sites project to regions of the central nervous system involved in processing visual, electrosensory, and mechanosensory stimuli, they are alleged to function in regulating the sensitivity of the eyes, ampullae of Lorenzini, and/or lateral line system during the copulatory period (Forlano et al., 2000). Such actions would significantly influence reproductive success in elasmobranchs, particularly rays, which are known to use electroreception to detect potential mates (Tricas et al., 1995). Wright and Demski (1993) also proposed that GnRH fibers in the midbrain may serve to regulate movement of the claspers, given that they project to regions in the spinal cord where motor neurons for these copulatory organs are located (Wright and Demski, 1991, 1993; Liu and Demski, 1993).

Like the TN, GnRH-producing neurons in the midbrain of certain elasmobranchs project to regions of the forebrain that appear to regulate ventral lobe and gonadal function (Forlano et al., 2000). Therefore, these cells have the potential to influence steroidogenesis and gametogenesis, in addition to their purported actions on sensory perception and/or locomotor activity. In fact, recent information suggests that the GnRH nucleus in the elasmobranch midbrain may be involved in conveying information regarding environmental cues that are believed to initiate cyclic activity of the BPG axis. Mandado et al. (2001) has reported the presence of neural projections from the pineal organ to GnRH-immunoreactive neurons in the dogfish midbrain, signifying that these cells may alter hormone production in relation to photoreceptive stimuli. Such findings add weight to the long-held, but largely unexplored premise that changes in day length, along with temperature or food availability, are the major environmental signals regulating elasmobranch reproduction.

11.7.3 Pituitary Gonadotropins

The presence of immunoreactive GTHs in the ventral lobe of the elasmobranch pituitary has been demonstrated by both immunocytochemistry (Mellinger and DuBois, 1973) and radioimmunoassay (Scanes et al., 1972). Furthermore, extracts of this organ have been shown to possess biologically active GTHs due to their ability to stimulate steroidogenesis in chondricthyan, reptilian, and avian testicular cells (Lance and Callard, 1978; Sumpter et al., 1980; Sourdaine et al., 1990), as well as follicular and luteal components of the elasmobranch ovary (Callard and Klosterman, 1988). Treatment of both male (Sumpter et al., 1978b) and female (Callard and Klosterman, 1988) elasmobranchs with ventral lobe extracts also may increase steroidogenesis *in vivo*, but these responses may vary depending on the stage of reproduction. Last, hypophysectomy or more selective removal of the ventral lobe has been shown to cause partial regression of the testis (Dodd et al., 1960; Dobson and Dodd, 1977a,b) and reduced androgen concentrations (Sumpter et al., 1978a; Fasano et al., 1989) in male elasmobranchs, as well as follicular atresia and impaired oviposition in female *Scyliorhinus canicula* (Norris, 1997). However, the nature of some of these responses also may depend on reproductive stage and/or environmental stimuli such as water temperature (Dobson and Dodd, 1977c). Although the effects of ventral lobectomy support a role for elasmobranch GTHs in regulating gonadal activity, this procedure is not capable of completely suppressing gametogenesis and steroidogenesis (Dobson and Dodd, 1977a,b; Sumpter et al., 1978b). Therefore, the direct actions of GnRH on elasmobranch gonadal function may represent a vital determinant of reproductive efficiency in these fishes.

Although a gonadotropic fraction has been purified from the ventral lobe of *S. canicula* (Sumpter et al., 1978a,b,c), the number and biochemical structure(s) of GTHs produced by the elasmobranch pituitary have long been unresolved. However, Quérat et al. (2001) recently have demonstrated the presence of two GTHs in the dogfish ventral lobe, which are structurally similar to paired gonadotropins from both tetrapods — i.e., follicle-stimulating hormone (FSH) and luteinizing hormone (LH) — and teleosts (GTH1 and GTH2). Given the discrete actions that FSH/LH and GTH1/GTH2 exert on gonadal function in these groups, future studies should investigate if the two elasmobranch GTHs serve to regulate dissimilar aspects of steroidogenesis and/or gametogenesis.

11.7.4 Gonadal Steroid Hormones in the Female

The ovary of female elasmobranchs produces three major gonadal steroids: 17β-estradiol (E_2), testosterone (T), and progesterone (P_4). As indicated by measurements of steroid production by *Squalus acanthias* ovarian subcomponents *in vitro*, the synthesis of these compounds appears to be a shared function of both granulosa and theca cells (Tsang and Callard, 1992; Callard et al., 1993). Granulosa cells from active ovarian follicles secrete low levels of T and E_2 in unstimulated cultures, and are capable of dramatically increasing production of E_2 and P_4 in response to stimulation by ventral lobe extracts. In contrast, isolated theca cells also synthesize T and E_2, but do not appear to contribute to the production of P_4 or increase steroidogenesis in response to gonadotropic stimulation. The cooperative nature of ovarian steroidogenesis was revealed through co-incubation of these cell layers, which resulted in a more modest increase in P_4 production, as well as a significant rise in T concentrations in stimulated cultures. Based on these results, it appears likely that stimulated theca cells utilize P_4 secreted by granulosa cells to produce heightened levels of T. The nature of these findings suggests that both of these cell layers contribute to total follicular steroidogenesis in intact animals.

Production of P_4 in female elasmobranchs also is a function of ovarian corpora lutea, which form primarily from granulosa cells either prior to or after ovulation. Evidence for expression of 3β-HSD, the key enzyme involved in P_4 synthesis, has been demonstrated in corpora lutea from *S. acanthias* (Callard et al., 1992). Also, luteal minces from both *S. acanthias* (Tsang and Callard, 1987a) and little skate, *Raja erinacea* (Fileti and Callard, 1988), have been shown to be capable of secreting substantial quantities of P_4 *in vitro*. As shown in these studies, production of P_4 increases with the maturation of the corpus luteum, but declines with its age.

As demonstrated in oviparous and viviparous species (Koob et al., 1986; Tsang and Callard, 1987b; Fasano et al., 1992; Manire et al., 1995; Snelson et al., 1997; Heupel et al., 1999, Rasmussen et al.,

1999; Tricas et al., 2000, Sulikowski, unpubl. data), circulating concentrations of E_2 in female elasmo-branchs generally peak during the period of follicular development. Such increases reflect the common role of E_2 on synthesis of vitellogenin, the precursor to egg yolk proteins (Koob and Callard, 1999). Production of vitellogenin occurs in the liver and is stimulated by E_2 through interactions with hepatic estrogen receptors (ER), which have been identified in at least one elasmobranch, *R. erinacea* (Paolucci and Callard, unpubl.). Afterward, it is transported to the ovary via systemic circulation and is sequestered by oocytes through receptor-mediated endocytosis. The stimulatory effect that E_2 exerts on vitellogenin production in elasmobranchs has been demonstrated in female *Scyliorhinus canicula* (Craik, 1978), *Squalus acanthais* (Ho et al., 1980), and *R. erinacea* (Perez and Callard, 1992; 1993) in response to hormone treatment. Furthermore, increased levels of circulating vitellogenin have been shown to correspond with the preovulatory rise in E_2 in female *Scyliorhinus canicula* (Craik, 1978, 1979) and *R. erinacea* (Perez and Callard, 1993). Male elasmobranchs also possess the ability to synthesize vitellogenin, but normally do not express this protein presumably due to lower levels of circulating E_2. However, treatment with E_2 can result in induction of vitellogenin synthesis in at least some male elasmobranchs, as demonstrated in *R. erinacea* (Perez and Callard, 1992; 1993) and *Sphyrna tiburo* (Gelsleichter, unpubl. data). Because the presence of vitellogenin in nonmammalian male vertebrates is commonly used as a tool for detecting exposure of these animals to estrogen or estrogen-like (e.g., organochlorine pesticides) compounds (see review by Denslow et al., 1999), similar use of this procedure for male elasmobranchs is currently being evaluated (Gelsleichter, unpubl. data).

Because elevated concentrations of E_2 during the follicular stage coincide with increased growth of the oviducal gland in female *R. erinacea* (Koob et al., 1986), *S. tiburo* (Manire, unpublished data), *Hemiscyllium ocellatum* (Heupel et al., 1999), and winter skate, *Leucoraja ocellata* (Sulikowski, unpubl. data), it is reasonable to consider that E_2 may regulate the development and functions of this organ. A similar relationship likely exists in all female elasmobranchs for which data regarding E_2 profiles are available (e.g., *D. sabina*, *Squalus acanthias*, *R. eglanteria*), but presumably was not apparent because changes in oviducal gland size generally were not measured. Presence of ER in the oviducal gland has been demonstrated in *R. erinacea* (Reese and Callard, 1991), and E_2 treatment has been shown to cause enlargement and increased protein secretion by this organ in female *Scyliorhinus canicula* (Dodd and Goddard, 1961). The response of the oviducal gland to hormone treatment, as well as the rise in E_2 observed in *R. eglanteria* specifically during egg capsulation (Rasmussen et al., 1999), would seem to indicate that E_2 functions to regulate this event. However, there are reasons to consider that E_2 also may influence other functions of the oviducal gland such as the storage of spermatozoa. In fact, recent examinations of female *Sphyrna tiburo* from populations exhibiting unusually high rates of infertility have linked a reduced peak in preovulatory E_2 levels with an apparent decline in the viability of stored spermatozoa (Manire et al., unpubl. data). Given that follicular development did not appear to be impaired in these animals, it is hypothesized that impaired sperm storage and perhaps infertility were directly associated with a disruption in E_2 production and/or effects. However, attempts to localize the specific presence of ERs in oviducal sperm storage follicles in *S. tiburo* have been largely unsuccessful, despite positive immunocytochemical detection of these receptors in other components (e.g., albumin-secreting cells) of this organ (Gelsleichter, unpubl. data).

In female *S. tiburo* (Manire et al., 1995), *Squalus acanthias* (Tsang and Callard, 1987b), and *D. sabina* (Snelson et al., 1997; Tricas et al., 2000), the rise in circulating E_2 concentrations beginning prior to ovulation overlaps to some extent with the passage of fertilized ova to the uterus. Because of this, it seems possible that E_2 may play a role in regulating uterine function in a manner that ensures the success of this transport process. This notion is supported by the observation that E_2 increases the compliance of the isthmus in female *S. acanthias* (Koob et al., 1983), the region of the reproductive tract lying between the oviducal gland and the uterus. As discussed by Koob and Callard (1999), increased extensibility of this tissue is probably necessary for permitting the movement of delicate ova without damage to its integrity.

A secondary rise in endogenous levels of E_2 has been observed to occur in female *S. acanthias*, during the latter stages of gestation (Tsang and Callard, 1987b). As this species is a continuous breeder, the change in E_2 concentrations at this time appears to regulate vitellogenesis and development of follicles for the succeeding reproductive cycle. However, there is also evidence to suggest that E_2 may potentiate

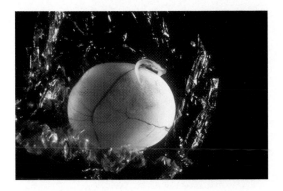

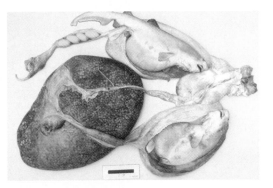

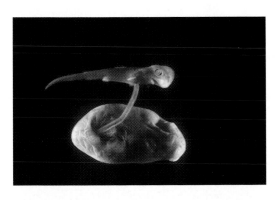

COLOR FIGURE 10.1 Finetooth shark (*Carcharhinus porosus*); 15 mm embryo attached to yolk sac. (Photo copyright José I. Castro. Used with permission.)

COLOR FIGURE 10.2 Reproductive tract of a sandtiger shark (*Carcharias taurus*) carrying one oophagous embryo in each uterus. The ovary can be seen on the left and several egg cases are being shunted to the embryos. (Photo copyright José I. Castro. Used with permission.)

COLOR FIGURE 10.3 Atlantic sharpnose shark (*Rhizoprionodon terranovae*); 43 mm embryo. Note yolk sac becoming flaccid as yolk is being consumed. (Photo copyright José I. Castro. Used with permission.)

COLOR FIGURE 10.4 Atlantic sharpnose shark (*Rhizoprionodon terranovae*); 250 mm embryo at midterm. (Photo copyright José I. Castro. Used with permission.)

COLOR FIGURE 10.8 Male nurse shark (*Ginglymostoma cirratum*) (left) gripping a female's pectoral fin as a prelude to copulation. (Photo copyright Jeffrey C. Carrier. Used with permission.)

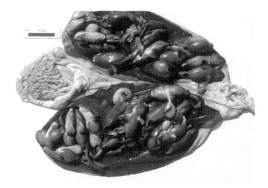

COLOR FIGURE 10.10 Entire reproductive system of pregnant nurse shark (*Ginglymostoma cirratum*) showing young in both uteri. (Photo copyright José I. Castro. Used with permission.)

COLOR FIGURE 14.3 Cross-sectioned spine from a 43-year-old female *Centrophorus squamosus*. M = mantle; OTL = outer trunk layer; ITL = inner trunk layer; TP = trunk primordium; L = lumen; A = annuli. Scale bar = 500 μm. (From Clarke, M. W. et al. 2002b. *J. Fish Biol.* 60:501–514. With permission.)

the effects of peptide hormones on uterine function in a manner that influences the maintenance of pregnancy (Koob and Callard, 1999). The actions of one of these hormones, relaxin, is discussed later in this chapter.

Post-oogenic elevations in circulating E_2 concentrations have been reported to also occur in seasonally breeding viviparous elasmobranchs during pregnancy. As follicular development for the subsequent year does not begin until after parturition in these species, increased levels of E_2 are believed to reflect a role for this hormone in the gestation process. In female *Sphyrna tiburo*, a rise in serum E_2 concentrations occurs coincident with the formation of placental connections between the gravid female and developing embryos (Manire et al., 1995). Similarly, in *D. sabina*, elevated E_2 concentrations coincide with secretion of uterine histotroph, which functions to nourish embryos between the middle and late stages of pregnancy (Snelson et al., 1997; Tricas et al., 2000). Because E_2 has well-characterized actions on uterine function in many vertebrates, it is reasonable to consider that it may be involved in modulating placental function in *S. tiburo* and/or the secretion of nutritive substances by the stingray uterus. Alternatively, Koob and Callard (1999) have postulated that E_2 may influence embryonic sustenance in these species by regulating nutrient availability in the pregnant female. In the only study that has addressed these hypotheses, Henningsen and Trant (unpubl. data) found no effect of E_2 on protein secretion by stingray uterine trophonemata *in vitro*. Nonetheless, since other factors (e.g., fatty acid profiles) are likely to be more valid indicators of the nutritive value of histotroph, the putative role of E_2 on uterine function in *D. sabina* (or *S. tiburo*) remains unresolved.

In viviparous elasmobranchs such as *S. tiburo* (Manire et al., 1995), *Squalus acanthias* (Tsang and Callard, 1987b), *D. sabina* (Snelson et al., 1997; Tricas et al., 2000), and *Torpedo marmorata* (Fasano et al., 1992), endogenous P_4 concentrations generally peak during and/or shortly after the ovulatory period. As detailed in Koob and Callard (1999), these changes likely reflect a role for P_4 in suppressing further production of vitellogenin, which, until this point, was stimulated by the actions of E_2. Receptors for P_4 have been detected in the liver of female *R. eglanteria* (Paolucci and Callard, 1998), and P_4 treatment is capable of blocking E_2-stimulated vitellogenesis (Perez and Callard, 1992, 1993) and overall follicular development (Koob and Callard, 1985) in this species. Furthermore, attempts to induce vitellogenin production in pregnant *S. acanthias* have been shown to be unsuccessful until later stages of pregnancy, when circulating P_4 concentrations decline (Ho et al., 1980). Thus, the reduction in P_4 levels in *S. acanthias* that occurs during this period (Tsang and Callard, 1987b) appears to permit the development of follicles for the subsequent pregnancy.

Because the strongest support for an inhibitory action of P_4 on elasmobranch vitellogenesis has been derived from studies on *R. erinacea* (Koob and Callard, 1985; Perez and Callard, 1992, 1993; Paolucci and Callard, 1998), this response likely occurs regularly in oviparous elasmobranchs. This appears to be the case for female *R. erinacea*, which experience an ephemeral surge in endogenous P_4 levels just prior to ovulation (Koob et al., 1986). However, occurrence of this process is less supported in female *R. eglanteria*, in which the only significant rise in serum P_4 concentrations during the egg-laying period occurs at the time of oviposition (Rasmussen et al., 1999). Although the reasons for these dissimilarities remain unclear, they warrant further investigation, as they are the most distinct difference in what is known regarding the reproductive endocrinology of these species. In total, these findings may reflect important roles for P_4 in both suppression of vitellogenesis and egg-laying in skates. The latter of these two proposed functions is supported by presence of P_4 receptors (PR) in the skate reproductive tract (Callard et al., 1993), as well as the observation that P_4 treatment can cause early oviposition in *R. erinacea* (Callard and Koob, 1993).

In virtually all female elasmobranchs for which data regarding steroid hormone profiles are available, endogenous concentrations of androgens rise specifically during the period of follicular development (Koob et al., 1986; Tsang and Callard, 1987b, Manire et al., 1995; Snelson et al., 1997; Rasmussen et al., 1999; Tricas et al., 2000; Sulikowski, unpubl. data). Because elevations in T in particular overlap with the preovulatory peak in E_2, it is possible that it may partially serve as a precursor for E_2 synthesis during this stage. Alternatively, because seasonal peaks in T and DHT (dihydrotestosterone) production slightly precede those for E_2 in *R. eglanteria* and *D. sabina*, Rasmussen et al. (1999) and Tricas et al. (2000) have suggested that androgens may play a role in modulating copulatory behavior. However, because elevated levels of T continue 6 months beyond the mating period in female *Sphyrna tiburo*,

Manire et al. (1995) hypothesized that T may be involved in the regulation of oviducal sperm storage, which has been shown to occur at this time. Finally, based on an increase in circulating T and DHT levels specifically during late stages of the egg-laying process in *R. eglanteria*, Rasmussen et al. (1999) suggested that androgens might function in regulating oviposition in oviparous elasmobranchs. However, this is not supported by observations on *R. erincea*, in which T levels are minimal during this same period (Koob et al., 1986). Notwithstanding this myriad of hypotheses, no published studies to date have described distribution of androgen receptors or effects of androgens in female elasmobranchs. Therefore, the role of T and DHT in these animals remains largely unresolved.

11.7.5 Gonadal Steroids, Sex Differentiation, and Puberty

Although sex differentiation in elasmobranchs has been poorly studied, it appears to progress in a manner similar to that in amphibians and amniotes (see Hayes, 1998, for review). As in all vertebrates, proper development of the gonads and secondary sex organs in sharks and their relatives appear to be sensitive to endogenous levels of steroid hormones. For example, in embryonic *Torpedo ocellata*, Chieffi (1967) observed feminization of embryonic gonads and accessory ducts following injections of E_2, P_4, T, and deoxycorticosterone into the external yolk supply. Thiebold (1953, 1954) observed similar effects of E_2 and T in embryonic *Scyliorhinus canicula*. Although limited in number, these studies underscore the need for a clearer understanding of levels of exposure and effects of steroid hormones in developing elasmobranchs. In a recent effort to partially address this topic, Manire et al. (unpubl. data) examined yolk concentrations of E_2, P_4, T, and DHT in preovulatory (i.e., ovarian), ovulatory (i.e., oviducal), and postovulatory (i.e., uterine) ova. The results from this study indicated that significant concentrations of E_2, P_4, and T are transferred from female *Sphyrna tiburo* to early-stage embryos via yolk provisions. Furthermore, reductions in yolk concentrations of E_2 and T during early development suggest the active utilization of these steroids. Interestingly, increased levels of all three steroids during the later stages of yolk dependency in these animals may reflect the period during which embryonic steroidogenesis is initiated.

In most vertebrates, the period of sexual maturation is associated with activation of the BPG axis, which results in heightened production and release of gonadal steroids (see Bourguignon and Plant, 2000; Okuzawa, 2002 for reviews). Increased concentrations of these hormones are believed to be essential in regulating the development of the gonads and secondary sex organs, in addition to influencing activity of the hypothalamus and pituitary gland via feedback mechanisms. Changes in gonadal steroidogenesis with maturity also appear to occur in elasmobranchs, based on comparisons of circulating steroid concentrations in immature and mature animals (e.g., Rasmussen and Gruber, 1993; Manire et al., 1999). More concrete evidence of these changes recently has been observed in serially examined male *S. tiburo*, which exhibit significant, but stage-specific increases in serum T, DHT, E_2, and P_4 concentrations during pubertal development (Gelsleichter et al., in press). Although increases in the concentrations of these hormones coincide with development of the testis and accessory sex organs (i.e., epididymis, seminal vesicle, clasper), their roles in such processes remain unclear. Nonetheless, since E_2 treatment has been shown to be capable of promoting maturation of the reproductive tract in immature female *Musteus canis* (Hisaw and Abramowitz, 1939) and *Scyliorhinus canicula* (Dodd and Goddard, 1961), it seems likely that the pubertal surge in steroid concentrations in maturing elasmobranchs has a functional significance.

11.7.6 Gonadal Steroids in the Male

The presence of numerous gonadal steroids has been reported in male elasmobranchs (Callard, 1988; Manire et al., 1999), but several of these observations require confirmation via analysis of compounds produced by testicular tissues and/or cells cultured in the presence of radiolabeled precursors. Unlike many other male vertebrates, in which gonadal steroids are largely produced by cells that lie between testicular spermatocysts (i.e., Leydig or interstitial cells), male elasmobranchs appear to synthesize the bulk of these compounds in Sertoli cells. This notion was first established by studies that demonstrated that these cells possess both the cytological (Holstein, 1969; Pudney and Callard, 1984a) and enzymatic

(Simpson and Wardle, 1967) characteristics of steroid producers, and has been further validated via direct measurement of steroids secreted by isolated Sertoli cell monolayers (DuBois et al., 1989). Although Leydig-like interstitial cells with steroidogenic features generally occur in the testis of male elasmobranchs, Pudney and Callard (1984b) reported that these cells are undifferentiated in appearance and do not undergo structural changes that occur in Sertoli cells in association with spermatogenic progression (Pudney and Callard, 1984a). Because of this, the involvement of these cells in testicular steroidogenesis in male elasmobranchs has long been questioned. However, recent observations on the testis of male *T. marmorata* have argued for the presence of true Leydig cells in this species, which exhibit both ultrastructural and enzymatic attributes of steroid-producing cells (Marina et al., 2002). Moreover, because these cells appear most active in regions bordering early-stage spermatocysts, they have been proposed to function in partially regulating the initial stages of spermatogenesis. Thus, it is reasonable to consider that Leydig-like cells supplement gonadal steroidogenesis in at least some male elasmobranchs, albeit far less than Sertoli cell components.

Despite the large number of gonadal steroids that have been detected in male elasmobranchs, patterns in the endogenous concentrations of only T, DHT, E_2, and P_4 have been well investigated in relation to the reproductive cycle. Associations between testicular and/or circulating levels of these hormones and certain breeding stages suggest that they function in regulating essential aspects of male reproduction. For example, in *Sphyrna tiburo* (Manire and Rasmussen, 1997), *D. sabina* (Snelson et al., 1997; Tricas et al., 2000), and *H. ocellatum* (Heupel et al., 1999), serum T and/or DHT concentrations significantly increase during the middle to late stages of spermatogenesis. In all these species, this period is characterized by a peak in gonadosomatic index (GSI), as well as an increase in the presence of mature spermatocysts in the testis. The increase in androgen concentrations experienced during this period likely reflects increased production of these compounds by late-stage, postmeiotic spermatocysts, which has been demonstrated to occur in the testis of both *Squalus acanthias* (Callard et al., 1985; Cuevas et al., 1993) and *Scyliorhinus canicula* (Sourdaine et al., 1990; Sourdaine and Garnier, 1993). Although this initially suggested that androgens directly regulate the final stages of sperm maturation, Cuevas and Callard (1992) have demonstrated that androgen receptors in the elasmobranch testis are primarily localized in early-stage (i.e., premeiotic and meiotic) spermatocysts. Therefore, T and/or DHT produced by Sertoli cells in mature spermatocysts more likely function to regulate the developmental advance of spermatogonia (Callard, 1992). This phenomenon appears to be made possible by the route of blood flow through the elasmobranch testis, which proceeds from more advanced to less advanced stages of spermatocyst differentiation (Cuevas et al., 1992).

Because serum androgen concentrations in male *Sphyrna tiburo* (Manire and Rasmussen, 1997), *D. sabina* (Snelson et al., 1997; Tricas et al., 2000), *H. ocellatum* (Heupel et al., 1999), and *Scyliorhinus canicula* (Garnier et al., 1999) are elevated during periods of increased semen transport, these compounds probably influence development and function of the gonaducts, and/or the maturation and viability of spermatozoa. Whereas such actions are likely based on the roles of these compounds in other vertebrates, no published studies have reported on the effects of androgens on these aspects of male elasmobranch reproduction. Nonetheless, multiple routes of steroid hormone transfer between the testis and urogenital system in male chondrichthyans exist. In addition to transport in the general circulation, steroid hormones appear capable of accessing putative binding sites in spermatozoa and/or the male reproductive tract through the occurrence of Sertoli cell cytoplasts and/or remnants (Pudney and Callard, 1986; Marina et al., 2002) and steroidogenic enzyme activity (Simpson et al., 1964) in elasmobranch semen. This mechanism may represent a significant contribution to the regulation of reproductive events occurring after spermiation because the seminal fluid of some sharks contains high concentrations of certain steroids (Simpson et al., 1963; Gottfried and Chieffi, 1967).

Although increased clasper size coincides with peak T concentrations in some mature male elasmobranchs (Garnier, 1999; Heupel et al., 1999), no studies have confirmed androgen sensitivity of this organ. Even during puberty, when growth of the clasper and other sexually dimorphic skeletal elements (i.e., the cephalofoil of male *Sphyrna tiburo*; Kajiura et al., unpubl. data) are at their maximum, Gelsleichter et al. (in press) found no direct relationship between circulating androgen concentrations and rates of clasper elongation in serially examined male *S. tiburo*. Furthermore, treatment of clasper cartilage explants from pubertal male *S. tiburo* with T was found to have no significant effects on the growth of this tissue (Gelsleichter, unpubl. data). Finally, both hypophysectomy and administration of T are incapable of altering clasper growth *in vivo* in immature male elasmobranchs (Wourms, 1977). Although these findings appear to argue against a function for androgens in the development and growth of the male elasmobranch copulatory organs, the effects of androgens on skeletal growth in mammals are currently believed to be largely mediated through estrogen-regulated (i.e., obtained via aromitization of T) increases in GH and IGF-I production (Grumbach, 2000). Androgens also may influence the pubertal growth of the mammalian skeletal system by stimulating increased production of IGF-I receptors (Phillip et al., 2001). Future studies should evaluate the possible links between the BPG and growth axes in the cartilaginous fishes to clarify the putative roles of T and DHT on the external genitalia.

Elevated serum androgen concentrations occur during copulatory activity in male *D. sabina* (Snelson et al., 1997; Tricas et al., 2000), *H. ocellatum* (Heupel et al., 1999), and *Negaprion brevirostris* (Rasmussen and Gruber, 1993), suggesting that these hormones may function in modulating certain aspects of reproductive behavior. Although this topic has not been extensively studied, recent evidence supports the notion that androgens are capable of influencing mating activity of elasmobranchs through effects on sensory organ responsiveness. Sisneros and Tricas (2000) have demonstrated that the electrosensory abilities of male *D. sabina* are significantly improved during the seasonal peak in circulating androgen levels. Similarly, increased sensitivity of ampullary electroreceptors in these animals occurred following treatment with DHT. Androgen-mediated changes in electroreception specifically improve the ability of male stingrays to detect low-frequency stimuli, such as those generated by conspecifics. Therefore, these changes are likely to influence both the detection of potential mates and overall reproductive success. As bioelectric information produced by the typical prey of this species is generally of a higher frequency, it is doubtful that the seasonal changes in electroreceptive ability are more associated with feeding behavior rather than reproduction.

Unlike those observed for androgens, seasonal patterns in circulating E_2 concentrations in most male elasmobranchs generally reveal little about the role of this hormone in reproduction. For example, in both male *S. tiburo* (Manire and Rasmussen, 1997) and *Scyliorhinus canicula* (Garnier et al., 1999), endogenous levels of E_2 vary irregularly during the reproductive cycle. However, in male *D. sabina* (Snelson et al., 1997; Tricas et al., 2000), serum E_2 concentrations exhibit a clear pattern of variation in which levels of this hormone rise specifically during the early to middle stages of spermatogenesis. These changes better reflect patterns in testicular E_2 production in *Squalus acanthias*, which appears to be highest in spermatocysts undergoing meiosis (Callard et al., 1985; Cuevas and Callard, 1992). Whereas evidence for peak E_2 production during meiosis suggests a role for this hormone during mid-spermatogenesis, ERs are primarily localized in regions of the testis containing pre-meiotic spermatocysts (Callard et al., 1985; Callard, 1992). Therefore, as suggested for androgens, testicular E_2 appears to have its greatest effect on downstream germ cells undergoing earlier stages of spermatogenesis. Because treatment of these pre-meiotic cells with E_2 results in a dose-dependent reduction in both cell proliferation and programmed cell death, this hormone appears to regulate spermatogenic progression through developmental arrest via a negative feedback system (Betka and Callard, 1998). As this effect is largely paracrine in nature, circulating levels of E_2 may not necessarily reflect its rate of production or role in the testis in certain male elasmobranchs (e.g., *Sphyrna tiburo*).

A function for E_2 in regulating the development and actions of the reproductive tract in male vertebrates is well supported by studies that have demonstrated presence of ER in both the epididymis and seminal vesicle of several taxa (e.g., Kwon et al., 1997; Misao et al., 1997). Recent experiments using the transgenic ER-alpha knockout mouse model have confirmed that estrogens play vital roles in maintaining virtually all aspects of genital tract function, particularly in the epididymis (Eddy et al., 1996). No published studies have investigated the presence or distribution of ER in the gonaducts of male elasmobranchs. However, since a peak in E_2 concentrations coincides with increased cell proliferation and growth in the epididymis and seminal vesicle of male *D. sabina* (Piercy et al., unpubl. data), this is a topic that should be addressed in future studies.

Because changes in circulating levels of P_4 mirror those of T and DHT in mature male *S. tiburo*, it may function as a substrate for androgen synthesis during the latter stages of spermatogenesis (Manire and Rasmusssen, 1997). However, in serially examined pubertal male *S. tiburo*, elevations in serum T concentrations precede those of P_4 by several months (Gelsleichter et al., in press). Similarly, in male

D. sabina, increased levels of P_4 both occur later and persist longer than the peak in serum androgen levels. Together, along with the lack of correlation between endogenous T and P_4 concentrations in *Scyliorhinus canicula* (Garnier et al., 1999), these findings suggest that P_4 functions as more than merely a precursor for other steroids in the elasmobranch testis. This notion is supported by the presence of PR in the testis of *Squalus acanthias*, which are primarily localized in late-stage (postmeiotic) spermatocysts (Cuevas and Callard, 1992). Such observations suggest that P_4 plays a role in regulating spermiogenesis and/or spermiation in male elasmobranchs. As testicular P_4 synthesis is greatest in postmeiotic spermatocysts (Callard et al., 1985), these actions may be regulated through an autocrine mechanism.

11.7.7 Other Hormones Involved in Reproduction: Relaxin

Relaxin, a 6-kDa polypeptide hormone best known for its ability to prepare the mammalian reproductive tract for successful parturition, has been detected in the ovaries of *S. acanthias* (Bullesbach et al., 1986), *R. erinacea* (Bullesbach et al., 1987), and the sandtiger shark, *Carcharias taurus* (Gowan et al., 1981; Reinig et al., 1981). Koob et al. (1984) determined that relaxin and its structural homologue insulin were capable of increasing cervical cross-sectional area in late-stage, pregnant *S. acanthias*, leading to the premature loss of developing fetuses. In similar studies, treatment of female *R. erincea* with homologues of porcine relaxin resulted in increased compliance of the cervix and other portions of the reproductive tract (Callard et al., 1993). As these effects mirror the ability of relaxin to increase circumference of the mammalian birth canal (Steinetz et al., 1983), it appears likely that this hormone may participate in pupping and/or egg laying in female elasmobranchs. Relaxin also has been shown to reduce the frequency of myometrial contractions *in vitro* and *in vivo* in uterus of pregnant *S. acanthias*, suggesting that it functions in the maintenance of pregnancy prior to parturition (Sorbera and Callard, 1995). This response is also analogous to one of the roles proposed for relaxin in certain mammals (Downing and Sherwood, 1985). Last, recent assessments of changes in serum relaxin concentrations associated with the reproductive cycle in female *Sphyrna tiburo* indicate that production of this hormone may be greatest just prior to ovulation (Gelsleichter et al., in press). Based on these findings and corresponding data in mammals (Bagnell et al., 1993; Song et al., 2001), relaxin may participate in the release of ova in *S. tiburo* via effects on ovarian connective tissue.

Although typically considered a "female hormone" due to the effects previously discussed, relaxin is also produced by the reproductive organs (e.g., testis, prostate, seminal vesicle) of some male vertebrates and is believed to play a role in regulating male fertility (Weiss, 1989). Relaxin has been purified from the testis of male *Squalus acanthias* (Steinetz et al., 1998) and, more recently, Gelsleichter et al. (2003) have demonstrated that serum relaxin concentrations in male *Sphyrna tiburo* are elevated specifically during late spermatogenesis and the copulatory period. These observations tentatively suggest that relaxin regulates certain aspects of semen quality, a hypothesis also proposed for its role in mammals (Weiss, 1989). However, as concentrations of relaxin in semen of male *S. tiburo* are approximately 1000 times greater than that in circulation (Gelsleichter and Steinetz, unpubl. data), it is also reasonable to consider that this hormone may facilitate insemination through regulating uterine contractibility in postmated females. A similar function has been proposed for a relaxin-like compound produced by the alkaline gland of skates and stingrays, a structure homologous with the mammalian prostrate gland (Bullesbach et al., 1997).

11.7.8 Other Hormones Involved in Reproduction: Thyroid Hormones

Thyroid hormones are believed to play a permissive role in regulating vertebrate reproduction, largely through interactions with the BPG axis (Karsch et al., 1995). Evidence for a similar function in elasmobranchs was first proposed in early studies that reported sexual dimorphism in this organ or its increased activity during oogenesis (see Dodd, 1975, for review). Further support for this premise was provided by the observation that thyroidectomy is capable of impairing follicular development in female *Scyliorhinus canicula* (Lewis and Dodd, 1974). However, despite these findings, the role of thyroid hormones in elasmobranch reproduction has long been unresolved.

Two recent studies have readdressed the relationship between thyroid activity and reproduction in the cartilaginous fishes. Volkoff et al. (1999) determined that both thyroid gland activity and circulating levels of triiodothyronine (T_3) were significantly elevated in female *D. sabina* during ovulation and throughout the period of gestation. These observations contradict the earlier notion that thyroid gland function in female elasmobranchs is greatest during follicular development. Gash (2000) observed an increase in thyroid gland activity and hormone production in two populations of female *Sphyrna tiburo* throughout pregnancy, with peak levels occurring during formation of the maternal–fetal placental connection. Because both species provide nourishment to developing embryos through energetically demanding processes, Gash (2000) hypothesized that increased production of thyroid hormones may address greater metabolic need during this period.

Gash (2000) also observed a seasonal pattern in thyroid gland activity of mature male *S. tiburo*, which was characterized by increased hormone production during both the spring and fall. Although it is tempting to consider that thyroid hormones may serve a function during the mating period, which occurs between September and November in the population in question, it is important to note that immature males exhibited a similar hormonal pattern. Because of this, Gash (2000) acknowledged the possibility that activity of the thyroid gland in male *S. tiburo* may be associated more with migratory activity, which increases dramatically during both of these periods due to changes in environmental stimuli. Similar caution in interpreting changes in thyroid gland activity of seasonally breeding elasmobranchs is stressed, because the reproductive cycle is more than likely associated with environmental cues.

Maternally provided thyroid hormones are critically important during early stages of development in many vertebrates. Because of this, McComb et al. (unpubl. data) recently have begun to examine the maternal transfer of thyroid hormones via yolk to developing elasmobranchs and the utilization of these compounds during gestation. Preliminary results from this study have characterized both the presence and abundance of T_3 and thyroxine (T_4) in yolk from preovulatory (i.e., ovarian) and postovulatory (i.e., uterine) ova of *S. tiburo* from two populations. Interestingly, levels of thyroid hormones in yolk were lower in sharks from the population exhibiting comparatively lower rates of embryonic development, size at birth, size at maturity, and maternal investment in reproduction. Therefore, these findings may reflect a role for thyroid hormones in dictating the rate of embryonic development of sharks and their relatives.

11.7.9   Other Hormones Involved in Reproduction: Calcitonin

In addition to its ability to regulate calcium balance in mammals through the inhibition of bone demineralization, the hormone calcitonin is generally believed to play a role in regulating vertebrate reproduction (Zaidi et al., 2002). Relationships between production of gonadal steroids and calcitonin have been observed in virtually all major vertebrate groups, and evidence has been presented to link calcitonin with a number of reproductive processes including pregnancy and lactation in mammals, follicular development in birds and teleosts, and embryonic development in all of these groups (e.g., Dacke et al., 1976; Björnsson et al., 1986; Lu et al., 1998). Although produced in parafollicular C-cells of the thyroid gland in mammals, calcitonin is largely produced in a separate organ, the ultimobranchial gland, in all other jawed vertebrates (Wendelaar Bonga and Pang, 1991). In elasmobranchs, this paired organ is embedded in the musculature lying between the pharynx and pericardial cavity. A role for calcitonin in regulating reproduction in cartilaginous fishes is supported by the presence of ERs in the ultimobranchial gland of the stingray *Dasyatis akajei* (Yamamoto et al., 1996), as well as the ability of E_2 to cause an increase in calcitonin production in the same species (Takagi et al., 1995). In contrast, a role for calcitonin in regulating calcium metabolism in elasmobranchs is less well supported (Wendelaar Bonga and Pang, 1991).

Recent studies have indicated that calcitonin may have important functions in viviparous elasmobranch during the period of gestation. Nichols et al. (2003) reported a temporal pattern in serum calcitonin concentrations in female *S. tiburo*, in which peak levels of this hormone were observed during the yolk-dependent stage of pregnancy. Although immunoreactive calcitonin was not detected in any reproductive or major nonreproductive tissues other than its site of production in pregnant females, it was localized in both the duodenum and pancreas of developing embryos during the same reproductive stage. These findings suggest that calcitonin may be involved in digestion of yolk and overall fetal nutrition in this species. However, this action appears to be limited to early stages of development, as calcitonin was not detected in the gastrointestinal system of late-stage, placental embryos. The ultimobranchial gland of embryonic

S. tiburo does not appear to be active during yolk dependency; thus, calcitonin present in the digestive tract of these animals reflects either *in situ* production or maternal transfer through the fetal egg capsule.

A relationship between ultimobranchial gland function and the reproductive cycle also has been identified in female *D. sabina* (Gelsleichter, unpubl. data). In this species, increased activity of the ultimobranchial gland was observed in females specifically during the stage in which embryos are nourished via maternal production of uterine histotroph. Similar changes were not detected in ultimobranchial glands of male *D. sabina* collected during the same time period, suggesting a relationship between calcitonin release and female reproduction. Based on the absence of immunoreactive calcitonin in the embryonic gut, these findings do not appear to reflect a role for this hormone in fetal nutrition, as observed in *S. tiburo*. Instead, the only tissue other than the ultimobranchial gland that was found to possess calcitonin immunoreactivity was the maternal gill. Based on these findings, calcitonin may aid in regulating calcium homeostasis in the gravid female, specifically during the demanding process of uterolactation. However, the actual mode of action of calcitonin on gill function in this species remains unclear.

11.8 Conclusions and Future Directions

Although discussions concerning elasmobranch endocrinology often bemoan the amount of available data on this topic, it is clear that there is a wealth of information regarding the manner in which hormones regulate the biology of sharks and their relatives. As demonstrated in this chapter, many of these regulatory mechanisms are strikingly similar to those in advanced vertebrates, indicating that they have been highly conserved throughout evolution. Thus, in addition to providing a better understanding of how elasmobranchs function, studies on hormonal regulation in these fishes contribute significantly to that which is known regarding vertebrate endocrinology as a whole.

Although recent years have proved to be fruitful with regard to uncovering the functions of certain hormones in elasmobranch physiology, the premise that less is known about the endocrine system in elasmobranchs than in any other vertebrate group except perhaps the jawless fish still rings true. With that in mind, future studies should address deficiencies regarding major aspects of endocrine regulation in these fishes that still remain largely unclear (e.g., the role of hormones in growth, stress, and development). In addition, as illustrated by studies regarding the roles for gonadal steroids in elasmobranch steroidogenesis, there is an important need to continue and expand investigations on the distribution of hormone receptors in order to truly understand the roles of certain hormones.

Acknowledgments

The author acknowledges the following individuals for their assistance in the preparation of this chapter: Charles A. Manire, D. Michelle McComb, Stephanie Nichols, Andrew Piercy, and John Tyminski. Illustrations for this chapter were prepared by Neil Aschliman. Portions of this research were funded by Mote Marine Laboratory, National Science Foundation Grant IBN-9986070 to the author, and Environmental Protection Agency Grant R826128-01-0 to C. A. Manire.

References

Acher, R., J. Chauvet, M.T. Chauvet, and Y. Rouille. 1999. Unique evolution of neurohypophysial hormones in cartilaginous fishes: possible implications for urea-based osmoregulation. *J. Exp. Zool.* 284:475–484.

Aldman, G., A.C. Jonsson, J. Jensen, and S. Holmgren. 1989. Gastrin/CCK-like peptides in the spiny dogfish, *Squalus acanthias*; concentrations and actions in the gut. *Comp. Biochem. Physiol.* C 92:103–108.

Amemiya, Y., A. Takahashi, N. Suzuki, Y. Sasayama, and H. Kawauchi. 2000. Molecular cloning of proopiomelanocortin cDNA from an elasmobranch, the stingray, *Dasyatis akajei. Gen. Comp. Endocrinol.* 118:105–112.

Amer, S. and J.A. Brown. 1995. Glomerular actions of arginine vasotocin in the *in situ* perfused trout kidney. *Am. J. Physiol.* 269:R775–R780.

Anderson, W.G., J.M. Conlon, and N. Hazon. 1995. Characterization of the endogenous intestinal peptide that stimulates the rectal gland of *Scyliorhinus canicula*. *Am. J. Physiol.* 268:R1359–R1364.

Anderson, W.G., M.C. Cerra, A. Wells, M.L. Tierney, B. Tota, Y. Takei, and N. Hazon. 2001a. Angiotensin and angiotensin receptors in cartilaginous fishes. *Comp. Biochem. Physiol. A Mol. Integr. Physiol.* 128:31–40.

Anderson, W.G., Y. Takei, and N. Hazon. 2001b. The dipsogenic effect of the renin-angiotensin system in elasmobranch fish. *Gen. Comp. Endocrinol.* 124:300–307.

Anderson, W.G., M.F. Ali, I.E. Einarsdottir, L. Schaffer, N. Hazon, and J.M. Conlon. 2002. Purification, characterization, and biological activity of insulins from the spotted dogfish, *Scyliorhinus canicula*, and the hammerhead shark, *Sphyrna lewini*. *Gen. Comp. Endocrinol.* 126:113–122.

Andrews, P.L. and J.Z. Young. 1988. The effect of peptides on the motility of the stomach, intestine and rectum in the skate (*Raja*). *Comp. Biochem. Physiol. C* 89:343–348.

Armour, K.J., L.B. O'Toole, and N. Hazon. 1993a. The effect of dietary protein restriction on the secretory dynamics of 1α hydroxycorticosterone and urea in the dogfish *Scyliorhinus canicula*: a possible role for 1α hydroxycorticosterone in sodium retention. *J. Endocrinol.* 138:275–282.

Armour, K.J., L.B. O'Toole, and N. Hazon. 1993b. Mechanisms of ACTH- and angiotensin II-stimulated 1 alpha-hydroxycorticosterone secretion in the dogfish, *Scyliorhinus canicula*. *J. Mol. Endocrinol.* 10:235–244.

Bagnell, C.A., Q. Zhang, B. Downey and L. Ainsworth. 1993. Sources and biological actions of relaxin in pigs. *J. Reprod. Fertil. Suppl.* 48:127–138.

Bajaj, M., T.L. Blundell, J.E. Pitts, S.P. Wood, M.A. Tatnell, S. Falkmer, S.O. Emdin, L.K. Gowan, H. Crow, and C. Schwabe. 1983. Dogfish insulin. Primary structure, conformation and biological properties of an elasmobranchial insulin. *Eur. J. Biochem.* 135:535–542.

Battersby, B.J., W.J. McFarlane, and J.S. Ballantyne. 1996. Short-term effects of 3,5,3'-triiodothyronine on the intermediary metabolism of the dogfish shark *Squalus acanthias*: evidence from enzyme activities. *J. Exp. Zool.* 274:157–162.

Bautista, C.M., S. Mohan, and D.J. Baylink. 1990. Insulin-like growth factors I and II are present in the skeletal tissues of ten vertebrates. *Metabolism* 39:96–100.

Bayliss, W.M. and E.H. Starling. 1903. On the uniformity of the pancreatic mechanisms in vertebrata. *J. Physiol.* (London) 29:174–180.

Bean, J.W. 1942. Specificity in the renin-hypertensinogen reaction. *Am. J. Physiol.* 136:731–742.

Belfry, C.S. and F.B. Cowan. 1995. Peptidergic and adrenergic innervation of the lachrymal gland in the euryhaline turtle, *Malaclemys terrapin*. *J. Exp. Zool.* 273:363–375.

Bennett, H.P., P.J. Lowry, and C. McMartin. 1974. Structural studies of alpha-melanocyte-stimulating hormone and a novel beta-melanocyte-stimulating hormone from the neurointermediate lobe of the pituitary of the dogfish *Squalus acanthias*. *Biochem. J.* 141:439–444.

Berglund, M.M., P.A. Hipskind, and D.R. Gehlert. 2003. Recent developments in our understanding of the physiological role of PP-fold peptide receptor subtypes. *Exp. Biol. Med.* (Maywood) 228:217–244.

Betka, M. and G.V. Callard. 1998. Negative feedback control of the spermatogenic progression by testicular oestrogen synthesis: insights from the shark testis model. *APMIS* 106:252–258.

Bjenning, C. and S. Holmgren. 1988. Neuropeptides in the fish gut. An immunohistochemical study of evolutionary patterns. *Histochemistry* 88:155–163.

Bjenning, C., A.C. Jonsson, and S. Holmgren. 1990. Bombesin-like immunoreactive material in the gut, and the effect of bombesin on the stomach circulatory system of an elasmobranch fish, *Squalus acanthias*. *Regul. Pept.* 28:57–69.

Bjenning, C., A.P. Farrell, and S. Holmgren. 1991. Bombesin-like immunoreactivity in skates and the *in vitro* effect of bombesin on coronary vessels from the longnose skate, *Raja rhina*. *Regul. Pept.* 35:207–219.

Bjenning, C., N. Hazon, A. Balasubramaniam, S. Holmgren, and J.M. Conlon. 1993. Distribution and activity of dogfish NPY and peptide YY in the cardiovascular system of the common dogfish. *Am. J. Physiol.* 264:R1119–R1124.

Björnsson, B.T., C. Haux, L. Forlin, and L.J. Deftos. 1986. The involvement of calcitonin in the reproductive physiology of the rainbow trout. *J. Endocrinol.* 108:17–23.

Bourguignon, J.P. and T.M. Plant, Eds. 2000. *The Onset of Puberty in Perspective*. Elsevier, Amsterdam.

Bullesbach, E.E., L.K. Gowan, C. Schwabe, B.G. Steinetz, E. O'Byrne, and I.P. Callard. 1986. Isolation, purification, and the sequence of relaxin from spiny dogfish (*Squalus acanthias*). *Eur. J. Biochem.* 161:335–341.

Bullesbach, E.E., C. Schwabe, and I.P. Callard. 1987. Relaxin from an oviparous species, the skate (*Raja erinacea*). *Biochem. Biophys. Res. Commun.* 143:273–280.

Bullesbach, E.E., C. Schwabe, and E.R. Lacy. 1997. Identification of a glycosylated relaxin-like molecule from the male Atlantic stingray, *Dasyatis sabina*. *Biochemistry* 36:10735–10741.

Bullock, T.H. and R.G. Northcutt. 1984. Nervus terminalis in dogfish (*Squalus acanthias*, Elasmobranchii) carries tonic efferent impulses. *Neurosci. Lett.* 44:155–160.

Burton, M. and D.R. Idler. 1986. The cellular location of 1 alpha-hydroxycorticosterone binding protein in skate. *Gen. Comp. Endocrinol.* 64:260–266.

Butler, P.J., J.D. Metcalfe, and S.A. Ginley. 1986. Plasma catecholamines in the lesser spotted dogfish and rainbow trout at rest and during different levels of exercise. *J. Exp. Biol.* 123:409–421.

Callard, G.V. 1988. Reproductive physiology (Part B: The male), in *Physiology of Elasmobranch Fishes*. T.J. Shuttleworth, Ed., Springer-Verlag, Heidelberg, 292–317.

Callard, G.V. 1992. Autocrine and paracrine role of steroids during spermatogenesis: studies in *Squalus acanthias* and *Necturus maculosus*. *J. Exp. Zool.* 261:132–142.

Callard, I.P. and L. Klosterman. 1988. Reproductive physiology (Part A: The female), in *Physiology of Elasmobranch Fishes*. T.J. Shuttleworth, Ed., Springer-Verlag, Heidelberg, 279–292.

Callard, I.P. and T.J. Koob. 1993. Endocrine regulation of the elasmobranch reproductive tract. *J. Exp. Zool.* 266:368–377.

Callard, G.V. and P. Mak. 1985. Exclusive nuclear location of estrogen receptors in *Squalus* testis. *Proc. Natl. Acad. Sci. U.S.A.* 82:1336–1340.

Callard, G.V., J.A. Pudney, P. Mak, and J.A. Canick. 1985. Stage-dependent changes in steroidogenic enzymes and estrogen receptors during spermatogenesis in the testis of the dogfish, *Squalus acanthias*. *Endocrinology* 117:1328–1335.

Callard, I.P., L.A. Fileti, L.E. Perez, L.A. Sorbera, G. Giannoukos, L.L. Klosterman, P. Tsang, and J.A. McCracken. 1992. Role of the corpus luteum and progesterone in the evolution of vertebrate viviparity. *Am. Zool.* 32:264–275.

Callard, I.P., L.A. Fileti, and T.J. Koob. 1993. Ovarian steroid synthesis and the hormonal control of the elasmobranch reproductive tract. *Environ. Biol. Fish.* 38:175–186.

Calvin, J.L., C.H. Slater, T.G. Bolduc, A.P. Laudano, and S.A. Sower. 1993. Multiple molecular forms of gonadotropin-releasing hormone in the brain of an elasmobranch: evidence for IR-lamprey GnRH. *Peptides* 14:725–729.

Camargo, C.R., M.A. Visconti, and A.M. Castrucci. 1999. Physiological color change in the bullfrog, *Rana catesbeiana*. *J. Exp. Zool.* 283:160–169.

Carroll, R.G., D.F. Opdyke, and N.E. Keller. 1984. Vascular recovery following hemorrhage in the dogfish shark *Squalus acanthias*. *Am. J. Physiol.* 246:R825–R828.

Chevins, P.F. and J.M. Dodd. 1970. Pituitary innervation and control of colour change in the skates *Raia naevus*, *R. clavata*, *R. montagui*, and *R. radiata*. *Gen. Comp. Endocrinol.* 15:232–241.

Chiba, A. 1998. Ontogeny of serotonin-immunoreactive cells in the gut epithelium of the cloudy dogfish, *Scyliorhinus torazame*, with reference to coexistence of serotonin and neuropeptide Y. *Gen. Comp. Endocrinol.* 111:290–298.

Chiba, A. 2000. Immunohistochemical cell types in the terminal nerve ganglion of the cloudy dogfish, *Scyliorhinus torazame*, with special regard to neuropeptide Y/FMRFamide-immunoreactive cells. *Neurosci. Lett.* 286:195–198.

Chiba, A. 2001. Marked distributional difference of alpha-melanocyte-stimulating hormone (alpha-MSH)-like immunoreactivity in the brain between two elasmobranchs (*Scyliorhinus torazame* and *Etmopterus brachyurus*): an immunohistochemical study. *Gen. Comp. Endocrinol.* 122:287–295.

Chiba, A. and Y. Honma. 1992. Distribution of neuropeptide Y-like immunoreactivity in the brain and hypophysis of the cloudy dogfish, *Scyliorhinus torazame*. *Cell Tissue Res.* 268:453–461.

Chiba, A., Y. Honma, and S. Oka. 1995. Ontogenetic development of neuropeptide Y-like-immunoreactive cells in the gastroenteropancreatic endocrine system of the dogfish. *Cell Tissue Res.* 282:33–40.

Chiba, A., S. Oka, and Y. Honma. 1996. Ontogenetic changes in neuropeptide Y-like-immunoreactivity in the terminal nerve of the chum salmon and the cloudy dogfish, with special reference to colocalization with gonadotropin-releasing hormone-immunoreactivity. *Neurosci. Lett.* 213:49–52.

Chiba, A., S. Oka, and E. Saitoh. 2002. Ontogenetic changes in neuropeptide Y-immunoreactive cerebrospinal fluid-contacting neurons in the hypothalamus of the cloudy dogfish, *Scyliorhinus torazame* (Elasmobranchii). *Neurosci. Lett.* 329:301–304.

Chieffi, G. 1967. The reproductive system of elasmobranchs: developmental and endocrinological aspects, in *Sharks, Skates, and Rays.* P.W. Gilbert, R.F. Mathewson, and D.P. Rall, Eds., Johns Hopkins University Press, Baltimore, MD, 553–580.

Chipkin, S.R., J.S. Stoff, and N. Aronin. 1988. Immunohistochemical evidence for neural mediation of VIP activity in the dogfish rectal gland. *Peptides* 9:119–124.

Cimini, V., S. van Noorden, G. Giordano-Lanza, V. Nardini, G.P. McGregor, S.R. Bloom, and J.M. Polak. 1985. Neuropeptides and 5-HT immunoreactivity in the gastric nerves of the dogfish (*Scyliorhinus stellaris*). *Peptides* 6(Suppl. 3):373–377.

Cimini, V., S. van Noorden, and J.M. Polak. 1989. Co-localisation of substance P-, bombesin- and peptide histidine isoleucine (PHI)-like peptides in gut endocrine cells of the dogfish *Scyliorhinus stellaris. Anat. Embryol.* (Berlin) 179:605–614.

Cimini, V., S. van Noorden, and M. Sansone. 1992. Neuropeptide Y-like immunoreactivity in the dogfish gastroenteropancreatic tract: light and electron microscopical study. *Gen. Comp. Endocrinol.* 86:413–423.

Conlon, J.M. and L. Thim. 1986. Primary structure of insulin and a truncated C-peptide from an elasmobranchian fish, *Torpedo marmorata. Gen. Comp. Endocrinol.* 64:199–205.

Conlon, J.M., D.V. Agoston, and L. Thim. 1985. An elasmobranchian somatostatin: primary structure and tissue distribution in *Torpedo marmorata. Gen. Comp. Endocrinol.* 60:406–413.

Conlon, J.M., I.W. Henderson, and L. Thim. 1987. Gastrin-releasing peptide from the intestine of the elasmobranch fish, *Scyliorhinus canicula* (common dogfish). *Gen. Comp. Endocrinol.* 68:415–420.

Conlon, J.M., C. Bjenning, and N. Hazon. 1992. Structural characterization of neuropeptide Y from the brain of the dogfish, *Scyliorhinus canicula. Peptides* 13:493–497.

Conlon, J.M., L. Agius, K. George, M.M. Alberti, and N. Hazon. 1994a. Effects of dogfish urotensin II on lipid mobilization in the fasted dogfish, *Scyliorhinus canicula. Gen. Comp. Endocrinol.* 93:177–180.

Conlon, J.M., N. Hazon, and L. Thim. 1994b. Primary structures of peptides derived from proglucagon isolated from the pancreas of the elasmobranch fish, *Scyliorhinus canicula. Peptides* 15:163–167.

Craik, J.C. 1978. The effects of oestrogen treatment on certain plasma constituents associated with vitellogenesis in the elasmobranch *Scyliorhinus canicula* L. *Gen. Comp. Endocrinol.* 35:455–464.

Craik, J.C. 1979. Simultaneous measurement of rates of vitellogenin synthesis and plasma levels of oestradiol in an elasmobranch. *Gen. Comp. Endocrinol.* 38:264–266.

Cuevas, M.E. and G. Callard. 1992. Androgen and progesterone receptors in shark (*Squalus*) testis: characteristics and stage-related distribution. *Endocrinology* 130:2173–2182.

Cuevas, M.E., W. Miller, and G. Callard. 1992. Sulfoconjugation of steroids and the vascular pathway of communication in dogfish testis. *J. Exp. Zool.* 264:119–129.

Cuevas, M.E., K. Collins, and G.V. Callard. 1993. Stage-related changes in steroid-converting enzyme activities in *Squalus* testis: synthesis of biologically active metabolites via 3 beta-hydroxysteroid dehydrogenase/isomerase and 5 alpha-reductase. *Steroids* 58:87–94.

Dacke, C.G., B.J. Furr, J.N. Boelkins, and A.D. Kenny. 1976. Sexually related changes in plasma calcitonin levels in Japanese quail. *Comp. Biochem. Physiol.* A 55:341–344.

D'Antonio, M., M. Vallarino, D.A. Lovejoy, F. Vandesande, J.A. King, R. Pierantoni, and R.E. Peter. 1995. Nature and distribution of gonadotropin-releasing hormone (GnRH) in the brain, and GnRH and GnRH binding activity in serum of the spotted dogfish *Scyliorhinus canicula. Gen. Comp. Endocrinol.* 98:35–49.

Demski, L.S. and R.G. Northcutt. 1983. The terminal nerve: a new chemosensory system in vertebrates? *Science* 22:435–437.

Demski, L.S., R.D. Fields, T.H. Bullock, M.P. Schriebman, and H. Margolis-Nunno. 1987. The terminal nerve of sharks and rays: EM, immunocytochemical and electrophysiological studies. *Ann. N.Y. Acad. Sci.* 519:15–32.

Denning-Kendall, P.A., J.P. Sumpter, and P.J. Lowry. 1982. Peptides derived from pro-opiocortin in the pituitary gland of the dogfish, *Squalus acanthias. J. Endocrinol.* 93:381–390.

Denslow, N.D., M.C. Chow, K.J. Kroll, and L. Green. 1999. Vitellogenin as a biomarker of exposure for estrogen or estrogen mimics. *Ecotoxicology* 8:385–398.

deRoos, R. and C.C. deRoos. 1973. Elevation of plasma glucose levels by mammalian ACTH in the spiny dogfish shark (*Squalus acanthias*). *Gen. Comp. Endocrinol.* 21:403–409.

deRoos, R. and C.C. deRoos. 1978. Elevation of plasma glucose levels by catecholamines in elasmobranch fish. *Gen. Comp. Endocrinol.* 34:447–452.

deRoos, R. and C.C. deRoos. 1979. Severe insulin-induced hypoglycemia in the spiny dogfish shark (*Squalus acanthias*). *Gen. Comp. Endocrinol.* 37:186–191.

deRoos, R. and C.C. deRoos. 1992. Effects of mammalian ACTH on potential fuels and gluconeogenic substrates in the plasma of the spiny dogfish shark (*Squalus acanthias*). *Gen. Comp. Endocrinol.* 87:149–158.

deRoos, R., C.C. deRoos, C.S. Werner, and H. Werner. 1985. Plasma levels of glucose, alanine, lactate, and beta-hydroxybutyrate in the unfed spiny dogfish shark (*Squalus acanthias*) after surgery and following mammalian insulin infusion. *Gen. Comp. Endocrinol.* 58:28–43.

Dimaline, R., M.C. Thorndyke, and J. Young. 1986. Isolation and partial sequence of elasmobranch VIP. *Regul. Pept.* 14:1–10.

Dimaline, R., J. Young, D.T. Thwaites, C.M. Lee, T.J. Shuttleworth, and M.C. Thorndyke. 1987. A novel vasoactive intestinal peptide (VIP) from elasmobranch intestine has full affinity for mammalian pancreatic VIP receptors. *Biochim. Biophys. Acta* 930:97–100.

Dobson, S. and J.M. Dodd. 1977a. Endocrine control of the testis in the dogfish *Scyliorhinus canicula* L. I. Effects of partial hypophysectomy on gravimetric, hormonal and biochemical aspects of testis function. *Gen. Comp. Endocrinol.* 32:41–52.

Dobson, S. and J.M. Dodd. 1977b. Endocrine control of the testis in the dogfish *Scyliorhinus canicula* L. II. Histological and ultrastructural changes in the testis after partial hypophysectomy (ventral lobectomy). *Gen. Comp. Endocrinol.* 32:53–71.

Dobson, S. and J.M. Dodd. 1977c. The roles of temperature and photoperiod in the response of the testis of the dogfish, *Scyliorhinus canicula* L. to partial hypophysectomy (ventral lobectomy). *Gen. Comp. Endocrinol.* 32:114–115.

Dodd, J.M. 1975. The hormones of sex and reproduction and their effects in fish and lower chordates: twenty years on. *Am. Zool.* 12:325–339.

Dodd, J.M. and C.K. Goddard. 1961. Some effects of oestradiol benzoate on the reproductive ducts of the female dogfish *Scyliorhinus canciulus*. *Proc. Zool. Soc. London* 137:325–331.

Dodd, J.M., P.J. Evennett, and C.K. Goddard. 1960. Reproductive endocrinology in cyclostomes and elasmobranchs. *Symp. Zool. Soc. London* 1:77–103.

Downing, S.J. and O.D. Sherwood. 1985. The physiological role of relaxin in the pregnant rat. I. The influence of relaxin on parturition. *Endocrinology* 116:1200–1205.

DuBois, W., P. Mak, and G.V. Callard. 1989. Sertoli cell functions during spermatogenesis: the shark testis model. *Fish Physiol. Biochem.* 7:221–227.

Eberle, A., Y.-S. Chang, and R. Schwyzer. 1978. Chemical synthesis and biological activity of the dogfish (*Squalus acanthias*) α-melanotropins I and II and of related peptides. *Helv. Chim. Acta* 61:2360–2374.

Ecay, T.W. and J.D. Valentich. 1991. Chloride secretagogues stimulate inositol phosphate formation in shark rectal gland tubules cultured in suspension. *J. Cell. Physiol.* 146:407–416.

Eddy, E.M., T.F. Washburn, D.O. Bunch, E.H. Goulding, B.C. Gladen, D.B. Lubahn, and K.S. Korach. 1996. Targeted disruption of the estrogen receptor gene in male mice causes alteration of spermatogenesis and infertility. *Endocrinology* 137:4796–4805.

El-Salhy, M. 1984. Immunocytochemical investigation of the gastro-entero-pancreatic (GEP) neurohormonal peptides in the pancreas and gastrointestinal tract of the dogfish *Squalus acanthias*. *Histochemistry* 80:193–205.

Epstein, F.H., J.S. Stoff, and P. Silva. 1981. Hormonal control of secretion in shark rectal gland. *Ann. N.Y. Acad. Sci.* 372:613–625.

Epstein, F.H., J.S. Stoff, and P. Silva. 1983. Mechanism and control of hyperosmotic NaCl-rich secretion by the rectal gland of *Squalus acanthias*. *J. Exp. Biol.* 106:25–41.

Evans, D.H. and P.M. Piermarini. 2001. Contractile properties of the elasmobranch rectal gland. *J. Exp. Biol.* 204:59–67.

Faraldi, G., E. Bonini, L. Farina, and G. Tagliafierro. 1988. Distribution and ontogeny of glucagon-like cells in the gastrointestinal tract of the cartilaginous fish *Scyliorhinus stellaris* (L.). *Acta. Histochem.* 83:57–64.

Fasano, S., R. Pierantoni, S. Minucci, L. Di Matteo, M. D'Antonio, and G. Chieffi. 1989. Effects of intrastes-ticular injections of estradiol and gonadotropin-releasing hormone (GnRHA, HOE 766) on plasma androgen levels in intact and hypophysectomized *Torpedo marmorata* and *Torpedo ocellata*. *Gen. Comp. Endocrinol.* 75:349–354.

Fasano, S., M. D'Antonio, R. Pierantoni, and G. Chieffi. 1992. Plasma and follicular tissue steroid levels in the elasmobranch fish, *Torpedo marmorata*. *Gen. Comp. Endocrinol.* 85:327–333.

Fileti, L.A. and I.P. Callard. 1988. Corpus luteum function and regulation in the skate, *Raja erinacea*. *Bull. Mt. Desert Isl. Biol. Lab.* 29:129–130.

Forlano, P.M., K.P. Maruska, S.A. Sower, J.A. King, and T.C. Tricas. 2000. Differential distribution of gonadotropin-releasing hormone-immunoreactive neurons in the stingray brain: functional and evolutionary considerations. *Gen. Comp. Endocrinol.* 118:226–248.

Franklin, C.E., S. Holmgren, and G.C. Taylor. 1996. A preliminary investigation of the effects of vasoactive intestinal peptide on secretion from the lingual salt glands of *Crocodylus porosus*. *Gen. Comp. Endocrinol.* 102:74–78.

Garnier, D.H., P. Sourdaine, and B. Jegou. 1999. Seasonal variations in sex steroids and male sexual characteristics in *Scyliorhinus canicula*. *Gen. Comp. Endocrinol.* 116:281–290.

Gash, T.A. 2000. Seasonal Thyroid Activity in the Bonnethead Shark, *Sphyrna tiburo*. M.S. thesis. Texas A&M University, College Station.

Gelsleichter, J. and J.A. Musick. 1999. Effects of insulin-like growth factor-I, corticosterone, and 3,3′, 5-tri-iodo-L-thyronine on glycosaminoglycan synthesis in vertebral cartilage of the clearnose skate, *Raja eglanteria*. *J. Exp. Zool.* 284:549–556.

Gelsleichter, J., B.G. Steinetz, C.A. Manire, and C. Ange. 2003. Serum relaxin concentrations and reproduction in male bonnethead sharks, *Sphyrna tiburo*. *Gen. Comp. Endocrinol.* 132:27–34.

Gelsleichter, J., L.E.L. Rasmussen, C.A. Manire, J. Tyminski, B. Chang, and L. Lombardi-Carson. In press. Serum steroid concentrations and development of reproductive organs during puberty in male bonnethead sharks, *Sphyrna tiburo*. *Fish Physiol. Biochem.*

Gerstberger, R. 1988. Functional vasoactive intestinal polypeptide (VIP)-system in salt glands of the Pekin duck. *Cell Tissue. Res.* 252:39–48.

Gerstberger, R., H. Sann, and E. Simon. 1988. Vasoactive intestinal peptide stimulates blood flow and secretion of avian salt glands. *Am. J. Physiol.* 255:R575–R582.

Gottfried, H. and G. Chieffi. 1967. The seminal steroids of the dogfish *Scylliorhinus stellaris*. *J. Endocrinol.* 37:99–100.

Gowan, L.K., J.W. Reinig, C. Schwabe, S. Bedarkar, and T.L. Blundell. 1981. On the primary and tertiary structure of relaxin from the sand tiger shark (*Odontaspis taurus*). *FEBS Lett.* 129: 80–82.

Grant, W.C., Jr., F.J. Hendler, and P.M. Banks. 1969. Studies on the blood-sugar regulation in the little skate *Raja erinacea*. *Physiol. Zool.* 42:231–247.

Grumbach, M.M. 2000. Estrogen, bone, growth and sex: a sea change in conventional wisdom. *J. Pediatr. Endocrinol. Metab.* 13(Suppl. 6):1439–1455.

Guibbolini, M.E., I.W. Henderson, W. Mosley, and B. Lahlou. 1988. Arginine vasotocin binding to isolated branchial cells of the eel: effect of salinity. *J. Mol. Endocrinol.* 1:125–130.

Gutiérrez, J., J. Fernandez, J. Blasco, J.M. Gesse, and J. Planas. 1986. Plasma glucagon levels in different species of fish. *Gen. Comp. Endocrinol.* 63:328–333.

Gutiérrez, J., J. Fernandez, and J. Planas. 1988. Seasonal variations of insulin and some metabolites in dogfish plasma, *Scyliorhinus canicula*, L. *Gen. Comp. Endocrinol.* 70:1–8.

Hansen, D. 1975. Evidence of a gastrin-like substance in *Rhinobatis productus*. *Comp. Biochem. Physiol. C* 52:61–63.

Hayashida, T. and U.J. Lewis. 1978. Immunochemical and biological studies with antiserum to shark growth hormone. *Gen. Comp. Endocrinol.* 36:530–542.

Hayes, T.B. 1998. Sex determination and primary sex differentiation in amphibians: genetic and developmental mechanisms. *J. Exp. Zool.* 281:373–399.

Hazon, N. and I.W. Henderson. 1984. Secretory dynamics of 1-alpha-hydroxycorticosterone in the elasmobranch fish, *Scyliorhinus canicula*. *J. Endocrinol.* 103:205–211.

Hazon, N. and I.W. Henderson. 1985. Factors affecting the secretory dynamics of 1 alpha-hydroxycorticosterone in the dogfish, *Scyliorhinus canicula*. *Gen. Comp. Endocrinol.* 59:50–55.

Hazon, N., R.J. Balment, M. Perrott, and L.B. O'Toole. 1989. The renin-angiotensin system and vascular and dipsogenic regulation in elasmobranchs. *Gen. Comp. Endocrinol.* 74:230–236.

Hazon, N., M.L. Tierney, W.G. Anderson, S. MacKenzie, C. Cutler, and G. Cramb. 1997. Ion and water balance in elasmobranch fish, in *Ionic Regulation in Animals*. N. Hazon, F.B. Eddy, and G. Flik, Eds., Springer-Verlag, Heidelberg, 70–86.

Henderson, I.W., J.A. Oliver, A. McKeever, and N. Hazon. 1981. Phylogenetic aspects of the renin–angiotensin system, in *Advances in Physiological Sciences*, Vol. 2: *Advances in Animal and Comparative Physiology*. G. Pethes and V.L. Frenyo, Eds., Pergamon Press, London, 355–363.

Heupel, M.R., J.M. Whittier, and M.B. Bennett. 1999. Plasma steroid hormone profiles and reproductive biology of the epaulette shark, *Hemiscyllium ocellatum. J. Exp. Zool.* 284:586–594.

Hisaw, L. and A. Abramowitz. 1939. Physiology of reproduction in the dogfishes *Mustelus canis* and *Squalus acanthias. Rep. Woods Hole Oceanogr. Inst.* 1938:22.

Ho, S.-M., G. Wulczyn, and I.P. Callard. 1980. Induction of vitellogenin synthesis in the spiny dogfish, *Squalus acanthias. Bull. Mt. Desert Isl. Biol. Lab.* 19:37–38.

Hoffmayer, E.R. and G.R. Parsons. 2001. The physiological response to capture and handling stress in the Atlantic sharpnose shark, *Rhizoprionodon terranovae. Fish Physiol. Biochem.* 25:277–285.

Hogben, L.T. 1942. Chromatic behaviour. *Proc. R. Soc. Lond.* B 131:111–136.

Holmgren, S. and S. Nilsson. 1983. Bombesin-, gastrin/CCK-, 5-hydroxytryptamine-, neurotensin-, soma-tostatin-, and VIP-like immunoreactivity and catecholamine fluorescence in the gut of the elasmobranch, *Squalus acanthias. Cell Tissue Res.* 234:595–618.

Holmgren, S. and S. Nilsson. 1999. Digestive system, in *Sharks, Skates and Rays. The Biology of Elasmobranch Fish*. W.C. Hamlett, Ed., John Hopkins University Press, Baltimore, MD, 144–173.

Holmgren, S., M. Axelsson, and A.P. Farrell. 1992a. The effects of catecholamines, substance P, and vasoactive intestinal polypeptide on blood flow to the gut in the dogfish *Squalus acanthias. J. Exp. Biol.* 168:161–175.

Holmgren, S., M. Axelsson, and A.P. Farrell. 1992b. The effects of neuropeptide Y and bombesin on blood flow to the gut in dogfish *Squalus acanthias. Regul. Pept.* 40:169.

Holstein, A.F. 1969. Zur Frange der lokalen Steuerung der Spermatogenese beim Dornhai (*Squalus acanthias* L.). *Z. Zellforsch.* 93:265–281.

Holt, W.F. and D.R. Idler. 1975. Influence of the interrenal gland on the rectal gland of a skate. *Comp. Biochem. Physiol.* C 50:111–119.

Idler, D.R. and K.M. Kane. 1976. Interrenalectomy and Na-K-ATPase activity in the rectal gland of the skate *Raja ocellata. Gen. Comp. Endocrinol.* 28:100–102.

Idler, D.R. and K.M. Kane. 1980. Cytosol receptor glycoprotein for 1α-hydroxycorticosterone in tissues of an elasmobranch fish (*Raja ocellata*). *Gen. Comp. Endocrinol.* 42:259–266.

Idler, D.R. and B. Truscott. 1966. 1α-hydroxycorticosterone from cartilaginous fish: a new adrenal steroid in blood. *J. Fish. Res. Board Can..* 23:615–619.

Jenkins, N., J.P. Joss, and J.M. Dodd. 1980. Biochemical and autoradiographic studies on the oestradiol-concentrating cells in the diencephalon and pituitary gland of the female dogfish (*Scyliorhinus canicula* L.). *Gen. Comp. Endocrinol.* 40:211–219.

Jensen, J. 2001. Regulatory peptides and control of food intake in non-mammalian vertebrates. *Comp. Biochem. Physiol.* A 128:471–479.

Johnsen, A.H., L. Jonson, I.J. Rourke, and J.F. Rehfeld. 1997. Elasmobranchs express separate cholecystokinin and gastrin genes. *Proc. Natl. Acad. Sci. U.S.A.* 94:10221–10226.

Jonsson, A.C. 1995. Endocrine cells with gastrin/cholecystokinin-like immunoreactivity in the pancreas of the spiny dogfish, *Squalus acanthias. Regul. Pept.* 59:67–78.

Kågstrom, J., M. Axelsson, J. Jensen, A.P. Farrell, and S. Holmgren. 1996. Vasoactivity and immunoreactivity of fish tachykinins in the vascular system of the spiny dogfish. *Am. J. Physiol.* 270:R585–R593.

Karsch, F.J., G.E. Dahl, T.M. Hachigian, and L.A. Thrun. 1995. Involvement of thyroid hormones in seasonal reproduction. *J. Reprod. Fertil. Suppl.* 49:409–422.

King, J.A. and R.P. Millar. 1979. Phylogenetic and anatomical distribution of somatostatin in vertebrates. *Endocrinology* 105:1322–1329.

King, J.A. and R.P. Millar. 1980. Comparative aspects of luteinizing hormone-releasing hormone structure and function in vertebrate phylogeny. *Endocrinology* 106:707–717.

King, J.A., A.A. Steneveld, R.P. Millar, S. Fasano, G. Romano, A. Spagnuolo, L. Zanetti, and R. Pierantoni. 1992. Gonadotropin-releasing hormone in elasmobranch (electric ray, *Torpedo marmorata*) brain and plasma: chromatographic and immunological evidence for chicken GnRH II and novel molecular forms. *Peptides* 13:27–35.

Koob, T.J. and I.P. Callard. 1985. Effect of progesterone on the ovulatory cycle of the little skate *Raja erinacea*. *Bull. Mt. Desert Isl. Biol. Lab.* 25:138–139.

Koob, T.J. and I.P. Callard. 1999. Reproductive endocrinology of female elasmobranchs: lessons from the little skate (*Raja erinacea*) and spiny dogfish (*Squalus acanthias*). *J. Exp. Zool.* 284:557–574.

Koob, T.J., J.J. Laffan, B. Elger, and I.P. Callard. 1983. Effects of estradiol on the Verschlussvorrichtung of *Squalus acanthias*. *Bull. Mt. Desert Isl. Biol. Lab.* 23:67–68.

Koob, T.J., J.J. Laffan, and I.P. Callard. 1984. Effects of relaxin and insulin on reproductive tract size and early fetal loss in *Squalus acanthias*. *Biol. Reprod.* 31:231–238.

Koob, T.J., P. Tsang, and I.P. Callard. 1986. Plasma estradiol, testosterone, and progesterone levels during the ovulatory cycle of the skate (*Raja erinacea*). *Biol. Reprod.* 35:267–275.

Kwon, S., R.A. Hess, D. Bunick, J.D. Kirby, and J.M. Bahr. 1997. Estrogen receptors are present in the epididymis of the rooster. *J. Androl.* 18:378–384.

Lance, V. and I.P. Callard. 1978. Gonadotrophic activity in pituitary extracts from a elasmobranch (*Squalus acanthias* L.). *J. Endocrinol.* 78:149–150.

Lehrich, R.W., S.G. Aller, P. Webster, C.R. Marino, and J.N. Forrest, Jr. 1998. Vasoactive intestinal peptide, forskolin, and genistein increase apical CFTR trafficking in the rectal gland of the spiny dogfish, *Squalus acanthias*. Acute regulation of CFTR trafficking in an intact epithelium. *J. Clin. Invest.* 101:737–745.

Leibson, L.G. and E.M. Plisetskaia. 1972. Hormones and their role in regulating metabolism in cold-blooded vertebrates. *Usp. Fiziol. Nauk* 3:26–44.

Le Roith D., C. Bondy, S. Yakar, J.-L. Liu, and A. Butler. 2001. The somatomedin hypothesis: 2001. *Endocr. Rev.* 22:53–74.

Lewis, M. and J.M. Dodd. 1974. Proceedings: thyroid function and the ovary in the spotted dogfish, *Scyliorhinus canicula*. *J. Endocrinol.* 63: 63P.

Lipshaw, L.A., G.J. Patent, and P.P. Foa. 1972. Effects of epinephrine and norepinephrine on the hepatic lipids of the nurse shark, *Ginglymostoma cirratum*. *Horm. Metab. Res.* 4:34–38.

Liu, Q. and L.S. Demski. 1993. Clasper control in the round stingray, *Urolophus halleri*: lower sensorimotor pathways. *Environ. Biol. Fish.* 38:219–232.

Love, R.M. and B.T. Pickering. 1974. A beta-MSH in the pituitary gland of the spotted dogfish (*Scyliorhinus canicula*): isolation and structure. *Gen. Comp. Endocrinol.* 24:398–404.

Lovejoy, D.A., N.M. Sherwood, W.H. Fischer, B.C. Jackson, J.E. Rivier, and T. Lee. 1991. Primary structure of gonadotropin-releasing hormone from the brain of a holocephalan (ratfish: *Hydrolagus colliei*). *Gen. Comp. Endocrinol.* 82:152–161.

Lovejoy, D.A., W.H. Fischer, S. Ngamvongchon, A.G. Craig, C.S. Nahorniak, R.E. Peter, J.E. Rivier, and N.M. Sherwood. 1992a. Distinct sequence of gonadotropin-releasing hormone (GnRH) in dogfish brain provides insight into GnRH evolution. *Proc. Natl. Acad. Sci. U.S.A.* 89:6373–6377.

Lovejoy, D.A., W.K. Stell, and N.M. Sherwood. 1992b. Partial characterization of four forms of immunoreactive gonadotropin-releasing hormone in the brain and terminal nerve of the spiny dogfish (Elasmobranchii; *Squalus acanthias*). *Regul. Pept.* 37:39–48.

Lowry, P.J. and A. Chadwick. 1970. Purification and amino acid sequence of melanocyte-stimulating hormone from the dogfish *Squalus acanthias*. *Biochem. J.* 118:713–718.

Lowry, P.J., H.P. Bennett, and C. McMartin. 1974. The isolation and amino acid sequence of an adrenocorticotrophin from the pars distalis and a corticotrophin-like intermediate-lobe peptide from the neurointermediate lobe of the pituitary of the dogfish *Squalus acanthias*. *Biochem. J.* 141:427–437.

Lu, C.C., S.C. Tsai, S.W. Wang, C.L. Tsai, C.P. Lau, H.C. Shih, Y.H. Chen, Y.C. Chiao, C. Liaw, and P.S. Wang. 1998. Effects of ovarian steroid hormones and thyroxine on calcitonin secretion in pregnant rats. *Am. J. Physiol.* 274:E246–E252.

Lundin, K., S. Holmgren, and S. Nilsson. 1984. Peptidergic functions in the dogfish rectum. *Acta Physiol. Scand.* 121:46A.

Lundstrom, H.M. and P. Bard. 1932. Hypophysial control of cutaneous pigmentation in an elasmobranch fish. *Biol. Bull.* 62:1–9.

MacKenzie, S., C.P. Cutler, N. Hazon, and G. Cramb. 2002. The effects of dietary sodium loading on the activity and expression of Na, K-ATPase in the rectal gland of the European dogfish (*Scyliorhinus canicula*). *Comp. Biochem. Physiol. B* 131:185–200.

Mandado, M., P. Molist, R. Anadon, and J. Yanez. 2001. A DiI-tracing study of the neural connections of the pineal organ in two elasmobranchs (*Scyliorhinus canicula* and *Raja montagui*) suggests a pineal projection to the midbrain GnRH-immunoreactive nucleus. *Cell Tissue Res.* 303:391–401.

Manire, C.A. and L.E. Rasmussen. 1997. Serum concentrations of steroid hormones in the mature male bonnethead shark, *Sphyrna tiburo*. *Gen. Comp. Endocrinol.* 107:414–420.

Manire, C.A., L.E. Rasmussen, D.L. Hess, and R.E. Hueter. 1995. Serum steroid hormones and the reproductive cycle of the female bonnethead shark, *Sphyrna tiburo*. *Gen. Comp. Endocrinol.* 97:366–376.

Manire C.A., L.E. Rasmussen, and T.S. Gross. 1999. Serum steroid hormones including 11-ketotestosterone, 11-ketoandrostenedione, and dihydroprogesterone in juvenile and adult bonnethead sharks, *Sphyrna tiburo*. *J. Exp. Zool.* 284:595–603.

Marina, P., L. Annamaria, D. Barbara, R. Loredana, A. Piero, and A. Francesco. 2002. Fine structure of Leydig and Sertoli cells in the testis of immature and mature spotted ray *Torpedo marmorata*. *Mol. Reprod. Dev.* 63:192–201.

Martin, S.C. and T.J. Shuttleworth. 1994. Vasoactive intestinal peptide stimulates a cAMP-mediated Cl$^-$ current in avian salt gland cells. *Regul. Pept.* 52:205–214.

McVey, D.C., D. Rittschof, P.J. Mannon, and S.R. Vigna. 1996. Localization and characterization of neuropeptide Y/peptide YY receptors in the brain of the smooth dogfish (*Mustelis canis*). *Regul. Pept.* 61:167–173.

Mellinger, J.C.A. and M.P. Dubois. 1973. Confirmation, par l'immunofluorescence, de la fonction corticotrope du lobe rostral et de la fonction gonadotrope du lobe ventral de l'hypophyse d'un poisson cartilagineux, la torpille marbree (*Torpedo marmorata*). *C. R. Acad. Sci.* 276:1979–1981.

Metcalfe, J.D. and P.J. Butler. 1984. Changes in activity and ventilation in response to hypoxia in unrestrained, unoperated dogfish (*Scyliorhinus canicula* L.). *J. Exp. Biol.* 108:411–418.

Misao, R., J. Fujimoto, K. Niwa, S. Morishita, Y. Nakanishi, and T. Tamaya. 1997. Immunohistochemical expressions of estrogen and progesterone receptors in human epididymis at different ages — a preliminary study. *Int. J. Fertil. Womens Med.* 42:39–42.

Moeller, J.F. and M. Meredith. 1998. Increase in gonadotropin-releasing hormone (GnRH) levels in CSF after stimulation of the nervus terminalis in Atlantic stingray, *Dasyatis sabina*. *Brain Res.* 806:104–107.

Mountjoy, K.G., L.S. Robbins, M.T. Mortrud, and R.D. Cone. 1992. The cloning of a family of genes that encode the melanocortin receptors. *Science* 257:1248–1251.

Nichols, S., J. Gelsleichter, C.A. Manire, and G.M. Caillet. 2003. Calcitonin-like immunoreactivity in serum and tissues of the bonnethead shark, *Sphyrna tiburo*. *J. Exp. Zool.* 298A:150–161.

Nishimura, H., M. Oguri, M. Ogawa, H. Sokabe, and M. Imai. 1970. Absence of renin in kidneys of elasmobranchs and cyclostomes. *Am. J. Physiol.* 218:911–915.

Norris, D.O. 1997. *Vertebrate Endocrinology*, 3rd ed. Academic Press, San Diego, CA.

Nozaki, M., T. Tsukahara, and H. Kobayashi. 1984. An immunocytological study of the distribution of neuropeptides in the brain of fish. *Biomed. Res.* (Suppl.) 4:135–145.

Okamoto, K., K. Yasumura, S. Shimamura, M. Nakamura, A. Tanaka, and H. Yajima. 1979. Synthesis of the nonatriacontapeptide corresponding to the entire amino acid sequence of dogfish adrenocorticotropic hormone (*Squalus acantias*). *Chem. Pharm. Bull.* (Tokyo) 27:499–507.

Okuzawa, K. 2002. Puberty in teleosts. *Fish Physiol. Biochem.* 26:31–41.

Oliver, A.S. and S.R. Vigna. 1996. CCK-X receptors in the endothermic mako shark (*Isurus oxyrinchus*). *Gen. Comp. Endocrinol.* 102:61–73.

Olson, K.R. 1999. Rectal gland and volume homeostasis, in *Sharks, Skates and Rays. The Biology of Elasmobranch Fish*. W.C. Hamlett, Ed., John Hopkins University Press, Baltimore, MD, 329–352.

Opdyke, D.F., R.G. Carroll, and N.E. Keller. 1982. Catecholamine release and blood pressure changes induced by exercise in dogfish. *Am. J. Physiol.* 242:R306–R310.

Opdyke, D.F., J. Bullock, N.E. Keller, and K. Holmes. 1983. Dual mechanism for catecholamine secretion in the dogfish shark *Squalus acanthias*. *Am. J. Physiol.* 244:R641–R645.

O'Toole, L.B., K.J. Armour, C. Decourt, N. Hazon, B. Lahlou, and I.W. Henderson. 1990. Secretory patterns of 1 alpha-hydroxycorticosterone in the isolated perifused interrenal gland of the dogfish, *Scyliorhinus canicula*. *J. Mol. Endocrinol.* 5:55–60.

Pan, J., N.V. McFerran, C. Shaw, and D.W. Halton. 1994. *Torpedo marmorata* gut PLY and the NPY/PP family: phylogenetic and structural considerations. *Biochem. Soc. Trans.* 22:6S.

Pan, J.Z., C. Shaw, D.W. Halton, L. Thim, C.F. Johnston, and K.D. Buchanan. 1992. The primary structure of peptide Y (PY) of the spiny dogfish, *Squalus acanthias*: immunocytochemical localisation and isolation from the pancreas. *Comp. Biochem. Physiol. B* 102:1–5.

Paolucci, M. and I.P. Callard. 1998. Characterization of progesterone-binding moieties in the little skate *Raja erinacea*. *Gen. Comp. Endocrinol.* 109:106–118.

Parker, G.H. 1936. Color change in elasmobranchs. *Proc. Natl. Acad. Sci. U.S.A.* 22:55–60.

Patent, G.J. 1970. Comparison of some hormonal effects on carbohydrate metabolism in an elasmobranch (*Squalus acanthias*) and a holocephalan (*Hydrolagus colliei*). *Gen. Comp. Endocrinol.* 14:215–42.

Perez, L.E. and I.P. Callard. 1992. Identification of vitellogenin in the little skate (*Raja erinacea*). *Comp. Biochem. Physiol. B* 103:699–705.

Perez, L.E. and I.P. Callard. 1993. Regulation of hepatic vitellogenin synthesis in the little skate (*Raja erinacea*): use of a homologous enzyme-linked immunosorbent assay. *J. Exp. Zool.* 266:31–39.

Phillip, M., G. Maor, S. Assa, A. Silbergeld, and Y. Segev. 2001. Testosterone stimulates growth of tibial epiphyseal growth plate and insulin-like growth factor-1 receptor abundance in hypophysectomized and castrated rats. *Endocrine* 16:1–6.

Pierantoni, R., M. D'Antonio, and S. Fasano. 1993. Morpho-functional aspects of the hypothalamus-pituitary-gonadal axis of elasmobranch fishes. *Environ. Biol. Fish.* 38:187–196.

Plesch, F., C. Scott, D. Calhoun, S. Aller, and J. Forrest. 2000. Cloning of the GHRH-like hormone/PACAP recursor polypeptide from the brain of the spiny dogfish shark, *Squalus acanthias*. *Bull. Mt. Desert Biol. Lab.* 39:127–129.

Powell, R.C., R.P. Millar, and J.A. King. 1986. Diverse molecular forms of gonadotropin-releasing hormone in an elasmobranch and a teleost fish. *Gen. Comp. Endocrinol.* 63:77–85.

Pudney, J. and G.V. Callard. 1984a. Development of agranular reticulum in Sertoli cells of the testis of the dogfish *Squalus acanthias* during spermatogenesis. *Anat. Rec.* 209:311–321.

Pudney, J. and G.V. Callard. 1984b. Identification of Leydig-like cells in the testis of the dogfish *Squalus acanthias*. *Anat. Rec.* 209:323–330.

Pudney, J. and G.V. Callard. 1986. Sertoli cell cytoplasts in the semen of the spiny dogfish *Squalus acanthias*. *Tissue Cell.* 18:375–382.

Quérat, B., C. Tonnerre-Doncarli, F. Genies, and C. Salmon. 2001. Duality of gonadotropins in gnathostomes. *Gen. Comp. Endocrinol.* 124:308–314.

Rasmussen, L.E. and S.H. Gruber. 1993. Serum concentrations of reproductively-related circulating steroid hormones in the free-ranging lemon shark, *Negaprion brevirostris*. *Environ. Biol. Fish.* 38:167–174.

Rasmussen, L.E., D.L. Hess, and C.A. Luer. 1999. Alterations in serum steroid concentrations in the clearnose skate, *Raja eglanteria*: correlations with season and reproductive status. *J. Exp. Zool.* 284:575–585.

Reese, J.C. and I.P. Callard. 1991. Characterization of a specific estrogen receptor in the oviduct of the little skate, *Raja erinacea*. *Gen. Comp. Endocrinol.* 84:170–181.

Reina, R.D. and P.D. Cooper. 2000. Control of salt gland activity in the hatchling green sea turtle, *Chelonia mydas*. *J. Comp. Physiol. B* 170:27–35.

Reinig, J.W., L.N. Daniel, C. Schwabe, L.K. Gowan, B.G. Steinetz, and E.M. O'Byrne. 1981. Isolation and characterization of relaxin from the sand tiger shark (*Odontaspis taurus*). *Endocrinology* 109:537–543.

Scanes, C.G., S. Dobson, B.K. Follett, and J.M. Dodd. 1972. Gonadotrophic activity in the pituitary gland of the dogfish (*Scyliorhinus canicula*). *J. Endocrinol.* 54:343–344.

Schaeffer, J.G. 1921. Beiträge zur Physiologie des Farbenwechsels der Fische. *Arch. Physiol.* 188:25–42.

Sekine, Y. and R. Yui. 1981. Immunohistochemical study of the pancreatic endocrine cells of the ray, *Dasyatis akajei*. *Arch. Histol. Jpn.* 44:95–101.

Shaw, C., V.P. Whittaker, and D.V. Agoston. 1990. Characterization of gastrin-releasing peptide immunoreactivity in distinct storage particles in guinea pig myenteric and *Torpedo* electromotor neurones. *Peptides* 11:69–74.

Sheikh, S.P. 1991. Neuropeptide Y and peptide YY: major modulators of gastrointestinal blood flow and function. *Am. J. Physiol.* 261:G701–G715.

Sherwood, N.M. and D.A. Lovejoy. 1993. Gonadotropin-releasing hormone in cartilaginous fishes: structure, location, and transport. *Environ. Biol. Fish.* 38:197–208.

Sherwood, N.M. and S.A. Sower. 1985. A new family member for gonadotropin-releasing hormone. *Neuropeptides* 6:205–214.

Shimamura, S., K. Okamoto, M. Nakamura, A. Tanaka, and H. Yajima. 1978. Synthesis of the nonatriacontapeptide corresponding to the entire amino acid sequence of dogfish adrenocorticotropin. *Int. J. Pept. Protein Res.* 12:170–172.

Silva, P., J.S. Stoff, D.R. Leone, and F.H. Epstein. 1985. Mode of action of somatostatin to inhibit secretion by shark rectal gland. *Am. J. Physiol.* 249:R329–R334.

Silva, P., S. Lear, S. Reichlin, and F.H. Epstein. 1990. Somatostatin mediates bombesin inhibition of chloride secretion by rectal gland. *Am. J. Physiol.* 258:R1459–R1463.

Silva, P., F.H. Epstein, K.J. Karnaky, Jr., S. Reichlin, and J.N. Forrest, Jr. 1993. Neuropeptide Y inhibits chloride secretion in the shark rectal gland. *Am. J. Physiol.* 265: R439–R446.

Silva, P., R.J. Solomon, and F.H. Epstein. 1999. Mode of activation of salt secretion by C-type natriuretic peptide in the shark rectal gland. *Am. J. Physiol.* 277:R1725–R1732.

Simpson, T.H. and C.S. Wardle. 1967. A seasonal cycle in the testis of the spurdog, *Squalus acanthias*, and the sites of 3β-hydroxysteroid dehydrogenase activity. *J. Mar. Biol. Assoc. U.K.* 47:699–708.

Simpson, T.H., R.S. Wright, and H. Gottfried. 1963. Steroids in the semen of dogfish *Squalus acanthias*. *J. Endocrinol.* 26:489–498.

Simpson, T.H., R.S. Wright, and J. Renfrew. 1964. Steroid biosynthesis in the semen of dogfish *Squalus acanthias*. *J. Endocrinol.* 31:11–20.

Sisneros, J.A. and T.C. Tricas. 2000. Androgen-induced changes in the response dynamics of ampullary electrosensory primary afferent neurons. *J. Neurosci.* 20:8586–8595.

Snelson, F.F., Jr., L.E. Rasmussen, M.R. Johnson, and D.L. Hess. 1997. Serum concentrations of steroid hormones during reproduction in the Atlantic stingray, *Dasyatis sabina*. *Gen. Comp. Endocrinol.* 108:67–79.

Song, L., P.L. Ryan, D.G. Porter, and B.L. Coomber. 2001. Effects of relaxin on matrix remodeling enzyme activity of cultured equine ovarian stromal cells. *Anim. Reprod. Sci.* 66:239–255.

Sorbera, L.A. and I.P. Callard. 1995. Myometrium of the spiny dogfish *Squalus acanthias*: peptide and steroid regulation. *Am. J. Physiol.* 269:R389–R397.

Sourdaine, P. and D.H. Garnier. 1993. Stage-dependent modulation of Sertoli cell steroid production in dogfish (*Scyliorhinus canicula*). *J. Reprod. Fertil.* 97:133–142.

Sourdaine, P., D.H. Garnier, and B. Jegou. 1990. The adult dogfish (*Scyliorhinus canicula* L.) testis: a model to study stage-dependent changes in steroid levels during spermatogenesis. *J. Endocrinol.* 127:451–460.

Steinetz, B.G., E.M. O'Byrne, M. Butler, and L. Hickman. 1983. Hormonal regulation of the connective tissue of the symphysis pubis, in *Biology of Relaxin and Its Role in the Human*. M. Bigazzi, F. Greenwood, and F. Gasparri, Eds., Excerpta Medica, Amsterdam, 71–92.

Steinetz, B.G., C. Schwabe, I.P. Callard, and L.T. Goldsmith. 1998. Dogfish shark (*Squalus acanthias*) testes contain a relaxin. *J. Androl.* 19:110–115.

Stell, W.K. 1984. Luteinizing hormone-releasing hormone (LHRH)- and pancreatic polypeptide (PP)-immunoreactive neurons in the terminal nerve of spiny dogfish, *Squalus acanthias*. *Anat. Rec.* 208:173A–174A.

Sumpter, J.P., B.K. Follett, N. Jenkins, and J.M. Dodd. 1978a. Studies on the purification and properties of gonadotrophin from ventral lobes of the pituitary gland of the dogfish (*Scyliorhinus canicula* L.). *Gen. Comp. Endocrinol.* 36:264–274.

Sumpter, J.P., N. Jenkins, and J.M. Dodd. 1978b. Gonadotrophic hormone in the pituitary gland of the dogfish (*Scyliorhinus canicula* L.): distribution and physiological significance. *Gen. Comp. Endocrinol.* 36:275–285.

Sumpter, J.P., N. Jenkins, and J.M. Dodd. 1978c. Hormonal control of steroidogenesis in an elasmobranch fish (*Scyliorhinus canicula* L.): studies using a specific anti-gonadotrophic antibody (proceedings). *J. Endocrinol.* 79:28P–29P.

Sumpter, J.P., N. Jenkins, R.T. Duggan, and J.M. Dodd. 1980. The steroidogenic effects of pituitary extracts from a range of elasmobranch species on isolated testicular cells of quail and their neutralisation by a dogfish anti-gonadotropin. *Gen. Comp. Endocrinol.* 40:331.

Sumpter, J.P., P.A. Denning-Kendall, and P.J. Lowry. 1984. The involvement of melanotrophins in physiological colour change in the dogfish *Scyliorhinus canicula*. *Gen. Comp. Endocrinol.* 56:360–367.

Tagliafierro, G., G. Faraldi, and M. Pestarino. 1985. Interrelationships between somatostatin-like cells and other endocrine cells in the pancreas of some cartilaginous fish. *Cell. Mol. Biol.* 31:201–207.

Tagliafierro, G., L. Farina, G. Faraldi, G.G. Rossi, and M. Vacchi. 1989. Distribution of somatostatin and glucagon immunoreactive cells in the gastric mucosa of some cartilaginous fishes. *Gen. Comp. Endocrinol.* 75:1–9.

Takagi, T., N. Suzuki, Y. Sasayama, and A. Kambegawa. 1995. Plasma calcitonin levels in the stingray (cartilaginous fish) *Dasyatis akajei. Zool. Sci.* 10:134.

Takei, Y. 2000. Structural and functional evolution of the natriuretic peptide system in vertebrates. *Int. Rev. Cytol.* 194:1–66.

Takei, Y., Y. Hasegawa, T.X. Watanabe, K. Nakajima, and N. Hazon. 1993. A novel angiotensin I isolated from an elasmobranch fish. *J. Endocrinol.* 139:281–285.

Thiebold, J.J. 1953. Action du benzoate d'oestradiol sur la differenciation sexuelle des embryons de *Scylliorhinus canicula. C. Compt. Rend. Soc. Biol.* 236:2174–2175.

Thiebold, J.J. 1954. Étude preliminaire de l'action des hormones sexuelles sur la morphogenese des voies genitales chez *Scylliorhinus canicula* L. *Bull. Biol. Fr. Belg.* 88:130–145.

Thorndyke, M.C. and T.J. Shuttleworth. 1985. Biochemical and physiological studies on peptides from the elasmobranch gut. *Peptides* 6(Suppl. 3):369–372.

Tierney, M., Y. Takei, and N. Hazon. 1997. The presence of angiotensin II receptors in elasmobranchs. *Gen. Comp. Endocrinol.* 105:9–17.

Torres, P., L. Tort, J. Planas, and R. Flos. 1986. Effect of confinement stress and additional zinc treatment on some blood parameters in the dogfish *Scyliorhinus canicula. Comp. Biochem. Physiol.* C 83:89–92.

Tort, L., F. Gonzalez-Arch, and J. Balasch. 1994. Plasma glucose and lactate and hematological changes after handling stresses in the dogfish. *Rev. Esp. Fisiol.* 50: 41–46.

Tricas, T.C., S.W. Michael, and J.A. Sisneros. 1995. Electrosensory optimization to conspecific phasic signals for mating. *Neurosci. Lett.* 202:129–132.

Tricas, T.C., K.P. Maruska, and L.E. Rasmussen. 2000. Annual cycles of steroid hormone production, gonad development, and reproductive behavior in the Atlantic stingray. *Gen. Comp. Endocrinol.* 118:209–225.

Tsang, P.C. and I.P. Callard. 1987a. Luteal progesterone production and regulation in the viviparous dogfish, *Squalus acanthias. J. Exp. Zool.* 241:377–382.

Tsang, P.C. and I.P. Callard. 1987b. Morphological and endocrine correlates of the reproductive cycle of the aplacental viviparous dogfish, *Squalus acanthias. Gen. Comp. Endocrinol.* 66:182–189.

Tsang, P.C. and I.P. Callard. 1992. Regulation of ovarian steroidogenesis *in vitro* in the viviparous shark, *Squalus acanthias. J. Exp. Zool.* 261:97–104.

Uva, B., M.A. Masini, N. Hazon, L.B. O'Toole, I.W. Henderson, and P. Ghiani. 1992. Renin and angiotensin converting enzyme in elasmobranchs. *Gen. Comp. Endocrinol.* 86:407–412.

Vallarino, M. and I. Ottonello. 1987. Neuronal localization of immunoreactive adrenocorticotropin-like substance in the hypothalamus of elasmobranch fishes. *Neurosci. Lett.* 80:1–6.

Vallarino, M., J.M. Danger, A. Fasolo, G. Pelletier, S. Saint-Pierre, and H. Vaudry. 1988a. Distribution and characterization of neuropeptide Y in the brain of an elasmobranch fish. *Brain. Res.* 448:67–76.

Vallarino, M., C. Delbende, S. Jegou, and H. Vaudry. 1988b. Alpha-melanocyte-stimulating hormone (alpha-MSH) in the brain of the cartilagenous fish. Immunohistochemical localization and biochemical characterization. *Peptides* 9:899–907.

Vallarino, M., A.C. Andersen, C. Delbende, I. Ottonello, A.N. Eberle, and H. Vaudry. 1989. Melanin-concentrating hormone (MCH) immunoreactivity in the brain and pituitary of the dogfish *Scyliorhinus canicula.* Colocalization with alpha-melanocyte-stimulating hormone (alpha-MSH) in hypothalamic neurons. *Peptides* 10:375–382.

Vallarino, M., M. Feuilloley, F. Vandesande, and H. Vaudry. 1991. Immunohistochemical mapping of galanin-like immunoreactivity in the brain of the dogfish *Scyliorhinus canicula. Peptides* 12:351–357.

Vigna, S.R. 1979. Distinction between cholecystokinin-like and gastrin-like biological activities extracted from gastrointestinal tissues of some lower vertebrates. *Gen. Comp. Endocrinol.* 39:512–520.

Vigna, S.R. 1983. Evolution of endocrine regulation of gastrointestinal function in lower vertebrates. *Am. Zool.* 23:729–738.

Visconti, M.A. and A.M. Castrucci. 1993. Melanotropin receptors in the lungfish, *Lepidosiren paradoxa,* and in the cartilaginous fish, *Potamotrygon reticularis. Comp. Biochem. Physiol.* C 106:523–528.

Visconti, M.A., G.C. Ramanzini, C.R. Camargo, and A.M. Castrucci. 1999. Elasmobranch color change: a short review and novel data on hormone regulation. *J. Exp. Zool.* 284:485–491.

Volkoff, H., J.P. Wourms, E. Amesbury, and F.F. Snelson. 1999. Structure of the thyroid gland, serum thyroid hormones, and the reproductive cycle of the Atlantic stingray, *Dasyatis sabina*. *J. Exp. Zool.* 284:505–516.

Waring, H. 1936. Colour change in the dogfish (*Scyliorhinus canicula*). *Proc. Liverpool Biol. Soc.* 49:17–68.

Watson, R.R. and K.A. Dickson. 2001. Enzyme activities support the use of liver lipid-derived ketone bodies as aerobic fuels in muscle tissues of active sharks. *Physiol. Biochem. Zool.* 74:273–282.

Weiss, G. 1989. Relaxin in the male. *Biol. Reprod.* 40:197–200.

Wells, A., W.G. Anderson, and N. Hazon. 2002. Development of an *in situ* perfused kidney preparation for elasmobranch fish: action of arginine vasotocin. *Am. J. Physiol. Regul. Integr. Comp. Physiol.* 282:R1636–R1642.

Wendelaar Bonga, S.E. and P.K.T. Pang. 1991. Control of calcium regulating hormones in the vertebrates: parathyroid hormone, calcitonin, prolactin, and stanniocalcin. *Int. Rev. Cytol.* 128:139–213.

White, J. and M. Meredith. 1995. Nervus terminalis ganglion of the bonnethead shark (*Sphyrna tiburo*): evidence for cholinergic and catecholaminergic influence on two cell types distinguished by peptide immunocytochemistry. *J. Comp. Neurol.* 351:385–403.

Wilson, J.F. and J.M. Dodd. 1973a. Effects of pharmacological agents on the *in vivo* release of melanophore-stimulating hormone in the dogfish, *Scyliorhinus canicula*. *Gen. Comp. Endocrinol.* 20:556–566.

Wilson, J.F. and J.M. Dodd. 1973b. The role of melanophore-stimulating hormone in melanogenesis in the dogfish, *Scyliorhinus canicula* L. *J. Endocrinol.* 58:685–686.

Wilson, J.F. and J.M. Dodd. 1973c. The role of the pineal complex and lateral eyes in the colour change response of the dogfish, *Scyliorhinus canicula* L. *J. Endocrinol.* 58:591–598.

Wourms, J.P. 1977. Reproduction and development of chondrichthyan fishes. *Am. Zool.* 17:379–410.

Wright, D.E. and L.S. Demski. 1991. Gonadotropin hormone-releasing hormone (GnRH) immunoreactivity in the mesencephalon of sharks and rays. *J. Comp. Neurol.* 307:49–56.

Wright, D.E. and L.S. Demski. 1993. Gonadotropin-releasing hormone (GnRH) pathways and reproductive control in elasmobranchs. *Environ. Biol. Fish.* 38:209–218.

Yamaguchi, K., A. Yasuda, U.J. Lewis, Y. Yokoo, and H. Kawauchi. 1989. The complete amino acid sequence of growth hormone of an elasmobranch, the blue shark (*Prionace glauca*). *Gen. Comp. Endocrinol.* 73:252–259.

Yamamoto, K., N. Suzuki, N. Takahashi, Y. Sasayama, and S. Kikuyama. 1999. Estrogen receptors in the stingray (*Dasyatis akajei*) ultimobranchial gland. *Gen. Comp. Endocrinol.* 110:107–114.

Zaidi, M., A.M. Inzerillo, B.S. Moonga, P.J. Bevis, and C.L. Huang. 2002. Forty years of calcitonin — where are we now? A tribute to the work of Iain Macintyre, FRS. *Bone* 30:655–663.

12

Sensory Biology of Elasmobranchs

Robert E. Hueter, David A. Mann, Karen P. Maruska, Joseph A. Sisneros,
and Leo S. Demski

CONTENTS

0-8493-1514-X/04/$0.00+$1.50
© 2004 by CRC Press LLC

12.1   Introduction

Sharks have become practically legendary for their sensory abilities. Some of the recognition is deserved, and some is often exaggerated. Accounts of sharks being able to smell or hear a single fish from miles away may be fish stories, but controlled measurements of elasmobranch sensory function have revealed that these animals possess an exquisite array of sensory systems for detecting prey and conspecifics, avoiding predators and obstacles, and orienting in the sea. This sensory array provides information to a central nervous system (CNS) that includes a relatively large brain, particularly in the rays and galeomorph sharks, whose brain-to-body weight ratios are comparable to those of birds and mammals (Northcutt, 1978).

Sensory system performance can be quantified in many ways. In the end, elasmobranch biologists wish to know, "How 'good' is elasmobranch hearing ... smell ... vision?" in a given behavioral or ecological context. To answer this basic question, sensory performance can be scaled in two general ways: *sensitivity*, which involves the minimum stimulus detectable by the system; and *acuity*, which is the ability of the system to discriminate stimulus characteristics, such as its location (direction of a sound or odor, resolution of a visual image, etc.) and type (frequency of sound, odor chemical, color of light, etc.). These parameters apply to all senses in one way or another and help to make comparisons across phylogenetic lines.

This chapter reviews the anatomy, physiology, and performance of elasmobranch senses within the context of sensory ecology and behavior. Special emphasis is placed on information that has come to light since publication of Hodgson and Mathewson's 1978 volume on elasmobranch senses (Hodgson and Mathewson, 1978a). Generalizations across all elasmobranch species are difficult and unwise, for with nearly 900 extant species, and only a fraction studied for their sensory capabilities, much still remains to be discovered about the diversity of sensory system function in elasmobranchs.

12.2   Vision

"My nose is sufficiently good. My eyes are large and gray; although, in fact, they are weak to a very inconvenient degree, still no defect in this regard would be suspected from their appearance."

— Edgar Allan Poe, "The Spectacles" (1844)

Poe could have been writing about the eyes and nose of a shark, for prior to the 1960s the perception, both scholarly and popular, was that vision in sharks was poor compared with the other senses, especially olfaction. Sharks' sense of smell was thought to be so much more important than vision and other senses that sharks were commonly called "swimming noses," even though visual scientists (e.g., Walls, 1942) recognized that elasmobranch ocular anatomy was highly developed. Sensory research in the 1960s and subsequent decades began to alter our understanding of shark visual capabilities. Several comprehensive reviews can be consulted for detailed research findings on elasmobranch vision (see Gilbert, 1963; Gruber and Cohen, 1978; Hueter and Cohen, 1991). This section summarizes what is known about the visual systems of sharks, skates, and rays with an emphasis on special adaptations for elasmobranch behavior and ecology.

12.2.1 Ocular Anatomy and Optics

Elasmobranch eyes are situated laterally on the head in the case of selachians and on the dorsal surface of the head in batoids, although the more benthic sharks (e.g., orectolobids, squatinids) have more dorsally positioned eyes and the less benthic rays (e.g., myliobatids, rhinopterids, mobulids) have more laterally positioned eyes, obvious adaptations for pelagic vs. benthic habits. Eye size in elasmobranchs is generally small in relation to body size but relatively larger in juveniles and in some notable species, such as the bigeye thresher shark, *Alopias superciliosus*.

In all elasmobranchs the two eyes oppose each other, which allows for a nearly 360° visual field, especially in the case of swimming sharks utilizing a laterally sinusoidal swimming pattern. Limited eye movements are observed in some species, primarily to compensate for swimming movements and stabilize the visual field (Harris, 1965). Binocular overlap is small, and blind areas exist directly in front of the snout or behind the head when the animal is still. The sizes of these blind areas depend on the configuration of the head and the separation of the eyes, but typically the forward blind area extends less than one body length in front of the rostrum.

The ocular adnexa are well developed and more elaborate than in most teleosts, although the upper and lower eyelids in most elasmobranchs do not move appreciably or cover the entire eyeball (Gilbert, 1963). Benthic shark species such as orectolobids have more mobile lids, which serve to protect the eyes while burrowing. Some shark species, especially the carcharhinids and sphyrnids, possess a third eyelid, the nictitating membrane, which can be extended from the lower nasal corner of the eye to cover the exposed portion of the eye (Gilbert, 1963). This membrane functions to protect the eye from damaging abrasion and may be extended when the shark feeds or comes into contact with another object. It does not naturally respond to bright light, although it can be conditioned to do so (Gruber and Schneiderman, 1975). Some other sharks not equipped with a nictitating membrane, including the white shark, *Carcharodon carcharias* (Tricas and McCosker, 1984) and the whale shark, *Rhincodon typus* (Hueter, pers. obs.), use the extraocular muscles to rotate the entire eye back into the orbit to protect it from abrasion during feeding and other activities.

The outer layer of the elasmobranch eye (Figure 12.1) comprises a thick cartilaginous sclera and a gently curving, transparent cornea, the fine structure of which includes sutural fibers that resist corneal

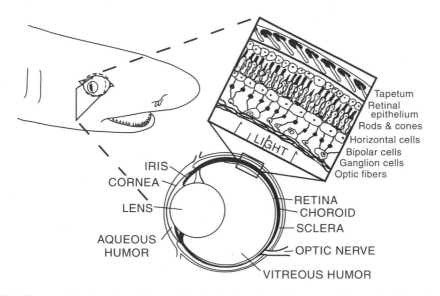

FIGURE 12.1 Cross section through a shark eye showing ocular and retinal anatomy. Tapetum lucidum shown in non-occluded state exposing reflective plates for greater visual sensitivity under scotopic conditions. (Modified from Hueter, R.E. and P.W. Gilbert. 1990. In *Discovering Sharks*. S.H. Gruber, Ed., American Littoral Society, Highlands, NJ, 48–55.)

swelling and loss of transparency in challenging chemical environments (Tolpin et al., 1969). Unlike teleosts, most elasmobranchs have a dynamic iris that can increase the size of the pupil in dim light or decrease it in bright light. Depending on species, the shape of the pupil can be circular (e.g., most deep-sea sharks, which have less mobile pupils for the more constant, low-light conditions), vertical slit (e.g., *Carcharhinus* spp., *Negaprion brevirostris*), horizontal slit (e.g., *Sphyrna tiburo*), oblique slit (e.g., *Scyliorhinus canicula, Ginglymostoma cirratum*), or crescent-shaped (e.g., many skates and rays). Mobile slit pupils are typically found in active predators with periods of activity in both photopic (bright light) and scotopic (dim light) conditions, such as the lemon shark, *N. brevirostris* (Gruber, 1967); a slit pupil that can be closed down to a pinhole is thought to be the most effective way to achieve the smallest aperture under photopic conditions, because a circular pupil is mechanically constrained from closing to a complete pinhole (Walls, 1942). In skates and rays, the combination of a U-shaped crescent pupil with multiple pupillary apertures under photopic conditions provides optical benefits including enhanced visual resolution, contrast, and focusing ability (Murphy and Howland, 1991).

The elasmobranch cornea is virtually optically absent underwater due to its similarity in refractive index to that of seawater (Hueter, 1991), leaving the crystalline lens to provide the total refractive power of the eye. Elasmobranch lenses are typically large, relatively free of optical aberration, and ellipsoidal in shape, although the spiny dogfish, *Squalus acanthias,* and clearnose skate, *Raja eglanteria,* have nearly spherical lenses (Sivak, 1978a; 1991). In the juvenile lemon shark, *N. brevirostris*, the principal power (D_p) of the lens is nearly +140 diopters (D), about seven times the optical power of the human lens (Hueter, 1991).

Some elasmobranch lenses contain yellowish pigments that are enzymatically formed oxidation products of tryptophan, similar to lens pigments found in many teleosts and diurnal terrestrial animals. These pigments filter near-ultraviolet (UV) light, which helps to minimize defocus of multiple wavelengths (chromatic aberration), enhance contrast sensitivity, and reduce light scatter and glare under conditions of bright sunlight (Zigman, 1991). They may also help to protect the retina from UV damage in shallow benthic and epipelagic species. Zigman (1991) found yellow lens pigments in coastal and surface-dwelling species such as the sandbar shark, *Carcharhinus plumbeus*, the dusky shark, *C. obscurus*, and the tiger shark, *Galeocerdo cuvier*, but interestingly not in another carcharhinid and shallow-water shark, the lemon shark, *N. brevirostris*, or in the shallow-dwelling nurse shark, *Ginglymostoma cirratum*. Both lemon and nurse sharks inhabit tropical waters where UV damage to the eye could be a problem, so the ecological correlations are unclear, and there may be other factors dictating the presence or absence of these lens filters. Nelson et al. (2003) described a related UV-filtering mechanism in the corneas of scalloped hammerhead sharks, *Sphyrna lewini*, in which the degree of UV protection by the cornea increased with duration of exposure to solar radiation.

Accommodation is the ability to change the refractive power of the eye to focus on objects at varying distances. Without accommodative ability, the focal plane of the eye is static, and in the absence of other optical adaptations, the image of any object in front of or behind that plane will be out of focus on the retina. Elasmobranchs that accommodate do not vary lens shape as humans do, but instead change the position of the lens by moving it toward the retina (for distant targets) or away from the retina (for near targets). The lens is supported dorsally by a suspensory ligament and ventrally by the pseudocampanule, a papilla with ostensibly contractile function (Sivak and Gilbert, 1976). Evidence of accommodation in elasmobranchs has been inconsistent across species, and many of the species studied have appeared to be hyperopic (far-sighted) in the resting state of the eye (Sivak, 1978b, 1991; Hueter, 1980; Hueter and Gruber, 1982; Spielman and Gruber, 1983), which is problematic.

Hueter et al. (2001), however, discovered that unrestrained, free-swimming lemon sharks (*N. brevirostris*) were not hyperopic and could accommodate, in contrast to previous findings for the same species under restraint (Hueter, 1980; Hueter and Gruber, 1982), suggesting that the hyperopia and absence of accommodation measured in many elasmobranchs under restraint is an induced, unnatural artifact resulting from handling stress. Eliminating this artifact, it is possible that most elasmobranchs would be emmetropic (neither far-sighted nor near-sighted) in the resting state and have accommodative ability. This complication aside, there is some indication that benthic elasmobranchs, such as the nurse shark *G. cirratum* and the bluntnose stingray, *Dasyatis sayi*, may have greater accommodative range than more active, mobile elasmobranchs (Sivak, 1978b). This may be attributable to the stability of the visual field in sedentary species, providing advantages for a more refined focusing mechanism, but more research into the interrelationship between vision and locomotion in elasmobranchs is needed.

At the back of the elasmobranch eye behind the retina and in front of the sclera lies the choroid, the only vascularized tissue within the adult elasmobranch eye. The elasmobranch retina itself is not vascularized and typically contains no obvious landmarks other than the optic disk (corresponding to a small blind spot in the visual field), which contains no photoreceptors and marks the exit of retinal ganglion cell fibers via the optic nerve from retina to CNS. The choroid in nearly all elasmobranchs contains a specialized reflective layer known as the tapetum lucidum, which consists of a series of parallel, platelike cells containing guanine crystals (Gilbert, 1963; Denton and Nicol, 1964). The function of this layer is to reflect back those photons that have passed through the retina and not been absorbed by the photoreceptor layer, allowing a second chance for detection of photons and thereby boosting sensitivity of the eye in dim light. The alignment of the tapetal cells provides for specular reflection; that is, photons are reflected back along the same path and are not scattered within the eye, which would blur the image.

Many elasmobranchs, furthermore, possess an occlusible tapetum, in which the reflective layer can be occluded by dark pigment granules that migrate under light-adapted conditions within tapetal melanophores to block the passage of photons (Nicol, 1964; Heath, 1991). Although there are exceptions, occlusible tapeta tend to be found in more surface-dwelling, arrhythmic species with both diurnal and nocturnal activity, which selects for visual adaptation to widely varying light levels. Non-occlusible tapeta in which the reflective layer is permanently exposed are found in sharks that inhabit the deep sea, where light levels are consistently dim (Nicol, 1964).

12.2.2 Retina and CNS

The largest impact on our understanding of visual capabilities in elasmobranchs came with the eventual finding that practically all elasmobranchs have duplex retinas containing both rod and cone photoreceptors (Gruber and Cohen, 1978), beginning with the unequivocal evidence of cones in the lemon shark (*N. brevirostris*) retina presented by Gruber et al. (1963). Cones subserve photopic and color vision and are responsible for higher visual acuity; rods subserve scotopic vision and are involved in setting the limits of visual sensitivity in the eye. Prior to 1963, elasmobranchs were thought to possess all-rod retinas, and thus were thought to have poor visual acuity and no capability for color vision, which we now know is untrue. The only elasmobranchs that appear to have no cone photoreceptors are skates (*Raja* spp.), but even their rods appear to have conelike functions under certain photic conditions (Ripps and Dowling, 1991; Dowling and Ripps, 1991).

Both rods and cones contain visual pigments that absorb photons and begin the process of vision. These pigments consist of a protein called opsin and a chromophore prosthetic group related to either vitamin A_1 or A_2, the former type called rhodopsins or chrysopsins and the latter called porphyropsins (Cohen, 1991). Rhodopsins are maximally sensitive to blue-green light, chrysopsins to deep-blue light, and porphyropsins to yellow-red light. Most elasmobranchs have been found to possess rhodopsin, which provides maximum sensitivity for clearer, shallow ocean waters associated with epipelagic environments (Cohen, 1991). Chrysopsin has been found in deep-sea squaliform sharks such as *Centrophorus, Centroscymnus,* and *Deania* (Denton and Shaw, 1963), which inhabit regions where the little available light is deep blue. Porphyropsin, which is common in freshwater teleosts and is more suited for turbid, yellowish photic conditions, is rare in elasmobranchs, even freshwater species.

However, Cohen et al. (1990) found a porphyropsin with maximum sensitivity (λ_{max}) of 522 nm (yellow-green) in the juvenile lemon shark, *N. brevirostris*, whereas adult lemon sharks have a rhodopsin with $\lambda_{max} = 501$ nm (blue-green). In this species, the visual pigment apparently changes over from a porphyropsin adapted for maximum sensitivity in inshore, shallow waters to a rhodopsin better suited for clearer, bluer oceanic waters (Figure 12.2). This visual adaptation matches a habitat shift from shallow to oceanic waters that occurs between juvenile and adult stages of the lemon shark (Cohen et al., 1990).

The density and spatial distribution of photoreceptors in the retina fundamentally affect visual acuity and sensitivity, as do the retinal interneurons (bipolar, amacrine, horizontal, ganglion cells), which transmit impulses ultimately to visual centers in the CNS. Elasmobranch retinas are rod-dominated,

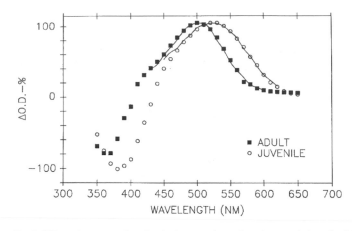

FIGURE 12.2 Normalized difference spectra for visual pigment absorption characteristics of adult vs. juvenile lemon sharks (*Negaprion brevirostris*). Peak absorption for the juvenile pigment is 522 nm whereas the adult peak is 501 nm, demonstrating a shift in this species from a more yellow-red-sensitive porphyropsin in the juvenile to a more blue-green-sensitive rhodopsin in the adult. (From Cohen, J.L. et al. 1990. *Vision Res.* 30:1949–1953. With permission.)

ranging from the skates with all-rod retinas (Dowling and Ripps, 1991), to species with apparently few cones such as *Mustelus* (Stell and Witkovsky, 1973; Sillman et al., 1996), to lamnid and carcharhinid sharks with perhaps as many as one cone for every 4 to 13 rods (Gruber et al., 1963; Gruber and Cohen, 1978). Some authors have suggested a correlation between greater rod-to-cone ratios and more scotopic habits (such as nocturnal behavior) or habitats (visually murky environments or deep-sea) in elasmobranch species. That sharks, skates, and rays have rod-dominated retinas does not inherently mean their vision is adapted primarily for low-light conditions, sensitivity to movement, and crude visual acuity; the human retina also has many more rods than cones, and our diurnal vision and acuity are among the best in the animal kingdom.

The spatial topography of retinal cells can, however, reveal much about the quality of vision in these animals. Although elasmobranchs do not have all-cone foveas, they do have retinal areas of higher cone and/or ganglion cell density, which indicate regional specializations for higher visual acuity (Hueter, 1991; Collin, 1999). Higher cone concentrations have been found in the "central" retina of the nurse shark, *Ginglymostoma cirratum* (Hamasaki and Gruber, 1965), white-spotted bamboo shark, *Chiloscyllium plagiosum* (Yew et al., 1984), and white shark, *Carcharodon carcharias* (Gruber and Cohen, 1985). Franz (1931) was the first to report horizontal streaks of higher ganglion cell density in the small-spotted catshark, *Scyliorhinus canicula,* and smooth hound, *Mustelus mustelus.*

More recently, Peterson and Rowe (1980), Hueter (1991), and Bozzano and Collin (2000) have used retinal whole-mount techniques to map the topographic distributions of retinal cells in nine elasmobranch species representing six families of sharks and skates. All of these species were found to have horizontal visual streaks of higher cell density, except for the cookie-cutter shark, *Isistius brasiliensis*, which has a specialized concentric area in the temporal retina (Bozzano and Collin, 2000). The horizontal visual streak is an adaptation for more or less two-dimensional terrain environments such as the sea bottom or sea surface, two environments commonly inhabited by many elasmobranch species, and species with prominent horizontal streaks include the horn shark, *Heterodontus francisci* (Peterson and Rowe, 1980), lemon shark, *N. brevirostris* (Hueter, 1991), small-spotted catshark, *Scyliorhinus canicula* (Bozzano and Collin, 2000), and tiger shark, *Galeocerdo cuvier* (Bozzano and Collin, 2000). The first three of these are benthically oriented species; the tiger shark feeds on prey such as birds, sea turtles, and marine mammals commonly found on or near the sea surface (Lowe et al., 1996).

Concentric retinal areas are more applicable for imaging a limited spot in the visual field or for operating in complex, three-dimensional visual environments, such as reefs. Interestingly, both the cookie-cutter shark and white shark are ambush predators in open water, and both appear to have retinal areas, not streaks. However, retinal topography in the white shark needs to be assessed more thoroughly before conclusions about this species' spatial vision can be made. In addition to habitat, locomotory style may influence the adaptiveness of visual streaks vs. areas (Hueter, 1991). A thoughtful discussion of the possible ecological and behavioral correlates with elasmobranch retinal topography has been presented by Bozzano and Collin (2000).

The elasmobranch retina projects via ganglion cell fibers in the optic nerve primarily to the mesencephalic optic tectum, but most species also possess at least ten other retinofugal targets in the brain in addition to the optic tectum, similar to the pattern in other vertebrates (Graeber and Ebbesson, 1972; Northcutt, 1979, 1991). These targets include the large elasmobranch telencephalon, once believed to be primarily an olfactory center but now thought to subserve the other senses as well, particularly for multimodal integration (Bodznick, 1991). In the lemon shark, *N. brevirostris*, the visual streak found in the cone and ganglion cell layers of the retina is preserved in the retinotectal projection to the surface of the optic tectum, where three times more tectal surface is dedicated proportionally to vision inside the streak than in the periphery of the visual field (Hueter, 1991).

A similar result was reported by Bodznick (1991) in the optic tectum of the skate *Raja erinacea*. The retinal topography of this skate is unknown but a related species (*R. bigelowi*) has a prominent visual streak (Bozzano and Collin, 2000). Bodznick (1991) furthermore found that a spatial map of electroreceptive input, aligned with the visual map, also overrepresented the animal's sensory horizon in the tectum. These findings give tantalizing insights into the coordination of multimodal sensory function in the elasmobranch brain, but much more work needs to be done in this area.

12.2.3 Visual Performance

Controlled experiments to test visual performance in sharks began in 1959 when Clark trained adult lemon sharks, *N. brevirostris,* to locate a square white target for food reward (Clark, 1959). Later, Clark (1963) trained lemon sharks to visually discriminate between a square vs. diamond and a white vs. black-and-white striped square. Parameters such as visual angle, contrast, and luminance of targets were not quantified, but the demonstration that sharks could learn certain visually mediated tasks was noteworthy at the time. Wright and Jackson (1964) and Aronson et al. (1967) added to Clark's findings with further conditioning experiments on lemon, bull (*Carcharhinus leucas*), and nurse sharks (*Ginglymostoma cirratum*), again without quantified visual parameters, but providing evidence that sharks can learn visual tasks about as quickly as teleosts (cichlids) and mammals (mice).

Rigorous methods of psychophysics, including both operant and classical conditioning techniques, were applied to the study of juvenile lemon shark vision by Gruber (reviewed in Gruber and Cohen, 1978). In a series of elegant behavioral experiments conducted over nearly two decades, Gruber elucidated many aspects of lemon shark visual performance, including brightness discrimination, dark adaptation, critical flicker fusion (CFF), and spectral (color) sensitivity. Among the many findings from this line of research were (1) lemon sharks can be trained to discriminate the brighter of two visual targets down to a 0.3 log unit difference (as opposed to a 0.2 log unit threshold in human subjects); (2) lemon sharks slowly dark-adapt to scotopic conditions over the course of about 1 h, eventually becoming more than 1 million times (6 log units) more sensitive to light than under photopic conditions (and more sensitive than dark-adapted human subjects); (3) a kink in the CFF vs. light intensity curve for the lemon shark demonstrates the rod–cone break characteristic of a duplex retina; and (4) a shift in the lemon shark's light-adapted vs. dark-adapted spectral sensitivity, also confirmed electrophysiologically by Cohen et al. (1977), provides further evidence of duplex visual function in this shark. The upshot of this work was the confirmation that sharks are capable of vision in extremely dim light and that they also are capable of color vision.

The ultimate test of whether elasmobranchs use color vision in the wild to discriminate visual targets has yet to be reported. Sharks can be attracted to bright colors, including the brilliant orange of life vests — a source of concern to the U.S. Navy, which funded many shark sensory studies in the 1960s and 1970s to understand shark behavior — but it is unclear whether the animals are visually cueing on color, brightness, or contrast. Similarly, the functional visual acuity of sharks in the wild is poorly known. Hueter (1991) calculated that the juvenile lemon shark has a theoretical resolving power of 4.5′ of arc, based on the closest separation of cones in the retina and the eye's optics. This acuity is about one ninth that of the human eye, which can resolve down to about 30″ of arc, but the prediction remains to be behaviorally tested.

The importance of vision in the daily lives of elasmobranchs certainly finds support in the complexity of their anatomical and physiological visual adaptations, many of which appear to be correlated with species behavior and ecology. Field reports of sharks appearing to use vision during the final approach to prey items are common, but controlled tests are not. One exception was a study of the Pacific angel shark, *Squatina californica,* by Fouts and Nelson (1999), in which chemical, mechanical, and electrical cues were eliminated to determine that visual stimuli released an ambush attack by these benthic sharks on nearby prey items. Based on their observations, the authors hypothesized that the angel shark visual system probably is specialized for anterodorsally directed vision. A study of retinal topography in this species would help to confirm this hypothesis. Strong (1996) tested behavioral preferences of white sharks, *Carcharodon carcharias,* approaching differently shaped visual targets. The sharks were attracted to the testing area with olfactory stimuli but they appeared to use vision as they approached the objects, which were ≥15-cm-diameter surface-borne targets to which the sharks appeared to visually orient from depths of ≥17 m. At that depth, a 15-cm target would subtend a visual angle of about 0.5°, or 30′ of arc, which is more than six times as large as the theoretical minimum separable angle of the juvenile lemon shark eye. This visual task should not be a problem for a white shark with a relatively large, cone-rich eye (Gruber and Cohen, 1985).

12.3 Hearing

Hearing in sharks is of great interest because sound in the ocean presents a directional signal that is capable of propagating over large distances. Sharks are not known to make sounds, so their hearing abilities have likely been shaped by the ambient noise (both physical and biological) in their environment. Hearing in sharks and rays has been reviewed by numerous authors (see Wisby et al., 1964; Popper and Fay, 1977; Corwin, 1981, 1989; Myrberg, 2001). These reviews provide both an excellent overview of shark hearing research and a historical perspective on the scientific approaches to studying shark hearing. The purpose of this section is to describe what is known about shark hearing with an emphasis on what remains to be learned.

12.3.1 Anatomy

12.3.1.1 Inner Ear — The inner ear of sharks, skates, and rays consists of a pair of membranous labyrinths with three semicircular canals and four sensory maculae each (Retzius, 1881; Maisey, 2001) (Figure 12.3). The semicircular canals are similar to those in other vertebrates, and are used to sense angular acceleration. They are not known to be involved in sound perception.

The saccule, lagena, and utricle are three sensory areas that are thought to be involved in both balance and sound perception. They consist of a patch of sensory hair cells on an epithelium overlain by an otoconial mass. The otoconia, made of calcium carbonate granules embedded in a mucopolysaccharide matrix, act as an inertial mass (Tester et al., 1972). As in fishes, these otolith organs are thought to be responsive to accelerations produced by a sound field, which accelerate the shark and the sensory macula relative to the otoconial mass. Some elasmobranchs, such as the spiny dogfish, *Squalus acanthias,* have been found to incorporate exogenous sand grains as a way to increase the endogenous otoconial mass (Lychakov et al., 2000).

12.3.1.2 Macula Neglecta — Sharks are unique among fishes in having a tympanic connection, the fenestra ovalis, to the posterior semicircular canal that enhances audition (Howes, 1883). The fenestra ovalis is located in the base of the parietal fossa, which makes a depression in the posterior portion of the skull. The fenestrae lead to the posterior canal ducts of the semicircular canals, each of which contains a sensory macula, the macula neglecta, that is not overlain by otoconia (Tester et al., 1972).

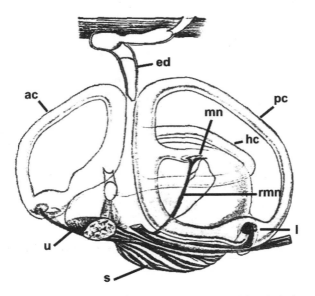

FIGURE 12.3 Anatomy of the ear of the thornback ray, *Raja clavata*. ed, endolymphatic duct; ac, anterior semicircular canal; pc, posterior semicircular canal; hc, horizontal semicircular canal; s, saccule; u, utricle; l, lagena; mn, macula neglecta; rmn, ramus of VIIIth nerve innervating macula neglecta. (Modified from Retzius, 1881.)

Elasmobranchs also have an endolymphatic duct that connects to the saccule and leads to a small opening on the dorsal surface of the shark. This connection has been hypothesized to act as a site of release of displacement waves (Tester et al., 1972), as any flow induced over the fenestrae ovalis would propagate down the posterior canal duct and into the sacculus.

Because of the specialization of the posterior canal in sharks, most hearing research has focused on the macula neglecta. The macula neglecta consists of one patch of sensory hair cells in rays, and two patches of sensory hair cells in carcharhinid sharks (Corwin, 1977, 1978). The macula neglecta lacks otoconia, but does have a crista like other hair cells in the semicircular canals. In rays, the hair cells show a variety of orientations. In carcharhinids, the hair cells are oriented in opposite directions in each sensory patch, and the orientation patterns are positioned so that fluid flows in the posterior canal would stimulate the hair cells. Variation of the structure of the macula neglecta has been hypothesized to be linked to the foraging behavior of different elasmobranchs (Corwin, 1978). However, until the function of the macula neglecta is determined, this hypothesis will be difficult to test.

The macula neglecta in rays has been shown to add hair cells continually as the fish grows (Corwin, 1983; Barber et al., 1985). Sex differences have also been found: females have been found to have more hair cells than males. The increase in hair cell number has been shown to increase vibrational sensitivity in neurons innervating the macula neglecta.

12.3.1.3 Central Pathways — As in other vertebrates, the ear of elasmobranchs is innervated by the VIIIth cranial (octaval) nerve. Studies of afferent connections and the physiology of the octaval nerve from individual end organs (saccule, lagena, utricle, and macula neglecta) show projections ipsilaterally to five primary octaval nuclei: magnocellular, descending, posterior, anterior, and periventricular. (Corwin and Northcutt, 1982; Barry, 1987). Much work remains to be done regarding both the anatomy and neurophysiology of the CNS.

12.3.2 Physiology

12.3.2.1 Audiograms — Audiograms are measures of hearing sensitivity to sounds of different frequencies. Audiograms are the most basic information that is collected about hearing systems in

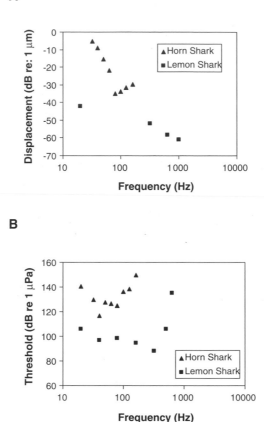

FIGURE 12.4 Elasmobranch audiograms: (A) Displacement audiograms, (B) pressure audiograms. (Redrawn from data presented in Fay, 1988; horn shark from Kelly and Nelson, 1975; lemon shark from Banner, 1967.)

animals. To date, there are only five published audiograms in elasmobranchs (summarized in Figure 12.4). Given the diversity of the group, more audiograms are warranted.

The greatest issue in measuring audiograms is what component of sound is relevant to acoustic detection in sharks. Fishes without swimbladders, such as flounders, detect the *particle displacement* component of sound. Fishes with swimbladders, especially those with connections between the swimbladder and ear like the goldfish, also detect the *pressure* component of sound. In these fishes, the swimbladder acts as a pressure-to-displacement transducer.

One way to determine the importance of particle displacement vs. pressure is to measure hearing sensitivity at different distances from a sound projector. The ratio of pressure to particle displacement changes as the distance from the sound changes. Measurements in the lemon shark, *Negaprion brevi-rostris*, and in the horn shark, *Heterodontus francisci*, show that sharks are sensitive to particle displacement rather than sound pressure at least at low frequencies (Banner, 1967; Kelly and Nelson, 1975). In both of these papers it was not clear that higher-frequency thresholds (640 Hz in Banner; 100 to 160 Hz in Kelly and Nelson) were dominated by either pressure or particle displacement sensitivity. This could be because of measurement errors in the setups or because the sharks are detecting some other measurement of the sound field, such as the pressure gradient.

Despite these issues, laboratory studies indicate that shark hearing is not as sensitive as that of some other fishes, especially those with hearing adaptations coupling a swimbladder to the inner ear. All the sharks tested show mainly low-frequency sensitivity, and there is no evidence that they are more sensitive at low frequencies than other fishes (Kritzler and Wood, 1961; Banner, 1967; Nelson, 1967; Kelly and Nelson, 1975; Casper et al., 2003).

Several papers show the importance of the macula neglecta in detecting sound and/or vibration (Lowenstein and Roberts, 1951). Fay et al. (1974) measured the response of the macula neglecta to vibrational stimuli applied to the parietal fossa. This showed that the parietal fossa is indeed in some way linked to hearing in the macula neglecta. Bullock and Corwin (1979) obtained similar results in finding that auditory evoked potentials were highest when a sound source was placed over the parietal fossa.

12.3.2.2 Pressure Sensitivity — Isolated preparations of dogfish, *Scyliorhinus canicula,* hair cells from the horizontal semicircular canals have recently been shown to respond to changes in ambient pressure (Fraser and Shelmerdine, 2002). Increased ambient pressure led to increased spike rates in response to an oscillation at 1 Hz. This result shows that sharks have a sensor that could be used to sense depth and atmospheric pressure, and recent studies by Heupel et al. (2003) demonstrate that blacktip sharks, *Carcharhinus limbatus,* behaviorally respond to decreases in atmospheric pressure associated with tropical storms. The physiological findings need to be pursued in other parts of the ear to determine whether responses to sound are modulated by pressure as well, and if shark hair cells could detect sound pressures directly. The ambient pressures tested were on the order of 200 dB re 1 μPa, which would be extremely loud for a sound.

12.3.3 Behavior

12.3.3.1 Attraction of Sharks with Sound — Several studies have shown that sharks can be attracted with low-frequency sounds in the field (Nelson and Gruber, 1963; Myrberg et al., 1969, 1972). In some of these tests, the received sound pressure levels were likely well below thresholds obtained from laboratory studies of shark hearing. This apparent disconnect between field and laboratory studies needs to be addressed. There are problems with each type of study. In the laboratory, sound fields are very complicated near-field stimuli that are rarely quantified. In the field, it is often difficult to know the distribution of sharks prior to playback and difficult to control for other stimuli, such as visual stimuli. The fact that sharks show a behavioral response to sound presentation should present a good system for testing theories about shark hearing abilities. New technology for tracking sharks should provide a means to monitor a shark's response to sound presentation in field situations.

12.3.3.2 Other Aspects of Hearing — There is more to hearing than just detection of sound. The ability to localize a sound source is just as an important as being able to hear the sound. The otolithic organs in other fishes have been shown to respond directionally to sound presentations due to the polarizations of the sensory hair cells (Lu and Popper, 2001). This is likely to be the case with sharks as well. One reason that the debate over the ability of sharks to detect sound pressure has been intense is that theoretical arguments have been made that sharks must be able to detect sound pressure to resolve a 180° ambiguity about the location of a source (see van den Berg and Schuijf, 1983; Kalmijn, 1988a). The acoustic attraction experiments show that sharks have the ability to localize a sound source, and laboratory experiments show that the lemon shark can localize a sound source to about 10° (Nelson, 1967).

There clearly needs to be more data collected about hearing sensitivity, masking by noise, frequency discrimination, intensity discrimination, and temporal sensitivity. Regardless of the actual mechanism of sound detection, data collected on these attributes of sound will be important for understanding the acoustic world of sharks.

12.4 Mechanosenses

The ability to detect water movements at multiple scales is essential in the lives of fishes. The detection of large tidal currents provides information important for orientation and navigation, and small-scale flows can reveal the location of prey, predators, and conspecifics during social behaviors. The mechanosensory lateral line system is stimulated by differential movement between the body and surrounding water, and is used by fishes to detect both dipole sources (e.g., prey) and uniform flow fields (e.g., currents). This sensory system functions to mediate behaviors such as rheotaxis (orientation to water currents), predator avoidance, hydrodynamic imaging to localize objects, prey detection, and social communication including schooling and mating (see Coombs and Montgomery, 1999, for review). In contrast to the amount of information available on lateral line morphology and function in bony fishes, relatively little is known about mechanosensory systems in elasmobranchs.

12.4.1   Peripheral Organization

The functional unit of all lateral line end organs is the mechanosensory neuromast, which is a group of sensory hair cells surrounded by support cells and covered by a gelatinous cupula (Figure 12.5A). Elasmobranch fishes have several different types of mechanosensory end organs that are classified by morphology and location: superficial neuromasts (pit organs or free neuromasts), pored and nonpored canals, spiracular organs, and vesicles of Savi. The variety of surrounding morphological structures and spatial distribution of these sensory neuromasts determine functional parameters such as response properties, receptive field area, distance range of the system, and which component of water motion (velocity or acceleration) is encoded (Denton and Gray, 1983, 1988; Münz, 1989; Kroese and Schellart, 1992).

Superficial neuromasts are distributed on the skin surface either in grooves positioned on raised papillae (skates, rays, and some sharks) or between modified placoid scales (sharks) with their cupulae directly exposed to the environment (Tester and Nelson, 1969; Peach and Marshall, 2000) (Figure 12.5B). Superficial neuromasts in the few batoids examined thus far are located in bilateral rows along the dorsal midline from the spiracle to the tip of the tail, a pair anterior to the endolymphatic pores, and a small group lateral to the eyes (Ewart and Mitchell, 1892; Maruska and Tricas, 1998; Maruska, 2001) (Figure 12.6A). In sharks, superficial neuromasts are positioned on the dorsolateral and lateral portions of the body and caudal fin (dorsolateral neuromasts), posterior to the mouth (mandibular row), between the

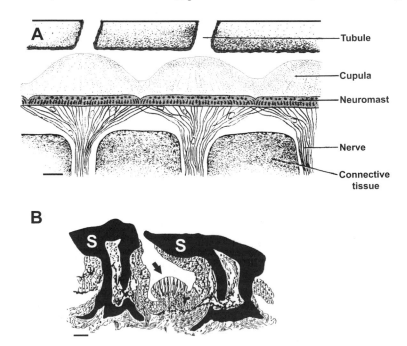

FIGURE 12.5 Morphology of the lateral line canal system and superficial neuromasts in elasmobranchs. (A) Diagrammatic longitudinal section of a pored canal from a juvenile grey reef shark, *Carcharhinus amblyrhynchos*. Innervated canal neuromasts are arranged in a nearly continuous sensory epithelium and covered by gelatinous cupulae. Pored canals are connected to the environment via tubules that terminate in openings on the skin surface. Scale bar = 150 μm. (Modified from Tester, A.L. and Kendall, J.I. *Pac. Sci.* 1969. With permission.) (B) Schematic transverse section of a single superficial neuromast (pit organ) in the nurse shark, *Ginglymostoma cirratum*. The sensory neuromast (arrow) is positioned between modified scales (S). Scale bar = 50 μm. Cupula is not shown. (Modified from Budker, 1958.)

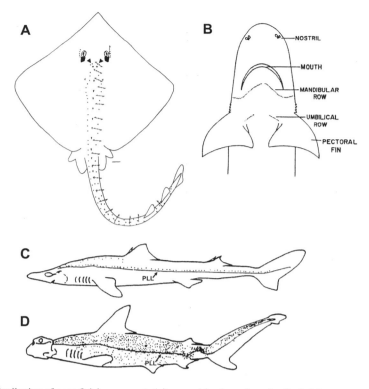

FIGURE 12.6 Distribution of superficial neuromasts (pit organs) in elasmobranchs. Each dot represents a single superficial neuromast. (A) Superficial neuromasts on the clearnose skate, *Raja eglanteria,* are located in bilateral rows along the dorsal midline to the end of the tail, a pair anterior to each endolymphatic pore (arrowheads), and a small group positioned lateral to each eye. Arrows indicate the groove orientation on every other neuromast. Scale bar = 1 cm. (B) Ventral surface of the lemon shark, *Negaprion brevirostris* (67 cm total length), shows the mandibular and umbilical rows of superficial neuromasts found on many shark species. (C) Superficial neuromasts on the spiny dogfish, *Squalus acanthias* (79 cm total length), are relatively few in number and positioned along the dorsal aspect of the posterior lateral line canal (PLL). (D) Superficial neuromasts on the scalloped hammerhead, *Sphyrna lewini* (61 cm total length), are more numerous (>600 per side) and located both dorsal and ventral to the posterior lateral line canal. (A, Modified from Maruska, K.P. 2001. *Environ. Biol. Fish.* 60, 47–75. With permission. B through D, Modified from Tester, A.L. and G.J. Nelson. 1969. In *Sharks, Skates, and Rays.* P.W. Gilbert, R.F. Mathewson, and D.P. Rall, Eds., Johns Hopkins University Press, Baltimore, 503–531. With permission.)

pectoral fins (umbilical row), and a pair anterior to each endolymphatic pore (Budker, 1958; Tester and Nelson, 1969; Peach and Marshall, 2000) (Figure 12.6B to D). However, the distribution pattern varies among taxa with one or more of the neuromast groups absent in some species. The number of superficial neuromasts ranges from less than 80 per side in the spiny dogfish, *Squalus acanthias,* to more than 600 per side in the scalloped hammerhead, *Sphyrna lewini* (Tester and Nelson, 1969) (Figure 12.6C and D). The position of the sensory epithelium within grooves or between scales differs from bony fishes and may enhance water flow parallel to the cupula to provide greater directional sensitivity. Superficial neuromasts encode the velocity of water motion and likely function to detect water movements generated by predators, conspecifics, or currents (Blaxter and Fuiman, 1989; Kroese and Schellart, 1992; Montgomery et al., 1997).

The most visible part of the mechanosensory system is the network of subepidermal fluid-filled canals distributed throughout the body. The main lateral line canals located on the head of elasmobranchs include the supraorbital, infraorbital, hyomandibular, and mandibular canals (Tester and Kendall, 1969; Boord and Campbell, 1977; Roberts, 1978; Chu and Wen, 1979; Maruska, 2001) (Figure 12.7). These canals show varying degrees of complex bifurcations on the head in sharks, or branching patterns that extend laterally onto the pectoral fins in skates and rays (Figure 12.7A). The principal canal on the remainder of the body is the posterior lateral line canal, which extends caudally from the endolymphatic pores on the dorsal surface of the head to the tip of the tail (Figure 12.7C). These lateral line canals all

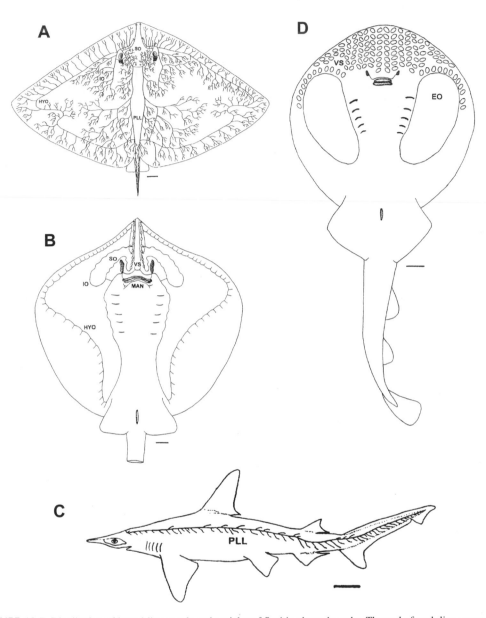

FIGURE 12.7 Distribution of lateral line canals and vesicles of Savi in elasmobranchs. The end of each line represents a pore opening on the skin surface. (A) Distribution of lateral line canals on the dorsal surface of the butterfly ray, *Gymnura micrura*. Canals are interconnected with extensive tubule branching that covers the majority of the disk surface. (B) Ventral lateral line system of the Atlantic stingray, *Dasyatis sabina*, contains pored canals along the disk margin, nonpored canals along the midline and around the mouth, and vesicles of Savi (ovals) on the rostral midline. (C) Lateral view of the posterior lateral line canal on the bonnethead shark, *Sphyrna tiburo*, which extends from the endolymphatic pores on the head to the upper lobe of the caudal fin. (D) Vesicles of Savi (ovals) on the ventral surface of the lesser electric ray, *Narcine brasiliensis*, are located in rows on the rostrum and along the anterior edge of the electric organ (EO). HYO = hyomandibular canal, IO = infraorbital canal, MAN = mandibular canal, PLL = posterior lateral line canal, SO = supraorbital canal, VS = vesicles of Savi. Scale bar = 1 cm in A, B, and D and 0.5 cm in C. (Modified from Maruska, K.P. 2001. *Environ. Biol. Fish.* 60, 47–75. With permission.)

contain between tens and thousands of neuromasts organized into an almost continuous sensory epithelium that results in multiple neuromasts between pores (Ewart and Mitchell, 1892; Johnson, 1917) (Figure 12.5A). This differs from bony fishes that have a single discrete neuromast positioned between adjacent pores, but the functional significance of this organization is unclear.

Elasmobranchs contain two different morphological classes of lateral line canals: pored and nonpored. Pored canals are in contact with the surrounding water via neuromast-free tubules that terminate in pores on the skin surface. These canals are abundant on the dorsal head of sharks and dorsal surface of batoids, where they often form complex branching patterns that increase the mechanosensory receptive field on the disk (Chu and Wen, 1979; Maruska, 2001) (Figure 12.7A). Pored canals encode water accelerations and are best positioned to detect water movements generated by prey, predators, conspecifics during social interactions or schooling, and distortions in the animal's own flow field to localize objects while swimming (Hassan, 1989; Kroese and Schellart, 1992; Montgomery et al., 1995; Coombs and Montgomery, 1999).

The presence of an extensive plexus of nonpored canals represents one of the most significant differences between teleost and elasmobranch lateral line systems. Nonpored canals are isolated from the environment and thus will not respond to pressure differences established across the skin surface. These canals are most common on the ventral surface of skates and rays, but are also found on the head of many shark species (Chu and Wen, 1979; Maruska and Tricas, 1998; Maruska, 2001). In the batoids, these nonpored canals have wide diameters, are located beneath compliant skin layers, and are concentrated along the midline, around the mouth, and on the rostrum (Maruska and Tricas, 1998; Maruska, 2001) (Figure 12.7B). These morphological characteristics indicate that nonpored canals may function as tactile receptors that encode the velocity of skin movements caused by contact with prey, the substrate, or conspecifics during social interactions (Maruska, 2001). The number and distribution of pored vs. nonpored canals differ widely among species and may be correlated with ecology and behavior, or explained by phylogeny.

Specialized mechanoreceptors in elasmobranchs are the spiracular organs and vesicles of Savi, both of which are isolated from the surrounding water. Spiracular organs are bilaterally associated with the first (spiracular) gill cleft and consist of a tube or pouch lined with sensory neuromasts and covered by a cupula (Barry and Bennett, 1989). This organ is found in both sharks and batoids, is stimulated by flexion of the cranial-hyomandibular joint, and although its biological role is unclear, morphological and physiological studies indicate it functions as a joint proprioceptor (Barry et al., 1988a,b; Barry and Bennett, 1989). Vesicles of Savi consist of neuromasts enclosed in sub-epidermal pouches, are most abundant on the ventral surface of the rostrum, and are thus far only found in some torpedinid, narcinid, and dasyatid batoids (Savi, 1844; Chu and Wen, 1979; Barry and Bennett, 1989; Maruska, 2001) (Figure 12.7B and D). Vesicular morphology differs slightly among these taxa and, although these mechanoreceptors are hypothesized to represent an obsolescent canal condition or serve as specialized touch or substrate-borne vibration receptors, their proper biological function also remains unclear (Norris, 1932; Nickel and Fuchs, 1974; Barry and Bennett, 1989; Maruska, 2001).

12.4.2   Adequate Stimulus and Processing

The necessary stimulus for the lateral line system is differential movement between the body surface and surrounding water. Because the flow amplitude of a dipole stimulus falls off rapidly with distance from the source (rate of $1/r^3$), the lateral line can only be stimulated within the inner regions of the so-called near-field (e.g., within one to two body lengths of a dipole source) (Denton and Gray, 1983; Kalmijn, 1989). Movement of the overlying cupula by viscous forces is coupled to stereocilia and kinocilia motions such that displacement of stereocilia toward the single kinocilium causes depolarization of the hair cell and an increase in the spontaneous discharge rate of the primary afferent neuron. Displacement in the opposite direction causes hyperpolarization of the hair cell and an inhibition or decrease in primary afferent firing. Thus, water motion stimuli effectively modulate the spontaneous primary afferent neuron discharges sent to the mechanosensory processing centers in the hindbrain. This modulation of neural activity from spatially distributed end organs throughout the body provides the animal with information about the frequency, intensity, and location of the stimulus source (Denton and

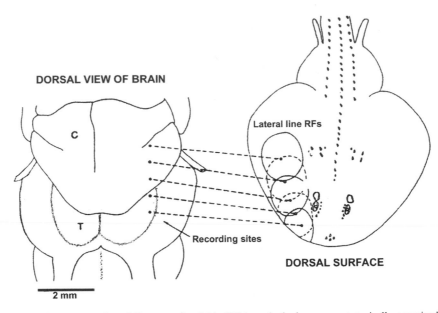

FIGURE 12.8 Mechanosensory lateral line receptive fields (RFs) on the body are somatotopically organized in a point-to-point rostrocaudal map in the midbrain of the thornback ray, *Platyrhinoidis triseriata*. Receptive fields on the anterior, mid, and posterior body are mapped onto the contralateral rostral, mid, and caudal dorsomedial nucleus of the midbrain. C = cerebellum, T = tectum. (Modified from Bleckmann, H. et al. 1987. *J. Comp. Physiol. A* 161:67–84. With permission.)

Gray, 1988; Kalmijn, 1989; Bleckmann et al., 1989). In general, neuromasts are sensitive to low-frequency stimuli (≤200 Hz), and neurophysiology studies indicate the lateral line system is sensitive to velocities in the μm s^{-1} range and accelerations in the mm s^{-2} range (Münz, 1985; Bleckmann et al., 1989; Coombs and Janssen, 1990).

Lateral line neuromasts are innervated by a distinct set of nerves separate from the traditional 11 to 12 cranial nerves described in most vertebrates (Northcutt, 1989a). The cephalic region of elasmobranchs is innervated by the ventral root of the anterior lateral line nerve complex and the body and tail by the posterior lateral line nerve complex (Koester, 1983). Both complexes contain efferents as well as afferent axons that enter the brain and terminate somatotopically within octavolateralis nuclei of the hindbrain (Bodznick and Northcutt, 1980; Koester, 1983; Bleckmann et al., 1987; Puzdrowski and Leonard, 1993). Ascending lateral line pathways continue to the lateral mesencephalic nucleus and tectum in the midbrain and to the thalamic and pallial nuclei in the forebrain (Bleckmann et al., 1987; Boord and Montgomery, 1989). Bleckmann et al. (1987) also demonstrated that mechanosensory receptive fields are somatotopically organized in a point-to-point rostrocaudal body map within the midbrain of the thornback ray (Figure 12.8). Further neurophysiological studies show bimodal and multimodal neurons within midbrain and forebrain centers that respond to hydrodynamic flow as well as to auditory, or visual, or electrosensory stimuli (Bleckmann and Bullock, 1989; Bleckmann et al., 1989). Thus, these processing regions can integrate information from several sensory systems to help mediate appropriate behavioral responses to complex biological stimuli.

12.4.3   Behavior

Among bony fishes, the lateral line system is known to function in schooling behavior, social communication, hydrodynamic imaging, predator avoidance, rheotaxis, and prey detection. However, behavioral experiments to demonstrate these lateral line–mediated behaviors in elasmobranch species are available only for prey detection and rheotaxis.

The best-known behavioral use of the lateral line system is in prey detection. The concentration of mechanoreceptors on the cephalic region of sharks and ventral surface of batoids, as well as the low-frequency, close range of the system, indicates an important role in the detection, localization, and capture of prey. Swimming and feeding movements of invertebrates and vortex trails behind swimming fish can produce water movements within the frequency and sensitivity range of the lateral line system (Montgomery et al., 1995). Montgomery and Skipworth (1997) showed that the ventral lateral line canal system of the short-tailed stingray, *Dasyatis brevicaudata*, could detect small transient water flows similar to those produced by the bivalves found in their diet. Furthermore, based on the peripheral morphology of the lateral line system and feeding behavior of the Atlantic stingray, *D. sabina*, Maruska and Tricas (1998) hypothesized that the nonpored canals on the ventral surface of the ray function as specialized tactile receptors that encode the velocity of skin movements caused by contact with small benthic prey. Neurophysiology experiments also demonstrate that touching the skin near the nonpored canals causes a transient stimulation of the neuromasts (Sand, 1937), which supports the hypothesized mechanotactile function. While prey detection is mediated by the integration of multiple sensory inputs (i.e., electroreception, olfaction, vision), the mechanosensory lateral line likely plays an important role in feeding behavior across elasmobranch taxa.

Recent evidence in sharks demonstrates that superficial neuromasts provide sensory information for rheotaxis, similar to that found in teleosts (Montgomery et al., 1997). Resting Port Jackson sharks, *Heterodontus portjacksoni,* with their dorsolateral superficial neuromasts (pit organs) ablated show a reduced ability to orient upstream in a flume when compared to intact individuals (Peach, 2001). Positive rheotaxis in sharks, skates, and rays may be important for species-specific behaviors and is hypothesized to facilitate water flow over the gills, to help maintain position on the substratum, to help orient to tidal currents, and to facilitate prey detection by enabling the animal to remain within an odor plume (see Peach, 2001).

The structure and function of the elasmobranch mechanosensory system are ripe for future study. For example, the variety of morphological specializations (e.g., nonpored canals, vesicles of Savi) found in elasmobranchs requires quantitative examinations of response properties among receptor types. Comparisons of specific mechanoreceptor distributions on the body are needed across elasmobranch taxa to test hypotheses on whether species-specific distributions have some ecological significance and represent specializations driven by evolutionary selective pressures. In addition, direct behavioral studies are sorely needed to clarify the many putative functions of the mechanosensory system in elasmobranch fishes such as schooling, object localization, predator avoidance, and social communication.

12.5 Electrosenses

All elasmobranch fishes possess an elaborate ampullary electroreceptor system that is exquisitely sensitive to low-frequency electric stimuli (see review by Bodznick and Boord, 1986; also see Montgomery, 1984; New, 1990; Tricas and New, 1998). The ampullary electroreceptor system consists of subdermal groups of electroreceptive units known as the ampullae of Lorenzini, which can detect weak extrinsic electric stimuli at intensities as low as 5 nV/cm (Kalmijn, 1982). The ampullae of Lorenzini were first recognized and described long ago by Stenonis (1664) and Lorenzini (1678), but their physiological and behavioral functions remained unknown for almost another three centuries. Initially, the ampullae of Lorenzini were thought to be mechanoreceptors (Parker, 1909; Dotterweich, 1932), but were then later shown to be also temperature sensitive (Sand, 1938; Hensel, 1955). A mechanoreceptive function was again proposed later (Murray, 1957, 1960a; Loewenstein, 1960) along with a proposed function as detectors for changes in salinity (Loewenstein and Ishiko, 1962) before current ideas about their use in electroreception were generally accepted. Murray (1960b) followed by Dijkgraaf and Kalmijn (1962) were the first to demonstrate the electrosensitivity of the ampullae of Lorenzini. Recently, the temperature sensitivity of ampullae was reconfirmed by Brown (2003), who demonstrated that the extracellular gel from the ampullae develops significant voltages in response to very small temperature gradients. Thus, temperature can be translated into electrical information by elasmobranchs without the need of cold-sensitive ion channels as used by mammals (Reid and Flonta, 2001, Viana et al., 2002). The extremely sensitive ampullary electroreceptor system of elasmobranchs is now known to mediate orientation to local inanimate electric fields (Kalmijn, 1974, 1982; Pals et al., 1982b), theorized to function in geomagnetic navigation (Kalmijn, 1974, 1988b, 2000; Paulin, 1995), and is known to be important for the detection of the bioelectric fields produced by prey (Kalmijn, 1971, 1982; Tricas, 1982; Blonder and Alevizon, 1988), potential predators (Sisneros et al., 1998), and conspecifics during social interactions (Tricas et al., 1995).

12.5.1   Anatomy

12.5.1.1   *Ampullae of Lorenzini* — Single ampullae of Lorenzini consist of a small chamber (the ampulla) and a subdermal canal about 1 mm wide that projects to the surface of the skin (Figure 12.9A) (Waltman, 1966). Small bulbous pouches known as alveoli form the ampulla chamber. Within each alveolus, hundreds of sensory hair-cell receptors and pyramidal support cells line the alveoli wall with only the apical surface of the sensory receptors and support cells exposed to the internal lumen of the ampulla chamber. Tight junctions unite the support cells and sensory receptors to create a high-resistance electrical barrier between the basal and apical surfaces of the sensory epithelium, which form the ampulla wall (Waltman, 1966; Sejnowski and Yodlowski, 1982). The basal surface of the sensory receptor cell is innervated by 5 to 12 primary afferents of the VIIIth cranial nerve with no efferents present (Kantner et al., 1962). The wall of the canal consists of a double layer of connective tissue fibers and squamous epithelial cells that are tightly joined together to form a high electrical resistance (6 MΩ-cm) between the outer and inner surface of the canal wall. In contrast, the canal and ampulla are filled with a high-potassium, low-resistance gel (25 to 31 Ω-cm) composed of mucopolysaccharides (Doyle, 1963) that form an electrical core conductor with a resistance equaling that of seawater, such that the ampullary chamber becomes isopotential with a charge at the skin pore (Murray and Potts, 1961; Waltman, 1966).

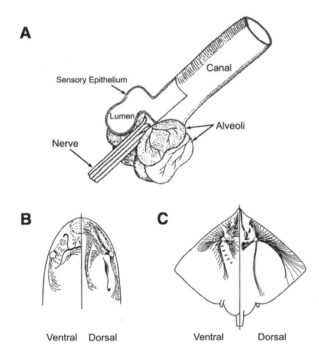

FIGURE 12.9 Ampullary electroreceptor organ of elasmobranchs. (A) The ampulla of Lorenzini consists of a small ampulla chamber composed of multiple alveoli that share a common lumen and a subdermal ampullary canal that projects to a pore on the surface of the skin. The sensory epithelium forms a high resistance ampulla wall composed of a single layer of sensory receptor cells and support cells. The basal surface of the sensory receptor cells is innervated by primary afferents of the VIIIth cranial nerve. (Modified from Waltman, B. 1966. *Acta Physiol. Scand.* 66(Suppl. 264):1–60. With permission.) (B) Diagrammatic representation of the horizontal distribution of the subdermal ampullary clusters and their radial canals that terminate at surface pores on the ventral and dorsal surfaces of the cat shark, *Scyliorhinus canicula.* (Modified from Dijkgraaf, S. and A.J. Kalmijn. 1963. *Z. Vergl. Physiol.* With permission.) (C) Horizontal distribution of the ampullae of Lorenzini in the skate, *Raja clavata.* (Modified from Murray, R.W. 1960. *J. Exp. Biol.* 37:417–424. With permission.)

In marine elasmobranchs, many individual ampullae are grouped into discrete, bilateral cephalic clusters from which project the subdermal canals that radiate in many directions and terminate at individual skin pores on the head of sharks (Figure 12.9B) and the head and pectoral fins of skates and rays (Figure 12.9C). The ampullary clusters, which usually vary in number (three to six per side of animal) and location depending on species, are innervated by different branches of the anterior lateral line nerve (VIII) (Norris, 1929). The special arrangement of the contiguously grouped ampullae within the cluster creates a common internal potential near the basal region of the sensory receptors within each cluster. The sensory receptor cells within individual ampullae detect potential differences between the animal's common internal potential at the ampullary cluster and seawater at the surface pore of the skin, which is isopotential with the subdermal canal and internal lumen of the ampulla (Bennett, 1971). In effect, electroreceptors measure the voltage drop of the electric field gradient along the length of the ampullary canal. Thus, ampullae with long canals sample across a greater distance within a uniform field, provide a larger potential difference for the sensory receptors, and thus have a greater sensitivity than do ampullae with short canals (Broun et al., 1979; Sisneros and Tricas, 2000). The morphological arrangement of the ampullary canals and clusters permits detection of both small local fields produced by small prey organisms and also the uniform electric fields of inanimate origins for possible use in orientation and navigation (Kalmijn, 1974; Tricas, 2001).

In contrast to marine species, freshwater elasmobranchs have a very different morphology and organization of the ampullary electroreceptors that are thought to reflect sensory adaptations to the highly resistive environment of freshwater (Kalmijn, 1974, 1982, 1988b; Raschi and Mackanos, 1989). One such adaptation is a thicker epidermis that functions to increase transcutaneous electrical resistance. In addition, the size of the ampullary electroreceptors in freshwater elasmobranchs is greatly reduced, and thus the ampullae are referred to as microampullae or miniampullae. Furthermore, the ampullary electroreceptors are distributed individually, rather than in clusters, over the head and pectoral fins and have very short subdermal canals (~0.3 to 2.1 mm long) that extend to the surface pores on the skin.

12.5.1.2 Central Pathways — The ampullae of Lorenzini are innervated by primary afferent neurons that convey sensory information to the brain via the dorsal root projections of the anterior lateral line (VIII). The electrosensory primary afferents from ipsilateral ampullae terminate in a somatotopic order within the central zone of the dorsal octavolateralis nucleus (DON), the first-order hindbrain electrosensory nucleus (Bodznick and Northcutt, 1980; Koester, 1983; Bodznick and Schmidt, 1984). The large electrosensory multipolar principal cells in the DON known as ascending efferent neurons (AENs) receive afferent input from the dorsal granular ridge and both the peripheral and central zones of the DON. AENs ascend to the midbrain via a lateral line lemniscus and terminate in somatotopic order in a part of the contralateral midbrain known as lateral mesencephalic nucleus (LMN) and in deep layers of the tectum (Bodznick and Boord, 1986). The LMN is one of the three elasmobranch midbrain nuclei that compose the lateral mesencephalic nuclear complex (Boord and Northcutt, 1982), which is a midbrain region considered to be homologous to the torus semicircularis in electrosensory teleost fishes (Platt et al., 1974; Northcutt, 1978). Electrosensory information processed in the LMN is sent to the posterior lateral nucleus of the thalamus, where it is then relayed to the medial pallium of the forebrain (Bullock, 1979; Bodznick and Northcutt, 1984; Schweitzer and Lowe, 1984). Some electrosensory information is also conveyed to the cerebellum (Tong and Bullock, 1982; Fiebig, 1988).

12.5.2 Physiology

12.5.2.1 Peripheral Physiology — Electrosensory primary afferent neurons that innervate the ampullae of Lorenzini exhibit a regular pattern of discharge activity in the absence of electrical stimulation. Average resting discharge rates of electrosensory afferents in batoid elasmobranchs range from 8.6 impulses/s at 7°C in the little skate, *Raja erinacea* (New, 1990), to 18.0 impulses/s at 16 to 18°C in the thornback guitarfish, *Platyrhinoidis triseriata* (Montgomery, 1984), 34.2 impulses/s at 18°C in the round stingray, *Urolophus halleri* (Tricas and New, 1998), 44.9 impulses/s at 20°C in the clearnose skate, *R. eglanteria* (Sisneros et al., 1998), and 52.1 impulses/s at 21 to 23°C in the Atlantic stingray,

Dasyatis sabina (Sisneros and Tricas, 2002a). These differences in resting discharge rates among batoids are most likely due to the influence of temperature, which in the case of higher temperatures can decrease the thresholds required for membrane depolarization of the sensory receptors and spike initiation of the electrosensory primary afferents (Carpenter, 1981; Montgomery and MacDonald, 1990). Resting discharge rates and discharge regularity of the electrosensory afferents are influenced by the animal's age. Both the rate and discharge regularity of electrosensory afferents increase during development from the neonate to the adult elasmobranch (Sisneros et al., 1998; Sisneros and Tricas, 2002a). The resting discharge rate and pattern of the electrosensory afferents are important determinants of the sensitivity and low-frequency information encoding of the electric sense (Stein, 1967; Ratnam and Nelson, 2000; Sisneros and Tricas, 2002a).

The resting discharge patterns of the electrosensory primary afferent neurons in all elasmobranch fishes are modulated by extrinsic electric fields as a function of stimulus polarity and intensity. Presentation of a cathodal (negative) stimulus at the ampullary pore increases the neural discharge activity of electrosensory afferents while an anodal (positive) stimulus decreases discharge activity (Murray, 1962, 1965). Stimulation of the electroreceptors with a sinusoidal electric field modulates the neural discharges of electrosensory afferents as a linear function of the stimulus intensity over the dynamic range of the peripheral electrosensory system, which is from 20 nV/cm to 25 μV/cm (Murray, 1965; Montgomery, 1984; Tricas and New, 1998). Electrosensory afferents are most responsive to electric fields oriented parallel to the vector between ampullary canal opening on the skin surface and the respective ampulla. Within the intensity range of natural biologically relevant electric fields, electroreceptors are broadly tuned to low-frequency electric stimuli and respond maximally to sinusoidal stimuli from approximately 0.1 to 15 Hz (Andrianov et al., 1984; Montgomery, 1984; Peters and Evers, 1985; New, 1990; Tricas et al., 1995; Tricas and New, 1998; Sisneros et al., 1998; Sisneros and Tricas, 2000). Sensitivity of the electrosensory afferents to a sinusoidal uniform electric field is 0.9 spikes/s per μV/cm for the little skate, *R. erinacea* (Montgomery and Bodznick, 1993), 4 spikes/s per μV/cm for thornback guitarfish, *P. triseriata* (Montgomery, 1984), 7.4 spikes/s per μV/cm average for the Atlantic stingray, *D. sabina* (Sisneros and Tricas, 2000, 2002a), 17.7 spikes/s per μV/cm average for the clearnose skate, *R. eglanteria* (Sisneros et al., 1998), and 24 spikes/s per μV/cm average for the round stingray, *U. halleri* (Tricas and New, 1998).

12.5.2.2   Central Physiology

— Although neurophysiological studies of the central electrosensory system in elasmobranchs are not very extensive, several features of electrosensory processing in the hindbrain and midbrain, and to a lesser extent in the thalamus and forebrain, are known. The principal cells of the DON known as AENs exhibit lower resting discharge rates and are more phasic in response than primary afferent neurons found in the peripheral electrosensory system (Bodznick and Schmidt, 1984; New, 1990). The resting discharge rates of AENs range from 0 to 5 spikes/s in the little skate, *R. erinacea* (Bodznick and Schmidt, 1984; New, 1990), to an average of 10 spikes/s in the thornback guitarfish, *P. triseriata* (Montgomery, 1984). However, AENs are similar to electrosensory primary afferents in that they are excited by cathodal stimuli and inhibited by anodal stimuli (New, 1990). Sensitivity to sinusoidal uniform electric fields is higher for second-order AENs than the primary afferent neurons. The sensitivity of AENs ranges from 2.2 spikes/sc per μV/cm for *R. erinacea* (Conley and Bodznick, 1994) to 32 spikes/s per μV/cm for *P. triseriata* (Montgomery, 1984). The increased gain of AENs is most likely due to the convergent input of multiple electrosensory primary afferents onto AENs, which have excitatory receptive fields that comprise two to five adjacent ampullary electroreceptor pores (Bodznick and Schmidt, 1984). AENs are also similar to electrosensory primary afferents in their frequency response with a maximum response in the range 0.5 to 10 Hz, followed by a sharp cutoff frequency between 10 and 15 Hz (Andrianov et al., 1984; Montgomery, 1984; New, 1990; Tricas and New, 1998).

One important function of the second-order AENs is to filter out unwanted noise or reafference created by the animal's own movements, which could interfere with the detection of biologically relevant signals (Montgomery and Bodznick, 1994). Electrosensory AENs show a greatly reduced response to sensory reafference that is essentially similar or common mode across all electrosensory primary afferents. An adaptive filter model was proposed by Montgomery and Bodznick (1994) to account for the ability of electrosensory AENs to suppress common mode reafference. The suppression of common mode signals by AENs is mediated by the balanced excitatory and inhibitory components of their spatial receptive fields (Bodznick and Montgomery, 1992; Bodznick et al., 1992; Montgomery and Bodznick, 1993).

The response properties of the central electrosensory system have also been studied in the midbrain of elasmobranchs. The midbrain electrosensory neurons of *P. triseriata* are usually "silent" and exhibit no resting discharge activity (Schweitzer, 1986). Midbrain unit thresholds range from less than 0.3 μV/cm, the lowest intensity tested, to 5 μV/cm in *P. triseriata* (Schweitzer, 1986) to even lower thresholds of 0.015 μV/cm measured with evoked potentials in the blacktip reef shark, *Carcharhinus melanopterus* (Bullock, 1979). Midbrain neurons respond maximally to frequency stimuli from 0.2 Hz (lowest frequency tested) to 4 Hz in *P. triseriata*, 10 to 15 Hz in the freshwater stingray, *Potamotrygon* sp., and at higher frequencies from 20 to 30 Hz in the blacktip reef shark, *C. melanopterus* (Bullock, 1979; Schweitzer, 1986). Such discrepancies in frequency sensitivity may be due to differences in methodology or to variation among species. Electrosensory neurons in the LMN of the midbrain may have small, well-defined minimum excitatory receptive fields that include 2 to 20 ampullary pores in *Platyrhinoidis triseriata* (Schweitzer, 1986) and 4 to 8 ampullary pores in the thorny skate, *R. radiata* (Andrianov et al., 1984). Electroreceptive fields are somatotopically mapped in the midbrain such that the anterior, middle, and posterior body surfaces are represented in the rostral, middle, and caudal levels of the contralateral midbrain. Like electrosensory primary afferents and AENs, the electrosensory midbrain neurons are also sensitive to the orientation of uniform electric fields with maximal response corresponding to the vector parallel to the length of the ampullary canal.

Neurophysiological recordings of electrosensory processing areas in the thalamus and forebrain have been limited at best. Multiunit and evoked potential recordings have localized electrosensory activity in the lateral posterior nucleus of the thalamus in *R. erinacea* (Bodznick and Northcutt, 1984) and in *P. triseriata* (Schweitzer, 1983). Bodznick and Northcutt (1984) also recorded electrosensory evoked potentials and multiple-unit activity throughout the central one third of the skate forebrain in a pallial area that corresponds to the medial pallium.

12.5.3 Behavior

12.5.3.1 Prey and Predator Detection — The first demonstrated use of the elasmobranch electric sense was for the detection of the bioelectric fields produced by prey organisms (Kalmijn, 1971). In laboratory behavioral experiments, Kalmijn (1971) demonstrated that both the catshark, *Scyliorhinus canicula*, and the skate, *Raja clavata*, executed well-aimed feeding responses to small, visually inconspicuous buried flounder (Figure 12.10A) and to flounder buried in a seawater agar-screened chamber that permitted the emission of the prey's bioelectric field but not its odor (Figure 12.10B). When the agar-screened prey was covered by a thin plastic film that insulated the prey electrically, the flounder remained undetected (Figure 12.10C). Feeding responses indistinguishable from those mediated by natural prey were observed again directed toward dipole electrodes that simulated bioelectric prey fields when buried under the sand or agar (Figure 12.10D). In later field experiments, Kalmijn (1982) also demonstrated that free-ranging sharks such as the smooth dogfish, *Mustelus canis*, and the blue shark, *Prionace glauca*, were attracted to an area by odor but preferentially attacked an active dipole source that simulated the prey's bioelectric field rather than the odor source of the prey. In addition, Tricas (1982) showed that the swell shark, *Cephaloscyllium ventriosum*, uses its electric sense to capture prey during nocturnal predation on small reef fish. More recent work with the Atlantic stingray (Blonder and Alevizon, 1988), sandbar shark, and scalloped hammerhead shark (Kajiura and Holland, 2002) also demonstrates well-aimed feeding responses at electrically simulated prey. Kajiura and Holland (2002) recently demonstrated that the "hammer" head morphology of sphyrnid sharks does not appear to confer a greater electroreceptive sensitivity to prey-simulating dipole electric fields than the "standard" head shark morphology, but it may provide a greater lateral search area to increase the probability of prey encounter and enhance maneuverability for prey capture.

Another important function of the elasmobranch electric sense is for use in predator detection and avoidance. Recent work on the clearnose skate, *R. eglanteria*, demonstrates that the electric sense of egg-encapsulated embryonic skates is well suited to detect potential egg predators (Sisneros et al., 1998),

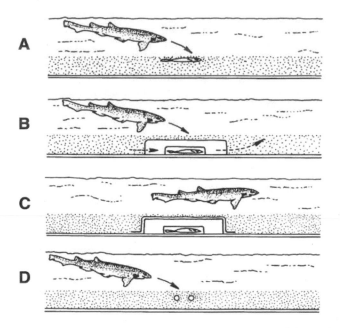

FIGURE 12.10 Use of the elasmobranch electric sense for the detection of electric fields produced by prey organisms. Behavioral responses of the catshark, *Scyliorhinus canicula*, to a small flounder buried in the sand (A), a flounder buried in a seawater agar-screened chamber permeable to bioelectric fields (B), a flounder in an agar chamber covered by a plastic film that insulates the prey electrically (C), and electrodes simulating the bioelectric fields produced by a flounder (D). Solid arrows indicate path of attack by the catshark; broken arrows indicate flow of seawater. (Modified from Kalmijn, A.J. 1971. *J. Exp. Biol.* 55:371–383. With permission.)

which include other elasmobranchs, teleost fishes, marine mammals, and molluscan gastropods (for review see Cox and Koob, 1993). Late-term embryonic skates circulate seawater within the egg case by undulating their tail in one corner of the egg near ventilation pores found in the horn of the egg case (Figure 12.11A). This action draws fresh seawater through pores on the opposite end of the egg case and creates a localized vortex near the exit pore by the tail, which can provide potential predators with olfactory, electrosensory, and mechanosensory cues needed for the detection and localization of the egg-encapsulated embryo. The peak frequency sensitivity of the peripheral electrosensory system in embryonic clearnose skates matches the frequency of phasic electric stimuli produced by large fish predators during ventilatory activity (0.5 to 2 Hz) and also corresponds to the same frequency of phasic electric stimuli that interrupts the respiratory movements of skate embryos and elicits an antipredator freeze behavior (Figure 12.11B and C) (Sisneros et al., 1998). This freeze response exhibited by embryonic skates stops the ventilatory streaming of seawater from the egg case and decreases the likelihood of sensory detection by predators. Phasic electric stimuli of 0.1 to 1 Hz are also known to interrupt the ventilatory activity of newly posthatched catsharks, *Scyliorhinus canicula* (Peters and Evers, 1985), and thus may represent an adaptive response in skates and other elasmobranchs to enhance survival during their early life history.

12.5.3.2 Orientation and Navigation — The electric sense of elasmobranchs is known to mediate orientation to local inanimate electric fields and in theory is sensitive enough to function in geomagnetic navigation. Pals et al. (1982b) showed via behavioral experiments that the catshark, *S. canicula*, could use electric DC fields for orientation in a captive environment. Furthermore, Kalmijn (1982) demonstrated that the round stingray, *Urolophus halleri*, can orient within a uniform electric DC field, discriminate the direction of the DC field based on its polarity, and detect voltage gradients as low as 5 nV/cm. The electric fields used in the behavioral experiments by Kalmijn (1982) were similar to those caused by both ocean and tidal currents, which can have peak amplitudes that range from 500 nV/cm (Kalmijn, 1984) to 8 μV/m (Pals et al., 1982a). Thus, in theory, elasmobranch fishes may be

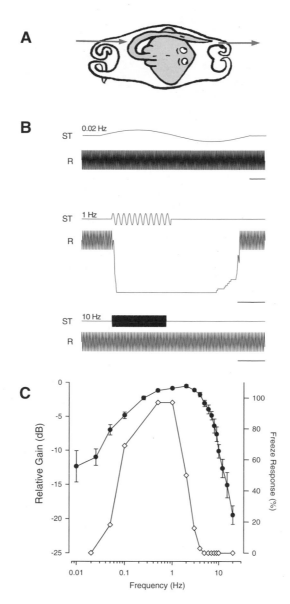

FIGURE 12.11 Behavioral response of embryonic clearnose skates, *Raja eglanteria*, to weak electric stimuli. (A) Ventilation behavior of embryonic skates. Diagram depicts a late-term embryonic skate circulating seawater within the egg case by undulating its tail in one corner of the egg near ventilation pores found in the horn of the egg case. The tail-beating action of the skate draws fresh seawater through pores on the opposite end of the case and creates a localized vortex near the exit pore by the tail. Arrow indicates flow of seawater. (B) Behavioral responses of skate embryos to sinusoidal uniform electric fields at stimulus (ST) frequencies of 0.02, 1, and 10 Hz. Stimuli were applied at an intensity of 0.56 $\mu V\ cm^{-1}$ across the longitudinal axis of the skate. The response (R) is expressed as a change in the peak-to-peak (PTP) tail displacement of the skate within the egg case. Prestimulus tail displacement for each record was 10 mm PTP. At 1 Hz, note the large tail displacement that occurs during coiling of the tail around the body after the onset of the electrical ST and a period of no tail movement during and after stimulation. Time bars = 5 s. (C) Freeze response of embryonic skates to weak electric stimuli. Behavioral responses (open diamonds) are shown as a percentage of total ST presentation to 0.02 to 20 Hz. Note that the peak frequency sensitivity of electrosensory primary afferent neurons (solid dots) for embryonic skates is at 1 to 2 Hz and is aligned with the freeze response peak of 0.5 to 1 Hz. (Modified from Sisneros, J.A. et al. 1998. *J. Comp. Physiol.* 183A:87–99. With permission.)

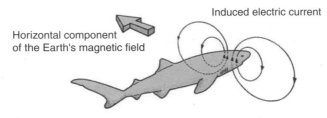

Shark heading east in the open ocean

FIGURE 12.12 Use of the elasmobranch electric sense in the active mode of navigation. Diagram depicts the induction of electric current induced in the head and body of the animal as the shark swims through the horizontal component of the Earth's geomagnetic field. (Modified from Kalmijn, A.J. 1988. In *Sensory Biology of Aquatic Animals*. J. Atema et al., Eds., Springer-Verlag, New York, 151–186. With permission.)

able to estimate their passive drift within the flow of tidal or ocean currents from the electric fields produced by the interaction of the water current moving through the Earth's magnetic field.

According to Kalmijn (1981, 1984), elasmobranchs can theoretically use the electric sense for two modes of navigation. In the passive mode, the elasmobranch simply measures the voltage gradients in the external environment. These electric fields are produced by the flow of ocean water through the Earth's magnetic field. In the active mode, the elasmobranch measures the voltage gradients that are induced through the animal's body due to its own swimming movements through the geomagnetic field (Figure 12.12). A different theory of active electronavigation proposed by Paulin (1995) maintains that directional information is acquired from the modulation of electrosensory inputs caused by head turning during swimming movements. Sufficient electrosensory information is obtained during head turns that allow the elasmobranch to extract directional cues from electroreceptor voltages induced in the animal as it swims in different directions. Thus, the comparison of electrosensory and vestibular inputs could then be used by the elasmobranch to determine a compass heading.

Evidence already exists to support the case that elasmobranchs use magnetic field information for orientation and navigation. Kalmijn (1982) showed that in the absence of an imposed electric field round stingrays, *U. halleri*, could be conditioned by food reward to locate and enter an enclosure in the magnetic east and to avoid a similar enclosure in the magnetic west. Kalmijn (1982) also showed that the stingrays could discriminate the direction and polarity of the magnetic field. More recently, Klimley (1993) showed that scalloped hammerhead sharks, *Sphyrna lewini*, seasonally aggregate near seamounts in the Gulf of California and follow daily routes to and from the seamounts, routes that correlate with the pattern of magnetic anomalies on the ocean floor. This suggests that under natural conditions elasmobranchs may use the geomagnetic field for navigation.

In contrast to the elasmobranch fishes, many other animals also use the Earth's magnetic field for navigation and homing. For these animals, many theories have been proposed that link magnetoreception to either the visual system or magnetite particles found in the head or body (Leask, 1977; Gould et al., 1978; Walcott et al., 1979; Phillips and Borland, 1992; Walker et al., 1997). Recently, Walker et al. (1997) were the first to discover, in any vertebrate, neurophysiologically identified magnetite-based magnetoreceptors, in the nasal region of the long-distance migrating rainbow trout, *Oncorhynchus mykiss*. Based on their behavioral, anatomical, and neurophysiological experiments, Walker et al. (1997) have provided the best evidence to date of a structure and function for a magnetite-based vertebrate magnetic sense. The identification of the key components of the magnetic sense in the rainbow trout will no doubt lead to new perspectives in the study of long-distance orientation and navigation in a variety of vertebrate groups.

12.5.3.3 Conspecific Detection — Work on non-electric stingrays demonstrates that the elasmobranch electric sense is used for conspecific detection and localization during social and reproductive behaviors (Tricas et al., 1995; Sisneros and Tricas, 2002b). Male and female round stingrays, *U. halleri*, use the electric sense to detect and locate the bioelectric fields of buried conspecifics during the mating season (Figure 12.13A). Stingrays produce a standing DC bioelectric field that is partially modulated by the ventilatory movements of the mouth, spiracles, and gill slits (Figure 12.13B) (Kalmijn, 1974; Tricas et al., 1995). Male rays use the electric sense to detect and locate females for mating, and females use their electric sense to locate and join other buried, less-receptive females for refuge (Tricas et al., 1995; Sisneros and Tricas, 2002b). The round stingray's peak frequency sensitivity of the peripheral electrosensory system matches the modulated frequency components of the bioelectric fields produced by conspecific stingrays (Figure 12.13C). Thus, the stingray's electric sense is "tuned" to social bioelectric stimuli and is used in a sex-dependent context for conspecific localization during the mating season.

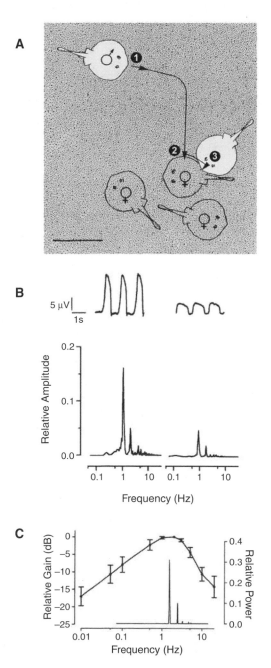

FIGURE 12.13 Detection of conspecific mates, bioelectric stimuli, and the frequency response of the peripheral electrosensory system in the round stingray, *Urolophus halleri*. (A) Orientation response by a male round stingray to cryptically buried conspecific females during the mating season. Males localize, orient toward, and inspect buried females buried in the sandy substrate. Search path of the male ray (1) changes abruptly after the detection of the female's bioelectric field. Males inspect buried females near the margins of her body disk (2) and pelvic fins (3). Active courtship and copulation begins after the male excavates the buried female and grasps the female's body disk with his mouth. Scale bar = 25 cm. (B) Bioelectric potentials recorded from a female stingray on the ventral surface near the gill slits (top, left record) and dorsal surface above the spiracle (top, right record). Recorded potentials are similar for both male (not shown) and female rays. Scales apply to both top records. Bottom graphs are Fourier transforms that show strong frequency components near 1 to 2 Hz that result from ventilatory movements. (C) Match between the peak frequency sensitivity of electrosensory primary afferent neurons and the frequency spectrum of the modulated bioelectric waveforms produced by round stingrays. The response dynamics of the electrosensory primary afferents in *U. halleri* show greatest frequency sensitivity at approximately 1 to 2 Hz with a 3 dB drop at approximately 0.5 and 4 Hz. Data are plotted as the relative gain of mean discharge peak (±1 SD). (Modified from Tricas, T.C. et al. 1995. *Neurosci. Lett.* 202:29–131. With permission.)

12.6 Olfaction and Other Chemosenses

Experimental studies in the first decades of the last century clearly identified olfaction as an important if not the primary means that sharks find food. The results provided well-founded starting points for later investigations. Interest in preventing shark attack on military personnel in World War II sparked a second generation of investigations on shark feeding and its olfactory control. This work continued into the mid-1970s (Hodgson and Mathewson, 1978b). More recent studies on olfaction in elasmobranchs have detailed aspects of the anatomy and physiology of olfactory systems, identified mechanisms of olfactory control of feeding, and suggested that female sex pheromones attract males and that predators may be detected by smell. Limited information on gustation and the common chemical sense or chemesthesis in elasmobranchs suggests similarities to their counterparts in other vertebrates.

12.6.1 Anatomy and Physiology of the Olfactory System

Information on the anatomical pathways for smell in elasmobranchs derives mostly from considerable work in comparative vertebrate neuroanatomy in the second half of the 20th century (Smeets, 1998). Physiological studies on elasmobranch olfaction, while limited, are consistent with the anatomical and behavioral data.

12.6.1.1 Peripheral Organ and Epithelium — The olfactory organs of elasmobranchs are situated in laterally placed cartilaginous capsules on the ventral aspect of the head well in front of the mouth. The ellipsoid saclike structures are typically divided by skin-covered flaps into a more lateral incurrent nostril (nares) and a more medial excurrent nostril (Tester, 1963a). A depression or groove helps to channel water into the incurrent opening where it traverses a rosette-like formation of plates or lamellae each with secondary folds that support the epithelium containing the primary olfactory receptors and supporting cells and tissues (Figure 12.14). The dynamics of the circulation path for the water movement have been analyzed in a series of detailed studies on several sharks (Theisen et al., 1986; Zeiske et al., 1986, 1987).

The epithelium is similar to that found in olfactory systems of most vertebrates with the major exception that the elasmobranch bipolar receptor cells are not ciliated but rather have a dendritic knob from which extends a tuft of microvilli (Reese and Brightman, 1970; Theisen et al., 1986; Zeiske et al., 1986, 1987). Similar microvillous receptors have been found along with the "typical" ciliated type in certain bony fishes. Cell surface lectin-binding patterns also differentiate the elasmobranch microvillous receptors (spotted dogfish, *Scyliorhinus canicula*) from the ciliated receptors of amphibians, rodents, and some bony fishes (Franceschini and Ciani, 1993). Studies on the clearnose skate, *Raja eglanteria*, identify two types of nonciliated olfactory receptor neurons (Takami et al., 1994). Type 1 is typical of those found in the other fishes (as above); the type 2 cell, so far unique to elasmobranchs, is distinguished from the type 1 by its thicker dendritic knob and microvilli that are shorter, thicker, and more regularly arranged. The functional meaning of the morphological differences in receptor types has yet to be determined.

The underwater electro-olfactogram (EOG) is a tool for recording the extracellular DC field potentials or analog of the summed electrical activity of the olfactory epithelium in response to chemical stimulation (Silver et al., 1976). EOG responses have been studied in two elasmobranchs, the Atlantic stingray, *Dasyatis sabina* (Silver et al., 1976; Silver, 1979) and the lemon shark, *Negaprion brevirostris* (Zeiske et al., 1986). Several amino acids, known to be effective stimuli for evoking EOGs in bony fishes and behavioral responses in both bony fishes and elasmobranchs, were tested in both species while extracts of squid muscle were also used in the lemon shark study. As expected, the squid extract evoked a significant response in the lemon shark. In both species, L isomers of the amino acids were highly stimulatory and for the most part their relative effectiveness was similar to that found in the teleosts. The EOG magnitude increased exponentially with the log of the stimulus concentration. Calculated thresholds ranged between 10^{-6} and 10^{-8} *M*. These levels are similar to those reported for bony fishes (teleosts) and for electroencephalographic (EEG) studies in nurse and lemon sharks (see above and

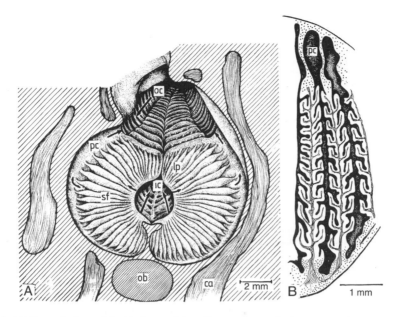

FIGURE 12.14 (A) Internal view of the hind part of the olfactory organ of a lemon shark (*Negaprion brevirostris*) as observed in an anterior-posterior direction with respect to the longitudinal axis of the organ. The olfactory cavity is divided into inlet and outlet chambers by lamellar protrusions of successive olfactory lamellae. Arrows indicate the calculated seawater flow direction. Ca, cartilage; ic, inlet chamber; lp, lamellar protrusion; ob, olfactory bulb; oc, outlet chamber; pc, peripheral channel; sf, secondary folds. (B) Lateral cross section through successive olfactory lamellae. Densely stippled areas represent the gap system between lamellae. The unstippled/white structures depict olfactory lamellae with secondary folds (the area covered by the sensory epithelium); fine lines indicate exiting olfactory nerve fibers (axons). Dark/shaded regions outline the peripheral channel (pc). (From Zeiske, E. et al. 1986. In *Indo-Pacific Fish Biology: Proceedings of the Second International Conference on Indo-Pacific Fishes*. T. Uyeno et al., Eds., Ichthyological Society of Japan, Tokyo, 381–391. With permission.)

Hodgson and Mathewson, 1978b). The similarity of detection abilities in the elasmobranchs and teleosts is surprising considering the far greater size of the olfactory organs in the former.

12.6.1.2   Olfactory Bulb — The first level of synaptic processing of olfactory information takes place in the olfactory bulb (OB), a part of the brain that receives the output from the olfactory receptors via their axons, which form the olfactory nerve. The olfactory bulbs of elasmobranchs are large structures that are closely applied to the olfactory epithelium or sac (Figure 12.15). The cytoarchitecture of the OB is conservative, and similar in elasmobranchs to other vertebrates (Andres, 1970; Smeets, 1998). Its concentric layers (from superficial to deep) include the olfactory nerve fibers; a layer of complex synaptic arrangements or glomeruli; a layer of large mitral cells, neurons functioning as the chief integrative units of the OB and, via their axons, the output pathway of the OB, the medial and lateral olfactory tracts; and a layer containing many small local circuit neurons, the granular cells. The olfactory tracts or peduncles travel to the cerebral hemispheres or telencephalon proper to make contact with secondary olfactory areas.

    Only fairly recently has information on the ultrastructure and electrophysiology of the OB of elasmobranchs become available. Studies on the topography of inputs and synaptic organization of the OB of bonnethead sharks, *Sphyrna tiburo* (Dryer and Graziadei, 1993, 1994a, 1996) and electrophysiology of the OB of the dogfish, *Scyliorhinus canicula* (Bruckmoser and Dieringer, 1973), and the little skate *Raja erinacea* (Cinelli and Salzberg, 1990), have greatly advanced the understanding of the structure in elasmobranchs and permit some useful comparisons to the OB of other better studied "model" species. Unlike other vertebrates, the OB of elasmobranchs is compartmentalized in a series of swellings or independent sub-bulbs each exclusively receiving input from the adjacent olfactory epithelium. The mitral cells in fishes (teleosts and elasmobranchs) lack the basal dendrites characteristic of mitral cells

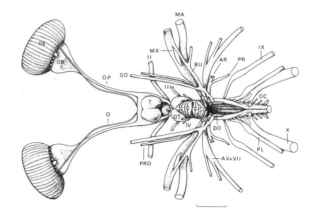

FIGURE 12.15 Dorsal view of the brain and olfactory system of the white shark, *Carcharodon carcharias*. The large partially divided olfactory bulb (OB) is closely applied to peripheral olfactory sac or epithelium (OE). Receptor cells in the epithelium project axons into the olfactory bulb (as the olfactory nerve) to make connections in complex synaptic arrangements. The mitral cells of the olfactory bulb distribute their axons to the secondary olfactory areas of the telencephalic hemisphere (T) via the elongated olfactory tracts or peduncles (OP). The terminal nerve or cranial nerve zero (O), which also extends from the olfactory epithelium to the hemisphere, may have chemosensory-related function(s) (see Demski and Schwanzel-Fukuda, 1987). Other abbreviations: AR, anterior ramus of the octaval nerve; AV, anteroventral lateral-line nerve; BU, buccal ramus of the anterodorsal lateral line nerve; DO, dorsal octavolateralis nucleus; MA, mandibular ramus of the trigeminal nerve; MX, maxillary ramus of the trigeminal nerve; OC, occipital nerves; OT, optic tectum; PL, posterior lateral line nerve; PR, posterior ramus of the octaval nerve; PRO, profundal nerve; SC, superficial ophthalmic ramus of the anterodorsal lateral line nerve; II, optic nerve; III, oculomotor nerve; IV, trochlear nerve; VII, facial nerve; IX, glossopharyngeal nerve; X, vagus nerve. Bar, 3 cm. (From Demski, L.S. and R.G. Northcutt. 1996. In *Great White Sharks: The Biology of Carcharodon carcharias*. A.P. Klimley and D.G. Ainley, Eds., Academic Press, San Diego, 121–130. With permission.)

of tetrapods, a finding that suggests differences in information processing, especially lateral inhibition (for details see Andres, 1970; Dryer and Graziadei, 1993, 1994a, 1996).

Species differences in the mass of the OB relative to total brain mass, as calculated in nine shark species (see Northcutt, 1978; Demski and Northcutt, 1996), suggest differences in reliance on smell in feeding and/or social behavior. The relative mass of the OB in the white shark, *Carcharodon carcharias* (Figure 12.15), at 18% is the highest followed by that of the smooth and spotted dogfishes (*Mustelus* and *Scyliorhinus*) at 14%. Intermediate in this regard are spiny dogfish, *Squalus acanthias*, deepwater dogfish, *Etmopterus* spp., and hammerhead sharks, *Sphyrna* spp., at 6, 7, and 9%, respectively. The lowest ratios are 3% for requiem sharks, *Carcharhinus* spp., blue sharks, *Prionace glauca*, and shortfin mako sharks, *Isurus oxyrinchus*. These figures must be interpreted cautiously as the percentages for the OB would be significantly higher in some sharks (e.g., *Carcharhinus* spp. and *Sphyrna* spp.) if their greatly enlarged telencephalic hemispheres are discounted.

The high ratio in the white shark is somewhat surprising, particularly compared with that of the closely related mako. The difference may reflect observations that, while both types consume fish, only the adult white sharks heavily prey on marine mammals including pinnipeds, the colonies of which introduce considerable odoriferous material into the water (Tricas and McCosker, 1984; Long et al., 1996; Strong et al., 1992; see also below).

12.6.1.3   Higher Level Systems — Projections from the OB to the telencephalic hemisphere have been mapped using contemporary neuroanatomical techniques in a variety of species (Ebbesson and Heimer, 1970; Ebbesson, 1972; 1980; Ebbesson and Northcutt, 1976; Northcutt, 1978; Smeets, 1983, 1998; Smeets et al., 1983; Dryer and Graziadei, 1994b). The results are in general agreement that the primary olfactory tract projection is to the lateral region of the ipsilateral hemisphere. Less well developed contralateral projections are reported in some species but not others. Spatial mapping of the projection of the medial and lateral olfactory tracts has been documented in the bonnethead shark, *Sphyrna tiburo* (Dryer and Graziadei, 1994b).

The findings refute earlier claims (see Aronson, 1963) that the entire hemisphere was dominated by the olfactory inputs and consequently that the enlarged hemispheres of sharks and rays could be attributed to their highly developed sense of smell. Other neuroanatomical, physiological, and behavioral studies have demonstrated that, other than the modest area of olfactory tract projection, most of the remainder of the hemisphere either receives specific inputs from other senses, including vision, hearing, mechanosenses, and electrosenses, or is multisensory in function (Ebbesson and Schroeder, 1971; Cohen et al., 1973; Graeber et al., 1973, 1978; Platt et al., 1974; Schroeder and Ebbesson, 1974; Graeber, 1978, 1980; Luiten, 1981a,b; Bleckmann et al., 1987; Smeets and Northcutt, 1987). This current view indicates that the elasmobranch telencephalon is similar in general organization and function to that of other vertebrates (see reviews by: Northcutt, 1978, 1989b; Demski and Northcutt, 1996).

There are few studies concerning the function of the olfactory areas in the elasmobranch hemisphere. Bruckmoser and Dieringer (1973) recorded evoked potentials from the surface of the hemisphere in response to electrical stimulation of the olfactory epithelium and OB in *Scyliorhinus canicula* and from electrical stimulation of the olfactory tracts in the torpedo ray, *Torpedo ocellata*. Short latency responses indicative of direct projections of the OB were observed only in the lateral olfactory area as defined by the anatomical studies.

Electrical stimulation of the lateral olfactory area in a free-swimming nurse shark (*Ginglymostoma cirratum*) evoked feeding-related responses of inconsistent mouthing or eating food (cut fish soaked to remove most of its juices) and a slow side-to-side head movement, which dragged the rostral sensory barbels across the substrate (Demski, 1977). The specific type of head movement was observed in unoperated sharks when colorless fish extracts were delivered to their home tank. Stimulation in the area also triggered circling toward the side of the electrode (ipsilateral). The latter result is consistent with Parker's observation that sharks with a unilateral occlusion of the nostril circle toward the side of the open nostril. Thus, the physiological and behavioral studies available are consistent with the anatomical projections and suggest that the olfactory area of the lateral hemisphere is involved in the arousal of feeding by olfactory stimulation.

Bruckmoser and Dieringer (1973) recorded potentials of longer latency (20 to 800 ms) including regular EEG-synchronous afterpotentials in other areas of the hemispheres. This secondary activity was more labile than the primary responses and differed in the two species. It is most likely indicative of areas involved in higher-level processing of the olfactory information and/or regions for multisensory or sensorimotor integration.

It should be noted that in bony fishes, the OBs project to the hypothalamus of the diencephalon (Finger, 1975; Bass, 1981; Murakami et al., 1983; Prasada Rao and Finger, 1984), an area from which feeding activity has been evoked by electrical stimulation (Demski, 1983) and potentials triggered by olfactory tract stimulation (Demski, 1981). Although a direct olfactory bulb projection to the hypothalamus has not been reported for elasmobranchs, projections from the lateral olfactory area of the hemisphere to the hypothalamus are suggested (Ebbesson, 1972; Smeets, 1998). Electrical stimulation of the hypothalamus in nurse sharks has evoked "feeding" as evidenced by relatively continuous swimming, consistent mouthing or eating food, and the barbel-dragging, side-to-side head movement (Demski, 1977). Based on the comparative data, a similar hypothalamic feeding area has been proposed for teleosts and sharks (Demski, 1982). Also in this regard, Tester (1963b) observed that thresholds for olfactory-triggered feeding in blacktip reef sharks, *Carcharhinus melanopterus*, are lowered by starvation (see below). Such increased sensitivity may have resulted from hypothalamic modulation of the olfactory system in response to changes in visceral sensory activity and/or blood-borne factors associated with the dietary conditions.

12.6.2 Olfactory Control of Feeding

12.6.2.1 *Studies of Sharks in Large Enclosures and Open Water* — The critical early studies on olfactory mediation of feeding in sharks were done by Sheldon and Parker (Sheldon, 1911; Parker and Sheldon, 1913; Parker, 1914) working with captive smooth dogfish, *Mustelus canis,* in large outdoor pens at Woods Hole, MA. The behavior patterns of normal animals with one or both nares blocked with cotton wool were described by Parker (1914). Normal animals and those with only one nostril blocked readily located a packet of crabmeat wrapped in cheesecloth to exclude visual

identification. Fish with both nostrils blocked totally ignored the bait. Normals essentially turned equally to either side and often made figure-8 movements while experimental sharks turned almost exclusively to the unblocked side, as if this was the direction of the odor corridor.

Olfactory involvement in elasmobranch feeding includes several phases, which can be roughly categorized as arousal; directed approach and attack; and if the prey or bait is not located or is lost, usually continued search. These components vary depending on circumstance and species. Notable studies on elasmobranch feeding and olfaction on several species of carcharhiniform sharks were carried out by Tester and his student Hobson (Tester, 1963a,b; Hobson, 1963). Tester's description of feeding in blind blacktip reef sharks, *Carcharhinus melanopterus,* is especially revealing. The blind sharks fed avidly on food that settled to the bottom of their tank, thusly: "The sharks would detect the odor while swimming in mid-water and would spiral down, converging on the food by swimming in a figure-8 pattern to the bottom." Tester commented on the similarity of the response to that reported by Parker (above) for *M. canis.* Indeed, the pattern may be typical of, at least, the carcharhiniform sharks.

Tester recorded responses of several shark species to a variety of extracts of fish and invertebrates as well as human urine, blood, and sweat. Essentially all food substance extracts were "attractive." Regarding responses to human materials, sharks demonstrated "attraction" to blood, "sensing" but otherwise indifference to urine, and although highly variable, "repulsion" to sweat. Blind reef blacktip sharks were more sensitive to odors than those with normal sight. In addition, starvation in these animals generally resulted in greater responses to food extracts. Sharks were "attracted" to introduction of water from containers with prey fish that were not stressed but the sharks soon adapted to the stimuli; in contrast, the sharks showed concerted "hunting reactions" to the test water when the prey fish were "frightened and excited by threatening them with a stick." The results strongly suggest that sharks can use odors to discriminate between stressed and unstressed prey fish.

Hobson's field studies complemented the laboratory tests reported above. Observations were made underwater with SCUBA and from the surface using a glass-bottomed boat. Three species were studied in the lagoon at Eniwetok: gray reef sharks; blacktip reef sharks; and whitetip reef sharks, *Triaenodon obesus.* He reported that the sharks used an olfactory corridor and local water currents to locate bait (a tethered but otherwise "uninjured" prey fish, or extract of grouper meat). The sharks could also accurately pinpoint a source of water flowing from tanks holding stressed but otherwise uninjured prey fish. The results were consistent with Tester's laboratory study.

Studies on two Atlantic species yielded similar results. Working at the Lerner Marine Laboratory in Bimini, Bahamas, Hodgson and Mathewson tested the responses of nurse sharks (*Ginglymostoma cirratum*) and lemon sharks (*Negaprion brevirostris*) to release of known chemical feeding attractants (glutamic acid and trimethylamine oxide, TMAO) in large outdoor pens (Hodgson and Mathewson, 1971, 1978b; Mathewson and Hodgson, 1972). The authors concluded that the two sharks use different strategies to localize the source of the odor. The nurse shark employs a true gradient search behavior (a klinotaxis or temporally based gradient sampling, i.e., sequential comparisons of concentrations from different points) as it scans across the olfactory corridor, whereas the lemon shark becomes aroused on contact with the stimuli and then turns in the direction of the greatest current and heads upstream. The pattern conforms to a rheotactic bias or release mechanism as suggested to account for odor tracking behavior of other sharks (Kleerekoper, 1969). Under most natural circumstances the lemon shark would find the source by moving upstream to its location. Indeed, from the studies cited above and others referred to by them, most advances by sharks on prey or artificial stimuli are from downstream locations. Field tests carried out in open water at Bimini indicated that several shark species use a rheotaxis-release mechanism in tracking a TMAO-glycine mixture.

Working at Dangerous Reef in South Australian waters, Strong and colleagues (1992, 1996) investigated olfactory orientation in tagged white sharks, *Carcharodon carcharias.* Female white sharks responded to baits of tuna and horsemeat by typically circling downstream of the olfactory corridor ("searching") for periods of up to 12 h. After several hours of circling, some sharks traveled among the nearby inshore islands ("island patrolling"). The authors indicated that considerable odoriferous material from nearby pinniped colonies is released into the sea and that white sharks cruise around these islands in search of such odors. The observations confirm for another shark species the use of downstream approach as a major strategy for at least general location of prey. Perhaps more important is the long-lasting effect of the stimulation and the complex behavior patterns it triggers. The movements among the islands could not be guided directly by odor trails but rather must represent innate and/or learned responses, which take control after the arousal by odors and the search for prey in the vicinity is not successful. Such behavior is most likely controlled by the highest integrative centers of the telencephalic hemispheres of the brain.

12.6.2.2 Laboratory Studies —

Kleerekoper (1978, 1982) analyzed the motor behavior of *Scyliorhinus stellaris* and *M. mustelus* in response to chemical stimulation in a circular arena. The preliminary studies revealed that the olfactory stimulus caused a decrease in the mean angle of turns made by the swimming sharks such that the area covered was greater during chemical tests vs. control stimulations. The behaviors, likened to "searching," occurred even in the absence of directional cues and thus can be considered arousal responses that trigger a stereotyped response leading in most cases to increased opportunity of locating the source of the odors.

Kleerekoper and colleagues (Kleerekoper et al., 1975; Kleerekoper, 1978, 1982) also studied the responses of nurse sharks to extracts of shrimp in a large enclosure (5 × 5 × 0.5 m). They concluded (Kleerekoper et al., 1975) that a precise localization of a chemical stimulus is dependent on both the flow of the medium and the stimulus gradient within it. In nonmoving water only generalized location of the release site is possible, whereas flowing water provides the nurse shark with a direction vector that is used to pinpoint the source of the stimulus. Thus, different mechanisms of location appear to be used in flow vs. stagnant conditions.

Johnsen and Teeter (1985) studied the reactions of bonnethead sharks, *Sphyrna tiburo*, to extracts of blue crabs injected either into the water of the study arena (a 1.8 m circular tank) or directly into the nares of one or both sides via tubes mounted to the shark's head. The animals were monitored under conditions of current and no current (Figure 12.16). The responses to direct and external presentation of the extract were essentially the same. In slack water the sharks reacted to the odor by suddenly stopping their typical circling around the outer limit of the tank and began moving in tight circles in the vicinity of the stimulation site. In contrast, when a current was present, the sharks, which typically swim against the current, reacted to the stimulation by reversing direction. The behavior was repeated "as sharks appeared to follow the stimulus bolus around the edge of the tank. The paths of the fish then had the appearance of connected loops." It is noteworthy that the responses to the direct stimulation of the nares continued past the point where calculations indicate that the stimuli would have been diluted below threshold levels. Such being the case, the responses are likely part of a neural program that is initiated by the odor but not dependent on it for continuation. The circling and looping would appear to be an effective means of guiding the animal back to the odor trail.

EEG studies in sharks carried out by Gilbert and colleagues (Gilbert et al., 1964; Hodgson et al., 1967; Hodgson and Mathewson, 1978b) at the Lerner Marine Laboratory were complementary to the field studies of the 1960s. The researchers studied electrical activity changes in the brains of nurse,

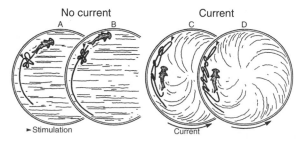

FIGURE 12.16 Responses of captive bonnethead sharks, *Sphyrna tiburo*, to chemical stimuli delivered either into the open water of a circular test tank or via a head mount directly into the shark's nares. Tests were conducted under two conditions of water current. A = Response to 10-ml sample of crab homogenate introduced in still water; B = Response to 0.5-ml sample delivered via head mount; C = Response to 10-ml sample delivered into 15 cm/s water current; D = Response to 0.5-ml sample delivered via the head mount in presence of 15 cm/s water current. (From Johnsen, P.B. and J.H. Teeter. 1985. *Mar. Behav. Physiol.* 11:283–291. With permission.)

bonnethead, and lemon sharks in response to stimulation with extracts of natural foods like crab and tuna as well as amino acids and amines known to stimulate feeding in fishes. Recording sites were located on the surface of the OBs, the telencephalic hemispheres, optic lobes or tectum, the cerebellum and medulla, and deep in the hemispheres. Sites in the tectum and cerebellum were unresponsive to chemical stimulation, whereas sites in the anterior telencephalon (an area later determined to be the primary olfactory center; see above) showed increases in both frequency and amplitude to most of the chemicals tested. Activity in the OB of a free-swimming lemon shark was greatly increased in response to a stimulus of 10^{-4} M glycine. These results and some additional preliminary findings of chemically evoked forebrain activity in anesthetized lemon sharks (Agalides, 1967) confirm the arousal function of food odors and artificial stimulants, and essentially mark the beginning of studies on elasmobranch brain systems that control both olfactory processing and feeding behavior.

12.6.3 Sex Pheromones in Mating

The evidence for use of olfactory cues in social-sexual behavior of elasmobranchs is indirect; nevertheless, it is consistent across several groups of sharks and batoids. The most compelling suggestion of olfactory sex attraction was reported by Johnson and Nelson (1978), who recounted an incident of "close following" behavior of blacktip reef sharks, *Carcharhinus melanopterus*, at Rangiroa Atoll in French Polynesia. One shark tracked down another, which was initially out of its view, and then followed it closely with its snout directed toward the leader's vent. The latter swam close to the substrate in an atypical slow, sinuous manner with its head inclined downward and its tail uplifted. The authors concluded that only an olfactory cue could have guided the second shark to the position of the other. While sex was not determined in this incidence, other observations indicated that unusual swimming and following behaviors appeared to be sex specific to the females and males, respectively.

There are scattered observations of males of other elasmobranch species following closely behind females, usually with their nose directed to the female's vent, sometimes pushing on it. This has been reported for the bonnethead shark, *Sphyrna tiburo* (Myrberg and Gruber, 1974), nurse shark, *Ginglymostoma cirratum* (Klimley, 1980; Carrier et al., 1994), spotted eagle ray, *Aetobatis narinari* (Tricas, 1980), clearnose skate, *Raja eglanteria* (Luer and Gilbert, 1985), and sand tiger shark, *Carcharias taurus* (Gordon, 1993); see also review by Demski (1991). Other indications of the sex-related nature of the encounters include the presence of scars on the females or swelling of the pelvic fins and cloacal area suggestive of recent mating; male attempts to mount the female; and in captive female sand tiger sharks, "cupping and flaring" of the pelvic fins in response to the close presence of the male. Thus, although there are no direct experimental findings to document female sex-attraction pheromones, behavioral observations in natural and captive environments strongly suggest their existence.

12.6.4 Olfaction and Predator Avoidance

Lemon sharks, *N. brevirostris,* and American crocodiles, *Crocodylus acutus*, overlap in their distributions, and where such is the case, the crocodiles will prey on the sharks. Rasmussen and Schmidt (1992) demonstrated that water samples taken from ponds holding *C. acutus* and delivered to the nares of juvenile lemon sharks consistently aroused the sharks from a state of tonic mobility (induced by inversion and restraint), an established bioassay for chemical awareness. Water from ponds containing alligators, *Alligator mississippiensis*, which have no substantial natural contact with lemon sharks, had no such effect. The authors identified three organic compounds produced by the crocodiles (2-ethyl-3-methyl maleimide; 2-ethyl-3-methyl succinimide; 2-ethylidene-3-methyl succinimide) that accounted for the positive results. Synthetic versions of the chemicals were also effective. The results strongly suggest that lemon sharks and perhaps other elasmobranchs use olfactory cues to avoid potential predators.

12.6.5 Gustation

Anatomical studies in elasmobranchs have identified receptors that closely resemble taste organs in other vertebrates. A few behavioral observations suggest gustation is important for the acceptance of food in sharks (see Sheldon, 1909, and review by Tester, 1963a). Cook and Neal (1921) mapped the distribution of taste buds in the oral-pharyngeal cavity of the spiny dogfish, *Squalus acanthias*. While located over the entire region, the receptor organs appear most numerous on the roof of the cavity. In microscopic section, the taste buds are characterized as small papillae covered with a multilayer epithelium that has a central cluster of elongate sensory receptor cells. Nerve fibers are associated with the base of the receptors. Older descriptive anatomical studies of several sharks indicate that the taste organs are supplied by branches of the facial, glossopharyngeal, and vagus nerves (Norris and Hughes, 1920; Herrick, 1924; Daniel, 1928; Aronson, 1963), as is the case with other vertebrates.

Whitear and Moate (1994) carried out a detailed ultrastructural analysis of the taste buds of the dogfish, *Scyliorhinus canicula*. The apical regions of the receptors with their protruding microvilli form pores, which are clearly visible in their scanning electron micrographs. Nerve fibers were associated with the receptors as well as possible free nerve endings. Part of a taste bud was reconstructed from serial transmission electron micrographs. In general, the organization of the peripheral gustatory system of sharks appears comparable with that of other vertebrates. Unfortunately, detailed physiological and behavioral studies are not available to further support this observation. It seems reasonable to assume that the gustatory apparatus in sharks functions primarily in the final determination of food vs. nonfood.

12.6.6   Common Chemical Sense

Studies in *Mustelus canis* demonstrated that sharks respond behaviorally to injections into the nostrils of certain chemicals (irritants) even with the olfactory tracts severed. In these cases, detection was through components of the maxillary branch of the trigeminal nerve (Sheldon, 1909). The animals reacted similarly to applications on the body surface. The latter responses were triggered via spinal nerves. Sheldon considered that this chemosensitivity was mediated by free nerve endings.

Studies in other vertebrates indicate that the nerves involved in such reactions are part of the somatosensory system and appear to represent a subset of temperature- and pain-sensitive fibers. The sense conveyed by these chemosensitive components has been renamed "chemesthesis" to reflect this relationship (Bryant and Silver, 2000). The function of this system in elasmobranchs, as in other vertebrates, appears to be protection from damaging chemicals. It seems likely that the adverse reactions certain sharks demonstrate to natural toxins, such as that produced by the skin of the Moses sole, *Pardachirus marmoratus* (Clark, 1974), is likely mediated by this category of unmyelinated somatosensory ending.

12.7   Conclusions

Are elasmobranchs sensory marvels, or not? There is no doubt that the combination of well-developed visual, acoustical, mechanical, electrical, and chemical sensing systems in sharks, skates, and rays distinguishes the group and makes them well-adapted for life in the sea. The sensory ecology of these fishes is complex. Depending on species and ambient conditions, elasmobranchs may utilize one or more of their senses to monitor their environment, detect and locate prey and mates, avoid predators, and find their way in the ocean.

The range at which each of the sensory modalities operates depends on the qualities of the particular sensory system, the strength of the stimulus, and the physical characteristics of the environment that affect transmission of the signal. For example, the range of effective vision in sharks is not a fixed distance but depends on the sensitivity and acuity of the shark's eye, the optical characteristics of the target, the ambient light levels, and the absorption and scattering of light by the surrounding seawater. Similarly, olfactory range depends on the sensitivity of the shark's nose for a given chemical scent, the concentration of odorant at the source, and water characteristics affecting the odorant's transmission, such as currents. In typical situations, it is clear that the senses of olfaction and hearing can orient an elasmobranch to a stimulus source from far distances, mechanosenses and vision typically can begin to operate at intermediate distances from the stimulus, and electrosenses and gustation operate very near to the stimulus. The sequence with which each of the sensory modalities comes into play depends on a multitude of factors, however, and there is no single sensory hierarchy that operates under all circumstances for all elasmobranch species.

Integration of this multimodal sensory information in the elasmobranch CNS ultimately leads to a behavioral response at the level of the whole animal. How sharks, skates, and rays integrate the complex input of environmental information through their various senses to form an adaptive response is among the most interesting questions in elasmobranch sensory biology, and a ripe area for further research.

Acknowledgments

The authors wish to thank Samuel Gruber, Joel Cohen, William Tavolga, and Timothy Tricas for many useful discussions and collaborations in elasmobranch sensory biology. R.E.H. was supported by the Perry W. Gilbert Chair in Shark Research at Mote Marine Laboratory. L.S.D.'s work on elasmobranchs has been supported by grants from the National Science Foundation and the Florsheim endowment to the New College Foundation.

References

Agalides, E. 1967. The electrical activity of the forebrain of sharks recorded with deep chronic implanted electrodes. *Trans. N.Y. Acad. Sci.* 29:378–389.

Andres, K. H. 1970. Anatomy and ultrastructure of the olfactory bulb in fish, amphibia, reptiles, birds and mammals, in *Taste and Smell in Vertebrates*. G.E.W. Wolstenhome and J. Knight, Eds., Ciba Foundation Symposium/J. and A. Churchill, London, 177–196.

Andrianov, G.N., G.R. Broun, O.B. Ilyinsky, and V.M. Muraveiko. 1984. Frequency characteristics of skate electroreceptive central neurons responding to electric and magnetic stimulation. *Neurophysiology* 16:365–376.

Aronson, L.R. 1963. The central nervous system of sharks and bony fishes with special reference to sensory and integrative mechanisms, in *Sharks and Survival*. P.W. Gilbert, Ed., D.C. Heath, Boston, 165–241.

Aronson, L.R., F.R. Aronson, and E. Clark. 1967. Instrumental conditioning and light-dark discrimination in young nurse sharks. *Bull. Mar. Sci.* 17:249–256.

Banner, A. 1967. Evidence of sensitivity to acoustic displacements in the lemon shark, *Negaprion brevirostris* (Poey), in *Lateral Line Detectors*. P. Cahn, Ed., Indiana University Press, Bloomington, IN, 265–273.

Barber, V.C., K.I. Yake, V.F. Clark, and J. Pungur. 1985. Quantitative analyses of sex and size differences in the macula neglecta and ramus neglectus in the inner ear of the skate, *Raja ocellata. Cell Tissue Res.* 241:597–605.

Barry, M.A. 1987. Afferent and efferent connections of the primary octaval nuclei in the clearnose skate, *Raja eglanteria. J. Comp. Neurol.* 266:457–477.

Barry, M.A. and M.V.L. Bennett. 1989. Specialized lateral line receptor systems in elasmobranchs: the spiracular organs and vesicles of Savi, in *The Mechanosensory Lateral Line: Neurobiology and Evolution.* S. Coombs, P. Görner, and H. Münz, Eds., Springer-Verlag, New York, 591–606.

Barry, M.A., D.H. Hall, and M.V.L. Bennett. 1988a. The elasmobranch spiracular organ I. Morphological studies. *J. Comp. Physiol.* A 163:85–92.

Barry, M.A., D.H. Hall, and M.V.L. Bennett. 1988b. The elasmobranch spiracular organ II. Physiological studies. *J. Comp. Physiol.* A 163:93–98.

Bass, A. 1981. Olfactory bulb efferents in the channel catfish, *Ictalurus punctatus. J. Comp. Morphol.* 169:91–111.

Bennett, M.V.L. 1971. Electroreception, in *Fish Physiology*. Vol. 5. W.S. Hoar and D.J. Randall, Eds., Academic Press, New York, 493–574.

Blaxter, J.H.S. and L.A. Fuiman. 1989. Function of the free neuromasts of marine teleost larvae, in *The Mechanosensory Lateral Line: Neurobiology and Evolution.* S. Coombs, P. Görner, and H. Münz, Eds., Springer-Verlag, New York, 481–499.

Bleckmann, H. and T.H. Bullock. 1989. Central nervous physiology of the lateral line, with special reference to cartilaginous fishes, in *The Mechanosensory Lateral Line: Neurobiology and Evolution.* S. Coombs, P. Görner, and H. Münz, Eds., Springer-Verlag, New York, 387–408.

Bleckmann, H., T.H. Bullock, and J.M. Jorgensen. 1987. The lateral line mechanoreceptive mesencephalic, diencephalic, and telencephalic regions in the thornback ray, *Platyrhinoidis triseriata* (Elasmobranchii). *J. Comp. Physiol.* A 161:67–84.

Bleckmann, H., O. Weiss, and T.H. Bullock. 1989. Physiology of lateral line mechanoreceptive regions in the elasmobranch brain. *J. Comp. Physiol.* A 164:459–474.

Blonder, B.I. and W.S. Alevizon. 1988. Prey discrimination and electroreception in the stingray *Dasyatis sabina. Copeia* 1988:33–36.

Bodznick, D. 1991. Elasmobranch vision: multimodal integration in the brain. *J. Exp. Zool.* Suppl. 5:108–116.

Bodznick, D. and R.L. Boord. 1986. Electroreception in Chondrichthyes, in *Electroreception.* T.H. Bullock and W. Heiligenberg, Eds., John Wiley & Sons, New York, 225–256.

Bodznick, D. and J.C. Montgomery. 1992. Suppression of ventilatory reafference in the elasmobranch electrosensory system: medullary neuron receptive fields support a common mode rejection mechanism. *J. Exp. Biol.* 171:127–137.

Bodznick, D. and R.G. Northcutt. 1980. Segregation of electro- and mechanoreceptive inputs to the elasmobranch medulla. *Brain Res.* 195:313–321.

Bodznick, D. and R.G. Northcutt. 1984. An electrosensory area in the telencephalon of the little skate, *Raja erinacea. Brain Res.* 298:117–124.

Bodznick, D. and A.W. Schmidt. 1984. Somatotopy within the medullary electrosensory nucleus of the skate, *Raja erinacea. J. Comp. Neurol.* 225:581–590.

Bodznick, D., J.C. Montgomery, and D.J. Bradley. 1992. Suppression of common mode signals within the electrosensory system of the little skate, *Raja erinacea. J. Exp. Biol.* 171:107–125.

Boord, R.L. and C.B.G. Campbell. 1977. Structural and functional organization of the lateral line system of sharks. *Am. Zool.* 17:431–441.

Boord, R.L. and J.C. Montgomery. 1989. Central mechanosensory lateral line centers and pathways among the elasmobranchs, in *The Mechanosensory Lateral Line: Neurobiology and Evolution.* S. Coombs, P. Görner, and H. Münz, Eds., Springer-Verlag, New York, 323–340.

Boord, R.L. and R.G. Northcutt. 1982. Ascending lateral line pathways to the midbrain of the clearnose skate, *Raja eglanteria. J. Comp. Neurol.* 207:274–282.

Bozzano, A. and S.P. Collin. 2000. Retinal ganglion cell topography in elasmobranchs. *Brain Behav. Evol.* 55:191–208.

Broun, G.R., O.B. Il'inskii, and B.V. Krylov. 1979. Responses of the ampullae of Lorenzini in a uniform electric field. *Neurophysiology* 11:118–124.

Brown, B.R. 2003. Sensing temperature with ion channels. *Nature* 421:495.

Bruckmoser, P. and N. Dieringer. 1973. Evoked potentials in the primary and secondary olfactory projection areas of the forebrain in Elasmobranchia. *J. Comp. Physiol.* 87: 65–74.

Bryant, B. and W. Silver. 2000. Chemesthesis: the common chemical sense, in *The Neurobiology of Taste and Smell,* 2nd ed. T.E. Finger, W.L. Silver, and D. Restrepo, Eds., Wiley-Liss, New York, 73–98.

Budker, P. 1958. Les organes sensoriels cutanes des selaciens, in *Traité de Zoologie.* Vol. 15, Library de l'Academie de Medicine. Masson et Cie, Paris, 1033–1062.

Bullock, T.H. 1979. Processing of ampullary input in the brain: comparisons of sensitivity and evoked responses among siluroids and elasmobranchs. *J. Physiol.* (Paris) 75:315–317.

Bullock, T.H. and J.T. Corwin. 1979. Acoustic evoked activity in the brain in sharks. *J. Comp. Physiol.* 129:223–234.

Carpenter, D.O. 1981. Ionic and metabolic bases of neuronal thermosensitivity. *Fed. Proc.* 40:2808–2813.

Carrier, J.C., H.L. Pratt, Jr., and L.K. Martin. 1994. Group reproductive behaviors in free-living nurse sharks, *Ginglymostoma cirratum. Copeia* 1994:646–656.

Casper, B.M., P.S. Lobel, and H.Y. Yan. 2003. The hearing sensitivity of the little skate, *Raja erinacea*: a comparison of two methods. *Environ. Biol. Fishes* 68:371–379.

Chu, Y.T. and M.C. Wen. 1979. A study of the lateral-line canal system and that of Lorenzini ampullae and tubules of elasmobranchiate fishes of China. *Monograph of Fishes of China.* Academic Press, Shanghai.

Cinelli, A.R. and B.M. Salzberg. 1990. Multiple optical recording of transmembrane voltage (MSORTV), single-unit recordings, and evoked potentials from the olfactory bulb of skate (*Raja erinacea*). *J. Neurophysiol.* 64:1767–1790.

Clark, E. 1959. Instrumental conditioning of sharks. *Science* 130:217–218.

Clark, E. 1963. Maintenance of sharks in captivity with a report on their instrumental conditioning, in *Sharks and Survival*. P.W. Gilbert, Ed., D.C. Heath, Boston, 115–149.

Clark, E. 1974. The red sea's sharkproof fish. *Natl. Geog.* 146:719–727.

Cohen, D.H., T.A. Duff, and S.O.E. Ebbesson. 1973. Electrophysiological identification of a visual area in the shark telencephalon. *Science* 182:492–494.

Cohen, J.L. 1991. Adaptations for scotopic vision in the lemon shark (*Negaprion brevirostris*). *J. Exp. Zool.* Suppl. 5:76–84.

Cohen, J.L., S.H. Gruber, and D.I. Hamasaki. 1977. Spectral sensitivity and Purkinje shift in the retina of the lemon shark, *Negaprion brevirostris* (Poey). *Vision Res.* 17:787–792.

Cohen, J.L., R.E. Hueter, and D.T. Organisciak. 1990. The presence of a porphyropsin-based visual pigment in the juvenile lemon shark (*Negaprion brevirostris*). *Vision Res.* 30:1949–1953.

Collin, S.P. 1999. Behavioural ecology and retinal cell topography, in *Adaptive Mechanisms in the Ecology of Vision*. S.N. Archer, M.B.A. Djamgoz, E. Loew, J.C. Partridge, and S. Vallerga, Eds., Kluwer, Dordrecht, 509–535.

Conley, R.A. and D. Bodznick. 1994. The cerebellar dorsal granular ridge in an elasmobranch has proprioceptive and electroreceptive representations and projects homotopically to the medullary electrosensory nucleus. *J. Comp. Physiol. A* 174:707–721.

Cook, M.H. and H.V. Neal. 1921. Are taste buds of elasmobranchs endodermal in origin? *J. Comp. Neurol.* 33:45–63.

Coombs, S. and J. Janssen. 1990. Behavioral and neurophysiological assessment of lateral line sensitivity in the mottled sculpin, *Cottus bairdi*. *J. Comp. Physiol. A* 167:557–567.

Coombs, S. and J.C. Montgomery. 1999. The enigmatic lateral line system, in *Comparative Hearing: Fish and Amphibians*. R.R. Fay and A.N. Popper, Eds., Springer-Verlag, New York, 319–362.

Corwin, J.T. 1977. Morphology of the macula neglecta in sharks of the genus *Carcharhinus*. *J. Morphol.* 152:341–362.

Corwin, J.T. 1978. The relation of inner ear structure to feeding behavior in sharks and rays, in *Scanning Electron Microscopy*. O. Johari, Ed., S.E.M., Inc., Chicago, 1105–1112.

Corwin, J.T. 1981. Audition in elasmobranchs, in *Hearing and Sound Communication in Fishes*. W.N. Tavolga, A.N. Popper, and R.R. Fay, Eds., Springer-Verlag, New York, 81–105.

Corwin, J.T. 1983. Postembryonic growth of the macula neglecta auditory detector in the ray, *Raja clavata*: continual increases in hair cell number, neural convergence, and physiological sensitivity. *J. Comp. Neurol.* 217:345–356.

Corwin, J.T. 1989. Functional anatomy of the auditory system in sharks and rays. *J. Exp. Zool.* Suppl. 2:62–74.

Corwin, J.T. and R.G. Northcutt. 1982. Auditory centers in the elasmobranch brain stem: deoxyglucose autoradiography and evoked potential recording. *Brain Res.* 236:261–273.

Cox, D.L. and T.J. Koob. 1993. Predation on elasmobranch eggs. *Environ. Biol. Fishes* 38:117–125.

Daniel, J.F. 1928. *The Elasmobranch Fishes*. University of California Press, Berkeley.

Demski, L.S. 1977. Electrical stimulation of the shark brain. *Am. Zool.* 17:487–500.

Demski, L.S. 1981. Hypothalamic mechanisms of feeding in fishes, in *Brain Mechanisms of Behaviour in Lower Vertebrates*. P.J. Laming, Ed., Cambridge University Press, Cambridge, U.K., 225–237.

Demski, L.S. 1982. A hypothalamic feeding area in the brains of sharks and teleosts. *Fla. Sci.* 45:34–40.

Demski, L.S. 1983. Behavioral effects of electrical stimulation of the brain, in *Fish Neurobiology*. Vol. 2. R.E. Davis and R.G. Northcutt, Eds., University of Michigan Press, Ann Arbor, 317–359.

Demski, L.S. 1991. Elasmobranch reproductive behavior: implications for captive breeding. *J. Aquaricult. Aquat. Sci.* 5:84–95.

Demski, L.S. and R.G. Northcutt. 1996. The brain and cranial nerves of the white shark: an evolutionary perspective, in *Great White Sharks: The Biology of Carcharodon carcharias*. A.P. Klimley and D.G. Ainley, Eds., Academic Press, San Diego, 121–130.

Demski, L.S. and M. Schwanzel-Fukuda, Eds. 1987. *The Terminal Nerve (Nervus Terminalis): Structure, Function and Evolution*. *Ann. N.Y. Acad. Sci.* 519.

Denton, E.J. and J.A.B. Gray. 1983. Mechanical factors in the excitation of clupeid lateral lines. *Proc. R. Soc. Lond.* B 218:1–26.

Denton, E.J. and J.A.B. Gray. 1988. Mechanical factors in the excitation of the lateral lines of fishes, in *Sensory Biology of Aquatic Animals.* J. Atema, R.R. Fay, A.N. Popper, and W.N. Tavolga, Eds., Springer-Verlag, New York, 595–617.

Denton, E.J. and J.A.C. Nicol. 1964. The chorioidal tapeta of some cartilaginous fishes (Chondrichthyes). *J. Mar. Biol. Assoc. U.K.* 44:219–258.

Denton, E.J. and T.I. Shaw. 1963. The visual pigments of some deep-sea elasmobranchs. *J. Mar. Biol. Assoc. U.K.* 43:65–70.

Dijkgraaf, S. and A.J. Kalmijn. 1962. Verhaltensversuche zur Funktion der Lorenzinischen Ampullen. *Naturwissenschaften* 49:400.

Dijkgraaf, S. and A.J. Kalmijn. 1963. Untersuchungen über die Funktion der Lorenzinischen Ampullen an Haifischen. *Z. Vergl. Physiol.* 47:438–456.

Dotterweich, H. 1932. Bau und Funktion der Lorenzinischen Ampullen. *Zool. Jahrb. Abt.* 3. 50:347–418.

Dowling, J.E. and H. Ripps. 1991. On the duplex nature of the skate retina. *J. Exp. Zool.* Suppl. 5:55–65.

Doyle, J. 1963. The acid mucopolysaccharides in the glands of Lorenzini of elasmobranch fish. *Biochem. J.* 88:7.

Dryer, L. and P.P.C. Graziadei. 1993. A pilot study on morphological compartmentalization and heterogeneity in the elasmobranch olfactory bulb. *Anat. Embryol.* 188:41–51.

Dryer, L. and P.P.C. Graziadei. 1994a. Mitral cell dendrites: a comparative approach. *Anat. Embryol.* 189:91–106.

Dryer, L. and P.P.C. Graziadei. 1994b. Projections of the olfactory bulb in an elasmobranch fish, *Sphyrna tiburo*: segregation of inputs to the telencephalon. *Anat. Embryol.* 190:563–572.

Dryer, L. and P.P.C. Graziadei. 1996. Synaptology of the olfactory bulb of an elasmobranch fish, *Sphyrna tiburo. Anat. Embryol.* 193:101–114.

Ebbesson, S.O.E. 1972. New insights into the organization of the shark brain. *Comp. Biochem. Physiol.* 42A:121–129.

Ebbesson, S.O.E. 1980. On the organization of the telencephalon in elasmobranchs, in *Comparative Neurology of the Telencephalon.* S.O.E. Ebbesson, Ed., Plenum Press, New York, 1–16.

Ebbesson, S.O.E. and L. Heimer. 1970. Projections of the olfactory tract fibers in the nurse shark (*Ginglymostoma cirratum*). *Brain Res.* 17:47–55.

Ebbesson, S.O.E. and R.G. Northcutt. 1976. Neurology of anamniotic vertebrates, in *Evolution of Brain and Behavior in Vertebrates.* R.B. Masterton, M.E. Bitterman, C.B.G. Campbell, and N. Hotton, Eds., Lawrence Erlbaum Associates, Hillsdale, NJ, 115–146.

Ebbesson, S.O.E. and D.M. Schroeder, 1971. Connections of the nurse shark's telencephalon. *Science* 173:254–256.

Ewart, J.C. and H.C. Mitchell. 1892. On the lateral sense organs of elasmobranchs. II. The sensory canals of the common skate (*Raja batis*). *Trans. R. Soc. Edinb.* 37: 87–105.

Fay, R.R. 1988. *Hearing in Vertebrates: A Psychophysical Databook.* Hill-Fay Associates, Winnetka, IL.

Fay, R.R., J.I. Kendall, A.N. Popper, and A.L. Tester. 1974. Vibration detection by the macula neglecta of sharks. *Comp. Biochem. Physiol.* 47A:1235–1240.

Fiebig, E. 1988. Connections of the corpus cerebelli in the thornback guitarfish, *Platyrhinoidis triseriata* (Elasmobranchii): a study with WGA-HRP and extracellular granule cell recording. *J. Comp. Neurol.* 268:567–583.

Finger, T.E. 1975. The distribution of the olfactory tracts in the bullhead catfish, *Ictalurus nebulosus. J. Comp. Neurol.* 161:125–142.

Fouts, W.R. and D.R. Nelson. 1999. Prey capture by the Pacific angel shark, *Squatina californica*: visually mediated strikes and ambush-site characteristics. *Copeia* 1999:304–312.

Franceschini, V. and F. Ciani. 1993. Lectin binding to the olfactory system in a shark, *Scyliorhinus canicula. Fol. Histochem. Cytobiol.* 31:133–137.

Franz, V. 1931. Die Akkommodation des Selachierauges und seine Abblendungsapparate, nebst Befunden an der Retina. *Zool. Jahrb. Abt. Allg. Zool. Physiol.* 49:323–462.

Fraser, P.J. and R.L. Shelmerdine. 2002. Dogfish hair cells sense hydrostatic pressure. *Nature* 415:495–496.

Gilbert, P.W. 1963. The visual apparatus of sharks, in *Sharks and Survival.* P.W. Gilbert, Ed., D.C. Heath, Boston, 283–326.

Gilbert, P.W., E.S. Hodgson, and R.F. Mathewson. 1964. Electroencephalograms of sharks. *Science* 145:949–951.

Gordon, I. 1993. Pre-copulatory behaviour of captive sandtiger sharks, *Carcharias taurus*. *Environ. Biol. Fishes* 38:159–164.

Gould, J.L., J.L. Kirschvink, and K.D. Deffeyes. 1978. Bees have magnetic remanence. *Science* 201:1026–1028.

Graeber, R.C. 1978. Behavioral studies correlated with central nervous system integration of vision in sharks, in *Sensory Biology of Sharks, Skates, and Rays*. E.S. Hodgson and R.F. Mathewson, Eds., U.S. Office of Naval Research, Arlington, VA, 195–225.

Graeber, R.C. 1980. Telencephalic function in elasmobranchs, a behavioral perspective, in *Comparative Neurology of the Telencephalon*. S.O.E. Ebbesson, Ed., Plenum Press, New York, 17–39.

Graeber, R.C. and S.O.E. Ebbesson. 1972. Retinal projections in the lemon shark (*Negaprion brevirostris*). *Brain Behav. Evol.* 5:461–477.

Graeber, C.R., S.O.E. Ebbesson, and J.A. Jane. 1973. Visual discrimination in sharks without optic tectum. *Science* 180:413–415.

Graeber, R.C., D.M. Schroeder, J.A. Jane, and S.O.E. Ebbesson. 1978. Visual discrimination following partial telencephalic ablations in nurse sharks (*Ginglymostoma cirratum*). *J. Comp. Neurol.* 180:325–344.

Gruber, S.H. 1967. A behavioral measurement of dark adaptation in the lemon shark, *Negaprion brevirostris*, in *Sharks, Skates, and Rays*. P.W. Gilbert, R.F. Mathewson, and D.P. Rall, Eds., Johns Hopkins University Press, Baltimore, 479–490.

Gruber, S.H. and J.L. Cohen. 1978. Visual system of the elasmobranchs: state of the art 1960–1975, in *Sensory Biology of Sharks, Skates, and Rays*. E.S. Hodgson and R.F. Mathewson, Eds., U.S. Office of Naval Research, Arlington, VA, 11–105.

Gruber, S.H. and J.L. Cohen. 1985. Visual system of the white shark, *Carcharodon carcharias*, with emphasis on retinal structure. *S. Calif. Acad. Sci. Mem.* 9:61–72.

Gruber, S.H. and N. Schneiderman. 1975. Classical conditioning of the nictitating membrane response of the lemon shark (*Negaprion brevirostris*). *Behav. Res. Methods Inst.* 7:430–434.

Gruber, S.H., D.I. Hamasaki, and C.D.B. Bridges. 1963. Cones in the retina of the lemon shark (*Negaprion brevirostris*). *Vision Res.* 3:397–399.

Hamasaki, D.I. and S.H. Gruber. 1965. The photoreceptors of the nurse shark, *Ginglymostoma cirratum* and the sting ray, *Dasyatis sayi*. *Bull. Mar. Sci.* 15:1051–1059.

Harris, A.J. 1965. Eye movements of the dogfish *Squalus acanthias* L. *J. Exp. Biol.* 43:107–130.

Hassan, E.S. 1989. Hydrodynamic imaging of the surroundings by the lateral line of the blind cave fish, *Anoptichthys jordani*, in *The Mechanosensory Lateral Line: Neurobiology and Evolution*. S. Coombs, P. Görner, and H. Münz, Eds., Springer-Verlag, New York, 217–227.

Heath, A.R. 1991. The ocular tapetum lucidum: a model system for interdisciplinary studies in elasmobranch biology. *J. Exp. Zool.* Suppl. 5:41–45.

Hensel, H. 1955. Quantitative Beziehungen zwischen Temperaturreiz und Aktionspotentialen der Lorenzinischen Ampullen. *Z. Vergl. Physiol.* 37:509–526.

Herrick, C.J. 1924. *Neurological Foundations of Animal Behavior*. Henry Holt and Company; reprint edition 1965 by Hafner, New York.

Heupel, M.R., C.A. Simpfendorfer, and R.E. Hueter. 2003. Running before the storm: sharks respond to falling barometric pressure associated with Tropical Storm Gabrielle. *J. Fish Biol.* 63:1357–1363.

Hobson, E.S. 1963. Feeding behavior in three species of sharks. *Pac. Sci.* 17:171–194.

Hodgson, E.S. and R.F. Mathewson. 1971. Chemosensory orientation in sharks. *Ann. N.Y. Acad. Sci.* 188:175–182.

Hodgson, E.S. and R.F. Mathewson, Eds. 1978a. *Sensory Biology of Sharks, Skates, and Rays*. U.S. Office of Naval Research, Arlington, VA.

Hodgson, E.S. and R.F. Mathewson. 1978b. Electrophysiological studies of chemoreception in elasmobranchs, in *Sensory Biology of Sharks, Skates, and Rays*. E.S. Hodgson and R.F. Mathewson, Eds., U.S. Office of Naval Research, Arlington, VA, 227–267.

Hodgson, E.S., R.F. Mathewson, and P.W. Gilbert. 1967. Electroencephalographic studies of chemoreception in sharks, in *Sharks, Skates, and Rays*. P.W. Gilbert, R.F., Mathewson, and D.P. Rall, Eds., Johns Hopkins University Press, Baltimore, 491–501.

Howes, G.B. 1883. The presence of a tympanum in the genus *Raja*. *J. Anat. Physiol.* 17:188–191.

Hueter, R.E. 1980. Physiological Optics of the Eye of the Juvenile Lemon Shark (*Negaprion brevirostris*). M.S. thesis. University of Miami, Coral Gables, FL.

Hueter, R.E. 1991. Adaptations for spatial vision in sharks. *J. Exp. Zool.* Suppl. 5:130–141.

Hueter, R.E. and J.L. Cohen, Eds. 1991. *Vision in Elasmobranchs: A Comparative and Ecological Perspective. J. Exp. Zool.* Suppl. 5.

Hueter, R.E. and P.W. Gilbert. 1990. The sensory world of sharks, in *Discovering Sharks*. S.H. Gruber, Ed., American Littoral Society, Highlands, NJ, 48–55.

Hueter, R.E. and S.H. Gruber. 1982. Recent advances in studies of the visual system of the juvenile lemon shark (*Negaprion brevirostris*). *Fla. Sci.* 45:11–25.

Hueter, R.E., C.J. Murphy, M. Howland, J.G. Sivak, J.R. Paul-Murphy, and H.C. Howland. 2001. Refractive state and accommodation in the eyes of free-swimming versus restrained juvenile lemon sharks (*Negaprion brevirostris*). *Vision Res.* 41:1885–1889.

Johnsen, P.B. and J.H. Teeter. 1985. Behavioral responses of bonnethead sharks (*Sphyrna tiburo*) to controlled olfactory stimuli. *Mar. Behav. Physiol.* 11:283–291.

Johnson, R.H. and D.R. Nelson. 1978. Copulation and possible olfaction-mediated pair formation in two species of carcharhinid sharks. *Copeia* 1978:539–542.

Johnson, S.E. 1917. Structure and development of the sense organs of the lateral canal system of selachians (*Mustelus canis* and *Squalus acanthias*). *J. Comp. Neurol.* 28:1–74.

Kajiura, S.M. and K.N. Holland. 2002. Electroreception in juvenile scalloped hammerhead and sandbar sharks. *J. Exp. Biol.* 205:3609–3621.

Kalmijn, A.J. 1971. The electric sense of sharks and rays. *J. Exp. Biol.* 55:371–383.

Kalmijn, A.J. 1974. The detection of electric fields from inanimate and animate sources other than electric organs, in *Handbook of Sensory Physiology*. Vol. 3. A. Fessard, Ed., Springer, Berlin, 147–200.

Kalmijn, A.J. 1981. Biophysics of geomagnetic field detection. *IEEE Trans. Magn.* MAG-17:1113–1124.

Kalmijn, A.J. 1982. Electric and magnetic field detection in elasmobranch fishes. *Science* 218:916–918.

Kalmijn, A.J. 1984. Theory of electromagnetic orientation: a further analysis, in *Comparative Physiology of Sensory Systems*. L. Bolis, R.D. Keynes, and S.H.P. Madrell, Eds., Cambridge University Press, Cambridge, U.K., 525–560.

Kalmijn, A. 1988a. Hydrodynamic and acoustic field detection, in *Sensory Biology of Aquatic Animals*. J. Atema, R.R. Fay, A.N. Popper, and W.N. Tavolga, Eds., Springer-Verlag, New York, 83–130.

Kalmijn, A.J. 1988b. Detection of weak electric fields, in *Sensory Biology of Aquatic Animals*. J. Atema, R.R. Fay, A.N. Popper, and W.N. Tavolga, Eds., Springer-Verlag, New York, 151–186.

Kalmijn, A.J. 1989. Functional evolution of lateral line and inner ear sensory systems, in *The Mechanosensory Lateral Line: Neurobiology and Evolution*. S. Coombs, P. Görner, and H. Münz, Eds., Springer-Verlag, New York, 187–215.

Kalmijn, A.J. 2000. Detection and processing of electromagnetic and near-field acoustic signals in elasmobranch fishes. *Philos. Trans. R. Soc. Lond.* 355:1135–1141.

Kantner, M., W.F. Konig, and W. Reinbach. 1962. Bau und Innervation der Lorenzinischen Ampullen und deren Bedeutung als niederes Sinnesorgan. *Z. Zellforsch.* 57:124–135.

Kelly, J.C. and D.R. Nelson. 1975. Hearing thresholds of the horn shark, *Heterodontus francisci*. *J. Acoust. Soc. Am.* 58:905–909.

Kleerekoper, H. 1969. *Olfaction in Fishes*. Indiana University Press, Bloomington.

Kleerekoper, H. 1978. Chemoreception and its interaction with flow and light perception in the locomotion and orientation of some elasmobranchs, in *Sensory Biology of Sharks, Skates, and Rays*. E.S. Hodgson and R.F. Mathewson, Eds., U.S. Office of Naval Research, Arlington, VA, 269–329.

Kleerekoper, H. 1982. The role of olfaction in the orientation of fishes, in *Chemoreception in Fishes: Developments in Aquaculture and Fisheries Sciences*, Vol. 8. T.J. Hara, Ed., Elsevier, Amsterdam, 201–225.

Kleerekoper, H., D. Gruber, and J. Matis. 1975. Accuracy of localization of a chemical stimulus in flowing and stagnant water by the nurse shark, *Ginglymostoma cirratum*. *J. Comp. Physiol.* 98:257–275.

Klimley, A.P. 1980. Observations of courtship and copulation in the nurse shark, *Ginglymostoma cirratum*. *Copeia* 1980:878–882.

Klimley, A.P. 1993. Highly directional swimming by scalloped hammerhead sharks, *Sphyrna lewini* and subsurface irradiance, temperature, bathymetry, and geomagnetic field. *Mar. Biol.* 117:1–22.

Koester, D.M. 1983. Central projections of the octavolateralis nerves of the clearnose skate, *Raja eglanteria*. *J. Comp. Neurol.* 221:199–215.

Kritzler, H. and L. Wood. 1961. Provisional audiogram for the shark, *Carcharhinus leucas*. *Science* 133:1480–1482.

Kroese, A.B. and N.A.M. Schellart. 1992. Velocity- and acceleration-sensitive units in the trunk lateral line of the trout. *J. Neurophysiol.* 68:2212–2221.

Leask, M.J.M. 1977. A physicochemical mechanism for magnetic field detection by migratory birds and homing pigeons. *Nature* 359:142–144.

Loewenstein, W.R. 1960. Mechanisms of nerve impulse initiation in a pressure receptor (Lorenzinian ampulla). *Nature* 188:1034–1035.

Loewenstein, W.R. and N. Ishiko. 1962. Sodium chloride sensitivity and electrochemical effects in a Lorenzinian ampulla. *Nature* 194:292–294.

Long, D.J., K.D. Hanni, P. Pyle, J. Roletto, R.E. Jones, and R. Bandar. 1996. White shark predation on four pinniped species in central California waters: geographic and temporal patterns inferred from wounded carcasses, in *Great White Sharks: The Biology of Carcharodon carcharias*. A.P. Klimley and D.G. Ainley, Eds., Academic Press, San Diego, 263–274.

Lorenzini, S. 1678. *Osservazioni intorno alle Torpedini*, Vol. 1. Florence. 136 pp.

Lowe, C.G., B.M. Wetherbee, G.L. Crow, and A.L. Tester. 1996. Ontogenetic dietary shifts and feeding behavior of the tiger shark, *Galeocerdo cuvier*, in Hawaiian waters. *Environ. Biol. Fishes* 47:203–211.

Lowenstein, O. and T.D.M. Roberts. 1951. The localization and analysis of the responses to vibration from the isolated elasmobranch labyrinth: a contribution to the problem of the evolution of hearing in vertebrates. *J. Physiol.* 114:471–489.

Lu, Z. and A.N. Popper. 2001. Neural response directionality correlates with hair cell orientation in a teleost fish. *J. Comp. Physiol.* A 187:453–465.

Luer, C.A. and P.W. Gilbert. 1985. Mating behavior, egg deposition, incubation period, and hatching in the clearnose skate,.*Raja eglanteria. Environ. Biol. Fishes* 13:161–171.

Luiten, P.G.M. 1981a. Two visual pathways to the telencephalon in the nurse shark (*Ginglymostoma cirratum*). 1. Retinal projections. *J. Comp. Neurol.* 196:531–538.

Luiten, P.G.M. 1981b. Two visual pathways to the telencephalon in the nurse shark (*Ginglymostoma cirratum*). 2. Ascending thalamo-telencephalic connections. *J. Comp. Neurol.* 196:539–548.

Lychakov, D.V., A. Boyadzhieva-Mikhailova, I. Christov, and I.I. Evdokimov. 2000. Otolithic apparatus in Black Sea elasmobranchs. *Fisheries Res.* 46: 27–38.

Maisey, J.G. 2001. Remarks on the inner ear of elasmobranchs and its interpretation from skeletal labyrinth morphology. *J. Morphol.* 250:236–264.

Maruska, K.P. 2001. Morphology of the mechanosensory lateral line system in elasmobranch fishes: ecological and behavioral considerations. *Environ. Biol. Fishes* 60:47–75.

Maruska, K.P. and T.C. Tricas. 1998. Morphology of the mechanosensory lateral line system in the Atlantic stingray, *Dasyatis sabina*: the mechanotactile hypothesis. *J. Morphol.* 238:1–22.

Mathewson, R.F. and E.S. Hodgson. 1972. Klinotaxis and rheotaxis in orientation of sharks toward chemical stimuli. *Comp. Biochem. Physiol.* 42A:79–84.

Montgomery, J.C. 1984. Frequency response characteristics of primary and secondary neurons in the electrosensory system of the thornback ray. *Comp. Biochem. Physiol.* 79A:189–195.

Montgomery, J.C. and D. Bodznick. 1993. Hindbrain circuitry mediating common mode suppression of ventilatory reafference in the electrosensory system of the little skate, *Raja erinacea. J. Exp. Biol.* 183:203–215.

Montgomery, J.C. and D. Bodznick. 1994. An adaptive filter that cancels self-induced noise in the electrosensory and lateral line mechanosensory systems of fish. *Neurosci. Lett.* 174:145–148.

Montgomery, J.C. and J.A. MacDonald. 1990. Effects of temperature on the nervous system: implications for behavioral performance. *Am. J. Physiol.* 259:191–196.

Montgomery, J.C. and E. Skipworth. 1997. Detection of weak water jets by the short-tailed stingray *Dasyatis brevicaudata* (Pisces: Dasyatidae). *Copeia* 1997:881–883.

Montgomery, J.C., S. Coombs, and M. Halstead. 1995. Biology of the mechanosensory lateral line in fishes. *Rev. Fish Biol. Fish.* 5:399–416.

Montgomery, J.C., C.F. Baker, and A.G. Carton. 1997. The lateral line can mediate rheotaxis in fish. *Nature* 389:960–963.

Münz, H. 1985. Single unit activity in the peripheral lateral line system of the cichlid fish *Sarotherodon niloticus* L. *J. Comp. Physiol.* A 157:555–568.

Münz, H. 1989. Functional organization of the lateral line periphery, in *The Mechanosensory Lateral Line: Neurobiology and Evolution.* S. Coombs, P. Görner, and H. Münz, Eds., Springer-Verlag, New York, 285–297.

Murakami, T., Y. Morita, and H. Ito. 1983. Extrinsic and intrinsic fiber connections of the telencephalon in a teleost, *Sebastiscus marmoratus. J. Comp. Neurol.* 216:115–131.

Murphy, C.J. and H.C. Howland. 1991. The functional significance of crescent-shaped pupils and multiple pupillary apertures. *J. Exp. Zool.* Suppl. 5:22–28.

Murray, R.W. 1957. Evidence for a mechanoreceptive function of the ampullae of Lorenzini. *Nature* 179:106–107.

Murray, R.W. 1960a. The response of ampullae of Lorenzini of elasmobranchs to mechanical stimulation. *J. Exp. Biol.* 37:417–424.

Murray, R.W. 1960b. Electrical sensitivity of the ampullae of Lorenzini. *Nature* 187:957.

Murray, R.W. 1962. The response of the ampullae of Lorenzini in elasmobranchs to electrical stimulation. *J. Exp. Biol.* 39:119–128.

Murray, R.W. 1965. Receptor mechanisms in the ampullae of Lorenzini of elasmobranch fishes. *Cold Spring Harbor Symp. Quant. Biol.* 30:235–262.

Murray, R.W. and T.W. Potts. 1961. The composition of the endolymph and other fluids of elasmobranchs. *Comp. Biochem. Physiol.* 2:65–75.

Myrberg, A.A., Jr. 2001. The acoustical biology of elasmobranchs. *Environ. Biol. Fishes* 60:31–45.

Myrberg, A.A., Jr. and S.H. Gruber. 1974. The behavior of the bonnethead shark, *Sphyrna tiburo. Copeia* 1974:358–374.

Myrberg, A.A., Jr., A. Banner, and J.D. Richard. 1969. Shark attraction using a video-acoustic system. *Mar. Biol.* 2:264–276.

Myrberg, A.A., Jr., S.J. Ha, S. Walewski, and J.C. Banbury. 1972. Effectiveness of acoustic signals in attracting epipelagic sharks to an underwater sound source. *Bull. Mar. Sci.* 22:926–949.

Nelson, D.R. 1967. Hearing thresholds, frequency discrimination, and acoustic orientation in the lemon shark, *Negaprion brevirostris* (Poey). *Bull. Mar. Sci.* 17:741–767.

Nelson, D.R. and S.H. Gruber. 1963. Sharks: attraction by low-frequency sound. *Science* 142:975–977.

Nelson, P.A., S.M. Kajiura, and G.S. Losey. 2003. Exposure to solar radiation may increase ocular UV-filtering in the juvenile scalloped hammerhead shark, *Sphyrna lewini. Mar. Biol.* 142:53–56.

New, J.G. 1990. Medullary electrosensory processing in the little skate. I. Response characteristics of neurons in the dorsal octavolateralis nucleus. *J. Comp. Physiol.* 167A:285–294.

Nickel, E. and S. Fuchs. 1974. Organization and ultrastructure of mechanoreceptors (Savi vesicles) in the elasmobranch *Torpedo. J. Neurocytol.* 3:161–177.

Nicol, J.A.C. 1964. Reflectivity of the chorioidal tapeta of selachians. *J. Fish. Res. Bd. Can.* 21:1089–1100.

Norris, H.W. 1929. The distribution and innervation of the ampullae of Lorenzini of the dogfish, *Squalus acanthias*: some comparisons with conditions in other plagiostomes and corrections of prevalent errors. *J. Comp. Neurol.* 47:449–465.

Norris, H.W. 1932. The laterosensory system of *Torpedo marmorata*, innervation and morphology. *J. Comp. Neurol.* 56:169–178.

Norris, H.W. and S.P. Hughes. 1920. The cranial, occipital, and anterior spinal nerves of the dogfish, *Squalus acanthias. J. Comp. Neurol.* 31:293–402.

Northcutt, R.G. 1978. Brain organization in the cartilaginous fishes, in *Sensory Biology of Sharks, Skates, and Rays.* E.S. Hodgson and R.F. Mathewson, Eds., U.S. Office of Naval Research, Arlington, VA, 117–193.

Northcutt, R.G. 1979. Retinofugal pathways in fetal and adult spiny dogfish, *Squalus acanthias. Brain Res.* 162:219–230.

Northcutt, R.G. 1989a. The phylogenetic distribution and innervation of craniate mechanoreceptive lateral lines, in *The Mechanosensory Lateral Line: Neurobiology and Evolution.* S. Coombs, P. Görner, and H. Münz, Eds., Springer-Verlag, New York, 17–78.

Northcutt, R.G. 1989b. Brain variation and phylogenetic trends in elasmobranch fishes. *J. Exp. Zool.* Suppl. 2:83–100.

Northcutt, R.G. 1991. Visual pathways in elasmobranchs: organization and phylogenetic implications. *J. Exp. Zool.* Suppl. 5:97–107.

Pals, N., R.C. Peters, and A.A.C. Schoenhage. 1982a. Local geo-electric fields at the bottom of the sea and their relevance for electrosensitive fish. *Neth. J. Zool.* 32:479–494.

Pals, N., P. Valentijn, and D. Verwey. 1982b. Orientation reactions of the dogfish, *Scyliorhinus canicula*, to local electric fields. *Neth. J. Zool.* 32:495–512.

Parker, G.H. 1909. The influence of eyes and ears and other allied sense organs on the movement of *Mustelus canis. Bull. U.S. Bureau Fisheries* 29:43–58.

Parker, G.H. 1914. The directive influence of the sense of smell in the dogfish. *Bull. U.S. Bur. Fisheries* 33:61–68.

Parker, G.H. and R.E. Sheldon. 1913. The sense of smell in fishes. *Bull. U.S. Bur. Fisheries* 32:33–46.

Paulin, M.G. 1995. Electroreception and the compass sense of sharks. *J. Theor. Biol.* 174:325–339.

Peach, M.B. 2001. The dorso-lateral pit organs of the Port Jackson shark contribute sensory information for rheotaxis. *J. Fish. Biol.* 59:696–704.

Peach, M.B. and N.J. Marshall. 2000. The pit organs of elasmobranchs: a review. *Philos. Trans. R. Soc. Lond. B* 355:1131–1134.

Peters, R.C. and H.P. Evers. 1985. Frequency selectivity in the ampullary system of an elasmobranch fish (*Scyliorhinus canicula*). *J. Exp. Biol.* 118:99–109.

Peterson, E.H. and M.H. Rowe. 1980. Different regional specializations of neurons in the ganglion cell layer and inner plexiform layer of the California horned shark, *Heterodontus francisci. Brain Res.* 201:195–201.

Phillips, J.B. and S.C. Borland. 1992. Behavioral evidence for use of a light-dependent magnetoreception mechanism in a vertebrate. *Nature* 359:142–144.

Platt, C.J., T.H. Bullock, G. Czéh, N. Kovačevic, D.J. Konjević, and M. Gojković. 1974. Comparison of electroreceptor, mechanoreceptor, and optic evoked potentials in the brain of some rays and sharks. *J. Comp. Physiol.* 95:323–355.

Popper, A.N. and R.R. Fay. 1977. Structure and function of the elasmobranch auditory system. *Am. Zool.* 17:443–452.

Prasada Rao, P.D. and T.E. Finger. 1984. Asymmetry of the olfactory system in the winter flounder, *Pseudopleuronectes americanus. J. Comp. Neurol.* 225:492–510.

Puzdrowski, R.L. and R.B. Leonard. 1993. The octavolateral systems in the stingray, *Dasyatis sabina*. I. Primary projections of the octaval and lateral line nerves. *J. Comp. Neurol.* 332:21–37.

Raschi, W. and L.A. Mackanos. 1989. The structure of the ampullae of Lorenzini in *Dasyatis garouaensis* and its implications on the evolution of freshwater electroreceptive systems. *J. Exp. Zool.* 2:101–111.

Rasmussen, L.E.L. and M.J. Schmidt. 1992. Are sharks chemically aware of crocodiles? in *Chemical Signals in Vertebrates*, Vol. IV. R.L. Doty and D. Müller-Schwarze, Eds., Plenum Press, New York, 335–342.

Ratnam, R. and M.E. Nelson. 2000. Nonrenewal statistics of electrosensory afferent spike trains: implications for the detection of weak sensory signals. *J. Neurosci.* 20:6672–6683.

Reese, T.S. and W.M. Brightman. 1970. Olfactory surface and central olfactory connections in some vertebrates, in *Taste and Smell in Vertebrates*. G.E.W. Wolstenhome and J. Knight, Eds., Ciba Foundation Symposium/J. and A. Churchill, London, 115–149.

Reid, G. and M.L. Flonta. 2001. Physiology: cold current in thermoreceptive neurons. *Nature* 413:480.

Retzius, G. 1881. *Das Gehörorgan der Wirbelthiere*, Vol. 1. Samson and Wallin, Stockholm.

Ripps, H. and J.E. Dowling. 1991. Structural features and adaptive properties of photoreceptors in the skate retina. *J. Exp. Zool.* Suppl. 5:46–54.

Roberts, B.L. 1978. Mechanoreceptors and the behavior of elasmobranch fishes with special reference to the acoustico-lateralis system, in *Sensory Biology of Sharks, Skates, and Rays*. E.S. Hodgson and R.F. Mathewson, Eds., U.S. Office of Naval Research, Arlington, VA, 331–390.

Sand, A. 1937. The mechanism of the lateral sense organs of fishes. *Proc. R. Soc. B* 123:472–495.

Sand, A. 1938. The function of the ampullae of Lorenzini, with some observations on the effect of temperature on sensory rhythms. *Proc. R. Soc. B* 125:524–553.

Savi, P. 1844. Etudes anatomiques sur le systeme nerveux et sur l'organe electrique de la *Torpille*, in *Traité des Phenomenes Electrophysiologiques des Animaux*. C. Matteucci, Ed., Chez L. Mechelsen, Paris, 272–348.

Schroeder, D.M. and S.O.E. Ebbesson. 1974. Nonolfactory telencephalic afferents in the nurse shark (*Ginglymostoma cirratum*). *Brain Behav. Evol.* 9:121–155.

Schweitzer, J. 1983. The physiological and anatomical localization of two electroreceptive diencephalic nuclei in the thornback ray, *Platyrhinoidis triseriata. J. Comp. Physiol.* 153A:331–341.

Schweitzer, J. 1986. Functional organization of the electroreceptive midbrain in an elasmobranch (*Platyrhinoidis triseriata*): a single unit study. *J. Comp. Physiol.* 158:43–48.

Schweitzer, J. and D.A. Lowe. 1984. Mesencephalic and diencephalic cobalt-lysine injections in an elasmobranch: evidence for two parallel electrosensory pathways. *Neurosci. Lett.* 44:317–322.

Sejnowski, T.J. and M.L. Yodlowski. 1982. A freeze fracture study of the skate electroreceptors. *J. Neurocytol.* 11:897–912.

Sheldon, R.E. 1909. The reactions of the dogfish to chemical stimuli. *J. Comp. Neurol.* 19:273–311.

Sheldon, R.E. 1911. The sense of smell in selachians. *J. Exp. Zool.* 10:51–62.

Sillman, A.J., G.A. Letsinger, S. Patel, E.R. Loew, and A.P. Klimley. 1996. Visual pigments and photoreceptors in two species of shark, *Triakis semifasciata* and *Mustelus henlei*. *J. Exp. Zool.* 276:1–10.

Silver, W.L. 1979. Olfactory responses from a marine elasmobranch, the Atlantic stingray, *Dasyatis sabina*. *Mar Behav. Physiol.* 6:297–305.

Silver, W.L., J. Caprio, J.F. Blackwell, and D. Tucker. 1976. The underwater electro-olfactogram: a tool for the study of the sense of smell of marine fishes. *Experientia* 32:1216–1217.

Sisneros, J.A. and T.C. Tricas. 2000. Androgen-induced changes in the response dynamics of ampullary electrosensory primary afferent neurons. *J. Neurosci.* 20:8586–8595.

Sisneros, J.A. and T.C. Tricas. 2002a. Ontogenetic changes in the response properties of the peripheral electrosensory system in the Atlantic stingray (*Dasyatis sabina*). *Brain Behav. Evol.* 59:130–140.

Sisneros, J.A. and T.C. Tricas. 2002b. Neuroethology and life history adaptations of the elasmobranch electric sense. *J. Physiol.* (Paris) 96:379–389.

Sisneros, J.A., T.C. Tricas, and C.A. Luer. 1998. Response properties and biological function of the skate electrosensory system during ontogeny. *J. Comp. Physiol.* 183A:87–99.

Sivak, J.G. 1978a. Optical characteristics of the eye of the spiny dogfish (*Squalus acanthias*). *Rev. Can. Biol.* 37:209–217.

Sivak, J.G. 1978b. Refraction and accommodation of the elasmobranch eye, in *Sensory Biology of Sharks, Skates, and Rays*. E.S. Hodgson and R.F. Mathewson, Eds., U.S. Office of Naval Research, Arlington, VA, 107–116.

Sivak, J.G. 1991. Elasmobranch visual optics. *J. Exp. Zool.* Suppl. 5:13–21.

Sivak, J.G. and P.W. Gilbert. 1976. Refractive and histological study of accommodation in two species of sharks (*Ginglymostoma cirratum* and *Carcharhinus milberti*). *Can. J. Zool.* 54:1811–1817.

Smeets, W.J.A.J. 1983. The secondary olfactory connections in two chondrichthians, the shark *Scyliorhinus canicula* and the ray *Raja clavata*. *J. Comp. Neurol.* 218:334–344.

Smeets, W.J.A.J. 1998. Cartilaginous fishes, in *The Central Nervous System of Vertebrates*, Vol. 1. R. Nieuwenhuys, H.J. ten Donkelaar, and C. Nicholson, Eds., Springer, Berlin, 551–654.

Smeets, W.J.A.J. and R.G. Northcutt. 1987. At least one thalamotelencephalic pathway in cartilaginous fishes projects to the medium pallium. *Neurosci. Lett.* 78: 277–282.

Smeets, W.J.A.J., R. Nieuwenhuys, and B.L Roberts. 1983. *The Central Nervous System of Cartilaginous Fishes: Structure and Functional Correlations*. Springer, Berlin.

Spielman, S.L. and S.H. Gruber. 1983. Development of a contact lens for refracting aquatic animals. *Ophthal. Physiol. Opt.* 3:255–260.

Stein, R.B. 1967. The information capacity of nerve cells using a frequency code. *Biophys. J.* 7:797–826.

Stell, W.K. and P. Witkovsky. 1973. Retinal structure in the smooth dogfish, *Mustelus canis*: light microscopy of photoreceptor and horizontal cells. *J. Comp. Neurol.* 148:33–46.

Stenonis, N. 1664. De musculis et glandulis observationum specimen cum duabus epistolis quarum una ad guil. Pisonum de anatome Rajae etc. Amstelodami.

Strong, W.R., Jr. 1996. Shape discrimination and visual predatory tactics in white sharks, in *Great White Sharks: The Biology of Carcharodon carcharias*. A.P. Klimley and D.G. Ainley, Eds., Academic Press, San Diego, 229–240.

Strong, W.R., Jr., R.C. Murphy, B.D. Bruce, and D.R. Nelson. 1992. Movements and associated observations of bait-attracted white sharks, *Carcharodon carcharias*: a preliminary report. *Aust. J. Mar. Freshwater Res.* 43:13–20.

Strong, W.R., Jr., B.D. Bruce, D.R. Nelson, and R.C. Murphy. 1996. Population dynamics of white sharks in Spencer Gulf, South Australia, in *Great White Sharks: The Biology of Carcharodon carcharias*. A.P. Klimley and D.G. Ainley, Eds., Academic Press, San Diego, 401–414.

Takami, S., C.A. Luer, and P.P.C. Graziadei. 1994. Microscopic structure of the olfactory organ of the clearnose skate, *Raja eglanteria*. *Anat. Embryol.* 190:211–230.

Tester, A.L. 1963a. Olfaction, gustation, and the common chemical sense in sharks, in *Sharks and Survival*. P.W Gilbert, Ed., D.C. Heath, Boston, 255–282.

Tester, A.L. 1963b. The role of olfaction in shark predation. *Pac. Sci.* 17:145–170.

Tester, A.L. and J.I. Kendall. 1969. Morphology of the lateralis canal system in the shark genus *Carcharhinus*. *Pac. Sci.* 23:1–16.

Tester, A.L. and G.J. Nelson. 1969. Free neuromasts (pit organs) in sharks, in *Sharks, Skates, and Rays*. P.W. Gilbert, R.F. Mathewson, and D.P. Rall, Eds., Johns Hopkins University Press, Baltimore, 503–531.

Tester, A.L., J.I. Kendall, and W.B. Milisen. 1972. Morphology of the ear of the shark genus *Carcharhinus*, with particular reference to the macula neglecta. *Pac. Sci.* 26:264–274.

Theisen, B., E. Zeiske, and H. Breucker. 1986. Functional morphology of the olfactory organs in the spiny dogfish (*Squalus acanthias* L.) and the small-spotted catshark (*Scyliorhinus canicula* L.). *Acta Zool.* (Stockholm) 67:73–86.

Tolpin, W., D. Klyce, and C.H. Dohlman. 1969. Swelling properties of dogfish cornea. *Exp. Eye Res.* 8:429–437.

Tong, S.L. and T.H. Bullock. 1982. The sensory functions of the cerebellum of the thornback ray, *Platyrhinoidis triseriata*. *J. Comp. Physiol.* 148A:399–410.

Tricas, T.C. 1980. Courtship and mating-related behaviors in myliobatid rays. *Copeia* 1980:553–556.

Tricas, T.C. 1982. Bioelectric-mediated predation by swell sharks, *Cephaloscyllium ventriosum*. *Copeia* 1982:948–952.

Tricas, T.C. 2001. The neuroecology of the elasmobranch electrosensory world: why peripheral morphology shapes behavior. *Environ. Biol. Fishes* 60:77–92.

Tricas, T.C. and J.E. McCosker. 1984. Predatory behavior of the white shark (*Carcharodon carcharias*), with notes on its biology. *Proc. Calif. Acad. Sci.* 43:221–238.

Tricas, T.C. and J.G. New. 1998. Sensitivity and response dynamics of electrosensory primary afferent neurons to near threshold fields in the round stingray. *J. Comp. Physiol.* 182A:89–101.

Tricas, T.C., S.W. Michael, and J.A. Sisneros. 1995. Electrosensory optimization to conspecific phasic signals for mating. *Neurosci. Lett.* 202:29–131.

van den Berg, A.V. and A. Schuijf. 1983. Discrimination of sounds based on the phase difference between particle motion and acoustic pressure in the shark *Chiloscyllium griseum*. *Proc. R. Soc. Lond. B* 218:127–134.

Viana, F., E. de la Pena, and C. Belmonte. 2002. Specificity of cold thermotransduction is determined by differential ionic channel expression. *Nature Neurosci.* 5:254–260.

Walcott, C., J.L. Gould, and J.L. Kirschvink. 1979. Pigeons have magnets. *Science* 205:1027–1029.

Walker, M.M., C.E. Diebel, C.V. Haugh, P.M. Pankhurst, J.C. Montgomery, and C.R. Green. 1997. Structure and function of the vertebrate magnetic sense. *Nature* 390:371–376.

Walls, G.L. 1942. *The Vertebrate Eye and Its Adaptive Radiation*. Cranbrook Institute of Science; reprint edition 1967 by Hafner, New York.

Waltman, B. 1966. Electrical properties and fine structure of the ampullary canals of Lorenzini. *Acta Physiol. Scand.* 66(Suppl. 264):1–60.

Whitear, M. and R.M. Moate. 1994. Microanatomy of the taste buds in the dogfish, *Scyliorhinus canicula*. *J. Submicrosc. Cytol. Pathol.* 29:357–367.

Wisby, W.J., J.D. Richard, D.R. Nelson, and S.H. Gruber. 1964. Sound perception in elasmobranchs, in *Marine Bio-Acoustics*. W.N. Tavolga, Ed., Pergamon Press, New York, 255–268.

Wright, T. and R. Jackson. 1964. Instrumental conditioning of young sharks. *Copeia* 1964:409–412.

Yew, D.T., Y.W. Chan, M. Lee, and S. Lam. 1984. A biophysical, morphological and morphometrical survey of the eye of the small shark (*Hemiscyllium plagiosum*). *Anat. Anz. Jena* 155:355–363.

Zeiske, E., J. Caprio, and S.H. Gruber. 1986. Morphological and electrophysiological studies on the olfactory organ of the lemon shark, *Negaprion brevirostris* (Poey), in *Indo-Pacific Fish Biology: Proceedings of the Second International Conference on Indo-Pacific Fishes*. T. Uyeno, R. Arai, T. Taniuchi, and K. Matsuura, Eds., Ichthyological Society of Japan, Tokyo, 381–391.

Zeiske, E., B. Theisen, and S.H. Gruber. 1987. Functional morphology of the olfactory organ of two carcharhinid shark species. *Can. J. Zool.* 65:2406–2412.

Zigman, S. 1991. Comparative biochemistry and biophysics of elasmobranch lenses. *J. Exp. Zool.* Suppl. 5:29–40.

13

The Immune System of Sharks, Skates, and Rays

Carl A. Luer, Catherine J. Walsh, and Ashby B. Bodine

CONTENTS

13.1 Introduction

The realization that present-day elasmobranch fishes are descendants of animals that diverged from the main line of vertebrate phylogeny some 400 million years ago has provided the opportunity for some unique glimpses into the evolution of functional systems that may have had their origins during the critical transition from jawless to jawed vertebrates. A prime example is the immune system, the complex scheme of recognition reactions that are responsible for distinguishing "self" from "non-self." In understanding the phylogeny of immunity, elasmobranch fishes fill a crucial niche because, in addition to utilizing basic nonspecific mechanisms of defense against invading elements, it has become apparent in recent years that sharks and their relatives are the first animal group to possess all the components necessary to perform the specific responses associated with adaptive immunity.

In this chapter, the immune system of sharks, skates, and rays is described in terms of the cellular components and tissue sites involved, the current understanding of how various nonspecific and specific immune responses are achieved and how they compare to immune responses in higher vertebrates, the ontogeny of immune tissues and cells, and examples of how experimental approaches using elasmobranch models are advancing our knowledge of comparative immunology.

13.2   Cells of the Elasmobranch Immune System

The primary cell types characteristic of peripheral blood in higher vertebrates are also found in elasmobranch blood (Hyder et al., 1983; Fänge, 1987). These include erythrocytes (red blood cells), leukocytes (white blood cells), and thrombocytes. Of these, the leukocytes (lymphocytes, granulocytes, monocytes, and macrophages) are typically responsible for vertebrate immune functions. In addition to fully differentiated cell types, elasmobranch blood includes cells at varying stages of mitosis as well as cells in "immature" stages (Hyder et al., 1983; Walsh and Luer, in press). Although the presence of differentiating cells complicates classification of cells into recognizable categories, many morphological similarities with higher vertebrate cells exist. Even so, attempts to correlate function with specific cell types in elasmobranchs are, in many cases, inconclusive.

13.2.1   Lymphocytes

The most common leukocyte in elasmobranch peripheral blood is the lymphocyte, accounting for approximately 40 to 60% of circulating white blood cells. Morphologically, elasmobranch lymphocytes are similar to lymphocytes from other vertebrates, and occur in varying sizes reflecting their degree of maturation (Figure 13.1). The majority of circulating lymphocytes are small (mature) or medium (maturing), but large (immature) lymphocytes are present as well. As lymphocytes mature, the nucleus occupies an increasingly greater proportion of the cytoplasm (Blaxhall and Daisley, 1973), so that in mature lymphocytes, the cytoplasm is often not clearly visible.

The two principal subsets of lymphocytes in vertebrates are B lymphocytes (bursa- or bone marrow-derived lymphocytes) and T lymphocytes (thymus-derived lymphocytes). Morphologically, B lymphocytes and T lymphocytes are indistinguishable. Because of the presence of circulating antibodies (see Section 13.5.1) and the identification of Ig-producing cells using immunocytochemical techniques and electron microscopy (Ellis and Parkhouse, 1975; Kobayashi et al., 1985; Tomonaga et al., 1992), the existence of B lymphocytes at the phylogenetic level of elasmobranchs has never been an issue. However, with uncertainty in the early literature that sharks might not possess a true thymus plus the lack of success demonstrating higher vertebrate T-lymphocyte functions (see Section 13.5.2), the existence of T lymphocytes in elasmobranchs has been debated. One hypothesis suggests that elasmobranchs may have a primitive equivalent of mammalian T lymphocytes that have not yet evolved the necessary accessory molecules and signals to respond in an acute fashion (McKinney, 1992a). Further, inability to demonstrate "T-cell help" could account for some of the features observed with elasmobranch B lymphocytes, which do not differentiate into plasma or memory cells as they do in higher vertebrates (McKinney, 1992a). The unequivocal identification of the anatomical location, organ arrangement, and cellular composition of thymus (Luer et al., 1995) (see Section 13.3.1), identification of genes coding for T-cell antigen receptors and major histocompatibility gene complexes (see Section 13.5.3), and expression of genes associated with T-lymphocyte function in higher vertebrates (see Sections 13.6.2 and 13.7.3) have provided convincing evidence that T lymphocytes do exist.

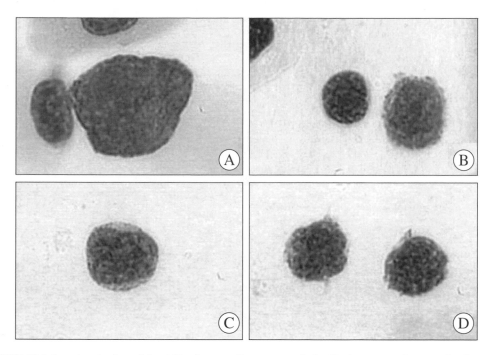

FIGURE 13.1 Lymphocytes in peripheral blood smears from a nurse shark, *Ginglymostoma cirratum*, showing a very large, immature lymphocyte (A), a small, mature lymphocyte adjacent to a medium-sized, maturing lymphocyte (B), a mature lymphocyte with a distinct nucleus and small rim of cytoplasm (C), and two mature lymphocytes (D) (stain: Wright-Giemsa; original magnification: 1000×).

13.2.2 Granulocytes

Granulocytes have been described in several species of elasmobranchs, but have been inconsistently identified and classified, probably as a result of great variability in size, shape, and staining properties of these cells (Rowley et al., 1988; Walsh and Luer, in press). Not all granulocytes found in elasmobranch blood have a clear mammalian counterpart and attempts to classify these cells have complicated the issue. As mentioned previously, immature cells are common in elasmobranch blood (Ellis, 1977) and cells with different morphologies or staining properties can be mistakenly considered to be different end cells when they actually may be different developmental stages from the same lineage (Hine and Wain, 1987).

The most common granulocyte in elasmobranch blood is referred to in nonmammalian hematology as the heterophil (analogous to the mammalian neutrophil) (Walsh and Luer, 2003). Heterophils have cytoplasmic granules of varying shapes, sizes, and staining intensity, all of which can vary among species as well as with maturity of the cell (Figure 13.2A through D). Although heterophils are the predominant granulocyte, their numbers vary widely among elasmobranch species, ranging from 20 to 50% of the total leukocytes. Infection, disease, and stressful conditions can result in even greater numbers of heterophils (Ellsaesser et al., 1985).

Eosinophilic granulocytes are also present in elasmobranch peripheral blood, although typically in much fewer numbers than heterophils. This type of granulocyte is referred to as an eosinophil and characteristically contains intensely staining granules. Eosinophils usually account for only 2 to 3% of total leukocytes, but can range from non-existent to more than 10% of the total leukocyte count. As with heterophils, granule shape in eosinophils varies with species (Walsh and Luer, in press), ranging from thin rods in the nurse shark to exceptionally large, spherical granules in the clearnose skate (Figure 13.2E and F). As in other vertebrates, eosinophils play a role in control of parasite infection and are involved in immune responses to a variety of antigens. Eosinophils have been observed to phagocytize bacteria and other foreign substances, but not with the efficiency of heterophils (Walsh and Luer, 1998). A third type of granulocyte is the basophil, which, as in higher vertebrates, is uncommon and accounts for less than 1% of the total leukocytes in elasmobranch peripheral blood (Figure 13.3A).

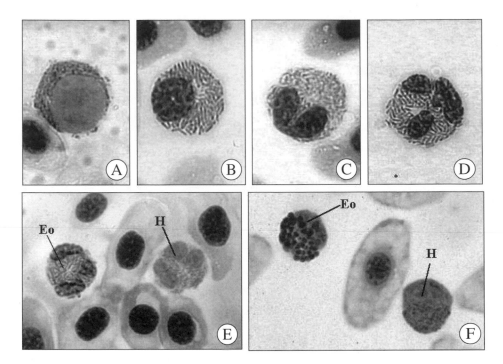

FIGURE 13.2 Representative granulocytes in peripheral blood smears showing a progression of maturing cells (A to D) and a comparison of eosinophilic and heterophilic granulocytes (E and F). Granuloblast from a blacktip shark, *Carcharhinus limbatus*, with its characteristically large nucleus (A), immature granulocyte from a nurse shark, *Ginglymostoma cirratum*, with a round, nonlobed, eccentrically positioned nucleus (B), granulocyte from a nurse shark, *G. cirratum*, with a "band"-shaped nucleus (C), and a mature granulocyte from a nurse shark, *G. cirratum*, with a segmented nucleus containing multiple lobes (D). Comparisons of granule morphology and staining intensity for representative eosinophils (Eo) and heterophils (H) are shown from a blacktip shark, *C. limbatus* (E) and a clearnose skate, *Raja eglanteria* (F) (stain: Wright-Giemsa; original magnification: 1000×).

13.2.3   Monocytes and Macrophages

Morphologically, monocytes and macrophages in the elasmobranch immune system resemble those of higher vertebrates. Monocytes are large, agranular cells with abundant cytoplasm and account for less than 3% of the leukocytes in elasmobranch peripheral blood (Walsh and Luer, in press). They are typically larger than lymphocytes and are often irregular in shape due to pseudopodial processes. The nucleus occupies less than half of the cell volume, is eccentric in location, and has a characteristic kidney-shape, often appearing to be bilobed or indented (Figure 13.3B). Nuclear chromatin in monocytes is less densely packed than in lymphocytes, and gives the nucleus a more lacelike and delicate appearance than the clumped chromatin in lymphocyte nuclei. A characteristic property of mammalian monocytes is adherence to plastic or glass, a feature also demonstrated with elasmobranch monocytes (Parish et al., 1986a; McKinney et al., 1986; Walsh and Luer, 1998).

Among higher vertebrates, the term *monocyte* typically refers to an immature, circulating cell, and the term *macrophage* describes a mature cell type found in tissues. In fish, however, a distinction is not often made, with this cell type being referred to as the monocyte/macrophage (Secombes, 1996). Hyder et al. (1983) suggested that in the nurse shark, differentiation of immature monocyte-like cells to fully differentiated macrophage-like cells takes place in the circulation, complicating the distinction between these cell types in the peripheral blood.

Functionally, monocytes are involved in nonspecific immune responses and are highly phagocytic (see Section 13.4.3). Macrophages participate in cellular defense against microbes and parasites by functioning as accessory cells for lymphocyte responses and produce reactive oxygen and nitrogen intermediates

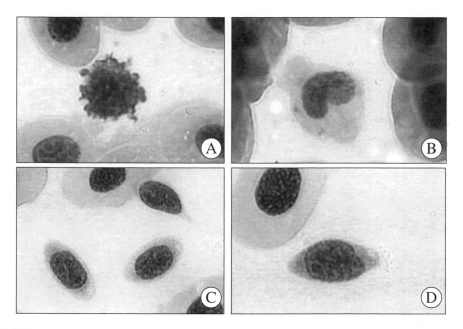

FIGURE 13.3 Miscellaneous cells in peripheral blood smears include a basophil from a blacktip shark, *Carcharhinus limbatus*, with dark granules that tend to obscure the nucleus (A), a monocyte from a cownose ray, *Rhinoptera bonasus*, depicting the characteristic indented or kidney-shaped nucleus (B), and thrombocytes from a nurse shark, *Ginglymostoma cirratum*, showing both elliptical (C) and spindle-shaped (D) forms (stain: Wright-Giemsa; original magnification: 1000×).

(see Section 13.4.1). Monocytes/macrophages also play a role in inflammation and accumulate at the site of injury or infection (Secombes, 1996).

13.2.4 Thrombocytes

A nonleukocyte cell type commonly found in elasmobranch peripheral blood is the thrombocyte. Although not routinely included as part of a differential cell count, thrombocytes can account for as much as 20% of the non-erythroid cells in the peripheral circulation (Walsh, unpubl.). In peripheral blood smears, thrombocytes can assume a variety of shapes, including spindle-shaped, elliptical, or round, probably varying with the stage of maturity or degree of reactivity (Figure 13.3C). An easily recognizable form of thrombocytes is an elongated spindle-shaped cell, with long spicules extending from either end (Figure 13.3D).

Although their role in blood clotting has not been experimentally demonstrated in elasmobranchs, thrombocytes are thought to play a role in coagulation comparable to platelets in mammals (Stokes and Firkin, 1971; Ellis, 1977). Unlike platelets, however, elasmobranch thrombocytes may have an immune function, based on observations that they can accumulate dyes and engulf latex beads and yeast cells (Stokes and Firkin, 1971; Walsh and Luer, 1998).

13.3 Lymphomyeloid Tissues

Tissue sites that provide the environments for immune cell production in elasmobranch fishes consist of both sites that are common to other vertebrate immune systems and sites that are unique to sharks, skates, and rays. Thymus and spleen, both vital to immune cell production in higher vertebrates, have their earliest phylogenetic appearance in the cartilaginous fish. In the absence of bone marrow and lymph nodes, however, alternative tissue sites, often referred to as "bone marrow equivalents" have evolved to serve remarkably similar functions.

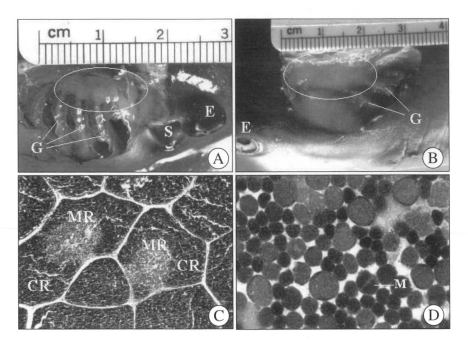

FIGURE 13.4 Dorsal views of dissections of a newborn Atlantic stingray, *Dasyatis sabina* (A) and a fetal blacknose shark, *Carcharhinus acronotus* (B), showing the anatomical location of the thymus, dorsomedial to the gill arches (G). Eyes (E) and spiracle (S) are labeled. Paraffin-embedded 10-μm section of thymus from a near-term fetal sandbar shark, *Carcharhinus plumbeus* (C), showing characteristic lobular architecture composed of cortical regions (CR) of tightly packed thymocytes and medullary regions (MR) of less densely populated thymocytes (stain: hematoxylin and eosin; original magnification: 100×). Tissue imprint of thymus from a near-term fetal southern stingray, *D. americana* (D), showing small, darkly staining mature thymocytes, larger immature thymocytes of varying sizes, and a thymocyte in the process of mitosis (M) (stain: methylene blue; original magnification: 1000×). (B, From Walsh, C.J. and C.A. Luer. 2003. In *Elasmobranch Husbandry Manual*. M. Smith et al., Eds., Special Publication of the Ohio Biological Survey 16, Columbus, OH. With permission.)

13.3.1   Thymus

The thymus is a paired organ situated dorsomedially to both gill regions (Figure 13.4A and B) (Luer et al., 1995). Because its mass and location relative to the surrounding musculature change with somatic growth and sexual maturation, the thymus is often extremely difficult to identify. As in higher vertebrates, the elasmobranch thymus is organized into distinct lobules, each lobule consisting of an outer cortex and an inner medulla (Figure 13.4C). Tissue imprints reveal that the cortex and medulla contain thymocytes at varying stages of maturation (Figure 13.4D), although only a small percentage will complete their maturation in the thymus prior to release into the peripheral circulation as thymus-derived lymphocytes (T lymphocytes).

13.3.2   Spleen

The spleen is easily recognized among elasmobranch visceral organs by its rich dark red to purplish color. In sharks, the spleen is elongate and positioned along the outer margin of the cardiac and pyloric regions of the stomach (Figure 13.5A). In batoids, however, with their relatively compressed peritoneal cavity, the spleen is more compact and situated along the inner margin of the stomach (Figure 13.5B). Histologically, the elasmobranch spleen is composed of regions of red and white pulp, giving it a structural organization that is surprisingly similar in appearance to that of higher vertebrates (Figure 13.5C). The scattered regions of white pulp are dense accumulations of small lymphocytes with asymmetrically placed central arteries. Areas of white pulp are surrounded by less dense areas of red pulp containing venous sinuses. Instead of being filled with lymph as in mammals, these sinuses are filled primarily with erythrocytes and to a lesser extent with lymphocytes (Andrew

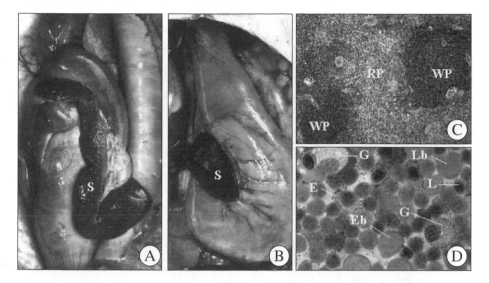

FIGURE 13.5 Ventral views of internal organs of a nurse shark, *Ginglymostoma cirratum* (A) and a clearnose skate, *Raja eglanteria* (B), showing the spleen, typically located along the outer curvature of the cardiac and pyloric regions of the stomach in sharks, and on the inner curvature of the stomach in batoids. Paraffin-embedded 10-μm section of spleen from a nurse shark, *G. cirratum* (C), showing characteristic red pulp (RP) composed of venous sinuses filled with red blood cells, and white pulp (WP) composed of dense accumulations of leukocytes (stain: hematoxylin and eosin; original magnification: 40×). Tissue imprint of spleen from a horn shark, *Heterodontus francisci* (D), showing granulocytes (G), lymphocytes (L), lymphoblasts (Lb), erythrocytes (E), and erythroblasts (Eb) (stain: methylene blue; original magnification: 1000×). (A and B, From Walsh, C.J. and C.A. Luer. 2003. In *Elasmobranch Husbandry Manual*. M. Smith et al., Eds., Special Publication of the Ohio Biological Survey 16, Columbus, OH. With permission.)

and Hickman, 1974). While the presence of mature, immature, and dividing cells in splenic imprints confirms this tissue as a site for lymphocyte production, granulocytes may also be produced in the spleen (Figure 13.5D).

13.3.3 Epigonal and Leydig Organs

The most conspicuous of the bone marrow equivalent tissues are the epigonal and Leydig organs. Both tissues, described by comparative anatomists long before their function was realized, are unique to the elasmobranch fishes (Fänge and Mattisson, 1981; Mattisson and Fänge, 1982; Honma et al., 1984). The epigonal organ continues caudally from the posterior margin of the gonads in all shark and batoid species (Figure 13.6A, B, and C). Its size and shape vary dramatically depending on the species. Because the postmortem deterioration of this tissue is extremely rapid, it is often not recognizable if an animal has been dead too long before examination. Histologically, the epigonal is composed of sinuses reminiscent of mammalian bone marrow (Figure 13.6D), except for the absence of adipose cells (fat cells). Tissue imprints demonstrate that epigonal sinuses are filled with leukocytes at various stages of maturation. Most of the cells are granulocytes, with lymphocytes present to a significant but lesser degree (Figure 13.6E).

Unlike the epigonal organ, the Leydig organ is not ubiquitous among elasmobranch species. Anecdotal observations support the notion that species possessing Leydig organs tend to have smaller epigonal organs, fueling speculation that Leydig tissue may compensate for the lack of lymphomyeloid tissue when epigonal tissue is limited. When present, Leydig organs can be visualized as whitish masses beneath the epithelium on both dorsal and ventral sides of the esophagus (Figure 13.7A). Leydig organ histology is virtually identical to that of the epigonal organ, composed of sinuses that again are reminiscent of mammalian bone marrow (Figure 13.7B and C). Tissue imprints are also similar to those of epigonal tissue, indicating leukocytes at various stages of maturation (Figure 13.7D). Again, cells are primarily granulocytes, although lymphocytes are also present.

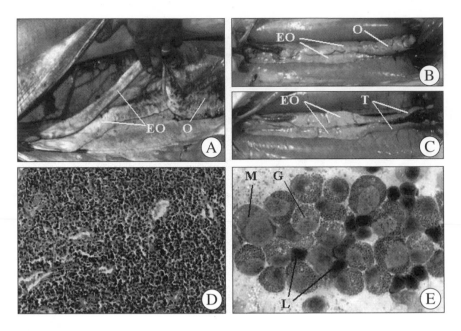

FIGURE 13.6 Dissections of a mature female blacktip shark, *Carcharhinus limbatus* (A), and immature female (B) and male (C) bonnethead sharks, *Sphyrna tiburo*, showing the anatomical location of the epigonal organ (EO), relative to the ovary (O) or testis (T). Paraffin-embedded 6-μm section of epigonal organ from clearnose skate, *Raja eglanteria* (D), showing leukocyte-filled sinuses reminiscent of mammalian bone marrow (stain: hematoxylin and eosin; original magnification: 100×). Tissue imprint of epigonal organ from a blacknose shark, *Carcharhinus acronotus* (E), showing the presence of granulocytes (G), myeloblasts (M), and lymphocytes (L) (stain: Wright-Giemsa; original magnification: 1000×).

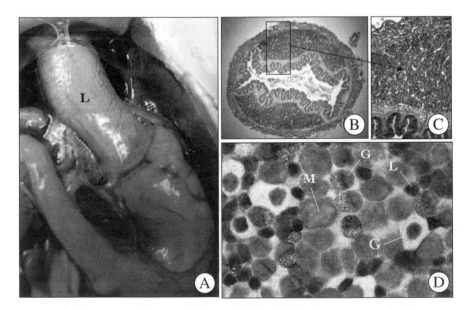

FIGURE 13.7 Ventral view of the esophagus and stomach of a clearnose skate, *Raja eglanteria* (A), showing the anatomical location of the Leydig organ (L). The organ has dorsal and ventral lobes on the respective surfaces of the esophagus (B), and consist of leukocyte-filled sinuses (C) much like the epigonal organ (stain: hematoxylin and eosin; original magnifications: 25× and 40×, respectively). Tissue imprint of Leydig organ from an Atlantic guitarfish, *Rhinobatos lentiginosus* (D), showing the presence of granulocytes (G), myeloblasts (M), and lymphocytes (L). In addition to granulocytes with darkly staining granules, granulocytes with neutrally staining granules are visible in this species (stain: Wright-Giemsa; original magnification: 1000×). (B, From Walsh, C.J. and C.A. Luer. 2003. In *Elasmobranch Husbandry Manual*. M. Smith et al., Eds., Special Publication of the Ohio Biological Survey 16, Columbus, OH. With permission.)

13.3.4 Miscellaneous Sites

In addition to the well-defined, encapsulated lymphomyeloid tissues described previously, pockets or aggregations of leukocytes can be found in various locations ranging from the intestinal mucosa to the meninges of the brain (Chiba et al., 1988; Zapata et al., 1996) and occasionally in the rectal gland (Luer and Walsh, unpubl.). Intestinal aggregations known as gut-associated lymphoid tissue, or GALT, can often be substantial (Tomonaga et al., 1986), but appear to be sites where immune cells accumulate rather than sites of immune cell production (Hart et al., 1988). The only site outside of the encapsulated lymphoid organs where cycling of leukocytes does appear to take place is the peripheral circulation. Although peripheral replication of leukocytes is not observed in higher vertebrates, flow cytometric analysis of elasmobranch peripheral blood leukocytes reveals as many as 20 to 23% of the circulating leukocytes can be actively synthesizing DNA (S-phase) and 2 to 7% can be undergoing mitosis (G_2/M phase) (Bodine et al., unpubl.).

13.4 Natural or Innate Immunity

Natural or innate immunity is common to all multicellular organisms and provides the first line of defense against invading pathogens. This type of immunity involves both preexisting and inducible defense mechanisms, and is often referred to as nonspecific since it does not depend on prior exposure or recognition of distinctive molecular structures. A distinct advantage of innate immune responses is that they typically have short lag times before reactions occur compared to specific immune responses that require considerably longer periods of activation. Even inducible innate defenses, such as inflammation, are initiated quickly and allow pathogens little time to become established. Unlike specific defense mechanisms, there is no memory component associated with innate responses and subsequent exposure to the same pathogen does not result in quicker and more intense secondary responses. Innate immune responses tend to be temperature independent, whereas specific immune defenses are typically sensitive to temperature. As a consequence, it has been postulated that innate responses play a prominent role in immune protection of poikilothermic vertebrates during periods of decreased environmental temperatures when specific immunity may be suppressed (Pettey and McKinney, 1983).

A variety of studies have clearly demonstrated that the elasmobranch immune system has the capacity to participate in innate immune responses. For example, elasmobranch immune cells demonstrate vigorous phagocytic activity (Hyder et al., 1983; Walsh and Luer, 1998) (see Section 13.4.3), strong nonspecific cytotoxic activity (Pettey and McKinney, 1981, 1983, 1988; McKinney et al., 1986), and chemotactic responses (Obenauf and Hyder Smith, 1985). In addition, elasmobranchs possess a well-characterized cascade of complement proteins (Nonaka and Smith, 2000) (see Section 13.4.2). Recently, the gene for the inflammatory cytokine interleukin-1β has been sequenced from *Scyliorhinus canicula* (Bird et al., 2002a,b), providing unequivocal evidence that inflammatory cytokines function in the elasmobranch immune system. A few of the innate responses, as they are currently understood for elasmobranchs, are described in further detail in the following subsections.

13.4.1 Nitric Oxide

Nitric oxide, a reactive nitrogen intermediate, is among the most potent host defense molecules utilized in the innate immune system. In addition to its role in combating bacterial and parasitic infections, nitric oxide also functions in intercellular signaling, vasodilation, and antitumor defense (Nathan and Hibbs, 1991; Nathan, 1992; Nussler and Billiar, 1993). Nitric oxide, released following the enzymatic conversion of arginine to citrulline, is produced in a variety of cell types by a group of enzymes referred to as nitric oxide synthases (NOS). Only the inducible form, iNOS, is transcriptionally upregulated through bacterial or cytokine challenge (Nathan, 1992).

Although nitric oxide production has been demonstrated in several species of teleost fishes (Neumann et al., 1995; Yin et al., 1997; Mulero and Meseguer, 1998; Laing et al., 1999; Campos-Perez et al., 2000; Saeij et al., 2000; Tafalla and Novoa, 2000), the recent study by Walsh et al. (unpubl.) represents the first evidence of its production by immune cells at the phylogenetic level of elasmobranchs. In this study, peripheral blood leukocytes from nurse sharks, *Ginglymostoma cirratum*, produced nitric oxide in response to stimulation with bacterial cell wall lipopolysaccharide (LPS). Maximal NO production occurred 72 h after stimulation and reached concentrations in the 5 to 20 μM range, similar to the kinetics of nitric oxide production reported for mammalian immune cells (McCarthy et al., 1995). When nurse shark leukocytes were stimulated with LPS in the presence of L-NIL, an arginine analogue and selective inhibitor of iNOS (Bryk and Wolff, 1998), nitric oxide production was significantly reduced, providing additional support that nitric oxide may be an important component of the innate defense mechanism in elasmobranchs.

13.4.2 Complement and Inflammation

The complement system in vertebrates comprises a multiprotein complex that evokes destructive activity against antigens by formation of an osmosis-disrupting membrane attack complex (MAC). In addition, the complement system provides important inflammatory mediator proteins that recruit and assist in eliciting effector functions of phagocytic cells. The complement system has both a classical pathway elicited by antibody/antigen complexes and an alternative pathway initiated by antigen surface-binding phenomena. Another pathway may be activated by carbohydrate-binding (lectin) proteins, such as the acute phase mannose-binding protein (Janeway et al., 1999).

The classical and alternative pathways of the complement cascade have been rigorously established in the nurse shark, *G. cirratum* (Jensen et al., 1981). In contrast to the nine-member mammalian complement pathway, the nurse shark classical complement pathway comprises only six functionally distinct proteins: C1n, C2n, C3n, C4n, C8n, and C9n. Early reports by Ross and Jensen (1973) revealed a strong hemolytic activity for nurse shark complement against shark antibody-sensitized sheep red blood cells. Given the wide diversity of shark antibody structures (Greenberg et al., 1996; Schluter et al., 1997), understanding the underlying mechanism of the shark complement cascade could reveal considerable knowledge about the evolution of serine protease-based amplification pathways. The cascade is activated by immune complexes and requires the presence of Ca^{2+} and Mg^{2+}. Smith et al. (1997) demonstrated significant spasmogenic and chemotactic activity in zymosan-activated pooled nurse shark sera, indicating the presence of analogues to higher vertebrate C3a and C5a.

Homologues of mammalian C1q, C3, and C4 have been isolated and shown to bear distinct similarities to those in higher vertebrates. Shark C1q has been reported to be composed of at least two chains, each bearing approximately 50% homology to mammalian C1q A and B chains. In addition, the N-terminal amino acid sequences of the α and β chains of nurse shark, *G. cirratum*, complement C3 and C4 proteins have considerable homology with the human C3 and C4 proteins (Dodds et al., 1998). The shark C3 protein has two chains while the C4 protein comprises three chains. It is apparent, therefore, that the earliest divergence of the C3 and C4 genes occurs in elasmobranchs. The structure/function of an MAC complex in the shark complement cascade is inferred based on isolation of C8 (MW = 185 kDa) and C9 (MW = 200 kDa) proteins (Smith, 1998).

Factor B/C2 cDNA clones have been isolated from two shark species, nurse shark, *G. cirratum*, and the banded houndshark, *Triakis scyllium* (Takemoto et al., 2000). This, in conjunction with the identification of a heat-labile 90-kDa factor B-like protein, provides evidence for an alternative complement pathway in sharks that may be similar to that found in mammals (Smith, 1998).

13.4.3 Phagocytosis and Pinocytosis

Phagocytosis, the internalization of cells or particles, and pinocytosis, the ingestion of nutrients and fluid, are essential components of innate immune defense (Silverstein et al., 1977). In addition to aiding in physical clearance of foreign material, phagocytic processes trigger the cascade of events involved in immune responses to foreign antigen through antigen uptake, presentation, and cytokine release (Hiemstra, 1993). In most vertebrates, neutrophils or heterophils are highly phagocytic. Eosinophils, although capable of ingesting and killing microorganisms, typically have little phagocytic activity (Hiemstra, 1993). Among elasmobranchs, phagocytic activity has been attributed to eosinophils in the lesser spotted

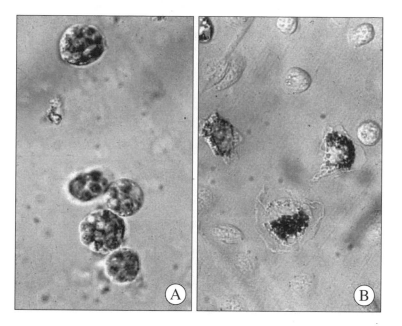

FIGURE 13.8 Light microscopy of immune cells from the clearnose skate, *Raja eglanteria*, demonstrating phagocytosis (A) and pinocytosis (B). Using differential interference contrast (DIC) optics, cells isolated from skate epigonal organ can be seen to engulf Congo Red stained yeast (A), while cells in the adherent cell population of skate peripheral blood can be seen to accumulate neutral red dye (B) following 24 h of coculture with the corresponding target (original magnification: 1000×).

dogfish (more commonly, small spotted catshark), *S. canicula* (Fänge and Pulsford, 1983), to neutrophils in the Port Jackson shark, *Heterodontus portusjacksoni* (Stokes and Firkin, 1971), and smooth dogfish, *Mustelus canis* (Weissman et al., 1978), and to neutrophils and macrophages in the nurse shark, *G. cirratum* (Hyder et al., 1983). Using the nomenclature described previously in Section 13.2.2, Walsh and Luer (1998) report that heterophils are the most actively phagocytic and pinocytic cells in the nurse shark and clearnose skate (Figure 13.8), although macrophages are also phagocytic. Thrombocytes may be involved in phagocytic processes to a limited extent (Stokes and Firkin, 1971; Walsh and Luer, 1998; Parish et al., 1986a,b), although it is not clear if they have the capacity for intracellular digestion and degradation (Secombes, 1996).

Phagocytic activity occurs without opsonization in elasmobranchs (Walsh and Luer, 1998), indicating that a direct cell particle interaction is involved. As a defense against fungi, vertebrates have developed various recognition mechanisms for β-glucan, the major structural component of yeast cell walls, including the activation of nonspecific defense mechanisms such as the alternative complement pathway (Czop and Kay, 1991). Consequently, phagocytosis of yeast by elasmobranch immune cells may occur through binding to β-glucan receptors and activating the alternative complement pathway as it does in other vertebrate species (Culbreath et al., 1991). Antibody-mediated target cell death through phagocytosis is another important receptor-mediated immune process. Throughout evolution, IgM reacts with particulate antigen and binds to Fcμ receptors, thus enhancing phagocytosis. Receptors for the Fcμ region of shark IgM are expressed on nurse shark neutrophils (Haynes et al., 1988; McKinney and Flajnik, 1997) and are likely comparable in function to the role of these receptors on the surface of immune cells in higher vertebrates.

13.4.4 Nonspecific Cytotoxic Cells

In sharks, both spontaneous and antibody-dependent cytotoxic activity have been observed (McKinney and Flajnik, 1997). Specifically, it was reported that shark neutrophils (heterophils), rather than macrophages, possess the Fc receptor for binding of IgM. It had been shown earlier (Pettey and McKinney,

1983; Haynes and McKinney, 1991) that a temperature-sensitive, non-adherent, nonphagocytic cell population was responsible for downregulation of the macrophage spontaneous cytotoxic activity at elevated environmental temperatures (>26°C). McKinney (1990) reported that target cells that had been treated with amino group reactive ligands were not recognized by shark cytotoxic macrophages. In contrast, cells treated with sulfhydryl reactive reagents were reactive toward macrophage-induced cytotoxicity. These data were interpreted to mean that shark cytotoxic macrophages interact with target cells via the target cell's surface amino groups.

13.5 Specific or Adaptive Immunity

It is widely viewed that members of Subclass Elasmobranchii represent the earliest phylogenetic group to possess all the components necessary for an adaptive immune system. These components include the presence of (1) immunoglobulin (Ig) molecules to mediate humoral immunity, (2) lymphocytes responsible for cellular immunity, and (3) rearranging immune receptor genes, including genes for heavy and light chain Ig, major histocompatibility complex (MHC), and T-cell receptor (TCR).

13.5.1   Humoral Immunity

Humoral immunity is mediated through molecules in the cell-free portion of the blood. The first true immunoglobulin to be identified in elasmobranchs was IgM (Marchalonis and Edelman, 1965, 1966), isolated initially from the smooth dogfish, *M. canis*, and confirmed soon after in lemon sharks, *Negaprion brevirostris* (Clem and Small, 1967) and nurse sharks, *G. cirratum* (Clem et al., 1967). As in mammals, elasmobranch IgM exists as a high-molecular-weight pentamer of monomeric subunits, each consisting of two heavy and two light chains covalently linked together by disulfide bonds. Unlike mammals, however, elasmobranch IgM is characterized by a marked lack of structural diversity (Rosenshein et al., 1986). Although secondary antibody responses consistent with some type of immunization have been reported (reviewed in Flajnik and Rumfelt, 2000), they require repeated monthly immunizations and occur over a much longer time interval than is seen in mammals (Mäkelä and Litman, 1980; Litman et al., 1982).

Although it was once thought that IgM was the only immunoglobulin circulating in elasmobranch blood, monomeric immunoglobulins unrelated to IgM, termed IgX or IgR, have been found in skates (Kobayashi et al., 1984; Kobayashi and Tomonaga, 1988; Harding et al., 1990b) and primitive sharks (Kobayashi et al., 1992). A long form of IgX in the clearnose skate, *Raja eglanteria* (Anderson et al., 1999), first detected in the little skate, *Leucoraja* (formerly *Raja*) *erinacea* (Harding et al., 1990a), was recently found to be orthologous to two other monomeric immunoglobulins, called IgW from sandbar sharks, *Carcharhinus plumbeus* (Bernstein et al., 1996a) and Ig new antigen receptor (NARC) from nurse sharks, *G. cirratum* (Greenberg et al., 1996). The functions of these monomeric Ig forms in humoral immune responses have yet to be fully understood.

13.5.2   Cell-Mediated Immunity

Specific immune responses that are independent of antibody are referred to as cell-mediated immunity — a term that originates from transfer of antigen-specific responses through live cells. Lymphocytes, specifically T lymphocytes, are the cells primarily responsible for cell-mediated immunity. T-lymphocyte function, at least as it occurs in higher vertebrates, has not been clearly demonstrated in cartilaginous fishes, even though TCR genes and MHC genes have been identified in several species of elasmobranchs (see Section 13.5.3), the most primitive vertebrate group possessing the MHC/TCR system. Activities and functions characteristic of allogeneic recognition responses in higher animals, such as a mixed leukocyte response (MLR) or typical graft vs. host reactions (GVHR), have yet to be demonstrated for elasmobranch immune cells. *In vitro* T-lymphocyte functions, such as cellular proliferation, are also indistinct in elasmobranchs. Sharks are capable of mounting a hapten-specific immune response, but it is associated with a lack of affinity maturation and minimal fine specificity (Mäkelä and Litman, 1980; Litman et al., 1982).

Mammalian transplantation experiments provide most of the initial understanding of cell-mediated immune processes in higher vertebrates. The cell-mediated rejection of allogeneic grafts involves antigen specificity and immunological memory, hallmark features of specific immunity, and is based on specific MHC restriction of foreign determinants. Transplantation studies in lower vertebrate groups have also been useful in investigations of cell-mediated immunity, and allograft rejection has been demonstrated in many teleost species. Transplantation studies in elasmobranchs, however, have been inconclusive, with the allograft response described as weak or chronic (Perey et al., 1968; Borysenko and Hildemann, 1970). The chronic nature of graft rejection in elasmobranchs occurs even though genes coding for both class I and II MHC have been identified in several species of elasmobranch (see Section 13.5.3). Although Nakanishi et al. (1999) demonstrated a strong relationship between MHC I sequences and intensity of skin allograft rejection in shark, indicating that classical MHC I was established at the level of elasmobranchs, Hashimoto et al. (1992) demonstrated the existence of a sequence resembling the MHC class Iα3 domains from banded houndshark, *T. scyllium*, a species that does not exhibit acute allograft rejection. Kasahara et al. (1992, 1993) also isolated cDNA clones encoding typical MHC class IIα chains from nurse sharks and demonstrated polymorphism of these genes. These observations suggest that lack of MHC polymorphism is not responsible for the absence of acute graft rejection in cartilaginous fishes (Manning and Nakanishi, 1996).

Even though genes encoding polypeptides homologous to TCR molecules have been reported in the horn shark, *H. francisci* (Rast and Litman, 1994), and four distinct classes of TCR genes occur in the clearnose skate, *R. eglanteria* (Rast et al., 1997), the possibility also exists that elasmobranchs utilize other antigen receptors besides immunoglobulin and TCR during specific immune responses (Greenberg et al., 1995).

In typical vertebrate cell-mediated responses, the first response to antigen is proliferation of the reactive cell — an important amplification stage not yet fully described in elasmobranchs. A number of substances, primarily plant lectins such as phytohemagglutinin (PHA) and concanavalin A (ConA), are specific for carbohydrate moieties on T lymphocytes and can be used to induce proliferation of T cells. Most mitogen-induced proliferation studies in lower vertebrates have been done with bony fish, as a result of availability of established techniques for good separation and optimal culture conditions (De Koning and Kaattari, 1992). These tools are not yet available for studying elasmobranch immune cells, although a few reports of T-lymphocyte-like responses in ConA- and PHA-stimulated shark leukocytes do exist (Lopez et al., 1974; Sigel et al., 1978; Pettey and McKinney, 1981). In these studies, the responding leukocyte did not express surface Ig, but also did not appear to involve MHC restriction (Haynes and McKinney, 1991). Although cells activated by T-lymphocyte mitogens apparently existed in the shark leukocyte population, a T-cell/B-cell-like heterogeneity was not clearly established.

13.5.3 Immune System Genes

In many cases the identification of genes coding for various elasmobranch immune function molecules has preceded the physical isolation of the transcript. To date, genes that have been isolated include genes coding for IgM heavy and light chains from sharks and skates (Hinds and Litman, 1986; Kokubu et al., 1988a; Shamblott and Litman, 1989; Harding et al., 1990a; Hohman et al., 1992, 1993; Anderson et al., 1995), IgX heavy chains from skates (Harding et al., 1990b; Anderson et al., 1994, 1999), IgNAR (Greenberg et al., 1995), IgW (Bernstein et al., 1996a), IgNARC (Greenberg et al., 1996) heavy chains from sharks, MHC I and MHC II genes from sharks (Hashimoto et al., 1992; Kasahara et al., 1992, 1993; Bartl and Weissman, 1994; Bartl et al., 1997; Okamura et al., 1997; Ohta et al., 2000; Bartl, 2001), all four T-cell receptor antigens from skates (Rast et al., 1997), lymphocyte-specific enzymes TdT (terminal deoxynucleotidyltransferase) and Rag-1 (recombination-activating gene) in sharks and skates (Hinds-Frey et al., 1993; Greenhalgh and Steiner, 1995; Bernstein et al., 1996b; Miracle et al., 2001), transcription factors in the Ikaros and PU.1 gene families in skates (Haire et al., 2000; Anderson et al., 2001), and novel immune receptors from skates with putative natural killer (NK) cell function (Litman et al., unpubl.).

Closer examination of how immune function genes in elasmobranchs are organized has revealed unique insights into the evolutionary origins and diversification of immunoglobulins and T-cell receptors. Unlike mammalian immunoglobulin genes, in which gene segments coding for V (variable), D (diverse), J (joining), and C (constant) regions of the antibody molecule are located in separate clusters on the same chromosome, elasmobranch immunoglobulin genes are arranged in more than 100 clusters distributed on several different chromosomes, with each cluster containing one V segment, two D segments (D_1 and D_2), and one J segment (linked to a C segment) (Kokubu et al., 1987). In mammals, antibodies result from the recombination of one gene segment from each cluster, whereas in elasmobranchs, only the four gene segments from a single cluster are recombined. This difference in gene arrangement means that the mammalian system has greater potential for antibody diversity through recombination (combinatorial diversity). The existence of a second D region in elasmobranchs, however, creates the potential for somewhat greater variation in elasmobranchs by providing an additional site for the random deletion or insertion of base pairs during recombination (junctional diversity) (Kokubu et al., 1988b; Hinds-Frey et al., 1993). Like mammals, elasmobranchs also utilize somatic mutation of rearranged genes to generate additional diversity (Hinds-Frey et al., 1993), but unlike mammals, elasmobranchs appear to rely heavily on a form of inherited diversity (called germline joining), where a large percentage (as many as half in some species) of the gene clusters in every cell is inherited with their V, D_1, D_2, and J gene segments entirely or partially prejoined (Harding et al., 1990a; Anderson et al., 1995). Further details of elasmobranch gene organization and phylogenetic relationships among the genes can be found in recent reviews (Warr, 1995; Schluter et al., 1997; Litman et al., 1999; Flajnik and Rumfelt, 2000).

Although phylogenetically "primitive" features appear to be maintained in Ig gene organization, the opposite is true for TCR genes, which closely resemble their higher vertebrate counterparts with respect to overall inferred structure (Rast and Litman, 1994) as well as diversity (Hawke et al., 1996). In the clearnose skate, *R. eglanteria,* all four TCR genes (α, β, γ, and δ) have been identified and are similar to the four mammalian gene types in comparisons of both V and C region sequences, junctional characteristics, absence of D regions in TCR α and γ genes, and presence of D regions in TCR β and δ genes (Rast et al., 1997). That the organization and diversity of TCRs have changed little over the course of vertebrate phylogeny suggests that these four TCR types were likely present in the common ancestor of the living jawed vertebrates.

13.5.4 Cytokine-Like Molecules

Cytokines are essential coordinators of the immune system and comprise a family of protein molecules that initiate and regulate both innate and adaptive immune processes. Interleukin-1 (IL-1), the first cytokine to be characterized from mammalian systems, is a major mediator of inflammation and other defense mechanisms. Although hundreds of cytokines are now well characterized in mammals, efforts to identify cytokine-like molecules in lower vertebrates are relatively few but interest in their phylogenetic significance is expanding rapidly. Tumor necrosis factor-α (TNF-α), IL-1β, several chemokines, and some cytokine receptors have now been cloned and sequenced from bony fish (Secombes et al., 2001). To date, the only cytokine gene sequenced from an elasmobranch fish is IL-1β, the complete coding sequence having been reported from the small spotted catshark, *S. canicula* (Bird et al., 2002a). Even though this gene does not have a high degree of homology with IL-1β genes from other vertebrates, expression studies in *S. canicula* indicate that IL-1β is produced in response to LPS stimulation (Bird et al., 2002b) as it is in higher vertebrate species.

Although no other cytokines or cytokine genes have been identified from elasmobranchs, cytokine-like activity has been documented from the media conditioned during short-term culture of immune cells (see Section 13.7.1). In a co-mitogenic assay, for example, the conditioned medium from nurse shark epigonal cells has been demonstrated to cause a 300-fold increase in the proliferative response of ConA-stimulated chick thymocytes compared to ConA-stimulated control thymocytes (Luer et al., unpubl.). Such a response is characteristic of IL-1-like activity.

13.6 Ontogeny of the Elasmobranch Immune System

Investigations into the ontogeny of the elasmobranch immune system are ideally suited to oviparous species, whose embryos are readily accessible and whose developmental ages can be determined if eggs are laid in captivity and can be maintained at a constant temperature that is optimal for the particular species. The clearnose skate, *R. eglanteria*, whose reproductive biology and embryonic development is well understood, serves as an excellent model (Luer and Gilbert, 1985; Luer, 1989). When eggs of this species are maintained at 20°C, embryonic development is completed in approximately 12 weeks.

13.6.1   Organogenesis of Lymphomyeloid Tissues

Examination of histological serial sections of embryos as a function of embryonic age has provided a way to determine the relative ontogenetic appearance of lymphomyeloid structures. Although the structures may be recognizable at certain ages, the expression of functional immune molecules may not actually occur until later in development (see Section 13.6.2). Figure 13.9 provides a representative glimpse of developing lymphomyeloid regions at the midpoint of embryonic development in *R. eglanteria*.

The first immune tissue to appear is the thymus, which derives from the dorsal walls of the second through the fifth pairs of pharyngeal pouches. Clusters of cells (lobes) that will become a functional thymus first appear around week 3 (by the end of the first quarter of development), but are not yet organized into cortical and medullary regions. During the time between weeks 4 and 7, lobes become separated from the pharyngeal epithelium and continue to grow laterally, eventually connecting to form a continuous mass of lobules. By the end of the second trimester (week 8) distinct cortex and medulla are apparent.

The spleen is present by the end of the first trimester (week 4), but lacks distinct regions of red and white pulp until after hatching. The Leydig organ begins to appear during week 5, visible as clusters of cells both dorsal and ventral to the developing esophagus. Tissue that will ultimately differentiate into epigonal organ is the last lymphomyeloid area to appear. Although a genital ridge is visible by the end of week 3, granulocytes do not appear until week 7, and it is not until the last quarter of development (weeks 9 through 12) that regions of both granulocytes and developing gonad are evident.

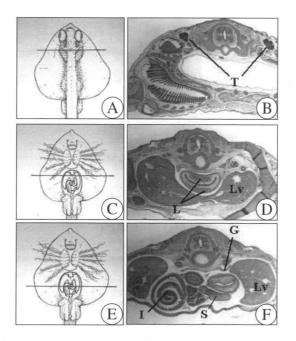

FIGURE 13.9 Paraffin-embedded cross sections of a midterm clearnose skate, *Raja eglanteria*, embryo (6 weeks into its 12-week developmental period). Figures depict the approximate location of each section and the corresponding histology through the gill region (A and B), anterior peritoneum (C and D), and mid-peritoneum (E and F), showing the developing thymus (T), Leydig organ (L), liver (Lv), gonad (G), intestine (I), and spleen (S).

13.6.2   Embryonic Expression of Immune Regulatory Genes

The most comprehensive study exploring the ontogenetic expression of genes related to immune function in elasmobranchs is the study of Miracle et al. (2001) in which a combination of real-time polymerase chain reaction (PCR) ribonuclease protection assays, and Northern blot analyses was used to evaluate expression of Ig, TCR, Rag-1, and TdT genes as a function of embryonic age in the clearnose skate, *R. eglanteria*. In this study, the ontogeny of mRNA expression and localization of gene transcripts appeared to share considerable similarity with higher vertebrate orthologues.

In characterizing the expression of TCR and Ig as a function of embryonic age, all four classes of TCR (α, β, γ, and δ) were found expressed in the skate thymus at the end of the second trimester of embryonic development (8 weeks into the 12-week developmental period). Although a dramatic upregulation of TCR gene expression occurred at this time, not all four TCR genes appeared coincidentally. TCR α, δ, and γ expression peaked at 8 weeks, while TCR β expression reached a maximum at 7 weeks and leveled off during later weeks. At this stage of embryogenesis, TCR gene expression was restricted to the thymus. Later in development, however, TCR gene expression also occurred in peripheral sites, suggesting that T lymphocytes originate in the elasmobranch thymus as they do in the thymus of higher vertebrates (Miracle et al., 2001). Interestingly, the onset of TCR genes in the skate is similar to the onset of TCR expression during embryonic development in the mouse (Fowlkes and Pardoll, 1989).

Even though TCR gene expression occurred primarily in the thymus, this tissue is not exclusively a T-lymphocyte organ during embryonic development of the skate. Several Ig genes, including IgX and IgM heavy chains and light chains I and II, were also expressed in the embryonic thymus, again with peak expression occurring 8 weeks into development as observed with TCR genes. This considerable expression of Ig genes in the developing thymus suggests this tissue may serve as a site for B-cell development as well during embryogenesis. Mature skates no longer exhibit B-cell gene expression in the thymus (Miracle et al., 2001).

In contrast to expression of TCR genes, tissue expression of Ig genes during ontogeny was far more complex and involved the spleen, Leydig, and epigonal organs. Ig and Rag-1 genes were abruptly expressed in various tissues in the 8-week embryo, suggesting B-cell development occurred at multiple sites in the developing embryo, in contrast to restriction of T-lymphocyte development in the thymus. Variation in sites of B-cell development also occurred in other vertebrates (Reynaud et al., 1987). The highest relative abundance of both IgM and IgX occurred at 8 weeks of embryonic development, but fell off dramatically by weeks 9 to 12. At 8 weeks, the highest relative abundance of IgM and IgX occurred in the spleen compared with other tissues, but relative abundance of IgX was greater than IgM in most tissues (Miracle et al., 2001).

In addition to lymphomyeloid tissues as a site of TCR and Ig gene expression during ontogeny, significant lymphoid gene expression was observed in embryonic skate liver, but not in adult liver. This observation is similar to the tissue-specific expression pattern occurring in mammals (Owen et al., 1974; Velardi and Cooper, 1984), suggesting that skate embryonic liver also may play a role in early lymphoid development (Miracle et al., 2001).

Expression of two important genes with major roles in generation of immune receptor repertoire, Rag-1 and TdT, was also found in embryonic skate thymus. Rag-1 is an integral component in the segmental rearrangement of Ig and TCR genes; TdT functions in junctional diversification of both Ig and TCR. Expression of these genes in the embryonic skate thymus, along with expression of Ig and TCR genes, suggests rearrangement and junctional diversification is occurring at this stage of development (Miracle et al., 2001).

Expression of genes for IgM and IgX heavy chains, IgM light chains I and II, and TCR α, β, γ, and δ was also detected in clearnose skate embryos using RNA *in situ* hybridization (RISH) with digoxigenin-labeled RNA probes (Walsh et al., unpubl.). Weak expression of TCR and Ig genes began to appear in the thymus at 7 weeks into development, with stronger hybridization signals at 8 weeks (i.e., by the end of the second trimester of development). Typically, TCR δ and γ chains were expressed more strongly than TCR α and β chains. Expression of Ig genes also appeared by week 8 of development in the spleen, Leydig organ, and liver. Although Miracle et al. (2001) reported a decrease in expression of some genes from the peak at 8 weeks until hatching at 12 weeks (corresponding to the final trimester of development), expression of genes detected using RISH persisted throughout development and a decrease in signal intensity was not observed.

13.7 Experimental Immunology

13.7.1 *In Vitro* Culture of Elasmobranch Immune Cells

The normal osmolarity of elasmobranch peripheral blood is approximately three times higher than that of mammalian blood (approximately 940 to 980 mOsm for elasmobranchs compared to 310 mOsm for mammals). To achieve the high osmolarity, commercially available mammalian cell culture media is adjusted to approximately 970 mOsm by increasing the salt concentration and adding urea. Elasmobranch immune cells have been successfully cultured using elasmobranch modified RPMI-1640 (E-RPMI) and elasmobranch buffered saline (E-PBS) for tissue culture and cell isolation procedures (Luer et al., 1995; Walsh and Luer, 1998; Walsh et al., 2002). E-RPMI is prepared by adding 360 mM urea and 188 mM NaCl to RPMI (Gibco, Grand Island, NY) with mammalian osmolarity. Routinely, penicillin (50 units/ml), streptomycin (50 μg/ml), neomycin (0.1 mg/ml), and amphotericin B (0.25 μg/ml) are added to control bacterial and fungal growth. Sodium bicarbonate (238 mM) is added for pH balance, and the pH of the medium adjusted to 7.2 to 7.4. E-PBS is prepared by increasing the salt concentration of normal phosphate buffered saline to achieve elasmobranch osmolarity. Reagents are filter sterilized using a 0.2 μm filter before use.

Adequate methods for separating cell populations from elasmobranch peripheral blood are not yet available. Consequently, culture of elasmobranch peripheral blood cells has primarily involved mixed cell populations, except for a few reports of cell culture separating adherent from non-adherent cell populations (Pettey and McKinney, 1983; McKinney, 1992b). Although methods for separating cell populations are desirable, the properties of elasmobranch peripheral blood cells complicate isolation and eliminate the effectiveness of traditional cell separation media. One method for achieving separation of leukocytes from erythrocytes in elasmobranch whole blood involves the use of repetitive slow speed centrifugation (10 to 20 min at $50 \times g$). Using this procedure, the denser erythrocytes will sink while PBL will remain suspended and can be aspirated and resuspended in E-PBS or E-RPMI. Before culturing, cells can be enumerated using a hemacytometer and viability assessed with trypan blue exclusion (Baur et al., 1975) using 0.2% trypan blue prepared in E-PBS. Elasmobranch cell cultures are incubated at 25°C in a humidified incubator containing 5% CO_2. Typically, cultures last for up to 96 h with greater than 90% viability, although cultures of up to 6 weeks have been maintained.

Successful culture of immune cells from elasmobranch lymphomyeloid tissues has been achieved by mincing fresh tissue into small pieces (2 to 5 mm²) using sterile scissors and forceps. Tissue pieces can then be cultured as described for PBL using serum-free E-RPMI and incubation at 25°C in a humidified incubator containing 5% CO_2. Although the length of time in culture varies, conditioned media dialyzed to remove urea and reduce salt have been shown to possess TNF-like and IL-1-like activity when cocultured against mammalian cell lines sensitive to the presence of the particular cytokine (Luer et al., unpubl.).

13.7.2 Experimental Induction of Apoptosis

Apoptosis, or programmed cell death, is an important component of the adaptive immune system and is essential in shaping the T- and B-lymphocyte repertoire during immune cell maturation. In the mammalian thymus, for example, autoreactive T-lymphocyte precursors are removed by negative selection, presumably as a result of apoptosis (Aguilar et al., 1994; Shortman and Scollay, 1994; Surh and Sprent, 1994). The removal of apoptotic cells via phagocytosis typically occurs through the activity of thymic nurse cells (De Waal Malefijt et al., 1986; Aguilar et al., 1994), a mechanism that may also be active in the elasmobranch thymus, since apoptotic cells can occasionally be observed within large cells thought to be phylogenetic counterparts of mammalian thymic nurse cells (Luer et al., 1995). Apoptosis also functions in removal of activated lymphocytes following an immune response (Fesus, 1992).

Experimental induction of apoptosis in response to glucocorticoids has been demonstrated in a variety of vertebrate groups, including mammals (Wyllie, 1980; Migliorati et al., 1994), birds (Compton et al., 1990; Machaca and Compton, 1993), amphibians (Ruben et al., 1994; Ducoroy et al., 1999), and teleost fish (Weyts et al., 1997). Recently, the first evidence for apoptosis occurring at the phylogenetic level of elasmobranchs was reported in the thymus, spleen, Leydig organ, gonad, and peripheral blood of clearnose skates, *R. eglanteria*, following *in vivo* administration of dexamethasone-21-phosphate (Walsh et al., 2002). In this study, cells undergoing apoptosis displayed similar morphologies and biochemical changes as apoptotic cells in higher vertebrates, including extensive DNA fragmentation, reduced cell volume, condensed chromatin, and plasma membrane blebbing (Fesus, 1992; Cohen et al., 1992).

13.7.3 RNA *in Situ* Hybridization

Although lymphocytes are easily recognized in blood smears and in imprints or histological sections of preserved or frozen immune tissues, the morphologies of the different lymphocyte subsets are similar enough to make them impossible to distinguish from each other using conventional staining techniques. In mammals, antibodies to cell surface antigens that are specific for either B or T lymphocytes have been linked to visible or fluorescent probes that allow the recognition of individual lymphocyte subsets. Unfortunately, mammalian antibody probes have not been useful in distinguishing lymphocyte subsets in elasmobranchs.

With the identification of heavy chain genes from IgM and IgX, variable region light chain genes (LC I and LC II), and TCR genes (α, β, γ, and δ), digoxigenin-labeled RNA probes prepared from these eight genes have been utilized in RNA *in situ* hybridization studies following procedures modified from Dijkman et al. (1995) and Panoskaltis-Mortori and Bucy (1995). Using frozen sections from a variety of tissues from adult clearnose skates, *R. eglanteria*, enzymatic detection of hybridized probes resulted in identification of lymphocytes expressing mRNA for the various B- or T-lymphocyte-specific genes (Figure 13.10). IgM and IgX heavy chain mRNA was expressed in all lymphomyeloid tissues, with IgM expression consistently greater in abundance and intensity than IgX. Hybridization of TCR probes was most intense in thymus, with expression to lesser and varying degrees in spleen, epigonal, Leydig, and intestine (Luer et al., unpubl.). None of the RNA probes demonstrated hybridization in either liver or kidney from adult specimens.

13.7.4 Transplantation Studies

Allogeneic transplantation studies (graft transplantation between two genetically different individuals of the same species) are valuable as experimental models for elucidating mechanisms of lymphocyte activation and have provided an initial understanding of cell-mediated immunity in fish. The basic phenomena appear to be similar throughout the vertebrate phyla, where recognition and rejection of grafts involves cellular infiltration with an accelerated response on secondary exposure to the same donor antigens (Manning and Nakanishi, 1996).

Allograft survival times for fish are variable and not always easy to compare, especially considering the variety of skin types and methods for determining end points of graft survival. The two existing reports in the literature investigating allograft rejection in elasmobranchs concluded that rejection is chronic with weak, if any, memory responses (Perey et al., 1968; Borysenko and Hildemann, 1970). Evaluating their results, however, is difficult because two very different skin types and drastically different criteria to determine end points were utilized. Using southern stingrays, *Dasyatis americana*, with a thin epidermis and a dermis with no protruding dermal denticles, Perey et al. (1968) used complete sloughing of the graft as the end point. Using hornsharks, *H. francisci*, with a very thick multilayered epidermis and a collagen-laden dermis containing dermal denticles projecting through the epidermis, Borysenko and Hildemann (1970) defined granulation or destruction of melanophores as a measure of rejection. In neither study did the elasmobranch skin respond satisfactorily to suturing methods used in mammals, resulting in irregularly shaped grafts that often did not adhere or that stretched or tore at the suture sites. Because of these technical difficulties, as well as high rates of mortality among experimental animals, results from these studies are difficult to interpret.

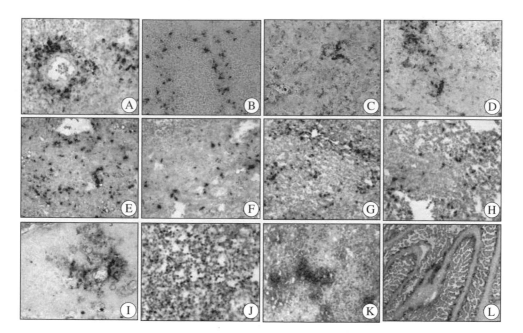

FIGURE 13.10 *In situ* hybridization of digoxigenin-labeled RNA probes. Probes specific for immunoglobulin (Ig) heavy chain (H) and light chain (LC) genes and T-cell receptor (TCR) genes (α, β, γ, and δ) identified from clearnose skate, *Raja eglanteria*, hybridize with cytoplasmic mRNA expressed in lymphocytes of various immune tissues. Cells in which hybridization occurs are visualized using an alkaline phosphatase-linked antibody to digoxigenin and appear dark against the methyl green background counterstain. (A to L) Examples of visualized RNA probes in 6- to 8-μm frozen sections of various immune tissues, described by the gene from which the probe was prepared, immune tissue examined, and original magnification. IgM H, spleen, 200× (A); IgX H, spleen, 200× (B); IgM LC I, spleen, 200× (C); IgM LC II, spleen, 200× (D); IgM H, epigonal, 200× (E); IgX H, epigonal, 200× (F); IgM H, Leydig, 200× (G); IgX H, Leydig, 200× (H); TCRα, thymus, 100× (I); TCRβ, spleen, 200× (J); TCRδ, epigonal, 100× (K); TCRδ, spiral intestine, 40× (L).

In currently ongoing transplantation studies, clearnose skate, *R. eglanteria*, adults and laboratory-raised offspring are being used (Luer et al., unpubl.). The advantage of laboratory-raised offspring is that, with known parental stock, it can be assured that donor and recipient are not related. In these studies, grafts are not sutured in place, but are rather inserted into "pockets" between the recipient skin and muscle layer. After allowing 1 week to adhere, the graft is exposed by trimming away the recipient skin. Visual examination of physical appearance and histological comparison of biopsied tissue reveal allografts that go through a period of melanocyte degranulation but are eventually accepted by the recipient. Healing at peripheral margins and infiltration of melanocytes appear to be important components of graft acceptance. At 2 to 4 weeks after transplantation, allografts appear to lose pigmentation and immune cells accumulate beneath the attachment site (Figure 13.11). However, the pale regions exist only temporarily, and pigmentation in allografts is eventually restored within 4 to 8 weeks. Healing of allografts at the peripheral margins precedes the eventual infiltration of melanocytes into the allograft surface (Figure 13.12). As expected, elapsed time before complete acceptance of allografts (5 to 12 weeks) is longer than the 2 to 4 weeks for the acceptance of control autografts (graft transplanted within the same individual).

13.7.5 Effects of Ionizing Radiation on Elasmobranch Immune System

Studies of stem cell origins in many animal species have been conducted using transplantation of suspected stem cell sources into a radiation–induced immune cell–depleted host. Little research, however, has been done on the effects of ionizing radiation on the immune system of elasmobranchs or on potential sites for stem cell recruitment and/or development (Egami et al., 1984; Pica et al., 2000).

In studies to examine the effect of radiation on the major immune organs of the clearnose skate, *R. eglanteria*, laboratory-bred hatchlings were exposed in a single dose to 0 to 75 Gy radiation delivered by

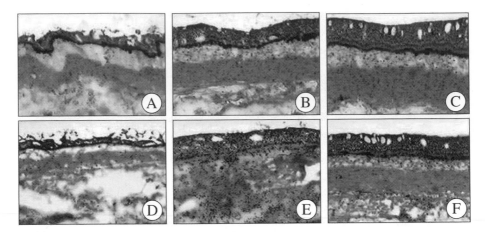

FIGURE 13.11 Paraffin-embedded cross sections of biopsied skin and underlying muscle comparing responses to autografts (A to C) and allografts (D to F) in clearnose skates, *Raja eglanteria*. Histology depicted is 1 week (A and D), 3 weeks (B and E), and 5 weeks (C and F) after transplantation of graft. Loss of pigmentation and accumulation of immune cells are visible in the 3-week allograft (E), with visual indications that normal architecture is beginning to recover by week 5 (F) (stain: hematoxylin and eosin; original magnification: 100×).

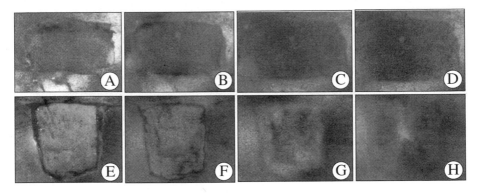

FIGURE 13.12 Visual documentation of chronic changes in autograft (A to D) and allograft (E to H) in clearnose skate, *Raja eglanteria*. Photographs were taken 1 week (A and E), 4 weeks (B and F), 8 weeks (C and G), and 12 weeks (D and H) after transplantation of grafts (original magnification, 25×).

a 6-Mv linear accelerator (Wyffels, 2001). Skates were sacrificed at 10, 12, 20, 30, or 40 days postexposure and formalin-fixed immune tissues and blood smears were examined histologically. Thymic area declined logarithmically as a function of dose beginning at 1.5 Gy and becoming asymptotic at 25 Gy, while the medulla was infiltrated with large cysts containing numerous apoptotic cells. Following a 9-Gy exposure, the thymus began to repopulate by day 30, but repopulation was incomplete by day 40. The spleen showed total destruction of the white and red pulp at exposures of 9 to 15 Gy, but at less than 9 Gy, the spleen revealed some recovery by day 30 but was still incomplete by day 40. In excess of 1.5 Gy, leukocytes decreased logarithmically as a function of radiation dose, becoming asymptotic at 24 Gy. In skates exposed to 9 Gy there was incomplete recovery of the leukocyte counts by day 40. At radiation exposures above 13.5 Gy, only the thymus revealed histological signs of recovery by day 40. For clearnose skate hatchlings, a 30-day postexposure LD_{50} for radiation was calculated to be between 9 and 15 Gy.

13.8 Future Directions

Even though considerable progress has been made in understanding the elasmobranch immune system, there are still many questions that remain unanswered. A fundamental issue yet to be resolved is the existence of a pluripotent stem cell. Established as the common precursor cell for all immune cell pathways in higher vertebrates, stem cells have not been identified from any elasmobranch species. One obstacle has been that antibodies to cell surface markers on mammalian immune cells do not recognize their elasmobranch counterparts, demonstrating the need to identify suitable elasmobranch-specific markers. The need to establish cell surface markers and the generation of antibodies to recognize them goes beyond stem cells, however, as such tools will ultimately be necessary to delineate the maturation pathways of the various blood cell lineages, to distinguish mature lymphocyte populations (such as helper T cells from cytotoxic T cells), to help define leukocyte microenvironments, and to assist in deciphering poorly understood immune functions. Cell surface markers will also be useful in developing methods to isolate individual cell types. Procedures that routinely separate blood cell types in other vertebrates have limited success when applied to elasmobranch blood, due in part to dissimilar cell sizes and densities, and drastically different osmotic conditions. The realization that experimental approaches to immune cell function have, to date, been achieved with mixed cultures underscores the importance of this future need.

For our knowledge of elasmobranch immune function to continue to advance, progress must also be made in understanding regulatory mechanisms. Critical to achieving this goal is the characterization (i.e., isolation, purification, and determination of activity profiles) of cytokines (see Section 13.5.4). In higher vertebrates, these molecules regulate not only the activity, growth, and differentiation of a variety of immune cells, including the cells from which they are secreted, but also the activity of other cytokines. Defining the sources and biological activities of elasmobranch cytokines is hoped to lead to the isolation and cloning of cytokine genes, evaluation of gene expression and function, and generation of recombinant cytokine molecules. The development and application of such molecular tools will be pivotal in the successful delineation of immunological structures, functions, and mechanisms that exist at this crucial stage of vertebrate phylogeny.

Acknowledgments

The authors gratefully acknowledge the use of the facilities at Mote Marine Laboratory (Sarasota, FL) and Clemson University (Clemson, SC). The authors wish to thank Jennifer Wyffels for sharing portions of her Ph.D. dissertation related to organogenesis and radiation exposure studies, David Noyes for technical assistance in the laboratory, and David Noyes and Tom Story for maintenance of experimental animals. We would also like to thank Gary Litman, Michele Anderson, and Jonathan Rast for providing RNA probes for the *in situ* hybridization studies. Portions of the research described in this chapter were funded by grants to C.A.L. from the Henry L. and Grace Doherty Charitable Foundation, the Melville and Sylvia Levi Fund of the Community Foundation of Sarasota County, the Vernal W. and Florence H. Bates Foundation, and the Disney Wildlife Conservation Fund, and by a grant to C.J.W. from the National Science Foundation (MCB95-09105). The authors wish to acknowledge the Ohio Biological Survey for granting permission to republish four photographs (Figures 13.4B, 13.5A, 13.5B, and 13.7B) that also appear in a chapter by C.J.W. and C.A.L. in the *Elasmobranch Husbandry Manual*.

References

Aguilar, L.K., E. Aguilar-Cordova, J. Cartwright, Jr., and J.W. Belmont. 1994. Thymic nurse cells are sites of thymocyte apoptosis. *J. Immunol.* 152:2645–2651.

Anderson, M., C. Amemiya, C. Luer, R. Litman, J. Rast, Y. Niimura, and G. Litman. 1994. Complete genomic sequence and patterns of transcription of a member of an unusual family of closely related, chromosomally dispersed immunoglobulin gene clusters in *Raja. Int. Immunol.* 6:1661–1670.

Anderson, M., M.J. Shamblott, R.T. Litman, and G.W. Litman. 1995. The generation of immunoglobulin light chain diversity in *Raja erinacea* is not associated with somatic rearrangement, an exception to a central paradigm of B cell immunity. *J. Exp. Med.* 181:109–119.

Anderson, M., S.J. Strong, R.T. Litman, C.A. Luer, C.T. Amemiya, J.P. Rast, and G.W. Litman. 1999. A long form of the skate IgX gene exhibits a striking resemblance to the new shark IgW and IgNARC genes. *Immunogenetics* 49:56–67.

Anderson, M., X. Sun, A.L. Miracle, G.W. Litman, and E.V. Rothenberg. 2001. Evolution of hematopoiesis: Three members of the PU.1 transcription factor family in a cartilaginous fish, *Raja eglanteria*. *Proc. Natl. Acad. Sci. U.S.A.* 98:553–558.

Andrew, W. and C.P. Hickman. 1974. Circulatory systems, in *Histology of the Vertebrates – A Comparative Text*. C.V. Mosby, St. Louis, MO, 133–165.

Bartl, S. 2001. New major histocompatibility complex class IIB genes from nurse shark. *Adv. Exp. Med. Biol.* 484:1–11.

Bartl, S. and I.L. Weissman. 1994. Isolation and characterization of major histocompatibility complex class IIB genes from the nurse shark. *Proc. Natl. Acad. Sci. U.S.A.* 91:262–266.

Bartl, S., M.A. Baish, M.F. Flajnik, and Y. Ohta. 1997. Identification of class I genes in cartilaginous fish, the most ancient group of vertebrates displaying an adaptive immune response. *J. Immunol.* 159:6097–6104.

Baur, H., S. Kasperek, and E. Pfaff. 1975. Criteria of viability of isolated liver cells. *Hoppe-Seylers Z. Physiol. Chem.* 356:827–838.

Bernstein, R.M., S.F. Schluter, S. Shen, and J.J. Marchalonis. 1996a. A new high molecular weight immunoglobulin class from the carcharhine shark: implications for the properties of the primordial immunoglobulin. *Proc. Natl. Acad. Sci. U.S.A.* 93:3289–3293.

Bernstein, R.M., S.F. Schluter, H. Bernstein, and J.J. Marchalonis. 1996b. Primordial emergence of the recombination activating gene 1 (RAG 1): sequence of the complete shark gene indicates homology to microbial integrases. *Proc. Natl. Acad. Sci. U.S.A.* 93:9454–9459.

Bird, S., T. Wang, J. Zou, C. Cunningham, and C.J. Secombes. 2002a. The first cytokine sequence within cartilaginous fish: IL-1 beta in the small spotted catshark (*Scyliorhinus canicula*). *J. Immunol.* 168:3329–3340.

Bird, S., J. Zou, T.B. Wang, B. Munday, C. Cunningham, and C.J. Secombes. 2002b. Evolution of interleukin-1 beta. *Cytokine Growth Factor Rev.* 13:483–502.

Blaxhall, P.C. and K.W. Daisley. 1973. Routine hematological methods for use with fish blood. *J. Fish Biol.* 5:771–782.

Borysenko, M. and W.H. Hildemann. 1970. Reactions to skin allografts in the horn shark, *Heterodontus francisci*. *Transplantation* 10:545–551.

Bryk, R. and D.J. Wolff. 1998. Mechanism of inducible nitric oxide synthase inactivation by aminoguanidine and L-N^6-(1-iminoethyl)lysine. *Biochemistry* 37:4844–4852.

Campos-Perez, J.J., M. Ward, P.S. Grabowski, A.E. Ellis, and C.J. Secombes. 2000. The gills are an important site of iNOS expression in rainbow trout *Oncorhynchus mykiss* after challenge with the gram-positive pathogen *Renibacterium salmoninarum*. *Immunology* 99:153–161.

Chiba, A., M. Torroba, Y. Honma, and A.G. Zapata. 1988. Occurrence of lymphohaemopoietic tissue in the meninges of the stingray *Dasyatis akajei* (Elasmobranchii, chondrichthyes). *Am. J. Anat.* 183:268–276.

Clem, L.W. and P.A. Small, Jr. 1967. Phylogeny of immunoglobulin structure and function. I. Immunoglobulins of the lemon shark. *J. Exp. Med.* 125:893–920.

Clem, L.W., F. DeBoutaud, and M.M. Sigel. 1967. Phylogeny of immunoglobulin structure and function. II. Immunoglobulins of the nurse shark. *J. Immunol.* 99:1226–1235.

Cohen, J.J., R.C. Duke, V.A. Fadok, and K.S. Sellins. 1992. Apoptosis and programmed cell death in immunity. *Annu. Rev. Immunol.* 10:267–293.

Compton, M.M., P.S. Gibbs, and L.R. Swicegood. 1990. Glucocorticoid-mediated activation of DNA degradation in avian lymphocytes. *Gen. Comp. Endocrinol.* 80:68–79.

Culbreath, L., S.L. Smith, and S.D. Obenauf. 1991. Alternative complement pathway activity in nurse shark serum. *Am. Zool.* 31:131A.

Czop, J.K. and J. Kay. 1991. Isolation and characterization of β-glucan receptors on human mononuclear phagocytes. *J. Exp. Med.* 173:1511–1520.

De Koning, J. and S.L. Kaattari. 1992. Use of homologous salmonid plasma for the improved responsiveness of salmonid leukocyte cultures, in *Techniques in Fish Immunology*. II. J.S. Stolen et al., Eds., SOS Publications, Fair Haven, NJ, 61–65.

De Waal Malefijt, R., W. Leene, P.J.M. Roholl, J. Wormmeester, and K.A. Hoeben. 1986. T cell differentiation within thymic nurse cells. *Lab. Invest.* 55:25–34.

Dijkman, H.B.P.M., S. Mentzel, A.S. DeJohng, and K.J.M. Assmann. 1995. RNA *in situ* hybridization using digoxigenin-labeled cRNA probes. *Biochemica* 2:23–27.

Dodds, A.W., S.L. Smith, R.P. Levine, and A.C. Willis. 1998. Isolation and initial characterization of complement components C3 and C4 of the nurse shark and the channel catfish. *Dev. Comp. Immunol.* 22:207–216.

Ducoroy, P., M. Lesourd, M.R. Padros, and A. Tournefier. 1999. Natural and induced apoptosis during lymphocyte development in the axolotl. *Dev. Comp. Immunol.* 23:241–252.

Egami, N., H. Mitani, Y. Shimada, J. Suzuki, Y. Akimoto, N. Onizuka, A. Kanamori, A. Shimada, S. Amemeya, and T. Satao. 1984. A note of the acute radiation death of sharks. *J. Fac. Sci. Univ. Tokyo* 15:363–365.

Ellis, A.E. 1977. The leucocytes of fish: a review. *J. Fish Biol.* 11:453–491.

Ellis, A.E. and R.M.E. Parkhouse. 1975. Surface immunoglobulins on the lymphocytes of the skate *Raja naevus. Eur. J. Immunol.* 5:726–728.

Ellsaesser, C.F., N.W. Miller, and M.A. Cuchens. 1985. Analysis of channel catfish peripheral blood leukocytes by bright-field microscopy and flow cytometry. *Trans. Am. Fish. Soc.* 114:279–285.

Fänge, R. 1987. Lymphomyeloid system and blood cell morphology in elasmobranchs. *Arch. Biol.* 98:187–208.

Fänge, R. and A. Mattisson. 1981. The lymphomyeloid (hemopoietic) system of the Atlantic nurse shark, *Ginglymostoma cirratum. Biol. Bull.* 160:240–249.

Fänge, R. and A. Pulsford. 1983. Structural studies on lymphomyeloid tissues of the dogfish, *Scyliorhinus canicula* L. *Cell Tissue Res.* 230:337–351.

Fesus, L. 1992. Apoptosis. *Immunol. Today* 13:A16–A17.

Flajnik, M.F. and L.L. Rumfelt. 2000. The immune system of cartilaginous fish. *Curr. Top. Microbiol. Immunol.* 248:249–270.

Fowlkes, B.J. and D.M. Pardoll. 1989. Molecular and cellular events of T cell development. *Adv. Immunol.* 44:207–264.

Greenberg, A.S., D. Avila, M. Hughes, A. Hughes, E.C. McKinney, and M.F. Flajnik. 1995. A new antigen receptor gene family that undergoes rearrangement and extensive somatic diversification in sharks. *Nature* 374:168–173.

Greenberg, A.S., A.L. Hughes, J. Guo, D. Avila, E.C. McKinney, and M.F. Flajnik. 1996. A novel "chimeric" antibody class in cartilaginous fish: IgM may not be the primordial immunoglobulin. *Eur. J. Immunol.* 26:1123–1129.

Greenhalgh, P. and L.A. Steiner. 1995. Recombination activating gene 1 (Rag1) in zebrafish and shark. *Immunogenetics* 41:54–55.

Haire, R.N., A.L. Miracle, J.P. Rast, and G.W. Litman. 2000. Members of the Ikaros gene family are present in early representative vertebrates. *J. Immunol.* 165:306–312.

Harding, F.A., N. Cohen, and G.W. Litman. 1990a. Immunoglobulin heavy chain gene organization and complexity in the skate, *Raja erinacea, Nucleic Acids Res.* 18:1015–1020.

Harding, F.A., C.T. Amemiya, R.T. Litman, N. Cohen, and G.W. Litman. 1990b. Two distinct immunoglobulin heavy chain isotypes in a primitive, cartilaginous fish, *Raja erinacea, Nucleic Acids Res.* 18:6369–6376.

Hart, S., A.B. Wrathmell, J.E. Harris, and T.H. Grayson. 1988. Gut immunology in fish: a review. *Dev. Comp. Immunol.* 12:453–480.

Hashimoto, K., T. Nakanishi, and Y. Kurosawa. 1992. Identification of a shark sequence resembling the major histocompatibility complex class I α3 domain. *Proc. Natl. Acad. Sci. U.S.A.* 89:2209–2212.

Hawke, N.A., J.P. Rast, and G.W. Litman. 1996. Extensive diversity of transcribed TCR-β in a phylogenetically primitive vertebrate. *J. Immunol.* 156:2458–2464.

Haynes, L. and E.C. McKinney. 1991. Shark spontaneous cytotoxicity: characterization of the regulatory cell. *Dev. Comp. Immunol.* 15:123–134.

Haynes, L., L. Fuller, and E.C. McKinney. 1988. Fc receptor for shark immunoglobulin. *Dev. Comp. Immunol.* 12:561–571.

Hiemstra, P.S. 1993. Role of neutrophils and mononuclear phagocytes in host defense and inflammation. *J. Int. Fed. Clin. Chem.* 5:94–99.

Hinds, K.R. and G.W. Litman. 1986. Major reorganization of immunoglobulin V_H segmental elements during vertebrate evolution. *Nature* 320:546–549.

Hinds-Frey, K.R., H. Nishikata, R.T. Litman, and G.W. Litman. 1993. Somatic variation precedes extensive diversification of germline sequences and combinatorial joining in the evolution of immunoglobulin heavy chain diversity. *J. Exp. Med.* 178:825–834.

Hine, P.M. and J.M. Wain. 1987. The enzyme cytochemistry and composition of elasmobranch granulocytes. *J. Fish Biol.* 30:465–476.

Hohman, V.S., S.F. Schluter, and J.J. Marchalonis. 1992. Complete sequence of a cDNA clone specifying sandbar shark immunoglobulin light chain: gene organization and implications for the evolution of light chains. *Proc. Natl. Acad. Sci. U.S.A.* 89:276–280.

Hohman, V.S., D.B. Schuchman, S.F. Schluter, and J.J. Marchalonis. 1993. Genomic clone for the sandbar shark lambda light chain: generation of diversity in the absence of gene rearrangement. *Proc. Natl. Acad. Sci. U.S.A.* 90:9882–9886.

Honma, Y., K. Okabe, and A. Chiba. 1984. Comparative histology of the Leydig and epigonal organs in some elasmobranchs. *Jpn. J. Ichthyol.* 31:47–54.

Hyder, S.L., M.L. Cayer, and C.L. Pettey. 1983. Cell types in peripheral blood of the nurse shark: an approach to structure and function. *Tissue Cell* 15:437–455.

Janeway, C.A., P. Travers, and M. Walport. 1999. The complement system in humoral immunity, in *Immuno- biology: The Immune System in Health and Disease.* Garland, New York, 339–359.

Jensen, J.A., E. Festa, D.S. Smith, and M. Cayer. 1981. The complement system of the nurse shark: hemolytic and comparative aspects. *Science* 214:566–569.

Kasahara, M., M. Vazquez, K. Sato, E.C. McKinney, and M.F. Flajnik. 1992. Evolution of the major histo- compatibility complex: Isolation of class II α cDNA clones from the cartilaginous fish. *Proc. Natl. Acad. Sci. U.S.A.* 89:6688–6692.

Kasahara, M., E.C. McKinney, M.F. Flajnik, and T. Ishibashi. 1993. The evolutionary origin of the major histocompatibility complex: polymorphism of class II α chain genes in the cartilaginous fish. *Eur. J. Immunol.* 23:2160–2165.

Kobayashi, K. and S. Tomonaga. 1988. The second immunoglobulin class is commonly present in cartilaginous fish belonging to the order Rajiformes. *Mol. Immunol.* 25:115–120.

Kobayashi, K., S. Tomonaga, and T. Kajii. 1984. A second class of immunoglobulin other than IgM present in the serum of a cartilaginous fish, the skate, *Raja kenojei*: isolation and characterization. *Mol. Immunol.* 21:397–404.

Kobayashi, K., S. Tomonaga, K. Teshima, and T. Kajii. 1985. Ontogenic studies on the appearance of two classes of immunoglobulin-forming cells in the spleen of the Aleutian skate, *Bathyraja aleutica*, a cartilaginous fish. *Eur. J. Immunol.* 15:952–956.

Kobayashi, K., S. Tomonaga, and S. Tanaka. 1992. Identification of a second immunoglobulin in the most primitive shark, the frill shark, *Chlamydoselachus anguineus. Dev. Comp. Immunol.* 16:295–299.

Kokubu, F., K. Hinds, R. Litman, M.J. Shamblott, and G.W. Litman. 1987. Extensive families of constant region genes in a phylogenetically primitive vertebrate indicate an additional level of immunoglobulin complexity. *Proc. Natl. Acad. Sci. U.S.A.* 84:5868–5872.

Kokubu, F., K. Hinds, R. Litman, M.J. Shamblott, and G.W. Litman, G.W. 1988a. Complete structure and organization of immunoglobulin heavy chain constant region genes in a phylogenetically primitive vertebrate. *EMBO J.* 7:1979–1988.

Kokubu, F., R. Litman, M.J. Shamblott, K. Hinds, and G.W. Litman. 1988b. Diverse organization of immu- noglobulin V_H gene loci in a primitive vertebrate. *EMBO J.* 7:3413–3422.

Laing, K.J., L.J. Hardie, W. Aartsen, P.S. Grabowski, and C.J. Secombes. 1999. Expression of an inducible nitric oxide synthase gene in rainbow trout, *Oncorhynchus mykiss. Dev. Comp. Immunol.* 23:71–85.

Litman, G.W., B.W. Erickson, L. Lederman, and O. Mäkelä. 1982. Antibody response in *Heterodontus. Mol. Cell Biochem.* 45:49–57.

Litman, G.W., M.K. Anderson, and J.P. Rast. 1999. Evolution of antigen binding receptors. *Annu. Rev. Immunol.* 17:109–147.

Lopez, D.M., M.M. Sigel, and J.C. Lee. 1974. Phylogenetic studies on T cells. I. Lymphocytes of the shark with differential response to phytohemagglutinin and concanavalin A. *Cell. Immunol.* 10:287–293.

Luer, C.A. 1989. Elasmobranchs (sharks, skates, and rays) as animal models for biomedical research, in *Nonmammalian Models for Biomedical Research.* A. Woodhead, Ed., CRC Press, Boca Raton, FL, 121–147.

Luer, C.A. and P.W. Gilbert. 1985. Mating behavior, egg deposition, incubation period, and hatching in the clearnose skate, *Raja eglanteria. Environ. Biol. Fish.* 13:161–171.

Luer, C.A., C.J. Walsh, A.B. Bodine, J.T. Wyffels, and T.R. Scott. 1995. The elasmobranch thymus: anatomical, histological, and preliminary functional characterization. *J. Exp. Zool.* 273:342–354.

Machaca, K. and M.M. Compton. 1993. Analysis of thymic lymphocyte apoptosis using *in vitro* techniques. *Dev. Comp. Immunol.* 17:263–276.

Mäkelä, O. and G.W. Litman. 1980. Lack of heterogeneity in anti-hapten antibodies of a phylogenetically primitive shark. *Nature* 287:639–640.

Manning, M.J. and T. Nakanishi. 1996. The specific immune system: cellular defenses, in *The Fish Immune System*. G. Iwama and T. Nakanishi, Eds., Academic Press, San Diego, 160–205.

Marchalonis, J.J. and G.M. Edelman, G.M. 1965. Phylogenetic origins of antibody structure. I. Multichain structure of immunoglobulins in the smooth dogfish (*Mustelus canis*). *J. Exp. Med.* 122:601–618.

Marchalonis, J.J. and G.M. Edelman. 1966. Polypeptide chains of immunoglobulins of the smooth dogfish (*Mustelus canis*). *Science* 154:1567–1568.

Mattison, A. and R. Fänge. 1982. The cellular structure of the Leydig organ in the shark, *Etmopterus spinax* (L.). *Biol. Bull.* 162:182–194.

McCarthy, J.E., P.H. Redmond, S.M. Duggan, R.W.G. Watson, C.M. Condron, J.R. O'Donnell, and D.J. Bouchier-Hayes. 1995. Characterization of the defects in murine peritoneal macrophage function in the early postsplenectomy period. *J. Immunol.* 155:387–396.

McKinney, E.C. 1990. Shark cytotoxic macrophages interact with target membrane amino groups. *Cell. Immunol.* 127:506–513.

McKinney, E.C. 1992a. Shark lymphocytes: primitive antigen reactive cells. *Annu. Rev. Fish Dis.* 2:43–51.

McKinney, E.C. 1992b. Proliferation of shark leukocytes. *In Vitro Cell. Dev. Biol.* 28A:303–305.

McKinney, E.C. and M.F. Flajnik. 1997. IgM-mediated opsonization and cytotoxicity in the shark. *J. Leukocyte Biol.* 61:141–146.

McKinney, E.C., L. Haynes, and A.L. Droese. 1986. Macrophage-like effector of spontaneous cytotoxicity from the shark. *Dev. Comp. Immunol.* 10:497–508.

Migliorati, G., I. Nicoletti, G. Nocentini, M.C. Pagliacci, and C. Roccordi. 1994. Dexamethasone and interleukins modulate apoptosis of murine thymocytes and peripheral T-lymphocytes. *Pharmacol. Res.* 30:43–52.

Miracle, A.L., M.K. Anderson, R.T. Litman, C.J. Walsh, C.A. Luer, E.V. Rothenberg, and G.W. Litman. 2001. Complex expression patterns of lymphocyte-specific genes during the development of cartilaginous fish implicate unique lymphoid tissues in generating an immune repertoire. *Int. Immunol.* 13:567–580.

Mulero, V. and J. Meseguer. 1998. Functional characterization of a macrophage-activating factor produced by leucocytes of gilthead seabram (*Sparus aurata* L.). *Fish Shellfish Immunol.* 8:143–156.

Nakanishi, T., K. Aoyagi, C. Xia, J.M. Dijkstra, and M. Ototake. 1999. Specific cell-mediated immunity in fish. *Vet. Immunol. Immunopathol.* 72:101–109.

Nathan, C. 1992. Nitric oxide as a secretory product of mammalian cells. *FASEB J.* 6:3051–3064.

Nathan, C. and J.B. Hibbs, Jr. 1991. Role of nitric oxide synthesis in macrophage antimicrobial activity. *Curr. Opin. Immunol.* 3:65–70.

Neumann, N.F., D. Faga, and M. Belosevic. 1995. Macrophage activating factor(s) secreted by mitogen stimulated goldfish kidney leukocytes synergize with bacterial lipopolysaccharide to induce nitric oxide production in teleost macrophages. *Dev. Comp. Immunol.* 19:473–482.

Nonaka, M. and S.L. Smith. 2000. Complement system of bony and cartilaginous fish. *Fish Shellfish Immunol.* 10:215–228.

Nussler, A. and T. Billiar. 1993. Inflammation, immunoregulation, and inducible nitric oxide synthase. *J. Leukocyte Biol.* 54:171–178.

Obenauf, S.D. and S. Hyder Smith. 1985. Chemotaxis of nurse shark leukocytes. *Dev. Comp. Immunol.* 9:221–230.

Ohta, Y., K. Okamura, E.C. McKinney, S. Bartl, K. Hashimoto, and M.F. Flajnik. 2000. Primitive synteny of vertebrate major histocompatibility complex class I and class II genes. *Proc. Natl. Acad. Sci. U.S.A.* 97:4712–4717.

Okamura, K., M. Ototake, T. Nakanishi, Y. Kurosawa, and K. Hashimoto. 1997. The most primitive vertebrates possess highly polymorphic class I genes comparable to those of humans. *Immunity* 7:777–790.

Owen, J.J., M.D. Cooper, and M.C. Raff. 1974. *In vitro* generation of B lymphocytes in mouse foetal liver, a mammalian "bursa equivalent." *Nature* 249:361–363.

Panoskaltis-Mortori, A. and R.P. Bucy. 1995. *In situ* hybridization with digoxigenin-labeled RNA probes: facts and artifacts. *BioTechniques* 18:300–307.

Parish, N., A. Wrathmell, S. Hart, and J.E. Harris. 1986a. Phagocytic cells in the dogfish, *Scyliorhinus canicula* L. I. *In vitro* studies. *Acta Zool.* (Stockholm) 67:215–224.

Parish, N., A. Wrathmell, S. Hart, and J.E. Harris. 1986b. Phagocytic cells in the peripheral blood of the dogfish, *Scyliorhinus canicula* L. II. *In vivo* studies. *Acta Zool.* (Stockholm) 67:225–234.

Perey, D.Y.E., J. Finstad, B. Pollara, and R.A. Good. 1968. Evolution of the immune response. VI. First and second set skin homograft rejections in primitive fishes. *Lab. Invest.* 19:591–597.

Pettey, C.L. and E.C. McKinney. 1981. Mitogen-induced cytotoxicity in the nurse shark. *Dev. Comp. Immunol.* 5:53–64.

Pettey, C.L. and E.C. McKinney. 1983. Temperature and cellular regulation in the shark. *Eur. J. Immunol.* 13:133–138.

Pettey, C.L. and E.C. McKinney. 1988. Induction of cell-mediated cytotoxicity by shark 19S IgM. *Cell. Immunol.* 111:28–38.

Pica, A., L. Cristino, F. Sasso, and P. Guerriero. 2000. Haemopoietic regeneration after autohaemotransplant in sub-lethal X-irradiated marbled electric rays. *Comp. Haematol. Int.* 10:43–49.

Rast, J.P. and G.W. Litman. 1994. T cell receptor gene homologs are present in the most primitive jawed vertebrates. *Proc. Natl. Acad. Sci. U.S.A.* 91:9248–9252.

Rast, J.P., M.K. Anderson, S.J. Strong, C. Luer, R.T. Litman, and G.W. Litman. 1997. α, β, γ, and δ T cell antigen receptor genes arose early in vertebrate phylogeny. *Immunity* 6:1–11.

Reynaud, C.-A., V. Anquez, H. Grimal, and J.-C. Weill. 1987. A hyperconversion mechanism generates the chicken light chain preimmune repertoire. *Cell* 48:379–388.

Rosenshein, I.L., S.F. Schluter, and J.J. Marchalonis. 1986. Conservation among the immunoglobulins of carcharhine sharks and phylogenetic conservation of variable region determinants. *Vet. Immunol. Immunopathol.* 12:13–20.

Ross, G. and J. Jensen. 1973. The first component (C1n) of the complement system of the nurse shark (*Ginglymostoma cirratum*). *J. Immunol.* 110:175–182.

Rowley, T.C., T.C. Hunt, M. Page, and G. Mainwaring. 1988. Fish, in *Vertebrate Blood Cells*. Cambridge University Press, Cambridge, U.K., 19–127.

Ruben, L.N., D.R. Buchholz, P. Ahmadi, R.O. Johnson, R.H. Clothier, and S. Shiigi. 1994. Apoptosis in thymus of adult *Xenopus laevis*. *Dev. Comp. Immunol.* 18:231–238.

Saeij, J.P., R.J. Stet, A. Groeneveld, L. Verburg van Kemenade, W. Muiswinkel, and G.F. Wiegertjes. 2000. Molecular and functional characterization of a fish inducible-type nitric oxide synthase. *Immunogenetics* 51:339–346.

Schluter, S.F., R.M. Bernstein, and J.J. Marchalonis. 1997. Molecular origins and evolution of immunoglobulin heavy-chain genes of jawed vertebrates. *Immunol. Today* 81:543–548.

Secombes, C.J. 1996. The nonspecific immune system: cellular defenses, in *The Fish Immune System*. G. Iwama and T. Nakanishi, Eds., Academic Press, San Diego, 63–103.

Secombes, C.J., T. Wang, S. Hong, S. Peddie, M. Crampe, K.J. Laing, C. Cunningham, and J. Zou. 2001. Cytokines and innate immunity of fish. *Dev. Comp. Immunol.* 25:713–723.

Shamblott, M.J. and G.W. Litman. 1989. Complete nucleotide sequence of primitive vertebrate immunoglobulin light chain genes. *Proc. Natl. Acad. Sci. U.S.A.* 86:4684–4688.

Shortman, K. and R. Scollay. 1994. Death in the thymus. *Nature* 372:44–45.

Sigel, M.M., J.C. Lee, E.C. McKinney, and D.M. Lopez. 1978. Cellular immunity in fish as measured by lymphocyte stimulation. *Mar. Fish Rev.* 40:6–11.

Silverstein, S.C., R.M. Steinman, and Z.A. Cohn. 1977. Endocytosis. *Annu. Rev. Biochem.* 46:669–722.

Smith, S. 1998. Shark complement: an assessment. *Immunol. Rev.* 166:67–78.

Smith, S., M. Riesgo, S. Obenauf, and C. Woody. 1997. Anaphylactic and chemotactic response of mammalian cells to zymosan- activated shark serum. *Fish Shellfish Immunol.* 7:503–514.

Stokes, E.E. and B.G. Firkin. 1971. Studies of the peripheral blood of the Port Jackson shark (*Heterodontus portusjacksoni*) with particular reference to the thrombocytes. *Br. J. Haematol.* 20:427–435.

Surh, C.D. and J. Sprent. 1994. T-cell apoptosis detected *in situ* during positive and negative selection in the thymus. *Nature* 372:101–103.

Tafalla, C. and B. Novoa. 2000. Requirements for nitric oxide production by turbot (*Scophthalmus maximus*) head kidney macrophages. *Dev. Comp. Immunol.* 24:623–631.

Takemoto, T., S. Smith, T. Terado, H. Kimura, and M. Nonaka. 2000. Molecular cloning of complement B/C2 and C3/C4 of a Japanese shark *Triakis scyllia*. *Dev. Comp. Immunol.* 24:S25.

Tomonaga, S., K. Kobayashi, K. Hagiwara, K. Yamaguchi, and K. Awaya. 1986. Gut-associated lymphoid tissue in the elasmobranchs. *Zool. Sci.* 3:453–458.

Tomonaga, S., H. Zhang, K. Kobayashi, R. Fujii, and K. Teshima. 1992. Plasma cells in the spleen of the Aleutian skate, *Bathyraja aleutica. Arch. Histol. Cytol.* 55:287–284.

Velardi, A. and M.D. Cooper. 1984. An immunofluorescence analysis of the ontogeny of myeloid, T, and B lineage cells in mouse hemopoietic tissues. *J. Immunol.* 133:672–677.

Walsh, C.J. and C.A. Luer. 1998. Comparative phagocytic and pinocytic activities of leucocytes from peripheral blood and lymphomyeloid tissues of the nurse shark (*Ginglymostoma cirratum* Bonaterre) and the clearnose skate (*Raja eglanteria* Bosc). *Fish Shellfish Immunol.* 8:197–215.

Walsh, C.J. and C.A. Luer. In press. Elasmobranch hematology: identification of cell types and practical applications, in *Elasmobranch Husbandry Manual.* M. Smith, D. Warmolts, D. Thoney, and R. Hueter, Eds., Special Publication of the Ohio Biological Survey 16, Columbus, OH.

Walsh, C.J., J.T. Wyffels, A.B. Bodine, and C.A. Luer. 2002. Dexamethasone-induced apoptosis in immune cells from peripheral circulation and lymphomyeloid tissues of juvenile clearnose skates, *Raja eglanteria. Dev. Comp. Immunol.* 26:623–633.

Warr, G.W. 1995. The immunoglobulin genes of fish. *Dev. Comp. Immunol.* 19:1–12.

Weissman, G., M.C. Finkelstein, J. Csernanasky, J.P. Quigley, R.S. Quinn, L. Techner, W. Troll, and P.B. Dunham. 1978. Attack of sea urchin eggs by dogfish phagocytes: model of phagocyte-mediated cellular toxicity. *Proc. Natl. Acad. Sci. U.S.A.* 75:1825–1829.

Weyts, F.A.A., B.M.L. Verburg-van Kemenade, G. Flik, J.G.D. Lambert, and S.E Wendelaar Bonga. 1997. Conservation of apoptosis as an immune regulatory mechanism: effects of cortisol and cortisone on carp lymphocytes. *Brain Behav. Immunol.* 11:95–105.

Wyffels, J.T. 2001. Acute Radiation Exposure in the Clearnose Skate, *Raja eglanteria*: A Histological Study of the Immune System. Ph.D. dissertation, Clemson University, Clemson, SC.

Wyllie, A.H. 1980. Glucocorticoid-induced thymocyte apoptosis is associated with endogenous endonuclease activation. *Nature* 284:555–556.

Yin, Z., T.J. Lam, and Y.M. Sin. 1997. Cytokine-mediated antimicrobial immune response of catfish, *Clarias gariepinus*, as a defense against *Aeromonas hydrophila. Fish Shellfish Immunol.* 7:93–104.

Zapata, A.G., M. Torroba, R. Sacedon, A. Varas, and A. Vicente. 1996. Structure of the lymphoid organs of elasmobranchs. *J. Exp. Zool.* 275:125–143.

Part III

Ecology and Life History

14

Age Determination and Validation in Chondrichthyan Fishes

Gregor M. Cailliet and Kenneth J. Goldman

CONTENTS

14.1 Introduction

It is important to understand the ages, growth characteristics, maturation processes, and longevity of fishes to assess their current population status and to predict how their populations will change in time (Ricker, 1975; Cailliet et al., 1986b). Fishery biologists have used age, length, and weight data as important tools for their age-based population models. Especially important are details about growth and mortality rates, age at maturity, and life span (Ricker, 1975; Cortés, 1997). Over the past several decades, it has become obvious that fisheries for chondrichthyan fishes have not been easily sustainable. In 1974, Holden suggested that these fishes had life histories that made them vulnerable to overfishing. Included in the characteristics he cited were slow growth, late age at maturity, few offspring, and lengthy gestation periods. Since then, fishing pressure on elasmobranchs, both as directed and as bycatch fisheries (and discards) has increased (Bonfil, 1994; Casey and Myers, 1998; Stevens et al., 2000; Baum et al., 2003), stimulating many studies on many important aspects of their life histories, such as age, growth, and reproduction.

In the first review of elasmobranch aging by Cailliet et al. (1986b), the age verification studies were relatively few, including statistical analyses, direct measurements of growth, marking anatomical features such as vertebrae or spines with tetracycline and then describing their location over time both in laboratory and field studies, and several relatively new chemical studies of calcified structures (Welden et al., 1987). In a second review, Cailliet (1990) updated progress made, stressing that many additional studies had derived age estimates based on opaque and translucent bands in calcified structures and that more studies were attempting to verify the periodicity with which these bandings were deposited. These included an additional 42 studies that addressed age verification in 39 species of elasmobranchs. For only six species was there sufficient information to validate banding patterns in chondrichthyan hard parts. In some of these studies, poor calcification and only partially verified band patterns prevented full understanding of growth patterns. In one species, the Pacific angel shark, *Squatina californica*, it was determined that they did not deposit any predictable growth bands in their vertebral centra (Natanson and Cailliet, 1990).

In this chapter, we review the studies that have been done on the age, growth, maturity, and longevity of chondrichthyan fishes since Cailliet et al.'s (1986b) and Cailliet's (1990) reviews, and summarize their approaches and findings. We briefly describe the methodology used to estimate, verify, and validate age estimates in chondrichthyan fishes and examine emerging technology now being applied to both age determination and validation studies. We then summarize the results of these recent chondrichthyan age and growth studies, providing an analysis of the range of asymptotic sizes (L_∞ or DW_∞) and growth coefficients (k) in the von Bertalanffy growth equation, plus a summary of the ages at maturity and longevities of chondrichthyan fishes studied recently. Finally, we briefly discuss the implications of growth rate, age at maturity, longevity, and the demographic traits of chondrichthyan fishes to their management and conservation.

14.2 Methodology

The age determination process consists of the following steps: collection of hard part samples, preparation of the hard part for age determination, examination (age reading), assessment of the validity and reliability of the resulting data, and interpretation (modeling growth). This section briefly discusses the hard parts that have been used to age chondrichthyan fishes and how to collect and prepare them for age determination. The examination of hard parts and assessment of validity and reliability of age estimates and modeling growth are discussed in Sections 14.3 and 14.4, respectively.

14.2.1 Structures

14.2.1.1 Vertebrae — Whole vertebral centra, as well as transverse and sagittally (i.e., longitudinally) sectioned centra, have been used for aging elasmobranches (Figure 14.1). Transverse sectioning will prevent bands on opposing halves from obscuring each other when illuminated from below. However, determining the age of older animals can still be problematic as bands become more tightly grouped at the outer edge of vertebrae, and may be inadvertently grouped and counted together if transverse sections or whole centra are used for aging, thereby causing underestimates of age (Cailliet et al., 1983a, 1986; Cailliet, 1990). As such, sagittally sectioned vertebrae should be used for aging unless it can be unequivocally demonstrated that identical ages can repeatedly be obtained from a given species using whole centra (Campana, 2001; Goldman, in press).

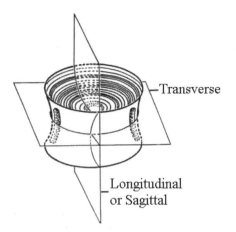

FIGURE 14.1 The two sectioning planes that can be used on vertebral centra. (Courtesy of G.M. Cailliet, Moss Landing Marine Laboratories.)

14.2.1.2   Spines — Dorsal fin spines (Figure 14.2) have been another useful hard part for aging some elasmobranchs, most notably dogfish sharks (Family Squalidae) (Ketchen, 1975; Nammack et al., 1985; McFarlane and Beamish, 1987a). Spines from the second dorsal fin are preferred for aging as the tips of first dorsal fin spines tend to be more worn down, which leads to an underestimation of age. Correction factors can be calculated to estimate ages of individuals with worn spines (Ketchen, 1975; Sullivan, 1977).

Spines can be read whole (without further preparation), by wet-sanding the enamel and pigment off the surface and polishing the spine, or from the exposed surface resulting from a longitudinal cut (Ketchen, 1975; McFarlane and Beamish, 1987a). Cross-sectioned dorsal fin spines have also proved useful in assessing ages in some squaloids (Figure 14.3) and chimaeras (Sullivan, 1977; Freer and Griffiths, 1993; Clarke et al., 2002a,b; Calis et al., in press).

14.2.1.3   Neural Arches — Calcium deposits have been documented in the neural arches of elasmobranch fishes (Peignoux-Deville et al., 1982; Cailliet, 1990), but they had not been used for

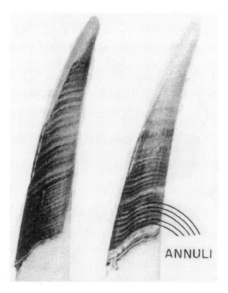

FIGURE 14.2 Spiny dogfish, *Squalus acanthias*, second dorsal fin spines showing annuli. First spine was aged at 42 years; second spine aged at 46 years. (Courtesy of G.A. McFarlane, Pacific Biological Station.)

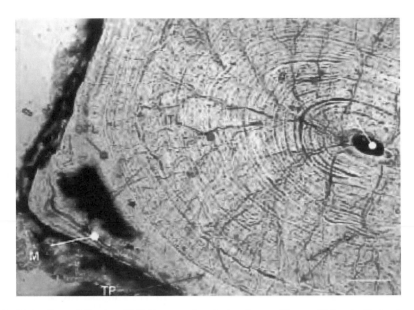

FIGURE 14.3 (Color figure follows p. 304.) Cross-sectioned spine from a 43-year-old female *Centrophorus squamosus*. M = mantle; OTL = outer trunk layer; ITL = inner trunk Layer; TP = trunk primordium; L = lumen; A = annuli. Scale bar = 500 μm. (From Clarke, M.W. et al. 2002b. *J. Fish Biol.* 60:501–514. With permission.)

aging. McFarlane et al. (2002) recently introduced the first attempt using this structure for aging elasmobranchs by silver nitrate staining the neural arches of sixgill sharks, *Hexancius griseus*. The results from this preliminary study indicate that neural arches may provide another aging structure for elasmobranch species with poorly calcified vertebral centra, but the method has not been validated (Figure 14.4).

14.2.1.4   Caudal Thorns and Other Structures

14.2.1.4   Caudal Thorns and Other Structures — Novel approaches to aging various elasmobranchs continue to arise, and researchers may want to begin collecting additional hard parts from specimens in the field to be experimented with in the laboratory. For example, Gallagher and Nolan (1999) used caudal thorns (Figure 14.5) along with vertebral centra to determine age in four bathyrajid species, demonstrating high precision in ages between the two parts. Gallagher et al. (in press) further elaborated on the structure and growth processes in caudal thorns. Comparing counts in more than one hard part is a common age verification technique used in teleost aging studies; however, it is not frequently conducted on cartilaginous fishes because of the lack of multiple hard parts available for comparison. The use of thorns as a reliable hard part for aging, where appropriate, has the potential to greatly aid our understanding of the life histories of several species of skate and ray.

Tanaka (1990) experimented with growth bands in the upper jaw of one specimen of the wobbegong, *Orectolobus japonicus*, kept in captivity, and found evidence for growth bands there. However, his search for nonvertebral cartilaginous tissues in this species and the swell shark, *Cephaloscyllium umbratile*, were not productive.

14.2.2   Sampling and Processing Specimens

14.2.2.1   Taking Samples

14.2.2.1   Taking Samples — The location in the vertebral column from which samples are taken for aging can have a statistically significant effect on increment counts (Officer et al., 1996). As such, it is important to use the larger, more anterior (thoracic) centra for age studies, as smaller centra from the caudal region may lack some bands (Cailliet et al., 1983b). This emphasizes the importance of standardizing the vertebral sampling region for all aging studies, allowing for precise, valid comparisons among individuals within a population and for more accurate comparisons between populations. All aging techniques require centra free of tissue; however, neural arches should be left on several centra

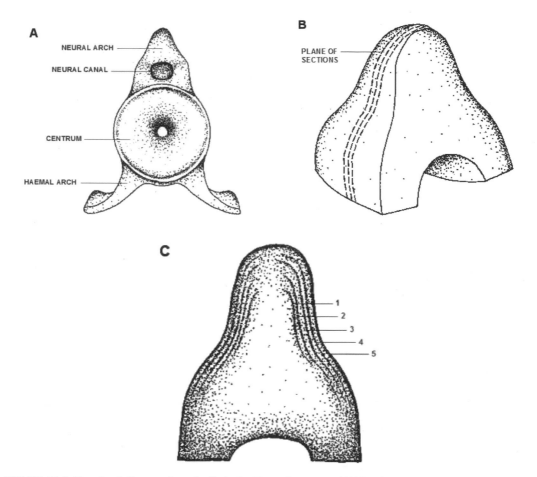

FIGURE 14.4 Neural arch diagrams from sixgill sharks, *Hexanchus griseus*. (A) The whole centra, (B) the planes at which sectioning took place, and (C) the resulting banding pattern after silver nitrate staining. (From McFarlane, G.A. et al. 2002. *Fish. Bull.* 100:861–864. With permission.)

based on their potential for use in aging if vertebral centra or spines show no banding pattern (McFarlane et al., 2002). Dorsal fin spines should be removed by cutting horizontally just above the notochord to ensure that the spine base and stem are intact.

Vertebral samples are typically individually bagged, labeled, and stored frozen until ready for preparation. If freezing is not an option, vertebrae can be fixed in 10% formalin for 24 h and then preserved in alcohol. Dorsal fin spines are typically bagged, labeled, and frozen until returned to the laboratory or placed immediately in 70 to 95% ethyl alcohol or 95% isopropyl alcohol.

14.2.2.2 *Centrum Cleaning and Sample Preparation* — Vertebral samples need to be thawed if frozen, or washed if preserved in alcohol, and cleaned of excess tissue and separated into individual centra. Tissue-removal techniques vary with species. For many, soaking the centrum in distilled water for 5 min followed by air-drying allows the connective tissue to be peeled away. Soaking in bleach may be required for other species. Bleaching time is proportional to centrum size and ranges from 5 to 30 min. After bleaching, the centrum is rinsed thoroughly in water. Another simple and effective method is to soak vertebral sections in a 5% sodium hypochlorite solution. Soak times can range from 5 min to 1 h depending on the size of the vertebrae and should be followed by soaking centra in distilled water for 30 to 45 min (Johnson, 1979; Schwartz, 1983). This method also assists in removal of the vertebral fascia between centra and does not affect the staining process. Centra are typically permanently stored in 70 to 95% ethyl alcohol or 95% isopropyl alcohol; however, a subsample of centra should be

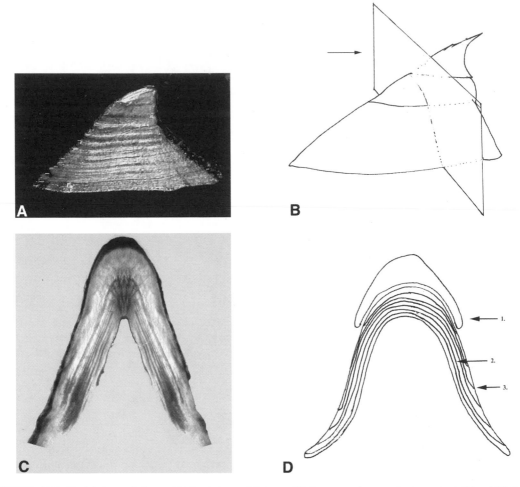

FIGURE 14.5 Caudal thorn of skates. (A) A whole caudal thorn, (B) the plane where sections were cut, (C and D) the observed banding patterns. (From Gallagher, M. et al. In press. With permission.)

permanently stored in a freezer in case it is needed for staining, and because long-term exposure to alcohol may reduce the resolution of the banding pattern (Wintner et al., 2002; Allen and Wintner, 2002).

Vertebrae may be analyzed whole or sectioned, but recent experience in many publications indicates that sectioning is ideal (Figure 14.6). Vertebral sectioning is typically done with a low-speed diamond-bladed saw (e.g., Isomet rotary diamond saw). However, sections can be cut with small handsaws and even scalpels when working with very small centra, or half of the centrum can be worn away with aluminum oxide wheel points and fine sandpaper attachments for the same tool (Cailliet et al., 1983a,b). Large vertebrae may be secured in a vise and cut with a small circular saw attachment on a jeweler's drill or even ground in half with a grinder. After mounting the sections to slides, they should be sanded with wet fine-grit sandpaper in a series (grades 320, 400, and finally 600 for polishing) to approximately 0.3 to 0.5 mm and air-dried. A binocular dissecting microscope with transmitted light is generally used for identification of growth rings and image analysis.

14.3 Age Determination

Although concentric growth bands have been documented in the vertebral centra of chondrichthyans for more than 80 years (Ridewood, 1921), aging these fishes has proved a slow and difficult process. Counts

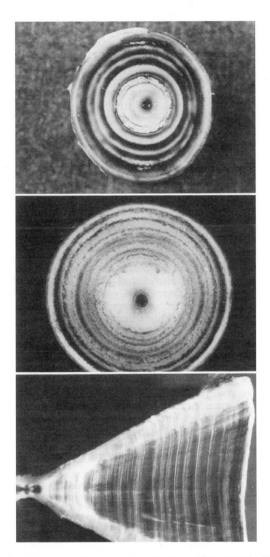

FIGURE 14.6 Vertebrae stained using the cobalt nitrate and ammonium sulfide method of Hoenig and Brown (1988). The top image is a smooth dogfish, *Mustelus canis*, centrum, the middle and bottom images are of lemon shark, *Negaprion brevirostris*, centra. (Courtesy of J.M. Hoenig, Virginia Institute of Marine Science.)

of opaque and translucent banding patterns in vertebrae, dorsal spines, caudal thorns, and neural arches have provided the only means of information on growth rates in these fishes as they lack the hard parts, such as otoliths, scales, and bones typically used in age and growth studies of teleost fishes (Cailliet et al., 1986a,b; Cailliet, 1990; Gallagher and Nolan, 1999; McFarlane et al., 2002). Unfortunately, the vertebral centra of many elasmobranch species (such as numerous deep-water species) are too poorly calcified to provide information on age, most species have no dorsal spines, and there may be no tangible relationship between observed banding patterns and growth (Caillet et al., 1986b; Cailliet, 1990; Natanson and Cailliet, 1990; McFarlane et al., 2002). These circumstances continue to cause difficulties in making age estimates for many species.

The most commonly distinguishable banding pattern in sectioned centra when viewed microscopically is one of wide bands separated by distinct narrow bands (Figure 14.7). The terms *opaque* and *translucent* are commonly used to describe these bands, and they tend to occur in summer and winter, respectively. However, the opacity and translucency of these bands vary considerably with species, light source, and methodology (Cailliet et al., 1986b, 1990; Wintner et al., 2002; Goldman 2002). It should not be assumed

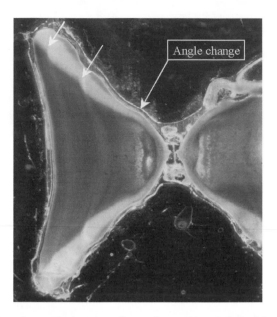

FIGURE 14.7 Sagittal section of a vertebral centrum from a 2-year-old smooth dogfish, *Mustelus canis*, showing the distinct notching pattern (white arrows) that accompanied the distinct banding pattern. (Courtesy of C. Conrath, Virginia Institute of Marine Science.)

that the opaque and translucent nature of vertebral bands in different species will be similar, however, the pattern of wide/narrow banding tends to be consistent. An annulus is usually defined as the winter band. The difference in appearance between summer (wide) and winter (narrow) growth bands provides the basis for age determinations. In many species, this so-called winter band actually forms in the spring (Sminkey and Musick, 1995a).

In elasmobranch vertebral sections, each pair of wide/narrow bands, which extends across one arm of the corpus calcareum, across the intermedialia, and across the opposing corpus calcareum arm, is considered to represent an annual growth cycle; the narrow bands, hereafter referred to as "rings or annuli," are what are counted (Figure 14.8). It must be noted that counting these rings, at this point in the process, carries with it the assumption that each one represents a year's growth; however, the validity

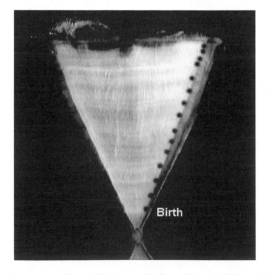

FIGURE 14.8 Blue shark vertebral section. Dots indicate annuli. (From Skomal, G.B. and L.J. Natanson. 2003. *Fish. Bull.* 101:627–639. With permission.)

of this assumption must be tested. (The term *annulus* is defined as a ringlike figure, part, structure, or marking, but they must be shown to be annual in their deposition). The age determination process for spines is virtually identical to that for vertebrae; however, Ketchen's (1975) method for calculating age from worn spines should be used instead of discarding them. This method uses an age to spine-base-diameter regression for unworn spines to allow an estimation of age for individuals with worn spines.

14.3.1 Centrum Staining

Numerous techniques have been used in attempts to enhance the visibility of growth bands in elasmobranch vertebral centra. Many are simply stained (Figure 14.9), but the list of techniques includes alcohol immersion (Richards et al., 1963), xylene impregnation (Daiber, 1960), histology (Ishiyama, 1951; Casey et al., 1985; Natanson and Cailliet, 1990), X-radiography (Aasen, 1963; Cailliet et al., 1983a,b; Martin and Cailliet, 1988; Natanson and Cailliet, 1990), X-ray spectrometry (Jones and Green, 1977), cedarwood oil (Cailliet et al., 1983a; Neer and Cailliet, 2001), alizarin red (LaMarca, 1966; Gruber and Stout, 1983; Cailliet et al., 1983a; Goosen and Smale, 1997), silver nitrate (Stevens, 1975; Schwartz, 1983; Cailliet et al., 1983a,b), crystal violet (Johnson, 1979; Schwartz, 1983; Anislado-Tolentino and Robinson-Mendoza, 2001; Carlson et al., 2003), graphite microtopography (Parsons, 1983, 1985; Neer and Cailliet, 2001), a combination of cobalt nitrate and ammonium sulfide (Hoenig and Brown, 1988), and the use of copper-, lead-, and iron-based salts (Gelsleichter et al., 1998a). Many of these studies used multiple techniques on a number of species for comparison, particularly Schwartz (1983) and Cailliet et al. (1983a). These studies show that the success of each technique is often species specific and that slight modifications in technique may enhance the results.

In addition to their effectiveness, the various techniques mentioned vary in their simplicity, cost, and technological requirements. Histological processes have proved useful, but require specialized equipment, a number of chemicals, and are relatively time-consuming. The resulting staining process is long lasting, with no color change in vertebral sections after 15 years (Casey et al., 1985). X-radiography has proved useful in many studies, but has the obvious necessity of an appropriate X-ray machine and film processing capabilities. While X-ray spectrometry may hold promise (Jones and Green, 1977; Casselman, 1983; Cailliet et al., 1983a, 1986b), it is also time-consuming and expensive. Simpler, less expensive, and more time-efficient staining techniques, such as crystal violet, silver nitrate, cedarwood oil, graphite microtopography, and alizarin red, should be used prior to considering other methods. Although these techniques have been tried, many have not yet been thoroughly evaluated. For example, the cobalt nitrate and ammonium sulfide stain suggested by Hoenig and Brown (1988) is easy to use, time efficient, and has provided quality results for two species, but has not been extensively applied. A microradiographic method using injected fluorochrome dyes to aid in resolving individual hyperminer-

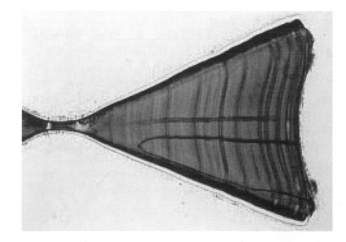

FIGURE 14.9 Vertebral section stained with haemotoxylin. (Staining by Sho Tanaka, photograph courtesy of K.G. Yudin and G.M. Cailliet.)

alized increments was applied to captive gummy sharks, *Mustelus antarcticus*, with success (Officer et al., 1997), but this method has not been extensively applied or thoroughly evaluated. The possibility that this method may also have application as a validation technique needs to be investigated.

14.3.2   Precision and Bias

The most commonly used methods for evaluating precision among age determinations have been the average percent error (APE) technique of Beamish and Fournier (1981) and the modification of their method by Chang (1982). However, Hoenig et al. (1995) and Evans and Hoenig (1998) have demonstrated that there may be differences in precision that these methods obscure because the APE assumes that the variability among observations of individual fish can be averaged over all age groups and that this variability can be expressed in relative terms. Also, APE does not result in values that are independent of the age estimates. APE indices do not test for systematic differences, do not distinguish all sources of variability (such as differences in precision with age), and do not take experimental design into account (i.e., number of times each sample was read in each study) (Hoenig et al., 1995). Within a given aging study, however, APE indices may serve as good relative indicators of precision within and between readers provided that each reader ages each vertebra the same number of times. However, even this appears to tell us only which reader was less variable, not which was better or if either was biased. Comparing precision between studies would seem to hold importance only if the study species is the same, but caution should be used if samples are from different geographic areas.

Goldman (in press) provided a simple and accurate approach to estimating precision, which is to: (1) calculate the percent reader agreement (PA = [No. agreed/No. read] × 100) within and between readers for all samples; (2) calculate the percent agreement plus or minus one year (PA ± 1 year) within and between readers for all samples; (3) calculate the percent agreement within and between readers, with individuals divided into appropriate length or disk width groups (e.g., 5 to 10 cm increments) as an estimate of precision (this should be done with sexes separate and together); and (4) test for bias using one or more of the methods discussed below. The criticism of percent agreement as a measure of precision has been that it varies widely among species and ages within a species (Beamish and Fournier, 1981; Campana, 2001). There is, however, validity in using percent agreement with individuals grouped by length as a test of precision because it does not rely on ages (which have been estimated), but rather on lengths, which are empirical values. Age could be used if, and only if, validation of absolute age for all available age classes had been achieved.

Several methods can be used to compare counts (ages) by multiple readers such as regression analysis of the first reader counts vs. the second reader counts, a paired *t*-test of the two reader's counts and a Wilcoxon matched pairs signed-ranks test (DeVries and Frie, 1996). Campana et al. (1995) state the importance of a separate measure for bias, and even that bias should be tested for prior to running any tests for precision. They suggest an age bias plot (Figure 14.10), graphing one reader vs. the other, which is interpreted by referencing the results to the equivalence line of the two readers (45° line through the origin). Similarly, Hoenig et al. (1995) and Evans and Hoenig (1998) state that comparisons of precision are only of interest if there is no evidence of systematic disagreement among readers or methods, and suggest testing for systematic differences between readers using chi-square tests of symmetry such as Bowker's (Bowker, 1948), McNemar's (McNemar, 1947), and their Evans–Hoenig test to determine whether differences between and within readers were systematic (biased) or due to random error. This is of particular importance if initial percent agreement and precision estimates are low. We recommend these tests of symmetry for testing for bias regardless of precision, as they place all age values in contingency tables and test the hypothesis that values in a given table are symmetrical about the main diagonal, and because they can be set up to test among all individual age classes or groups of age classes. The test statistic (the chi-square variable) will tend to be large if a systematic difference exists between the two readers.

14.3.3   Back-Calculation

Back-calculation is a method for describing the growth history of each individual sampled, and numerous variations in methodology exist (see Francis, 1990, for a thorough review and Goldman, in press, for

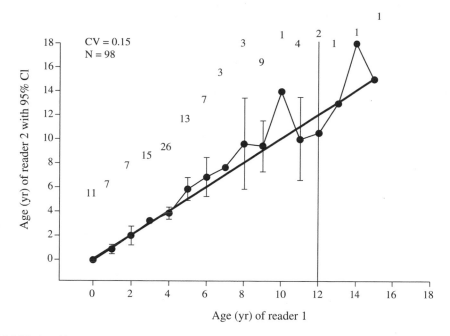

FIGURE 14.10 Age–bias curve. (From Skomal, G.B. and L.J. Natanson. 2003. *Fish. Bull.* 101:627–639. With permission.)

description and application to elasmobranchs). Back-calculations estimate lengths-at-previous-ages for each individual and should be used if sample sizes are small and if samples have not been obtained from each month. Back-calculation formulas that follow a hard part or body proportion hypothesis are recommended (Francis, 1990; Campana, 1990; Ricker, 1992). The proportional relationship between animal length or disk width and the radius of the vertebral centrum among different length animals within a population is used as a basis for empirical relationships regarding population and individual growth, as is the distance from the focus to each annulus within a given centrum. Centrum radius (CR) and distance to each ring should be measured as a straight line from the central focus to the outer margin of the corpus calcareum to the finest scale possible. Lengths or disk widths should then be plotted against CR to determine the proportional relationship between somatic and vertebral growth, which will assist in determining the most appropriate back-calculation method.

Providing biological and statistical reasoning behind the choice of a back-calculation method is extremely important for obtaining accurate life history parameter estimates from a growth function (e.g., Gompertz) when using back-calculated data. Although one method may prove to be more statistically appropriate for back-calculation, researchers should conduct several methods for comparison to available sample length-at-age data to verify that statistical significance equates to biological accuracy. Biological accuracy can be determined by plotting the sample mean length-at-age data against the difference between mean back-calculated length-at-age estimates and the sample mean length-at-age data to see which method provides results that most accurately reflect sample data (Goldman, 2002, in press). Although the most commonly used back-calculation method has been the Dahl–Lea direct proportions method (Carlander, 1969), linear and quadratic modified Dahl–Lea methods (Francis, 1990) and the Frazer–Lee birth-modified back-calculation method (Campana, 1990; Ricker, 1992) should be conducted, where appropriate, and compared to sample length-at-age data (Goldman, 2002, in press).

14.4 Verification and Validation

Cailliet (1990) stated that the process of evaluating growth zone deposition in fishes can be categorized into the terms *verification* and *validation*. Verification is defined as "confirming an age estimate by comparison with other indeterminate methods," and validation as "proving the accuracy of age estimates by comparison with a determinate method." These definitions are used throughout this discussion.

Estimates of age, growth rate, and longevity in chondrichthyans assume that the growth rings are an accurate indicator of age. Although this is probably true for most species, few studies on elasmobranch growth have validated the temporal periodicity of band deposition in vertebral centra and even fewer have validated the absolute age (Cailliet et al., 1986b; Cailliet, 1990; Campana, 2001).

Validation can be achieved via several methods such as chemically tagging wild fish, mark–recapture studies of known-age individuals, and bomb carbon dating (the latter two can also be used to validate absolute age). A combination of using known-aged individuals, tag and recapture, and chemical marking is probably the most robust method for achieving complete validation (Beamish and McFarlane, 1983; Cailliet, 1990; Campana, 2001; Natanson et al., 2002). Although this is a rather daunting task to accomplish with most elasmobranch species, the current necessity to obtain age-growth data for fisheries management purposes dictates that it be attempted. The most frequently applied method used with elasmobranchs has been chemical marking of wild fish even though recaptures can be difficult to obtain for many species. As validation has proved difficult in elasmobranchs, verification methods such as centrum edge analysis and relative marginal increment analysis are frequently employed.

Obtaining the absolute age of individual fish (complete validation) is the ultimate goal of every aging study, yet it is the frequency of growth ring formation for which validation is typically attempted. The distinction between validating absolute age and validating the periodicity of growth ring formation is important (Beamish and McFarlane, 1983; Cailliet, 1990; Campana, 2001). Validation of the frequency of growth ring formation must prove that the mark being considered an annulus forms once a year (Beamish and McFarlane, 1983). However, it is the consistency of the marks in "number per year" that really matters, be it one or more than one. Two or more marks (rings) may make up an "annulus" if, and only if, consistent multiple marks per year can be proved. Strictly speaking, validation of absolute age is only complete when it has been done for all age classes available, with validation of the first growth ring the critical component for obtaining absolute ages (Beamish and McFarlane, 1983; Cailliet, 1990; Campana, 2001).

In the following sections, both verification and validation are discussed. It is important to remember that some techniques, especially if used in conjunction with others, can be verification and/or validation.

14.4.1   Size Mode Analysis

This technique monitors the progression of discrete length modes of fish over time. Although commonly considered either a basic approach to studying age composition and even growth, its use as a growth tool is primarily verification. That is, if the size modes seen in data from a presumed random sample of all sizes of fish in a population appear to coincide with the mean or median sizes in an age class (as determined by aging studies or other means), then this lends support to the contention that these age classes are real. For example, Kusher et al. (1992) used this method to show that young leopard sharks, *Triakis semifasciata*, in Elkhorn Slough, CA followed growth patterns that would have been predicted by the von Bertalanffy growth function determined by size at age patterns from vertebral sections. Similarly, Natanson et al. (2002) determined growth rates for age 0 and 1 porbeagle sharks, *Lamna nasus*, by monitoring the progression of those two discrete length modes of across months within a year.

14.4.2   Tag–Recapture

In addition to size mode analysis, tag–recapture data are often used to produce growth curves. This usually involves capturing, measuring, weighing, and tagging specimens in the field, which are released. Through recaptures obtained from either dedicated surveys or from recreational or commercial fishers, tagged specimens provide information on growth (length and or weight) over a distinct period of time. This has been done in many studies, including the Pacific angel shark, *Squatina californica*, in which the von Bertalanffy growth functions are based on size-at-capture and recapture, but not oxytetracycline (OTC) (Cailliet et al., 1992). A significant amount of literature exists on the procedures of estimating growth parameters from tag–recapture data (Gulland and Holt, 1959; Fabens, 1965; Cailliet et al., 1992).

The method developed by Gulland and Holt (1959) is fairly straightforward; however, efforts should be made to use several methodologies, such as GROTAG (Francis, 1988; Natanson et al., 2002) when analyzing growth increment data.

14.4.3 Marking, Field Tag–Recapture, and Laboratory Studies

Validation of absolute age is extremely difficult to achieve with elasmobranch fishes; hence the (few) studies that have attempted validation in these fishes have focused on validating the temporal periodicity of ring (growth increment) formation. The tetracycline validation method is a standard among fisheries biologists for marking free-swimming individuals (Cailliet, 1990; DeVries and Frie, 1996; Campana, 2001; Smith et al., 2003) to test the assumption of annual periodicity of growth rings. OTC, a general antibiotic that can be purchased through veterinary catalogs, binds to calcium and is subsequently deposited at sites of active calcification. It is typically injected intramuscularly at a dose of 25 mg kg^{-1} body weight (Tanaka, 1990; Gelsleichter et al., 1998b) and an external identification tag is simultaneously attached to each injected animal. OTC produces highly visible marks in vertebral centra and dorsal fin spines of recaptured sharks when viewed under ultraviolet light (Holden and Vince, 1973; Gruber and Stout, 1983; Smith, 1984; Beamish and McFarlane, 1985; McFarlane and Beamish, 1987a,b; Branstetter, 1987; Brown and Gruber, 1988; Natanson and Cailliet, 1990; Tanaka, 1990; Kusher et al., 1992; Natanson, 1993; Gelsleichter et al., 1998b; Wintner and Cliff, 1999; Natanson et al., 2002; Simpfendorfer et al., 2002; Goldman, 2002b; Skomal and Natanson, 2003).

The combination of body growth information and a discrete mark in the calcified structure permit direct comparison of time at liberty with growth band deposition, such that the number of rings deposited in the vertebra or spine since the OTC injection can be counted and related to the time at liberty. Although there may be problems associated with using captive growth as a surrogate to growth in the wild and with recapturing animals that have been at large for a sufficiently long period of time, this method has been used on a number of species in the laboratory and field (Cailliet et al., 1986b; Cailliet, 1990).

Nevertheless, this technique, when successful, has proved to be invaluable at validating growth characteristics of chondrichthyans. The best recent examples are tag–recaptures, some with OTC, of the blue shark, *Prionace glauca*, by Skomal and Natanson (2003), shown in Figure 14.11, and the 20 year tag return of a leopard shark, *Triakis semifasciata*, by Smith et al. (2003), shown in Figure 14.12. In both cases, it was possible to define birth years and to identify individual growth characteristics for individual years from zones on sections of the vertebrae, relative to the OTC mark from the original release.

Several other chemical markers such as fluorescein and calcein have been used to validate growth ring periodicity in teleost otoliths, but very few studies have evaluated these in elasmobranchs (Gelsleichter et al., 1997; Officer et al., 1997). Gelsleichter et al. (1997) found that while doses of 25 mg kg^{-1}

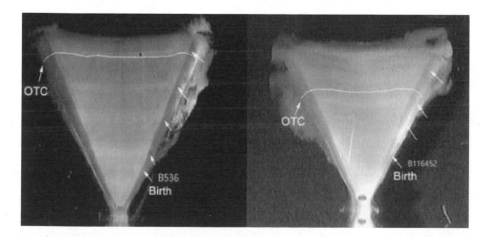

FIGURE 14.11 Sagittally cut vertebral section from tagged and recaptured OTC injected blue sharks. (From Skomal, G.B. and L.J. Natanson. 2003. *Fish. Bull.* 101:627–639. With permission.)

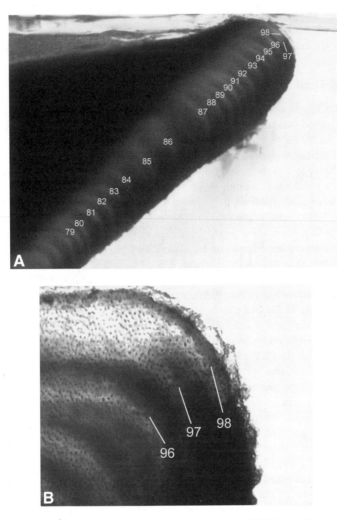

FIGURE 14.12 Sagittally cut vertebral section from tagged and recaptured OTC injected leopard shark after 20 years at large. (A) 20 years of annual ring formation from 1978 to 1998; and (B) the last 3 years at the edge of the section. (From Smith, S.E. et al. 2003. *Fish. Bull.* 101:194–198. With permission.)

body weight (typical dose for teleosts) induced physiological stress and mortality in elasmobranchs, doses of 5 to 10 mg kg^{-1} body weight produced suitable marks without causing physiological trauma or death. Based on this evaluation, any alternative chemical markers tested should consider that doses for teleosts might be too high for elasmobranchs.

14.4.4 Centrum Edge and Relative Marginal Increment Analysis

Centrum edge analysis compares the opacity and translucency (width and/or density) of the centrum edge over time in many different individuals to discern seasonal changes in growth. The centrum edge is categorized as opaque or translucent, and the bandwidth is measured or graded, then compared to season or time of year (Kusher et al., 1992; Wintner and Dudley, 2000; Wintner et al., 2002). A more detailed centrum edge analysis can be conducted by analyzing the levels of calcium and phosphorus at the centrum edge using X-ray or electron microprobe spectrometry (Cailliet et al., 1986b; Cailliet and Radtke, 1987). This technique has only been applied in a single study on recaptured nurse sharks that had been injected with tetracycline (Carrier and Radtke, 1988, cited in Cailliet, 1990).

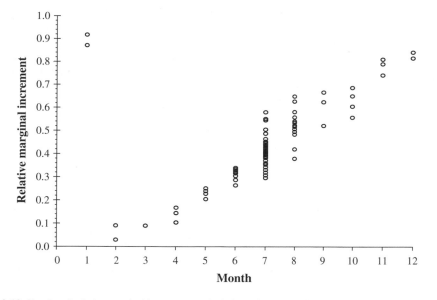

FIGURE 14.13 Results of relative marginal increment analysis from Goldman (2002) showing annuli formation in salmon sharks occurs between January and March.

Relative marginal increment analysis (RMI; sometimes referred to as MIR, or marginal increment ratio) is a useful, direct technique with which to assess seasonal band and ring deposition (Figure 14.13). The margin, or growth area of a centrum from the last growth ring to the centrum edge, is divided by the width of the last (previously) fully formed annulus (Branstetter and Musick, 1994; Natanson et al., 1995; Goldman, 2002; Wintner et al., 2002). Resulting RMI values are then plotted against month of capture to determine temporal periodicity of band formation. Age 0 animals cannot be used in this analysis since they have no fully formed increments.

Recently, ecologists have employed stable isotope composition to trace the early life histories of fishes, including analyses of habitats and environments occupied, as well as biochronologies (Campana and Thorrold, 2001). This approach has not been used to study either process in sharks and rays, but is certainly a field open to provide useful, additional information on the ecology of chondrichthyans.

14.4.5   Captive Rearing

The operation, often seen in public aquaria but also in research laboratories, of keeping chondrichthyans alive in captivity can produce some very useful growth information. Van Dykhuizen and Mollet (1992) analyzed growth of captive sevengill sharks, *Notorynchus cepedianus*, and provided the first estimates of growth for this species, which has poorly calcified vertebrae that cannot easily be analyzed for growth characteristics (Cailliet et al., 1983a). Similar studies have been done for such open-water species as the pelagic stingray, *Dasyatis violacea* (Mollet et al., 2002). Laboratory growth has also been used as a way of determining the periodicity of growth zone formation in the vertebral centra of several species, such as the Atlantic sharpnose shark, *Rhizoprionodon terraenovae* (Branstetter, 1987), and the wobbegong, *Orectolobus japonicus* (Tanaka, 1990).

The public aquarium trade is beginning to emphasize research as part of its husbandry practices, especially on topics such as the relationship between food intake and growth in chondrichthyan fishes. In a recent review on age and growth of captive sharks, Mohan et al. (in press) pointed out how carefully taken morphometric data on captive sharks can result in useful information on their age and growth. In addition, such growth information can provide data on their life histories that could not be obtained from animals in the wild. Unfortunately, there is also the caveat that captivity in itself can influence growth in a way that does not reflect what might occur in the wild.

One of the biggest problems with captive growth is accurately measuring individual specimens without harming them for display (Mohan et al., in press). However, recent handling techniques and the advent of remote measuring techniques has made this less of a problem. Mohan et al. (in press) provide very useful weight–length relationships for 17 species of sharks. The study also provides a detailed life history summary of ten species (eight kept in captivity, ranging from lamnoids and charharinids to the angel shark; the other two not kept in captivity, including *Carcharodon carcharias* and *Isurus oxyrinchus*). In addition, von Bertalanffy growth curves are presented as a result of this summary of life histories for 16 species.

14.4.6   Bomb Carbon

Bomb carbon dating is a technique recently applied to age validation in elasmobranchs. A rapid increase in radiocarbon (^{14}C) occurred in the world's oceans due to atmospheric atom bomb testing in the 1950s and 1960s (Druffel and Linick, 1978). The uptake of bomb-produced ^{14}C was virtually synchronous in marine carbonates, including corals and fish hard parts. This period of increase serves as a dated marker in calcified structures exhibiting growth bands (Druffel and Linick, 1978; Weidman and Jones, 1993; Kalish, 1995; Campana, 1997, 1999). Hence, all fishes born prior to ~1956 contain relatively low, naturally occurring levels of ^{14}C, and all those born after possess elevated levels of ^{14}C until the mid-1960s, after which they decline. Agreement in the period of ^{14}C chronology in fish hard parts with published ^{14}C chronometers for a region allows validation of annulus formation and absolute age of individual fish.

While this method has been used for aging several teleost fishes, Campana et al. (2002) reported the first application of bomb ^{14}C to validate ages in sharks, specifically the porbeagle, *Lamna nasus,* and also gave preliminary results for the shortfin mako, *Isurus oxyrinchus* (Figure 14.14). This method may be one of the best approaches to age validation of long-lived fishes; however, it does have some requirements and drawbacks. Bomb ^{14}C validation requires fish born during the period of ^{14}C increase (~1955?–1970), making this technique applicable to long-lived fishes and shorter-lived fishes with archived vertebral collections. In addition, the species must have resided in the mixed layer of the ocean, at least during a portion of its life history. The main drawback of this technique is high cost. The required use of high technology equipment, such as an accelerator mass spectrometer (AMS), may make this method unavailable for many researchers. Bomb ^{14}C age validation may be a key technique in resolving certain aging discrepancies, such as periodicity of ring formation in some species.

14.5 Review and Status of Chondrichthyan Age, Growth, Verification, and Validation Studies

Since the last review of age determination and validation in chondrichthyan fishes by Cailliet (1990), 115 publications on at least 91 species of chondrichthyans have been produced using some form of verification or validation (Table 14.1). Of the species studied, ~68 are new to the list. This chapter includes 3 species of chimaeras, 3 sawfishes, 18 guitarfish, torpedo rays, and stingrays, 15 skates, 1 angel shark, 2 cow sharks, 3 carpet sharks, 7 dog sharks, 8 mackerel sharks, and 32 species of ground sharks. The increase in diversity of taxa studied compares favorably with those in Cailliet et al. (1986b), who reviewed verification studies for approximately 30 species, and Cailliet (1990), who reviewed 42 studies on 39 species of chondrichthyans, some of which were the same as in the previous review. Thus, the total number of species of chondrichthyans for which there is age and growth information is now well over 110.

Most of these studies have used calcified structures, primarily vertebral centra (Table 14.1). Approximately 70% of the studies reviewed in Table 14.1 used vertebrae, either whole or sectioned, some of which were stained in one way or another. However, other structures are now also being used. For example, dorsal spines were used in 7% of the studies, and jaws and neural arches were used in one study each. Other techniques, not necessarily involving structures, have also been employed to calculate growth coefficients or annual increments of growth. These include captive growth (9%), field tag–recapture (7%), and embryonic growth (4%) methodologies. In many cases, combinations of techniques were used.

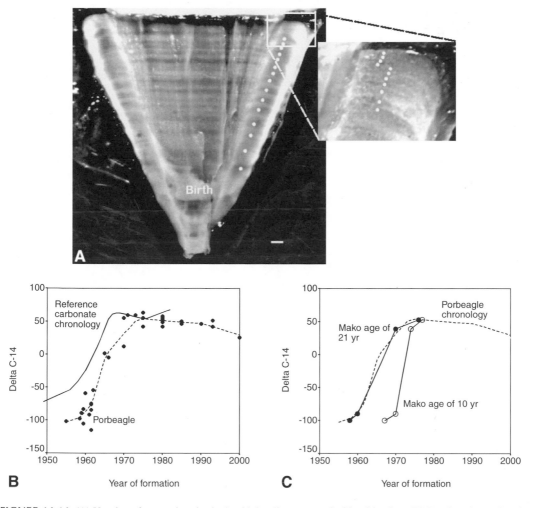

FIGURE 14.14 (A) Vertebrae from porbeagle shark with banding pattern elucidated by dots, (B) Bomb carbon values in different annuli from porbeagle (*Lamna* nasus) vertebrae, and (C) bomb carbon values for the shortfin mako along with porbeagle chronology. (From Campana, S.E. et al. 2002. *Can. J. Fish. Aquat. Sci.* 59:450–455. With permission.)

Precision analyses have started to become more common, with 21 studies calculating an APE, D, PA, or CV value among readers. However, this is still not a high proportion of the studies being reviewed.

There has been an increase in the use of both verification and validation methodology in chondrichthyan growth studies (Table 14.1). The most common method employed has been some form of marginal increment analysis or ratio (41 studies, 17 of which were not statistically significant; comprising 50% of the studies) and centrum edge analysis (12 studies). Even though they are not very robust methods, authors retained the use of size frequency modal analysis (25 studies on 22 species) and back-calculation (15 studies, 13 species). Because other techniques are more difficult, costly, or time-consuming, they were more infrequently utilized. Tag–recapture analysis provided growth estimates in 20 studies (19 species), whereas laboratory growth was employed in 21 studies. In 18 of these studies, OTC injections were involved to mark growth zones. Perhaps the most exciting study was the use of bomb carbon by Campana et al. (2002). This technique was successfully used, along with size frequency analysis and a previous OTC/tag–recapture study (Natanson et al., 2002), to validate the growth of the porbeagle, *Lamna nasus*, and to provide preliminary results from one vertebral centrum from the shortfin mako (*Isurus oxyrinchus*), suggesting that the one-band pair per year hypothesis is correct for this species.

Again, using combinations of verification and/or validation approaches is most likely to produce convincing results. For example, validation methods involving both captive growth and OTC marking

TABLE 14.1

Summary of the Age Verification and Validation Studies Completed since the Cailliet (1990) Review

Taxonomic Group and Species	Author(s) and Date of Study	Structure, Treatment or Method	Precision Analysis	Size Frequency Analysis	Back Calculation	Marginal Increment Analysis	Centrum Edge Analysis	Tag/Recapture Analysis	Laboratory Growth	OTC Injection	Bomb Carbon
Chimaeras											
Callorhynchus capensis	Freer and Griffiths (1993)	Sectioned spines				(X)					
Callorhynchus milii	Sullivan (1977)	Sectioned spine		X							
	Francis (1997)	Length frequency and tagging		X				X			
Chimaera monstrosa	Calis et al. (in press)	Sectioned spine									
Sawfishes											
Pristis microdon	Tanaka (1991)	Vertebrae/sectioned/stained hematoxylin									
Pristis pteroteti	Simpfendorfer (2000a)	Tagging/demography						X			
Pristis pectinata	Simpfendorfer (2000a)	Tagging/demography						X			
Guitarfishes, Torpedo Rays, and Stingrays Marine and Brackish											
Dasyatis americanus	Henningson (pers. comm.)	Whole vertebrae							X	X	
Dasyatis chrysonata	Cowley (1997)	Sectioned vertebrae							X		
Dasyatis pastinaca	Ismen (2003)	Whole vertebrae		X							
Dasyatis violacea	Mollet et al. (2002)	Captive growth							X		
	Neer (in press)	Vertebrae/sectioned/stained									
Gymnura altavela	Henningsen (1996)	Captive growth							(X)		
Rhinobatus productus	Timmons and Bray (1997)	Whole vertebrae and captive growth	PA				X		X	X	
Torpedo californica	Neer and Cailliet (2001)	Whole vertebrae	APE, D								

Species	Reference	Method						
Torpedo marmorata	Mellinger (1971)	Whole vertebrae, embryonic growth		X				
	Aloj Totaro et al. (1985)	Lipofuscin						
Trygonoptera mucosa	White et al. (2002a)	Verts/sectioned/ embryonic growth		X		X		
Trygonoptera personalis	White et al. (2002a)	Verts/sectioned/ embryonic growth		X		X		
Urolophus lobatus	White et al. (2001)	Sectioned vertebrae				X		
Freshwater								
Himantura chaophraya	White et al. (2001)	Sectioned vertebrae						
Himantura gerrardi	White et al. (2001)	Sectioned vertebrae						
Himantura imbricata	Tanaka and Ohnishi (1998)	Sectioned stained vertebrae						
Himantura laosensis	Tanaka and Ohnishi (1998)	Sectioned stained vertebrae						
Himantura signifer	Tanaka and Ohnishi (1998)	Sectioned stained vertebrae						
Himantura uarnak	Tanaka and Ohnishi (1998)	Sectioned stained vertebrae						
Pastinachus sephen	Tanaka and Ohnishi (1998)	Sectioned stained vertebrae						
Skates								
Bathyraja albomaculata	Gallagher and Nolan (1999)	Caudal thorns (whole and sectioned)					X	X
Bathyraja brachyurops	Gallagher and Nolan (1999)	Caudal thorns (whole and sectioned)						
Bathyraja griseocauda	Gallagher and Nolan (1999)	Caudal thorns (whole and sectioned)						
Bathyraja scaphiops	Gallagher and Nolan (1999)	Caudal thorns (whole and sectioned)						
Dipturus batis	DuBuit (1972)	Sectioned vertebrae						
Dipturus innominatus	Francis et al. (2001)	Sectioned X-rayed vertebrae						
Dipturus nasutus	Francis et al. (2001)	Sectioned X-rayed vertebrae						
Dipturus pullopunctata	Walmsley-Hart et al. (1999)	Vertebrae/sectioned/ stained	APE		X			

TABLE 14.1 (Continued)

Summary of the Age Verification and Validation Studies Completed since the Cailliet (1990) Review

Taxonomic Group and Species	Author(s) and Date of Study	Structure, Treatment or Method	Precision Analysis	Verification Techniques					Laboratory Growth	Validation Techniques	
				Size Frequency Analysis	Back Calculation	Marginal Increment Analysis	Centrum Edge Analysis	Tag/Recapture Analysis		OTC Injection	Bomb Carbon
Leucoraja erinacea	Natanson (1993)	Lab growth									
Leucoraja naevus	DuBuit (1972)	Sectioned vertebrae									
Leucoraja ocellata	Sulikowski et al. (2003)	Sectioned vertebrae	CV vs. age plots			X					
Leucoraja wallacei	Walmsley-Hart et al. (1999)	Vertebrae/sectioned/stained	APE		X	X					
Raja binoculata	Zeiner and Wolf (1993)	Sectioned vertebrae					X				
Raja miraletus	Abdel-Aziz (1992)	Whole vertebrae		X		X					
Raja rhina	Zeiner and Wolf (1993)	Sectioned vertebrae			X		(X)				
Angel Sharks											
Squatina californica	Cailliet et al. (1992)	Tag–recapture						X			
Cow Sharks											
Hexanchus griseus	McFarlane et al. (2002)	Sectioned neural arches									
Notorynchus cepedianus	Van Dykhuizen and Mollet (1992)	Captive lab growth							X		
Carpet Sharks											
Gingylomostoma cirratum	Carrier and Luer (1990)	Lab growth							X		
	Carrier and Luer (1990)	Tag–recapture						X			
	Schmid et al. (1990)	Captive lab growth						X	X		
	Robinson and Motta (2002)	Captive lab growth							X		
Orectolobus japonicus	Tanaka (1990)	Lab growth and sectioned vertebrae/jaws							X	X	

Species	Reference	Method	Code						
Rhincodon typus	Chang et al. (1997)	Embryonic growth							
	Wintner (2000)	Whole vertebrae	APE, D	X					
Dog Sharks									
Centrophorus squamosus	Clarke et al. (2002b)	Second dorsal spines							
Cephaloscyllium umbratile	Tanaka (1990)	Sectioned vertebrae			X			X	
Chiloscyllium plagiosum	Tullis and Peterson (2000)	Embryonic lab growth							
Deania calcea	Clarke et al. (2002a)	Sectioned first dorsal spines							
Squalus acanthias	Avsar (2001)	Sectioned spines	APE, D			X			
	Stenberg (in press)	Sectioned spines							
Squalus megalops	Watson and Smale (1999)	Sectioned spines							
Squalus mitsukurii	Wilson and Seki (1994)	Whole spines					X		
	Taniuchi and Tachikawa (1999)	Sectioned second dorsal spines							
	Taniuchi and Tachikawa (1999)	Sectioned second dorsal spines							
	Taniuchi and Tachikawa (1999)	Sectioned second dorsal spines							
Mackerel Sharks									
Alopias pelagicus	Liu et al. (1999)	Whole vertebrae					X	X	
Alopias superciliosus	Liu et al. (1998)	Whole vertebrae					X	X	
Alopias vulpinus	Smith et al. (in press)	Whole vertebrae							
Carcharias taurus	Schmid et al. (1990)	Captive lab growth				X			X
	Govender et al. (1991)	Captive growth				X			X
	Branstetter and Musick (1994)	Sectioned vertebrae				X	(X)		
Carcharodon carcharias	Goldman (2002)	Sectioned vertebrae	PA			X	(X)	X	X
	Wintner and Cliff (1999)	Whole vertebrae					X		X
Isurus oxyrinchus	Campana et al. (2002)	Sectioned vertebrae	APE, D		X		X		
	Ribot-Carballal (2003)	Whole vertebrae	APE, D				(X)		
Lamna ditropis	Tanaka (1980)	Sectioned vertebrae			X		X		
	Goldman (2002)	Sectioned vertebrae	PA				X		

TABLE 14.1 (Continued)

Summary of the Age Verification and Validation Studies Completed since the Cailliet (1990) Review

Taxonomic Group and Species	Author(s) and Date of Study	Structure, Treatment or Method	Precision Analysis	Size Frequency Analysis	Back Calculation	Verification Techniques — Marginal Increment Analysis	Verification Techniques — Centrum Edge Analysis	Verification Techniques — Tag/Recapture Analysis	Validation Techniques — Laboratory Growth	Validation Techniques — OTC Injection	Validation Techniques — Bomb Carbon
Lamna nasus	Francis and Stevens (2000)	Embryonic development & sizes		X							
	Campana et al. (2002)	Sectioned vertebrae									X
	Natanson et al. (2002)	Sectioned vertebrae						X		X	
Ground Sharks											
Carcharhinus acronotus	Carlson et al. (1999)	Sectioned vertebrae		X	X	(X)					
	Carlson et al. (1999)	Sectioned vertebrae		X	X	(X)					
	Carlson et al. (1999)	Sectioned vertebrae		X	X	(X)					
	Driggers et al. (in press)	Sectioned vertebrae				(X)			X		
Carcharhinus brachyurus	Walter and Ebert (1991)	Sectioned vertebrae			X						
Carcharhinus brevipinna	Allen and Wintner (2002)	Sectioned vertebrae				(X)					
Carcharhinus caatus	White et al. (2002b)	Sectioned vertebrae		X		X					
Carcharhinus falciformis	Bonfil et al. (1993)	Sectioned vertebrae		X		(X)					
Carcharhinus isodon	Carlson et al. (2003)	Sectioned vertebrae				X					
Carcharhinus leucas	Schmid et al. (1990)	Captive growth							X		
	Wintner et al. (2002)	Sectioned vertebrae					X	X			
	Tanaka (1991)	Vertebrae/sectioned/stained hematoxylin									
Carcharhinus longimanus	Seki et al. (1998)	Sectioned vertebrae			X	(X)					
	Lessa et al. (1999b)	Sectioned vertebrae			X	X					
Carcharhinus obscurus	Natanson et al. (1995)	Sectioned vertebrae		X		(X)					
	Simpfendorfer (2000b)	Tagging (OTC vs. not)						X			
	Simpfendorfer (2002b)	Sectioned vertebrae						X		X	

Species	Reference	Method	Notes						
Carcharhinus plumbeus	Schmid et al. (1990)	Captive growth						X	
	Casey and Natanson (1992)	Captive/tag–recapture (26 yr)			X				
	Casey and Natanson (1992)	Sectioned vertebrae							
	Sminkey and Musick (1995a)	Sectioned vertebrae				X			
	Sminkey and Musick (1995a)	Sectioned vertebrae				X			
	Sminkey and Musick (1995a)	Sectioned vertebrae				X			
	Merson and Pratt (2001)	Nursery ground growth — FL by month		X					
Carcharhinus porosus	Lessa and Santana (1998)	Sectioned vertebrae				(X)			
Carcharhinus sorrah	Davenport and Stevens (1998)	Whole vertebrae						X	X
Carcharhinus tilstoni	Davenport and Stevens (1998)	Whole vertebrae						X	X
Furgaleus macki	Simpfendorfer et al. (2000)	Sectioned vertebrae				(X)		X	
Galeocerdo cuvier	Natanson et al. (1999)	Tag returns (Gulland-Holt 1959)		X				X	
	Wintner and Dudley (2000)	Sectioned vertebrae				(X)	X		X
Galeorhinus galeus	Ferreira and Vooren (1991)	Sectioned vertebrae		X		(X)	X		
	Francis and Mulligan (1998)	Sectioned vertebrae	D, CV	X					
Galeus melastomus	Correia and Figueiredo (1997)	Sectioned vertebrae	PA						
Glyphis sp.	Tanaka (1991)	Vertebra/sectioned/stained hematoxylin							
Isogomphodon oxyrhynchus	Lessa et al. (2000)	Sectioned vertebrae		X		(X)			
Mustelus antarcticus	Officer et al. (1996)	Sectioned vertebrae							
	Officer et al. (1997)	Sectioned vertebrae		X					
	Walker et al. (1998)	Sectioned vertebrae							
	Walker et al. (1998)	Sectioned vertebrae							

TABLE 14.1 (Continued)

Summary of the Age Verification and Validation Studies Completed since the Cailliet (1990) Review

Taxonomic Group and Species	Author(s) and Date of Study	Structure, Treatment or Method	Precision Analysis	Verification Techniques						Validation Techniques	
				Size Frequency Analysis	Back Calculation	Marginal Increment Analysis	Centrum Edge Analysis	Tag/Recapture Analysis	Laboratory Growth	OTC Injection	Bomb Carbon
	Troynikov and Walker (1999)	Sectioned vertebrae					(X)				
Mustelus canis	Conrath et al. (2002)	Sectioned vertebrae	PA			X					
Mustelus henlei	Yudin and Cailliet (1990)	Sectioned vertebrae			X						
Mustelus lenticulatus	Francis and Francis (1992)	Whole vertebrae		X							
	Francis and Maolagain (2000)	Whole vertebrae									
Mustelus manazo	Cailliet et al. (1990)	Sectioned vertebrae	APE, D, PA								
	Cailliet et al. (1990)	Sectioned vertebrae	APE, D, PA								
	Yamaguchi et al. (1996)	Sectioned vertebrae					X				
	Yamaguchi et al. (1999)	Sectioned vertebrae					(X)				
Mustelus mustelus	Goosen and Smale (1997)	Sectioned vertebrae					X				
Prionace glauca	Tanaka et al. (1990)	Sectioned vertebrae	APE, D, PA	X							
	Nakano (1994)	Whole vertebrae									
	Skomal and Natanson (2003)	Sectioned vertebrae	APE, D, PA		X			X		X	
Rhizoprionodon taylori	Simpfendorfer (1993, 1999)	Whole vertebrae	APE, D		X	X					
Rhizoprionodon terraenovae	Loefer and Sedberry (2002)	Sectioned vertebrae		X		X					
	Carlson and Baremore (2003)	Vertebrae/sectioned/ crystal violet	APE, D								

Species	Reference	Method	Precision					
Scyliorhinus canicula	Thomason et al. (1996)	Embryonic growth					X	
	Rodriguez-Cabella et al. (1998)	Length frequency and tagging			X			
Sphyrna lewini	Chen et al. (1990)	Sectioned vertebrae		(X)				
	Anislado-Tolentino and Robinson-Mendoza (2001)	Sectioned vertebrae	APE, D, PA	X				
Spyrna tiburo	Parsons (1993)	Whole vertebrae		X		X		X
	Parsons (1993)	Whole vertebrae		X		X		X
	Carlson and Parsons (1997)	Whole vertebrae		X		X		
Triakis semifasciata	Kusher et al. (1992)	Sectioned vertebrae	APE, D, PA	X		X		X
	Smith et al. (2003)	Sectioned vertebrae		X		X		X

Note: (X) = not statistically significant; PA = percentage agreement; APE = average percent agreement; D = Chang's 1982 index of precision; CV = coefficient of variation.

were used in eight studies (eight species), while the combination of tag–recapture analysis with OTC marking was used in seven studies (seven species) to validate growth.

14.6 Growth Models

A number of models and variations of models exist for estimating growth parameters in fishes, of which the von Bertalanffy and Gompertz growth models are the most commonly applied (see Ricker, 1979; Haddon, 2001; Summerfelt and Hall, 1987, for thorough reviews). The von Bertalanffy growth function (VBGF) has most often been used to describe fish growth, and the Gompertz curve is often used to describe larval and early life growth of fishes and growth in many invertebrates (Zweifel and Lasker, 1976; Ricker, 1979). However, weight can be used in place of length in the von Bertalanffy model and length may be substituted for weight in the Gompertz model. Many statistical packages include modules (i.e., functions) that can be used to calculate the best fitting growth parameters for the available length-at-age or weight-at-age data pairs. For example, a nonlinear least-squares regression algorithm (e.g., 'nls' in S-Plus, Mathsoft, Inc., 2000), a maximum likelihood function or the PROC NLIN function in SAS can be used to fit the von Bertalanffy and Gompertz curves to the data (SAS Institute, Inc., 1999), and programs such as PC-YIELD (Punt and Hughes, 1989) can calculate a wide range of growth models for comparison (Wintner et al., 2002). Additionally, FISHPARM (Prager et al., 1987), a fishery-based statistics program, is simple to use and provides quality statistical results for the two models presented here. Both models can also be fit to data on a spreadsheet via a nonlinear regression using the "solver" function in Microsoft Excel.

The von Bertalanffy growth function has been widely used since its introduction into fisheries by Beverton and Holt (1957), and although it has received much criticism over the years (Roff, 1982), it is the most widely used growth function in fisheries biology today (Haddon, 2001). It maintains its attractiveness, in part, because its approach to modeling growth is based on the biological premise that the size of an organism at any moment depends on the resultant of two opposing forces: anabolism and catabolism. Additionally, it is convenient to use and allows for much easier comparison between populations, but several alternative forms of the model can be fit to the age–length data (Haddon, 2001, presents a variety of growth models including generalized models as possible alternatives).

Small sample size, particularly of small or large individuals can cause poor parameter estimates using the von Bertalanffy model (Cailliet and Tanaka, 1990; Francis and Francis, 1992). In lieu of using t_0 (due to its lack of biological meaning) some researchers suggest using an estimate of length at birth (L_0) as a more robust method (Goosen and Smale, 1997; Carlson et al., 2003, H. Mollet, pers. comm.). This method was first introduced by Fabens (1965) as an alternative equation to the von Bertalanffy growth model. Although only a few studies have used Faben's (1965) equation to estimate growth parameters in elasmobranchs, it has provided more realistic parameter estimates for some species when the sample size was small (Goosen and Smale, 1997), and extremely similar results to the von Bertalanffy model when sample size was adequate (Carlson et al., 2003). This appears to be an excellent alternative to the von Bertalanffy model and should be applied where appropriate for comparison to other models.

From a mathematical point of view, it does not matter whether L_0 or t_0 is used as the third parameter, and for mathematical manipulations it may even be advantageous to use t_0 as the third parameter (H. Mollet, pers. comm.). However, from a biological point of view and in particular for elasmobranchs, the size at birth (L_0) is often well defined, whereas t_0 has no biological meaning. Holden (1974) originally proposed t_0 to be an estimate of gestation time, which unreasonably implies that embryonic growth is governed by the same growth parameters as postnatal or posthatch growth. The use of L_0 as the third parameter allows an easy evaluation of the growth curve. If t_0 is needed for comparison with published data, it can easily be calculated from reported L_∞, k, and t_0 using the formula $L_0 = L_\infty (1 - \exp(k \times t_0))$. This would indicate whether previously reported results were reasonable, because L_0 is often the best-known parameter.

The Gompertz growth function is an S-shaped model function (similar to the logistic function; for use of the logistic function and several alternatives to the Gompertz function, see Ricker, 1975, 1979). The estimated instantaneous growth rate in the Gompertz function is proportional to the difference between the logarithms of the asymptotic disk width or length and the actual disk width or length (Ricker,

1975, 1979). This growth function has been used most often for skates and stingrays (Mollet et al., 2002) and may be better suited to elasmobranchs that hatch from eggs, but it may also be the most appropriate model for some shark species (Wintner et al., 2002). This model may offer a better option when the volume of an organism greatly expands with age, such as myliobatiform rays where considerable thickness is added to the animal over time, but not so much disk width or length (W. Smith, pers. comm.). The body mass may be distributed differently than would be readily detectable by length measurement and by the von Bertalanffy model. Additionally, captive growth rates (particularly when starting with young, small animals) may be better estimated by this function as newly captured specimens may not grow in their typical fashion due to physiological stress or a reduction in feeding that often accompanies that stress, which may cause growth rates to slow (Mollet et al., 2002).

It is interesting to note how prevalent the use of VBGF models is in recent chondrichthyan studies (Table 14.2). Almost 100 new sets of growth parameters from VBGF equations covering 88 species were generated during this time period. In ~20% of these studies, these VBGF models were for both sexes combined, whereas in the other 80% male and female growth models were produced separately, and in some cases both sexes were combined and also presented.

We decided not to plot or statistically analyze the VBGF parameters in Table 14.1 in a synoptic fashion because these data do not represent either a random sample or all of the papers published in which VBGFs are presented. Rather, we thought it would be more beneficial to the reader to summarize the distribution and ranges in these values in the following sections.

14.7 Comparative Growth Coefficients (k in the von Bertalanffy)

The growth coefficient. k, from the VBGF describes the average rate at which an organism in the population achieves its maximum length (or size) from its length at birth.

An analysis of these growth coefficients in Table 14.2 indicates that there is a considerable range in k values, both among all chondricthyans and even within separate taxonomic groups. In most cases, studies fit their growth data to the von Bertalanffy growth model, yielding estimates of asymptotic size (L_∞), a growth coefficient (k), and either a size at birth (L_0) or a mathematically fit estimate of the age at theoretical size = 0 (t_0).

Even though it is commonly thought that chondrichthyans all have relatively slow growth (i.e., low k values), Table 14.2 indicates that these values vary a great deal. In the chimaeras, k values ranged from 0.05 to 0.47, while in the sawfishes, all three species studied produced low values, ranging from 0.07 to 0.08. Therefore, from this limited data set, it appears that the sawfishes are, indeed, slow-growing fishes. The guitarfishes, torpedo rays, and stingrays had k values ranging from 0.2 to 0.5, while the skates had a bit broader range of 0.05 to 0.5. The angel shark, *Squatina californica,* and sevengill shark, *Notorynchus cepedianus,* had relatively high k values, ranging from 0.15 to 0.29. The carpet sharks, however, represented here only as the whale shark, *Rhincodon typus,* appear to be slower growing, with k values estimated to be between 0.03 and 0.05.

The remaining three major groups of sharks also had widely ranging growth coefficients (Table 14.2). The dog shark k values ranged from a very low 0.03 to 0.25, and the mackerel sharks 0.03 to 0.23. The more diverse ground sharks, represented by 31 species, also had a very wide range in k values, from 0.06 for the sandbar shark, *Carcharhinus plumbeus,* to 1.013 and 1.337 for female and male Australian longnose shark, *Rhizoprionodon taylori,* respectively. It must be noted, however, that this broad difference in k values can also be related to sample size, aging methodology, verification, validation, and growth model fitting techniques.

14.8 Comparative Ages of Maturity and Longevity among Studies

14.8.1 Age of Maturity Estimates

As with growth coefficients, age of maturity (T_{mat}) estimates varied a great deal, both within and among chondrichthyan groups, and they appeared to be related to longevity estimates (Table 14.2; Section

TABLE 14.2

Summary of Chondrichthyan Age and Growth Studies since the Cailliet (1990) Review (including species, authors, locations, and von Bertalanffy growth function parameters)

Species	Author and Date of Study	Location	Sex	von Bertalanffy Growth Parameters					Age at Maturity (T_{mat})	Max Age (T_{max})
				L_∞	k	t_0	L_0	n		
Chimaeras										
Callorhynchus capensis	Freer and Griffiths (1993)	South Africa	Female	108.9 cm FL	0.052	−0.606	20.0 cm FL	94	4.6 yr	10 yr
	Freer and Griffiths (1993)	South Africa	Male	65.9 cm FL	0.17	−0.721	20.0 cm FL	61	3.3 yr	5+ yr
	Freer and Griffiths (1993)	South Africa	Both sexes	68.6 mm FL	0.171	−0.721	20.0 cm FL	155		10 yr
Callorhynchus milli	Francis (1997)	New Zealand (Pegasus Bay 1966–68)	Female	156.9 cm FL	0.096	−0.87			5.5 yr	8 yr
	Francis (1997)	New Zealand (Pegasus Bay 1966–68)	Male	74.7 cm FL	0.231	−0.78			4.5 yr	8 yr
	Francis (1997)	New Zealand (Canterbury Bight 1966–68)	Female	203.6 cm FL	0.06	−1.06			6.0 yr	8 yr
	Francis (1997)	New Zealand (Canterbury Bight 1966–68)	Male	141.5 cm FL	0.089	−0.96			5.5 yr	6 yr
	Francis (1997)	New Zealand (Pegasus Bay 1983–1984)	Female	113.9 cm FL	0.195	−0.53			4.6 yr	8 yr
	Francis (1997)	New Zealand (Pegasus Bay 1983–1984)	Male	66.9 cm FL	0.473	−0.24			2.9 yr	7 yr
	Francis (1997)	New Zealand (Canterbury Bight 1988)	Female	94.1 cm FL	0.224	−0.69			5.6 yr	9 yr
	Francis (1997)	New Zealand (Canterbury Bight 1988)	Male	62.7 cm FL	0.466	−0.38			4.6 yr	5 yr
Chimaera monstrosa	Calis et al. (in press)	Ireland, west coast	Both sexes	78.87 cm PSCFL	0.067	−2.513	~10 cm PSCFL	58 (48F, 7M)	11.2–13.4 yr (F–M)	5–29 yr

Sawfishes

Species	Reference	Location	Sex	L_∞	k	t_0	Size at birth	n	Age at maturity	Max age
Pristis microdon	Tanaka (1991)	Northern Australia and Papua New Guinea	Both sexes	362.5 (397.9) cm TL	0.07 (0.05)	−4.07 (−5.54)	~80 mm TL	36	?	44 yr
Pristis perotteti	Simpfendorfer (2000a)	Southeastern U.S.	Both sexes	450 cm TL	0.08	−1.98			10 yr	30 yr
Pristis pectinata	Simpfendorfer (2000a)	Southeastern U.S.	Both sexes	450 cm TL	0.08	−1.98			10 yr	30 yr

Guitarfishes, Torpedo Rays, and Stingrays Marine and Brackish

Species	Reference	Location	Sex	L_∞	k	t_0	Size at birth	n	Age at maturity	Max age
Dasyatis americana	Henningsen (2002) Abstr.	Southeastern U.S.	Female	150 cm DW?	0.539				5–6 yr	26 yr
Dasyatis americana	Henningsen (2002) Abstr.	Southeastern U.S.	Male	112.5 cm DW?	0.206				3–4 yr	28 yr
Dasyatis chrysonata	Cowley (1997)	South Africa	Female	91.3 cm DW	0.07	−4.48		165	~4 yr	14 yr
Dasyatis chrysonata	Cowley (1997)	South Africa	Male	53.2 cm DW	0.17	−3.65		105	~4 yr	9 yr
Dasyatis pastinaca	Ismen (2003)	Eastern Mediterranean	Both sexes	121.5 cm TL	0.089	−1.615		256	4–5 yr	~10 yr
Dasyatis violacea	Mollet et al. (2002)	California, NE Pacific	Female	103.0 cm DW	0.32	−8.2		4	3.0 yr	8.5–8.7
Dasyatis violacea	Mollet et al. (2002)	California, NE Pacific	Male	67.0 cm DW	0.8	−5.6		2	3 yr	7.2–8.3
	Neer (in press)	California	Both sexes					45	3 yr	12 yr
Gymnura altavela	Henningsen (1996)	Baltimore Aquarium	Female		0.303 mm d^{-1}			1		
	Henningsen (1996)	Baltimore Aquarium	Male		0.063 mm d^{-1}			1		
Rhinobatus productos	Timmons and Bray (1997)	Southern California	Female	594 cm TL	0.016	−3.8	23 cm TL	19	8 yr	11 yr
	Timmons and Bray (1997)	Southern California	Male	142 cm TL	0.095	−3.942	23 cm TL	24	8 yr (50%)	11 yr
	Timmons and Bray (1997)	Southern California	Both sexes	228 cm TL	0.047	−4.03	23 cm TL	43	8 yr	11 yr
Torpedo californica	Neer and Cailliet (2001)	California, NE Pacific	Female	137.3	0.078	−1.934		81	9 yr	16 yr
	Neer and Cailliet (2001)	California, NE Pacific	Male	92.1	0.137	−1.483		116	6 yr	14 yr
Torpedo marmorata	Mellinger (1971)	Bay of Biscay, France	Female						~12 yr	20 yr
	Mellinger (1971)	Bay of Biscay, France	Male						5 yr	12–13 yr
Trygonoptera mucosa	White et al. (2002b)	Western Australia	Female	308.1 cm DW	0.241	−2.517	11.3 cm DW?	324	5 yr (50%)	17 yr
	White et al. (2002b)	Western Australia	Male	261.2 cm DW	0.493	−1.362	11.3 cm DW	400	2 yr	12 yr
Trygonoptera personalis	White et al. (2002b)	Western Australia	Female	302.8 cm DW	0.143	−3.858	12.5 cm DW?	352	4 yr (50%)	16 yr
Urolophus lobatus	White et al. (2002b)	Western Australia	Male	269.1 cm DW	0.203	−3.09		303	4 yr (50%)	10 yr
	White et al. (2001)	New Zealand?	Female	24.8 cm DW?	0.369	−1.35		388	~4 yr	14 yr
	White et al. (2001)	New Zealand?	Male	21.0 cm DW?	0.514	−1.18		428	~2 yr	13 yr

TABLE 14.2 (Continued)

Summary of Chondrichthyan Age and Growth Studies since the Cailliet (1990) Review (including species, authors, locations, and von Bertalanffy growth function parameters)

Species	Author and Date of Study	Location	Sex	von Bertalanffy Growth Parameters					Age at Maturity (T_{mat})	Max Age (T_{max})
				L_∞	k	t_0	L_0	n		
Freshwater										
Himantura chaophraya	White et al. (2001)	Rivers in South Asia	Both sexes				33.2 cm DW	1,3	<7 yr	
Himantura gerrardi	White et al. (2001)	Rivers in South Asia	Both sexes					3,3	3~7 yr	7+ yr
Himantura imbricata	Tanaka and Ohnishi (1998)	Rivers in South Asia	Both sexes					3	2+ yr	3+ yr
Himantura laosensis	Tanaka and Ohnishi (1998)	Rivers in South Asia	Both sexes					1,3	~1–4 yr	
Himantura signifer	Tanaka and Ohnishi (1998)	Rivers in South Asia	Female				15.0 cm DW	17	2~3 yr	5+ yr
	Tanaka and Ohnishi (1998)	Rivers in South Asia	Male				15.0 cm DW	26	1~2 yr	5+ yr
	Tanaka and Ohnishi (1998)	Rivers in South Asia	Both sexes				~35.0 cm DW	43		4 yr?
Himantura uarnak	Tanaka and Ohnishi (1998)	Rivers in South Asia	Female					1		
Pastinachus sephen	Tanaka and Ohnishi (1998)	Rivers in South Asia	Female					1		
	Tanaka and Ohnishi (1998)	Rivers in South Asia	Male					4	4–6 yr	7+ yr
Skates										
Dipturus batis	DuBuit (1972)	Celtic Sea	Both sexes	253.73 cm TL	0.057	−1.629		82	11 yr	50 yr
Dipturus innominatus	Francis et al. (2001)	New Zealand	Female					50	13 yr (50%)	24 yr
	Francis et al. (2001)	New Zealand	Male					46	8 yr (50%)	
Dipturus nasutus	Francis et al. (2001)	New Zealand	Both sexes	150.5 cm PL	0.095	−1.06		98		24 yr
Dipturus nasutus	Francis et al. (2001)	New Zealand	Female					75	6 yr (50%)	9 yr
	Francis et al. (2001)	New Zealand	Male					59	4 yr (50%)	
Dipturus nasutus	Francis et al. (2001)	New Zealand	Both sexes	91.3 cm PL	0.16	−1.2		134		9 yr
Dipturus pullopunctata	Walmsley-Hart et al. (1999)	South Africa	Female	132.68 cm DW	0.05	−2.2		51	~5 yr	14 yr

Species	Reference	Location	Sex	Size	K	t_0		n		
Leucoraja naevus	DuBuit (1972)	Celtic Sea	Both sexes	91.64 cm TL	0.019	−0.465		50	~9 yr	~13–14 yr
Leucoraja ocellata	Sulikowski et al. (2002)	Gulf of Maine/NWAtl	Female	137.4 cm TL	0.059	−1.609		121		18 yr
	Sulikowski et al. (2002)	Gulf of Maine/NWAtl	Male	121.8 cm TL	0.074	−1.418		88		19 yr
Leucoraja wallacei	Walmsley-Hart et al. (1999)	South Africa	Female	43.52 cm DW	0.26	−0.21		74	7 yr	12 yr
Raja binoculata	Zeiner and Wolf (1993)	Monterey Bay, CA	Female	167.9 cm DW	0.37	Logistic		68	10–12 yr	12 yr
	Zeiner and Wolf (1993)	Monterey Bay, CA	Male	139.3 cm TL	0.43	Logistic		103	7–8 yr	11 yr
Raja miraletus	Abdel-Aziz (1992)	Off Egypt	Female	91.92 cm TL	0.25	−0.172		219		17.2 yr
	Abdel-Aziz (1992)	Off Egypt	Male	87.87 cm TL	0.502	−0.193		318		15 yr
	Abdel-Aziz (1992)	Off Egypt	Male	77.05 cm DW	0.1	−2.37		56	~5 yr	18 yr
	Abdel-Aziz (1992)	Off Egypt	Both sexes	87.32 cm DW	0.08	−1.95		107	~5 yr	18 yr
Raja rhina	Zeiner and Wolf (1993)	Monterey Bay, CA	Female	106.9 cm TL	0.16	−0.3		68	10–11 yr	12 yr
	Zeiner and Wolf (1993)	Monterey Bay, CA	Male	96.7 cm TL	0.25	−0.73		64	7 yr	13 yr
	Zeiner and Wolf (1993)	Monterey Bay, CA	Male	40.54 cm DW	0.27	−0.08		65	7 yr	17 yr
	Zeiner and Wolf (1993)	Monterey Bay, CA	Both sexes	42.19 cm DW	0.26	−0.17		139	7 yr	17 yr
Angel Sharks										
Squatina californica	Cailliet et al. (1992)	California	Female	126.0 cm TL	0.162	NA	~25 cm TL		~11 yr?	35 yr
	Cailliet et al. (1992)	California	Male	125.9	0.152	NA			~11 yr	35 yr
	Cailliet et al. (1992)	California	Both sexes	127	0.146	NA			~11 yr?	35 yr
Cow Sharks										
Notorynchus cepedianus	Van Dykhuizen and Mollet (1992)	California	Female	192 cm TL	0.291			4	11–21 yr	
	Van Dykhuizen and Mollet (1992)	California	Male	239	0.2			1	5–15 yr	
Carpet Sharks										
Ginglymostoma cirratum	Carrier and Luer (1990)	Florida	Female						25 yr (50%)	35 yr
	Carrier and Luer (1990)	Florida	Male						16 yr (50%)	29 yr
Orectolobus japonicus	Tanaka (1990)	Japan	Both sexes					6,1		
Rhinocodon typus	Wintner (2000)	South Africa	Female						~22 yr	19–27 yr
	Wintner (2000)	South Africa	Male						~20 yr	20–31 yr

TABLE 14.2 (Continued)

Summary of Chondrichthyan Age and Growth Studies since the Cailliet (1990) Review (including species, authors, locations, and von Bertalanffy growth function parameters)

Species	Author and Date of Study	Location	Sex	von Bertalanffy Growth Parameters					Age at Maturity (T_{mat})	Max Age (T_{max})
				L_∞	k	t_0	L_0	n		
	Wintner (2000) with Pauly (1997) T_{max} ranges	South Africa	Both sexes (Pauly T_{max} = 60 yr)	1,179.0 cm PCL	0.031	−0.85		23 & 2		
	Wintner (2000) with Pauly (1997) T_{max} ranges	South Africa	Both sexes (Pauly T_{max} = 100 yr)	1,554.0 cm PCL	0.051	−1.03		23 & 2		
Dog Sharks										
Centrophorus squamosus	Clarke et al. (2002b)	NE Atlantic	Female					66	45 yr	70 yr
Deania calcea	Clarke et al. (2002b)	NE Atlantic	Male	119.3 cm TL	0.077	−0.933		61	30 yr	55 yr
	Clarke et al. (2002a)	NE Atlantic	Female	93.52 cm TL	0.135	0.165		?		?
	Clarke et al. (2002a)	NE Atlantic	Male					?		?
Squalus acanthias	Avsar (2001)	Black Sea	Female	145.0 cm TL	0.17	−0.73		160		13 yr
	Avsar (2001)	Black Sea	Male	128.2 cm TL	0.2	−0.29		168		14 yr
Squalus acanthias	Avsar (2001)	Black Sea	Both sexes	157.0 cm TL	0.12	−1.3		328		13–14 yr
Squalus megalops	Watson and Smale (1999)	South Africa	Female	93.2 cm TL	0.033	−8.12		509	15 yr (50%)	32 yr
Squalus megalops	Watson and Smale (1999)	South Africa	Male	52.6 yr	0.089	−6.94		255	9 yr (50%)	29 yr
Squalus mitsukurii	Wilson and Seki (1994)	North Pacific Ocean	Female	107 cm TL	0.041	−10.09	21–26 cm TL	105	15 yr (50%)	27 yr
	Wilson and Seki (1994)	North Pacific Ocean	Male	66 cm TL	0.155	−4.64	21–26 cm TL	102	4 yr (50%)	18 yr
	Taniuchi and Tachikawa (1999)	Japan, Choshi	Female	162.8 cm TL	0.039	−5.21		38	9 yr	21 yr
	Taniuchi and Tachikawa (1999)	Japan, Choshi	Male	109.3 cm TL	0.066	−5.03		85	10–11 yr	20 yr
	Taniuchi and Tachikawa (1999)	Japan, Osasawara	Female	111.2 cm TL	0.051	−5.12		130	9 yr	27 yr?
	Taniuchi and Tachikawa (1999)	Japan, Osasawara	Male	88.0 cm TL	0.06	−5.51		54	5 yr	21 yr?

Species	Reference	Location	Sex	L_∞	K	t_0	n	Age at maturity	Max age
	Taniuchi and Tachikawa (1999)	Japan, SE Harbor	Female	83.1 cm TL	0.103	−2.94	36	14–16 yr	17 yr
	Taniuchi and Tachikawa (1999)	Japan, SE Harbor	Male	64.5 cm TL	0.252	−0.43	28	6–7 yr	12 yr
Mackerel Sharks									
Alopias pelagicus	Liu et al. (1999)	Taiwan	Female	197.2	0.085	−7.67	508	9–9.2 yr	16 yr
	Liu et al. (1999)	Taiwan	Male	182.2	0.118	−5.48	323	7–8 yr	14 yr
Alopias superciliosus	Liu et al. (1998)	Taiwan	Female	224.6 cm TL	0.092	−4.21	214	12.3–13.4 yr	20 yr
	Liu et al. (1998)	Taiwan	Male	218.8	0.088	−4.224	107	9–10 yr	14 yr
Alopias vulpinus	Smith et al. (in press)	California	Female	464.3 cm TL	0.124	−3.35	129	5.8 yr	22 yr
	Smith et al. (in press)	California	Male	416.2 cm TL	0.184	−2.08	83	5 yr	10 yr
Carcharias taurus	Govender et al. (1991)	South Africa	Both sexes	249.8 cm FL	0.233	−2.238	3.5		
	Branstetter and Musick (1994)	Gulf of Mexico and Atlantic	Both sexes	303.0 cm TL	0.18	−2.09	55		
Carcharodon carcharias	Goldman (2002)	Atlantic	Female	295.8 cm PCL?	0.11	−4.2	48	9–10 yr	17 yr
	Goldman (2002)	Atlantic	Male	249.5 cm PCL	0.16	−3.4	48	6–7 yr	16 yr
	Wintner and Cliff (1999)	South Africa	Both sexes	544.0 cm TL	0.065	−4.4		8–13	13 (35?)
Isurus oxyrinchus	Campana et al. (2002)	NW Atlantic	Both sexes				1		10 yr
	Ribot-Carballal (2003)	Baja California, Mexico	Both sexes	359 cm TL	0.05	−4.96			25, 10 yr
Lamna ditropis	Tanaka (1980)	NW Pacific	Female	203.8 cm PCL	0.136	−3.946		8–10 yr	17 yr
	Tanaka (1980)	NW Pacific	Male	180.3 cm PCL	0.171	−3.628		5 yr	25 yr
	Goldman (2002)	NE Pacific	Female	207.4 cm PCL	0.17	−2.3	166	6–9 yr	20 yr
	Goldman (2002)	NE Pacific	Male	182.8 cm PCL	0.23	−1	16	3–5 yr	17 yr
Lamna nasus	Francis and Stevens (2000)	Australia and New Zealand	Both sexes (embryo and adult SFA)						
	Francis and Stevens (2000)	SW New Zealand	Linear: FL = 66.5 + 19.8 (Age)						
	Francis and Stevens (2000)	Australia	Linear: FL = 65.4 + 16.1 (Age)						
	Campana et al. (2002)	NW Atlantic	Both sexes						26 yr
	Natanson et al. (2002)	NW Atlantic	Female	369.8 cm TL	0.061	−5.9	291	13 yr	24 yr
	Natanson et al. (2002)	NW Atlantic	Male	257.7 cm TL	0.03	−5.87	283	8 yr	25 yr

TABLE 14.2 (Continued)

Summary of Chondrichthyan Age and Growth Studies since the Cailliet (1990) Review (including species, authors, locations, and von Bertalanffy growth function parameters)

Species	Author and Date of Study	Location	Sex	von Bertalanffy Growth Parameters				n	Age at Maturity (T_{mat})	Max Age (T_{max})
				L_∞	k	t_0	L_0			
Ground Sharks										
Carcharhinus acronotus	Carlson et al. (1999)	NW Florida	Female	113.7 cm FL	0.352	−1.2		23		10–16 yr
	Carlson et al. (1999)	NW Florida	Male	96.3 cm TL	0.59	−0.754		44		4.5–9 yr
	Carlson et al. (1999)	Tampa Bay, Florida	Female	124.13 cm FL	0.237	−1.54		26		10–16 yr
	Carlson et al. (1999)	Tampa Bay, Florida	Male	80.1 cm TL	0.771	−0.797		30		4.5–9 yr
	Carlson et al. (1999)	North Carolina	Female	165.0 cm FL	0.138	−2.68		42		10–16 yr
	Carlson et al. (1999)	North Carolina	Male	188.7 cm TL	0.117	−2.01		30		4.5–9 yr
	Driggers et al. (in press)	NW Atlantic (So. Carolina)	Female	113.5 cm FL	0.18	−4.07		117	4.5 yr	12.5 yr
	Driggers et al. (in press)	NW Atlantic (So. Carolina)	Male	105.8 cm TL	0.21	−3.9		109	4.3 yr	10.5 yr
Carcharhinus brachyurus	Walter and Ebert (1991)	South Africa	Female						19–20 yr	25 yr
	Walter and Ebert (1991)	South Africa	Male						13–19 yr	30 yr
	Walter and Ebert (1991)	South Africa	Both sexes	384.8 cm TL	0.0385	−3.477		61	13–20 yr	25–30 yr
Carcharhinus brevipinna	Allen and Wintner (2002)	South Africa	Female	232.8 cm PCL	0.1	−2.9		38	8–10 yr	17 yr
	Allen and Wintner (2002)	South Africa	Male	196.3 cm PCL	0.146	−2.3		29	8–10 yr	19 yr
Carcharhinus cautus	White et al. (2002a)	Western Australia	Female	123.8 cm TL	0.198	−2.52	~35 cm TL	171	6 yr (50%)	16 yr
	White et al. (2002a)	Western Australia	Male	110.5 cm TL	0.287	−1.75	~35 cm TL	57	4 yr (50%)	12 yr
Carcharhinus falciformis	Bonfil et al. (1993)	Western Gulf of Mexico	Female					40	12 yr	22 yr
	Bonfil et al. (1993)	Western Gulf of Mexico	Male					43	10 yr	20 yr
	Bonfil et al. (1993)	Western Gulf of Mexico	Both sexes	311.0 cm TL	0.101	−2.718		83		
Carcharhinus isodon	Carlson et al. (2003)	NE Gulf of Mexico	Female	155.9 cm TL	0.244	−2.07		117	3.9 yr (50%)	8.0 yr

Species	Reference	Location	Sex	Max length	K	t₀	n	Age at maturity	Max age
	Carlson et al. (2003)	Northeastern Gulf of Mexico	Male	133.8 cm TL	0.412	−1.39	123	4.3 yr (50%)	8.1 yr
Carcharhinus leucas	Wintner et al. (2002)	South Africa	Female				69	21 yr (50%)	32 yr
	Wintner et al. (2002)	South Africa	Male	230.0 cm TL	0.071	−5.12	123	20 yr (50%)	29 yr
	Wintner et al. (2002)	South Africa	Both sexes				123		
	Tanaka (1991)	Northern Australia and Papua New Guinea	Both sexes?	79.1 and 88.8 cm TL			2		
Carcharhinus longimanus	Seki et al. (1998)	Central Pacific	Female				114	4–5 yr	11+ yr
	Seki et al. (1998)	Central Pacific	Male				111	4–5 yr	11 yr
	Seki et al. (1998)	Central Pacific	Both sexes	244.58 cm PCL	0.103	−2.698	126	4–5 yr	11+ yr
	Lessa et al. (1999b)	South Atlantic	Female				60	6–7 yr	17 yr
	Lessa et al. (1999b)	South Atlantic	Male				44	6–7 yr	13 yr
	Lessa et al. (1999b)	South Atlantic	Both sexes	284.9 cm TL	0.0996	−3.391	104 + 6 = 110	6–7 yr	13–17 yr
Carcharhinus obscurus	Natanson et al. (1995)	NW Atlantic	Female	359 cm FL	0.039	−7.04	676	21 yr (50%)	33 yr
	Natanson et al. (1995)	NW Atlantic	Male	373 cm FL	0.038	−6.28	47	19 yr	25+ yr
	Simpfendorfer (2000b)	Western Australia	Female	354.4 cm FL	0.043	None	127	17–22 yr	32 yr
	Simpfendorfer (2000b)	Western Australia	Male	336.5 FL	0.045	None	111	20–22 yr	23 yr
Carcharhinus plumbeus	Casey and Natanson (1992)	NW Atlantic	Both sexes	186 cm TL	0.046	−6.45	33		23 yr
	Casey and Natanson (1992)	NW Atlantic	Female	303 cm FL	0.039	−3.92	447		
	Sminkey and Musick (1995a)	NW Atlantic	Female	197.0 cm TL	0.059	−4.8	150	15–16 cm TL	25 yr
	Sminkey and Musick (1995a)	NW Atlantic	Male	184.0 cm TL	0.059	−5.4	38	15–16 yr	18 yr
	Sminkey and Musick (1995a)	NW Atlantic	Female	165.0 cm TL	0.086	−3	191	15–16 cm TL	25 yr
	Sminkey and Musick (1995a)	NW Atlantic	Male	166.0 cm TL	0.087	−3.8	223	15–16 yr	18 yr
Carcharhinus porosus	Lessa and Santana (1998)	South Atlantic	Female					6 yr	12 yr
	Lessa and Santana (1998)	South Atlantic	Male					6 yr	?
	Lessa and Santana (1998)	South Atlantic	Both sexes	125.1 cm TL	0.101	−2.9	504	6 yr	12 yr
Carcharhinus sorrah	Davenport and Stevens (1998)	Australia	Female	133.9 cm TL	0.34	−1.9	133	2–3 yr	7 yr
	Davenport and Stevens (1998)	Australia	Male	98.4 cm TL	1.17	−0.6	80	3 yr?	7 yr

TABLE 14.2 (Continued)

Summary of Chondrichthyan Age and Growth Studies since the Cailliet (1990) Review (including species, authors, locations, and von Bertalanffy growth function parameters)

Species	Author and Date of Study	Location	Sex	von Bertalanffy Growth Parameters				n	Age at Maturity (T_{mat})	Max Age (T_{max})
				L_∞	k	t_0	L_0			
Carcharhinus tilsoni	Davenport and Stevens (1998)	Australia	Female	194.2 cm TL	0.14	−2.8		257	3–4 yr	12 yr
	Davenport and Stevens (1998)	Australia	Male	165.4 cm TL	0.19	−2.6		132	3–4 yr	8 yr
Furgaleus macki	Simpfendorfer et al. (2000)	Australia	Female	120.7	0.369	−0.544		112	6.5 yr	11.5 yr
	Simpfendorfer et al. (2000)	Australia	Male	121.5 cm TL	0.423	−0.472		67	4.5 yr	10.5 yr
Galeocerdo cuvier	Natanson et al. (1999)	NW Atlantic	Female						7 yr	27.3 yr
	Natanson et al. (1999)	NW Atlantic	Male						7 yr	27.3 yr
	Natanson et al. (1999)	NW Atlantic	Both sexes (TRA)	337.0 cm TL	0.178	−1.12		42 (TRA)	7 yr	27.3 yr?
	Wintner and Dudley (2000)	South Africa	Female						11 yr	11 yr
	Wintner and Dudley (2000)	South Africa	Male						8 yr	8 yr
	Wintner and Dudley (2000)	South Africa	Both sexes	365 cm PCL	0.117	−2.34		90	8–11 yr	8–11 yr
Galeorhinus galeus	Ferreira and Vooren (1991)	South Brazilian Coast	Female	163.0 cm TL	0.075	−3		26		43 yr
	Ferreira and Vooren (1991)	South Brazilian Coast	Male	152.0 cm TL	0.092	−2.69		33		35 yr
	Francis and Mulligan (1998)	New Zealand	Female	179.2 cm TL	0.086	−2.68		137	13–15 yr	25 yr
	Francis and Mulligan (1998)	New Zealand	Male	142.9 cm TL	0.154	−1.64		127	12–17 yr	25 yr
	Francis and Mulligan (1998)	New Zealand	Both sexes	165.8–180.4 cm TL	0.086–0.109	−2.37 to −2.41		264	25 yr	12–17 yr
Galeus melastomus	Correia and Figueiredo (1997)	NE Atlantic	Both sexes?					477		8 yr
Glyphis sp.	Tanaka (1991)	Northern Australia and Papua New Guinea	Both sexes?				~49.7 cm?	1		

Species	Reference	Locality	Group	L_∞	k	t_0	n	Age	Age
Isogomphodon oxyrhynchus	Lessa et al. (2000)	NE Atlantic	Female				52	6–7 yr	12 yr
	Lessa et al. (2000)	NE Atlantic	Male				46	5–6 yr	7 yr
	Lessa et al. (2000)	NE Atlantic	Both sexes	171.4 cm TL	0.121	-2.612	105	5–7 yr	12 yr
Mustelus antarcticus	Walker et al. (1998)	Australia	Male (constant variance)	137.2 cm TL	0.266	-0.8	95		
	Walker et al. (1998)	Australia	Female (constant variance)	201.9 cm TL	0.074	-2.99	134		
	Walker et al. (1998)	Australia	Male (proportional variance)	134.6 cm TL	0.288	-0.64	95		
	Walker et al. (1998)	Australia	Female (proportional variance)	250.9 cm TL	0.122	-1.55	134		
	Troynikov and Walker (1999)	Australia	Female (range: two locations and years)	(CD = 11–18 cm)	0.037–0.071	-2.80 to -2.43	69, 60, 61	11–18 yr	11–18 yr
	Troynikov and Walker (1999)	Australia	Male (range: two locations and years)	(CD = 9 cm)	0.096–0.100	-2.11 to -2.04	50, 51, 63	9 yr?	9 yr?
Mustelus canis	Conrath et al. (2002)	NW Atlantic	Female	123.5 cm TL	0.292	-1.94	531	4–7 yr	16 yr
	Conrath et al. (2002)	NW Atlantic	Male	105.2 cm TL	0.439	-1.52	363	2–3 yr	10 yr
Mustelus henlei	Yudin and Cailliet (1990)	California	Female	976 mm TL	0.225	-1.375	56	1–4 yr	13 yr
	Yudin and Cailliet (1990)	California	Male	861 mm TL	0.285	-1.086	15	1–4 yr	13 yr
	Yudin and Cailliet (1990)	California	Both sexes	977 mm TL	0.244	-1.296	71	1–4 yr	13 yr
Mustelus lenticulatus	Francis and Francis (1992)	New Zealand	Female				134	7.5 yr	
	Francis and Francis (1992)	New Zealand	Male				110	5.5 yr	
	Francis and Francis (1992)	New Zealand	Both sexes	147.2 cm TL	0.119	-2.35	244	5.5–7.5 yr	
Mustelus manazo	Cailliet et al. (1990)	Japan (Choshi)	Female	176.5 cm TL	0.07	-3.24	30		
	Cailliet et al. (1990)	Japan (Choshi)	Male	133.4 cm TL	0.1	-3.42	30		
	Cailliet et al. (1990)	Japan (Nagasaki)	Female	99.9 cm TL	0.2	-2.88	30		
	Cailliet et al. (1990)	Japan (Nagasaki)	Male	84.6 cm TL	0.22	-3.69	30		
	Yamaguchi et al. (1996)	Japan (3 areas)	Female	134.1 cm TL	0.113	-2.65	226		

TABLE 14.2 (Continued)

Summary of Chondrichthyan Age and Growth Studies since the Cailliet (1990) Review (including species, authors, locations, and von Bertalanffy growth function parameters)

Species	Author and Date of Study	Location	Sex	L_∞	k	t_0	L_0	n	Age at Maturity (T_{mat})	Max Age (T_{max})
	Yamaguchi et al. (1996)	Japan (3 areas)	Male	124.1 cm TL	0.12	−2.59		180		
	Yamaguchi et al. (1999)	Japan (3 areas)	Female	82.9 cm TL	0.233	−2.16		71	2–5 yr	9 yr
	Yamaguchi et al. (1999)	Japan (3 areas)	Male	113.7 cm TL	0.124	−2.78		115	2–4 yr	5 yr
Mustelus mustelus	Goosen and Smale (1997)	South Africa	Female	204.9 cm TL	0.03	−3.55		68	12–15 yr	24 yr
	Goosen and Smale (1997)	South Africa	Male	145.1 cm TL	0.12	−2.14		68	6–9 yr	17 yr
Prionace glauca	Tanaka et al. (1990)	NE and West Pacific	Female	304.0 cm TL	0.16	−1.01		152		
	Tanaka et al. (1990)	NE and West Pacific	Male	369.0 cm TL	0.1	−1.38		43		
	Nakano (1994)	Central Pacific	Female	243.3 cm PCL	0.144	−0.849		123	5–6 yr	15–16 yr
	Nakano (1994)	Central Pacific	Male	289.7 cm PCL	0.129	−0.756		148	4–5 yr	15–16 yr
	Skomal and Natanson (2003)	NW Atlantic	Female	310.0 cm FL	0.13	−1.77		119	5 yr	15 yr
	Skomal and Natanson (2003)	NW Atlantic	Male	282.0 cm TL	0.18	−1.35		287	5 yr	16 yr
Rhizoprionodon taylori	Simpfendorfer (1993, 1999)	Western Australia	Female	73.25 cm TL	1.0123	0.0455		85	3 yr	7 yr
	Simpfendorfer (1993, 1999)	Western Australia	Male	65.22 cm TL	1.337	0.41		52	3 yr	6 yr
Rhizoprionodon terraenovae	Loefer and Sedberry (2003)	East Coast, U.S.	Female	74.9 cm PCL	0.49	−0.94		433	2–4 yr	10 yr
	Loefer and Sedberry (2003)	East Coast, U.S.	Male	74.5 cm TL	0.5	−0.91		379	3 yr	9 yr
	Carlson and Baremore (2003)	Gulf of Mexico	Female	95.62 cm TL	0.63	−1.03		143	1.6 yr	5.5 yr
	Carlson and Baremore (2003)	Gulf of Mexico	Male	91.95 cm TL	0.85	−0.73		161	1.3 yr	4.0 yr

Species	Reference	Location	Sex	L_∞ or D_∞	k	t_0	n	T_{mat}	T_{max}
Scyliorhinus canicula	Carlson and Baremore (2003)	Gulf of Mexico	Both sexes	94.02 cm TL	0.73	−0.88	304	?	10 yr
	Rodriguez-Cabella et al. (1998)	Cantabrian Sea	Both sexes	88.8–98.8 cm TL	0.09–0.13	−1.33	2,153 SFA; 19 TRA	9–11 cm TL	~30 yr?
Sphyrna lewini	Chen et al. (1990)	Taiwan	Female	319.72 cm TL	0.249	−0.413		4 yr	14 yr
	Chen et al. (1990)	Taiwan	Male	320.59 cm TL	0.222	−0.746		3.8 yr	10.6 yr
	Anislado-Tolentino and Robinson-Mendoza (2001)	Mexico	Female	353.3 cm TL	0.153	−0.633	51	8.8 yr	19 yr?
	Anislado-Tolentino and Robinson-Mendoza (2001)	Mexico	Male	336.4 cm TL	0.131	−1.09	50		
Spyrna tiburo	Parsons (1993)	Gulf of Mexico (Tampa Bay)	Female	115.0 cm TL	0.34	−1.11	46	2.2 yr	7 yr
	Parsons (1993)	Gulf of Mexico (Tampa Bay)	Male	88.8 cm TL	0.58	−0.77	48	2.0 yr	6 yr
	Parsons (1993)	Gulf of Mexico (Florida Bay)	Female	103.3 cm TL	0.37	−0.6	45	2.3 yr	7 yr
	Parsons (1993)	Gulf of Mexico (Florida Bay)	Male	81.5 cm TL	0.5	−0.64	44	2.0 yr	6 yr
	Carlson and Parsons (1997)	Gulf of Mexico	Female	122.6 cm TL	0.28	−0.79	65	2.4 yr	6 yr
	Carlson and Parsons (1997)	Gulf of Mexico	Male	89.7 cm TL	0.69	−0.04	50	2.0 yr	5 yr
Triakis semifasciata	Kusher et al. (1992)	California	Female	160.2 cm TL	0.073	−2.74	77	10 yr	24 yr
	Kusher et al. (1992)	California	Male	149.9 cm TL	0.089	−2.03	85	7 yr	24 yr

Note: See Table 14.1 for verification and validation techniques utilized.

Legend: L_∞ or D_∞ = asymptotic length or disk width; k = von Bertalanffy growth coefficient; t_0 = theoretical age at which the organism was 0 length; n = sample size; TRA = tag–recapture analysis; T_{mat} = estimated age at maturity, with (50%) indicating 50% of organisms are mature at that age; T_{max} = oldest aged hard part, not estimated longevity.

Measurements: TL = total length; PL = pelvic length; PCL = pre-caudal length; PSCFL = pre-supra caudal fin length; DW = disk width; and CD = centrum diameter.

14.8.2). It was difficult in some papers to distinguish between ages at first maturity vs. age at 50% or some other cumulative probability function. Therefore, we tended to assume it was the former, unless indicated otherwise in the text. Chimaeras had a relatively narrow range of T_{mat} values, mainly in the 2.9 to 6.0 year range, but *Chimaera monstrosa* was estimated to mature at between 11.2 and 13.4 years. The sawfishes had T_{mat} values of 10 years for the two species studied, while the guitarfishes, torpedo rays, and stingrays had a wide range from 1 or 2 to 9 years. Skates matured, on the average, a bit later than the rays just mentioned, ranging from 4 to 13 years, while the angel shark matured at 11 and the sixgill shark at 5 to 21 years, depending on the sex and the estimate method.

The remaining shark groups also had wide-ranging age at maturity estimates (Table 14.2). The carpet sharks were generally late to mature, ranging from T_{mat} values of 16 to 25 years. The dog sharks ranged from 4 to 30 or 45 years, depending on the species and the study, but most were close to or above 10 years, thus relatively late-maturing. Mackerel sharks had a similar low end of T_{mat} values (5 years), but the upper end (the bigeye thresher, *Alopias superciliosus*) was 13.4 years. Again, the quite diverse ground sharks also had a wide range of T_{mat} values, starting at ~1 year but peaking at 25 years in one study of the soupfin shark, *Galeorhinus galeus*. If one were to estimate a mean or median value of T_{mat} for this group, it would be at about 5 to 6 years, with a small group of late-maturing species clumped around 15 to 25 years.

14.8.2 Longevity Estimates

Longevity (T_{max}) estimates also had quite a wide range both within and among chondrichthyan groups (Table 14.2), and there appears to be a strong relationship between the longevity and age at maturity estimates (see Section 14.8.1). Most chimaeras had T_{max} estimates between 5 and 10 years, but again *C. monstrosa* had estimates up to 29 years. The sawfishes appear to be both long-lived, from 30 to 44 years, and late-maturing (~10 years). The guitarfishes, torpedo rays, and stingrays had a wide range of T_{max} values, starting at 3 years for one of the freshwater stingrays to 28 years for the southern stingray, *Dasyatis americana*. The skates had an even wider range in longevities, ranging from T_{max} estimates of 9 to 24 years and even 50 years for two species of the genus *Dipturus*. The angel shark was estimated to live to 35 years, while no estimate of longevity is available for the sevengill shark.

The remainder of the shark groups had wide ranges, but tended to be relatively long-lived (Table 14.2). The carpet sharks ranged from 19 to 35 years, while the dog sharks were estimated to have T_{max} values from 12 to 70 years; the maxima (55 and 70 years for males and females) were for the deep-water *Centrophorus squamosus*. This species now competes for the longest-lived chondrichthyan with previously published estimates of longevity for the spiny dogfish, *Squalus acanthias* (see Cailliet, 1990, for a review). Although there remain some controversies over aging techniques and interpretations, and in general a lack of verification or validation, the mackerel sharks appear to have T_{max} values ranging from 10 to 25 years. The ground sharks have a wider range in longevity estimates, ranging from 4 to 32 years.

14.9 Implication of Growth, Longevity, and Demography to Fisheries Management

As noted in the introduction, a better understanding of age and growth processes will lead to better estimates of the potential for chondrichthyan populations to grow, especially in response to additional sources of mortality from fisheries. Likewise, understanding the ages of maturity and longevities of these organisms will make management strategies more effective. For example, organisms that have high population growth rates often also have early ages of maturity and low longevities, resulting in population turnover times that may be able to respond to fishing mortality better than those with low population growth rates, late maturity, and high longevity.

The demographic consequences of these age and growth studies are, therefore, very important and have stimulated numerous authors to apply the results of these studies to their demographic analyses. This has been done with specific species, such as the leopard shark, *Triakis semifasciata* (Au and Smith,

1997), and the Atlantic sharpnose shark, *Rhizoprionodon terraenovae* (Cortés, 1994). It also has been done with different populations of the same species, such as the bonnethead shark, *Sphyrna tiburo*, by Cortés and Parsons (1996) and Carlson and Parsons (1997), and sandbar sharks, *Carcharhinius plumbeus*, before and after fishery "depletion" (Sminkey and Musick, 1995a,b). It also has been more broadly applied to many species of sharks, using matrix and life table demographic approaches (Cortés, 1997, 2000, 2002; Frisk et al., 2001; Goldman, 2002) and a related approach, called the intrinsic rebound potential by Smith et al. (1998). This subject is further considered in Chapter 15 of this book.

Acknowledgments

We thank all the authors who published papers in the past two decades for doing such a good job of starting to fill the gaps in knowledge about the life histories of chondrichthyan fishes. We especially thank Enric Cortés, John Carlson, Jeff Carrier, Dave Ebert, Lisa Natanson, Julie Neer, and Wade Smith for providing references we had missed and for comments on both the summary tables and text. We thank Steve Campana, Susan Smith, and their co-authors for letting us use the figures illustrating bomb carbon and tag–recapture age validation, and John Carlson, Michael Gallagher, Gordon (Sandy) McFarlane, John Hoenig, and Christina Conrath for other images. Thanks to Consuelo Goldman for assisting with reference and table checking.

References

Aasen, O. 1963. Length and growth of the porbeagle (*Lamna nasus* Bonnaterre) in the North West Atlantic. *Fiskeridir. Skr. Ser. Havunders.* 13(6):20–37.

Abdel-Aziz, S. H. 1992. The use of vertebral rings of the brown ray *Raja miraletus* (Linnaeus, 1758) off Egyptian Mediterranean coast for estimation of age and growth. *Cybium* 16(2):121–132.

Allen, B. R. and S. P. Wintner. 2002. Age and growth of the spinner shark *Carcharhinus brevipinna* (Muller and Henle, 1839) off the KwaZulu-Natal coast, South Africa. *S. Afr. J. Mar. Sci.* 24:1–8.

Aloj Totaro, A., F. A. Pisanti, P. Russo, and P. Brunetti. 1985. Evaluation of aging parameters in *Torpedo marmorata*. *Ann. Soc. R. Zool. Belg.* 115(2):203–209.

Anislado-Tolentino, V. and C. Robinson-Mendoza. 2001. Age and growth for the scalloped hammerhead shark, *Sphyrna lewini* (Griffith and Smith, 1834), along the central Pacific coast of Mexico. *Cien. Mar.* 27:501–520.

Au, D. W. and S. E. Smith. 1997. A demographic method with population density compensation for estimating productivity and yield per recruit of the leopard shark (*Triakis semifasciata*). *Can. J. Fish. Aquat. Sci.* 54:415–420.

Avsar, D. 2001. Age, growth, reproduction and feeding of the spurdog (*Squalus acanthias* Linnaeus, 1758) in the South-eastern Black Sea. *Estuarine Coastal Shelf Sci.* 52:269–278.

Baum, J. K., R. A. Myers, D. G. Kehler, B. Worm, S. J. Harley, and P. A. Doherty. 2003. Collapse and conservation of shark populations in the northwest Atlantic. *Science* 299:389–392.

Beamish, R. J. and D. A. Fournier. 1981. A method for comparing the precision of a set of age determinations. *Can. J. Fish. Aquat. Sci.* 38:982–983.

Beamish, R. J. and G. McFarlane. 1983. The forgotten requirement for validation in fisheries biology. *Trans. Am. Fish. Soc.* 112:735–743.

Beamish, R. J. and G. McFarlane. 1985. Annulus development on the second dorsal spines of the spiny dogfish (*Squalus acanthias*) and its validity for age determinations. *Can. J. Fish. Aquat. Sci.* 42:1799–1805.

Beverton, R. J. H. and S. J. Holt. 1957. On the dynamics of exploited fish populations. *U.K. Min. Agric. Fish. Fish. Invest.* (Ser. 2) 19, 533 pp.

Bonfil, R. 1994. Overview of world elasmobranch fisheries. FOA Fisheries Technical Paper 341:119.

Bonfil, R., R. Mena, and D. De Anda. 1993. Biological parameters of commercially exploited silky sharks, *Carcharhinus falciformis*, from the Campeche Bank, Mexico. U.S. Department of Commerce, NOAA Technical Report 115:73–86.

Bowker, A. H. 1948. A test for symmetry in contingency tables. *J. Am. Stat. Assoc.* 43:572–574.

Branstetter, S. 1987. Age and growth validation of newborn sharks held in laboratory aquaria, with comments on the life history of the Atlantic sharpnose shark, *Rhizoprionodon terraenovae*. *Copeia* 1987:291–300.

Branstetter, S. and J. A. Musick. 1994. Age and growth estimates for the sand tiger in the Northwestern Atlantic Ocean. *Trans. Am. Fish. Soc.* 123:242–254.

Brown, C. A. and S. H. Gruber. 1988. Age assessment of the lemon shark, *Negaprion brevirostris*, using tetracycline validated vertebral centra. *Copeia* 1988:747–753.

Cailliet, G. M. 1990. Elasmobranch age determination and verification: an updated review, in Elasmobranchs as Living Resources: Advances in the Biology, Ecology, Systematics, and the Status of the Fisheries, W. S. Pratt, Jr., S. H. Gruber, and T. Taniuchi, Eds., NOAA Tech. Rep. 90:157–165.

Cailliet, G. M. and R. L. Radtke. 1987. A progress report on the electron microprobe analysis technique for age determination and verification in elasmobranchs, in *The Age and Growth of Fish*, R. C. Summerfelt and G. E. Hall, Eds., Iowa State University Press, Ames, 359–369.

Cailliet, G. M. and S. Tanaka. 1990. Recommendations for research needed to better understand the age and growth of elasmobranchs, in Elasmobranchs as Living Resources: Advances in the Biology, Ecology, Systematics, and the Status of the Fisheries, W.S. Pratt, Jr., S. H. Gruber, and T. Taniuchi, Eds., NOAA Tech. Rep. 90:505–507.

Cailliet, G. M., L. K. Martin, J. T. Harvey, D. Kusher, and B. A. Welden. 1983a. Preliminary studies on the age and growth of blue (*Prionace glauca*), common thresher (*Alopias vulpinus*), and shortfin mako (*Isurus oxyrinchus*) sharks from California waters, in Proceedings International Workshop on Age Determination of Oceanic Pelagic Fishes: Tunas, Billfishes, Sharks, E. D. Prince and L. M. Pulos, Eds., NOAA Tech. Rep. NMFS 8:179–188.

Cailliet, G. M., L. K. Martin, D. Kusher, P. Wolf, and B. A. Welden. 1983b. Techniques for enhancing vertebral bands in age estimation of California elasmobranchs, in Proceedings International Workshop on Age Determination of Oceanic Pelagic Fishes: Tunas, Billfishes, Sharks, E. D. Prince and L. M. Pulos, Eds., NOAA Tech. Rep. NMFS 8:157–165.

Cailliet, G. M., M. S. Love, and A. W. Ebeling. 1986a. *Fishes: A Field and Laboratory Manual on Their Structure, Identification, and Natural History*. Original publisher: Wadsworth Publishing Company, Belmont, CA; present publisher: Waveland Press, Prospect Heights, IL.

Cailliet, G. M., R. L. Radtke, and B. A. Welden. 1986b. Elasmobranch age determination and verification: a review, in *Indo-Pacific Fish Biology: Proceedings of the Second International Conference on Indo-Pacific Fishes*, T. Uyeno, R. Arai, T. Taniuchi, and K. Matsuura, Eds., Ichthyological Society of Japan, Tokyo, 345–359.

Cailliet, G. M., K. G. Yudin, S. Tanaka, and T. Taniuchi. 1990. Growth characteristics of two populations of *Mustelus manazo* from Japan based upon cross-readings of vertebral bands, in Elasmobranchs as Living Resources: Advances in the Biology, Ecology, Systematics, and the Status of the Fisheries, W. S. Pratt, Jr., S. H. Gruber, and T. Taniuchi, Eds., NOAA Tech. Rep. 90:167–176.

Cailliet, G. M., H. F. Mollet, G. G. Pittenger, D. Bedford, and L. J. Natanson. 1992. Growth and demography of the Pacific angel shark (*Squatina californica*), based on tag returns off California. *Aust. J. Mar. Freshwater Res.* 43:1313–1330.

Calis, E., E. H. Jackson, C. P. Nolan, and F. Jeal. In press. An insight into the life history of the rabbitfish, *Chimaera monstrosa*, with implications for future resource management. MS J-501, NAFO Elasmobranch Symposium, Spain, September 2002.

Campana, S. E. 1990. How reliable are growth back-calculations based on otoliths? *Can. J. Fish. Aquat. Sci.* 47:2219–2227.

Campana, S. E. 1997. Use of radiocarbon from nuclear fallout as a dated marker in the otoliths of haddock, *Melanogrammus aeglefinus*. *Mar. Ecol. Prog. Ser.* 150:49–56.

Campana, S. E. 1999. Chemistry and composition of fish otoliths: pathways, mechanisms and applications. *Mar. Ecol. Prog. Ser.* 188:263–297.

Campana, S. E. 2001. Accuracy, precision and quality control in age determination, including a review of the use and abuse of age validation methods. *J. Fish Biol.* 59:197–242.

Campana, S. E. and S. R. Thorrold. 2001. Otoliths, increments, and elements: keys to a comprehensive understanding of fish populations? *Can. J. Fish. Aquat. Sci.* 58:30–38.

Campana, S. E., M. C. Annand, and J. I. McMillan. 1995. Graphical and statistical methods for determining the consistency of age determinations. *Trans. Am. Fish. Soc.* 124:131–138.

Campana, S. E., L. J. Naatanson, and S. Myklevoll. 2002. Bomb dating and age determination of large pelagic sharks. *Can. J. Fish. Aquat. Sci.* 59:450–455.

Carlander, K. D. 1969. *Handbook of Freshwater Fishery Biology,* Vol. 1. Iowa University Press, Ames.

Carlson, J. K. and I. E. Baremore. 2003. Changes in biological parameters of Atlantic sharpnose shark *Rhizoprionodon terraenovae* in the Gulf of Mexico: evidence for density-dependent growth and maturity? *Mar. Freshwater Res.* 54:1–8.

Carlson, J. K. and G. R. Parsons. 1997. Age and growth of the bonnethead shark, *Sphyrna tiburo*, from northwest Florida, with comments on clinal variation. *Environ. Biol. Fish.* 50:331–341.

Carlson, J. K., E. Cortés, and A. G. Johnson. 1999. Age and growth of the blacknose shark, *Carcharhinus acronotus*, in the eastern Gulf of Mexico. *Copeia* 3:684–691.

Carlson, J. K., E. Cortés, and D. M. Bethea. 2003. Life history and population dynamics of the finetooth shark (*Carcharhinus isodon*) in the northeastern Gulf of Mexico. *Fish. Bull.* 101:281–292.

Carrier, J. C. and C. A. Luer. 1990. Growth rates in the nurse shark, *Ginglymostoma cirratum. Copeia* 1990:686–692.

Carrier, J. C. and R. Radtke. 1988. Preliminary evaluation of age and growth in juvenile nurse sharks (*Ginglymostoma cirratum*) using visual and electron microprobe assessment of tetracycline-labeled vertebral centra. (Abstr. unpubl. pap., Ann Arbor, MI.) Available from author, Biology Department, Albion College, Albion, MI 49224.

Casey, J. G. and L. J. Natanson. 1992. Revised estimates of age and growth of the sandbar shark (*Carcharhinus plumbeus*) from the Western North Atlantic. *Can. J. Fish. Aquat. Sci.* 49(7):1474–1477.

Casey, J. G., H. L. Pratt, Jr., and C. Stillwell. 1985. Age and growth of the sandbar shark (*Carcharhinus plumbeus*) from the western North Atlantic. *Can. J. Fish. Aquat. Sci.* 42 (5):963–975.

Casey, J. M. and R. A. Myers. 1998. Near extinction of a large, widely distributed fish. *Science* 281:690–692.

Casselman, J. M. 1983. Age and growth assessments of fish from their calcified structures — techniques and tools, in Proceedings of the International Workshop on Age Determination of Oceanic Pelagic Fishes: Tunas, Billfishes, and Sharks, E. D. Prince and L. M. Pulos, Eds., NOAA Tech. Rep. NMFS 8:1–17.

Chang, W. Y. B. 1982. A statistical method for evaluating the reproducibility of age determination. *Can. J. Fish. Aquat. Sci.* 39:1208–1210.

Chang, W. B., M. Y. Leu, and L. S. Fang. 1997. Embryos of the whale shark, *Rhincodon typus*: early growth and size distribution. *Copeia* 1997:444–446.

Chen, C. T., T. C. Leu, S. J. Joung, and N. C. H. Lo. 1990. Age and growth of the scalloped hammerhead, *Sphyrna lewini*, in Northeastern Taiwan waters. *Pac. Sci.* 44:156–170.

Clarke, M. W., P. L. Connolly, and J. J. Bracken. 2002a. Catch, discarding, age estimation, growth and maturity of the squalid shark *Deania calceus* west and north of Ireland. *Fish. Res.*, 56:139–153.

Clarke, M. W., P. L. Connolly, and J. J. Bracken. 2002b. Age estimation of the exploited deepwater shark *Centrophorus squamosus* from the continental slopes of the Rockall Trough and Porcupine Bank. *J. Fish Biol.* 60:501–514.

Conrath, C. L., J. Gelsleichter, and J. A. Musick. 2002. Age and growth of the smooth dogfish (*Mustelus canis*) in the northwest Atlantic Ocean. *Fish. Bull.* 100:674–682.

Correia, J. P. and I. M. Figueiredo. 1997. A modified decalcification technique for enhancing growth bands in deep-coned vertebrae of elasmobranchs. *Environ. Biol. Fish.* 50:225–230.

Cortés, E. 1994. Demographic analysis of the Atlantic sharpnose shark, *Rhizoprionodon terraenovae*, in the Gulf of Mexico. *Fish. Bull.* 93:57–66.

Cortés, E. 1997. Demographic analysis as an aid in shark stock assessment and management. *Fish. Res.* 39:199–208.

Cortés, E. 2000. Life history patterns and correlations in sharks. *Rev. Fish. Sci.* 8(4):299–344.

Cortés, E. 2002. Incorporating uncertainty into demographic modeling: application to shark populations. *Conserv. Biol.* 16(4):1048–1062.

Cortés, E. and G. R. Parsons. 1996. Comparative demography of two populations of the bonnethead shark (*Sphyrna tiburo*). *Can. J. Fish. Aquat. Sci.* 53:709–718.

Cowley, P. D. 1997. Age and growth of the blue stingray *Dasyatis chrysonota chrysonota* from the South-Eastern Cape coast of South Africa. *S. Afr. J. Mar. Sci.* 18:31–38.

Daiber, F. C. 1960. A technique for age determination in the skate *Raja eglantaria. Copeia* 1960:258–260.

Davenport, S. and J. D. Stevens. 1998. Age and growth of two commercially important sharks (*Carcharhinus tilsoni* and *C. sorrah*) from northern Australia. *Aust. J. Mar. Freshwater Res.* 39:417–433.

DeVries, D. R. and R. V. Frie. 1996. Determination of age and growth, in *Fisheries Techniques*, 2nd ed., Murphy, B.R. and D.W. Willis, Eds., American Fisheries Society, Bethesda, MD, 483–512.

Driggers, W. B., III, J. K. Carlson, D. Oakley, G. Ulrich, B. Cullum, and J. M. Dean. In press. Life history of the blacknose shark (*Carcharhinus acronotus*) in the western North Atlantic Ocean. *Environ. Biol. Fish.*

Druffel, E. M. and T. W. Linick. 1978. Radiocarbon in annual coral rings of Florida. *Geophys. Res. Lett.* 5:913–916.

DuBuit, M. H. 1972. Age et croissance de *Raja batis* et de *Raja naevus* en Mer Celtique. *J. Cons. Int. Explor. Mer* 37:261–265.

Evans, G. T. and J. M. Hoenig. 1998. Testing and viewing symmetry in contingency tables, with application to readers of fish ages. *Biometrics* 54:620–629.

Fabens, A. J. 1965. Properties and fitting of the von Bertalanffy growth curve. *Growth* 29:265–289.

Ferreira, B. P. and C. M. Vooren. 1991. Age, growth, and structure of vertebra in the school shark, *Galeorhinus galeus* (Linnaeus, 1758) from southern Brazil. *Fish. Bull.* 89(1):19–31.

Francis, M. P. 1997. Spatial and temporal variation in the growth rate of elephantfish (*Callorhinchus milii*). *N.Z. J. Mar. Freshwater Res.* 31:9–23.

Francis, M. P. and R. I. C. C. Francis. 1992. Growth rate estimates for New Zealand rig (*Mustelus lenticulatus*). *Aust. J. Mar. Freshwater Res.*43:1157–1176.

Francis, M. P. and C. O. Maolagain. 2000. Age growth and maturity of a New Zealand endemic shark (*Mustelus lenticulatus*) estimated from vertebral bands. *Mar. Freshwater Res.* 51:35–42.

Francis, M. P. and K. P. Mulligan. 1998. Age and growth of the New Zealand school shark, *Galeorhinus galeus*. *N.Z. J. Mar. Freshwater Res.* 32:427–440.

Francis, M. P. and J. D. Stevens. 2000. Reproduction, embryonic development, and growth of the porbeagle shark, *Lamna nasus*, in the southwest Pacific Ocean. *Fish. Bull.* 98:41–63.

Francis, M. P., C. O. Maolagain, and J. D. Stevens. 2001. Age, growth, and sexual maturity of two New Zealand endemic skates, *Dipturus nasutus* and *D. innominatus*. *N.Z. J. Mar. Freshwater Res.* 35:831–842.

Francis, R. I. C. C. 1988. Maximum likelihood estimation of growth and growth variability from tagging data. *N.Z. J. Mar. Freshwater Res.* 22:43–51.

Francis, R. I. C. C. 1990. Back-calculation of fish length: a critical review. *J. Fish Biol.* 36:883–902.

Freer, D. W. L. and C. L. Griffiths. 1993. Estimation of age and growth in the St. Joseph *Callorhinchus capensis* (Dumeril). *S. Afr. J. Mar. Sci.* 13:75–82.

Frisk, M. G., T. J. Miller, and M. J. Fogarty. 2001. Estimation and analysis of biological parameters in elasmobranch fishes: a comparative life history study. *Can. J. Fish. Aquat. Sci.* 58:969–981.

Gallagher, M. and C. P. Nolan. 1999. A novel method for the estimation of age and growth in rajids using caudal thorns. *Can. J. Fish. Aquat. Sci.* 56:1590–1599.

Gallagher, M.J., C.P. Nolan, and F. Jeal. In press. The structure and growth processes of caudal thorns. MS J-529, NAFO Elasmobranch Symposium, Spain, September 2002.

Gelsleichter, J., J. A. Musick, and P. Van Veld. 1995. Proteoglycans from vertebral cartilage of the clearnose skate, *Raja eglentaria*: inhibition of hydroxyapetite formation. *Fish. Physiol. Biochem.* 14:247–251.

Gelsleichter, J., E. Cortés, C. A. Manire, R. E. Hueter, and J. A. Musick. 1997. Use of calcein as a fluorescent marker for elasmobranch vertebral cartilage. *Trans. Am. Fish. Soc.* 126:862–865.

Gelsleichter, J., E. Cortés, C. A. Manire, R. E. Hueter, and J. A. Musick. 1998a. Evaluation of toxicity of oxytetracycline on growth of captive nurse sharks, *Ginglymostoma cirratum*. *Fish. Bull.* 96:624–627.

Gelsleichter, J., A. Piercy, and J. A. Musick. 1998b. Evaluation of copper, iron, and lead substitution techniques in elasmobranch age determination. *J. Fish Biol.* 53:465–470.

Goldman, K. J. 2002. Aspects of Age, Growth, Demographics and Thermal Biology of Two Lamniform Shark Species. Ph.D. dissertation, College of William and Mary, School of Marine Science, Virginia Institute of Marine Science, Williamsburg, VA, 220 pp.

Goldman, K. J. In press. Age and growth of elasmobranch fishes, in Technical Manual for the Management of Elasmobranchs, J. A. Musick and R. Bonfil, Eds., Asia Pacific Economic Cooperation and IUCN Shark Specialist Group Publication.

Goosen, A. J. J. and M. J. Smale. 1997. A preliminary study of age and growth of the smoothhound shark *Mustelus mustelus* (Triakidae). *S. Afr. J. Mar. Sci.* 18:85–92.

Govender, A., N. Kistnasamy, and R. P. Van Der Elst. 1991.Growth of spotted ragged-tooth sharks *Carcharias taurus* (Rafinesque) in captivity. *S. Afr. J. Mar. Sci.* 11:15–19.

Gruber, S.H. and R.G. Stout. 1983. Biological materials for the study of age and growth in a tropical marine elasmobranch, the lemon shark, *Negaprion brevirostris* (Poey), in Proceedings of the International Workshop on Age Determination of Oceanic Pelagic Fishes: Tunas, Billfishes, and Sharks, E. D. Prince and L. M. Pulos, Eds., NOAA Tech. Rep. NMFS 8:193–205.

Gulland, J. A. and S. J. Holt. 1959. Estimation of growth parameters for data at unequal time intervals. *J. Cons. Int. Explor. Mer* 25:47–49.

Haddon, M. 2001. Growth of Individuals, in *Modelling and Quantitative Measures in Fisheries*, Chapman & Hall/CRC Press, Boca Raton, FL, 187–246.

Henningsen, A. D. 1996. Captive husbandry and bioenergetics of the spiny butterfly ray, *Gymnura altavela* (Linnaeus). *Zoo Biol.* 15:135–142.

Henningsen, A. D. 2002. Age and growth in captive southern stingrays, *Dasyatis americana*. Abstract, American Elasmobranch Society, 2002 Annual Meeting, Kansas City, MO.

Hoenig, J. M. and C. A. Brown. 1988. A simple technique for staining growth bands in elasmobranch vertebrae. *Bull. Mar. Sci.* 42(2):334–337.

Hoenig, J. M., M. J. Morgan, and C. A. Brown. 1995. Analyzing differences between two age determination methods by tests of symmetry. *Can. J. Fish. Aquat. Sci.* 52:364–368.

Holden, M. J. 1974. Problems in the rational exploitation of elasmobranch populations and some suggested solutions, in *Sea Fisheries Research*, F.R. Jones, Ed., Halstead Press/John Wiley & Sons, New York, 117–138.

Holden, M.J. and M.R. Vince. 1973. Age validation studies on the centra of *Raja clavata* using tetracycline. *J. Cons. Int. Explor. Mer.* 35:13–17.

Ishiyama, R. 1951. Studies on the rays and skates belonging to the family *Rajidae*, found in Japan and adjacent regions. 2. On the age determination of Japanese black-skate *Raja fusca* Garman (preliminary report) [in English]. *Bull. Jpn. Soc. Sci. Fish.* 16(12):112–118.

Ismen, A. 2003. Age, growth, reproduction and food of common stingray (*Dasyatis pastinaca* L., 1758) in Ikenderun Bay, the eastern Mediterranean. *Fish. Res.* 60:169–176.

Johnson, A.G. 1979. A simple method for staining the centra of teleosts vertebrae. *Northeast. Gulf Sci.* 3:113–115.

Jones, B.C. and G.H. Green. 1977. Age determination of an elasmobranch (*Squalus acanthias*) by X-ray spectrometry. *J. Fish. Res. Board Can.* 34:44–48.

Kalish, J. M. 1995. Radiocarbon and fish biology, in *Recent Developments in Fish Otolith Research*, D. H. Secor, J. M. Dean, and S. E. Campana, Eds., University of South Carolina Press, Columbia, 637–653.

Ketchen, K. S. 1975. Age and growth of dogfish (*Squalus acanthias*) in British Columbia waters. *J. Fish. Res. Board Can.* 32:13–59.

Kusher, D. I., S. E. Smith, and G. M. Cailliet. 1992. Validated age and growth of the leopard shark, *Triakis semifasciata*, from central California. *Environ. Biol. Fish.* 35:187–203.

LaMarca, M. J. 1966. A simple technique for demonstrating calcified annuli in the vertebrae of large elasmobranchs. *Copeia* 1966:351–352.

Lessa, R. and F. M. Santana. 1998. Age determination and growth of the smalltail shark, *Carcharhinus porosus*, from northern Brazil. *Mar. Freshwater Res.* 49:705–111.

Lessa, R., F. M. Santana, R. Menni, and Z. Almeida. 1999a. Population structure and reproductive biology of the smalltail shark (*Carcharhinus porosus*) off Maranhno (Brazil). *Mar. Freshwater Res.* 50:383–388.

Lessa, R., F. M. Santana, and R. Paglerani. 1999b. Age, growth and stock structure of the oceanic whitetip shark, *Carcharhinus longimanus*, from the southwestern equatorial Atlantic. *Fish. Res.* 42:21–30.

Lessa, R., F. M. Santana, V. Batista, and Z. Almeida. 2000. Age and growth of the daggernose shark, *Isogomphodon oxyrhynchus*, from northern Brazil. *Mar. Freshwater Res.* 51:339–347.

Liu, K. M., P. J. Chiang, and C. T. Chen. 1998. Age and growth estimates of the bigeye thresher shark, *Alopias superciliosus*, in northeastern Taiwan waters. *Fish. Bull.* 96:482–491.

Liu, K. M., C. T. Chen, T. H. Liao, and S. J. Joung 1999. Age, growth, and reproduction of the pelagic thresher shark, *Alopias pelagicus* in the Northwestern Pacific. *Copeia* 1999:68–74.

Loefer, J. K. and G. R. Sedberry. 2003. Life history of the Atlantic sharpnose shark (*Rhizoprionodon terraenovae*) (Richardson, 1836) off the southeastern United States. *Fish. Bull.* 101:75–88.

Martin, L. K. and G. M. Cailliet. 1988. Age and growth of the bat ray, *Myliobatis californica*, off central California. *Copeia* 1988:762–773.

Mathsoft, Inc. 2000. S-Plus 2000 Professional Release 1. Mathsoft, Inc., Seattle, WA.

McFarlane, G. A. and R. J. Beamish. 1987a. Validation of the dorsal spine method of age determination for spiny dogfish, in *The Age and Growth of Fish*, R. C. Summerfelt and G. E. Hall, Eds., Iowa State University Press, Ames, 287–300.

McFarlane, G. A. and R. J. Beamish. 1987b. Selection of dosages of oxytetracycline for age validation studies. *Can. J. Fish. Aquat. Sci.* 44:905–909.

McFarlane, G. A., J. R. King, and M. W. Saunders. 2002. Preliminary study on the use of neural arches in the age determination of bluntnose sixgill sharks (*Hexanchus griseus*). *Fish. Bull.* 100:861–864.

McNemar, Q. 1947. Note on the sampling error of the difference between correlated proportions or percentages. *Psychometrika* 12:153–157.

Mellinger, J. 1971. Croissance et reproduction de la torpille (*Torpedo marmorata*). I. Introduction, ecologie, croissance generale et dimorphisme sexual, cycle, fecondite [Growth and reproduction of the electric ray (*Torpedo marmorata*). I. Introduction. ecology, body growth and sexual dimorphism, cycle, fecundity]. *Bull. Biol. Fr. Belg.* 105:167–215.

Merson, R. R. and H. L. Pratt, Jr. 2001. Distribution, movements and growth of young sandbar sharks, *Carcharhinus plumbeus*, in the nursery grounds of Delaware Bay. *Environ. Biol. Fish.* 61:13–24.

Mohan, P. J., S. T. Clar, and T. H. Schmid. In press. Age and growth of captive sharks, in *Elasmobranch Husbandry Manual*, M. F. L. Smith, D. I. Warmolts, D. A. Thoney, and R. E. Hueter, Eds., Special Publication of the Ohio Biological Survey No. 16, Columbus, chap. 5.6.

Mollet, H. F., J. M. Ezcurra, and J. B. O'Sullivan. 2002. Captive biology of the pelagic stingray, *Dasyatis violacea* (Bonaparte, 1830). *Mar. Freshwater Res.* 53:531–541.

Nakano, H. 1994. Age, reproduction and migration of blue shark in the North Pacific Ocean. *Bull. Nat. Res. Inst. Far Seas Fish.*, 31:141–256.

Nammack, M.F., J.A. Musick, and J.A. Colvocoresses. 1985. Life history of spiny dogfish off the Northeastern United States. *Trans. Am. Fish. Soc.* 114:367–376.

Natanson, L. J. 1993. Effect of temperature on band deposition in the little skate, *Raja erinacea*. *Copeia* 1993:199–206.

Natanson, L. J. and G. M. Cailliet. 1990. Vertebral growth zone deposition in angel sharks. *Copeia* 1990:1133–1145.

Natanson, L. J., J. G. Casey, and N. E. Kohler. 1995. Age and growth estimates for the dusky sharks, *Carcharhinus obscurus*, in the western North Atlantic Ocean. *Fish. Bull.* 93:116–126.

Natanson, L. J., J. G. Casey, N. E. Kohler, and T. Colket IV. 1999. Growth of the tiger shark, *Galeocerdo cuvier*, in the western North Atlantic based on tag returns and length frequencies; and a note on the effects of tagging. *Fish. Bull.* 97:944–953.

Natanson, L. J., J. J. Mello, and S. E. Campana. 2002. Validated age and growth of the porbeagle shark (*Lamna nasus*) in the western North Atlantic Ocean. *Fish. Bull.* 100:266–278.

Neer, J. A. and G. M. Cailliet. 2001. Aspects of the life history of the Pacific electric ray, *Torpedo californica* (Ayers). *Copeia* 2001:842–847.

Neer, J. A. In press. Ecology of the pelagic stingray, *Dasyatis violacea* (Bonaparte, 1832), in *Pelagic Sharks*. E. K. Pikitch and M. Camhi, Eds., Blackwell Scientific, Oxford.

Officer, R. A., A. S. Gason, T. I. Walker, and J. G. Clement. 1996. Sources of variation in counts of growth increments in vertebrae from gummy shark, *Mustelus antarcticus*, and school shark, *Galeorhinus galeus*: implications for age determination. *Can. J. Fish. Aquat. Sci.* 53:1765–1777.

Officer, R. A., R. W. Day, J. G. Clement, and L. P. Brown. 1997. Captive gummy sharks, *Mustelus antarcticus*, for hypermineralized bands in their vertebrae during winter. *Can. J. Fish. Aquat. Sci.* 54:2677–2683.

Parsons, G. R. 1983. An examination of the vertebral rings of the Atlantic sharpnose shark, *Rhizopriodon terraenovae*. *Northeast. Gulf Sci.* 6:63–66.

Parsons, G. R. 1985. Growth and age estimation of the Atlantic sharpnose shark, *Rhizopriodon terraenovae*: a comparison of techniques. *Copeia* 1985:80–85.

Parsons, G. R. 1993. Age determination and growth of the bonnethead shark *Sphyrna tiburo*: a comparison of two populations. *Mar. Biol.* 117:23–31.

Pauly, D. 1997. Growth and mortality of the basking shark *Cetorhinus maximus* and their implications for management of whale sharks *Rhincodon typus*. International Seminar on Shark and Ray Biodiversity, Conservation and Management, Kota Kinabalu, Sabah, Malaysia.

Peignoux-Deville, J., F. Lallier, and B. Vidal. 1982. Evidence for the presence of osseous tissue in dogfish vertebrae. *Cell Tissue Res.* 222:605–614.

Prager, M. H., S. B. Saila, and C. W. Recksick. 1987. FISHPARM: a microcomputer program for parameter estimation of nonlinear models in fishery science. Old Dominion University, Department of Oceanography, Tech. Rep. 87-10, Norfolk, VA.

Punt, A. E. and G. S. Hughes. 1989. PC-YIELD II user's guide. Benguela Ecology Programme Report 18. Foundation for Research and Development: South Africa, 55 pp.

Ribot-Carballal, C. 2003. Age and Growth of the Shortfin Mako, *Isurus oxyrinchus*, from Baja California Sur. M.S. thesis, CICIMAR, La Paz, Mexico (Abstr. ASIH/AES Manaus, Brazil, Mexico).

Richards, S. W., D. Merriman, and L. H. Calhoun. 1963. Studies of the marine resources of southern New England. IX. The biology of the little skate, *Raja erinacea* Mitchell. *Bull. Bingham Oceanogr. Collect. Yale Univ.* 18(3):5–67.

Ricker, W. E. 1975. Computation and interpretation of biological statistics of fish populations. *Bull. Fish. Res. Board Can.* 191:1–382.

Ricker, W. E. 1979. *Growth Rates and Models*. Fish Physiology VIII series, Academic Press, San Diego, CA, 677–743.

Ricker, W. E. 1992. Back-calculation of fish lengths based on proportionality between scale and length increments. *Can. J. Fish. Aquat. Sci.* 49:1018–1026.

Ridewood, W. G. 1921. On the calcification of the vertebral centra in sharks and rays. *Physiol. Trans. R. Soc. Ser. B Biol.* 210:311–407.

Robinson, M. P. and P. J. Motta. 2002. Patterns of growth and the effects of scale on the feeding kinematics of the nurse shark (*Ginglymostoma cirratum*). *J. Zool.* (London) 256(4):449–462.

Rodriguez-Cabella, C., F. De La Gandata, and F. Sanchez. 1998. Preliminary results on growth and movements of dogfish *Scyliorhinus canicula* (Linnaeus, 1958) in the Cantabrian Sea. *Oceanol. Acta* 21(2):363–370.

Roff, D. A. 1982. *The Evolution of Life Histories: Theory and Analysis*. Chapman & Hall, New York.

SAS Institute, Inc. 1999. Version 8.0. Statistical Analysis Systems Institute, Inc., Cary, NC.

Schmid, T. H., F. L. Murru, and F. McDonald. 1990. Feeding habits and growth rates of bull (*Carcharhinus leucas* (Valenciennes)), sandbar (*Carcharhinus plumbeus* (Nardo)), sandtiger (*Eugomophodus taurus* (Rafinesque)) and nurse (*Ginglymostoma cirratum* (Bonnaterre)) sharks maintained in captivity. *J. Aquaricult. Aquat. Sci.* 5:100–105.

Schwartz, F. J. 1983. Shark aging methods and age estimates of scalloped hammerhead, *Sphyrna lewini*, and dusky, *Carcharhinus obscurus*, sharks based on vertebral ring counts, in Proceedings of the International Workshop on Age Determination of Oceanic Pelagic Fishes: Tunas, Billfishes, and Sharks, E. D. Prince and L. M. Pulos, Eds., NOAA Tech. Rep. NMFS 8:167–174.

Seki, T., T. Taniuchi, H. Nakano, and M. Shimizu. 1998. Age, growth, and reproduction of the oceanic whitetip shark from the Pacific Ocean. *Fish. Sci.* 64:14–20.

Simpfendorfer, C. A. 1993. Age and growth of the Australian sharpnose shark, *Rhizoprionodon taylori*, from north Queensland, Australia. *Environ. Biol. Fish.* 36:233–241.

Simpfendorfer, C. A. 1999. Mortality estimates and demographic analysis for the Australian sharpnose shark, *Rhizoprionodon taylori*, from northern Australia. *Fish. Bull.* 97:978–986.

Simpfendorfer, C. A. 2000a. Predicting population recovery rates for endangered western Atlantic sawfishes using demographic analysis. *Environ. Biol. Fish.* 58:371–377.

Simpfendorfer, C. A. 2000b. Growth rates of juvenile dusky sharks, *Carcharhinus obscurus* (Lesueur, 1818), from southwestern Australia estimated from tag–recapture data. *Fish. Bull.* 98:811–822.

Simpfendorfer, C. A., J. Chidlow, and R. McAuley. 2000. Age and growth of the whiskery shark, *Furgaleus macki*, from southwestern Australia. *Environ. Biol. Fish.* 58:335–343.

Simpfendorfer, C. A., R. E. Hueter, U. Bergman, and S. M. H. Connett. 2002a. Results of a fishery-independent survey for pelagic sharks in the western North Atlantic, 1977–1994. *Fish. Res.* 55:175–192.

Simpfendorfer, C. A., R. B. McAuley, J. Chidlow, and P. Unsworth. 2002b. Validated age and growth of the dusky shark, *Carcharhinus obscurus*, from Western Australia waters. *Mar. Freshwater Res.* 53:567–573.

Skomal, G. B. and L. J. Natanson. 2003. Age and growth of the blue shark, *Prionace glauca*, in the North Atlantic Ocean. *Fish. Bull.* 101:627–639.

Sminkey, T. R. and J. A. Musick. 1995a. Age and growth of the sandbar shark, *Carcharhinus plumbeus*, before and after population depletion. *Copeia* 1995:871–883.

Sminkey, T. R. and J. A. Musick. 1995b. Demographic analysis of the sandbar shark, *Carcharhinus plumbeus*, in the western North Atlantic. *Fish. Bull.* 94:341–347.

Smith, J. W. and J. V. Merriner. 1987. Age and growth, movements and distribution of the cownose ray, *Rhinoptera bonasus*, in Chesapeake Bay. *Estuaries* 10:153–164.

Smith, S. E. 1984. Timing of vertebral band deposition in tetracycline injected leopard sharks. *Trans. Am. Fish. Soc.* 113:308–313.

Smith, S. E., D. W. Au, and C. Show. 1998. Intrinsic rebound potentials of 26 species of Pacific sharks. *Mar. Freshwater Res.* 49:663–678.

Smith, S. E., R. A. Mitchell, and D. Fuller. 2003. Age, validation of a leopard shark (*Triakis semifasciata*) recaptured after 20 years. *Fish. Bull.* 101:194–198.

Smith, S. E., R. C. Rasmussen, D. A. Ramon, and G. M. Cailliet. In press. Biology and ecology of thresher sharks (Family: Alopiidae), in *Pelagic Sharks*. E. K. Pikitch and M. Camhi, Eds., Blackwell Scientific, Oxford.

Stenberg, C. In press. Life history of the piked dogfish (*Squalus acanthias* L.) in Swedish waters. MS J-525, NAFO Elasmobranch Symposium, Spain, September 2002.

Stevens, J. D. 1975. Vertebral rings as a means of age determination in the blue shark (*Prionace glauca*). *J. Mar. Biol. Assoc. U.K.* 55:657–665.

Stevens, J. D., R. Bonfil, N. K. Dulvey, and P. A. Walker. 2000. The effects of fish on sharks, rays, and chimaeras (chondrichthyans), and the implications for marine ecosystems. *ICES J. Mar. Sci.* 57:476–494.

Sulikowski, J. A., M. D. Morin, S. H. Suk, and W. H. Howell. 2003. Age and growth estimate of the winter skate (*Leuoraja ocellata*) in the western Gulf of Maine. *Fish. Bull.* 101:405–413.

Sullivan, K. J. 1977. Age and growth of the elephant fish *Callorhinchus milii* (Elasmobranchii: Callorhynchidae). *N.Z. J. Mar. Freshwater Res.* 11(4):745–753.

Summerfelt, R. C. and G. E. Hall. 1987. *Age and Growth of Fish.* Iowa State University Press, Ames.

Tanaka, S. 1980. Biological investigation of *Lamna ditropis* in the north-western waters of the North Pacific, in Report of Investigation on Sharks as a New Marine Resource (1979), Japan Marine Fishery Resource Research Center, Tokyo [English abstract, translation by Nakaya].

Tanaka, S. 1990. Age and growth studies on the calcified structures of newborn sharks in laboratory aquaria using tetracycline, in Elasmobranchs as Living Resources: Advances in the Biology, Ecology, Systematics, and the Status of the Fisheries, H.L. Pratt, Jr., S.H. Gruber, and T. Taniuchi, Eds., NOAA Tech. Rep. NMFS 90:189–202.

Tanaka, S. 1991. Age estimation of freshwater sawfish and sharks in northern Australia and Papua New Guinea. *Univ. Mus. Univ. Tokyo, Nat. Cult.* 3:71–82.

Tanaka, S. and S. Ohnishi. 1998. Some biological aspects of freshwater stingrays collected from Chao Phraya, Mekong, and Ganges River systems, in Adaptability and Conservation of Freshwater Elasmobranchs, Report of Research Project, Grant-in-Aid for International Scientific Research (Field Research) in the financial year of 1996 and 1997, S. Tanaka, Ed., pp. 102–119.

Tanaka, S., G. M. Cailliet, and K. G. Yudin. 1990. Differences in growth of the blue shark, *Prionace glauca*: technique or population? in Elasmobranchs as Living Resources: Advances in the Biology, Ecology, Systematics, and the Status of the Fisheries, H. L. Pratt, Jr., S. H. Gruber, and T. Taniuchi, Eds., NOAA Tech. Rep. NMFS 90:177–187.

Taniuchi, T. and H. Tachikawa. 1999. Geographical variation in age and growth of *Squalus mitsukurii* (Elasmobranchii: Squalidae) in the North Pacific, in *Proc. 5th Indo-Pac. Fish Conf.*, Noumea, 1997, B. Séret and J.-Y. Sire, Eds., Society of French Ichthyologists, Paris, 321–328.

Thomason, J. C., W. Conn, E. Le Comte, and J. Davenport. 1996. Effect of temperature and photoperiod on the growth of the embryonic dogfish, *Scyliorhinus canicula*. *J. Fish Biol.* 49:739–742.

Timmons, M. and R. N. Bray. 1997. Age, growth, and sexual maturity of shovelnose guitarfish, *Rhinobatos productus* (Ayres). *Fish. Bull.* 95:349–359.

Troynikov, V. S. and T. I. Walker. 1999. Vertebral size-at-age heterogeneity in gummy shark harvested off southern Australia. *J. Fish Biol.* 54:863–877.

Tullis, A. and G. Peterson. 2000. Growth and metabolism in the embryonic white-spotted bamboo shark, *Chiloscyllium plagiosum*: comparison with embryonic birds and reptiles. *Physiol. Biochem. Zool.* 73:271–282.

Van Dykhuizen, G. and H. F. Mollet. 1992. Growth, age estimation and feeding of captive sevengill sharks, *Notorynchus cepedianus*, at the Monterey Bay Aquarium. *Aust. J. Mar. Freshwater Res.* 43:297–318.

Walker, T. I., B. L. Taylor, R. J. Hudson, and J. P. Cottier. 1998. The phenomenon of apparent change of growth rate in gummy shark (*Mustelus antarcticus*) harvested off southern Australia. *Fish. Res.* 39:139–163.

Walmsley-Hart, S. A., W. H. H. Sauer, and C. D. Buxton. 1999. The biology of the skates *Raja wallacei* and *R. Pullopunctata* (Batoidea: Rajidae) on the Agulhas Bank, South Africa. *S. Afr. J. Mar. Sci.* 21:165–179.

Walter, J. P. and D. A. Ebert. 1991. Preliminary estimates of age of the bronze whaler *Carcharhinus brachyurus* (Chondrichthyes: Carcharhinidae) from southern Africa, with a review of some life history parameters. *S. Afr. J. Mar. Sci.* 10:37–44.

Wang, Y. G. and D. A. Milton. 2000. On comparison of growth curves: how do we test whether growth rates differ? *Fish. Bull.* 98:874–880.

Watson, G. and M. J. Smale. 1999. Age and growth of the shortnose spiny dogfish, *Squalus megalops*, from the Agulhas Bank, South Africa. *S. Afr. J. Mar. Sci.* 21:9–18.

Weidman, C. R. and G. A. Jones. 1993. A shell-derived time history of bomb C-14 on Georges Bank and its Labrador Sea implications. *J. Geophys. Res.* 98:14,577–14,588.

Welden, B. A., G. M. Cailliet, and A. R. Flegal. 1987. Comparison of radiometric with vertebral band age estimates in four California elasmobranchs, in *The Age and Growth of Fish*, R.C. Summerfelt and G.E. Hall, Eds., Iowa State University Press, Ames, 301–315.

White, W. T., M. E. Platell, and I. C. Potter. 2001. Relationship between reproductive biology and age composition and growth in *Urolophus lobatus* (Batoidea: Urolophidae). *Mar. Biol.* 138:135–147.

White, W. T., N. G. Hall, and I. C. Potter. 2002a. Size and age compositions and reproductive biology of the nervous shark, *Carcharhinus cautus*, in a large subtropical embayment, including an analysis of growth during pre- and postnatal life. *Mar. Biol.* 141:1153–1166.

White, W. T., N. G. Hall, and I. C. Potter. 2002b. Reproductive biology and growth during pre- and postnatal life of *Trygonoptera personata* and *T. mucosa* (Batoidea: Urolophidae). *Mar. Biol.* 140:699–712.

Wilson, C. D. and M. P. Seki. 1994. Biology and population characteristics of *Squalis mitzukuii* from a seamount in the central North Pacific Ocean. *Fish. Bull.* 92:851–864.

Wintner, S. P. 2000. Preliminary study of vertebral growth rings in the whale shark, *Rhincodon typus*, from the east coast of South Africa. *Environ. Biol. Fish.* 59:441–451.

Wintner, S. B. and G. Cliff. 1999. Age and growth determination of the white shark, *Carcharodon carcharias*, from the east coast of South Africa. *Fish. Bull.* 97:153–169.

Wintner, S. P. and S. F. J. Dudley. 2000. Age and growth estimates for the tiger shark, *Galeocerdo cuvier*, from the east coast of South Africa. *Mar. Freshwater Res.* 51:43–53.

Wintner, S. P., S. F. J. Dudley, N. Kistnasamy, and B. Everett. 2002. Age and growth estimates for the Zambezi shark, *Carcharhinus leucas*, from the east coast of South Africa. *Mar. Freshwater Res.* 53:557–566.

Yamaguchi, A., T. Taniuchi, and M. Shimuzu. 1996. Age and growth of the starspotted dogfish, *Mustelus manazo* from Tokyo Bay, Japan. *Fish. Sci.* 62:919–922.

Yamaguchi, A., S. Y. Huang, C. T. Chen, and T. Taniuchi. 1999. Age and growth of the starspotted smooth-hound, *Mustelus manazo* (Chondrichthyes: Triakidae) in the waters of north-eastern Taiwan, in *Proc. 5th Indo-Pac. Fish Conf.*, Noumea, 1997, B. Séret and J.-Y. Sire, Eds. Society of French Ichthyologists, Paris, 1999:505–513.

Yudin, K. G. and G. M. Cailliet. 1990. Age and growth of the gray smoothhound, *Mustelus californicus*, and the brown smoothhound, *M. henlei*, sharks from central California. *Copeia* 1990:191–204.

Zeiner, S. J. and P. Wolf. 1993. Growth characteristics and estimates of age at maturity of two species of skate (*Raja binoculata and Raja rhina*) from Monterey, California. NOAA Tech. Rep. 115:87–99.

Zweifel, J. R. and R. Lasker. 1976. Prehatch and posthatch growth of fishes — a general model. *Fish. Bull.* 74:609–621.

15

Life History Patterns, Demography, and Population Dynamics

Enric Cortés

CONTENTS

15.1 Introduction

There is mounting evidence of recent declines in a number of elasmobranch populations as a result of overharvesting (Campana et al., 1999, 2001, 2002; Simpfendorfer, 2000; Cortés et al., 2002; Baum et al., 2003), and two species of skate have even become locally extirpated or almost extinct (Brander, 1981; Casey and Myers, 1998). Yet our knowledge of life history traits of most species is still limited and we are just beginning to gain insight into the life history patterns shared by some species and the relationships among life history traits (Compagno, 1990; Cortés, 2000; Frisk et al., 2001). Within the past two decades, our scant but increasing knowledge of the life history of numerous species (Compagno, 1984) has given rise to the development of demographic (life table and matrix population) models for elasmobranchs that attempt to characterize the vulnerability to exploitation of the populations under study. Increased fishing pressure on some species (Hoff and Musick, 1990), largely due to an increase in demand for shark fins (Bonfil, 1994), also prompted the emergence of population models to assess stock status.

With that in mind, I start by reviewing the progress that has been made in our understanding of life history patterns in elasmobranchs, with emphasis on sharks. Then I introduce the frameworks used to incorporate our knowledge of the biology of each species into population models. The first step is to present an overview of methodological issues relevant to the study of demography and dynamics of elasmobranch populations, which is critical to understanding the data requirements, limitations, and advantages of different population modeling approaches. After setting the methodological background, I critically review the complementary approaches used to model elasmobranch populations and arrange the individual studies in a summary table. I conclude with a synthesis of the review and recommendations for future work.

15.2 Life History Patterns

15.2.1 Comparative Life History Patterns

Life history strategies can be interpreted using three basic frameworks: (1) *r-K* theory, (2) bet-hedging theory, and (3) age-specific models that focus on optimal reproductive effort (Stearns, 1992). The *r-K* theory is the simplest scheme in that it is deterministic and assumes environmental stability, and it is the most common paradigm used in elasmobranch life history studies. Indeed, the vulnerability of sharks to fishing pressure is almost invariably attributed to their *K*-selected life history strategies. In contrast, almost no reference exists in the literature to the stochastic bet-hedging theory or age-specific models. This is in part because vital rates of elasmobranchs are believed to be less susceptible to environmental variability than those of teleosts, for example, which generally produce planktonic larvae (Stevens, 1999). Meanwhile, there have been no comparative tests of these theories, making our knowledge of the selective pressures operating on life histories of sharks very limited and speculative.

Despite the heavy criticism received by the *r-K* theory, one appealing aspect of it is that it provides a framework for explaining the observed variability in life history traits of species by predicting that certain traits will generally tend to be found in *r*-selected species, whereas others will tend to occur in *K*-selected species. Hoenig and Gruber (1990) recognized this feature and advocated the use of *r-K* selection theory as a tool to classify elasmobranch species according to their relative abilities to withstand exploitation.

Several attempts have been made at distinguishing separate life history strategies or patterns in elasmobranchs. Compagno (1990) qualitatively classified the life history styles of chondrichthyans into at least 18 groupings, which he termed ecomorphotypes, based on ecomorphological factors such as habitat, morphology, feeding preferences, and behavior. Branstetter (1990) used relative and absolute size at birth, litter size, growth during the first year of life, and the growth completion rate (*k*) from the von Bertalanffy growth (VBG) equation generally used to describe growth in elasmobranchs, to classify several species of carcharhinoid and lamnoid sharks into broad categories. Cortés (2000) identified at least three separate groupings among 40 populations of 34 shark species using principal component analysis and cluster analysis of adult maximum size, offspring size, fecundity, *k*, and longevity. The groups identified by Cortés (2000) using statistical ordination techniques generally agreed with Branstetter's (1990) *ad hoc* classification. Cortés (2000) argued that the alternative life history groupings he identified could be used to explain how different species may cope with juvenile mortality. Species such as the blue shark, *Prionace glauca*, would exemplify a first group characterized by large litter size, variable but generally long lifespan, intermediate to large body length, small offspring, and fairly low *k*. Species in this group would invest in many small offspring, with high vulnerability to predators, which they would compensate by growing rapidly during the early life stages. In contrast, species such as the dusky shark, *Carcharhinus obscurus*, would typify a second group characterized by large size, large offspring, small litter size, low *k*, and generally long lifespan. Species in this group would produce fewer, larger offspring less vulnerable to predation, not requiring growth to be as rapid as in the blue shark. A small species such as the Atlantic sharpnose shark, *Rhizoprionodon terraenovae*, would exemplify a third group characterized by small litter size, small to moderate body length, short to moderate lifespan, small offspring, and generally high *k*. Species in this group would allocate reproductive effort differently, by producing a few, small offspring, born at a higher proportion of maximum adult size and growing faster than their counterparts in the other groups to overcome mortality in the early life stages.

In all, it is difficult to explain the observed life history traits of elasmobranchs using a single theory. This is partly because what is often observed is a collection of selected life history traits rather than the whole set of biological events that make up a life history pattern (Hoenig and Gruber, 1990) or the coordinated evolution of all life history traits (Stearns, 1992). It is too simplistic to talk about life history patterns and strategies without taking account of spatial factors such as movement and dispersal, or even morphological, physiological, or behavioral aspects. Despite these caveats, there is some evidence that mortality, expressed through predation or competition rather than environmental variability, may be the main selective force in sharks (Stevens, 1999). Most adult sharks reach a large size, suggesting low mortality from predation once adulthood is reached (Roff, 1992), and implying that mortality primarily affects the juvenile stages.

According to the *r-K* theory, if a population is under stable or predictable environmental conditions, nearing its carrying capacity, and with strong intraspecific competition, then natural selection will favor *K*-selection, with delayed reproduction and high longevity to allow for protracted reproductive output (Stearns, 1992). The bet-hedging theory predicts that environmental variability causes relatively high and variable juvenile mortality, and thus *K*-selected traits are also favored because a long reproductive life is needed to offset years of high juvenile mortality (Stearns, 1992). In contrast, this theory also suggests that in more stable environments where juvenile mortality may be more constant, *r*-selected traits would be favored because predictable juvenile mortality does not require a long reproductive life to counteract juvenile mortality.

Stevens (1999) attempted to describe the different life history "strategies" of the school shark, *Galeorhinus galeus*, and gummy shark, *Mustelus antarcticus*, off Australia through these two competing theories, concluding that, if driven by juvenile mortality, they would be better explained by the *r-K* theory than by the bet-hedging theory. Using these theories to explain the life history patterns of the species most representative of the three groups identified by Cortés (2000) yields inconclusive results and underscores the limitations of theories that link habitats to life histories (Stearns, 1992). The life history of the Atlantic sharpnose shark seems to adhere to the *r-K* theory because it is more *r*-selected, and one may contend that the shallow nursery areas where individuals spend the first few years of life and the coastal habitats where adults mostly occur represent a more unstable and unpredictable environment than the open ocean, for example. In contrast, the blue shark life history can perhaps be better explained by the bet-hedging theory in that the pelagic environment where blue sharks occur is a more stable environment, and juvenile survival is likely to be relatively constant, favoring *r*-selected species such as the blue shark. The life history of dusky sharks does not appear to conform to either of these two schemes because they occur mostly in what can be considered unstable coastal habitats; yet they are believed to have low juvenile mortality and to be *K*-selected.

15.2.2 Life History Relationships

Examining correlations between life history traits is useful for comparisons among different taxonomic groups, and developing empirical relationships between life history parameters is also useful because it allows estimation of parameters that are difficult to measure or estimate using more readily available parameters. Two recent studies were aimed at providing these kinds of analyses for elasmobranchs. Cortés (2000) provided a compendium of life history traits for 230 shark populations encompassing 164 species, 19 families, and 7 orders, and examined correlations between pairs of traits and the effect of body size on some of the relationships. Frisk et al. (2001) developed regressions between pairs of vital parameters and estimated invariant life history ratios for several species of sharks, skates, and rays.

Cortés (2000) found that several life history traits related to reproduction, growth, and age of sharks varied with body size and that controlling the effect of body size changed the nature of some of the relationships between traits. He reported that interspecifically maternal length positively correlated with litter size and offspring length, and litter size negatively correlated with offspring size only when the latter was expressed as a proportion of parental size. Garrick (1982) previously described this trade-off predicted by life history theory for sharks of the genus *Carcharhinus*. The relationship between offspring length and the growth coefficient *k* was negative, but became weakly positive after expressing offspring length as a proportion of parental length. This pattern, in conjunction with the negative correlation observed between *k* and parental size, suggested to Cortés (2000) that the smaller species with generally higher values of k are born at a higher proportion of their maximum size than larger, slower-growing species, supporting previous findings by Pratt and Casey (1990).

Cortés (2000) also reported differences between males and females in traits related to body size, growth, and age. He found that, in general, females of the populations he examined reached maturity at a larger size and older age than males (bimaturism), attained a larger maximum size and older age than males, and took longer to complete their growth than males. He attributed bimaturism to the need for females to reach a larger size than males to carry pups, and to a smaller proportional partitioning of energy for growth in favor of reproduction, which would be ultimately reflected in a delayed onset of sexual maturity in females. Stearns (1992) ascribed this pattern, common in many taxa, to a continuous gain in fecundity for females after males reach a size of "diminishing returns." However, Cortés (2000) found that both males and females reach maturity on average at 75% of their maximum size, supporting similar observations by Holden (1972) and Garrick (1982). Frisk et al. (2001) found a value of 73% in dogfishes, skates, and rays, and indicated that this life history ratio remains relatively invariant among taxonomic groups, as first pointed out by Beverton and Holt (1959). Cortés (2000) also found that the ratio of age at maturity to maximum age was similar in both sexes (48% in males, 54% in females), whereas Frisk et al. (2001) found an average value of 38% in their analysis, a value in the upper range of those found for other fish groups by Beverton (1992). The lower value found by Frisk et al. (2001) may possibly be attributed to their use of extrapolations from the age–length curve to estimate theoretical lifespan in some cases, yielding almost invariably higher values of lifespan than empirical observations (Cortés, 2000) and thus lower ratios of age at maturity to lifespan.

Cortés (2000) also found a strong positive correlation between size at maturity and maximum size in both sexes, as did Frisk et al. (2001) for sexes combined. Cortés (2000) found a weaker correlation between body size and lifespan, especially in females, and a negative correlation between k and lifespan, supporting the life history prediction that long-lived species tend to complete their growth at a slower rate than short-lived species. Frisk et al. (2001) reported that another invariant ratio, the M/k ratio (M, instantaneous rate of natural mortality), for the 30 elasmobranch species they examined, was significantly different from those of other taxa. However, it was unclear whether this difference was real or a result of limited sample size and the way in which M was estimated.

Body size has been identified as an indicator of vulnerability to exploitation in skates and rays (Walker and Hislop, 1998; Dulvy and Reynolds, 2002; Frisk et al., 2002). In a literature review of information on body size and latitudinal and depth ranges for a large number of species, Dulvy and Reynolds (2002) found that locally extinct species tended to have larger body size and that geographic range size was not a good predictor of extinction vulnerability in skates. While there are other life history traits not examined by these authors that are related to body size and that may be better predictors of vulnerability, using this trait for prediction is appealing because of the simplicity with which it can be obtained.

Other evidence linking body size to measures of population productivity is weaker. Walker and Hislop (1998) and Frisk et al. (2002) found a decreasing trend in productivity measured by the intrinsic rate of population increase, r, with increasing body length in analyses of five species of skates and rays and three species of skates, respectively. Frisk et al. (2001) included 36 elasmobranch species in their analysis, and were ambiguous in their interpretation of the value of total length as an indicator of resilience, but recommended that large species (>200 cm total length) be subjected only to conservative fishing limits. They based the value of maximum length as an indicator of resilience to exploitation on its negative correlation with a calculated potential rate of increase proposed by Jennings et al. (1999). Mollet and Cailliet (2002) indicated that incorrect values of annual fecundity had been used in Frisk et al.'s (2001) calculations of productivity, making it unclear how this may have affected the trends observed by these authors. Smith et al. (1998) also found that, of the 28 species they analyzed, those with the lowest rebound potentials generally tended to be larger. However, both Frisk et al. (2001) and Smith et al. (1998) included mostly large species, which have received more attention and been the focus of more research than, for example, many small squaliform sharks, which are probably very long-lived and have low productivity.

In contrast to these findings, Cortés (2002a) found no correlation between population growth rates (λ, finite rate) and maximum length in a study of 41 populations from 38 species of sharks. Furthermore, Cortés found that some small or relatively small species perceived to be fairly productive had very low

λ values, leading to the proposal that, at least for sharks, elasticities (proportional matrix sensitivities; De Kroon et al., 1986) might be better predictors of resilience to exploitation than population growth rates. Cited as an example was the blacknose shark, *Carcharhinus acronotus*, a small species estimated to have low λ values, but that still showed an elasticity pattern consistent with those of other small and more productive species characterized by early age at maturity, fast growth, and short lifespan.

Calculation of population growth rates or elasticities requires multiple estimates of life history traits, which are often not available. A single life history trait, such as age at maturity, may instead be a good indicator of vulnerability because this trait is negatively correlated with population growth rate (Smith et al., 1998; Musick, 1999; Cortés, 2002a). Use of a more easily observed trait, such as maximum body size, is obviously preferable to provide practical management advice, but using it as the sole indicator of resilience to exploitation is potentially misleading, especially for sharks, since the evidence is still equivocal.

15.3 Population Dynamics

Populations are made up of individuals with a life cycle consisting of a series of sequential and recognizable states of development that can be described by age, stage, or size (cohorts). Population dynamics attempts to describe changes in the cohort-specific abundance of a population in space and with time as a result of various sources of variability. In general terms, the sources of variability governing population dynamics are both ecological and genetic processes (Cortés, 1999). The cohort-specific abundance of individuals over time and space is determined by three basic vital rates (birth, growth, and death) and the demographic processes of emigration and immigration, which are subject to genetic, demographic, environmental, sampling, and human-induced stochasticity. The effect that these sources of variability have on vital rates and demographic processes ultimately determines the fate of the population. Ideally, a population dynamics model should thus capture the interaction of vital rates and demographic processes with all sources of variability to provide knowledge on population abundance in time and space.

The reality for elasmobranch population modeling is quite different, however. Our knowledge of vital rates and demographic processes is still fragmentary for most species, let alone our grasp on the spatial distribution of populations, stock-recruitment dynamics, and the effect of most sources of stochasticity on elasmobranch populations. Despite this state of affairs, considerable progress has been made in the recent past in the fields of demographic analysis and population modeling of elasmobranchs. Two main approaches with separate philosophies and purposes have emerged. Life tables and population matrix models have been developed to gain a basic understanding of the population ecology of some species while assessing their vulnerability to fishing, and to address conservation issues by producing population metrics that can be used to generate mostly qualitative management measures. In contrast, stock assessment models traditionally used in fisheries research have been applied to several stocks to produce estimates of population status that can be used for implementing quantitative management measures. Table 15.1 summarizes all known elasmobranch population models arranged into several groups according to the following factors: (1) whether the model was cohort-structured or considered lumped biomass only, (2) whether the model was static or dynamic, (3) whether the cohort structure of the population was classified as age or stage, (4) whether the model dealt with uncertainty or not (deterministic vs. stochastic), and (5) whether the model was linear or nonlinear (with density dependence; see Chaloupka and Musick, 1997). Table 15.1 also includes the modeling approach, species, geographic location, purpose of the study, and citation.

15.3.1 Methodological Background

Before describing the various population modeling approaches, it is convenient to define some terms and describe the limitations of sampling design in relation to the data requirements of the different methods.

TABLE 15.1

Summary of Elasmobranch Demography and Population Dynamics Studies

Structure	Time	Cohort Type	Mode	Shape	Model Type(s)	Species	Aim	Area	Ref.
Biomass	Dyn	—	Det	NL	Schaefer	Spiny dogfish	Sa/Ma	NEA	Aasen (1964)
Biomass	Dyn	—	Det	NL	Schaefer	Large sharks	Sa/Ma	NWA	Otto et al. (1977)
Biomass	Dyn	—	Det	NL	Fox, Pella-Tomlinson	Pelagic sharks	Sa/Ma	NWA	Anderson (1980)
Biomass	Dyn	—	Det	NL	Fox	Kitefin shark	Sa/Ma	Azores	Silva (1983, 1987)
Biomass	Dyn	—	Det	NL	Schaefer, Fox, Pella-Tomlinson	Rajid assemblage	Sa/Ma	Falkland Islands	Agnew et al. (2000)
Biomass	Dyn	—	Stoch	NL	Schaefer, Fox	—	Sa/Ma	—	Bonfil (1996)
Biomass	Dyn	—	Stoch	NL	Schaefer (Bayesian)	Sandbar and blacktip sharks	Sa/Ma	NWA	McAllister et al. (2001)
Biomass	Dyn	—	Stoch	NL	Schaefer (Bayesian)	Small coastal sharks	Sa/Ma	NWA	Cortés (2002b)
Biomass	Dyn	—	Stoch	NL	Schaefer (Bayesian)	Large coastal sharks	Sa/Ma	NWA	Cortés et al. (2002)
Cohort	Static	Age	Det	Linear	Life table	Sandbar shark	Da/Ma	NWA	Hoff (1990)
Cohort	Static	Age	Det	Linear	Life table	Leopard shark	Da/Ma	California	Cailliet (1992)
Cohort	Static	Age	Det	Linear	Life table	Angel shark	Da/Ma	California	Cailliet et al. (1992)
Cohort	Static	Age	Det	Linear	Life table	Atlantic sharpnose shark	Da/Ma	NWA	Cortés (1995)
Cohort	Static	Age	Det	Linear	Life table	Bonnethead	Da	EGM	Cortés and Parsons (1996)
Cohort	Static	Age	Det	Linear	Life table	Sandbar shark	Da/Ma	NWA	Sminkey and Musick (1996)
Cohort	Static	Age	Det	Linear	Life table	Atlantic sharpnose shark	Da/Ma	SEGM	Márquez and Castillo (1998)
Cohort	Static	Age	Det	Linear	Life table	Bonnethead	Da/Ma	SEGM	Márquez et al. (1998)
Cohort	Static	Age	Det	Linear	Life table	Lemon, sandbar, dusky, blacktip, bonnethead, and Atlantic sharpnose sharks	Da/Ma	NWA	Cortés (1998)
Cohort	Static	Age	Det	Linear	Life table	Scalloped hammerhead	Da/Ma	NWP	Liu and Chen (1999)
Cohort	Static	Age	Det	Linear	Life table	Australian sharpnose shark	Da/Ma	Northern Australia	Simpfendorfer (1999a)
Cohort	Static	Age	Det	Linear	Life table	Dusky shark	Da/Ma	Southwest Australia	Simpfendorfer (1999b)
Cohort	Static	Age	Det	Linear	Life table	Pacific electric ray	Da	California	Neer and Cailliet (2001)
Cohort	Static	Age	Det	Linear	Life table	Two species of sawfish	Da/Ma	WA	Simpfendorfer (2000)
Cohort	Static	Age	Det	Linear	Life table	Porbeagle	Da/Ma	NWA	Campana et al. (2002)
Cohort	Static	Age	Det	Linear	Modified Euler–Lotka equation	Up to 31 species of shark and 1 species of ray	Da/Ma	Multiple locations	Smith et al. (1998, in press), Au et al. (in press)

Cohort	Static	Age	Det	Linear	Modified "dual" Euler-Lotka equation	Gummy and school sharks	Southern Australia	Da	Xiao and Walker (2000)
Cohort	Static	Age	Det	Linear	BLL matrix	Lemon shark	NWA	Da/Ma	Hoenig and Gruber (1990)
Cohort	Static	Age	Det	Linear	BLL matrix	One species of skate and four species of ray	North Sea	Da/Ma	Walker and Hislop (1998)
Cohort	Static	Age	Det	Linear	BLL matrix	Leopard and angel sharks	California	Da/Ma	Heppell et al. (1999)
Cohort	Static	Age	Det	Linear	BLL matrix	Pelagic stingray, white, pelagic thresher, and sandtiger sharks	Multiple locations	Da	Mollet and Cailliet (2002)
Cohort	Static	Age	Det	NL	BLL matrix	Spiny dogfish	NWA	Sa/Ma	Silva (1993)
Cohort	Static	Age	Det, Stoch	Linear	BLL matrix	Little and winter skates	NWA	Sa/Ma	Frisk et al. (2002)
Cohort	Static	Age	Stoch	Linear	BLL matrix, life table	Sandbar and blacktip sharks	NWA	Input to Sa	McAllister et al. (2001)
Cohort	Static	Age	Stoch	Linear	BLL matrix, life table	Small coastal sharks	NWA	Input to Sa	Cortés (2002b)
Cohort	Static	Age	Stoch	Linear	BLL matrix, life table	41 shark species	Multiple locations	Da/Ma	Cortés (2002a)
Cohort	Static	Age	Stoch	Linear	Life table	Silky shark	NWA	Da	Beerkircher et al. (2003)
Cohort	Static	Stage	Det	Linear	Usher matrix	Sandbar shark	NWA	Da/Ma	Brewster-Geisz and Miller (2000)
Cohort	Static	Stage	Det	Linear	Usher matrix	Barndoor skate	NWA	Da/Ma	Frisk et al. (2002)
Cohort	Static	Stage	Det	Linear	Usher matrix	Pelagic stingray, white, pelagic, thresher, and sandtiger sharks	Multiple locations	Da	Mollet and Cailliet (2002)
Cohort	Static	Stage	Stoch	Linear	Usher matrix, life table	Sandbar shark	NWA	Da/Ma	Cortés (1999)
Cohort	Static	Age	Det	Linear	Yield per recruit, Cohort analysis	School shark	Australia	Sa/Ma	Grant et al. (1979)
Cohort	Static	Age	Det	Linear	Yield per recruit	Little skate	NWA	Ma	Waring (1984)
Cohort	Static	Age	Det	Linear	Yield per recruit, VPA	Leopard shark	California	Sa/Ma	Smith and Abramson (1990)
Cohort	Static	Age	Det	Linear	Recruitment-adjusted yield per recruit	Leopard shark	California	Ma	Au and Smith (1997)
Cohort	Static	Age	Det	Linear	Yield per recruit	Sandbar shark	NWA	Ma	Cortés (1998)
Cohort	Static	Age	Det	Linear	Yield per recruit	Porbeagle	NWA	Sa/Ma	Campana et al. (1999, 2001, 2002)
Cohort	Static	Age	Det	Linear	Age-structured	Spiny dogfish	NWA	Sa/Ma	Rago et al. (1998)
Cohort	Dyn	Age	Det	NL	Dynamic pool	Gummy shark	Southern Australia	Sa/Ma	Walker (1992, 1994a,b)
Cohort	Dyn	Age	Stoch	NL	Fully age-structured (Bayesian)	School shark	Southern Australia	Sa/Ma	Punt and Walker (1998), Punt et al. (2000)

TABLE 15.1 (Continued)

Summary of Elasmobranch Demography and Population Dynamics Studies

Structure	Time	Cohort Type	Mode	Shape	Model Type(s)	Species	Area	Aim	Ref.
Cohort	Dyn	Age	Det	NL	Fully age-structured (maximum likelihood)	Whiskery shark	Southwest Australia	Sa/Ma	Simpfendorfer et al. (2000)
Cohort	Dyn	Age	Stoch	NL	Fully age-structured (Bayesian)	Blacktip shark	NWA	Sa/Ma	Apostolaki et al. (2002)
Cohort	Dyn	Age	Stoch	NL	Fully age-structured (Bayesian)	Porbeagle	NWA	Sa/Ma	Harley (2002)
Cohort	Dyn	Age	Stoch	NL	Fully age-structured (Bayesian and maximum likelihood)	Blacktip and sandbar sharks	NWA	Sa/Ma	Brooks et al. (2002); Cortés et al. (2002)
Delay difference	Dyn	Age	Det	NL	Deriso–Schnute	School shark	Southern Australia	Sa/Ma	Walker (1995)
Delay difference	Dyn	Age	Stoch	NL	Deriso–Schnute	—	—	Da/Ma	Bonfil (1996)
Delay difference	Dyn	Age	Stoch	NL	Lagged recruitment, survival and growth (Bayesian)	Small coastal sharks	NWA	Sa/Ma	Cortés (2002b)
Delay difference	Dyn	Age	Stoch	NL	Lagged recruitment, survival and growth (Bayesian)	Large coastal sharks	NWA	Sa/Ma	Cortés et al. (2002)

Abbreviations: Dyn, dynamic; Det, deterministic; Stoch, stochastic; NL, nonlinear; NEA, Northeastern Atlantic; NWA, Northwestern Atlantic; WA, Western Atlantic; SEGM, Southeastern Gulf of Mexico; NWP, Northwestern Pacific; EGM, Eastern Gulf of Mexico; Sa, stock assessment; Ma, management advice; Da, demographic analysis.

15.3.1.1   Demographic Unit or Stock — One of the main assumptions of a population dynamics model is that the stock, population, or demographic unit under study can be distinguished in time and space from other similar units. Although movement, migratory patterns, and genetic stock identification of elasmobranchs are starting to be better understood (see Musick et al., Chapter 2, and Heist, Chapter 16, this volume), identifying discrete demographic units or stocks still remains a major challenge in the study of elasmobranch populations. Many shark species, for example, are widely distributed and highly migratory, posing an especially difficult problem because individuals from potentially different stocks are likely to co-occur in some areas or habitats. In some other cases, as with the spiny dogfish, *Squalus acanthias*, and school shark, genetically separate stocks have been identified and little mixing is believed to occur (Walker, 1998). Ideally, demographic and population modeling of elasmobranchs should focus on genetically distinct stocks. In practice, the transboundary nature of many populations or stocks poses a practical problem for management, which is generally restricted geographically because of jurisdictional issues.

15.3.1.2   Population Sampling Design — Vital rates and demographic processes are affected by three separate, yet often confounded, time effects (Chaloupka and Musick, 1997). Indeed, demographic rates may vary from year to year due to external factors, may differ among cohorts due to genetic factors, and are also age-specific. A realistic population dynamics model thus needs to uncouple the effects of year, age, and cohort factors. However, it is not always possible to separate these time effects because of shortcomings in the modeling framework or, more often, owing to sampling limitations. This is the case with elasmobranch population modeling studies, which usually rely on only one set of estimates of demographic rates that are often not age specific. These models thus do not consider year effects, let alone cohort effects.

At present we simply do not know how these confounding time effects may bias estimates of population parameters for elasmobranchs. Given the life histories of elasmobranchs, it is reasonable to assume that year factors will not have the pronounced effect they can have on other fishes because vital rates of elasmobranchs are believed to be less sensitive to environmental influences and therefore more stable and predictable (Stevens, 1999). It is unknown how genetic influences, expressed through cohort factors, affect vital rates of elasmobranchs. In terms of age factors, we know from life history theory that natural mortality, for example, varies with age (Roff, 1992). In sharks, it is believed that intraspecific mortality generally remains fairly low and stable once individuals attain a certain size, but that juvenile mortality decreases from birth to adulthood as individuals grow and predation risk decreases (Cortés and Parsons, 1996).

There are only a few direct estimates of instantaneous natural mortality rate (M) or instantaneous total mortality rate (Z) for elasmobranchs based on mark–recapture techniques or catch curves. Direct estimates of natural mortality were obtained only in the mark–depletion experiments conducted for age-0 (Manire and Gruber, 1993) and juvenile (Gruber et al., 2001) lemon sharks, *Negaprion brevirostris*. Estimates of natural mortality derived from Z were obtained in mark–recapture studies for school shark (Grant et al., 1979), little skate, *Raja erinacea* (Waring, 1984), and juvenile blacktip sharks, *Carcharhinus limbatus* (Heupel and Simpfendorfer, 2002), and from length-converted catch curves for bonnetheads, *Sphyrna tiburo* (Cortés and Parsons, 1996), rays, *R. clavata* and *R. radiata* (Walker and Hislop, 1998), and porbeagle, *Lamna nasus* (Campana et al., 2001).

The majority of population modeling studies for elasmobranchs has relied, however, on indirect estimates of mortality obtained through methods based on predictive equations of life history traits. Most of these methods make use of parameters estimated from the VBG function, including those of Pauly (1980), Hoenig (1983), Chen and Watanabe (1989), and Jensen (1996) (see Roff, 1992; Cortés, 1998, 1999; and Simpfendorfer, 1999a for reviews of these methods). These equations do not yield age-specific estimates of natural mortality except in part for the Chen and Watanabe (1989) method. In contrast, a method proposed by Peterson and Wroblewski (1984) that has generated considerable debate (Cortés, 2002a; Mollet and Cailliet, 2002), allows estimation of size-specific natural mortality, which can then be transformed into age-specific estimates through the VBG function.

Back-transformation of lengths into ages through the VBG function is the usual method for estimating age-specific life history traits in elasmobranchs, because determining age of individuals is much more difficult than simply measuring their lengths. Thus, very few studies have determined age at maturity directly. Use of ages at maturity or age-specific fecundity estimates derived in this way can result in biased estimates of population metrics because this procedure does not account for variability in age at length, and vice versa. Many elasmobranch population models also describe maturity as a knife-edge process in which it is assumed that 100% of females reach maturity at the same size (age). This assumption is a direct consequence of reproductive studies that do not attempt to fit an ogive (logistic function) to describe the proportion of mature females at size or age in a population.

The distinction between static and dynamic population models is arbitrary because in a strict sense only models that incorporate temporal variation in demographic rates and allow for feedback mechanisms such as potential density-dependent responses reflect the dynamics of a population (Chaloupka and Musick, 1997). In studies of elasmobranch populations, the year, age, and cohort effects are often confounded because a year-specific state space vector (Getz and Haight, 1989) of absolute abundance is not available and thus the transient or time-dependent behavior of the population is being modeled in relative, rather than absolute, terms. For this review, only models that include year-specific vectors of absolute abundance (with or without varying demographic rates) are considered dynamic.

15.3.1.3 Stock–Recruitment Curve

15.3.1.3   Stock–Recruitment Curve — Knowledge of the relationship between stock and recruitment is central to the understanding of the population dynamics of marine organisms. No empirical data on this relationship have been published for any species of elasmobranch, but because of their reproductive limitations it is generally assumed that recruitment is directly related to spawning (pupping) stock size (Holden, 1977).

Walker (1994a) first produced some indirect support for a Beverton–Holt-type of stock–recruitment curve. By assuming that a density-dependent response was elicited through natural mortality of pre-recruit ages, he found that the number of gummy shark recruits off southeastern Australia predicted by an age-structured model remained relatively constant over a fairly wide range of high stock biomass levels. More recently, several stock assessments of elasmobranchs have also used the Beverton–Holt stock–recruitment curve, or a reparameterization that uses a steepness parameter, defined simply as the recruitment occurring at 20% of virgin biomass. A steepness of 0.2 indicates that recruitment is directly proportional to spawning stock and 1 is the theoretical maximum (Hilborn and Mangel, 1997). Simpfendorfer et al. (2000) constrained steepness between 0.205 and a maximum given by recruitment at virgin biomass and unexploited egg production in an age-structured model for whiskery shark, *Furgaleus macki*, off southwestern Australia. Harley (2002) estimated steepness values ranging from 0.25 to 0.67 for porbeagle through a relationship between steepness and maximum reproductive rate proposed by Myers et al. (1999). Apostolaki et al. (2002) estimated pup survival at low densities, a function of steepness and pup production and recruitment under virgin conditions, in an age-structured model application to blacktip shark. Brooks et al. (2002) also estimated steepness in an age-structured model application to sandbar, *Carcharhinus plumbeus*, and blacktip sharks. Cortés (2002b) and Cortés et al. (2002) assigned uninformative, uniform prior distributions for steepness ranging from 0.2 to 0.9, in Bayesian lagged recruitment, survival, and growth models for small and large coastal sharks, respectively.

15.3.2   Biomass Dynamic Models

Biomass dynamic models, also known as (surplus) production models, are widely used in the assessment of teleost stocks. Use of these models in assessment of elasmobranch stocks, however, has been criticized because of invalid assumptions, notably the presupposition that r responds immediately to changes in stock density and that it is independent of the age structure of the stock (Holden, 1977; Walker, 1998). In general, production models trade biological realism for mathematical simplicity, combining growth, recruitment, and mortality into one single "surplus production" term. However, they are useful in situations where only catch and effort data on the stock are available and for practical stock assessments because they are easy to implement and provide management parameters, such as maximum sustainable yield (MSY) and virgin biomass (Meyer and Millar, 1999a).

Walker (1998) cited some of the early assessment work on elasmobranchs (Aasen, 1964; Holden, 1974; Otto et al., 1977; Anderson, 1980; Silva 1983, 1987), which was based on application of production models, and therefore thought to produce questionable results. But the lack of quality data for many species of elasmobranchs and the need for management benchmarks have prompted the resurgence of this methodology more recently. Bonfil (1996) used simulation to compare the performance of several dynamic production models and a delay difference model in estimating assessment and management parameters of elasmobranchs, concluding that only the Schaefer (1954) model gave acceptable results. Agnew et al. (2000) used what they called a constant recruitment model, a Schaefer production model, a Fox (1970) model, and a Pella–Tomlinson (1969) model to assess the multispecies skate and ray fishery off the Falkland Islands. They were able to demonstrate that there are two distinct rajid communities off the islands, with different sustainable yields, and that species composition was affected by fishing, such that smaller and earlier-maturing species took over larger and slower-maturing species. More sophisticated applications of surplus production models have been used for assessment of large coastal (McAllister et al., 2001; Cortés et al., 2002) and small coastal (Cortés, 2002b) sharks off the United States. These will be described in a later section because they are dynamic models that incorporate uncertainty and stochasticity.

15.3.3 Cohort-Structured Models

15.3.3.1 Static Models —

15.3.3.1.1 Age-Structured Models. Demographic studies of elasmobranchs are typically based on deterministic, density-independent population growth theory, whereby populations grow at an exponential rate r and converge to a stable age distribution. Indeed, most of the age-structured life tables and matrix population models reviewed here assumed time-invariant (stationary with respect to time) and density-independent demographic rates; i.e., the estimates of demographic rates were generally collected from a single point in time and thus they provide only a snapshot of the population.

The majority of demographic analyses of elasmobranch populations are (1) deterministic life tables based on a discrete implementation of the Euler–Lotka equation (Euler, 1760; Lotka, 1907) or (2) age-based Leslie or Bernardelli-Leslie-Lewis (BLL; Manly, 1990) matrix population models. Hoff (1990) and Cailliet (1992), and Hoenig and Gruber (1990), respectively, pioneered the use of these two analogous methods (Table 15.1), with the aim of producing basic population statistics, measuring the sensitivity of r to variation in some demographic rates, and assessing the vulnerability of each population to fishing. The latter is generally accomplished by adding a constant instantaneous fishing mortality (F) term to M starting at a given age and thereafter, and recalculating r while still assuming fixed demographic rates with time and exponential population growth. This approach is straightforward, but has obvious limitations given the numerous implicit assumptions (Cortés, 1998). Nevertheless, it has become a common framework for evaluating the effect of harvesting on population growth of elasmobranchs, having been used for leopard shark, *Triakis semifasciata* (Cailliet, 1992), Pacific angel shark, *Squatina californica* (Mollet et al. 1992), Atlantic sharpnose shark (Cortés, 1995), sandbar shark (Sminkey and Musick, 1996), bonnethead (Márquez and Castillo, 1998), Australian sharpnose shark, *Rhizoprionodon taylori* (Simpfendorfer 1999a), dusky shark (Simpfendorfer, 1999b), scalloped hammerhead, *Sphyrna lewini* (Liu and Chen, 1999), Pacific electric ray, *Torpedo californica* (Neer and Cailliet, 2001), and porbeagle (Campana et al., 2002).

Deterministic, age-structured BLL matrices have also been used in a number of studies of elasmobranch populations. Walker and Hislop (1998) compared the demography of four *Raja* species; Heppell et al. (1999) compared the demography of several long-lived marine vertebrates, including the leopard and angel sharks; Mollet and Cailliet (2002) modeled the demography of the pelagic stingray, *Dasyatis violacea*, pelagic thresher, *Alopias pelagicus*, white shark, *Carcharodon carcharias*, and sandtiger, *Carcharias taurus*; and Frisk et al. (2002) compared the demography of two *Leucoraja* species. Elasticities were also calculated in these studies, leading to the almost unanimous conclusion that juvenile survival was the vital rate that had the largest effect on population growth rate.

Two modifications of the horizontal life table approach involving the Euler–Lotka equation have been proposed. Au and Smith (1997) introduced a demographic technique applied to leopard shark that combines the traditional Euler–Lotka equation with concepts of density dependence from standard fisheries models. The density-dependent compensation is manifested in preadult survival as a result of increased mortality in the adult ages. These so-called rebound potentials were later calculated for a suite of shark species (Smith et al., 1998, in press; Au et al., in press) and were found to be strongly affected by age at maturity. Xiao and Walker (2000) developed another modification of the Lotka equation that allowed calculation of the intrinsic rate of increase with time and the intrinsic rate of decrease with age and applied it to gummy and school sharks. They concluded that the intrinsic rate of increase with time is a function of the reproductive and total mortality schedules, but that the intrinsic rate of decrease with age is a function of the reproductive schedules only.

Walker (1998) stated that, because life tables or Leslie matrix models do not account for density dependence, they always produce pessimistic outlooks for shark exploitation. However, results from both deterministic and stochastic simulations also include very optimistic prognoses. We must not forget that population growth rates obtained through density-independent approaches imply exponential population growth, and as such, we may also argue that they are unrealistically optimistic, contrary to Walker's (1998) interpretation.

15.3.3.1.2 Stage-Structured Models.   Stage-structured analogs of the age-based BLL matrix models, referred to as Lefkovitch or Usher models (see Getz and Haight, 1989, and Manly, 1990, for details), have been applied in deterministic analyses of some elasmobranch populations. Brewster-Geisz and Miller (2000) used this approach in combination with stage-based matrix elasticity analysis to examine management implications for the sandbar shark. They concluded that of the five stages they considered (neonate, juvenile, subadult, pregnant adult, and resting adult), juveniles and subadults affected λ the most. Frisk et al. (2002) also applied a stage-based matrix model and elasticity analysis to the barndoor skate, *Dipturus laevis*, but found that adult survival contributed the most to λ. Mollet and Cailliet (2002) applied life tables, and age- and stage-based matrix models to the pelagic stingray, sandtiger, pelagic thresher, and white shark to demonstrate the effect of various methodological issues on population statistics. When using stage-based models, they found that if stage duration was fixed, population growth rates were identical to those obtained with the other methods, but net reproductive rates and generation times differed.

15.3.3.1.3 Yield-per-Recruit Models.   Yield-per-recruit (YPR) models are a form of age-structured analysis that takes account of age-specific weight and survival, but does not include fecundity rates and assumes constant and density-independent recruitment. As originally devised by Beverton and Holt (1957), the main application of this model in elasmobranchs has been to determine the fishing mortality rate (F) that maximizes the yield per recruit when considering different ages of entry into the fishery (age at first capture). It is often applied in combination with methods that analyze tag–recapture or length–frequency information to estimate mortality, which is then used in the YPR model.

Most researchers who have used YPR analysis to model elasmobranch populations have concluded that the predicted maximum YPR is likely not to be sustainable. Grant et al. (1979) first applied this methodology to the school shark in Australia after estimating natural and fishing mortality rates through cohort analysis (Pope, 1972) and found that to achieve the maximum YPR the fishery should be expanded, but they cautioned that such action could reduce the breeding stock. Waring (1984) used catch curves to estimate Z, which he then used in a YPR analysis of little skate off the northeastern United States, also concluding that the value of F that maximized yield per recruit could result in overexploitation given the low fecundity of little skate. Smith and Abramson (1990) used YPR analysis in combination with backward virtual population analysis (VPA) to estimate population replacement of leopard sharks off California, and concluded that imposition of a 100-cm total length size limit would allow the stock to be maintained while providing a yield per recruit close to the predicted maximum. Au and Smith (1997) used their modified demographic method described earlier to adjust the estimates of YPR obtained by Smith and Abramson (1990) for the effects of reduction in recruitment as a result of fishing. Their results showed that the leopard shark is much easier to overfish than originally thought when the adjustment for reduced recruitment is introduced. Cortés (1998) used estimates of M and Z from life table analysis in a YPR analysis of the sandbar shark in the northwestern Atlantic, and estimated that the maximum YPR when using the value of F that results in MSY would be attained at an age of 22 years. He also concluded that sustainable YPR values for this population could be reached only with ages of entry into the fishery of 15+ years and at low values of F. Finally, Campana et al. (1999, 2001, 2002) used F estimates from Petersen analysis of tag–recaptures (Ricker, 1975), Paloheimo Zs (Paloheimo, 1961), and M from catch curves in a YPR analysis of the porbeagle in the northwestern Atlantic, concluding that the fishing mortality that would result in MSY is very low for this stock.

15.3.3.2 *Dynamic Age-Structured Models* — Deterministic models described under this section incorporate time explicitly in the equations describing the population dynamics, and many include nonlinear terms to account for density dependence in the three main components: growth, recruitment, and mortality. Stochastic age-structured models or models that incorporate uncertainty are treated in the next section. While the structure of age-based dynamic models is biologically more realistic than that of biomass dynamic models, for example, it comes at the price of having to provide or estimate values for an increased number of parameters. Age-structured models are thus more sophisticated, but also more assumption laden (Chaloupka and Musick, 1997). Some of the major assumptions of a typical fully age-structured model are that (1) growth is described adequately by a VBG function; (2) catch-at-age can be obtained by back-transforming catch-at-length through the VBG function in the absence of an age–length key, but even if an age–length key is available, it is still year and cohort invariant; (3) age at maturity and lifespan are fixed, year- and cohort-invariant parameters; (4) recruitment is constant from year to year (although this can be modified in nonlinear models); (5) all members of a cohort become vulnerable to the fishing gear at the same age and size; (6) natural and fishing mortality are time invariant (also modifiable in nonlinear models); and (7) removals are adequately described by a constant, time-invariant Baranov-type catch equation (Quinn and Deriso, 1999).

Wood et al. (1979) developed the first dynamic pool (or age-structured; Quinn and Deriso, 1999) model to describe the population dynamics of spiny dogfish off western Canada. Their model simulated the effects of assumptions on density-dependent regulation of mortality, reproduction, and growth, leading them to conclude that adult natural mortality was the compensatory mechanism regulating stock abundance in this species. Walker (1992) applied an age-structured simulation model to gummy shark off southern Australia that was sex specific, included terms to account for selectivity of the fishing gear, and assumed that density-dependent regulation operated through pre-recruit natural mortality. He subsequently refined the model for gummy shark with updated data and the ability to estimate some parameters, such as catchability and natural mortality (Walker, 1994a), and replaced the assumption of constant natural mortality for sharks recruited to the fishery with an asymmetric U-shaped function that varied with age (Walker, 1994b). Silva (1993) developed an analogous approach using a BLL nonlinear model for spiny dogfish in the Northwest Atlantic Ocean, which incorporated density-dependent terms for growth, fecundity, and recruitment. He concluded that the observed increase in abundance of spiny dogfish in the late 1980s was due at least in part to an increase in juvenile growth rate during the early 1970s.

Delay difference models bridge the gap between the simple, but biologically unrealistic production models and the more complex age-structured population models (Quinn and Deriso, 1999). Unlike production models, delay difference models consider the age-specific structure of the population, including the lag that exists between spawning and recruitment, and consider separately growth, recruitment, and natural mortality processes. Unlike fully age-structured models, no age data are required for fitting delay difference models because the age-specific equations are collapsed into a single equation for the entire population (Meyer and Millar, 1999a). Walker (1995) applied a Deriso–Schnute delay difference model (Quinn and Deriso, 1999) to the school shark off southern Australia using a Beverton–Holt (1957) curve to describe the stock–recruitment relationship. The model estimated the catchability coefficient (q) and the stock–recruitment parameters through maximum likelihood (ML) estimation techniques, but assumed knife-edge selectivity and did not fully utilize all available information on reproduction.

15.3.4 Models Incorporating Uncertainty and Stochasticity

Uncertainty in estimates of demographic rates has been incorporated into various forms of demographic analysis of elasmobranchs using Monte Carlo simulation. Cortés (1999) used life tables and stage-based matrix population models to incorporate uncertainty in size-specific estimates of fecundity and survivorship for sandbar shark, but fixed the values of age at maturity and maximum age. Cortés (1999) added a constant exploitation vector separately to each of the six stages identified and considered three fixed-quota harvesting strategies to simulate the effect of fishing on population abundance 20 years into the future. The model was dynamic in that it included a vector of stage-specific abundance that was updated at each time step (year), and the transition matrix varied yearly as a result of different values being drawn randomly from the distributions describing fecundity and survivorship. This author found that removal of large juveniles resulted in the greatest population declines, whereas removal of age-0 individuals at low values of fishing ($F = 0.1$) could be sustainable. These results were in agreement with findings from a deterministic stage-structured matrix population model by Brewster-Geisz and Miller (2000), who found that population growth rates of sandbar sharks were most sensitive to variations in the juvenile and subadult stages.

Cortés (2002a) used Monte Carlo simulation applied to age-structured life tables and BLL matrices to reflect uncertainty in estimates of demographic rates and to calculate population statistics and elasticities in a comparative analysis of 41 shark populations. He also used correlation analysis to identify the demographic rates that explained most of the variance in population growth rates. He reported that the populations examined fell along a continuum of life history characteristics that could be linked to elasticity patterns. Early maturing, short-lived, and fecund sharks that generally had high values of λ and short generation times were at the fast end of the spectrum, whereas late-maturing, long-lived, and less fecund sharks that had low values of λ and long generation times were placed at the slow end of the spectrum. "Fast" sharks tended to have comparable adult and juvenile survival elasticities, whereas "slow" sharks had high juvenile survival elasticity and low age-0 survival (or fertility) elasticity. Ratios of adult survival to fertility elasticities and juvenile survival to fertility elasticities suggested that many of the 41 populations considered were biologically incapable of withstanding even moderate levels of exploitation. While elasticity analysis suggested that changes in juvenile survival would have the greatest effect on λ, correlation analysis indicated that variation in juvenile survival, age at maturity, and reproduction accounted for most of the variance in λ. Combined results from the application of elasticity and correlation analyses in tandem led Cortés (2002a) to recommend that research, conservation, and management efforts be focused on those demographic traits.

Monte Carlo simulation of demographic rates has also been used to generate statistical distributions of the intrinsic rate of increase for use as informative prior distributions (priors) in Bayesian stock assessments. Both McAllister et al. (2001) and Cortés (2002b) used a variety of statistical distributions to describe vital rates of sandbar and blacktip sharks and four species of small coastal shark, respectively, in the northwestern Atlantic, producing probability density functions for r that were subsequently used in Bayesian stock assessments of these species.

An increasing number of models used to describe the population dynamics of elasmobranchs for stock assessment purposes have started to incorporate sources of stochasticity. Typically, in stock assessment work two stochastic components must be taken into consideration (Hilborn and Mangel, 1997): natural variability affecting the annual change in population biomass (also known as process error) and uncertainty in the observed indices of relative abundance owing to sampling and measurement error (observation error).

Punt and Walker (1998) and Simpfendorfer et al. (2000) developed age- and sex-structured population dynamics models for school and whiskery shark, respectively, off southern Australia, and used probabilistic risk analysis to predict stock status under several harvesting strategies. Both studies incorporated catch-at-age estimates and accounted for the effect of gear selectivity. Punt and Walker (1998) used a Bayesian statistical framework in which they incorporated an observation error component in the catch rate series and a process error term to account for recruitment variability under virgin conditions, both of which were assumed normally distributed. These authors incorporated two forms of assumed density dependence: in pup production, which the model related to the number of breeding females and their fecundity, and in natural mortality, which they described with a decreasing exponential function for ages 0 to 2, a constant value for adults, and with values increasing toward an asymptote for old ages (30+ years). Simpfendorfer et al. (2000) used a likelihood approach, fixed the value of the process error term based on Punt and Walker (1998), estimated the observation error, assumed that the stock–recruitment relationship was described by a Beverton–Holt curve, and fixed the value of natural mortality.

Punt et al. (2000) later refined their model to consider explicitly the spatial structure of multiple stocks of school shark obtained from extensive tagging studies. They identified two sources of uncertainty in their study: uncertainty in the model structural assumptions, and statistical uncertainty in the variability of parameter estimates. McAllister et al. (2001) and Cortés (2002b) used a Bayesian Schaefer production model to describe the dynamics of large and small coastal sharks, respectively, in the northwestern Atlantic. Both studies considered observation error only, which was integrated along with q from the joint posterior distribution using the analytical approach described by Walters and Ludwig (1994). All Bayesian studies described here used the sampling/importance resampling (SIR) algorithm as the method of numerical integration (see McAllister et al., 2001, and references therein for details).

Both process and observation errors can be incorporated easily when using a dynamic state-space modeling framework of time series (Meyer and Millar, 1999b). This approach relates observed states (catch per unit of effort, or CPUE, observations) to unobserved states (biomasses) through a stochastic model. State-space models allow for stochasticity in population dynamics because they treat the annual biomasses as unknown states, which are a function of previous states, other unknown model parameters, and explanatory variables (e.g., catch). The observed states are in turn linked to the biomasses in a way that includes observation error by specifying the distribution of each observed CPUE index given the biomass of the stock in that year. A Bayesian approach to state-space modeling has only been applied very recently to fisheries (Meyer and Millar, 1999a). One advantage of using a Bayesian approach is that it allows fitting nonlinear and highly parameterized models that are more likely to capture the complex dynamics of natural populations. Meyer and Millar (1999a,b) advocated the use of the Gibbs sampler, a special Markov chain Monte Carlo (MCMC) method, to compute posterior distributions in nonlinear state-space models.

Cortés (2002b) and Cortés et al. (2002) applied the Bayesian nonlinear, nonnormal state-space surplus production model developed by Meyer and Millar (1999b) to small and large coastal sharks, respectively, in the northwestern Atlantic. Cortés (2002b) and Cortés et al. (2002) also applied a simplified version of the delay difference model developed by Meyer and Millar (1999a) to the same two shark complexes using the Gibbs sampler for numerical integration. The lagged recruitment, survival and growth model (Hilborn and Mangel, 1997) is an approximation of the Deriso (1980) delay difference model that describes annual changes in biomass through a parameter combining natural mortality and growth, incorporates a lag phase to account for the time elapsed between reproduction and recruitment to the fishery, and describes the stock–recruitment relationship through a Beverton–Holt curve. The model assumes that fish reach sexual maturity and recruit to the fishery at the same age, although some alternative models that alleviate this assumption have been developed (Mangel, 1992).

Apostolaki et al. (2002), Harley (2002), Brooks et al. (2002), and Cortés et al. (2002) presented detailed models with the ability to incorporate fleet-disaggregated, fully explicit age- and sex-structured population dynamics based on Bayesian inference for parameter estimation. The model of Apostolaki et al. (2002), applied to blacktip shark in the northwestern Atlantic as an example, used a Beverton–Holt stock–recruitment curve in its baseline application, but also investigated the effects of considering a generalized hockey stick model (Barrowman and Myers, 2000) and a Ricker (1954) function. Apostolaki et al. (2002) reported the somewhat surprising finding that stock depletion was essentially unaffected by the form of the stock–recruitment curve. Their model also allowed for incorporation of separate spatial areas, considered observation uncertainty only, and used the SIR algorithm for numerical integration. Harley (2002) used a statistical catch-at-length approach applied to the porbeagle in the northwestern Atlantic. The model assumed that the stock–recruitment relationship could be described by a Beverton–Holt curve, allowed for interannual recruitment variability, considered observation error, and allowed for incorporation of mark–recapture data. The application to blacktip and sandbar sharks from the northwestern Atlantic by Brooks et al. (2002) and Cortés et al. (2002) was based on a model developed by Porch (2003). The model was a state-space implementation of an age-structured production model, a step-up in complexity with respect to a production model, which can incorporate age-specific vectors for fecundity, maturity, and fleet-specific gear selectivity while considering both observation error and process error for several parameters. The model assumed that the stock–recruitment relationship is described by a Beverton–Holt curve and allowed specification of either ML or Bayesian techniques for parameter estimation.

15.4 Conclusions

Although *r-K* theory may still be a more or less adequate categorization tool, future efforts should focus on identifying the causes (selection pressures) for different life history patterns and understanding the evolution of individual life history traits in elasmobranchs. Of particular importance is an understanding of the role of density dependence in the evolution of life history traits (Stearns, 1992).

Using life history correlates and predictive equations to estimate values of life history parameters and provide conservation and management advice is a useful shortcut, but should be applied cautiously, especially when based on limited data. To gain a good understanding of elasmobranch population dynamics, we should invest in obtaining empirical estimates of vital rates and demographic processes. Uncritical use of some measures of productivity alone to assess vulnerability to exploitation is also potentially dangerous because these measures are correlated with population size. This is problematic because calculation of productivity measures requires extensive biological data while assessment of absolute population abundance in elasmobranchs is particularly difficult. Other measures of vulnerability, such as elasticity analysis or similar approaches, hold promise but must be thoroughly evaluated before using them as the sole basis for conservation and management actions. Integrated approaches that provide both qualitative and quantitative conservation and management advice likely should be pursued.

Despite significant development of population models of elasmobranchs for conservation and stock assessment purposes in the recent past, empirical research is still limited. Highly sophisticated age-structured population dynamics models describe reality better by incorporating a large number of parameters, but their greater realism is also their pitfall in that they require many parameter estimates. There may be greater predictive return from investing in increased data quality rather than model sophistication.

In all, much remains to be done in the field of elasmobranch population modeling. In addition to validation of ages for the majority of species, very little is known of crucial vital rates such as mortality or of the relationship between parental stock and recruitment. Implicitly related to the latter is also an understanding of the density-dependent mechanisms that control the size of elasmobranch populations. Very little is still known of the temporal and spatial structure of populations, but there is hope that the increased number of mark–recapture programs and telemetry studies in existence will provide insight in the years to come. Even less is known of competitive intrapopulation, intraspecific, and interspecific processes or ecological interactions with other species in the marine ecosystem. Indeed, we are only starting to gain an understanding of these processes and interactions through emerging food web studies.

Acknowledgments

The views expressed in this chapter are solely those of the author and do not imply endorsement by NOAA Fisheries. I thank J. Neer for her comments on an earlier version of this manuscript.

References

Aasen, O. 1964. The exploitation of the spiny dogfish (*Squalus acanthias* L.) in European waters. *Fiskeridir. Skr. Ser. Havunders.* 13:5–16.

Agnew, D. J., C. P. Nolan, J. R. Beddington, and R. Baranowski. 2000. Approaches to the assessment and management of multispecies skate and ray fisheries using the Falkland Islands fishery as an example. *Can. J. Fish. Aquat. Sci.* 57:429–440.

Anderson, E. D. 1980. MSY estimate of pelagic sharks in the western North Atlantic. Woods Hole Lab. Ref. Doc. 80-18, U.S. Department of Commerce.

Apostolaki, P., M. K. McAllister, E. A. Babcock, and R. Bonfil. 2002. Use of a generalized stage-based, age-, and sex-structured model for shark stock assessment. *Col. Vol. Sci. Pap. Int. Comm. Conserv. Atl. Tunas* 54:1182–1198.

Au, D. W. and S. E. Smith. 1997. A demographic method with population density compensation for estimating productivity and yield per recruit. *Can. J. Fish. Aquat. Sci.* 54:415–420.

Au, D. W., S. E. Smith, and C. Show. In press. Estimating productivity and fishery-entry ages that guard reproductive potential and collapse thresholds of sharks, in *Pelagic Sharks*. E. K. Pikitch and M. Camhi, Eds., Blackwell Scientific, Oxford.

Barrowman, N. J. and R. A. Myers. 2000. Still more spawner-recruitment curves: the hockey stick and its generalizations. *Can. J. Fish. Aquat. Sci.* 57:665–676.

Baum, J. K., R. A. Myers, D. G. Kehler, B. Worm, S. J. Harley, and P. A. Doherty. 2003. Collapse and conservation of shark populations in the northwest Atlantic. *Science* 299:389–392.

Beerkircher, L., M. Shivji, and E. Cortés. 2003. A Monte Carlo demographic analysis of the silky shark (*Carcharhinus falciformis*): implications of gear selectivity. *Fish. Bull.* 101:168–174.

Beverton, R. J. H. 1992. Patterns of reproductive strategy parameters in some marine teleost fishes. *J. Fish Biol.* 41(Suppl. B):137–160.

Beverton, R. J. H. and S. J. Holt. 1957. *On the Dynamics of Exploited Fish Populations*. Chapman & Hall, New York.

Beverton, R. J. H. and S. J. Holt. 1959. A review of the life-spans and mortality rates of fish in nature, and their relationship to growth and other physiological characteristics. *Ciba Found. Colloq. Ageing* 54:142–180.

Bonfil, R. 1994. Overview of world elasmobranch fisheries. FAO Fish. Tech. Paper 341. FAO, Rome.

Bonfil, R. 1996. Elasmobranch Fisheries: Status, Assessment and Management. Ph.D. dissertation, University of British Columbia, Vancouver, Canada.

Brander, K. 1981. Disappearance of common skate *Raia batis* from Irish Sea. *Nature* 290:48–49.

Branstetter, S. 1990. Early life history implications of selected carcharhinoid and lamnoid sharks of the northwest Atlantic, in Elasmobranchs as Living Resources: Advances in the Biology, Ecology, Systematics, and the Status of the Fisheries. H. L. Pratt, Jr., S. H. Gruber, and T. Taniuchi, Eds., NOAA Tech. Rep. NMFS 90, U.S. Department of Commerce, Washington, D.C., 17–28.

Brewster-Geisz, K. K. and T. J. Miller. 2000. Management of the sandbar shark, *Carcharhinus plumbeus*: implications of a stage-based model. *Fish. Bull.* 98:236–249.

Brooks, E., E. Cortés, and C. Porch. 2002. An age-structured production model (ASPM) for application to large coastal sharks. Sust. Fish. Div. Contrib. SFD-01/02-166. NOAA Fisheries, Miami, FL.

Cailliet, G. M. 1992. Demography of the Central California population of the leopard shark (*Triakis semifasciata*). *Aust. J. Mar. Freshwater Res.* 43:183–193.

Cailliet, G. M., H. F. Mollet, G. G. Pittinger, D. Bedford, and L. J. Natanson. 1992. Growth and demography of the Pacific angel shark (*Squatina californica*), based upon tag returns off California. *Aust. J. Mar. Freshwater Res.* 43:1313–1330.

Campana, S., L. Marks, W. Joyce, P. Hurley, M. Showell, and D. Kulka. 1999. An analytical assessment of the porbeagle shark (*Lamna nasus*) population in the northwest Atlantic. Res. Doc. 99/158, Canadian Stock Assessment Secretariat, Ottawa, Canada.

Campana, S., L. Marks, W. Joyce, and S. Harley. 2001. Analytical assessment of the porbeagle shark (*Lamna nasus*) population in the northwest Atlantic, with estimates of long-term sustainable yield. Res. Doc. 2001/067. Canadian Science Advisory Secretariat, Ottawa, Canada.

Campana, S. E., W. Joyce, L. Marks, L J. Natanson, N. E. Kohler, C. F. Jensen, J. J. Mello, H. L. Pratt, Jr., and S. Myklevoll. 2002. Population dynamics of the porbeagle in the Northwest Atlantic Ocean. *North Am. J. Fish. Manage.* 22:106–121.

Casey, J. M. and R. A. Myers. 1998. Near extinction of a large, widely distributed fish. *Science* 281:690–692.

Chaloupka, M.Y. and J. A. Musick. 1997. Age, growth, and population dynamics, in *Biology of Sea Turtles*. P. L. Lutz and J. A. Musick, Eds., CRC Press, Boca Raton, FL, 233–276.

Chen, S. B. and S. Watanabe. 1989. Age dependence of natural mortality coefficient in fish population dynamics. *Nip. Suisan Gak.* 55:205–208.

Compagno, L. J. V. 1984. *Sharks of the World. An Annotated and Illustrated Catalogue of Shark Species Known to Date*. FAO Species Catalogue. Vol. 4, Parts 1 and 2. FAO Fish. Synopsis 125, FAO, Rome.

Compagno, L. J. V. 1990. Alternative life history styles of cartilaginous fishes in time and space. *Environ. Biol. Fish.* 28:33–75.

Cortés, E. 1995. Demographic analysis of the Atlantic sharpnose, *Rhizoprionodon terraenovae*, in the Gulf of Mexico. *Fish. Bull.* 93:57–66.

Cortés, E. 1998. Demographic analysis as an aid in shark stock assessment and management. *Fish. Res.* 39:199–208.

Cortés, E. 1999. A stochastic stage-based population model of the sandbar shark in the western North Atlantic, in *Life in the Slow Lane: Ecology and Conservation of Long-Lived Marine Animals.* J. A. Musick, Ed., Symposium 23. American Fisheries Society, Bethesda, MD, 115–136.

Cortés, E. 2000. Life history patterns and correlations in sharks. *Rev. Fish. Sci.* 8:299–344.

Cortés, E. 2002a. Incorporating uncertainty into demographic modeling: application to shark populations and their conservation. *Conserv. Biol.* 16:1048–1062

Cortés, E. 2002b. Stock assessment of small coastal sharks in the U.S. Atlantic and Gulf of Mexico. Sust. Fish. Div. Contrib. SFD-01/02-152. NOAA Fisheries, Panama City, FL.

Cortés, E. and G. R. Parsons. 1996. Comparative demography of two populations of the bonnethead shark (*Sphyrna tiburo*). *Can. J. Fish. Aquat. Sci.* 53:709–718.

Cortés, E., E. Brooks, and G. Scott. 2002. Stock assessment of large coastal sharks in the U.S. Atlantic and Gulf of Mexico. Sust. Fish. Div. Contrib. SFD-02/03-177. NOAA Fisheries, Panama City, FL.

De Kroon, H., A. Plaisier, J. van Groenendael, and H. Caswell. 1986. Elasticity: the relative contribution of demographic parameters to population growth rate. *Ecology* 67:1427–1431.

Deriso, R. B. 1980. Harvesting strategies and parameter estimation for an age-structured model. *Can. J. Fish. Aquat. Sci.* 37:268–282.

Dulvy, N. K., and J. D. Reynolds. 2002. Predicting extinction vulnerability in skates. *Conserv. Biol.* 16:440–450.

Euler, L. 1760. Recherches générales sur la mortalité et la multiplication du genre humain. *Mem. Acad. R. Sci. Belles Lett.* (Belgique) 16:144–164.

Fox, W. W. 1970. An exponential surplus-yield model for optimizing exploited fish populations. *Trans. Am. Fish. Soc.* 99:80–88.

Frisk, M. G., T. J. Miller, and M. J. Fogarty. 2001. Estimation and analysis of biological parameters in elasmobranch fishes: a comparative life history study. *Can. J. Fish. Aquat. Sci.* 58:969–981.

Frisk, M. G., T. J. Miller, and M. J. Fogarty. 2002. The population dynamics of little skate *Leucoraja erinacea*, winter skate *Leucoraja ocellata*, and barndoor skate *Dipturus laevis*: predicting exploitation limits using matrix analyses. *ICES J. Mar. Sci.* 59:576–586.

Garrick, J. A. F. 1982. Sharks of the Genus *Carcharhinus*. NOAA Tech. Rep. NMFS Circ. 445, U.S. Department of Commerce, Washington, D.C.

Getz, W. N. and R. G. Haight. 1989. *Population Harvesting.* Princeton University Press, Princeton, NJ.

Grant, C. J., R. L. Sandland, and A. M. Olsen. 1979. Estimation of growth, mortality and yield per recruit of the Australian school shark, *Galeorhinus galeus* (Macleay), from tag recoveries. *Aust. J. Mar. Freshwater Res.* 30:625–637.

Gruber, S. H., J. R. C. de Marignac, and J. M. Hoenig. 2001. Survival of juvenile lemon sharks at Bimini, Bahamas, estimated by mark-depletion experiments. *Trans. Am. Fish. Soc.* 130:376–384.

Harley, S. J. 2002. Statistical catch-at-length model for porbeagle shark (*Lamna nasus*) in the Northwest Atlantic. *Col. Vol. Sci. Pap. Int. Comm. Conserv. Atl. Tunas* 54:1314–1332.

Heppell, S. S., L. B. Crowder, and T. R. Menzel. 1999. Life table analysis of long-lived marine species, with implications for conservation and management, in *Life in the Slow Lane: Ecology and Conservation of Long-Lived Marine Animals.* J. A. Musick, Ed., Symposium 23, American Fisheries Society, Bethesda, MD, 137–148.

Heupel, M. R. and C. A. Simpfendorfer. 2002. Estimation of mortality of juvenile blacktip sharks, *Carcharhinus limbatus*, within a nursery area using telemetry data. *Can. J. Fish. Aquat. Sci.* 59:624–632.

Hilborn, R. and M. Mangel. 1997. *The Ecological Detective.* Princeton University Press, Princeton, NJ.

Hoenig, J. M. 1983. Empirical use of longevity data to estimate mortality rates. *Fish. Bull.* 82:898–903.

Hoenig, J. M. and S. H. Gruber. 1990. Life history patterns in the elasmobranchs: implications for fisheries management, in Elasmobranchs as Living Resources: Advances in the Biology, Ecology, Systematics, and the Status of the Fisheries. H. L. Pratt, Jr., S. H. Gruber, and T. Taniuchi, Eds., NOAA Tech. Rep. NMFS 90, U.S. Department of Commerce, Washington, D.C., 1–16.

Hoff, T. B. 1990. Conservation and Management of the Western North Atlantic Shark Resource Based on the Life History Strategy Limitations of the Sandbar Shark. Ph.D. dissertation, University of Delaware, Newark.

Hoff, T. B. and J. A. Musick. 1990. Western North Atlantic shark-fishery management problems and international requirements, in Elasmobranchs as Living Resources: Advances in the Biology, Ecology, Systematics, and the Status of the Fisheries. H. L. Pratt, Jr., S. H. Gruber, and T. Taniuchi, Eds., NOAA Tech. Rep. NMFS 90, U.S. Department of Commerce, Washington, D.C., 455–472.

Holden, M. J. 1972. Are long-term sustainable fisheries for elasmobranchs possible? *J. Cons. Int. Explor. Mer* 164:360–367.

Holden, M. J. 1974. Problems in the rational exploitation of elasmobranch populations and some suggested solutions, in *Sea Fisheries Research.* F. R. Harden-Jones, Ed., Halsted Press, New York, 117–137.

Holden, M. J. 1977. Elasmobranchs, in *Fish Population Dynamics.* J. A. Gulland, Ed., John Wiley & Sons, New York, 187–214.

Jennings, S., S. R. P. Greenstreet, and J. D. Reynolds. 1999. Structural change in an exploited fish community: a consequence of differential fishing effects on species with contrasting life histories. *J. Anim. Ecol.* 68:617–627.

Jensen, A. L. 1996. Beverton and Holt life history invariants result from optimal trade-off of reproduction and survival. *Can. J. Fish. Aquat. Sci.* 53:820–822.

Liu, K. M. and C. T. Chen. 1999. Demographic analysis of the scalloped hammerhead, *Sphyrna lewini*, in the northwestern Pacific. *Fish. Sci.* 65:218–223.

Lotka, A. J. 1907. Studies on the mode of growth of material aggregates. *Am. J. Sci.* 24:199–216.

Mangel, M. 1992. Comparative analyses of the effects of high seas driftnets on the northern right whale dolphin *Lissodelphus borealis. Ecol. Appl.* 3:221–229.

Manire, C. A. and S. H. Gruber. 1993. A preliminary estimate of natural mortality of age-0 lemon sharks, *Negaprion brevirostris*, in Conservation Biology of Elasmobranchs. S. Branstetter, Ed., NOAA Tech. Rep. NMFS 115, U.S. Department of Commerce, Washington, D.C., 65–71.

Manly, B. J. F. 1990. *Stage-Structured Populations: Sampling, Analysis, and Simulation.* Chapman & Hall, London.

Márquez, J. F. and J. L. Castillo. 1998. Fishery biology and demography of the Atlantic sharpnose shark, *Rhizoprionodon terraenovae*, in the southern Gulf of Mexico. *Fish. Res.* 39:183–198.

Márquez, J. F., J. L. Castillo, and M. C. Rodríguez de la Cruz. 1998. Demography of the bonnethead shark, *Sphyrna tiburo* (Linnaeus, 1758), in the southeastern Gulf of Mexico. *Cien. Mar.* 24:13–34.

McAllister, M. K., E. K. Pikitch, and E. A. Babcock. 2001. Using demographic methods to construct Bayesian priors for the intrinsic rate of increase in the Schaefer model and implications for stock rebuilding. *Can. J. Fish. Aquat. Sci.* 58:1871–1890.

Meyer, R. and R. B. Millar. 1999a. Bayesian stock assessment using a state-space implementation of the delay difference model. *Can. J. Fish. Aquat. Sci.* 56:37–52.

Meyer, R. and R. B. Millar. 1999b. BUGS in Bayesian stock assessments. *Can. J. Fish. Aquat. Sci.* 56:1078–1086.

Mollet, H. F. and G. M. Cailliet. 2002. Comparative population demography of elasmobranchs using life history tables, Leslie matrices and stage-based matrix models. *Mar. Freshwater Res.* 53:503–516.

Musick, J. A. 1999. Ecology and conservation of long-lived marine animals, in *Life in the Slow Lane: Ecology and Conservation of Long-Lived Marine Animals.* J. A. Musick, Ed., Symposium 23, American Fisheries Society, Bethesda, MD, 1–10.

Myers, R. A., K. G. Bowen, and N. J. Barrowman. 1999. Maximum reproductive rate of fish at low population sizes. *Can. J. Fish. Aquat. Sci.* 56:2404–2419.

Neer, J. A. and G. M. Cailliet. 2001. Aspects of the life history of the Pacific electric ray, *Torpedo californica* (Ayres). *Copeia* 2001:842–847.

Otto, R. S., J. R. Zuboy, and G. T. Sakagawa. 1977. Status of northwest Atlantic billfish and shark stocks, in Report of the La Jolla Working Group, 28 March–8 April 1977.

Paloheimo, J. E. 1961. Studies on estimation of mortalities. I. Comparison of a method described by Beverton and Holt and a new linear formula. *J. Fish. Res. Board Can.* 18:645–662.

Pauly, D. 1980. On the interrelationship between natural mortality, growth parameters, and mean environmental temperature in 175 fish stocks. *J. Cons. Int. Explor. Mer* 39:175–192.

Pella, J. J. and P. K. Tomlinson. 1969. A generalized stock production model. *Inter-Am. Trop. Tuna Comm. Bull.* 13:419–496.

Peterson, I. and J. S. Wroblewski. 1984. Mortality rates of fishes in the pelagic ecosystem. *Can. J. Fish. Aquat. Sci.* 41:1117–1120.

Pope, J. G. 1972. An investigation of the accuracy of virtual population analysis using cohort analysis. *Res. Bull. Int. Comm. Northw. Atl. Fish.* 9:65–74.

Porch, C. E. 2003. A preliminary assessment of Atlantic white marlin (*Tetrapturus albidus*) using a state-space implementation of an age-structured production model. *Col. Vol. Sci. Pap. Int. Comm. Conserv. Atl. Tunas* 55:559–527.

Pratt, H. L. and J. G. Casey. 1990. Shark reproductive strategies as a limiting factor in directed fisheries, with a review of Holden's method of estimating growth-parameters, in Elasmobranchs as Living Resources: Advances in the Biology, Ecology, Systematics, and the Status of the Fisheries. H. L. Pratt, Jr., S. H. Gruber, and T. Taniuchi, Eds., NOAA Tech. Rep. NMFS 90, U.S. Department of Commerce, Washington, D.C., 97–109.

Punt, A. E. and T. I. Walker. 1998. Stock assessment and risk analysis for the school shark *Galeorhinus galeus* (Linnaeus) off southern Australia. *Mar. Freshwater Res.* 49:719–731.

Punt, A. E., F. Pribac, T. I. Walker, B. L. Taylor, and J. D. Prince. 2000. Stock assessment of school shark *Galeorhinus galeus*, based on a spatially explicit population dynamics model. *Mar. Freshwater Res.* 51:205–220.

Quinn, T. J. and R. B. Deriso. 1999. *Quantitative Fish Dynamics.* Oxford University Press, New York.

Rago, P. J., K. A. Sosebee, J. K. T. Brodziak, S. A. Murawski, and E. D. Anderson. 1998. Implications of recent increases in catches on the dynamics of northwest Atlantic spiny dogfish (*Squalus acanthias*). *Fish. Res.* 39:165–181.

Ricker, W. E. 1954. Stock and recruitment. *J. Fish. Res. Board Can.* 11:559–623.

Ricker, W. E. 1975. Computation and interpretation of biological statistics of fish populations. *Bull. Fish. Res. Board Can.* 191.

Roff, D. A. 1992. *The Evolution of Life Histories: Theory and Analysis.* Chapman & Hall, New York.

Schaefer, M. B. 1954. Some aspects of the dynamics of populations important to the management of commercial marine fisheries. *Inter-Am. Trop. Tuna Comm. Bull.* 2:247–285.

Silva, H. M. 1983. Preliminary studies of the exploited stock of kitefin shark *Scymnorhinus licha* (Bonnaterre, 1788) in the Azores. ICES Council Meeting Papers No. ICES CM 1983/G:18, International Council for the Exploration of the Sea, Copenhagen, Denmark.

Silva, H. M. 1987. An assessment of the Azorean stock of kitefin shark *Scymnorhinus licha* (Bonnaterre, 1788) in the Azores. ICES Council Meeting Papers No. ICES CM 1987/G:66, International Council for the Exploration of the Sea, Copenhagen, Denmark.

Silva, H. M. 1993. A density-dependent Leslie matrix-based population model of spiny dogfish, *Squalus acanthias*, in the NW Atlantic. ICES Council Meeting Papers No. ICES CM 1993/G:54, International Council for the Exploration of the Sea, Copenhagen, Denmark.

Simpfendorfer, C. A. 1999a. Mortality estimates and demographic analysis for the Australian sharpnose shark, *Rhizoprionodon taylori*, from northern Australia. *Fish. Bull.* 97:978–986.

Simpfendorfer, C. A. 1999b. Demographic analysis of the dusky shark fishery in southwestern Australia, in *Life in the Slow Lane: Ecology and Conservation of Long-Lived Marine Animals.* J. A. Musick, Ed., Symposium 23, American Fisheries Society, Bethesda, MD, 149–160.

Simpfendorfer, C. A. 2000. Predicting population recovery rates for endangered western Atlantic sawfishes using demographic analysis. *Environ. Biol. Fish.* 58:371–377.

Simpfendorfer, C. A., K. Donohue, and N. G. Hall. 2000. Stock assessment and risk analysis for the whiskery shark (*Furgaleus macki* (Whitley)) in south-western Australia. *Fish. Res.* 47:1–17.

Sminkey, T. R. and J. A. Musick. 1996. Demographic analysis of the sandbar shark, *Carcharhinus plumbeus*, in the western North Atlantic. *Fish. Bull.* 94:341–347.

Smith, S. E. and N. J. Abramson. 1990. Leopard shark *Triakis semifasciata* distribution, mortality rate, yield and stock replenishment estimates based on a tagging study in San Francisco Bay. *Fish. Bull.* 88:371–381.

Smith, S. E., D. W. Au, and C. Show. 1998. Intrinsic rebound potentials of 26 species of Pacific sharks. *Mar. Freshwater Res.* 49:663–678.

Smith, S. E., D. W. Au, and C. Show. In press. Shark intrinsic rates of increase with emphasis on pelagic species, in *Pelagic Sharks.* E. K. Pikitch and M. Camhi, Eds., Blackwell Scientific, Oxford.

Stearns, S. C. 1992. *The Evolution of Life Histories.* Oxford University Press, Oxford.

Stevens, J. D. 1999. Variable resilience to fishing pressure in two sharks: the significance of different ecological and life history parameters, in *Life in the Slow Lane: Ecology and Conservation of Long-Lived Marine Animals*. J. A. Musick, Ed., Symposium 23, American Fisheries Society, Bethesda, MD, 11–15.

Walker, P. A. and J. R. G. Hislop. 1998. Sensitive skates or resilient rays? Spatial and temporal shifts in ray species composition in the central and north-western North Sea between 1930 and the present day. *ICES J. Mar. Sci.* 55:392–402.

Walker, T. I. 1992. A fishery simulation model for sharks applied to the gummy shark, *Mustelus antarcticus* Günther, from southern Australian waters. *Aust. J. Mar. Freshwater Res.* 43:195–212.

Walker, T. I. 1994a. Fishery model of gummy shark, *Mustelus antarcticus*, for Bass Strait, in *Proceedings of Resource Technology '94 New Opportunities Best Practice*. I. Bishop, Ed., Centre for Geographic Information Systems and Modelling, University of Melbourne, Melbourne, Australia, 422–438.

Walker, T. I. 1994b. Stock assessments of the gummy shark, *Mustelus antarcticus* Günther, in Bass Strait and off South Australia, in *Population Dynamics for Fisheries Management*. D. A. Hancock, Ed., Australian Society for Fish Biology Workshop Proceedings 1, Australian Government Printing Service, Canberra, 173–187.

Walker, T. I. 1995. Stock assessment of the school shark, *Galeorhinus galeus* (Linnaeus), off southern Australia by applying a delay-difference model. Report to Southern Shark Fishery Assessment Group, Marine and Freshwater Resources Institute, Queenscliff, Victoria, Australia.

Walker, T. I. 1998. Can shark resources be harvested sustainably? A question revisited with a review of shark fisheries. *Mar. Freshwater Res.* 49:553–572.

Walters, C. J. and D. Ludwig. 1994. Calculation of Bayes posterior probability distributions for key population parameters: a simplified approach. *Can. J. Fish. Aquat. Sci.* 51:713–722.

Waring, G. T. 1984. Age, growth, and mortality of the little skate off the northeast coast of the United States. *Trans. Am. Fish. Soc.* 113:314–321.

Wood, C. C., K. S. Ketchen, and R. J. Beamish. 1979. Population dynamics of spiny dogfish (*Squalus acanthias*) in British Columbia waters. *J. Fish. Res. Board Can.* 36:647–656.

Xiao, Y. and T. I. Walker. 2000. Demographic analysis of gummy shark (*Mustelus antarcticus*) and school shark (*Galeorhinus galeus*) off southern Australia by applying a generalized Lotka equation and its dual equation. *Can. J. Fish. Aquat. Sci.* 57:214–222.

16

Genetics of Sharks, Skates, and Rays

Edward J. Heist

CONTENTS

16.1 Elasmobranch Cytogenetics

The genetic code of animals including elasmobranchs is compartmentalized into two cellular organelles: the nucleus and the mitochondrion. The vast majority of DNA is found in the nucleus where it is packaged into discrete chromosomes (Futuyma, 1998). Chromosomes segregate during meiosis in germ cells, a process that ultimately leads to the formation of haploid gametes (sperm and egg). Thus, nuclear DNA exhibits biparental inheritance; each diploid parent contributes a haploid chromosome complement to form a new diploid offspring. Mitochondrial (mt) DNA is a haploid code that typically exhibits maternal inheritance in vertebrates (Futuyma, 1998) including, presumably, elasmobranchs. Every cell contains numerous mitochondria each with multiple copies of the mitochondrial genome. When an egg is fertilized, only the mtDNA derived from the female parent is retained in the developing embryo. Thus, all mtDNA in an elasmobranch is derived from a small number of copies present in the ovum.

16.1.1 Genome Sizes

The size of a nuclear genome, measured in picograms of DNA per haploid nucleus, is directly proportional to the number of base pairs in the genetic code of the organism. Shark genomes measured so far range from 3 to 34 pg (Stingo and Rocco, 2001) compared to 3.4 pg found in the human genome. With the exception of dipnoans (lungfishes) and urodeles (salamanders and allies), elasmobranchs possess the largest vertebrate genomes. Genome size varies widely among elasmobranch species, and the size of the genome does not seem to exhibit an evolutionary trend among primitive and derived forms (Stingo and Rocco, 2001; Schwartz and Maddock, 2002).

Complete mtDNA sequences have been published for small-spotted catshark, *Scyliorhinus canicula* (Delarbre et al., 1998), starspotted smooth-hound, *Mustelus manazo* (Cao et al., 1998), spiny dogfish, *Squalus acanthias* (Rasmussen and Arnason, 1999b), starry skate, *Raja radiata* (Rasmussen and Arnason, 1999a), and horn shark, *Heterodontus franscisci* (Arnason et al., 2001). The sizes of the published mitochondrial genomes are similar to those of other vertebrates, ranging from 16,707 base pairs (bp) to 16,783 bp. Gene order and arrangement of RNAs and noncoding regions is identical to that of mammals and bony fishes but differs slightly from that of sea lamprey (Lee and Kocher, 1995). Consistent with other vertebrates, elasmobranch mtDNA contains 13 uninterrupted protein-coding genes (12 of which are found on the "heavy" strand), 22 tRNAs, 2 rRNAs, and a noncoding control region or D-loop approximately 1000 to 1100 bp in length. Interestingly, the largest vertebrate mtDNA known (18,580 bp) is that of the rabbit fish, *Chimaera monstrosa*, the only holocephalan for which a complete mtDNA genome has been published (Arnason et al., 2001). The reason for the large genome is an unusually large (1672 bp) control region (Arnason et al., 2001).

16.1.2   Chromosome Complements

Among the vertebrates, fish have the least-studied chromosome complements and the chromosomes of cartilaginous fishes are not as well studied as those of bony fishes (Solari, 1994). Stingo and Rocco (2001) reported that of approximately 1100 species of Selachii (elasmobranchs and holocephalans) karyotypes have been described for only 63 species. The limited data present indicate that relative to other vertebrates, elasmobranches possess large genomes comprising a large number of chromosomes, some of which are very small in size. Chromosome counts in elasmobranchs range from 28 to 106 chromosomes in a full diploid complement (Stingo and Rocco, 2001), and some elasmobranch chromosomes are so small that they are near the limit of resolution of the light microscope (Maddock and Schwartz, 1996). Thus, discrepancies among authors in the chromosome counts for particular species can arise through differences in the ability to resolve the presence of tiny "microchromosomes." Stingo and Rocco (2001) surmised that poyploidy played an important role in the evolution of elasmobranchs and that the evolutionary trend from primitive (e.g., Hexanchidae) to more derived (e.g., Carcharhinidae) forms was a reduction in the number of telocentric chromosomes with fusion into a smaller number of metacentric chromosomes accompanied by a loss of microchromosomes. This trend is apparent both among superorders and within superorders; for example, Galeomorphs tend to have fewer larger chromosomes than Squalomorphs while within the galeomorphs the primitive horn sharks (Heterodontidae) have a larger number of shorter chromosomes than do more advanced requiem sharks (Carcharhinidae) and hammerheads (Sphyrnidae). There is also an evolutionary trend toward a reduction in the quantity of AT-rich DNA, which is presumably associated with repetitive noncoding regions in more advanced forms (Stingo et al., 1989; Stingo and Rocco, 2001).

16.1.3   Sex-Determination

In many gonochoristic species the separate sexes have morphologically distinguishable chromosome complements that can be used to infer the genetic mechanism of sex determination. Often one sex possesses a matched set of chromosomes (i.e., it is homogametic) whereas the other has one single chromosome or one pair of unmatched chromosomes (heterogametic). In mammals males are heterogametic (XY) whereas female birds are the heterogametic (WZ) sex. Fishes exhibit XY, WZ, and at least six other chromosomal sex-determining systems as well as several varieties of hermaphroditism (Tave, 1993). To date, there has been very little investigation into the sex-determining mechanisms in elasmobranchs. Maddock and Schwartz (1996) determined that two species of guitarfish (*Rhinobatus* sp.) exhibited XY sex determination and found evidence of male heterogamy in white shark, *Carcharodon carcharias*, Atlantic sharpnose shark, *Rhizoprionodon terranovae*, blacknose shark, *Carcharhinus acronotus,* and blacktip shark, *C. limbatus,* and evidence of female heterogamy in southern stingray, *Dasyatis americana.* They suggested that male heterogamy is the predominate sex-determining mechanism in elasmobranchs. Castro (1996) described a hermaphroditic blacktip shark and stated that hermaphroditism in elasmobranchs is very rare.

16.1.4 Genetic Oddities

Albinism has been reported in at least nine species of elasmobranch (Smale and Heemstra, 1997, and references therein). To date, there have been no documented cases of hybridization or parthenogenesis in elasmobranchs, although the morphological similarity of many species that occur in sympatry might make recognition of hybrids in the field problematic. Recent tales of captive birth by female elasmobranchs housed alone or with other species are almost certainly due to long-term sperm storage. To date, the only record of triploidy in an elasmobranch is for a nurse shark by Kendall et al. (1994).

16.2 Population Genetics, Stock Structure, and Forensics

16.2.1 Molecular Markers

Since the development of isozyme electrophoresis in the 1960s, molecular markers have increasingly been used to partition genetic variation among species and to define the presence of multiple discrete units (stocks) within species (for a historical review see Utter, 1991). During the 1980s and 1990s analysis of mtDNA restriction fragment length polymorphisms (RFLP) was popular, fueled by the compact size and therefore manageability of the mitochondrial genome coupled with the ability to isolate mtDNA away from the much larger and more complex nuclear genome (Avise, 1994). The revolution of polymerase chain reaction (PCR), which started in the late 1980s and has continued to this day, has provided access to specific segments of DNA. As more is learned about the nuclear genomes of organisms, techniques that explore nuclear DNA, including analysis of highly polymorphic major histocompatibility complex (MHC) genes and DNA microsatellites, have provided the resolution to go beyond the species and population level and to examine genetic traits at the level of the family and the individual.

16.2.1.1 Isozymes and Allozymes — Isozymes are enzymes with similar catalytic properties that differ in the rate of migration in an electric field and can thus be resolved as discrete zones of activity on an electrophoretic medium (e.g., starch gel or cellulose acetate plate; Murphy et al., 1996). Although some isozymes are the result of products at different gene loci, allozymes are a subset of isozymes that possess allelic variation at a single locus. During the 1970s studies of allozymes in *Drosophila* and other organisms demonstrated that natural populations contained far more genetic variation than was previously assumed and ultimately led to the development of the neutral theory of molecular evolution, which stated, briefly, that the majority of genetic variation found at the molecular level is selectively neutral and thus subject to such random forces as genetic drift (Futuyma, 1998).

In the first published study of allozymes in elasmobranchs, Smith (1986) reported that variation in allozymes (as indicated by mean heterozygosity and the percentage of loci that are polymorphic) is low in sharks. In that study mean heterozygosity ranged from 0.001 in spotted estuary smoothhound, *Mustelus lenticulatus,* to 0.037 in blue shark, *Prionace glauca.* In a review paper Ward et al. (1994) reported a mean heterozygosity of 0.064 for marine fishes. Low levels of allozyme heterozygosity were subsequently reported in gummy shark, *M. antarcticus* (mean heterozygosity = 0.006), by MacDonald (1988), and in sandbar shark by Heist et al. (1995) (mean heterozygosity = 0.005). Larger amounts of intraspecific variation were observed in two species of *Carcharhinus* (*C. tilstoni* and *C. sorrah*) by Lavery and Shaklee (1989) (mean heterozygosity 0.037 and 0.035, respectively), and in Pacific angel sharks (*Squatina* sp.) (mean heterozygosity 0.056) by Gaida (1997).

The amount of genetic variation present within a species is a function of the mutation rate, the long-term effective population size of the organism, natural selection, and with some markers (e.g., allozymes) the ability of different research protocols (e.g., number of buffer systems employed, separatory media employed, experience and skills of researchers). For example, MacDonald (1988) detected variation in only 1 of 32 presumed allozyme loci with a mean heterozygosity of 0.006 in gummy shark from the waters of southern Australia. Gardner and Ward (1998) found variation in 7 of 28 presumed loci with a mean heterozygosity of 0.099 in the same species from many of the same locations. Although the latter study scored polymorphism at four loci not surveyed by MacDonald, they also detected variation at two loci MacDonald scored as monomorphic and suggested that their use of additional buffer systems afforded them greater resolution.

16.2.1.2   *Mitochondrial DNA* — After allozymes, the next type of molecular marker to be widely used for determining stock structure of fishes was mtDNA. The reasons for the use of mtDNA rather than nuclear DNA have to do with the compact size of mtDNA relative to nuclear DNA (see above) coupled with the ability to isolate mtDNA from nuclear DNA. Although mitochondrial genes tend to evolve more rapidly than nuclear genes (Brown, 1979), there is evidence that mtDNA evolves more slowly in sharks than in mammals (Martin et al., 1992; Martin, 1995). The first studies of mtDNA employed whole-molecule analysis of RFLPs. As "universal" PCR primers for mtDNA genes were developed (Kocher et al., 1989), smaller fragments of PCR-amplified mtDNA were analyzed using either RFLP or direct sequencing. Studies that employed RFLP analysis of whole-molecule mtDNA include those of Heist et al. (1995; 1996a,b). Sandbar shark, *Carcharhinus plumbeus* (Heist et al., 1995), and Atlantic sharpnose shark (Heist et al., 1996b) did not exhibit significant stock structure between the Gulf of Mexico and U.S. Atlantic coasts. Mitochondrial haplotypes in shortfin mako, *Isurus oxyrinchus*, showed significant frequency differences between ocean basins (Heist et al., 1996a; Schrey and Heist, in press); however, there was no strong phylogeographic signal to indicate that stocks were completely isolated.

Another way to analyze mtDNA variation is to use the PCR to amplify a segment of mtDNA and use either direct sequencing or RFLP to score variation. Pardini et al. (2001) found that white shark, *Carcharodon carcharias*, mtDNA haplotypes from South Africa, Australia, and New Zealand clustered into two highly divergent clades. One clade was found only in 48 of 49 individuals surveyed in Australia and New Zealand while the other clade was found in 39 of 40 individuals from South Africa and in 1 of the 49 individuals surveyed in the Australia/New Zealand sample. These data indicate that in terms of female migration white shark populations are highly structured and stand in strong contrast to the higher levels of gene flow in shortfin mako (Heist et al., 1996a). Current work based on sequencing the mitochondrial D-loop in blacktip shark young-of-the-year from continental nursery areas in the Atlantic, Gulf of Mexico, and Caribbean show significant mtDNA haplotype frequency differences among widespread nurseries (Figure 16.1), indicating that at least females of this species segregate into reproductive stocks (Keeney et al., 2002). Kitamura et al. (1996) examined mtDNA sequence variation in ten bull sharks, *Carcharhinus leucas*, collected from widespread locations including some freshwater specimens from Mexico, Nicaragua, and Northern Australia. Eight haplotypes were detected but sample sizes were too small to estimate stock structure or gene flow.

16.2.1.3   *Nuclear DNA* — The nuclear genomes of vertebrates are far larger and more complex than the mitochondrial genomes. Most segments of nuclear DNA evolve very slowly and thus exhibit very little intra- and interspecific variation. Because of the combination of size, complexity, and low variation in the nuclear genome, most studies of population genetics and systematics in elasmobranchs have utilized mitochondrial data. However, as more is learned about the makeup of vertebrate nuclear genomes and as nuclear entities with higher levels of variation are characterized, more studies are employing nuclear data. Types of nuclear markers that have been employed to study elasmobranchs include ribosomal internal transcribed spacers (ITS), microsatellites, and MHC genes.

Among the most conserved nuclear genes in vertebrates are the 5.8S, 18S, and 28S, ribosomal RNA (rRNA) genes, which are found as multiple copies of a single long transcript of all three conserved genes separated by more polymorphic ITS segments. Because rRNA gene sequences are so highly conserved, PCR primers developed in one species have very broad taxonomic utility and can be used to amplify the more variable ITS regions. Using a combination of conserved PCR primers located in the 5.8S and

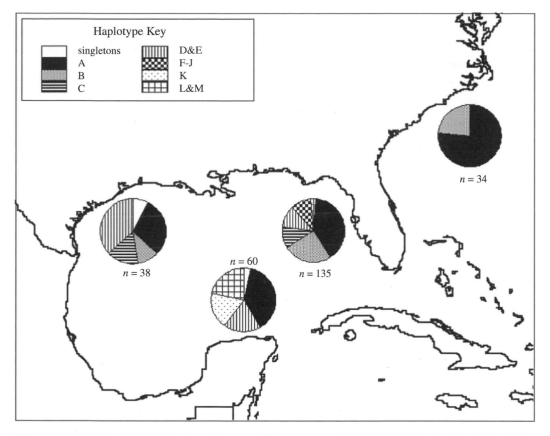

FIGURE 16.1 Mitochondrial DNA haplotype frequency differences among blacktip sharks (*Carcharhinus limbatus*) from continental nursery areas in South Carolina, Florida, Texas, and Yucatan indicating biologically and statistically significant partitioning of genetic variation ($F_{ST} = 0.152$, $p < 0.0001$).

28S rRNA genes and species-specific primers in the ITS2 region, Pank et al. (2001) and Shivji et al. (2002) produced forensic tools for the identification of shark species (see Section 16.2.3).

Microsatellites are short repetitive segments of DNA, e.g., $(GT)_n$ and $(GA)_n$, where n refers to the number of repeats of the core motif, that are highly variable for the number of repeats and hence the size of a PCR fragment that is produced by primers flanking the specific repeat (Ashley and Dow, 1994; Wright and Bentzen, 1994; O'Connell and Wright, 1997). Microsatellites are among the most polymorphic markers yet developed, with many loci possessing more than 20 alleles and heterozygosities exceeding 95%. Microsatellites are useful for studies of population genetics; however, they tend to underestimate genetic divergence among populations because of the large amount of variation and high rate of homoplasy (Balloux et al., 2000). To date, polymorphic microsatellite loci have been developed in sandbar shark (Heist and Gold, 1999b), white shark (Pardini et al., 2000), lemon shark, *Negaprion brevirostris* (Feldheim et al., 2001a,b), shortfin mako (Schrey and Heist, 2002), and nurse shark, *Ginglymostoma cirratum* (Heist et al., 2003). Development of microsatellite loci can be a difficult and time-consuming process, but once a set of primers is developed in one species they often retain utility in closely related species. For example, of the five polymorphic microsatellite loci developed in shortfin mako by Schrey and Heist (2002) all were polymorphic in porbeagle, *Lamna nasus*, and salmon shark, *L. ditropis*, and two were polymorphic in white shark and common thresher, *Alopias vulpinus*.

MHC genes are, with more than 100 alleles in some species, the most highly polymorphic markers known in vertebrates (Potts and Wakeland, 1990). MHC genes have received considerable study in elasmobranchs owing to the presumed basal location of elasmobranchs in the lineage that includes bony fishes and tetrapods coupled with the lack of MHC genes in jawless fishes (Bartl, 1998; Flajnik et al.,

1999). Thus, elasmobranchs are an important group for studying the evolution of immunity in vertebrates. Variation at MHC loci is typically scored via amplification of a particular locus using primers designed in conserved regions flanking the highly polymorphic antigen-binding cleft. This is followed by digesting with a restriction enzyme that cuts the products of both alleles into a population of DNA fragments, the sizes of which are resolved via gel electrophoresis. By comparing the patterns produced by parents and offspring the DNA restriction fragments associated with individual alleles can be resolved. The high allelic diversity makes each MHC locus potentially more powerful than a single microsatellite locus for studies of relatedness and paternity (see Section 16.3). The first documentation of multiple paternity in elasmobranchs was an unexpected outcome of a study of gene linkage of MHC loci in a litter of nurse sharks (Ohta et al., 2000). However, given the limited number of loci available and the difficulty in resolving individual alleles, microsatellites are ultimately the more powerful marker for many applications. Variation at MHC loci appear to be maintained by balancing selection (Edwards and Hedrick, 1998). Thus, population genetics models that assume neutrality may not be suitable for analysis of MHC data.

16.2.2 Measuring Stock Structure with Molecules

Since the development of allozyme electrophoresis in the 1960s, molecular markers have been increasingly used to determine stock structure in fishes including elasmobranchs (Utter, 1991). The three most commonly applied markers for estimating stock structure are allozymes, mtDNA, and microsatellites (Ward, 2000). When a species is divided into multiple reproductively isolated populations, the evolutionary forces of mutation and genetic drift cause frequencies of neutral alleles to change such that over time significant differences in gene frequencies develop. These disruptive forces are countered by migration, which has a tendency to homogenize allele frequencies throughout the range of the species. Once equilibrium has been achieved between the disruptive forces of mutation and drift and the homogenizing force of migration, the magnitude of the variance in allele frequencies among geographic units, as determined by various estimators of Wright's (1969) F_{ST}, is indicative of the reproductive isolation, and hence stock structure, of the units involved. If we can assume that the rate at which new mutations spread by migration is large relative to the rate at which new mutations arise in isolated populations, mutation can be effectively ignored and F_{ST} is a function of migration and drift (Ward and Grewe, 1994). Generally F_{ST} values of less than 0.05 indicate little genetic differentiation whereas F_{ST} values greater than 0.15 indicate great genetic differentiation (Hartl and Clark, 1997).

Although it may seem counterintuitive, the magnitude of F_{ST} among locations is determined not by the rate of migration among regions but by the absolute number of migrants, abbreviated by $N_e m$, which stands for the product of the effective population size (N_e) and the migration rate (m). The reason for this relationship is that the rate at which populations diverge due to genetic drift is inversely proportional to population size (N_e), and therefore smaller populations require a larger migration rate to arrive at the same F_{ST} as larger populations with a smaller migration rate.

Under the island model of migration, which assumes that multiple same-sized populations exist with an equal rate of exchange among all populations, the relationship between F_{ST} and $N_e m$ is

$$F_{ST} \approx \frac{1}{4N_e m + 1} \tag{16.1}$$

This relationship has been widely used (and abused) (Neigel, 2002) in estimating the degree of reproductive isolation among fishery stocks. Given the unrealistic assumptions that accompany this model (large number of populations with constant equal migration among the populations, equilibrium between migration and drift, etc.), this equation should really be considered an approximation rather than an absolute measure of migration. The above relationship holds only for nuclear genes, which are diploid and biparentally inherited. For the haploid maternally inherited mitochondrial DNA the relationship is

$$F_{ST} \approx \frac{1}{2N_e m_f + 1} \tag{16.2}$$

where $N_e m_f$ refers to the number of female migrants. Thus, estimates of gene flow based on nuclear markers indicate movement by both (or either) sex and will tend to indicate the pattern of gene flow caused by the most dispersive sex. Conversely, mitochondrial DNA reflects only the movements of females and, in situations in which females are philopatric while males are more likely to roam, there can be large discrepancies in the estimates of gene flow and stock structure based on nuclear and mitochondrial markers (Pardini et al., 2001).

Mitochondrial markers have some decided advantages over nuclear markers in estimating stock structure. In species with equal levels of male and female-mediated gene flow the magnitude of F_{ST} for mitochondrial markers is larger than that of nuclear markers (Figure 16.2); thus, there can be greater statistical power for detecting nonzero F_{ST} values (see below). Furthermore, it is easier to unambiguously interpret the magnitude of genetic divergence among mtDNA haplotypes based on the number of nucleotide substitutions between haplotypes than it is for many kinds of nuclear DNA data (e.g., allozymes and microsatellites) where two alleles that appear very different may differ by only a single mutation.

There are some significant difficulties in the use of gene frequencies to estimate stock structure in highly motile species (such as many elasmobranchs) that live in environments with few barriers to migrations (i.e., the seas) (Waples, 1998). Figure 16.2 demonstrates the relationship between F_{ST} and $N_e m$ (or $N_e m_f$) for nuclear and mitochondrial markers. At levels of low gene flow (i.e., less than one individual per generation) the relationship between a measured estimate of F_{ST} and an inferred level of gene flow is robust, as F_{ST} values are large and modest errors in the measurement of F_{ST} result in only small changes in the estimate of $N_e m$. However, at greater levels of gene flow F_{ST} approaches zero and small errors in the measurement of F_{ST} produce large errors in the estimate of $N_e m$. Furthermore, it is very difficult to interpret the meaning of a small but statistically significant (nonzero) F_{ST}. Small factors in the sampling regimen (e.g., collection of related individuals within samples or variation in gene frequencies among sampling years) coupled with statistically powerful tests of genetic homogeneity can produce significant F_{ST} values that are not representative of the long-term genetic structure of populations. Highly polymorphic microsatellite loci tend to underestimate F_{ST} because of the high degree of homoplasy (i.e., two alleles that are identical in state but derived from different ancestors). The maximum value that F_{ST} can assume is equal to homozygosity (Hedrick, 1999), which for many highly polymorphic microsatellite loci is less than 0.05.

The inability to resolve stock structure in the presence of moderate amounts of gene flow in the marine environment may lead to improper management and conservation practices. Studies that employ small sample sizes and/or markers with low levels of variation may result in a failure to reject the null hypothesis of a single stock (type II error) when in fact multiple stocks do exist. Managing multiple, largely independent stocks as a single stock may be more injurious to the resource than managing for multiple stocks. Thus, Dizon et al. (1995) recommend evaluating the consequences of type I and type II errors and perhaps lowering the rejection criterion (alpha) to balance the risks associated with both types of

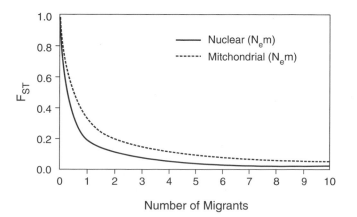

FIGURE 16.2 Relationship between F_{ST} and $N_e m$ for nuclear loci and $N_e m_f$ for mitochondrial loci.

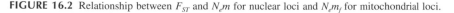

error. For such a strategy to be effective, calculations of statistical power associated with hypothesis testing should be employed.

16.2.3 Forensic Identification and Cryptic Species

Many elasmobranch genera include multiple species that are morphologically similar to one another. This has caused considerable confusion in species identification for management and scientific purposes. Thus, new species of elasmobranchs are continually being described as subtle differences in morphology and/or genetic differences are detected within nominal species. The tools of molecular genetics, including allozymes and DNA sequencing, have provided aids for identifying specimens (Shivji et al., 2002) and for identifying the presence of cryptic species (Lavery and Shaklee, 1991).

In diploid organisms, allozymes exhibit disomic inheritance, meaning that each individual inherits two alleles, one from each parent. When a locus exhibits multiple alleles in a large outbreeding population, the mixture of homozygotes and heterozygotes typically conform to Hardy–Weinberg equilibrium (i.e., the frequency of heterozygotes is equal to twice the product of the individual allele frequencies) (Futuyma, 1998). A deficit of heterozygotes indicates that two or more groups within the population are reproductively isolated. Thus, it is easier to identify the presence of cryptic species when two similar species occur in sympatry by using molecular genetics to identify the presence of reproductive isolation. It is more difficult to decide whether allopatric forms that are morphologically similar constitute discrete species, although molecular genetics can provide estimates of the genetic divergence that can be used in concert with morphological data to assign species status.

Deviations from Hardy–Weinberg expectations and especially a complete lack of heterozygotes has been used to detect the presence of cryptic species of elasmobranchs. Solé-Cava et al. (1983) confirmed that two morphotypes of Brazilian angel shark (*Squatina* sp.) were distinct species based on fixed allelic differences at two esterase loci. Later, Solé-Cava and Levy (1987) identified a third, less common species. Similarly, Lavery and Shaklee (1991) showed that two morphs of "blacktip shark" in the waters of northern Australia were heterospecific. Although it had previously been assumed that all blacktip sharks in northern Australia were the widely distributed *Carcharhinus limbatus*, Lavery and Shaklee (1991) determined that *C. limbatus* was rare in northern Australia and that the common species was actually *C. tilstoni*, as suggested from a previous morphological study (Stevens and Wiley, 1986). Gardner and Ward (2002) examined mtDNA RFLPs, allozymes, and morphology in 550 specimens of *Mustelus* from Australia and concluded that in addition to the two named species (*M. lenticulatus* and *M. antarcticus*) two additional species were present. In a study that included all known species of thresher sharks (*Alopias*), Eitner (1995) suggested the presence of an unrecognized species based on fixed allelic differences among individuals in a group of specimens identified as *A. superciliosis* by fishers in Baja California. Unfortunately, the specimens were not retained so no morphological comparison could be made, nor are there any published DNA sequence data that would support the presence of an undescribed thresher shark. Thus, the status of this "unrecognized species" remains unsettled. Undoubtedly, additional species of elasmobranchs will continue to be identified and molecular genetics will play a large role in providing evidence for species discrimination. However, given the propensity of misidentification of elasmobranch specimens, accurate collection of morphological data and retention of voucher specimens are crucial.

DNA sequences have proved useful for forensically identifying elasmobranchs and may become useful tools for identifying fins, carcasses, and other shark parts. Because heads and fins are typically removed at sea by commercial fishers, it is very difficult to identify whether prohibited species are present in the landed catch. Smith and Benson (2001) used isoelectric focusing, a protein-based technique, to demonstrate that 40% of the shark filets labeled as *M. lenticulatus* in New Zealand were actually other species and in some cases prohibited species. In the United States, landings of dusky shark, *Carcharhinus obscurus*, are currently prohibited while landings of very similar species (e.g., sandbar and bignose, *C. altimus,* sharks) are allowed. Heist and Gold (1999a) provided a diagnostic means of discriminating among the most commonly utilized species of *Carcharhinus* in the U.S. Atlantic large coastal shark fishery through the use of PCR RFLP. A segment of the mitochondrial cytochrome *b* gene was amplified and digested with a panel of seven restriction enzymes. A unique restriction profile was generated for

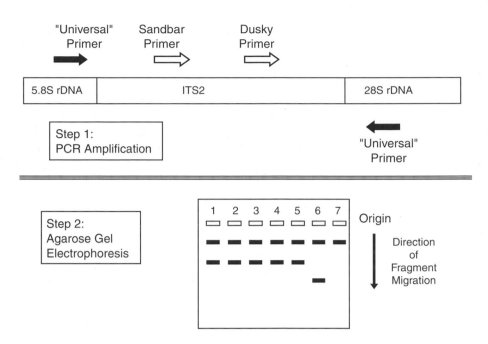

FIGURE 16.3 Protocol employed by Pank et al. (2001) for producing species-specific fragment profiles to distinguish between sandbar shark (*Carcharhinus plumbeus*) and dusky shark (*C. obscurus*). Results in lanes 1 to 5 indicate that the unknown tissue came from sandbar shark, lane 6 is a dusky shark, lane 7 is tissue from another species.

each species. Pank et al. (2001) described an innovative approach for distinguishing among sandbar and dusky sharks without the added time and expense of digesting with restriction enzymes (Figure 16.3). By using a species-specific primer that matches the DNA sequence of only one species between two "universal" primers that amplify a larger "positive control" fragment in many species, a species-specific PCR fragment profile is produced. Shivji et al. (2002) expanded this approach to show that six species of sharks (shortfin mako; longfin mako, *I. paucus*; porbeagle; dusky; silky, *C. falciformis*; and blue, *Prionace glauca*) could be distinguished from each other and from all but one other species likely to be encountered in North Atlantic fisheries (dusky sharks could not be distinguished from oceanic whitetip, *C. longimanus*).

16.3 Molecular Ecology

The development of highly polymorphic molecular markers has fostered a revolution in studies of ecology and evolutionary biology. The high resolution provided by these markers allows for assessments of relatedness among individuals to determine familial relationships within populations to determine, for example, how many fathers sired a litter of offspring, whether related individuals associate or cooperate, and which adults in a population are successful breeders. Results of various studies have either confirmed or contradicted data from field observations of mating behavior concerning fidelity to mates and reproductive success of dominant individuals.

16.3.1 Philopatry and Sex-Biased Dispersal

Philopatry, or the tendency of an animal to return to or stay in a particular location, has been confirmed in several elasmobranchs based on tracking and telemetry data (Hueter et al., in press). Strong reproductive philopatry has been demonstrated based on tagging and field observations in lemon (Feldheim et al., 2002) and nurse sharks (Pratt and Carrier, 2001). Heupel and Hueter (2001) found that after juvenile blacktip sharks from southwest Florida left their nursery area for the winter they faithfully returned each of the following two summers. Whether either of these three species returns to its natal nursery area for mating or parturition has yet to be established. The best-studied examples of reproductive philopatry in animals involve various species of anadromous salmonids. The strong natal philopatry exhibited by many salmonid fishes is accompanied by significant differences in gene frequencies among drainages and even among seasonal runs within the same tributary (Allendorf and Waples, 1996). To date, no genetic studies of elasmobranchs have indicated significant gene frequency differences across such small scales. Feldheim et al. (2001b) found only slight differences in microsatellite allele frequencies in lemon sharks between the Bahamas and Brazil, indicating a significant amount of gene flow in this philopatric species. Keeney et al. (2002) found significant differences in mtDNA haplotype frequencies among blacktip shark nursery areas from South Carolina, Florida, Texas, and the Yucatan (see Figure 16.1). However, nursery areas separated by tens of kilometers in Florida did not exhibit significant differences. Perhaps the smallest scale over which stock structure was indicated in elasmobranchs is the study of Gaida (1997) of Pacific angel sharks. The mean F_{ST} value of 0.085 among California's Channel Islands separated by a distance of less than 100 km was attributed to the tendency of angel sharks to remain in less than 100 m water depth and the deep (greater than 500 m) channel between islands.

Differences between maternally inherited mitochondrial markers and nuclear encoded loci have been used to demonstrate sex-biased dispersal and female philopatry in a variety of marine taxa including marine mammals (Palumbi and Baker, 1994; Lyrholm et al., 1999; Gladden et al., 1999) and sea turtles (Karl et al., 1992; Bowen and Karl, 1997). Several species of sea turtles exhibit significant mtDNA differences among nesting beaches, indicating strong female natal philopatry, accompanied by much lower levels of divergence in nuclear markers, indicating considerable male-mediated gene flow (Bowen and Karl, 1997). Humpback, *Megaptera novaengliae* (Palumbi and Baker, 1994), beluga, *Delphinapterus leucas* (Gladden et al., 1999), and sperm, *Physeter macrocephalus* (Lyrholm et al., 1999), whale populations also exhibit higher levels of genetic structure in mitochondrial than nuclear markers, indicating female natal philopatry and male dispersal.

Several recent studies of elasmobranchs indicate sex-biased dispersal. The high degree of mtDNA divergence among white sharks between Australia/New Zealand and South Africa detected by Pardini et al. (2001) (see above) was accompanied by data from five microsatellite loci that exhibited no significant difference in allele frequency. The most obvious explanation for this discrepancy is that female white sharks do not travel far; thus, genetic drift operating on maternally inherited mtDNA has resulted in significant structure, while males do occasionally move great distances and in doing so homogenize allele frequencies at nuclear microsatellite (loci). Similarly the moderate levels of mtDNA divergence in shortfin mako reported by Heist et al. (1996a) are accompanied by a lack of divergence at microsatellite loci (Schrey and Heist, 2003) and the significant difference in mtDNA haplotype frequency among blacktip shark nursery areas are 50 times as large as the differences at microsatellite loci (Keeney et al., 2002). Differences between estimates of gene flow based on nuclear and mitochondrial markers do not necessarily imply sex-specific dispersal as the differences in the rate of genetic drift and the high mutation rate and high degree of homoplasy in microsatellite data can produce very different values for F_{ST} under several scenarios (Buonaccorsi et al., 2001). However, given the magnitude of the differences in the studies referenced above, it appears that female fidelity accompanied by male dispersal is emerging as a common pattern of elasmobranch population structure. Perhaps the similarities in reproductive biology between viviparous sharks, sea turtles, and whales, notably internal fertilization followed by parturition or egg-laying in nursery areas temporally and spatially removed from mating areas, are responsible for the similar patterns in genetic structure. Conclusions based on nuclear and mitochondrial data do not always disagree. Gardner and Ward (1998) found concordant differences in mtDNA and allozymes frequencies in gummy sharks from Australia, consistent with the segregation of both males and females into multiple stocks.

16.3.2 Parentage and Multiple Paternity

The extremely high levels of genetic variation provided by DNA microsatellite and MHC loci make it possible to determine whether a single clutch of offspring was sired by one or more fathers. In a study of linkage relationships among MHC loci in sharks, Ohta et al. (2000) observed that at least four fathers must have sired a litter of 17 nurse sharks. Saville et al. (2002) independently discovered that in another clutch of 32 nurse shark pups the number of MHC genotypes required a minimum of four fathers. The high resolution of multiple microsatellite loci has the potential to provide even greater power for determining the number of sires. Using six microsatellite loci, Heist et al. (2002) increased the minimum number of fathers in the clutch examined by Saville et al. (2002) to five and found that two additional litters had a minimum of five and six sires. Estimates of the number of sires in all three litters based on a simulation method of DeWoody et al. (2000), which assumes equal contributions from each father, ranged from 9 to 15 sires per clutch. With a sufficiently large number of microsatellite loci it will be possible to assign all of the pups to full-sib (same mother and father) and half-sib (same mother, different father) groups. Microsatellites have also been used to demonstrate philopatry and multiple paternity in lemon sharks. Feldheim et al. (2001a) detected the presence of multiple paternity in a single clutch of lemon sharks, and later Feldheim et al. (2002) showed that between two and four fathers sired two additional litters of lemon sharks.

Just as multiple microsatellite loci can be used to determine relatedness of pups within a litter, they are very powerful for matching offspring with parents. Feldheim et al. (2002) were able to identify the female parent of 89 subadult lemon sharks in Bimini Lagoon and the male parent of an additional 15 lemon sharks by matching the sharing of alleles at nine highly polymorphic microsatellite loci. The presence of sharks of different age classes with the same maternal parent indicated that female lemon sharks return to Bimini in alternate years to pup, a confirmation of previous observational data.

16.4 Summary

The following points summarize the state of our understanding of elasmobranch genetics and serve to identify critical areas for future research:

- Elasmobranchs possess large nuclear genomes comprising a large number of small chromosomes. There is an evolutionary trend to reduction in chromosome number and increase in chromosome size in more derived forms. Elasmobranch mitochondrial genomes are similar to those of fish and mammals.

- Elasmobranchs exhibit both XY and WZ sex-determining systems, and perhaps other mechanisms as well.

- The most common molecular markers used to study elasmobranchs are allozymes, mtDNA RFLP and sequencing, and nuclear MHC, ITS, and microsatellites. Sharks tend to have relatively low levels of allozyme variation and a reduced rate of mtDNA evolution compared to mammals.

- Because many elasmobranchs live in environments with few barriers, migration is expected to homogenize gene frequencies across vast distances. High levels of gene flow make detection of stock structure a challenge. Nevertheless, there is evidence of significant stock structure in several species.

- Molecular markers are useful tools for forensic identification of elasmobranch tissues and for detection and confirmation of previously unrecognized species.

- Several species of elasmobranchs show evidence of sex-biased dispersal, with males moving more than females.

- Highly polymorphic molecular markers indicate the presence of multiple paternity in elasmobranchs and can be used to determine paternity and relatedness among individuals.

References

Allendorf, F. W., and R. S. Waples. 1996. Conservation and genetics of salmonid fishes, in *Conservation Genetics: Case Histories from Nature.* J. C. Avise, Ed., Chapman & Hall, New York, 238–280.

Arnason, U., A. Gullberg, and A. Janke. 2001. Molecular phylogenetics of gnathostomous (jawed) fishes: old bones, new cartilage. *Zool. Scr.* 30:249–255.

Ashley, M. V. and B. D. Dow. 1994. The use of microsatellite analysis in population biology: background, methods and potential applications, in *Molecular Ecology and Evolution: Approaches and Applications.* B. Schierwater, B. Streit, and R. DeSalle, Eds., Birkhauser Verlag, Basel, Switzerland, 185–201.

Avise, J. C. 1994. *Molecular Markers, Natural History and Evolution.* Chapman & Hall, New York.

Balloux, F., H. Brunner, N. Lugon-Moulin, J. Hausser, and J. Goudet. 2000. Microsatellites can be misleading: An empirical and simulation study. *Evolution* 54:1414–1422.

Bartl, S. 1998. What sharks can tell us about the evolution of MHC genes. *Immunol. Rev.* 166:317–331.

Bowen, B. W. and S. A. Karl. 1997. Population genetics, phylogeography, and molecular evolution, in *The Biology of Sea Turtles.* P. L. Lutz and J. A. Musick, Eds., CRC Press, Boca Raton, FL, 29–50.

Brown, W. M. 1979. Rapid evolution of animal mitochondrial DNA. *Proc. Natl. Acad. Sci. U.S.A.* 78:1967–1971.

Buonaccorsi, V. P., J. R. McDowell, and J. E. Graves. 2001. Reconciling patterns of inter-ocean molecular variance from four classes of molecular markers in blue marlin (Makaira nigricans). *Mol. Ecol.* 10:1179–1196.

Cao, Y., P. J. Waddell, N. Okada, and M. Hasegawa. 1998. The complete mitochondrial DNA sequence of the shark *Mustelus manazo*: evaluating rooting contradictions to living bony vertebrates. *Mol. Biol. Evol.* 15:1637–1646.

Castro, J. I. 1996. Biology of the blacktip shark, *Carcharhinus limbatus*, off the southeastern United States. *Bull. Mar. Sci.* 59:508–522.

Delarbre, C., N. Spruyt, C. Delmarre, C. Gallut, V. Barriel, P. Janvier, V. Laudet, and G. Gachelin. 1998. The complete nucleotide sequence of the mitochondrial DNA of the dogfish, *Scyliorhinus canicula.* *Genetics* 150:331–344.

DeWoody, J. A., Y. D. DeWoody, A. C. Fiumera, and J. C. Avise. 2000. On the number of reproductives contributing to a half-sib progeny array. *Genet. Res.* 75:95–105.

Dizon, A. E., B. L. Taylor, and G. M. O'Corry-Crowe. 1995. Why statistical power is necessary to link analyses of molecular variation to decisions about population structure, in *Evolution and the Aquatic Ecosystem.* J. L. Neilson and D. A. Powers, Eds., American Fisheries Society, Bethesda, MD.

Edwards, S. V. and P. W. Hedrick. 1998. Evolution and ecology of MHC molecules: from genomics to sexual selection. *Trends Ecol. Evol.* 13:305–311.

Eitner, B. J. 1995. Systematics of the genus *Alopias* (Lamniformes: Alopiidae) with evidence for the existence of an unrecognized species. *Copeia* 1995:562–571.

Feldheim, K. A., S. H. Gruber, and M. V. Ashley. 2001a. Multiple paternity of a lemon shark litter (Chondrichthyes : Carcharhinidae). *Copeia* 2001:781–786.

Feldheim, K. A., S. H. Gruber, and M. V. Ashley. 2001b. Population genetic structure of the lemon shark (*Negaprion brevirostris*) in the western Atlantic: DNA microsatellite variation. *Mol. Ecol.* 10:295–303.

Feldheim, K. A., S. H. Gruber, and M. V. Ashley. 2002. The breeding biology of lemon sharks at a tropical nursery lagoon. *Proc. R. Soc. Lond. B Biol. Sci.* 269:1655–1661.

Flajnik, M. F., Y. Ohta, C. Namikawa-Yamada, and M. Nonaka. 1999. Insight into the primordial MHC from studies in ectothermic vertebrates. *Immunol. Rev.* 167:59–67.

Futuyma, D. J. 1998. *Evolutionary Biology.* Sinauer Associates, Inc., Sunderland, MA.

Gaida, I. H. 1997. Population structure of the Pacific angel shark, *Squatina californica* (Squatiniformes : Squatinidae), around the California Channel Islands. *Copeia* 1997:738–744.

Gardner, M. G. and R. D. Ward. 1998. Population structure of the Australian gummy shark (*Mustelus antarcticus* Gunther) inferred from allozymes, mitochondrial DNA and vertebrae counts. *Mar. Freshwater Res.* 49:733–745.

Gardner, M. G. and R. D. Ward. 2002. Taxonomic affinities within Australian and New Zealand *Mustelus* sharks (Chondrichthyes : Triakidae) inferred from allozymes, mitochondrial DNA and precaudal vertebrae counts. *Copeia* 2002:356–363.

Gladden, J. G. B., M. M. Ferguson, M. K. Friesen, and J. W. Clayton. 1999. Population structure of North American beluga whales (*Delphinapterus leucas*) based on nuclear DNA microsatellite variation and contrasted with the population structure revealed by mitochondrial DNA variation. *Mol. Ecol.* 8:347–363.

Hartl, D. L. and A. G. Clark. 1997. *Principles of Population Genetics.* Sinauer Associates, Sunderland, MA.

Hedrick, P. W. 1999. Perspective: highly variable loci and their interpretation in evolution and conservation. *Evolution* 53:313–318.

Heist, E. J. and J. R. Gold. 1999a. Genetic identification of sharks in the US Atlantic large coastal shark fishery. *Fish. Bull.* 97:53–61.

Heist, E. J. and J. R. Gold. 1999b. Microsatellite DNA variation in sandbar sharks (*Carcharhinus plumbeus*) from the Gulf of Mexico and mid-Atlantic Bight. *Copeia* 5:182–186.

Heist, E. J., J. E. Graves, and J. A. Musick. 1995. Population genetics of the sandbar shark (*Carcharhinus plumbeus*) in the gulf of Mexico and mid-Atlantic bight. *Copeia* 18:555–562.

Heist, E. J., J. A. Musick, and J. E. Graves. 1996a. Genetic population structure of the shortfin mako (*Isurus oxyrinchus*) inferred from restriction fragment length polymorphism analysis of mitochondrial DNA. *Can. J. Fish. Aquat. Sci.* 53:583–588.

Heist, E. J., J. A. Musick, and J. E. Graves. 1996b. Mitochondrial DNA diversity and divergence among sharpnose sharks, *Rhizoprionodon terraenovae*, from the Gulf of Mexico and Mid-Atlantic Bight. *Fish. Bull.* 94:664–668.

Heist, E. J., J. C. Carrier, and H. L. J. Pratt. 2002. Unpublished data.

Heist, E. J., J. L. Jenkot, D. B. Keeney, R. L. Lane, G. R. Moyer, B. J. Reading, and N. L. Smith. 2003. Isolation and characterization of polymorphic microsatellite loci in nurse shark (*Ginglymostoma cirratum*). *Mol. Ecol. Notes* 3:59–61.

Heupel, M. R. and R. E. Hueter. 2001. Use of an automated acoustic telemetry system to passively track juvenile blacktip shark movements, in *Electronic Tagging and Tracking in Marine Fisheries.* J. R. Silbert and J. L. Nielsen, Eds., Kluwer Academic, Amsterdam, the Netherlands, 217–236.

Hueter, R. E., M. R. Heupel, E. J. Heist, and D. B. Keeney. In press. The implications of philopatry in sharks for the management of shark fisheries. *J. Northw. Atl. Fish. Sci.*

Karl, S. A., B. W. Bowen, and J. C. Avise. 1992. Global population genetic-structure and male-mediated gene flow in the green turtle (*Chelonia-Mydas*) — RFLP analyses of anonymous nuclear loci. *Genetics* 131:163–173.

Keeney, D. B., M. R. Heupel, R. E. Hueter, and E. J. Heist. 2002. Unpublished data.

Kendall, C., S. Valentino, A. B. Bodine, and C. A. Luer. 1994. Triploidy in a nurse shark, *Ginglymostoma-Cirratum. Copeia* 1994:825–827.

Kitamura, T., A. Takemura, S. Watabe, T. Taniuchi, and M. Shimizu. 1996. Mitochondrial DNA analysis for the cytochrome *b* gene and D-loop region from the bull shark *Carcharhinus leucas. Fish. Sci.* 62:21–27.

Kocher, T. D., W. K. Thomas, A. Meyer, S. V. Edwards, A. Paabo, F. X. Villablanca, and A. C. Wilson. 1989. Dynamics of mitochondrial DNA evolution in animals: amplification and sequencing with conserved primers. *Proc. Natl. Acad. Sci. U.S.A.* 86:6196–6200.

Lavery, S. and J. B. Shaklee. 1989. Population genetics of two tropical sharks *Carcharhinus tilstoni* and *C. sorah*, in Northern Australia. *Aust. J. Mar. Freshwater Res.* 40:541–557.

Lavery, S. and J. B. Shaklee. 1991. Genetic evidence for separation of two sharks, *Carcharhinus limbatus* and *C. tilstoni*, from Northern Australia. *Mar. Biol.* 108:1–4.

Lee, W. J. and T. D. Kocher. 1995. Complete sequence of a sea lamprey (*Petromyzon marinus*) mitochondrial genome — early establishment of the vertebrate genome organization. *Genetics* 139:873–887.

Lyrholm, T., O. Leimar, B. Johanneson, and U. Gyllensten. 1999. Sex-biased dispersal in sperm whales: contrasting mitochondrial and nuclear genetic structure of global populations. *Proc. R. Soc. Lond. B Biol. Sci.* 266:347–354.

MacDonald, C. M. 1988. Genetic variation, breeding structure and taxonomic status of the gumy shark *Mustelus antarcticus* in southern Australian waters. *Aust. J. Mar. Freshwater Res.* 39:641–648.

Maddock, M. B. and F. J. Schwartz. 1996. Elasmobranch cytogenetics: methods and sex chromosomes. *Bull. Mar. Sci.* 58:147–155.

Martin, A. P. 1995. Mitochondrial DNA sequence evolution in sharks — rates, patterns, and phylogenetic inferences. *Mol. Biol. Evol.* 12:1114–1123.

Martin, A. P., G. J. P. Naylor, and S. R. Palumbi. 1992. Rates of mitochondrial DNA evolution in sharks are slow compared with mammals. *Nature* 357:153–155.

Murphy, R. W., J. M. J. Sites, D. G. Buth, and C. H. Haufler. 1996. Proteins: isozyme electrophoresis, in *Molecular Systematics.* D. M. Hillis, C. Moritz, and B. K. Mable, Eds., Sinauer Associates, Sunderland, MA, 51–120.

Neigel, J. E. 2002. Is F_{ST} obsolete? *Conserv. Genet.* 3:167–173.

O'Connell, M. and J. M. Wright. 1997. Microsatellite DNA in fishes. *Rev. Fish Biol. Fish.* 7:331–363.

Ohta, Y., K. Okamura, E. C. McKinney, S. Bartl, K. Hashimoto, and M. F. Flajnik. 2000. Primitive synteny of vertebrate major histocompatibility complex class I and class II genes. *Proc. Natl. Acad. Sci. U.S.A.* 97:4712–4717.

Palumbi, S. R. and C. S. Baker. 1994. Contrasting population structure from nuclear intron sequences and mtDNA in humpback whales. *Mol. Biol. Evol.* 11:426–435.

Pank, M., M. Stanhope, L. Natanson, N. Kohler, and M. Shivji. 2001. Rapid and simultaneous identification of body parts from the morphologically similar sharks *Carcharhinus obscurus* and *Carcharhinus plumbeus* (Carcharhinidae) using multiplex PCR. *Mar. Biotechnol.* 3:231–240.

Pardini, A. T., C. S. Jones, M. C. Scholl, and L. R. Noble. 2000. Isolation and characterization of dinucleotide microsatellite loci in the Great White Shark, *Carcharodon carcharias. Mol. Ecol.* 9:1176–1178.

Pardini, A. T., C. S. Jones, L. R. Noble, B. Kreiser, H. Malcolm, B. D. Bruce, J. D. Stevens, G. Cliff, M. C. Scholl, M. Francis, C. A. J. Duffy, and A. P. Martin. 2001. Sex-biased dispersal of great white sharks — in some respects, these sharks behave more like whales and dolphins than other fish. *Nature* 412:139–140.

Potts, W. K. and E. K. Wakeland. 1990. Evolution of diversity at the major histocompatibility complex. *Trends Ecol. Evol.* 5:181–187.

Pratt, H. L. and J. C. Carrier. 2001. A review of elasmobranch reproductive behavior with a case study on the nurse shark, *Ginglymostoma cirratum. Environ. Biol. Fish.* 60:157–188.

Rasmussen, A. S. and U. Arnason. 1999a. Molecular studies suggest that cartilaginous fishes have a terminal position in the piscine tree. *Proc. Natl. Acad. Sci. U.S.A.* 96:2177–2182.

Rasmussen, A. S. and U. Arnason. 1999b. Phylogenetic studies of complete mitochondrial DNA molecules place cartilaginous fishes within the tree of bony fishes. *J. Mol. Evol.* 48:118–123.

Saville, K. J., A. M. Lindley, E. G. Maries, J. C. Carrier, and H. L. Pratt. 2002. Multiple paternity in the nurse shark, *Ginglymostoma cirratum. Environ. Biol. Fish.* 63:347–351.

Schrey, A. W. and E. J. Heist. 2002. Microsatellite markers for the shortfin mako and cross-species amplification in Lamniformes. *Conserv. Genet.* 3:459–461.

Schrey, A. W. and E. J. Heist. 2003. Microsatellite analysis of population structure in the shortfin mako (*Isurus oxyrinchus*). *Can. J. Fish. Aquat. Sci.* 60:670–675.

Schwartz, F. J. and M. B. Maddock. 2002. Cytogenetics of the elasmobranchs: genome evolution and phylogenetic implications. *Mar. Freshwater Res.* 53:491–502.

Shivji, M., S. Clarke, M. Pank, L. Natanson, N. Kohler, and M. Stanhope. 2002. Genetic identification of pelagic shark body parts for conservation and trade monitoring. *Conserv. Biol.* 16:1036–1047.

Smale, M. J. and P. C. Heemstra. 1997. First record of albinism in the great white shark, *Carcharodon carcharias* (Linnaeus, 1758). *S. Afr. J. Sci.* 93:243–245.

Smith, P. J. 1986. Low genetic variation in sharks (Chondrichthyes). *Copeia* 1986:202–207.

Smith, P. J. and P. G. Benson. 2001. Biochemical identification of shark fins and fillets from the coastal fisheries in New Zealand. *Fish. Bull.* 99:351–355.

Solari, A. J. 1994. Sex chromosomes and sex determination in fishes, in *Sex Chromosomes and Sex Determination in Vertebrates.* A. J. Solari, Ed., CRC Press, Boca Raton, FL, 308.

Solé-Cava, A. M. and J. A. Levy. 1987. Biochemical evidence for a third species of angel shark off the east coast of South America. *Biochem. Syst. Ecol.* 15:139–144.

Solé-Cava, A. M., C. M. Voreen, and J. A. Levy. 1983. Isozymic differentiation of two sibling species of *Squatina* (Chondrichthyes) in south Brazil. *Comp. Biochem. Physiol.* 75B:355–358.

Stevens, J. D. and P. D. Wiley. 1986. Biology of two commercially important carcharhinid sharks from Northern Australia. *Aust. J. Mar. Freshwater Res.* 37:671–688.

Stingo, V. and L. Rocco. 2001. Selachian cytogenetics: a review. *Genetica* 111:329–347.

Stingo, V., T. Capriglione, L. Rocco, R. Improta, and A. Morescalchi. 1989. Genome size and A-T rich DNA in Selachians. *Genetica* 79:197–205.

Tave, D. 1993. *Genetics for Fish Hatchery Managers.* Van Nostrand Reinhold, New York.

Utter, F. M. 1991. Biochemical genetics and fishery management — an historical perspective. *J. Fish Biol.* 39:1–20.

Waples, R. S. 1998. Separating the wheat from the chaff — patterns of genetic differentiation in high gene flow species. *J. Hered.* 89:438–450.

Ward, R. D. 2000. Genetics in fisheries management. *Hydrobiologia* 420:191–201.

Ward, R. D. and P. M. Grewe. 1994. Appraisal of molecular-genetic techniques in fisheries. *Rev. Fish Biol. Fish.* 4:300–325.

Ward, R. D., M. Woodwark, and D. O. F. Skibinski. 1994. A comparison of genetic diversity levels in marine, freshwater, and anadromous fishes. *J. Fish Biol.* 44:213–232.

Wright, J. M. and P. Bentzen. 1994. Microsatellites — genetic markers for the future. *Rev. Fish Biol. Fish.* 4:384–388.

Wright, S. 1969. *Evolution and the Genetics of Populations,* Vol. 2, *The Theory of Gene Frequencies.* University of Chicago Press, Chicago.

17

Predator–Prey Interactions

Michael R. Heithaus

CONTENTS

17.1 Introduction

Predator–prey interactions play a central role in the behavior, ecology, and population biology of most taxa and are critical in community dynamics. For elasmobranchs, most studies focus only on their role as a predator. This is an important oversight as most species are both predator and prey, at least for periods of their life history. In this chapter I place predator–prey interactions in the rich theoretical framework that has developed over the last several decades, from both a behavioral and a trophic perspective. Rather than compiling an exhaustive list of predator–prey interactions, I develop a framework for these interactions highlighting relevant elasmobranch examples. While there have been intriguing studies, there is much work to do and I hope that this chapter will help stimulate studies based on hypothesis testing that will help answer unresolved issues in the behavior and ecology of elasmobranchs.

In this chapter I consider predator–prey interactions at a variety of levels. Initially, I investigate the behavioral strategies and tactics used by elasmobranchs to capture food and to avoid becoming another predator's meal. Then, I consider the implications of predation and competition or prey availability on elasmobranch populations. Finally, I attempt to synthesize available studies to help understand the role of elasmobranchs in shaping the population and community dynamics of their prey and other species in the marine environment.

Before investigating the behavioral ecology of predator–prey interactions, it is important to distinguish between a strategy and a tactic. A *strategy* is a genetically based decision rule, or set of rules, that results in the use of particular tactics. Animals use *tactics* (which include behaviors) to pursue a strategy (Gross, 1996). Tactics may be fixed or flexible, and may depend on the condition of the individual or environmental conditions (including predators and prey). For example, a juvenile shark's strategy may be to use that tactic that will optimize energy intake and survival probability. The shark may pursue this strategy by switching between habitat use tactics that place the shark in dangerous but productive areas for some time periods and low-risk and low-food habitats during others.

17.2 Elasmobranchs as Prey

17.2.1 Predators of Elasmobranchs

Although we generally do not think of sharks as prey, they are often included in the diet of other species, as are skates and rays. Large teleosts have been found with small sharks in their stomachs (e.g., Jones, 1971; Randall, 1977) and odontocete cetaceans are also occasional elasmobranch predators (see Heithaus, 2001a, for a review). Small sharks and rays have been found in a number of dolphins and small toothed whales, but killer whales, *Orcinus orca,* are the only cetaceans that will regularly take elasmobranchs or take large species. Killer whales have been noted taking several species of carcharhinid sharks and species as large as a 3- to 4-m great white shark, *Carcharodon carcharius,* and even larger basking, *Cetorhinus maximus,* and whale, *Rhiniodon typus,* sharks (Fertl et al., 1996; Pyle et al., 1999). In New Zealand, killer whales prey upon eagle rays and stingrays (Visser, 1999).

Large sharks are the most important predators of other sharks and rays. Many species of large sharks include sharks or rays in their diets and some, like white, bull, *Carcharhinus leucas,* great hammerhead, *Sphyrna mokarran,* and sevengill sharks, *Notorynchus cepidanus,* regularly consume other elasmobranchs (Tricas and McCosker, 1984; Cliff et al., 1989; Cliff and Dudley, 1991; Cliff, 1995; Ebert, 2002). Cannibalism has been recorded in a number of species of large sharks, and in some areas, adult conspecifics may actually be the most important predators of pups and juveniles (e.g., scalloped hammerhead sharks, *S. lewini,* Clarke, 1971; bull sharks, Snelson et al., 1984).

17.2.2 Avoiding Predators

There are many tactics that elasmobranchs can use to avoid being killed by predators. These range from immediate responses to a threat such as flight or defense to longer-term tactics like habitat use and group formation (Figure 17.1).

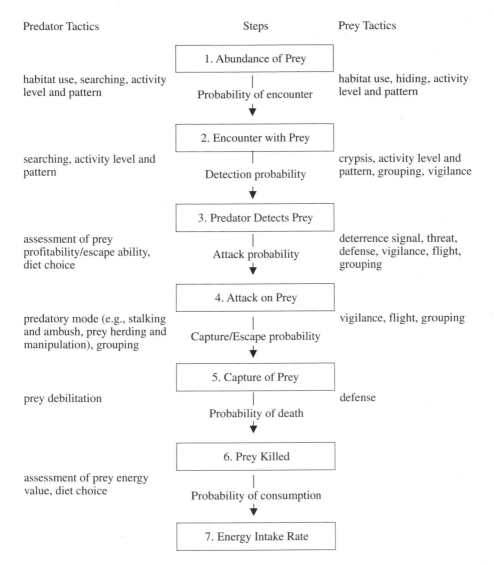

| Predator Tactics | Steps | Prey Tactics |

habitat use, searching, activity level and pattern

1. Abundance of Prey

Probability of encounter

habitat use, hiding, activity level and pattern

searching, activity level and pattern

2. Encounter with Prey

Detection probability

crypsis, activity level and pattern, grouping, vigilance

assessment of prey profitability/escape ability, diet choice

3. Predator Detects Prey

Attack probability

deterrence signal, threat, defense, vigilance, flight, grouping

predatory mode (e.g., stalking and ambush, prey herding and manipulation), grouping

4. Attack on Prey

Capture/Escape probability

vigilance, flight, grouping

prey debilitation

5. Capture of Prey

Probability of death

defense

6. Prey Killed

Probability of consumption

assessment of prey energy value, diet choice

7. Energy Intake Rate

FIGURE 17.1 Steps in a predator–prey interaction leading to energy intake for the predator. There are many tactics that may be used by both predator and prey throughout the predatory process and the energy intake of the predator (foraging success) and survival probability of prey will depend on tactics of the other. (Modified from Lima and Dill, 1990; Sih and Christensen, 2001; Heithaus et al., 2002b.)

17.2.2.1 Habitat Use — By selecting habitats where predators are relatively rare or absent, prey can greatly reduce their probability of being killed. The risk faced in a habitat is determined by two primary factors: the density of predators in the habitat and the intrinsic habitat risk. Intrinsic habitat risk is determined by habitat attributes that influence the probability of prey encounters with predators and the probability of death in an encounter situation (e.g., Hugie and Dill, 1994; Heithaus, 2001b). Intrinsic risk can be influenced by the presence of cover (habitat complexity), substrate color, light level, water depth, and water turbidity (Werner and Hall, 1988; Gotceitas and Colgan, 1989; Lima and Dill, 1990; Hugie and Dill, 1994; Miner and Stein, 1996), and in some situations intrinsic habitat risk may be the primary determinant of habitat use by prey species (e.g., Hugie and Dill, 1994; Heithaus, 2001b; see Section 17.3.5.1).

The use of shallow-water nursery habitats by juvenile sharks of many species is hypothesized to be driven largely by predator avoidance (e.g., Springer, 1967; Castro, 1987; Morrissey and Gruber, 1993a; Simpfendorfer and Milward, 1993; Castro, 1993). For example, Morrissey and Gruber (1993a) found that juvenile lemon sharks, *Negaprion brevirostris*, selected warm, shallow waters with sand or rock bottoms. They suggested that this habitat choice was driven largely by the need to avoid larger sharks. Similarly, juvenile blacktip shark, *Carcharhinus limbatus*, habitat use within their nursery may be driven largely by the need to avoid predators (Heupel and Heuter, 2002), and juvenile scalloped hammerhead, *S. lewini*, pups select the most turbid waters of their nursery area during the day, potentially as a refuge from predators (Clarke, 1971). However, it is possible that other demands (e.g., energy intake) may also be an important factor in juvenile shark habitat use (see Section 17.3.5.1). Large sharks are the primary predator of small sharks and tend to be more abundant in deeper waters (e.g., Morrissey and Gruber, 1993a; but see Heithaus et al., 2002a). By selecting shallow waters, juveniles probably reduce both the probability of encounter with predators as well as increase their likelihood of detecting a predator (less directions for a predator approach) and avoiding an attack (movement into waters too shallow for predators). Although this hypothesis remains the most plausible explanation for the use of shallow-water nurseries, it has yet to be experimentally tested.

Adult elasmobranchs also appear to select predator-free habitats, at least during some activities. For example, cowtail rays, *Pastinachus sephen*, in Shark Bay, Western Australia may rest in extremely shallow waters with sand bottoms that are free of predatory hammerhead, *Sphyrna* spp., and tiger, *Galeocerdo cuvier*, sharks (M. Heithaus, pers. obs.; C. Semenuick, unpubl. data).

17.2.2.2 *Activity Levels and Patterns* — By reducing their activity level (i.e., movement speed or duration), prey can often reduce their probability of being attacked by predators (Taylor, 1984; Gerritsen and Strickler, 1997; Werner and Anholt, 1993; Anholt and Werner, 1995). This can operate through reduced encounter rates with predators as well as reduced probability of being detected by a predator (Lima, 1998a,b). It is currently unknown if elasmobranchs use this antipredator tactic. Choosing an appropriate time of day to be active can greatly influence predation risk, and changes in light level can cause prey to modify their activities because of increased susceptibility to predators (Lima and Dill, 1990). For example, nocturnal foragers may reduce activity on nights with bright moonlight and diurnal foragers reduce activity during crepuscular periods because predators enjoy detection advantages at these times (Lima and Dill, 1990). Many species of elasmobranchs show distinct differences in movement patterns and rates between diurnal and nocturnal periods (Section 17.3.2.3). These patterns have generally been interpreted as preferred foraging times, but it is also possible that the diel patterns of activity level are in response to predation risk. For example, Hawaiian stingrays, *Dasyatis lata*, exhibit low activity rates during the day, which may be a mechanism for avoiding predators (Cartamil et al., 2003).

17.2.2.3 *Hiding and Crypsis* — Hiding behavior and cryptic coloration can reduce the probability of being killed by a predator, and examples of both are found in elasmobranchs. Skates and rays are the most obvious as they bury themselves in the substrate, thereby reducing the probability of being detected by a predator. Some small sharks, skates, and rays hide in structures such as reefs and rocks where predators are unlikely to detect and capture them. Cryptic coloration is found in a number of species. Benthic species may have dorsal surfaces that are the color of the substrate, and the complex body color patterns of reef-dwelling sharks probably help them blend into their surroundings. Sometimes, coloration changes through ontogeny. For example, nurse sharks are born with black spots that would help them blend into their reef and sponge habitat. As body size increases, they slowly develop the brown dorsal coloration common in adults (Castro, 2000). Finally, some species of elasmobranchs are able to change color to mimic those of their surroundings (physiological color change, see Chapter 11), which would help them avoid detection by predators.

17.2.2.4 *Group Formation* — Group formation can reduce the risk of being killed by a predator, and predation has been suggested as the selective force leading to sociality in many taxa (Bertram, 1978; Pulliam and Caraco, 1984). Many species of small sharks and rays occur in groups, but there have been almost no attempts to understand the factors leading to group formation and the dynamics of group living in elasmobranchs. There are four major ways in which groups can reduce the risk of predation. First, they may increase the probability that predators are detected. Second, fleeing groups may confuse

predators. Third, groups may be able to defend against attacking predators, but I am aware of no elasmobranch examples. A final benefit of group living is dilution whereby increasing the number of individuals in a group reduces an individual's probability of being captured and killed by a predator. Although there is still no experimental or other evidence to demonstrate a dilution effect, it is likely a benefit to elasmobranch groups. Scalloped hammerhead pups in nurseries of Kaneohe Bay, Oahu, HI form aggregations during the day, which may provide protection from their predators in the bay (Clark, 1971; Holland et al., 1993). Simpfendorfer and Milward (1993) suggested that multispecies nurseries may provide additional benefits for pups in the form of reduced predation risk, probably due to dilution. It is important to note that the antipredator benefits of grouping are often not mutually exclusive and sometimes inseparable. For example, it is not possible to separate dilution and detection benefits when group members do not share information about predatory attacks perfectly (Bednekoff and Lima, 1998).

Cowtail rays resting in shallow waters of Shark Bay, Western Australia show facultative group formation, which is apparently driven by predation risk (Semeniuk, 2003). When visibility is good, rays generally rest alone, but form groups when light levels are low and water is turbid. Experimental evidence suggests that grouped rays benefit from increased detection of predators. First, rays preferentially settled with model rays that had abnormally long tails (which would presumably mechanically detect predators sooner) over those with normal and shortened tails. Second, ray groups exhibited greater flight initiation distances (FID, Section 17.2.2.7) in response to the approach of a model predator than solitary rays, suggesting that groups detected predators sooner. Rays in these groups also tended to adopt specific geometries to increase detection area. Escape responses of the rays suggest that they may also benefit from predator confusion and dilution. Group formation in good visibility conditions, however, may actually increase predation risk. Although not fully resolved, groups show slower escape velocities and escape impedance (rays may have to move toward a predator before initiating flight), which may reduce the benefits of grouping if predators can be detected at sufficient distance by solitary rays in good visibility conditions (Semeniuk, 2003).

17.2.2.5 Vigilance

17.2.2.5   Vigilance — Vigilance, where prey stop other activities and watch for predators, is a common behavior used to reduce predation risk. Optimal vigilance level depends on both the risk of attack and the size of the group an animal is in, with higher levels when predation risk is high and when group size is low (Lima and Dill, 1990; Lima, 1995; Brown et al., 1999). I am not aware of any studies of vigilance in elasmobranchs and it is difficult to determine what behaviors may be considered vigilance. However, species that can remain stationary may stop searching for food to scan for predators. Vigilance may overlap substantially with searching behavior, and may be difficult to operationally isolate in elasmobranchs.

17.2.2.6   Deterrence and Defense — Pursuit-deterrence signals, where prey signal to predators that they have been observed, can result in a reduced probability of predatory attack when signals are honest (Caro, 1995). There are currently no studies of such signaling in elasmobranchs; however, they are likely to be effective. Some elasmobranch predators are unlikely to initiate attacks on potential prey that are vigilant (e.g., tiger sharks; Heithaus et al., 2002a). Thus, prey behaviors that signal their readiness to flee are likely to reduce attack probability.

Behavioral threat displays may also serve to thwart potential predators. Threat displays and subsequent attacks on divers and submersibles by gray reef sharks, *Carcharhinus amblyrhynchos*, may have an antipredator function. Nelson et al. (1986) approached gray reef sharks with submersibles and elicited threat displays and attacks. They suggested that the behavior served an antipredator function rather than to defend a territory or resources because (1) both solitary and grouped sharks attacked, (2) attacks were elicited primarily when sharks were pursued, (3) sharks fled after attacks rather than continue an attack as would be expected if trying to drive off an intruder or competitor, (4) cornering a shark against a barrier increased the likelihood of attack, and (5) there were no threat displays directed at, or attacks observed on, conspecifics.

A number of defensive morphological characters have evolved in elasmobranchs. These include the stinger found on the tail of many ray species, which can be used to inflict a painful wound on a potential attacker, as well as spines found anterior to the dorsal fins of horn sharks (Heterodontiformes), which may result in them being ejected from a predator's mouth after being engulfed. Electric rays can use electric discharges to thwart predatory attacks (Belbenoit, 1986).

17.2.2.7   Flight — Fleeing is an obvious reaction to the immediate threat of a predator and involves decisions of escape direction, at what distance from the predator to initiate flight (FID) and escape velocity. Because flight is costly (in terms of both energetic expenditure and lost foraging or mating opportunities), animals should not necessarily flee as soon as a predator is detected; instead, there should be an optimal flight response (Ydenberg and Dill, 1986; Dill, 1990). The cost of flight, the speed and angle of approach of the predator (i.e., the loom rate), and distance to safety (which is often influenced by habitat characteristics) influence escape responses (e.g., Ydenberg and Dill, 1986; Dill, 1990; Lima and Dill, 1990; Bonenfant and Kramer, 1994). FIDs will be shorter (predators allowed to approach closer) when costs of flight are high, the loom rate is lower, and habitat characteristics favor easier escape (e.g., distance to cover is shorter). Often, distance or time to safety is equivalent to the distance to a physical refuge, but in marine systems time to safety might be the amount of time it takes a prey animal to reach a critical velocity or maneuvering ability such that they are inaccessible to a predator (Heithaus et al., 2002b). Dill (1990) proposed that animals would maintain a consistent "margin of safety" by selecting a combination of escape velocity and FID. However, the margin of safety may increase as distance from cover increases, as some species appear to flee at well below their maximum speed. This choice of escape velocity would allow prey to respond to acceleration by the predator, and having such flexibility may be relatively more advantageous as distance from safety increases (Bonenfant and Kramer, 1994). For cryptic species, the decision to flee will also be based on the probability that a predator will detect and capture the prey if it flees relative to the probability of detection and capture if the prey remains motionless and cryptic; cryptic individuals show shorter FIDs than those that are not as well camouflaged (Lima and Dill, 1990).

There are few studies of FIDs in elasmobranchs, but solitary cowtail rays in Shark Bay, Western Australia initiate flight when approached by a model predator at a significantly greater distance when water visibility conditions are relatively good (C. Semeniuck and L. Dill, in prep.). This can be interpreted in several ways. First, it is possible that rays are able to detect the model predator at a greater distance in clear conditions and initiate flight as soon as a predator is detected. Alternatively, rays detected models at a similar distance in good and poor visibility conditions, but waited to initiate flight in poor-visibility conditions because remaining motionless during these conditions had a lower probability of being detected than if flight had been initiated.

17.2.3   Apparent Competition

Apparent competition occurs when two (or more) species share a common predator and high productivity by one prey species supports predators at a population level sufficient to eliminate another prey species (Holt, 1977, 1984). Apparent competition may also be manifested through behavioral mechanisms where predator abundance in a habitat is driven by one prey species and leads to habitat abandonment by another species (Heithaus and Dill, 2002a; Dill et al., 2003). I am not aware of apparent competition among elasmobranchs, but tiger sharks regulate apparent competition between their primary prey and Indian Ocean bottlenose dolphins, *Tursiops aduncus* (Heithaus and Dill, 2002a), and sharks evidently mediate apparent competition between gray seals, *Halichoerus grypus*, and harbor seals, *Phoca vitulina*, on Sable Island, Canada (Bowen et al., 2003; Section 17.5.1).

17.3   Elasmobranchs as Predators

Elasmobranchs feed on an amazing array of species from plankton and benthic invertebrates to marine mammals and other large vertebrates. Species also vary from batch-feeding filter feeders and scavengers to active predators and from opportunists with catholic diets to highly specialized feeders. Although a review of elasmobranch diets is beyond the scope of this chapter (see Chapter 8), in this section I review optimal diet theory and the predatory tactics of elasmobranchs.

17.3.1 Diets of Elasmobranchs and Optimal Diet Theory

Diet composition is the result of behavioral decisions associated with locating and capturing or rejecting potential prey items (Stephens and Krebs, 1986). Optimal diet theory (ODT) has a long history but has not been widely applied to studies of elasmobranchs. Indeed, most studies have been limited to catalogs of prey items found in a particular location. Although there is likely to be some consistency in the diets of a particular species throughout its range, there is certain to be a high degree of flexibility as ecological conditions vary. Bonnethead sharks, *Sphyrna tiburo*, in Florida, tiger sharks in Western Australia, mako sharks, *Isurus oxyrhinchus*, in the northwest Atlantic, leopard sharks, *Triakis semifasciata*, in Tomales Bay, CA, and scalloped hammerhead pups within a Hawaiian nursery ground show geographic variation in their diets (Clarke, 1971; Stillwell and Kohler, 1982; Cortés et al., 1996; Webber and Cech, 1998; Simpfendorfer et al., 2001). Despite the lack of data addressing questions of optimal diet choice in elasmobranchs, I briefly review relevant theory in hopes of stimulating future research.

The most basic form of optimal diet theory describes when a prey item should be accepted or rejected (Stephens and Krebs, 1986). If prey items vary in their net energy gain (energy content minus the energy expended in capture and handling), handling time, and encounter rate, then it is possible to make simple predictions about which prey should be eaten. In this simple situation, predators are predicted to rank prey items in order of their profitability; they either always consume or always reject a particular prey item (Stephens and Krebs, 1986). It is important to note that the predictions based on this theory do not specifically state whether a predator should diversify or narrow its diet choice as more prey types become available to the predator (e.g., seasonal changes in prey availability, ontogenetic changes in foraging abilities). Instead, the nature of prey selection will be based on the relative profitability of each potential prey item. The general theory is upheld in some situations (e.g., Werner and Hall, 1974), but animals may not conform to the prediction of always accepting or rejecting particular prey items (e.g., Hart and Ison, 1991; see Clark and Mangel, 2000). This partial preference, where individuals sometimes accept prey items of lower profitability, can be explained by differences in the state of the individual (e.g., body condition or energetic reserves). For example, three-spined stickleback, *Gasterosteus aculeatus*, feeding on larger marine invertebrates (*Asellus*) sometimes accept prey items that are not predicted to be taken (Hart and Ison, 1991). When guts were relatively empty, large prey items with very long handling times were accepted, but as gut fullness increased, the acceptance of these relatively unprofitable prey items decreased. Although traditional diet theory fails to explain such results, dynamic state variable models that include gut fullness and prey catchability show a good match with data (Hart and Gill, 1992a), and predictions from these types of model suggest that individuals in good condition or with relatively full stomachs should be the most selective for high-quality prey items. Dynamic state variable models (see Clark and Mangel, 2000) allow predictions to be made about prey choice when both ecological and internal conditions are variable and are likely to benefit studies of elasmobranch diets.

The diets of elasmobranchs are almost certainly a result of the decision processes described above, and there is evidence for selection of relatively profitable prey items. Southern stingrays, *Dasyatis americana*, preferentially ingest only large size classes of lancelets (Stokes and Holland, 1992), and mako sharks appear to selectively feed on large bluefish, *Pomatomus saltatrix*, to maximize energy intake (Stillwell and Kohler, 1982). A predator selecting an optimal diet may not stop the decision-making process with what type of prey to capture, but may also decide what portions of a prey item to consume to maximize energy intake rate. White sharks, blue sharks, and Greenland sharks, *Somniosus microcephalus*, seem to selectively feed on the blubber layer of marine mammal carcasses, thus maximizing their energy intake through ingestion of only high-quality food (Beck and Mansfield, 1969; Carey et al., 1982; Klimley, 1994; Klimley et al., 1996a; Long and Jones, 1996). Rays foraging on thin-shelled bivalves usually consume the entire shell, but when feeding on thick-shelled bivalves they consume the body of the animal and largely avoid ingesting the shell (e.g., cownose rays; Smith and Merriner, 1985).

Recent reviews of ODT have shown that while it is very good at predicting the diet of foragers consuming immobile prey, predictions are often not supported for species consuming mobile prey (Sih and Christensen, 2001). The reason for failure of ODT in the mobile prey scenario is likely to lie in two factors. First, diet preferences of a predator (i.e., prey are consumed disproportionately relative to their abundance in the environment) can be the result of both active predator choice (unequal attack probabilities in encounter situations with prey; attack probability, Figure 17.1) and antipredator tactics that result in differences among prey in their probability of detection or probability of escape (see Figure 17.1). ODT only considers active predator choice and does not account for antipredator behavior (Sih and Christensen, 2001). Second, most studies of ODT do not include information on all aspects of prey profitability, which would include probability of capture, handling time, and probability of prey escape after capture (Sih and Christensen, 2001). Even under ODT predictions, predators might not show preferential attacks on the prey species with the highest energy content (Sih and Christensen, 2001). For example, if energy-rich prey are rarely encountered or usually escape, predators may show no prey preferences at all. Therefore, to truly understand the diet selection process of predators it is important to understand the antipredatory tactics of their prey. This aspect of predator–prey interactions has largely been ignored in studies of elasmobranchs with mobile prey.

17.3.2   Finding Prey

17.3.2.1   *Habitat Use* — A predator can increase its probability of encountering and capturing prey through optimal selection of a forging habitat. In the absence of other individuals, this may entail selecting the habitat with the highest prey density. However, selecting the optimal patch to maximize energy intake does not necessarily equate to selecting the habitat with the highest density (i.e., abundance) of prey. Prey in highly structured environments (e.g., reefs or seagrass beds) may escape easily and thus become unavailable (e.g., Gotceitas and Colgan, 1989). Also, when other factors such as the presence of competitors or a third trophic level (the prey's food) are considered, then there may not be a good relationship between the abundance of prey and the distribution of predators even if energy intake is the only factor determining the distribution of the predator.

If a predator is pursuing a strategy to maximize net energetic gains, it should select the habitat with the highest energetic return rate (energy gained per unit time minus energetic costs of being in the habitat and capturing and consuming prey). Some elasmobranchs appear to conform to this prediction. Basking sharks are found in highest abundance in warm-water oceanic fronts that concentrate their planktonic prey, and appear to select higher-density plankton patches with abundant large copepods (Sims and Merrett, 1997; Sims and Quayle, 1998). Thus, these sharks appear to preferentially select habitat patches based on their potential energetic intake rate. Bat rays, *Myliobatis californica*, in Tomales Bay, CA appear to adopt foraging tactics to pursue a strategy that maximizes energy intake rate. Rays move into shallow waters of the inner bay from 2:50 to 14:50, then return to the cooler waters of the outer bay (Matern et al., 2000). The infaunal prey of the rays may be buried 0.5 to 1.0 m deep and require substantial energy to excavate, and thus a high metabolic rate during foraging would be beneficial (Matern et al., 2000). Over the temperature range observed in the bay, bat rays have an extremely high Q_{10} (the increase in metabolic rate with a 10°C change in temperature) of 6.8 (Hopkins and Chech, 1994). Thus, moving into the foraging areas during the day likely increases the foraging efficiency of the rays (Matern et al., 2000). Moving into cooler waters to rest would reduce metabolic rate (less energy wasted compared to warm water) without major effects on the rate of digestion and assimilation (Matern et al., 2000; see Chapter 7).

In the absence of other factors, foragers should begin using a patch only when the energy intake rate available in that habitat is above that, or equal to, the energy intake rate available in other areas within the environment. In some cases, this results in threshold foraging responses. For example, basking sharks will only forage on zooplankton concentrations above 0.48 to 0.70 g/m^3, which is close to the theoretically predicted threshold at which sharks would be foraging at a net energetic loss (Sims, 1999). Threshold foraging responses have also been observed in rays. Cownose rays will consume all bay scallops in some high-density patches but not forage in areas with low bay scallop density (Peterson et al., 2001). Similarly, infaunal bivalves, *Macomona lilliana*, have a refuge in low densities as eagle rays, *Myliobatis tenuicaudatus*, show a threshold foraging response at about 44 *Macomona*/0.25 m^2 (Hines et al., 1997). Furthermore, the rays forage on a relatively large spatial scale and respond to prey patches on a scale of 75 to 100 m, so rays did not forage on small but dense aggregations of bivalves (Hines et al., 1997). It is unclear how eagle rays avoid unprofitable prey patches, but they may be able to detect water jets from bivalves (Montgomery and Skipworth, 1997).

Many studies suggest that habitat use by elasmobranchs is influenced by the distribution of their prey, but the studies are not based on rigorous measurements of predator habitat selection and prey densities or availability. Morrissey and Gruber (1993a) suggested that juvenile lemon sharks may show a preference for sand and rock bottoms over seagrass areas because of the decreased availability of prey over seagrass, as the complex habitat serves as a prey refuge. The distribution of large white sharks along the California coast may be influenced by the distribution of their pinniped prey (Klimley, 1985), and individuals show site fidelity to particular seal colonies across years (e.g., Strong et al., 1992, 1996; Klimley and Anderson, 1996; Goldman and Anderson, 1999). In southern Australia, white sharks may select habitats where dolphin densities are high (Bruce, 1992), and Pacific angel sharks, *Squatina calicornica*, appear to select ambush sites in response to prey availability (Fouts and Nelson, 1999). Some habitats become available only periodically. For example, tidal mudflats and sandflats are often inaccessible except at high tide, and leopard sharks move into these muddy littoral zones on incoming tides, presumably to feed on worms and clam siphons (Ackerman et al., 2000).

Predators that forage on sessile prey and are in low densities may be able to make decisions in the manner described above, but often a predator's decision on which habitat to select will be influenced by antipredator decisions and behaviors of their prey, or by foraging decisions made by other predators, or by both. For example, as a predator spends more time in a particular habitat or more predators accumulate in a given habitat, prey should become more vigilant or leave a particular habitat, and thus prey availability will decline (see Brown et al., 1999, for a discussion of the "ecology of fear"). Therefore, there may be selection for a predator to switch among habitats as prey increase their investment in antipredator behavior. In such circumstances a tactic that involves covering an extremely large area may be optimal, as is often observed in large sharks that consume highly mobile prey not tied to a specific haul-out site. In Hawaii, tiger sharks make large moves between different areas of what appears to be a large home range (Holland et al., 1999). Sharks move in very straight lines across deep oceanic waters but then begin to make more turns when they encounter shallower banks. In Shark Bay, tiger sharks appear to move within extremely large home ranges, and although they preferentially select habitats that are rich in prey, they do not remain within a habitat patch for extended periods (Heithaus, 2001c; Heithaus et al., 2002a). Even within nursery areas, where they have relatively small home ranges, elasmobranchs may use a habitat-shifting tactic. Juvenile lemon sharks shift locations within their home ranges from day to day, which may allow prey availability to rebound (presumably through habitat-use shifts in prey or changes in prey vigilance) (Morrissey and Gruber, 1993b), but may also be in response to temporarily reduced availability in the area.

Often the intake rate of an individual predator will be influenced by the number of other predators in a habitat (i.e., frequency-dependent energy intake). A vast array of models have been developed to describe such situations. The most basic is the ideal free distribution (IFD) (Fretwell and Lucas, 1970). This model assumes that (1) prey distribution is fixed, (2) animals forage to maximize energy intake rate, (3) resources are split evenly among predators, and (4) energy intake of each predator is reduced with the addition of another predator. Under this model, predators are expected to be distributed across habitats proportional to food resources available there. This results in the basic prediction that intake rates of individuals in patches with high and low availability are identical due to the higher density of foragers in more productive patches. Although there is empirical support for this model in some non-elasmobranch situations (e.g., Tregenza, 1995), many factors may cause deviations from this distribution, including differences in competitive ability among foragers (see Tregenza, 1995) and predation risk (see Section 17.3.5.1). Studies of elasmobranch foraging are inconclusive. If habitat patches of all densities are considered, eagle rays foraging on benthic invertebrates do not conform to IFD predictions, but may conform to the predictions of the IFD at prey densities above a threshold level (Hines et al., 1997). Juvenile blacktip sharks, *Carcharhinus limbatus*, within a coastal nursery in Tampa Bay, FL did not appear to distribute themselves across the nursery area in relation to prey density (Heupel and Hueter, 2002). However, a mismatch between prey distribution and predator distribution does not necessarily indicate that factors other than food are important in habitat use decisions (see below). Finally, the IFD would not predict complete depletion of prey patches, as is observed in cownose rays foraging on bay scallops (Peterson et al., 2001). However, this result may be due to (1) the energy gained by completely depleting bay scallops is greater than the energy gained by leaving early and searching for new prey patches (i.e., high search costs) or (2) although all bay scallops were removed, rays were still harvesting resources in the patch (e.g., infaunal invertebrates), but bay scallop profitability was higher than other prey so they were consumed first, before the profitability of the habitat had dropped below that available in other habitats.

Antipredator decisions by prey are another important factor in habitat use decisions made by predators. This is especially likely for elasmobranchs that consume mobile prey, since these prey species show a diverse array of antipredator behaviors and can adaptively modify these in response to predation risk (see Lima and Dill, 1990; Lima, 1998a, for reviews). When decisions made by predators are influenced by decisions made by prey, and vice versa, game theory can provide insights into optimal decisions of both players. Game-theoretic models of habitat use by predators and prey in systems with three trophic levels, predators (e.g., a shark), their prey (e.g., teleost fish), and food of the prey (e.g., benthic invertebrates) have revealed counterintuitive results (Hugie and Dill, 1994; Sih, 1998; Heithaus, 2001b). In these situations, predators are actually predicted to be distributed across habitats in relation to the food of their prey (i.e., resources one trophic level removed) while prey are distributed across habitats relative to intrinsic habitat risk. This can result in predator distributions that do not relate well to those of their prey. Therefore, even if the fitness of a predator is determined solely by energy intake rate, field observations may not find a match between predator distribution and prey density.

Tiger shark habitat use in Shark Bay, Western Australia appears to be influenced primarily by the distribution of their prey species. Sharks showed a strong preference for shallow seagrass habitats (<4 m) over surrounding deeper waters covered by sand (6 to 15 m). These seagrass habitats have higher densities of potential prey including dugongs, *Dugong dugon*, sea snakes, sea turtles, and birds (Heithaus et al., 2002a). These results qualitatively match theoretical predictions of several foraging models including the IFD, a predator–prey game, or optimal foraging when frequency-dependent selection does not occur (Heithaus et al., 2002a).

Vertical migration can be considered a type of habitat use (Iwasa, 1982; Hugie and Dill, 1994; Heithaus, 2001a). For example, deep scattering layer organisms undergo daily vertical migrations from deeper waters during the day to shallower depths at night. Megamouth sharks, *Megachasma pelagios*, migrate with this layer, effectively keeping within the prey band (Nelson et al., 1997). These sharks appeared to migrate according to light level, following the 0.4 lux isolum (Nelson et al., 1997), but this is probably a proximate response that allows the sharks to maintain contact with prey patches. Other, deep-water pelagic sharks like cookie-cutter sharks, *Isistius* sp., may also migrate vertically with the deep scattering layer (Jones, 1971), but not all deep-water sharks migrate vertically. Depth telemetry of sixgill sharks, *Hexanchus griseus*, and gulper sharks, *Centrophorus acus*, have not indicated any diel vertical migrations. Instead, both of these benthopelagic species appeared to stay relatively close to the bottom (Yano, 1986; Carey and Clark, 1995).

Migration, where individuals move extremely large distances seasonally, can be considered an extreme case of habitat selection with preferred habitats widely separated. Seasonal migrations are common in elasmobranchs, which in many cases may be driven solely by temperature (climate) (Musick et al., 1993), by both reproductive needs and foraging needs (Section 17.3.5.2), or may be primarily food-driven. Whale sharks, *Rhincodon typus*, respond to seasonal changes in resources, and seasonally congregate in areas with abundant prey. Whale shark densities peak at Ningaloo Reef, Western Australia during coral spawning (Gunn et al., 1999; Wilson et al., 2001) and off the coast of Belize during spawning aggregations of snappers, *Lutjanus cyanopterus* and *L. jocu* (Heyman et al., 2001). Tiger sharks appear to congregate at islands offshore of the Western Australian coast when food abundance (fishing industry discards) peaks (Simpfendorfer et al., 2001). Movements of blue sharks from inshore to offshore waters at Catalina Island, CA may be due to movements of their prey with sharks moving inshore in winter to feed on spawning schools of market squid, *Loligo opalescens* (Sciarrotta and Nelson, 1977; Tricas, 1979). Finally, seasonal changes in the abundance of tiger sharks in Shark Bay coincide with changes in the abundance of high-quality prey items: dugongs and sea snakes (Heithaus, 2001c). Detailed studies of elasmobranch prey are needed to fully understand elasmobranch migrations. In general, the influence of spatial scale on predator–prey behavioral interactions has been overlooked (Lima, 2002), and this is especially true of elasmobranchs. Given the high mobility of both elasmobranchs and their prey in many situations, this is likely to be a challenging but exciting avenue of research.

17.3.2.2   Search Patterns — The probability of prey encounter and capture can be further increased within a habitat by adopting an optimal searching strategy. In some cases this may include remaining in one location, as in the case of ambush predators (Section 17.3.3.1), or it may involve roving movements within a single location. For example, tiger sharks in Shark Bay do not remain in one high-productivity habitat patch for an extended period, but instead move among productive shallow seagrass patches (Heithaus et al., 2002a). White sharks have several searching tactics within and among seal colonies. Sharks make relatively directional travels between islands that might have pinniped prey, then more restricted movements close to the shore of islands that have high pinniped prey densities (Strong et al., 1992, 1996; Goldman and Anderson, 1999; Klimley et al., 2001). Goldman and Anderson (1999) suggested that larger white sharks may have more restricted patrolling areas than smaller sharks, but their sample size was insufficient to conclusively test this hypothesis.

Vertical movements within a habitat may also be a form of searching behavior. Many species of sharks make regular movements up and down through the water column. The reason for these movements is still uncertain, and there are probably different factors that apply in certain situations. Oscillatory swimming appears to be a general rule in pelagic species (e.g., Carey and Scharold, 1990; Holts and Bedford, 1993; Klimley et al., 2002). Oscillatory swimming in pelagic habitats may be used for (1) conserving energy, (2) thermoregulation, (3) obtaining olfactory information for foraging or navigation, or (4) detecting magnetic gradients for navigation (see Klimley et al., 2002, for a discussion of these mechanisms). In some species, this behavior appears to be associated directly with foraging. Whale sharks at Ningaloo Reef, Western Australia make regular oscillations between the surface and the bottom, and are probably searching for food throughout the water column (Gunn et al., 1999). Tiger sharks in Shark Bay exhibit oscillatory swimming even when in shallow waters (3 to 10 m) (Heithaus et al., 2002a). This behavior appears to be a foraging tactic that aids in visual detection of air-breathing and benthic prey (Heithaus et al., 2002a), but may also aid in olfactory detection. Carey and Scharold (1990) suggested that blue shark dives were for both behavioral thermoregulation and foraging with oscillations aiding in both visual and olfactory detection of prey. Finally, white sharks making movements above and below the thermocline may increase olfactory detection of whale carcasses (Carey et al., 1982).

17.3.2.3   Activity Levels and Patterns — For roving predators, increased activity levels can result in increased access to food through increased encounter rates with prey. Elasmobranchs can also increase the probability of both encountering and capturing prey by selecting an appropriate time of day to feed. Diel patterns of foraging have often been inferred by increases in the rate of movement (ROM) as measured by a tracking boat (Sundström et al., 2001) or swimming speed; however, these results must be viewed with caution. First, ROM does not necessarily reflect the swimming speeds of the elasmobranch (Sundström et al., 2001). Also, although increased rates of movement and swimming speeds may be indicative of foraging behavior, it may also represent transit movements between areas of a home range or between foraging sites. For example, during summer, blue sharks patrol inshore waters during the night, then move in a relatively straight line to offshore waters during the day (Sciarrotta and Nelson, 1977). Juvenile lemon sharks also migrate between daytime and nighttime activity centers (Sundström et al., 2001). Furthermore, species feeding on prey with extensive handling times are likely to show lower rates of movement when foraging than when traveling. Also, there is not always agreement between ROM data and feeding data obtained from stomach contents analysis. For example, although juvenile lemon sharks appear to become more active at night (Morrissey and Gruber, 1993a) there are no diel patterns in feeding (Cortés and Gruber, 1990). Thus, increased movement rates should not be automatically interpreted as indicative of foraging activity, and the interpretation of ROM and swimming speeds should be done with caution unless combined with other technology to assess feeding behavior (e.g., stomach temperature pills, Klimley et al., 2001; animal-borne video devices, Heithaus et al., 2001, 2002a).

Elasmobranchs may have a large sensory advantage over their prey during crepuscular and nocturnal periods (see Chapter 12), and many studies have found evidence for increased foraging activity during these times. Scalloped hammerhead sharks in the Gulf of California spend their days in large groups swimming around seamounts and no feeding occurs. At dusk, these sharks disperse from the seamount into pelagic waters to feed (Klimley and Nelson, 1981, 1984; Klimley et al., 1988), apparently navigating along geomagnetic fields (Klimley, 1993). They appear to use the same seamounts repeatedly, probably as a reference point for their foraging excursions (Klimley et al., 1988) and would conform to a central-place foraging system. However, the sharks do not return to the seamount every day, which may be dictated largely by water temperature fluctuations (Klimley and Butler, 1988). Many gray reef sharks, *Carcharhinus amblyrhynchos*, form daytime aggregations that occupy relatively limited areas then disperse at night, presumably to forage (Nelson and Johnson, 1980; McKibben and Nelson, 1986; Economakis and Lobel, 1998). Scalloped hammerhead pups have been observed forming groups during the day and making larger excursions at night (Holland et al., 1993). However, many prey species of the pups are diurnal species that hide at night and it is unclear when and how they capture these prey (Clarke, 1971). Several species of rays have been shown to become more active at night. For example, Hawaiian stingrays, *Dasyatis lata*, have higher ROM and cover larger activity spaces at night (Cartamil et al., 2003). Similarly, Pacific electric rays, *Torpedo californica*, are buried during the day but appear to actively seek prey at night (Bray and Hixon, 1978; Lowe et al., 1994), and attack potential prey items that are presented to them significantly faster at night (Lowe et al., 1994).

For some visual predators, diurnal foraging may be the most common tactic. For example, white sharks arrived at bait stations off southern Australia primarily during daylight hours, and sharks foraging on pinnipeds are probably diurnal because of their reliance on vision during prey detection and capture (Strong, 1996). However, white sharks at Año Nuevo, CA patrolled near seal haul-outs during both diurnal and nocturnal periods with equal frequency (Klimley et al., 2001). Catch rates of tiger sharks were substantially higher during daylight hours in Shark Bay (Heithaus, 2001c) despite previous suggestions that tiger sharks were nocturnal (Randall, 1992). Given the flexibility observed in foraging tactics of many elasmobranchs, it is likely that there is geographic variation in diel patterns of foraging and movement, so generalizations from one or a few studies should be viewed with caution. The use of bioluminescence during prey capture at night, or at depth, has been suggested for a number of species, including Pacific angel sharks (Fouts and Nelson, 1999) and blue sharks (Tricas, 1979). Also, based on its diet and visual capabilities, which are tuned to bioluminescent light, the deep-water blackmouth dogfish, *Galeus melastomus*, likely uses bioluminescence to capture prey while the sympatric small spotted dogfish, *Sclyiorhinus canicula*, does not (Bozzano et al., 2001).

Some species do not show obvious diel patterns. Juvenile lemon sharks did not show a diel pattern in the number of burst-speed events that may be associated with prey capture or possibly predator avoidance (Sundström et al., 2001), and there are no diel patterns in the swimming behavior of whale sharks (Gunn et al., 1999) and mako sharks (Holts and Bedford, 1993).

17.3.3   Capturing and Consuming Prey

17.3.3.1   Stalking and Ambush — Stalking and ambush are two tactics elasmobranchs may use to capture swift or large prey. Stalking is the process by which a predator attempts to reduce the distance between itself and its intended prey without being seen (stealth) before a rapid chase over the final distance. Ambush predators conceal themselves and allow the prey to approach before a rapid attack. For both stalking and ambush tactics, a close, undetected approach to prey is critical to success.

Sevengill sharks, *Notorynchus cepedianus*, have been observed swimming slowly near prey and then suddenly making a speed burst to capture small leopard sharks that had not reacted to the predator (Ebert, 1991). Other ambush tactics include selecting turbid and low-light waters and slowly gliding up from below prey so they are not detected (Ebert, 1991). In Shark Bay, tiger sharks may use oscillatory swimming as a stalking tactic, as they are able to closely approach benthic prey items when descending from above (Heithaus et al., 2002a). When swimming along the bottom, sharks would have better chance of surprising prey at the surface. White sharks also use a stalking approach, staying close to the bottom, then rushing to the surface to capture unsuspecting prey silhouetted above (Tricas and McCosker, 1984; Strong, 1996). The use of this tactic has also been proposed for sevengill sharks (Ebert, 1991), tiger sharks (Heithaus et al., 2002a), and blue sharks (Carey and Scharold, 1990).

Staying close to the bottom, or below prey, is an effective stalking tactic for several reasons. First, approaching from below gives a predator a detection advantage over prey where prey have limited visibility into the water due to light attenuation and predators have better visual detection capabilities of silhouettes (Strong, 1996). An approach from below surface prey also is advantageous by limiting escape routes of prey (Strong, 1996). Finally, by staying close to the bottom, an elasmobranch may be camouflaged against the substrate or darker, deeper waters (Klimley, 1994; Goldman and Anderson, 1999; Heithaus et al., 2002a). One interesting aspect of white shark attacks is their tendency to attack floating objects that are not prey. Hunting success of stalking predators like white and tiger sharks relies on a close approach and attack before being detected because of these species' limited speed and maneuverability relative to their prey (Strong, 1996; Heithaus et al., 2002a). Therefore, the costs of waiting to gather more information about prey identity (in terms of probability of being detected) likely outweigh the costs of attacking nonprey items at the surface.

Ambush predation is widespread in elasmobranchs. Crypsis, including hiding as well as body coloration, can enhance the efficiency of ambush predators. Some elasmobranchs lie in wait buried in the sand or concealed in coral or rock caves and crevices. Pyjama sharks, *Poroderma africanum*, have been observed ambushing squid by concealing themselves among squid eggs (Smale et al., 1995). Once squid had habituated to the shark's presence and returned to the site, the shark would attack squid as they approach the seafloor to lay eggs. Diamond rays, *Gymnura matalensis*, ambush spawning squid by burying themselves in the sand near egg beds (Smale et al., 2001), and torpedo rays ambush their prey by jumping from the bottom (Bray and Hixon, 1978; Michaelson et al., 1979; Belbenoit, 1986; Lowe et al., 1994). Pacific angel sharks lie in body pits to ambush prey and appear to use the same ambush sites repeatedly, occasionally making longer movements that may be to select new ambush sites (Fouts and Nelson, 1999). Ambush sites were primarily adjacent to rock–sand interfaces or patch reefs, which may serve as refugia for prey species and thus may maximize encounter rates with prey.

17.3.3.2 Prey Herding and Manipulation

While some swift species like mako and salmon sharks, *Lamna ditropis*, may be capable of swimming down a meal, others rely on manipulating prey behavior to make capture more energetically efficient. Some species of elasmobranchs are thought to lure prey. Megamouth sharks might be able to attract prey toward the mouth with luminescent tissue along the upper jaw (Diamond, 1985; Compagno, 1990). Cookie-cutter sharks, *Isistius brasiliensis*, may be squid mimics (Jones, 1971) and lure a host close enough to attack. Widder (1998) expanded this hypothesis further suggesting that the "collar" of nonluminescent tissue on the underside of the cookie-cutter would stand out from the general luminescence of deep-sea waters, which would mimic the search image of pelagic predators approaching from below. Myberg (1991) suggested that the white tips on the fins of oceanic whitetip sharks, *Carcharhinus longimanus*, function to lure fast-swimming prey close enough to be successfully captured. Given the generally good water visibility in which oceanic whitetips are found, potential prey might be able to distinguish a predator from a great distance, which would make lures less advantageous. Although untested, these luring hypotheses are worthy of experimental tests. A first step would be to determine if prey species (such as tuna and mackerel for whitetips) will respond to shark models in the ways predicted by these hypotheses.

Some elasmobranchs may take advantage of antipredatory behavior of prey in order to increase feeding efficiency. Blacktip reef sharks, *C. melanopterus*, have been observed in groups chasing small teleosts up onto the shoreline, then beaching themselves to feed on the fish (see Wetherbee et al., 1990). Sharks in pelagic environments feed on baitfish herded into tight schools. For example, silky sharks, *C. falciformes*, will feed from schools herded together by bottlenose dolphins, *Tursiops truncatus* (Acevedo-Gutiérrez, 2002).

17.3.3.3 Prey Debilitation

Rather than attempting to capture prey immediately, some species debilitate it before they consume it. Thresher sharks apparently use the greatly exaggerated upper lobe of their caudal fin to hit small fish, killing or stunning them (Allen, 1923). A great hammerhead shark was observed using its head to hit a fleeing stingray into the bottom, then pinned the ray to the bottom with its head so it could debilitate it with bites to both of the ray's pectoral fins (Strong et al., 1990).

White sharks may also use a prey debilitation tactic, which involves the shark making a first bite followed by a release of the prey until it bleeds to death (Tricas and McCosker, 1984; Klimley et al., 1996a). This "bite and spit" tactic has been interpreted as sharks reducing their probability of being injured by their prey (Tricas and McCosker, 1984; McCosker, 1985), but it is still unclear whether this tactic is widely used (Klimley, 1994; Klimley et al., 1996a). Electric rays may either swim over their prey when actively searching or jump from the bottom when they are concealed and envelop their prey within their disk. Then, they discharge an electric current that debilitates the prey, after which they can orient the prey and ingest it (e.g., Bray and Hixon, 1978; Belbenoit, 1986; Lowe et al., 1994). The electric ray, *Torpedo ocellata,* is able to emit electric discharges immediately after birth but the voltage increases substantially in the first 3 weeks (Michaelson et al., 1979).

17.3.3.4   Benthic Foraging — Benthic foraging is found in a diverse array of elasmobranchs. While some merely capture prey along the bottom, others dig or excavate the bottom to capture infaunal prey. Epaulette sharks, *Hemiscyllium ocellatum*, have been observed burying their bodies up to the first gill slit to capture prey (Heupel and Bennett, 1998), and some rays are able to forage on infauna buried deep in the substrate (e.g., Smith and Merriner, 1985; Matern et al., 2000). When foraging on infaunal prey, rays produce pits that leave a record of their foraging activity. See Motta (Chapter 6, this volume) for details of benthic foraging tactics by rays and sharks.

17.3.3.5   Batch Feeding — While most species of elasmobranchs are raptorial predators, consuming a single prey item at a time, a few species have evolved into batch feeders. These include the whale, basking, and megamouth sharks and manta rays. There are several tactics used in batch feeding. The most common is ram-ventilation filter feeding where the predator swims through prey patches with its mouth open, straining prey from the water. This tactic is observed in whale sharks, basking sharks, and manta rays. Megamouth sharks probably gulp concentrations of their prey (Compagno, 1990), and whale sharks sometimes use a gulping tactic (Heyman et al., 2001).

For years, it was thought that basking sharks hibernated over winter because of observations of sharks without gill rakers (see Francis and Duffy, 2002). Recent findings have cast considerable doubt on this hypothesis including observations of basking sharks with gill rakers during winter, winter captures of sharks in midwater, and basking sharks foraging at plankton densities well below that previously thought to be profitable (Sims, 1999; Francis and Duffy, 2002). Basking sharks are selective filter feeders (Sims and Quayle, 1998), but they forage at swimming speeds slower than predicted by optimal filter-feeding models (Sims, 2000). The mismatch between theoretical predictions and observed swimming speed is likely due to higher drag incurred by basking sharks than the small teleost fishes for which model predictions were developed (Sims, 2000).

17.3.3.6   Ectoparasitism — Although we typically think of parasites as small organisms living in or on the bodies of another organism (see Chapter 18), there is an elasmobranch feeding tactic that is more similar to parasitism than it is to predation. This tactic, sometimes called ectoparasitism (Long and Jones, 1996; Heithaus, 2001a; Heithaus and Dill, 2002b), involves a shark gouging a mouthful of tissue from a "host." Cookie-cutter sharks are ectoparasites of many large teleosts and marine mammals and even other sharks (Jones, 1971; Diamond, 1985; Hiruki et al., 1993; Heithaus, 2001a). A shark is able to create a vacuum with its tongue and fleshy lips, then spin around using its teeth to remove a plug of flesh (Jones, 1971). Portuguese dogfish, *Centroscymmus coelophis*, may also parasitize marine mammals in this manner (e.g., Ebert et al., 1992), and the dentition, tongue, and fleshy lips of the kitefin shark, *Dalatias licha*, suggest that this species may also use an ectoparasitic tactic (Clark and Kristof, 1990). Small carcharhinid sharks may also use an ecotoparasitic strategy. Indian Ocean bottlenose dolphins, *Tursiops aduncus*, in Shark Bay have been observed with scars inflicted from small shallow-water sharks (Heithaus, 2001d). However, the generality of this foraging tactic in such species is unknown.

17.3.3.7   Scavenging — Scavenging, at least opportunistically, is probably one of the most common feeding tactics of elasmobranchs (e.g., Compagno, 1984a,b; Smith and Merriner, 1985), and for

some species or age classes it is probably the primary feeding tactic. For example, large great white sharks are thought to adopt a diet composed largely of cetacean carcasses in the Atlantic Ocean (Carey et al., 1982; Pratt et al., 1982). Some observations suggest that white sharks may defend carcasses from both conspecifics and other species (e.g., Pratt et al., 1982; McCosker, 1985; Long and Jones, 1996), but white sharks have been observed scavenging a whale carcass concurrently with tiger sharks (Dudley et al., 2000).

17.3.3.8   Group Foraging — Group foraging may increase an elasmobranch's ability to gather resources. Groups may form for reasons other than foraging, such as reproduction (see Chapter 10) or reducing predation risk (Section 17.2.2.4), forcing individuals to forage in close proximity to each other and resulting in intraspecific competition (see Section 17.3.4). For example, cownose rays forage in groups (Smith and Merriner, 1985), but it is unclear why these groups form. Not all groups of foraging elasmobranchs are necessarily beneficial to the individuals in the group. In most elasmobranchs, foraging groups are likely a result of mutual attraction to prey resources and are intensely competitive in nature. Groups of both shortnose spiny dogfish, *Squalus megalops*, and smoothhounds, *Mustelus mustelus*, have been recorded at spawning aggregations of chokka squid, *Loligo vulgaris reynaudii*, but little foraging was observed, and it is possible that groups formed for nonforaging reasons as well (Smale et al., 2001). Sevengill sharks will aggregate in large groups around potential food (Ebert, 1991). Finally, white sharks at Año Nuevo, CA appeared to hunt in relatively close proximity. Although movements are most consistent with an individual foraging tactic, sharks probably remained in relatively close proximity to take advantage of kills by other individuals (Klimley et al., 2001).

Some groups may increase the *per capita* intake of the individuals in the group through increased detection of prey or increased probability of prey capture and death. Cooperative foraging groups have not been conclusively shown in elasmobranchs (also see Motta, Chapter 6 of this volume), but sevengill sharks appear to cooperatively forage on seals, cetaceans, and large rays. These groups surround a potential prey item, and then slowly circle until they all converge for the kill (Ebert, 1991). Cooperation is defined as "an outcome that — despite individual costs — is 'good' in some appropriate sense for the members of the group, and whose achievement requires collective action" (Mesterson-Gibbons and Dugatkin, 1992). Sevengill sharks appear to conform to this definition as the outcome is "good" for the group in that larger prey is taken than would normally be available. The cost is somewhat unclear, but may involve lost foraging opportunities on other prey while large prey are subdued (see Heithaus and Dill, 2002b).

Shark groups are often found with groups of tuna and dolphins in pelagic waters (Leatherwood, 1977; Au, 1991). In the eastern tropical Pacific sharks are very common around tuna schools that have congregated around logs (Au, 1991). In the Gulf of Mexico, many pods of dolphins are also followed by sharks (Leatherwood, 1977), and oceanic whitetip sharks are found in association with short finned pilot whales, *Globicephala macrorhynchus*, off Hawaii (pers. obs.). In the first two cases, sharks may attack tuna or dolphins or they may be following to take advantage of prey detection capabilities of dolphins and tuna. Because of the body size differences between pilot whales and whitetip sharks, predation is extremely unlikely and the sharks may follow whales for foraging reasons.

17.3.4   Competition

Competition is an almost ubiquitous aspect of elasmobranch life. This competition may be intra- or interspecific and may take the form of exploitative or interference competition. In exploitative competition, the consumption of a prey item by one individual removes it from possible consumption by another. Interference competition may take several forms, where individuals actively exclude others from prey resources (contest competition) or merely get in the way of other foragers, reducing foraging efficiency.

Competition among elasmobranchs has been quantified in the North Sea (Ellis et al., 1996) and among sharks and between sharks and dolphins off the coast of South Africa (Heithaus, 2001a), and significant dietary overlaps have been found in both locations. Qualitative comparisons also suggest competition is common. Gray reef sharks engage in competition for bait, but there is no intraspecific aggression (Nelson and Johnson, 1980). Silky sharks have been observed competing with bottlenose dolphins over schooling fish (Acevedo-Gutiérrez, 2002), and as the number of sharks increased, the intake rate of dolphins appeared to decrease. Also, large beaked skates, *Dipturus chilensis*, compete with southern sea lions, *Otaria flavescens* (Koen Alonso et al., 2001). In some situations, competition may be relatively weak. For example, there does not appear to be interference competition among eagle rays foraging on bivalves even at the highest foraging densities, but this may be due to generally low densities of rays (Hines et al., 1997).

The implications of competition within and among elasmobranch species and between elasmobranchs and other taxa are largely unknown. However, competition appears to be important in determining the species composition and abundance of several elasmobranch communities (see Section 17.4). We know little about how elasmobranchs respond to competition behaviorally or over short timescales, but it is probably important in determining habitat use patterns and structuring inter- and intraspecific interactions. Theoretical extensions of the IFD have shown that if individuals differ in their competitive ability (i.e., the division of resources is not equal among individuals), then the distribution of animals may deviate substantially from that of their prey with final distributions dictated partially by individuals' abilities to monopolize resources (see Tregenza, 1995).

Territoriality is one way that competition can be manifested, with individuals defending food resources. Currently, there is no evidence for territoriality in elasmobranchs (e.g., Nelson et al., 1986; Klimley et al., 2001), which may be due to the indefensibility of most food resources. However, size-based dominance hierarchies may exist in many shark species. For example, large sevengill sharks displace smaller conspecifics from baited situations (Ebert, 1991), and large white sharks may displace smaller ones from whale carcasses (e.g., Pratt et al., 1982; Long and Jones, 1996).

Kleptoparasitism, or food stealing, is an extreme case of interference competition, which appears to be widespread in sharks. Sevengill sharks that have captured relatively large bodied prey (e.g., smaller sharks) and not consumed it quickly will have some of their prey taken by conspecifics (Ebert, 1991). White sharks will compete over pinniped prey, and the shark that makes a kill may be driven away from it (e.g., Klimley et al., 1996b, 2001). Klimley et al. (1996b) have argued that tail-slapping behavior, and in some cases breaches, observed near kills are displays directed at conspecifics that are competing for the carcass. Sharks may also kleptoparasitize other species. For example, Hawaiian monk seals, *Monachus schauinslandi*, foraging in open habitats are often followed by both whitetip reef sharks and gray reef sharks that attempt to steal food captured by the seal (F. Parrish, pers. comm.).

17.3.5 Foraging Trade-Offs

17.3.5.1 Foraging–Safety Trade-Offs — All organisms face trade-offs, which are inevitable as time and resources are limited and demands are often in conflict. One trade-off faced by many taxa is that between foraging and avoiding predators (see Lima and Dill, 1990; Lima, 1998a, for reviews). This is because the most energetically productive habitats are often the most dangerous, and behaviors that increase foraging efficiency often increase the risk of being killed by a predator (Lima and Dill, 1990; Lima 1998a,b). Because most species of elasmobranchs are both predators and prey, an energy intake–predation risk trade-off is certainly important for many species but has yet to be specifically studied in any species.

One way to test for the existence of food–risk trade-offs is to measure the giving up density (GUD) of foraging individuals. The GUD is the density of food remaining in a patch at the time an individual, or group of individuals, ceases foraging and abandons the patch (Brown, 1988, 1992b, 1999). GUDs should be greater in habitats with higher risk than those with lower risk because the marginal gain of continued foraging in a high-risk habitat does not outweigh the benefits of continued foraging at low food densities. However, the exact GUD in a patch will also be influenced by food availability in other patches, as time spent foraging represents lost foraging opportunities in other patches (Brown, 1988, 1992b, 1999). This prediction has been supported in a number of species foraging on immobile prey (e.g., granivorous rodents, squirrels; Brown 1988, 1992b). There are currently no studies of GUDs in elasmobranchs. Although measuring GUDs is likely to be very difficult for species that consume mobile prey, studying this parameter may be useful in studies of benthic foragers.

It is possible to make predictions about how predation risk should modify habitat use decisions of foraging animals. First, there are a variety of strategies that elasmobranchs might reasonably use to trade off risk with foraging (see review in Brown, 1992a). First, it is possible that animals must attain only some minimum energy (maintenance costs) to survive and any additional energy does not result in fitness increases. In this situation, they maximize their safety value as long as they meet this minimum energy requirement. When reproduction is relatively infrequent, individuals must survive for extended periods to realize increased reproductive output. In this situation, animals may try to maximize the product of safety (probability of survival) and the number of surviving descendants (which may be approximated by energy intake is some cases) (Brown, 1992a). Some studies suggest that animals may adopt a strategy of maximizing energy intake over their lifetimes by minimizing the risk of predation in a habitat (μ) divided by the energy intake in that habitat (f) (μ/f rule; Gilliam and Fraser, 1987). There is empirical support for this prediction (Gilliam and Fraser, 1987); it is likely to be a good predictor of the behavior of juvenile animals that are faced primarily with the challenges of growth, and survival and reproductive decisions are unlikely to cause deviations from such optimal habitat use. The μ/f rule as well as incorporating predation risk into IFD-based models (e.g., Moody et al., 1996) suggests that animals should forage in relatively less productive habitats if they are safer. However, some individuals may select higher-risk habitats to take advantage of greater growth options there (e.g., Gilliam and Fraser, 1987; Abrahams and Dill, 1989; Lima and Dill, 1990; Lima, 1998a).

Most models of food–risk trade-offs assume that predators are behaviorally inert and cannot modify their distributions in accordance with decisions made by their prey, which has been a major oversight only now beginning to be addressed (Lima, 2002). When both predators and prey can move freely, a predator–prey game ensues, and game theoretical modeling can help predict optimal behaviors of both predators and prey (see Dugatkin and Reeve, 1998). When predator–prey games are considered, habitat selection by the middle predator (e.g., small sharks or rays) may not be influenced by the amount of food in a habitat at all, as they should distribute themselves across habitats proportional to the intrinsic habitat risk (e.g., Hugie and Dill, 1994; Sih, 1998; Heithaus, 2001b).

A trade-off between food availability and predation risk may be important in habitat use decisions of juvenile sharks at multiple spatial scales including the use of nursery areas and movement patterns within these nurseries. Reducing predation risk seems to be of primary importance at both of these spatial scales (Section 17.2.2.1). For example, a study of juvenile blacktip shark movements within a Florida nursery failed to find a link between shark movements and prey abundance (Heupel and Hueter, 2002). However, with current techniques it is difficult to measure the relative importance of food and safety in determining habitat use of elasmobranchs because of their relatively low energetic requirements and the possibility that some species confine feeding activity to short time periods followed by long periods without foraging (see Chapter 8). Individuals are likely to modify their habitat use decisions depending on which behavioral state they are in, and most studies of elasmobranch habitat use employ techniques that rely on determining average habitat use of individuals over relatively long time intervals and are not able to assess an animal's behavior at any particular time (see Chapter 19). Therefore, our knowledge of how elasmobranchs might trade off food and safety is still limited.

The activity level of forager is also subject to a trade-off between energy intake and risk of death. This occurs because increased activity rates are generally associated with both increased food intake and increased probability of predation. Theoretical models suggest that the optimal activity level depends on the relationship between activity rate and feeding rate and predation risk, the density of foragers relative to their prey, and the state of the individual (e.g., Werner and Anholt, 1993; Walters and Juanes, 1993). Game theoretical models of activity levels of predators and prey reveal results strikingly similar to habitat use games. Predator activity level is predicted to parallel that of the availability of *their prey's food* while prey should maintain constant activity levels (Brown et al., 2001).

Foraging under predation risk may lead to changes in diet selectivity, but the nature of this change depends on how prey items influence the risk of death (Houtman and Dill, 1998). In general, animals will increase their acceptance of food items associated with lower predation risk (e.g., items with relatively low handling times), which may lead to increased diet selectivity (if profitable prey are safer), reduced selectivity (if less profitable prey are safer), or no change in selectivity (if prey are of similar risk) (Houtman and Dill, 1998).

Vigilance and other antipredator behaviors, like hiding, are often mutually exclusive with foraging. Optimal vigilance levels will vary with risk because individuals that overinvest in vigilance are likely to realize lower fitness than those that do not because of reduced energy intake, whereas those that underinvest in vigilance are more likely to be killed by predators (Brown et al., 1999). Vigilance levels may also be influenced by temporal variation in the risk of predation (i.e., pulses of high and low risk). Counterintuitively, vigilance during high-risk periods will be *lower* when the proportion of time spent at high risk is greater (Lima and Bednekoff, 1999; Hamilton and Heithaus, 2001). This is because an animal often at high risk cannot afford to invest heavily in antipredator behavior during periods of high risk without seriously compromising energy intake. Such a system may occur for elasmobranchs in shallow waters as periods of higher water may allow increased access to predators.

Finally, food–risk trade-offs likely influence elasmobranch group sizes. Generally, predation risk selects for larger groups while increased foraging competition tends to select for smaller groups, and observed group sizes often reflect a balance of these conflicting selective pressures (Bertram, 1978; Lima and Dill, 1990).

17.3.5.2 Foraging–Reproduction Trade-Offs

17.3.5.2 Foraging–Reproduction Trade-Offs — Sometimes animals must trade off gathering energy efficiently or growing with securing mates or investing in reproduction, and it is possible that mating systems are actually the result of tactical decisions in response to ecological conditions rather than simple fixed strategies (e.g., Lott, 1984; Siems and Sikes, 1998). Dental sexual dimorphism appears to be one example. While differences in male and female dentition are observed in a number of species, there are seasonal changes in this dimorphism within Atlantic stingrays, *Dasyatis sabina*. Female dentition is stable year-round, but males possess recurved cuspidiform dentition during the breeding season and female-like molariform dentition outside of the breeding season (Kajiura and Tricas, 1996). The molariform dentition is likely the most efficient for feeding, but is not well suited for reproduction by males, which must use their teeth to grasp females during reproduction (Kajiura and Tricas, 1996). Therefore, male changes in dentition represent a trade-off between foraging and reproductive success. Further studies are required to determine whether seasonal dental sexual dimorphism is common in elasmobranchs with well-defined mating seasons.

Reproductive effort may be reduced when predation risk is high, and the reproductive habitat use and reproductive tactics used may be influenced by risk (Lima and Dill, 1990). Sex-specific habitat use may be the result of foraging–reproduction trade-offs. Klimley has suggested that female scalloped hammerhead sharks move offshore and adopt a pelagic diet at a smaller size than males, which results in higher growth rates for females (Klimley, 1987). Cailliet and Goldman (Chapter 14 of this volume), however, suggest that growth rates for male and females are intrinsically different for many species. This raises the question of why juvenile males would not make the shift to a pelagic lifestyle at a smaller size to take advantage of higher growth rates. Klimley (1987) suggests that small individuals shifting to pelagic habitats incur a higher risk of predation. Females, however, are willing to accept this higher risk because large body size is important to reproductive success. Males likely do not take the risks because the benefits of increased growth are not high enough. Other species of sharks show similar sex differences in habitat use and growth rates (Klimley, 1987), raising the possibility that foraging–reproduction trade-offs are common in elasmobranchs. For example, sex differences in foraging habitat use and activity patterns of lesser spotted dogfish, *Scyliorhinus canicula* (see Section 17.3.6.2), probably represent females making a trade-off between foraging and male harassment (Sims et al., 2001).

The need to deliver pups in nursery areas with low predation risk (Castro, 1987) may cause females to abandon more productive foraging areas to migrate to nursery areas. Early work on sharks suggested another foraging–reproduction trade-off: that females might fast when they enter nursery areas in order to protect their young (e.g., Springer, 1960; Olsen, 1984). However, this hypothesis has yet to be verified, and there is mounting evidence that this is not the case (see Wetherbee et al., 1990).

17.3.5.3 Intraguild Predation

17.3.5.3 Intraguild Predation — Intraguild predation (IGP), where competitors are also predator and prey (Polis et al., 1989), creates special trade-offs for the intraguild prey. Intraguild predation may be symmetrical where both species eat each other or asymmetrical where only one species eats its competitor. It may also be age structured, where only certain age–sex classes are engaged in IGP. Cannibalism is an example of asymmetrical age-structured IGP within a species. IGP appears to be common among sharks and also between sharks and other taxa like cetaceans. For example, IGP occurs among sharks and between sharks and dolphins off South Africa (see Heithaus, 2001a, and references therein). Intraguild predation can have dramatic consequences for the coexistence and spatial distribution of intraguild predators and prey (e.g., Holt and Polis, 1997; Heithaus, 2001b), and it is possible that IGP between killer whales and white sharks was responsible for the displacement of white sharks from the Farallon Islands during a season when killer whales were present (Pyle et al., 1999; Heithaus, 2001a).

17.3.6 Variation in Feeding Strategies and Tactics

17.3.6.1 Ontogenetic Variation — Ontogenetic changes in diet or prey size are found within many species (see Chapter 8) and may result in changes in foraging tactics. For example, young white sharks are more agile than larger ones and, thus, are able to capture fast-swimming teleost prey while large sharks must rely on stealth to capture large mammalian prey (Tricas and McCosker, 1984). In cownose rays, the shift from nonburying and shallow-burying bivalves to deep-burrowing ones (Smith and Merriner, 1985) would result in a shift in foraging tactics from collecting benthic prey to excavation.

Ontogenetic changes in foraging tactics and habitat use can also result from changes in sensitivity to predation risk, with more susceptible juveniles selecting safe habitats and shifting into more productive but dangerous areas as their susceptibility to predators decreases (e.g., Werner and Hall, 1988; Bouskila et al., 1998). Juvenile elasmobranchs inhabiting nursery areas face this decision of when to shift from safe nurseries with relatively high intraspecific competition into more productive areas that have more predators. In general, juveniles should delay switching habitats as predation risk outside nurseries increases (Bouskila et al., 1998). However, if many juveniles synchronize their departure, risk can be reduced through dilution and switching may occur sooner (Bouskila et al., 1998).

17.3.6.2 Inter- and Intraindividual Variation — With their large ranges, large-scale seasonal movements, and sometimes diverse diets, it is not surprising that elasmobranchs show considerable variation in foraging tactics. White sharks vary their feeding tactics with pinniped prey type. White sharks attack sea lions with greater force than elephant seals, which may be due to sea lions' better escape abilities and probability of wounding a shark (Klimley et al., 1996a). It also appears that white sharks do not use a bite and spit tactic with elephant seals while they may with sea lions (Klimley et al., 1996a). Torpedo rays appear to change their foraging tactics from a sit-in-wait ambush predator during the day to actively searching for prey at night (Bray and Hixon, 1978; Lowe et al., 1994). Tiger sharks in Shark Bay show individual variation in habitat preference (Heithaus et al., 2002a), which may represent differences in foraging tactics. Changes in prey availability may cause individuals to modify their foraging tactics. Pyjama sharks modify their daily hunting rhythm to take advantage of diurnally spawning squid (Smale et al., 1995), and a number of elasmobranchs aggregate at these spawning sites to take advantage of this seasonally abundant prey (Smale et al., 2001). Although not yet studied in elasmobranches, variation in body condition within and among individuals may cause differences in diet selection, foraging tactics, and risk-taking behavior (e.g., Houston et al., 1993; Bouskila et al., 1998). Hungry animals tend to spend less time in refuges and engage in more risk-prone behaviors like foraging in high-productivity and high-risk habitats (e.g., Houston et al., 1993; Lima, 1998b).

Sex differences in foraging tactics have been observed in several different species of elasmobranchs. For example, male lesser spotted dogfish in Lough Hyne off southwest Ireland rest in deep waters during the day, then move into shallow waters to feed on crustacean prey throughout crepuscular and nocturnal hours. Female dogfish, however, refuge in shallow-water caves during the day and through some nights, only emerging at night to forage in deeper waters (Sims et al., 2001). These differences appear to be due to female avoidance of males. Other sex differences in foraging tactics may be due to variation in selective pressures between males and females (Section 17.3.5.2).

17.4 Regulation of Elasmobranch Populations

The above sections largely investigated the behavioral mechanisms that elasmobranchs use to capture prey and to avoid predators. The end result of these interactions can have profound consequences on the equilibrium populations of both predators and prey. There have been few studies that have identified density dependence, and none of these has addressed the mechanisms for this density dependence. Juvenile lemon sharks in North Sound, Bimini, Bahamas, show density-dependent survival, with a carrying capacity of about 30 pups in the sound (Gruber et al., 2001). However, it is not known what sets this level. Similarly, Walker (1998) argues that density-dependent natural mortality occurs in young age classes of gummy sharks, *Mustelus antarcticus*, presumably largely due to predation.

Competition and/or food availability and predation may play important roles in regulating population sizes of some elasmobranchs. While fishing pressure has decreased populations of large-bodied skates (with some species disappearing) in the Irish Sea, small-bodied species have increased in abundance and biomass resulting in stable aggregate catch trends (Dulvy et al., 2000). Previous studies have shown significant dietary overlaps among species (Ellis et al., 1996), leading to the suggestion that small species have increased due to competitive release (Dulvy et al., 2000). Similarly, Walker and Heesen (1996) have suggested that competitive release or increased food availability from fishery discards has led to increases in starry ray, *Raja radiata*, populations in the North Sea. Small shark populations off South Africa may have been regulated to some degree by competition with or predation by large sharks as reductions in catches of large sharks by recreational fishers were linked with increases in small shark catches (van der Elst, 1979).

Both competition and predation may be important factors influencing population sizes of animals, but the effects of these two are often inseparable (e.g., Sih et al., 1985; Walters and Juanes, 1993; Werner and Anholt, 1996; Walters, 2000). For example, there is empirical support for the "predation-sensitive food" (PSF) hypothesis, which states that food and risk both act to limit populations for species that are both predators and prey (e.g., McNamara and Houston, 1987; Sinclair and Arcese, 1995). This arises through animals taking larger risks as food becomes limited and risk-taking individuals are killed. Antipredator behavior that limits foraging can also cause prey populations to be limited by a combination of both food and predators (Walters and Juanes, 1993; Lima, 1998b; Walters, 2000) or may stabilize otherwise oscillatory predator–prey dynamics (Lima, 1998b). Spatial and activity level components of antipredator behavior can influence population dynamics of both predators and prey in several ways (see Walters, 2000, for a detailed description). If prey restrict their movements to areas that are relatively safe from predators (e.g., small shark nursery areas), there is a limited "foraging arena" that is generally much smaller than the range of the prey's food. Because of this restriction in foraging area, it may appear that food is the limiting factor for populations even though larger population size would be possible if predators were not present (Walters and Juanes, 1993). The restriction in prey distribution may actually allow coexistence of prey species with similar diets because neither species exploits the full range of the prey species (Walters, 2000). Antipredator behavior by the prey will also influence predator populations as energy flow rates will be restricted relative to situations that ignore prey behavior, and give the appearance of bottom-up control of predator populations (Walters, 2000). One important insight from these dynamics is the importance of the spatial scale of sampling of prey food as prey surveys at too large a scale may miss the importance of intraspecific competition within restricted foraging arenas. Also, this view of population regulation challenges the traditional view that increasing predation risk acts to lower intraspecific competition because prey are kept well below the carrying capacity set by food resources. Instead, reduced activity levels or restricted foraging areas may increase intraspecific competition within these areas or during safe times and therefore increase the limiting effects of food (Walters and Juanes, 1993). Castro (1987) suggested that many shark populations may be limited by nursery area availability. This situation would perfectly fit the "foraging arena" scenario (Walters and Juanes, 1993; Walters, 2000), where populations would be limited by the presence of predators and antipredator behaviors of juvenile sharks and intraspecific competition among these sharks within the foraging arena.

17.5 Role of Elasmobranchs in Marine Ecosystems

As top predators, elasmobranchs are generally thought of as critical components of marine ecosystems, perhaps regulating prey populations and even community structure. However, a detailed analysis of their role in marine environments has not been undertaken and information on their importance has been largely based on supposition rather than data. This section begins with a consideration of the theoretical framework for top-down regulation of prey populations and community structure and then evaluates the likely role of elasmobranchs in marine ecosystems.

There are two ways in which predators may affect equilibrium population levels of prey populations. First, they may exert a regulatory force through density mechanisms, either through mortality inflicted on prey items (density-mediated direct interactions, DMDI) or through indirect density effects (density-mediated indirect interactions, DMII) where a predator reduces populations of a competitor or predator species with resulting effects on the species in question. While the majority of ecological literature has been focused on these density-mediated interactions, it is becoming increasingly apparent that predators may have a profound influence on prey through behavioral interactions, which may be either direct (BMDI) or indirect (BMII). Until recently, the influence of antipredator behavior on prey growth rates or population size had been largely overlooked. However, recent work suggests that behavioral effects of predators on prey are extremely important (e.g., Lima 1998b) and may actually be of comparable or even greater magnitude than density effects in some situations (Schmitz et al., 1997; Peacor and Werner, 2000, 2001). While somewhat counterintuitive, such results occur because direct mortality usually removes a limited number of individuals from a population, which may result in decreased intraspecific feeding or reproductive competition. This, in turn, can result in increased reproduction or growth among remaining prey individuals (compensatory reproduction or growth) with an end result of no reduction in population size. In contrast, antipredator behaviors, which may include leaving high-risk but high-productivity habitats or reduced foraging rates, are generally performed by all (or most) individuals in a population and can result in lower access to food and a resulting reduction in the population's reproductive potential. Antipredator behaviors can also reverse competitive asymmetries between prey species (e.g., Lima, 1998b; Relyea, 2000) and allow coexistence of competitor species (Lima, 1998b).

17.5.1 Density-Mediated Effects

Many studies of elasmobranch feeding comment that elasmobranchs, especially sharks, are responsible for regulating prey populations through density mechanisms, and this claim is often made simply because numerous prey individuals are killed. However, because we do not know where density dependence operates in these prey species, it is currently not possible to evaluate these hypotheses. Even high rates of predation on a species may not affect equilibrium population sizes if density dependence operates at a life history stage different from that where most predation occurs (e.g., Piraino et al., 2002).

Sharks have played an important role in the population decline of harbor seals on Sable Island (Stobo and Lucas, 2000; Bowen et al., 2003). Based on carcasses washing ashore with shark bites that were obviously not scavenged, shark attacks have been steadily increasing and sharks now regularly kill all age–sex classes. Predation, especially on adult females, influenced the substantial decline in pup production between 1980 and 1997. In fact, observed shark mortality from 1994 to 1996 accounted for around 50% of the decline in pup production from 1995 to 1997 (Lucas and Stobo, 2000; Bowen et al., 2003). It is still unclear why there has been an increase in apparent shark-inflicted mortality (Lucas and Stobo, 2000), but it is likely that shark abundance has increased in response to the substantial population increases in gray seals on Sable (Bowen et al., 2003). Although they are killed by sharks (Brodie and Beck, 1983; Bowen et al., 2003), gray seal populations have not been affected by shark predation because predation is extremely low relative to pup production, and adults likely face much lower risk than harbor seals (Bowen et al., 2003).

Predatory attacks by sharks are not always successful, and often leave injured individuals in prey populations. Some studies have attempted to estimate the effects of shark predation on prey populations by measuring the rate of scarring or injury in the population. Such methods are fraught with biases. For example, differences in wounding rates may reflect the probability of escape after capture rather than differences in attack and death rate. Therefore, it is difficult to make comparisons among populations or species that either face different sizes of predators or differ substantially in body size, antipredator behavior, or escape abilities. Despite these biases, such studies may provide some useful insights into the importance of shark predation on the populations and behavior of their prey. The rate of scars and wounds from white shark attacks found on pinnipeds along the California coast has led some investigators to suggest top-down control of pinniped populations (e.g., McCosker 1985). However, further work is needed to verify this hypothesis. Nonetheless, even nonlethal white shark attacks have substantial reproductive consequences for female elephant seals, *Mirounga angustirostris*. At Año Nuevo, CA only 8 of 11 adult females with fresh bites successfully weaned their pups, and the successful seals had the least severe injuries (LeBoeuf et al., 1982). Furthermore, none of the injured females was observed copulating before returning to sea, resulting in a probable loss of 2 years of reproduction. A similar result was found at the Farallon Islands (Ainley et al., 1981).

Large shark injury rates on Hawaiian monk seals in the northwest Hawaiian Islands are relatively low, generally less than 3.5% injured annually (Bertilsson-Friedman, 2002). However, there are large differences in age–sex classes attacked throughout the chain. At French Frigate Shoals, the largest subpopulation, pups are attacked most frequently, while attacks on pups are relatively few at Laysan and Lisianski Islands, where juveniles are attacked most frequently (Bertilsson-Friedman, 2002). These differences may be due to variation in the physical habitats and accessibility for large sharks. At French Frigate, it appears that Galapagos sharks, *Carcharhinus galapegensis*, may be responsible for a large number of attacks on pups, before or near the time of weaning, with most attacks observed in the very shallow waters of a small sand island where the density of pups is quite high (Bertilsson-Friedman, 2002; B. Antonelis, pers. comm.). No such attacks have been recorded elsewhere in the archipelago. At Lisianski and Laysan Islands, the opportunities for such predation attempts appear to be quite low and attacks are made in different habitats and by other species of large shark such as tiger sharks (Bertilsson-Friedman, 2002). The population consequences of shark predation are unknown.

Off Natal, South Africa, between 10 and 19% of bottlenose dolphins exhibit bite scars, and an estimated 2.2% of the population is killed annually by sharks (Cockcroft et al., 1989). In other areas, dolphins with smaller body sizes facing predation risk from the same shark species have much higher rates of wounds and scars, and thus sharks probably kill a higher proportion of these populations each year (Heithaus, 2001d). In Moreton Bay, Queensland 36.6% of dolphins bear wounds (Corkeron et al., 1987) and 74.2% of dolphins in Shark Bay, Western Australia have been attacked at least once in their lives with at least 10% of the population attacked unsuccessfully each year (Heithaus, 2001d). The mortality rate of dolphins in these locations and, thus, the direct effects of predation are unknown.

Studies of wounding have shown that prey age–sex classes may be affected differentially by shark predators. Male dolphins in several locations have higher rates of scarring or multiple scarring than do females (Heithaus, 2001d). In Shark Bay, male loggerhead turtles have significantly higher rates of major shark-inflicted injuries (58%) than females (12%) while there are no sex differences in wounding rates of sympatric green turtles (<6% injured) (Heithaus et al., 2002b). It is likely that inter- and intraspecific variation in wounding is the result of different attack rates (Heithaus et al., 2002b).

Wounds from cookie-cutter sharks have been found on a diverse array of species (e.g., Hiruki et al., 1993; Heithaus, 2001a). Almost every adult spinner dolphin observed off Hawaii shows signs of attacks from these sharks (Norris and Dohl, 1980). The implications of these attacks for their prey species are unknown, and although they are certainly less detrimental than predatory attacks they may have fitness consequences as energy must be used for recuperation that could have been invested in growth or reproduction.

Understanding the role of sharks in regulating prey populations can be very difficult because of the mobility of both predators and prey. Benthic foraging rays, however, offer an opportunity for experimental studies, and it has been shown that rays can have a large impact on their prey. Exclusion experiments have shown that ray predation and disturbance of sediments can have a negative effect on a number of invertebrate species in soft-bottom communities (Thrush et al., 1994). Cownose rays, *Rhinoptera bonasus*, have been observed to completely remove bay scallops, *Argopecten irradians concentricus*, from the most productive habitat patches in the Cape Lookout lagoonal system in North Carolina, causing a population sink (Peterson et al., 2001). Rays removed scallops before reproduction occurred causing the individuals in the most productive habitats to not contribute to future generations. However, bay scallops are an annual species and individuals remaining in habitats with low initial densities produce enough offspring to maintain population levels (Peterson et al., 2001). Thus, rays do not appear to regulate equilibrium population sizes, but are an important factor in population dynamics of their prey.

In some situations, there appears to be little effect of elasmobranchs on populations of their prey. Mako sharks consume between 4 and 14% of the available bluefish biomass between Cape Hatteras, NC and Georges Bank (Stilwell and Kohler, 1982). However, although the bluefish is a very important prey item of the sharks, there does not appear to be a significant impact on bluefish populations (Stillwell and Kohler, 1982).

Our understanding of DMII in elasmobranchs is even less developed, but such interactions may be important in community dynamics (see Section 17.5.3). For example, bat rays and round stingrays, *Urolophus halleri*, have a positive DMII with sand dabs, *Citharichthys stigmaeus*, and possibly seastars, *Astropecten verrilli*. While excavating prey, rays eject infaunal species that are normally unavailable to sand dabs and seastars. Thus, ray foraging increases food availability for these species (VanBlaricom, 1982). Understanding density-mediated indirect interactions involving elasmobranchs can be important to commercial operations. For example, oyster growers in Humboldt Bay, CA have tried to reduce populations of the bat ray because of its supposed role in destroying oysters. However, this may have negative consequences for oyster farms as rays do not appear to regularly feed on oysters, and instead are major predators on the primary oyster predator, red rock crabs, *Cancer productus* (Gray et al., 1997). Therefore, a reduction in ray populations may actually result in increased losses of oysters.

17.5.2 Behavior-Mediated Effects

For some prey species, the probability of being killed by an elasmobranch predator is quite low (e.g., dolphins; Heithaus, 2001a; Simpfendorfer et al., 2001). However, this does not mean that these prey are unlikely to be influenced by the risk of predation from elasmobranchs. Especially in long-lived species with slow reproductive rates, even a low risk of predation can lead to extreme antipredator behaviors as longevity can be a major determinant of fitness (Lima, 1998b). For example, bottlenose dolphins in Shark Bay are rarely found in the stomach contents of tiger sharks (Simpfendorfer et al., 2001; Heithaus, 2001c), but dolphin habitat use is greatly influenced by the presence of tiger sharks. When sharks are absent from the bay in winter months, foraging bottlenose dolphins distribute themselves between shallow seagrass habitats and deeper waters proportional to the food available in each as predicted by the IFD (Heithaus and Dill, 2002a). However, when tiger sharks move into the bay in warmer months, sharks prefer shallow seagrass habitats, which contain high densities of prey and are also the most productive for dolphins. This results in dolphins largely avoiding the productive shallow habitats, and instead they forage in the lower productivity, but safer deep habitats (Heithaus and Dill, 2002a). Theoretically, the habitat change by dolphins could reduce equilibrium population size of dolphins through reduced access to food during summer. Therefore, it is possible that tiger sharks are important in determining dolphin population size through behavioral effects. However, this hypothesis requires further testing.

In a behaviorally mediated indirect interaction (BMII), a change in one species (the "initiator") causes a behavioral response in a second species (the "transmitter"), which in turn influences a third species (the "receiver") (Dill et al., 2003). BMII may create, enhance, ameliorate, or even reverse the sign (e.g., a species actually has a positive effect on its competitor) of interactions between species, and thus understanding the dynamics of BMII is important in understanding community dynamics and conservation biology (see Dill et al., 2003, for a review of BMII in marine communities). In Shark Bay, tiger sharks are an important *transmitter* of a BMII between their primary prey, dugongs and sea snakes, and less common prey, bottlenose dolphins. In this interaction, the seasonal occurrence and habitat use of dugongs and sea snakes result in tiger sharks selecting shallow seagrass habitats during warm months and being largely absent during winter months. This causes dolphins to switch from using high-productivity shallow waters for foraging in the winter to the less-productive deeper waters in the summer (Heithaus and Dill, 2002a). This BMII may be further transmitted through the ecosystem as teleost populations in shallow seagrass habitats would enjoy reduced predation pressure from dolphins during warm months (Dill et al., 2003). Although this is the only BMII of which I am aware that includes elasmobranchs, it is likely that BMII are common in marine communities (Dill et al., 2003), including those with elasmobranchs, which probably play the role of initiator, transmitter, and receiver.

17.5.3   Community Consequences of Elasmobranch Predation

Elasmobranchs have often been cited as top predators that almost certainly play an important role in the dynamics of marine communities. However, being at a high trophic level does not necessarily equate with having a critical role in an ecosystem (e.g., Kitchell et al., 2002; see below). A predator plays a "keystone" role in a community when it has a large effect on its community through indirect effects that are disproportionately large compared to their abundance (Paine, 1966; Power et al., 1996; Piraino et al., 2002) and may lead to trophic cascades (Pace et al., 1999). This may occur because a predator limits the population size of competitively dominant prey species such that poor competitors can coexist in the community (Paine, 1966; Sih et al., 1985). However, density effects are not the only mechanism for trophic cascades, as antipredator behavior may cause a similar effect (e.g., Power et al., 1985; Schmitz et al., 1997; Lima et al., 1998b; BMII, Section 17.5.2). For example, if a prey species switches habitats to avoid a predator, the prey's food experiences relaxed predation pressure (e.g., Turner and Mittlebach, 1990; Schmitz et al., 1997).

It is usually impossible to test the community consequences of elasmobranch predation, as manipulative experiments would be necessary. Benthic foraging rays provide a system where such experiments are possible, and rays may influence marine ecosystems through disturbance of bottom sediments. Within sandflats in New Zealand, a particular patch of sediment may be excavated once every 70 days (Thrush et al., 1991), and similar turnover rates have been noted elsewhere (e.g., Reidenauer and Thistle, 1981; Hines et al., 1997). Off La Jolla, CA, 5.4% of transect areas were disturbed by rays *every day* during warm months (when ray abundance is high) and between 25 and 100% of the bottom area may be in some stage of infaunal recolonization from this disturbance (VanBlaricom, 1982). In some cases, invertebrate communities show very little response to ray disturbance of sediments. Harpacticoid copepod communities within pits excavated by stingrays, *Dasyatus sabina*, off the Gulf coast of Florida were indistinguishable from those in undisturbed areas within 29 h of ray excavations (Reidenauer and Thistle, 1981). Similarly, bivalves disturbed by eagle rays, *Myliobatis tenuicaudatus*, recolonized pits quickly with densities returning to predisturbance levels within 2 days (Thrush et al., 1991). However, polychetes were much slower to recover; after 12 days their densities were still reduced. In general, it is difficult to detect a substantial impact of individual pits on marine invertebrate communities, but ray disturbance may still be important in structuring these communities. With the relatively high turnover rate of sediments, ray impacts may best be understood as large-scale habitat modification and may smooth distribution patterns of infaunal invertebrates (Thrush et al., 1991; Hines et al., 1997). Also, while eagle rays consumed only 1.6% of the *Macomona* population and disturbed 5% of a study site, the rate of removal was as high as 8 to 12% of bivalves and 36 to 39% disturbance over 2 to 3 months in high-density patches (Hines et al., 1997). Off La Jolla, round stingray and bat ray foraging disturbance, unlike other predators in the system, structures the infaunal community. Initially, feeding pits clear all infauna, but rapid accumulation of organic material results in elevated food availability, which attracts certain invertebrate species. Over a period of weeks, infaunal species recolonize pits at variable rates leading to a succession in the community such that ray disturbance causes a mosaic of infaunal communities throughout the benthic habitat in different stages of recolonization (VanBlaricom, 1982). Digging rays are abundant throughout tropical and temperate areas, and VanBlaricom (1982) has suggested that they are probably important in structuring benthic communities in many locations.

Ecosystem models have been used to gain insights into the role of elasmobranchs in marine communities, but, not surprisingly, the results are mixed and are highly dependent on the community being modeled. Ecopath with Ecosim is based on a mass-balance assumption where energy flow must balance within the community. These models provide working hypotheses of community dynamics and allow simulations of the likely effects of perturbations in the environment and analyses of the role of individual species. These models, however, make a number of simplifying assumptions. First, they assume that the system is closed and that all energy consumed is from within the system. Thus, if species move outside the modeled system, the model may not make accurate predictions. Second, the models assume that the diets of individual species are fixed. Therefore, seasonal changes in diets, which are observed in elasmobranchs, may cause problems for these models. Inaccurate diet information may also be problematic. This is especially likely for large sharks, where sample sizes may be limited and there is often geographic variation in diets. Third, these models do not account for many behavioral interactions that may lead to predictions that are opposite from those that are actually observed (see Dill et al., 2003). Although Ecosim models may be able to mimic antipredator decisions by designating proportions of prey populations that are vulnerable and invulnerable to predation (Walters, 2000), for accurate predictions to be made it is critical to understand the actual behavioral dynamics between predator and prey and how tactics of both predator and prey shape the trophic interaction. Furthermore, the actual impact of predators on prey may be far greater than would be suggested by energy flow alone (e.g., Peacor and Werner, 2000, 2001). Therefore, future studies must begin to explore behavioral interactions between elasmobranchs and their predators and prey. Fourth, results may be highly dependent on model assumptions and parameter estimates. Finally, these models are very difficult to test, so results must be viewed with caution. To fully establish the role of elasmobranchs in marine ecosystems, an experimental approach is much preferred. However, this is extremely difficult in many systems, and models such as Ecopath and Ecosim provide an excellent starting point for investigations of community dynamics.

Kitchell et al. (2002) created a model of the Central North Pacific ecosystem, which suggested that sharks, including the blue shark, lamnid sharks, and carcharinid sharks like oceanic whitetip and silky sharks, are not keystone species. A simulated reduction of shark populations did not result in profound ecological effects primarily because of their relatively low feeding rates and slow rates of population turnover. Billfishes and tunas are able to reproduce quickly and show rapid compensatory responses to reductions in shark populations, making them the most important apex predators in the Central North Pacific, and probably other similar areas (Kitchell et al., 2002). Similarly, although sharks occupied the highest trophic level in a model of the southwestern Gulf of Mexico, there was little effect of increases in shark populations (Manichchand-Heileman et al., 1998), but no information was given on the effects of shark population declines. In contrast, within systems where sharks are the dominant apex predator, and billfishes and tuna are rare, sharks may be keystone predators.

Stevens et al. (2000) used Ecosim to model three marine ecosystems in which sharks are a major component: the northeast Venezuelan shelf ecosystem, the Alaska Gyre oceanic ecosystem, and French Frigate shoals in the northwestern Hawaiian Islands. The Venezuelan model includes small sharks (e.g., *Mustelus canis*), and suggests that the depression of small shark populations could cause major changes in other ecosystem components, which are not necessarily common prey of sharks. Salmon sharks, *Lamna ditropis*, are the primary elasmobranch in the Alaskan model, and also appear to have widespread effects in the ecosystem. Tiger sharks and reef sharks (*Carcharhinus* sp.) were included as separate components of the French Frigate Shoals model. Removal of reef sharks had almost no effect on other species in the ecosystem, but removing tiger sharks caused major changes in biomass throughout the ecosystem (Stevens et al., 2000). The results for tiger sharks must be viewed with some caution as the diet of tiger sharks in the area include prey from outside the modeled system (e.g., Tricas et al., 1981), and the diet likely changes substantially with changes in prey abundance such as albatross chicks (e.g., DeCrosta et al., 1984). Also, the importance of tiger sharks was linked to predation on seabirds, which are also foraging largely outside of the atoll (Fernandez et al., 2001). One interesting prediction of these models is that the largest responses to shark removal are not always major prey items and that elasmobranch's role in regulating a particular prey group is not necessarily related to the relative contribution of that species to the elasmobranch's diet (Stevens et al., 2000).

It may be possible to identify habitats and shark species that are most likely to play a keystone role in marine ecosystems. In general, elasmobranchs that prey upon species with slow life history strategies and have few other predators (e.g., marine mammals) are likely to be the most important in influencing their prey's population size (through density and behavioral mechanisms) as well as community structure and dynamics. Also, elasmobranchs feeding on species that are also known to affect community structure will play a disproportionate role in structuring particular communities. For example, sea turtles may be important in structuring benthic communities and detrital cycles (Bjorndal, 1997), so sharks feeding on sea turtles may have a disproportionate effect on community dynamics. The French Frigate Shoals model provides support for this hypothesis by suggesting that reef sharks play very little role in structuring the marine community compared to tiger sharks. This effect is observed because the main prey of reef sharks (reef fishes) are self-regulating rather than regulated by predation (Stevens et al., 2000). In contrast, tiger sharks in the seagrass ecosystem of Shark Bay, Western Australia are likely to have large impacts on community dynamics because they feed on the major grazers in the system, dugongs and sea turtles, which appear to invest heavily in antipredator behavior (M. Heithaus and A. Wirsing, unpubl. data). Another factor that will influence the importance of an elasmobranch in a marine community is the presence of close competitors. When such competitors are present, it is likely that these species can fill the role of the elasmobranch and thus a loss of such species is unlikely to have cascading influences through an ecosystem.

An interesting result of ecosystem models is the importance of intraguild predation among or cannibalism within shark species. Even if sharks are 1% of the diet of other sharks, the effects of shark population decline may spread to both prey and competitors of sharks (Kitchell et al., 2002), and it is obvious that greater attention must be paid to the potential roles of cannibalism and intraguild predation among sharks. Both IGP and cannibalism are commonly found within elasmobranchs, especially sharks (Section 17.3.5.3), so further studies of IGP in a community context are warranted. It is also essential that behavior not be ignored in these circumstances. Indeed, antipredator behavior may allow co-occurrence of intraguild predators and prey even with strong IGP even though it is not predicted in models that ignore behavior (Heithaus, 2001b).

It is clear that the ecological role of elasmobranchs is still poorly understood and a subject that is currently in need of serious research efforts, as some populations of elasmobranchs, especially large-bodied species, are declining around the world. Future studies, using an experimental framework or using natural fluctuations in ecological conditions as a natural experiment, will be of the most value, as will more detailed theoretical studies.

17.6 Summary

This review of elasmobranch predator–prey interactions shows that although there have been many interesting studies, there is yet much to be done to gain a more general understanding of the tactics of both predators and prey and how these interactions shape elasmobranch populations and the communities of which they are a part. One of the emerging themes in current studies of predators and their prey is the importance not only of behaviors, but of the interplay of predator tactics and resulting prey tactics. Therefore, future studies will be greatly improved by both incorporating game theoretical ideas and field studies that simultaneously study both elasmobranchs and their predators and prey. These studies should not be limited to coarse-scale surveys of predator and prey distribution, but should endeavor to understand underlying mechanisms and tactics that cause these distributions. The hope is that such studies will allow us to gain a functional understanding of elasmobranch behavior and the ability to make predictions about how changes in ecological conditions will affect them, and how changes to their populations are likely to influence marine communities.

References

Abrahams, M. V. and L. M. Dill. 1989. A determination of the energetic equivalence of the risk of predation. *Ecology* 70:999–1007.

Acevedo-Gutiérrez, A. 2002. Interactions between marine predators: dolphin food intake is related to the number of sharks. *Mar. Ecol. Prog. Ser.* 240:267–271.

Ackerman, J. T., M. C. Kondratieff, S. A. Matern, and J. J. Cech, Jr. 2000. Tidal influences on spatial dynamics of leopard sharks, *Triakis semifasciata*, in Tomales Bay, California. *Environ. Biol. Fish.* 58:33–43.

Ainley, D. G., C. S. Strong, H. R. Huber, T. J. Lewis, and S. H. Morrell. 1981. Predation by white sharks on pinnipeds at the Farallon Islands. *Fish. Bull.* 78:941–945.

Allen, W. E. 1923. Behavior of the thresher shark. *Science* 58(1489):31–32.

Anholt, B. R. and E. E. Werner. 1995. Interaction between food availability and predation mortality mediated by adaptive behavior. *Ecology* 76:2230–2234.

Au, D. W. 1991. Polyspecific nature of tuna schools: shark, dolphin, and seabird associates. *Fish. Bull.* 89:343–354.

Beck, B. and A. W. Mansfield. 1969. Observations on the Greenland shark, *Somniosus microcephalus*, in northern Baffin Island. *J. Fish. Res. Board Can.* 26:143–145.

Bednekoff, P. A. and S. L. Lima. 1998. Re-examining safety in numbers: interactions between risk dilution and collective detection depend upon predator targeting behaviour. *Proc. R. Soc. Lond. B* 265:2021–2026.

Belbenoit, P. 1986. Fine analysis of predatory and defensive motor events in *Torpedo marmorata* (Pices). *J. Exp. Biol.* 121:197–226.

Bertilsson-Friedman, A.K. 2002. Shark Inflicted Injuries to the Endangered Hawaiian Monk Seal, *Monachus schauinslandi*. M.S. thesis, University of New Hampshire, Durham.

Bertram, B. C. R. 1978. Living in groups: predators and prey, in *Behavioural Ecology: An Evolutionary Approach*. J. R. Krebs and N. B. Davies, Eds., Blackwell Press, Oxford, 64–96.

Bjorndal, K. A. 1997. Foraging ecology and nutrition of sea turtles, in *The Biology of Sea Turtles*. P. L. Lutz and J. A. Musick, Eds., CRC Press, Boca Raton, FL, 199–231.

Bonenfant, M. and D. L. Kramer. 1994. The influence of distance to burrow on flight initiation distance in the woodchuck, *Marmota monax. Behav. Ecol.* 7:299–303.

Bouskila, A., M. E. Robinson, B. D. Roitberg, and B. Tenhumberg. 1998. Life-history decisions under predation risk: importance of a game perspective. *Evol. Ecol.* 12:701–715.

Bowen, W. D., S. L. Ellis, S. J. Iverson, and D. J. Boness. 2003. Maternal and newborn life-history traits during periods of contrasting population trends: implications for explaining the decline of harbour seals, *Phoca vitulina*, on Sable Island. *J. Zool. Lond.* 261:155–163.

Bozzano, A., R. Murgia, S. Vallerga, J. Hirano, and S. Archer. 2001. The photoreceptor system in the retinae of two dogfishes, *Sclyliorhinus canicula* and *Galeus melastomus*: possible relationship with depth distribution and predatory lifestyle. *J. Fish Biol.* 59:1258–1278

Bray, R. N. and M. A. Hixon. 1978. Night shocker: predatory behavior of the Pacific electric ray (*Torpedo californica*). *Science* 200:333–334.

Brodie, P. and B. Beck. 1983. Predation by sharks on the grey seal (*Halichoerus grypus*) in eastern Canada. *Can. J. Fish. Aquat. Sci.* 40:267–271.

Brown, J. S. 1988. Patch use as an indicator of habitat preference, predation risk, and competition. *Behav. Ecol. Sociobiol.* 22:37–47.

Brown, J. S. 1992a. Patch use under predation risk: I. Models and prediction. *Ann. Zool. Fen.* 29:301–309.

Brown, J. S. 1992b. Patch use under predation risk: II. A test with fox squirrels, *Sciurus niger. Ann. Zool. Fen.* 29:311–318.

Brown, J. S. 1999. Vigilance, patch use and habitat selection: foraging under predation risk. *Evol. Ecol. Res.* 1:49–71.

Brown, J. S., J. W. Laundré, and M. Gurung. 1999. The ecology of fear: optimal foraging, game theory, and trophic interactions. *J. Mamm.* 80:385–399.

Brown, J. S., B. P. Kotler, and A. Bouskila. 2001. Ecology of fear: foraging games between predators and prey with pulsed resources. *Ann. Zool. Fen.* 38:55–70.

Bruce, B. D. 1992. Preliminary observations on the biology of the white shark, *Carcharodon carcharias*, in South Australian waters. *Aust. J. Mar. Freshwater Res.* 43:1–11.

Carey, F. G. and E. Clark. 1995. Depth telemetry from the sixgill shark, *Hexanchus griseus*, at Bermuda. *Environ. Biol. Fish.* 42:7–14.

Carey, F. G. and J. V. Scharold. 1990. Movements of blue sharks (*Prionace glauca*) in depth and course. *Mar. Biol.* 106:329–342.

Carey, F. G., G. Gabrielson, J. W. Kanwisher, and O. Brazier. 1982. The white shark, *Carcharodon carcharias*, is warm-bodied. *Copeia* 1982:254–260.

Caro, T. M. 1995. Pursuit-deterrence revisited. *Trends Ecol. Evol.* 10:500–503.

Cartamil, D. P., J. J. Vaudo, C. G. Lowe, B. M. Wetherbee, and K. N. Holland. 2003. Diel movement patterns of the Hawaiian stingray *Dasyatis lata*: implications for ecological interactions between sympatric elasmobranch species. *Mar. Biol.* 142:841–847.

Castro, J. I. 1987. The position of sharks in marine biological communities: an overview, in *Sharks, An Inquiry into Biology, Behavior, Fisheries, and Use*. S. Cook, Ed., Oregon State University Extension Service, Corvallis, 11–17.

Castro, J. I. 1993. The shark nursery area of Bulls Bay, South Carolina, with a review of the shark nurseries of the southeastern coast of the United States. *Environ. Biol. Fish.* 38:37–48.

Castro, J. I. 2000. The biology of the nurse shark, *Ginglymostoma cirratum*, off the Florida east coast and the Bahama Islands. *Environ. Biol. Fish.* 58:1–22.

Clark, C. W. and M. Mangel. 2000. *Dynamic State Variable Models in Ecology: Methods and Application.* Oxford University Press, Oxford.

Clark, E. and E. Kristof. 1990. Deep-sea elasmobranchs observed from submersibles off Bermuda, Grand Cayman, and Freeport, Bahamas, in Elasmobranchs as Living Resources. H. L. Pratt, Jr., S. H. Gruber, and T. Taniuchi, Eds., NOAA Tech. Rep. 90, 269–284.

Clarke, T. A. 1971. The ecology of the scalloped hammerhead shark, *Sphyrna lewini*, in Hawaii. *Pac. Sci.* 25:133–144.

Cliff, G. 1995. Sharks caught in the protective gill nets off Kwazulu-Natal, South Africa. 8. The great hammerhead shark *Sphyrna mokarran* (Rüppell). *S. Afr. J. Mar. Sci.* 15:105–114.

Cliff, G. and S. F. J. Dudley. 1991. Sharks caught in the protective gill nets off Natal, South Africa. 4. The bull shark *Carcharhinus leucas* Valenciennes. *S. Afr. J. Mar. Sci.* 10:253–270.

Cliff, G., S. F. J. Dudley, and B. Davis. 1989. Sharks caught in the protective gill nets off Natal, South Africa. 2. The great white shark *Carcharodon carcharias* (Linnaeus).). *S. Afr. J. Mar. Sci.* 8:131–144.

Cockcroft, V. G., G. Cliff, and G. J. B. Ross. 1989. Shark predation on Indian Ocean bottlenose dolphins *Tursiops truncatus* off Natal, South Africa. *S. Afr. J. Zool.* 24:305–310.

Compagno, L. J. V. 1984a. FAO Species Catalogue. Vol. 4. *Sharks of the World.* Part 1. *Hexanchiformes to Lamniformes.* FAO Fish. Synop. 125:1–249.

Compagno, L. J. V. 1984b. FAO Species Catalogue. Vol. 4. *Sharks of the World.* Part 2. *Carchariniformes.* FAO Fish. Synop. 125:250–655.

Compagno, L. J. V. 1990. Relationships of the megamouth shark, *Megachasma pelagios* (Lamniformes: Megachasmidae), with comments on its feeding habits, in Elasmobranchs as Living Resources. H. L. Pratt, Jr., S. H. Gruber, and T. Taniuchi, Eds., NOAA Tech. Rep. 90, 357–379.

Corkeron, P. J., R. J. Morris, and M. M. Bryden. 1987. Interactions between bottlenose dolphins and sharks in Moreton Bay, Queensland. *Aquat. Mamm.* 13:109–113.

Cortés, E. 1999. Standardized diet compositions and trophic levels of sharks. *ICES J. Mar. Sci.* 56:707–717

Cortés, E. and S. H. Gruber. 1990. Diet, feeding habits and estimation of daily ration of young lemon sharks, *Negaprion brevirostris* (Poey). *Copeia* 1990:204–218.

Cortés, E., C. A. Manire, and R. E. Hueter. 1996. Diet, feeding habits, and diel feeding chronology of the bonnethead shark, *Sphyrna tiburo*, in southwest Florida. *Bull. Mar. Sci.* 58:353–367.

DeCrosta, M. A., L. R. Taylor, Jr., and J. D. Parrish. 1984. Age determination, growth, and energetics of three species of carcharhinid sharks in Hawaii, in *Proceedings of the Second Symposium on Resource Investigations in the Northwest Hawaiian Islands,* Vol. 2, May 25–27, 1983. University of Hawaii Sea Grant MR-84-01, 75–95.

Diamond, J. M. 1985. Filter-feeding on a grand scale. *Nature* 316:679–680.

Dill, L. M. 1990. Distance-to-cover and the escape decisions of an Africa cichlid fish, *Melanochromis chipokae*. *Environ. Biol. Fish.* 27:147–152.

Dill, L. M., M. R. Heithaus, and C. J. Walters. 2003. Behaviorally-mediated indirect species interactions in marine communities and their importance to conservation and management. *Ecology* 84:1151–1157.

Dudley, S. F. J., M. D. Anderson-Reade, G. S. Thompson, and P. B. McMullen. 2000. Concurrent scavenging off a whale carcass by great white sharks, *Carcharodon carcharias*, and tiger sharks, *Galeocerdo cuvier*. *Fish. Bull.* 98:646–649.

Dugatkin, L. A. and H. K. Reeve. 1998. *Game Theory and Animal Behavior.* Oxford University Press, Oxford.

Dulvy, N. K., J. D. Metcalfe, J. Glanville, M. G. Pawson, and J. D. Reynolds. 2000. Fishery stability, local extinctions, and shifts in community structure. *Conserv. Biol.* 14:283–293.

Ebert, D. A. 1991. Observations on the predatory behavior of the sevengill shark *Notorynchus cepedianus*. *S. Afr. J. Mar. Sci.* 11:455–465.

Ebert, D. A. 2002. Ontogenetic changes in the diet of the sevengill shark (*Notorynchus cepedianus*). *Mar. Freshwater Res.* 53:517–523.

Ebert, D. A., P. D. Cowley, and L. J. V. Compagno. 1991. A preliminary investigation of the feeding ecology of skates (Batoidea: Rajidae) off the west coast of southern Africa. *S. Afr. J. Mar. Sci.* 10:71–81.

Ebert, D. A., L. J. V. Compagno, and P. D. Cowley. 1992. A preliminary investigation of the feeding ecology of squaloid sharks off the west coast of southern Africa. *S. Afr. J. Mar. Sci.* 12:601–609.

Economakis, A. E. and P. S. Lobel. 1998. Aggregation behavior of grey reef sharks, *Carcharhinus amblry-hynchos*, at Johnson Atoll, Central Pacific Ocean. *Environ. Biol. Fish.* 51:129–139.

Ellis, J. R., M. G. Pawson, and S. E. Shackley. 1996. The comparative feeding ecology of six species of shark and four species of ray (Elasmobranchii) in the north-east Atlantic. *J. Mar. Biol. Assoc. U.K.* 76:89–106.

Fernandez, P., D. J. Anderson, P. R. Sievert, and K. P. Huyvaert. 2001. Foraging destinations of three low-latitude albatross (Phoebastria) species. *J. Zool.* (Lond.) 254:391–404.

Fertl, D., A. Acevedo-Guiterrez, and F. L. Darby. 1996. A report of killer whales (*Orcinus orca*) feeding on a carcharhinid shark in Costa Rica. *Mar. Mamm. Sci.* 12:606–611.

Fouts, W. R. and D. R. Nelson. 1999. Prey capture by the Pacific angel shark, *Squatina californica*: visually mediated strikes and ambush-site characteristics. *Copeia* 1999:304–312.

Francis, M. P. and C. Duffy. 2002. Distribution, seasonal abundance and bycatch of basking sharks (*Cetorhinus maximus*) in New Zealand, with observations on their winter habitat. *Mar. Biol.* 140:831–842.

Fretwell, S. D. and H. L. Lucas. 1970. On territorial behavior and other factors influencing habitat distribution in birds. *Acta Biotheor.* 19:16–36.

Gerritsen, J. and J. R. Strickler. 1977. Encounter probabilities and community structure in zooplankton: a mathematical model. *J. Fish. Res. Board Can.* 34:73–82.

Gilliam, J. F. and D. F. Fraser. 1987. Habitat selection under predation hazard: test of a model with foraging minnows. *Ecology* 68:1856–1862.

Goldman, K. J. and S. D. Anderson. 1999. Space utilization and swimming depth of white sharks, *Carcharodon carcharias*, at the South Farallon Islands, central California. *Environ. Biol. Fish.* 56:351–364.

Gotceitas, V. and P. Colgan. 1989. Predator foraging success and habitat complexity: quantitative test of the threshold hypothesis. *Oecologia* 80:158–166.

Gray, A. E., T. J. Mulligan, and R. W. Hannah. 1997. Food habits, occurrence, and population structure of the bat ray, *Myliobatis californica*, in Humboldt Bay, California. *Environ. Biol. Fish.* 49:227–238.

Gross, M. R. 1996. Alternative reproductive strategies and tactics: diversity within sexes. *Trends Ecol. Evol.* 11:92–98.

Gruber, S. H., J. R. C. de Marignac, and J. M. Hoenig. 2001. Survival of juvenile lemon sharks at Bimini, Bahamas, estimated by mark-depletion experiments. *Trans. Am. Fish. Soc.* 130:376–384.

Gunn, J. S., J. D. Stevens, T. L. O. Davis, and B. M. Norman. 1999. Observations on the short- term movements and behaviour of whale sharks (*Rhincodon typus*) at Ningaloo Reef, Western Australia. *Mar. Biol.* 135:553–559.

Hamilton, I. M. and M. R. Heithaus. 2001. The effects of temporal variation in predation risk on anti-predator behaviour: an empirical test using marine snails. *Proc. R. Soc. Lond.* B 268:2585–2588.

Hart, P. J. B. and A. B. Gill. 1992a. Choosing prey size: a comparison of static and dynamic foraging models for predicting prey choice by fish. *Mar. Behav. Phys.* 22:93–106.

Hart, P. J. B. and A. B. Gill. 1992b. Constraints on prey size selection by three-spined stickleback: energy requirements and the capacity and fullness of the gut. *J. Fish. Biol.* 40:205–218.

Hart, P. J. B. and S. Ison. 1991. The influence of prey size and abundance, and individual phenotype on prey choice by three-spined stickleback, *Gasterosteus aculeatus* L. *J. Fish. Biol.* 38:359–372.

Heithaus, M.R. 2001a. Predator–prey and competitive interactions between sharks (order Selachii) and dolphins (suborder Odontoceti): a review. *J. Zool. Lond.* 253:53–68.

Heithaus, M. R. 2001b. Habitat selection by predators and prey in communities with asymmetrical intraguild predation. *Oikos* 92:542–554.

Heithaus, M. R. 2001c. The biology of tiger sharks (*Galeocerdo cuvier*) in Shark Bay, Western Australia: sex ratio, size distribution, diet, and seasonal changes in catch rates. *Environ. Biol. Fish.* 61:25–36.

Heithaus, M. R. 2001d. Shark attacks on bottlenose dolphins (*Tursiops aduncus*) in Shark Bay, Western Australia: attack rate, bite scar frequencies, and attack seasonality. *Mar. Mamm. Sci.* 17:526–539.

Heithaus, M. R. and L. M. Dill. 2002a. Food availability and tiger shark predation risk influence bottlenose dolphin habitat use. *Ecology* 83:480–491.

Heithaus, M. R. and L. M. Dill. 2002b. Feeding strategies and tactics, in *Encyclopedia of Marine Mammals*. W. F. Perrin, B. Würsig, and J. G. M. Thewissen, Eds., Academic Press, New York, 412–422.

Heithaus, M. R., G. J. Marshall, B. M. Buhleier, and L. M. Dill. 2001. Employing CritterCam to study the behavior and habitat use of large sharks. *Mar. Ecol. Prog. Ser.* 209:307–310.

Heithaus, M. R., L. M. Dill, G. J. Marshall, and B. Buhleier. 2002a. Habitat use and foraging behavior of tiger sharks (*Galeocerdo cuvier*) in a seagrass ecosystem. *Mar. Biol.* 140:237–248.

Heithaus, M. R., A. Frid, and L. M. Dill. 2002b. Shark-inflicted injury frequencies, escape ability, and habitat use of green and loggerhead turtles. *Mar. Biol.* 140:229–236.

Heupel, M. R. and M. B. Bennett. 1998. Observations on the diet and feeding habits of the epaulette shark, *Hemiscyllium ocellatum* (Bonnaterre), on Heron Island Reef, Great Barrier Reef, Australia. *Mar. Freshwater Res.* 49:753–756.

Heupel, M. R. and R. E. Hueter. 2002. The importance of prey density in relation to the movement patterns of juvenile sharks within a coastal nursery area. *Mar. Freshwater Res.* 53:543–550.

Heyman, W. D., R. T. Graham, B. Kjerfve, and R. E. Johannes. 2001. Whale sharks *Rhincodon typus* aggregate to feed on fish spawn in Belize. *Mar. Ecol. Prog. Ser.* 215:275–282.

Hines, A. H., R. B. Whitlatch, S. F. Thrush, J. E. Hewitt, V. J. Cummings, P. K. Dayton, and P. Legendre. 1997. Nonlinear foraging response of a large marine predator to benthic prey: eagle ray pits and bivalves in a New Zealand sandflat. *J. Exp. Mar. Biol. Ecol.* 216:191–201.

Hiruki, L. M., W. G. Gilmartin, B. L. Becker, and I. Stirling. 1993. Wounding in Hawaiian monk seals (*Monachus schauinslandi*). *Can. J. Zool.* 71:458–468.

Holland, K. N., B. M. Wetherbee, J. D. Peterson, and C. G. Lowe. 1993. Movements and distribution of hammerhead shark pups on their natal grounds. *Copeia* 1993:495–502.

Holland, K. N., B. M. Wetherbee, C. G. Lowe, and C. G. Meyer. 1999. Movements of tiger sharks (*Galeocerdo cuvier*) in coastal Hawaiian waters. *Mar. Biol.* 134:665–673.

Holt, R. D. 1977. Predation, apparent competition and the structure of prey communities. *Theor. Pop. Biol.* 12:197–229.

Holt, R. D. 1984. Spatial heterogeneity, indirect interactions, and the coexistence of prey species. *Am. Nat.* 124:377–406.

Holt, R. D. and G. A. Polis. 1997. A theoretical framework for intraguild predation. *Am. Nat.* 149:745–764.

Holts, D. B. and D. W. Bedford. 1993. Horizontal and vertical movements of shortfin mako shark, *Isurus oxyrinchus*, in the southern California Bight. *Aust. J. Mar. Freshwater Res.* 44:901–909

Hopkins, T. E. and J. J. Cech, Jr. 1994. Effect of temperature on oxygen consumption of the bat ray, *Myliobatis californica* (Chondrichthyes, Myliobatidae). *Copeia* 1994:529–532.

Houston, A. I., J. M. McNamara, and J. M. C. Hutchinson. 1993. General results concerning the trade-off between gaining energy and avoiding predation. *Philos. Trans. R. Soc. Lond. B*: 341:375–397.

Houtman, R. and L. M. Dill. 1998. The influence of predation risk on diet selectivity: a theoretical analysis. *Evol. Ecol.* 12:251–262.

Hugie, D. M. and L. M. Dill. 1994. Fish and game: a game theoretic approach to habitat selection by predators and prey. *J. Fish Biol.* 45(Suppl. A):151–169.

Iwasa, Y. 1982. Vertical migration of zooplankton: a game between predator and prey. *Am. Nat.* 120:171–180.

Jones, E. C. 1971. *Isistius brasiliensis*, a squaloid shark, the probable cause of crater wounds on fishes and cetaceans. *Fish. Bull.* 69:791–798.

Kajiura, S. M. and T. C. Tricas. 1996. Seasonal dynamics of dental sexual dimorphism in the Atlantic stingray (*Dasyatis sabina*). *J. Exp. Biol.* 199:2297–2306.

Kitchell, J. F., T. E. Essington, C. H. Boggs, D. E. Schindler, and C. J. Walters. 2002. The role of sharks and longline fisheries in a pelagic ecosystem of the central Pacific. *Ecosystems* 5:2002–2016.

Klimley, A. P. 1985. The areal distribution and autoecology of the white shark, *Carcharodon carcharias*, off the west coast of North America. *Mem. South. Calif. Acad. Sci.* 9:15–40.

Klimley, A. P. 1987. The determinants of sexual selection in the scalloped hammerhead shark, *Sphyrna lewini*. *Environ. Biol. Fish.* 18:27–40.

Klimley, A. P. 1993. Highly directional swimming by scalloped hammerhead sharks, *Sphyrna lewini*, and subsurface irradiance, temperature, bathymetry, and geomagnetic field. *Mar. Biol.* 117:1–22.

Klimley, A. P. 1994. The predatory behavior of the white shark. *Am. Sci.* 52:122–133.

Klimley, A. P. and S. D. Anderson. 1996. Residency patterns of white sharks at the South Farallon Islands, California, in *Great White Sharks: The Biology of Carcharodon carcharias*. A. P. Klimley and D. G. Ainley, Eds., Academic Press, New York, 365–374.

Klimley, A. P. and S. B. Butler. 1988. Immigration and emigration of a pelagic fish assemblage to seamounts in the Gulf of California related to water mass movements using satellite imagery. *Mar. Ecol. Prog. Ser.* 49:11–20.

Klimley, A. P. and D. R. Nelson. 1981. Schooling of the scalloped hammerhead shark, *Sphyrna lewini*, in the Gulf of California. *Fish. Bull.* 79:356–360.

Klimley, A. P. and D. R. Nelson. 1984. Diel movement patterns of the scalloped hammerhead shark (*Sphyrna lewini*) in relation to El Bajo Espiritu Santo: a refuging central-position social system. *Behav. Ecol. Sociobiol.* 15:45–54.

Klimley, A. P., S. B. Butler, D. R. Nelson, and A. T. Stull. 1988. Diel movements of scalloped hammerhead sharks, *Sphyrna lewini* Griffith and Smith, to and from a seamount in the Gulf of California. *J. Fish. Biol.* 33:751–761.

Klimley, A. P., S. D. Anderson, P. Pyle, and R. P. Henderson. 1992. Spatiotemporal patterns of white shark (*Carcharodon carcharias*) predation at the South Farallon Islands, California. *Copeia* 1992:680–690.

Klimley, A. P., P. Pyle, and S. D. Anderson. 1996a. The behavior of white sharks and their pinniped prey during predatory attacks, in *Great White Sharks: The Biology of Carcharodon carcharias*. A. P. Klimley and D. G. Ainley, Eds., Academic Press, New York, 175–191.

Klimley, A. P., P. Pyle, and S. D. Anderson. 1996b. Tail slap and breach: agonistic displays among white sharks? in *Great White Sharks: The Biology of Carcharodon carcharias*. A. P. Klimley and D. G. Ainley, Eds., Academic Press, New York, 241–255.

Klimley, A. P., B. J. Le Boeuf, K. M. Cantara, J. E. Richert, S. F. Davis, S. Van Sommeran, and J. T. Kelly. 2001. The hunting strategy of white sharks (*Carcharodon carcharias*) near a seal colony. *Mar. Biol.* 138:617–636.

Klimley, A. P., S. C. Beavers, T. H. Curtis, and S. J. Jorgensen. 2002. Movements and swimming behavior of three species of sharks in La Jolla Canyon, California. *Environ. Biol. Fish.* 63:117–135.

Koen Alonso, M., E. A. Crespo, N. A. García, S. N. Pedraza, P. A. Mariotti, B. Berón Vera, and N. J. Mora. 2001. Food habits of *Dipturus chilensis* (Pisces: Rajidae) off Patagonia, Argentina. *ICES J. Mar. Sci.* 58:288–297.

Leatherwood, S. 1977. Some preliminary impressions of the numbers and social behavior of free-swimming bottlenosed dolphin calves (*Tursiops truncatus*) in the northeastern Gulf of Mexico, in Breeding Dolphins: Present Status, Suggestions for the Future. S. H. Ridgway, Ed., Report to the U.S. Marine Mammal Commission MM6AC009, 143–167.

LeBoeuf, B. J., M. Reidman, and R. S. Keyes. 1982. White shark predation on pinnipeds in California coastal waters. *Fish. Bull.* 80:891–895.

Lima, S. L. 1995. Back to the basics of anti-predatory vigilance: the group size effect. *Anim. Behav.* 49:11–20.

Lima, S. L. 1998a. Stress and decision making under the risk of predation: recent developments from behavioral, reproductive, and ecological perspectives. *Adv. Stud. Behav.* 27:215–290.

Lima, S. L. 1998b. Nonlethal effects in the ecology of predator–prey interactions. *BioScience* 48:25–34.

Lima, S. L. 2002. Putting predators back into behavioral predator–prey interactions. *Trends Ecol. Evol.* 17:70–75.

Lima, S. L. and P. A. Bednekoff. 1999. Temporal variation in danger drives antipredator behavior: the predation risk allocation hypothesis. *Am. Nat.* 153:649–659.

Lima, S. L. and L. M. Dill. 1990. Behavioral decisions made under the risk of predation: a review and prospectus. *Can. J. Zool.* 68:619–640.

Long, D. J. and R. E. Jones. 1996. White shark predation and scavenging on cetaceans in the eastern north Pacific Ocean, in *Great White Sharks: The Biology of Carcharodon carcharias*. A. P. Klimley and D. G. Ainley, Eds., Academic Press, New York, 293–307.

Lott, D. F. 1984. Intraspecific variation in the social system of wild vertebrates. *Behaviour* 88:266–325.

Lowe, C. G., R. N. Bray, and D. R. Nelson. 1994. Feeding and associated electrical behavior of the Pacific electric ray *Torpedo californica* in the field. *Mar. Biol.* 120:161–169.

Lucas, Z. and W. T. Stobo. 2000. Shark-inflicted mortality on a population of harbour seals (*Phoca vitulina*) at Sable Island, Nova Scotia. *J. Zool. Lond.* 252:405–414.

Manichchand-Heileman, S., L. A. Soto, and E. Escobar. 1998. A preliminary trophic model of the continental shelf, south-western Gulf of Mexico. *Estuarine Coastal Shelf Sci.* 46:885–899.

Matern, S. A., J. J. Cech, Jr., and T. E. Hopkins. 2000. Diel movements of bat rays, *Myliobatis californica*, in Tomales Bay, California: evidence for behavioral thermoregulation? *Environ. Biol. Fish.* 58:173–182.

McCosker, J. E. 1985. White shark attack behavior: observation of and speculations about predator–prey strategies. *Mem. South. Calif. Acad. Sci.* 9:123–135.

McKibben, J. N. and D. R. Nelson. 1986. Patterns of movement and grouping of gray reef sharks, *Carcharhinus amblyrhynchos*, at Enewetak, Marshall Islands. *Bull. Mar. Sci.* 38: 89–110.

McNamara, J. M. and A. I. Houston. 1987. Starvation and predation as factors limiting population size. *Ecology* 68:1515–1519.

Mesterson-Gibbons, M. and Dugatkin, L. A. 1992. Cooperation among unrelated individuals: evolutionary factors. *Q. Rev. Biol.* 67:267–281.

Michaelson, D. M., D. Sternberg, and L. Fishelson. 1979. Observations on feeding, growth, and electric discharge of newborn *Torpedo ocellata* (Chondrichthyes, Batoidei). *J. Fish. Biol.* 15:159–163.

Miner, J. G., and R. A. Stein. 1996. Detection of predators and habitat choice by small bluegills: effects of turbidity and alternative prey. *Trans. Am. Fish. Soc.* 125:97–103.

Montgomery, J. and E. Skipworth. 1997. Detection of weak water jets by the short-tailed stingray *Dasyatus brevicaudata. Copeia* 1997:881–883.

Moody, A. L., A. I. Houston, and J. M. McNamara. 1996. Ideal free distributions under predation risk. *Behav. Ecol. Sociobiol.* 38:131–143.

Morrissey, J. F. and S. H. Gruber. 1993a. Habitat selection by juvenile lemon sharks, *Negaprion brevirostris. Environ. Biol. Fish.* 38:311–319.

Morrissey, J. F. and S. H. Gruber. 1993b. Home range of juvenile lemon sharks, *Negaprion brevirostris. Copeia* 1993:425–434.

Musick, J.A., S. Branstetter, and J.A. Colvocoresses. 1993. Trends in shark abundance from 1974 to 1991 for the Chesapeake Bight region of the U.S. Mid-Atlantic coast, in Conservation Biology of Elasmobranchs, S. Branstetter, Ed., NOAA Tech Rep. NMFS 115, 1–18.

Myberg, A. A., Jr. 1991. Distinctive markings of sharks: ethological considerations of visual function. *J. Exp. Zool. Suppl.* 5:156–166.

Nelson, D. R. and R. H. Johnson. 1980. Behavior of the reef sharks of Rangiroa, French Polynesia. *Natl. Geogr. Soc. Res. Rep.* 12:479–499.

Nelson, D. R., R. H. Johnson, J. N. McKibben, and G. G. Pittenger. 1986. Agonistic attacks on divers and submersibles by gray reef sharks, *Carcharhinus amblyrhynchos*: antipredatory or competitive? *Bull. Mar. Sci.* 28:68–88.

Nelson, D. R., J. N. McKibben, W. R. Strong, Jr., C. G. Lowe, J. A. Sisneros, D. M. Schroeder, and R. J. Lavenberg. 1997. An acoustic tracking of a megamouth shark, *Megachasma pelagios*: a crepuscular vertical migrator. *Environ. Biol. Fish.* 49:389–399.

Norris, K. S. and T. P. Dohl. 1980. Behavior of the Hawaiian spinner dolphin, *Stenella longirostris. Fish. Bull.* 77:821–849.

Olsen, A. M. 1984. Synopsis of biological data on the school shark, *Galworhinus australis* (Maleay 1881). FAO Fish. Synop 139.

Pace, M. L., J. J. Cole, S. R. Carpenter, and J. F. Kitchell. 1999. Trophic cascades in diverse ecosystems. *Trends Ecol. Evol.* 14:483–488.

Paine, R. T. 1966. Food web complexity and species diversity. *Am. Nat.* 100:65–75.

Peacor, S. D. and E. E. Werner. 2000. Predator effects on an assemblage of consumers through induced changes in consumer foraging behavior. *Ecology* 81:1998–2010.

Peacor, S. D. and E. E. Werner. 2001. The contribution of trait-mediated indirect effects to the net effects of a predator. *Proc. Natl. Acad. Sci. U.S.A.* 98:3904–3908.

Peterson, C. H., F. J. Fodrie, H. C. Summerson, and S. P. Powers. 2001. Site-specific and density-dependent extinction of prey by schooling rays: generation of a population sink in top-quality habitat for bay scallops. *Oecologia* 129:349–356.

Piraino, S., G. Fanelli, and F. Boero. 2002. Variability of species' roles in marine communities: change of paradigms for conservation priorities. *Mar. Biol.* 140:1067–1074.

Platell, M. E., I. C. Potter, and K. R. Clarke. 1998. Resource partitioning by four species of elasmobranchs (Batoidea: Urolophidae) in coastal waters of temperate Australia. *Mar. Biol.* 131:719–734.

Polis, G. A., C. A. Meyers, and R. D. Holt. 1989. The ecology and evolution of intraguild predation: potential competitors that eat each other. *Annu. Rev. Ecol. Syst.* 20:297–330.

Power, M. E., W. J. Matthews, and A. J. Stewart. 1985. Grazing minnows, piscivorous bass, and stream algae: dynamics of a strong interaction. *Ecology* 66:1448–1456.

Power, M. E., D. Tilman, J. A. Estes, B. A. Menge, W. J. Bond, L. S. Mills, D. Gretchen, J. C. Castilla, J. Lubchenco, and R. T. Paine. 1996. Challenges in the quest for keystones. *BioScience* 46:1415–1427.

Pratt, H.L., Jr., J. G. Casey, and R. B. Conklin. 1982. Observations on large white sharks, *Carcharodon carcharias*, off Long Island, New York. *Fish. Bull.* 80:153–156.

Pulliam, H. R. and T. Caraco. 1984. Living in groups: is there an optimal group size? in *Behavioural Ecology: An Evolutionary Approach*. 2nd ed. J. R. Krebs and N. B Davies, Eds., Blackwell Scientific, Oxford, 122–147.

Pyle, P., M. J. Schramm, C. Keiper, and S. D. Anderson. 1999. Predation on a white shark (*Carcharodon carcharias*) by a killer whale (*Orcinus orca*) and a possible case of competitive displacement. *Mar. Mamm. Sci.* 15:563–568.

Randall, J. E. 1977. Contribution to the biology of the whitetip reef shark (*Triaenodon obesus*). *Pac. Sci.* 31:143–164.

Randall, J. E. 1992. Review of the biology of the tiger shark (*Galeocerdo cuvier*). *Aust. J. Mar. Freshwater Res.* 43:21–31.

Reidenauer, J. A. and D. Thistle. 1981. Response of a soft-bottom harpacticoid community to stingray (*Dasyatis sabina*) disturbance. *Mar. Biol.* 65:261–267.

Relyea, R. A. 2000. Trait-mediated indirect effects in larval anurans: reversing competition with the threat of predation. *Ecology* 81:2278–2289.

Schmitz, O. J., A. P. Beckerman, and K. M. O'Brien. 1997. Behaviorally mediated trophic cascades: effects of predation risk on food web interactions. *Ecology* 78:1388–1399.

Sciarrotta, T. C. and D. R. Nelson. 1977. Diel behavior of the blue shark, *Prionace glauca*, near Santa Catalina Island, California. *Fish. Bull.* 75:519–528.

Semeniuk, A. D. 2003. Resting Behaviour of the Cowtail Stingray (*Pastinachus sephen*) in Shark Bay, Western Australia. Master's thesis. Simon Fraser University, Burnaby, B.C., Canada.

Siems, D. P. and R. S. Sikes. 1998. Trade-offs between growth and reproduction in response to temporal variation in food supply. *Environ. Biol. Fish.* 53:319–329.

Sih, A. 1998. Game theory and predator–prey response races, in *Game Theory and Animal Behavior*. L. A. Dugatkin and H. K. Reeve, Eds., Oxford University Press, Oxford, 221–238.

Sih, A. and B. Christensen. 2001. Optimal diet theory: when does it work, and when and why does it fail? *Anim. Behav.* 61:379–390.

Sih, A., P. Crowley, M. McPeek, J. Petranka, and K. Strohmeier. 1985. Predation, competition, and prey communities: a review of field experiments. *Annu. Rev. Ecol. Syst.* 16:269–311.

Simpfendorfer, C. A. and N. E. Milward. 1993. Utilisation of a tropical bay as a nursery area by sharks of the families Carcharhinidae and Sphyrnidae. *Environ. Biol. Fish.* 37:337–345.

Simpfendorfer, C. A., A. B. Goodreid, and R. B. McAuley. 2001. Size, sex, and geographic variation in the diet of the tiger shark, *Galeocerdo cuvier*, from Western Australian waters. *Environ. Biol. Fish.* 61:37–46.

Sims, D. W. 1999. Threshold foraging behavior of basking sharks on zooplankton: life on an energetic knife-edge? *Proc. R. Soc. Lond.* B 266:1437–1443.

Sims, D. W. 2000. Filter-feeding and cruising swimming speeds of basking sharks compared with optimal models: they filter-feed slower than predicted for their size. *J. Exp. Mar. Biol. Ecol.* 249:65–76.

Sims, D. W. and D. A. Merrett. 1997. Determination of zooplankton characteristics in the presence of surface feeding basking sharks *Cetorhinus maximus*. *Mar. Ecol. Prog. Ser.* 158:297–302.

Sims, D. W. and V. A. Quayle. 1998. Selective foraging behaviour of basking sharks on zooplankton in a small-scale front. *Nature* 393:460–464

Sims, D. W., J. P. Nash, and D. Morritt. 2001. Movements and activity of male and female dogfish in a tidal sea lough: alternative behavioral strategies and apparent sexual selection. *Mar. Biol.* 139:1165–1175.

Sinclair, A. R. E. and P. Arcese. 1995. Population consequences of predation-sensitive foraging: the Serengeti wildebeest. *Ecology* 76:882–891.

Smale, M. J., W. H. H. Sauer, and R. T. Hanlon. 1995. Attempted ambush predation on spawning squids *Loligo vulgaris reynaudii* by benthic pyjama sharks, *Poroderma africanum*, off South Africa. *J. Mar. Biol. Assoc. U.K.* 75:739–742.

Smale, M. J., W. H. H. Sauer, and M. J. Roberts. 2001. Behavioural interactions of predators and spawning chokka squid of South Africa: towards quantification. *Mar. Biol.* 139:1095–1105.

Smith, J. W. and J. V. Merriner. 1985. Food habits and feeding behavior of the cownose ray, *Rhinoptera bonasus*, in lower Chesapeake Bay. *Estuaries* 8:305–310.

Snelson, F. F., Jr., T. J. Mulligan, and S. E. Williams. 1984. Food habits, occurrence, and population structure of the bull shark, *Carcharhinus leucas*, in Florida coastal lagoons. *Bull. Mar. Sci.* 34:71–80.

Springer, S. 1960. Natural history of the sandbar shark, *Eulamia milberti*. *Fish. Bull.* 61:1–38.

Springer, S. 1967. Social organization of shark populations, in *Sharks, Skates, and Rays*. P. W. Gilbert, R. F. Matheson, and D. P. Rall, Eds., Johns Hopkins University Press, Baltimore, 149–174.

Stephens, D. W. and J. R. Krebs. 1986. *Foraging Theory*. Princeton University Press, Princeton, NJ.

Stevens, J. D., R. Bonfil, N. K. Dulvy, and P. A. Walker. 2000. The effects of fishing on sharks, rays, and chimeras (chondrichthyans), and the implications for marine ecosystems. *ICES J. Mar. Sci.* 57:476–494.

Stillwell, C. E. and N. E. Kohler. 1982. Food, feeding habits, and estimates of daily ration of the shortfin mako (*Isurus oxyrinchus*) in the Northwest Atlantic. *Can. J. Fish. Aquat. Sci.* 39:407–414.

Stokes, M. D. and N. D. Holland. 1992. Southern stingray (*Dasyatis Americana*) feeding on lancelets (*Branchiostoma floridae*). *J. Fish. Biol.* 41:1043–1044.

Strong, W. R., Jr. 1991. Instruments of natural selection: how important are sharks? in *Discovering Sharks*. S. H. Gruber, Ed., American Littoral Society, Highlands, NJ, 70–73.

Strong, W. R., Jr. 1996. Shape discrimination and visual predatory tactics in white sharks, in *Great White Sharks: The Biology of Carcharodon carcharias*. A. P. Klimley and D. G. Ainley, Eds., Academic Press, New York, 229–240.

Strong, W. R., Jr., F. F. Snelson, and S. H. Gruber. 1990. Hammerhead shark predation on stingrays: an observation of prey handling by *Spyrna mokarran*. *Copeia* 1990:836–840.

Strong, W. R., Jr., R. D. Murphy, B. D. Bruce, and D. R. Nelson. 1992. Movements and associated observations of bait-attracted white sharks, *Carcharodon carcharias*: a preliminary report. *Aust. J. Mar. Freshwater Res.* 43:13–20.

Strong, W. R., Jr., B. D. Bruce, D. R. Nelson, and R. D. Murphy. 1996. Population dynamics of white sharks in Spencer Gulf, South Australia, in *Great White Sharks: The Biology of Carcharodon carcharias*. A. P. Klimley and D. G. Ainley, Eds., Academic Press, New York, 401–414.

Sundström, L. F., S. H. Gruber, S. M. Clermont, J. P. S. Correia, J. R. C. de Marignac, J. F. Morrissey, C. R. Lowrance, L. Thomassen, and M. T. Oliveira. 2001. Review of elasmobranch behavioral studies using ultrasonic telemetry with special reference to the lemon shark, *Negaprion brevirostris*, around Bimini Islands, Bahamas. *Environ. Biol. Fish.* 60:225–250.

Taylor, R. J. 1984. *Predation*. Chapman & Hall, London.

Thrush, S. F., R. D. Pridmore, J. E. Hewitt, and V. J. Cummings. 1991. Impact of ray feeding disturbances on sandflat macrobenthos: do communities dominated by polychaetes or shellfish respond differently. *Mar. Ecol. Prog. Ser.* 69:245–252.

Thrush, S. F., R. D. Pridmore, J. E. Hewitt, and V. J. Cummings. 1994. The importance of predators on a sandflat: interplay between seasonal changes in prey densities and predator effects. *Mar. Ecol. Prog. Ser.* 107:211–222.

Tregenza, T. 1995. Building on the ideal free distribution. *Adv. Ecol. Res.* 26:253–307.

Tricas, T. C. 1979. Relationships of the blue shark, *Prionace glauca*, and its prey species near Santa Catalina Island, California. *Fish. Bull.* 77:175–182.

Tricas, T. C. and J. E. McCosker. 1984. Predatory behavior of the white shark (*Carcharodon carcharias*), with notes on its biology. *Proc. Calif. Acad. Sci.* 43:221–238.

Tricas, T. C., L. R. Taylor, and G. Naftel. 1981. Diel behavior of the tiger shark, *Galeocerdo cuvier*, at French Frigate Shoals, Hawaiian Islands. *Copeia* 1981:904–908.

Turner, A. M. and G. G. Mittlebach. 1990. Predator avoidance and community structure: interactions among piscivores, planktivores, and plankton. *Ecology* 71:2241–2254.

VanBlaricom, G. R. 1982. Experimental analyses of structural regulation in a marine sand community exposed to oceanic swell. *Ecol. Monogr.* 52:283–305.

van der Elst, R. P. 1979. A proliferation of small sharks in the shore-based Natal sports fishery. *Environ. Biol. Fish.* 4:349–362.

Visser, I. 1999. Benthic foraging on stingrays by killer whales (*Orcinus orca*) in New Zealand waters. *Mar. Mamm. Sci.* 15:220–227.

Walker, P. A. and H. J. L. Heesen. 1996. Long-term changes in ray populations in the North Sea. *ICES J. Mar. Sci.* 53:1085–1093

Walker, T. I. 1998. Can shark resources be harvested sustain ably? A question revisited with a review of shark fisheries. *Mar. Freshwater Res.* 49:553–572.

Walters, C. 2000. Natural selection for predation avoidance tactics: implications for marine population and community dynamics. *Mar. Ecol. Prog. Ser.* 208:309–313.

Walters, C. and F. Juanes. 1993. Recruitment limitation as a consequence of natural selection for use of restricted feeding habitats and predation risk taking by juvenile fishes. *Can. J. Fish. Aquat. Sci.* 50:2058–2070.

Webber, J. D. and J. J. Cech, Jr. 1998. Nondestructive diet analysis of the leopard shark from two sites in Tomales Bay, California. *Calif. Fish. Game* 84:18–24.

Werner, E. E. and B. R. Anholt. 1993. Ecological consequences of the trade-off between growth and mortality rates mediated by foraging activity. *Am. Nat.* 142:242–272.

Werner, E. E. and B. R. Anholt. 1996. Predator-induced behavioral indirect effects: consequences to competitive interactions in anuran larvae. *Ecology* 77:157–169.

Werner, E. E. and D. J. Hall. 1974. Optimal foraging and size selection of prey by bluegill sunfish (*Lepomis macrochirus*). *Ecology* 55:1042–1052.

Werner, E. E. and D. J. Hall. 1988. Ontogenetic habitat shifts in bluegill: the foraging rate-predation risk trade-off. *Ecology* 69:1352–1366.

Wetherbee, B. M., S. H. Gruber, and E. Cortés. 1990. Diet, feeding habits, digestion, and consumption in sharks with special reference to the lemon shark, *Negaprion brevirostris*, in Elasmobranchs as Living Resources. H. L. Pratt, Jr., S. H. Gruber, and T. Taniuchi, Eds., NOAA Tech. Rep. 90, 29–47.

Widder, E. A. 1998. A predatory use of counterillumination by the squaloid shark, *Isistius brasiliensis. Environ. Biol. Fish.* 53:267–273.

Wilson, S. G., J. G. Taylor, and A. F. Pearce. 2001. The seasonal aggregation of whale sharks at Ningaloo Reef, Western Australia: currents, migrations and the El Niño/Southern Oscillation. *Environ. Biol. Fish.* 61:1–11.

Yano, K. 1986. A telemetric study on the movements of the deep sea squaloid shark, *Centrophorus actus*, in *Indo-Pacific Fish Biology: Proceedings of the Second International Conference of Indo-Pacific Fishes.* Ichthyological Society of Japan, Tokyo, 372–380.

Ydenberg, R. C. and L. M. Dill. 1986. The economics of fleeing from predators. *Adv. Stud. Behav.* 16:229–249.

18

Elasmobranchs as Hosts of Metazoan Parasites

Janine N. Caira and Claire J. Healy

CONTENTS

18.1 Introduction

The body of an elasmobranch offers a diversity of sites that can be, and often are, occupied by other animals. Indeed, essentially no organ system of elasmobranchs has escaped the attention of one or more groups of parasites (Figure 18.1). That is not to say that all sites of the body of an elasmobranch are equally parasitized. Certain organs and organ systems, such as the skin, digestive system, and gills, for example, tend to host particularly diverse faunas of parasites. And, as elasmobranch parasite survey work continues globally, new sites occupied by parasites continue to be discovered. It is our intention to provide a synthesis of the invertebrate metazoan parasites of elasmobranchs. Caira (1990), Cheung (1993), and Benz and Bullard (in review) have treated the parasites of these hosts within the last dozen or so years. All three of these works provided overviews of the taxonomic diversity of these parasites. In addition, Caira (1990) provided an overview of the life cycles and utility of metazoan parasites as indicators of elasmobranch biology, Cheung (1993) included an extensive list of many of the parasite species reported from elasmobranchs, and Benz and Bullard (in review) provided extensive information on the pathology caused by, and treatment of, each major parasite group. In this chapter, we have taken a slightly different approach and have treated the various parasites based on the sites they occupy within their elasmobranch hosts. We have concentrated on the parasitic metazoans, or multicellular parasites, because we did not feel we could do justice to the diversity of protistan (i.e., unicellular) taxa parasitizing elasmobranchs at this time. Readers interested in the protists are directed to Cheung (1993) for a list of the approximately 30 species reported from elasmobranchs. Also omitted from discussion here are the vertebrate associates of elasmobranchs such as hagfish, lampreys, eels, etc., which, although intriguing (e.g., Caira et al., 1997a), generally exhibit interactions with elasmobranchs that are more consistent with predation than parasitism. Finally, we wish to note that we have restricted the host groups under consideration to elasmobranchs; parasites of the holocephalans are not addressed.

The invertebrate metazoans parasitizing elasmobranchs belong to six phyla. In ascending order of their diversity in elasmobranchs, these are Mollusca, Acanthocephala, Annelida, Nematoda, Arthropoda, and Platyhelminthes. To date, only one species of mollusc and approximately eight species of acanthocephalans have been reported from elasmobranchs. The annelids and nematodes of elasmobranchs are somewhat more diverse. Recent counts suggest that approximately 20 species of annelids (all leeches; Burreson, pers. comm.) and perhaps as many as 80 species of nematodes are known to associate with elasmobranchs. By far the greatest diversity of elasmobranch parasites, however, is found among the arthropods and platyhelminths. Each of these phyla includes several major groups that are worthy of individual note. The major arthropod taxa parasitizing elasmobranchs, again in ascending order of their diversity in elasmobranchs, are mites, barnacles (Cirripedia), ostracods, amphipods, branchiurans (i.e., fish lice), isopods, and copepods. Significant differences in diversity exist among these taxa in elasmobranchs. For example, whereas there is a single record of a mite from an elasmobranch (see Benz and Bullard, in review), the copepods of elasmobranchs number in the hundreds and are encountered with regularity. The major groups of parasitic platyhelminths, in ascending order of their diversity in elasmobranchs, are triclads, aspidogastrids, digeneans, monogeneans, and cestodes. Whereas only a single triclad and two aspidogastrids are known from elasmobranchs, approximately 40 to 50 digeneans (see Cheung, 1993; Bray and Cribb, 2003), 193 monogeneans (Whittington and Chisholm, 2003), and we estimate well over 800 cestode species are known to parasitize elasmobranchs. In fact, elasmobranch cestode diversity exceeds that of all of the other metazoan groups parasitizing elasmobranchs combined.

Collectively, the metazoan parasites of elasmobranchs represent a total of approximately 113 families. These families are presented in Table 18.1 according to their higher classification in the six phyla listed above. To illustrate the spectacular diversity of morphologies exhibited by these parasites, we present scanning electron micrographs or light micrographs of a representative of most of these families in Figure 18.2 through Figure 18.85. Although these images certainly do not substitute for descriptions of the distinguishing features of each family, they do serve to provide readers with some idea of the morphological variation found among major taxa in each phylum of parasites. In several cases we have included images of representatives of taxa below the level of family, either because family level taxonomy is unstable (e.g., tetraphyllidean and lecanicephalidean cestodes) or because we felt family level diversity

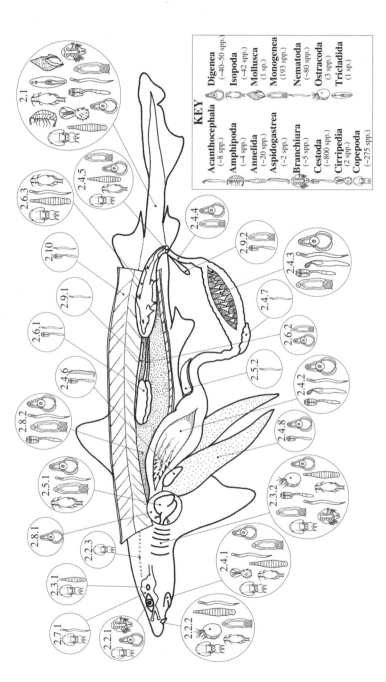

FIGURE 18.1 Overview of sites occupied by metazoan parasites of elasmobranchs, indicating text section of this chapter treating each site and approximate number of species of each parasite group found in elasmobranchs.

TABLE 18.1

Classification of Families of Metazoan Invertebrates Parasitic in Elasmobranchs

Phylum Mollusca[a]
  Class Gastropoda
    Subclass Prosobranchia
      F. Cancellariidae[b] (Figure 18.2)
Phylum Acanthocephala
  Class Paelaeacanthocephala
    Order Echinorhynchida
      F. Cavisomidae (Figure 18.3)
      F. Echinorhynchidae
      F. Illiosentidae (Figure 18.4)
      F. Rhadinorhynchidae
    Order Polymorphida
      F. Polymorphidae
Phylum Annelida
  Class Hirudinea
    Order Rhynchobdellida
      F. Piscicolidae
        SubF. Piscicolinae[c] (Figure 18.5)
        SubF. Pontobdellinae (Figure 18.6)
        SubF. Platybdellinae
        Unassigned (Curran et al., pers. comm.) (Figure 18.7)
Phylum Nematoda
  Class Rhabditea
    Order Ascaridida
      F. Acanthocheilidae (Figure 18.8)
      F. Anisakidae (Figure 18.9)
      F. Ascaridae (Figure 18.10)
    Order Spirurida
      F. Cucullanidae (Figure 18.11)
      F. Cystidicolidae
      F. Gnathostomatidae (Figure 18.12)
      F. Philometridae
      F. Physalopteridae (Figure 18.13)
      F. Rhabdochonidae
    Order Dracunculoidea
      F. Guyanemidae
      F. Micropleuridae
      Unassigned (Adamson et al., pers. comm.)
  Class Enoplea
    Order Trichinelloidea
      F. Capillariidae (Figure 18.14)
      F. Trichosomoididae (Figure 18.15)
Phylum Arthropoda
  Class Arachnida
    Order Acari
      Unidentified (Benz and Bullard, in review)
  Class Crustacea
    Subclass Malacostraca
    Order Amphipoda
      F. Lafystiidae (Figure 18.16)
      F. Lysianassidae
      F. Trischizostomatidae
    Order Isopoda
      Suborder Flabellifera
        F. Aegidae (Figure 18.17)
        F. Cirolanidae (Figure 18.18)
        F. Corallanidae (Figure 18.19)
        F. Cymothoidae

      Suborder Gnathiidea
        F. Gnathiidae (Figure 18.20)
    Subclass Maxillopoda
      Superorder Ostracoda
      Order Myodocopa
        F. Cypridinidae (Figure 18.21)
      Superorder Branchiura
        F. Argulidae (Figure 18.22)
      Superorder Cirripedia
      Order Thoracica
        F. Conchodermidae
        F. Anelasmatidae
      Superorder Copepoda
      Order Siphonostomatoida
        F. Caligidae (Figure 18.23)
        F. Cecropidae (Figure 18.24)
        F. Dichelesthiidae (Figure 18.25)
        F. Dissonidae (Figure 18.26)
        F. Eudactylinidae (Figure 18.27)
        F. Euryphoridae (Figure 18.28)
        F. Kroyeriidae (Figure 18.29)
        F. Lernaeopodidae (Figure 18.30)
        F. Pandaridae (Figure 18.31)
        F. Pennellidae (Figure 18.32)
        F. Sphyriidae (Figure 18.33)
        F. Trebiidae
      Order Poecilostomatoida
        F. Chondracanthidae
        F. Ergasilidae (Figure 18.34)
        F. Philichthyidae
        F. Taeniacanthidae
Phylum Platyhelminthes
  Order Tricladida
      F. Procerodidae (Figure 18.35)
  Subphylum Neodermata
    Class Trematoda
      Subclass Aspidogastrea
        F. Stichocotylidae
        F. Multicalycidae (Figure 18.36)
      Subclass Digenea
        F. Acanthocolpidae
        F. Azygiidae (Figure 18.37)
        F. Bucephalidae
        F. Campulidae
        F. Derogenidae (Figure 18.38)
        F. Didymozoidae
        F. Gorgoderidae (Figure 18.39)
        F. Hemiuridae
        F. Hirudinellidae
        F. Leptocreadiidae
        F. Opecoelidae
        F. Ptychogonimidae
        F. Sanguinicolidae
        F. Syncoeliidae (Figure 18.40)
        F. Tandanicolidae
        F. Zoogonidae (Figure 18.41)
    Class Cercomeromorpha
      Subclass Monogenea
      Superorder Monopisthocotylea
        F. Acanthocotylidae (Figure 18.42)
        F. Amphibdellidae (Figure 18.43)
        F. Capsalidae (Figure 18.44)
        F. Dioncidae

        F. Loimoidae (Figure 18.45)
        F. Microbothriidae (Figure 18.46)
        F. Monocotylidae (Figure 18.47)
        F. Udonellidae (Figure 18.48)
      Superorder Polyopisthocotylea
        F. Hexabothriidae (Figure 18.49)
    Subclass Cestoda
      Order Diphyllidea
        F. Ditrachyobothridiidae (Figure 18.50)
        F. Echinobothriidae (Figure 18.51)
      Order Lecanicephalidea
        F. Anteroporidae (Figure 18.52)
        F. Lecanicephalidae (Figure 18.53)
        F. Polypocephalidae (Figure 18.54)
        F. Tetragonocephalidae (Figure 18.55)
      Order Trypanorhyncha
        SuperF. Homeacanthoidea[d]
          F. Aporhynchidae
          F. Hepatoxylidae (Figure 18.60)
          F. Paranybeliniidae
          F. Pseudotobothriidae (Figure 18.61)
          F. Sphyriocephalidae
          F. Tentacularidae (Figure 18.62)
          F. Tetrarhynchobothriidae
        SuperF. Heteracanthoidea
          F. Eutetrarhynchidae (Figure 18.63)
          F. Gilquiniidae (Figure 18.64)
          F. Shirleyrhynchidae (Figure 18.65)
        SuperF. Otobothrioidea
          F. Grillotiidae (Figure 18.66)
          F. Molicolidae
          F. Otobothriidae (Figure 18.67)
          F. Pterobothriidae
          F. Rhinoptericolidae (Figure 18.68)
        SuperF. Poecilacanthoidea
          F. Dasyrhynchidae (Figure 18.69)
          F. Gymnorhynchidae (Figure 18.70)
          F. Hornelliellidae (Figure 18.71)
          F. Lacistorhynchidae (Figure 18.72)
          F. Mixodigmatidae (Figure 18.73)
          F. Mustelicolidae (Figure 18.74)
      Order Tetraphyllidea
        F. Cathetocephalidae (Figure 18.75)
        F. Dioecotaeniidae (Figure 18.76)
        F. Disculicepitidae (Figure 18.77)
        F. Litobothriidae (Figure 18.78)
        F. Onchobothriidae (Figure 18.79)
        F. Phyllobothriidae
          SubF. Echeneibothriinae (Figure 18.80)
          SubF. Phyllobothriinae (Figure 18.81)
          SubF. Rhinebothriinae (Figure 18.82)
          SubF. Thysanocephalinae (Figure 18.83)
          SubF. Trilocularinae (Figure 18.84)
        F. Prosobothriidae (Figure 18.85)

---

[a] Phyla are in order of their increasing diversity in elasmobranchs.
[b] Family.
[c] Subfamily.
[d] Superfamily.

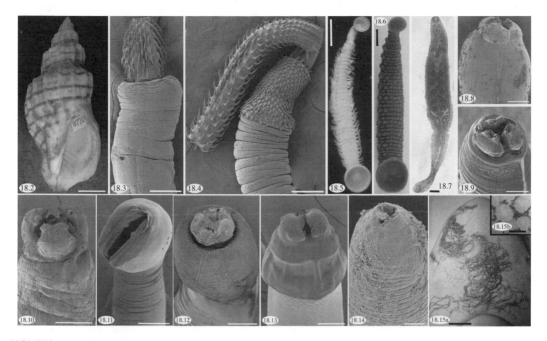

FIGURES 18.2 TO 18.15 Micrographs of Mollusca, Acanthocephala, Annelida, and Nematoda. 18.2. Mollusca: Cancellariidae: *Cancellaria cooperi* (USNM No. 877074). 18.3. Acanthocephala: Cavisomidae: *Megapriapus* sp. ex *Potamotrygon* sp. 18.4. Acanthocephala: Illiosentidae: *Tegorhynchus* sp. ex *Rhinoptera bonasus*. 18.5. Annelida: Piscicolidae: Piscicolinae: *Branchellion* sp. ex *Raja nasuta*. 18.6. Annelida: Piscicolidae: Pontobdellinae: *Stibarobdella* sp. ex *Carcharhinus plumbeus*. 18.7. Annelida: Piscicolidae ex *Zapteryx exasperata*. 18.8. Nematoda: Acanthocheilidae ex *Pristis zijsron*. 18.9. Nematoda: Anisakidae ex *Galeocerdo cuvier*. 18.10. Nematoda: Ascaridae ex *Potamotrygon* sp. 18.11. Nematoda: Cucullanidae: *Cucullanus* sp. ex *Heterodontus franscisi*. 18.12. Nematoda: Gnathostomatidae: *Echinocephalus* sp. ex *Himantura granulata*. 18.13. Nematoda: Physalopteridae: *Paraleptus* sp. ex *Hemiscyllium ocellatum*. 18.14. Nematoda: Capillariidae: *Piscicapillaria* sp. ex *Rhina ancylostoma*. 18.15. Nematoda: Trichosomoididae: *Huffmanela* sp. (a). Egg trail on ventral surface of head of *Carcharhinus plumbeus*. (b). Enlarged view of characteristic pigmented, bipolar eggs around bases of placoid scales of *C. sorrah*. Scale bars: Figure 18.2, 1 cm; Figures 18.3 and 18.4, 200 µm; Figure 18.5, 6 mm; Figure 18.6, 5.5 mm; Figure 18.7, 165 µm; Figures 18.8, 18.10, and 18.13, 50 µm; Figures 18.9, 18.11, and 18.12, 100 µm; Figure 18.14, 2 µm; Figure 18.15a, 1 cm; Figure 18.15b, 280 µm.

did not do sufficient justice to the morphological variation seen in a group (e.g., leeches). The families for which illustrations are provided are indicated in Table 18.1.

It is common to categorize parasites as either ectoparasitic or endoparasitic. Ectoparasites inhabit any exterior site and/or orifice of their host. Leeches, arthropods, and molluscs typically occupy such sites on elasmobranchs. Endoparasites generally inhabit sites associated with the interior cavities, organs, ducts (both between and within organs), and musculature of their host. With a few exceptions, the acanthocephalans, nematodes, and most of the major groups of platyhelminths (except most monogeneans) are endoparasitic in elasmobranchs. Some authors (e.g., Kabata, 1979; Benz, 1993) recognize a third category, the mesoparasites. This term is applied to those organisms that normally live with a significant portion of their body embedded within the host while a significant portion of their body extends outside of the host. Most mesoparasites of elasmobranchs are copepods, but one of the parasitic barnacles that associates with elasmobranchs (*Anelasma squalicola*) also exhibits this lifestyle.

It is feasible to treat the metazoan parasites of elasmobranchs based on the sites they occupy in or on their hosts because most of the major parasite groups exhibit remarkable specificity for particular organs or organ systems. In many cases this specificity is extreme. The various sites and the major groups of metazoans that occupy them are summarized in Figure 18.1. This figure also serves as a guide to the sections within this chapter that follow.

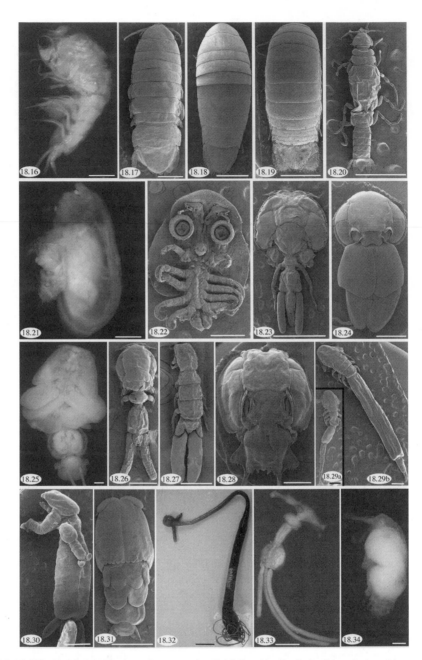

FIGURES 18.16 TO 18.34 Micrographs of Arthropoda. 18.16. Amphipoda: Lafystiidae: *Opisa tridentata* (USNM No. 127598). 18.17. Isopoda: Aegidae: *Rocinela* sp. ex *Dasyatis* sp. 18.18. Isopoda: Cirolanidae ex *Paragaleus pectoralis*. 18.19. Isopoda: Corallanidae ex *Chiloscyllium punctatum*. 18.20. Isopoda: Gnathiidae: *Gnathia* sp. ex *Centrolophus niger*. 18.21. Ostracoda: Cypridinidae: *Sheina orri* (USNM No. 112675) ex *Hemiscyllium ocellatum*. 18.22. Branchiura: Argulidae: *Argulus* sp. ex *Potamotrygon magdalenae*. 18.23. Copepoda: Caligidae ex *Himantura* c.f. *uarnacoides*. 18.24. Copepoda: Cecropidae ex *Prionace glauca*. 18.25. Copepoda: Dichelesthiidae ex *Prionace glauca*. 18.26. Copepoda: Dissonidae ex *Chiloscyllium punctatum*. 18.27. Copepoda: Eudactylinidae ex *Himantura* c.f. *pastinacoides*. 18.28. Copepoda: Euryphoridae ex *Sphyrna lewini*. 18.29. Copepoda: Kroyeriidae ex *Prionace glauca*: (a) male, (b) female. 18.30. Copepoda: Lernaeopodidae ex *Galeorhinus australis*, female with small parasitic male. 18.31. Copepoda: Pandaridae: *Pandarus* sp. ex *Squalus acanthias*. 18.32. Copepoda: Pennellidae: *Pennella filosa* (USNM No. 92174) ex *Makaira nigricans*. 18.33. Copepoda: Sphyriidae: *Norkus cladocephalus* (USNM No. 229971) ex *Rhinobatos productus*. 18.34. Copepoda: Ergasilidae: *Ergasilus myctarothes* (USNM No. 42255) ex *Sphyrna zygaena*. Scale bars: Figures 18.17, 18.22, 18.26, and 18.29a and b, 500 µm; Figures 18.18, 18.19, 18.23, 18.24, 18.28, and 18.31, 2 mm; Figures 18.20 and 18.30, 1 mm; Figure 18.27, 200 µm; Figure 18.32, 1 cm; Figure 18.33, 3 mm.

18.2 Sites Parasitized

18.2.1 Skin

Parasites that attach to the skin of elasmobranchs are often specific for the skin on a particular region of the body. This site specificity is most marked in the copepods and monogeneans. For example, females of the copepod *Echthrogaleus coleopterus* are highly specific for the surfaces of the pelvic fins of their blue shark hosts (Benz, 1986); the monogenean *Acanthocotyle greeni* is found only on the ventral surfaces of *Raja* species (MacDonald and Llewellyn, 1980). Unfortunately, a detailed treatment of parasites associated with the skin of each of the various regions of the body is beyond the scope of this chapter. We have instead treated the parasites of any region of the skin, including external surfaces of the body proper, fins, and claspers, together in this single section. Organisms parasitizing this site share the ability to attach below, around, or on top of the placoid scales of elasmobranchs. Many possess appendages and/or attachment structures useful for this purpose.

The prosobranch snail *Cancellaria cooperi* is the only species of mollusc known from elasmobranchs. It has been reported by O'Sullivan et al. (1987), apparently feeding on blood, on the dorsal surface of Pacific torpedo rays, *Torpedo californica*.

The only class of annelids known to include species that parasitize elasmobranchs is the Hirudinea, or leeches. Records to date suggest that leeches associated with elasmobranchs belong to at least three subfamilies of the family Piscicolidae (Burreson, pers. comm.): Piscicolinae, Pontobdellinae, and Platybdellinae (see Burreson and Kearn, 2000). However, Curran et al. (pers. comm.) discovered a piscicolid leech on the external surfaces of *Zapteryx exasperata* that they were unable to place into any of these three subfamilies. Approximately 18 of the 20 or so known species of elasmobranch leeches have been reported from the skin of their hosts (Burreson, pers. comm.). Several of these species are also known from other sites (see below), but the skin appears to be the preferred region of attachment for most leeches.

Although the skin is an unusual site in which to encounter evidence of nematodes, given that most nematode species are normally endoparasitic, at least two species of spirurids in the family Philometridae and one species of trichinelloid in the family Trichosomoididae have been found in this region of the elasmobranch body. Larvae of a philometrid similar to *Phlyctainophora lamnae* were reported by Ruyck and Chabaud (1960) from tumors at the bases of the fins of *Mustelus mustelus*. Adamson et al. (1987) reported adults of *P. squali* from lesions in the skin of several species of sharks. The trichosomoidid *Huffmanela carcharhini* is particularly interesting because, to date, it is known only from its darkly pigmented eggs, which are deposited by the female nematode in meandering trails around the bases of the placoid scales of the head and fins of a diversity of carcharhinid shark species. This nematode was originally described by MacCallum (1925) from *Carcharhinus plumbeus* (as *C. commersoni*). A detailed discussion of this genus is provided by Moravec (2001), who noted that the genus also includes seven species that parasitize teleosts, six of which are also known only from eggs.

A diversity of minor arthropod groups parasitize the skin of elasmobranchs. These include the barnacles, amphipods, branchiurans, and isopods. The mesoparasitic barnacle *Anelasma squalicola* has been found parasitizing several species of squaliform sharks, in which they are associated with, for example, the dorsal spines and the pectoral and pelvic fins of their hosts (e.g., Kabata, 1970; Long and Waggoner, 1993; Yano and Musick, 2000). An interesting, indirect association apparently exists between a second barnacle species, *Conchoderma virgatum*, and certain copepods found on the body surfaces of elasmobranchs. This species has been reported attached to members of two different families of copepods parasitizing the skin of large, pelagic sharks (e.g., Williams, 1978; Benz, 1984).

Although most amphipods are free-living, there exist a number of species that are normally found associated with hosts. Some of these, for example, the lysianassid *Opisa tridentata* and the lafystiid *Lafystius morhuanus*, have been reported from the skin of sharks and rays. However, these species have also been reported from a diversity of teleosts (Bousfield, 1987). A few amphipod species, such as the trischizostomatid *Trischizostoma raschi*, are known only from the body surfaces of certain squaliform sharks (Bousfield, 1987).

The branchiurans, or fish lice, number approximately 150 species (Kabata, 1988); all are parasitic on fishes. Only a small number of these are known from elasmobranchs. The majority of the elasmobranch

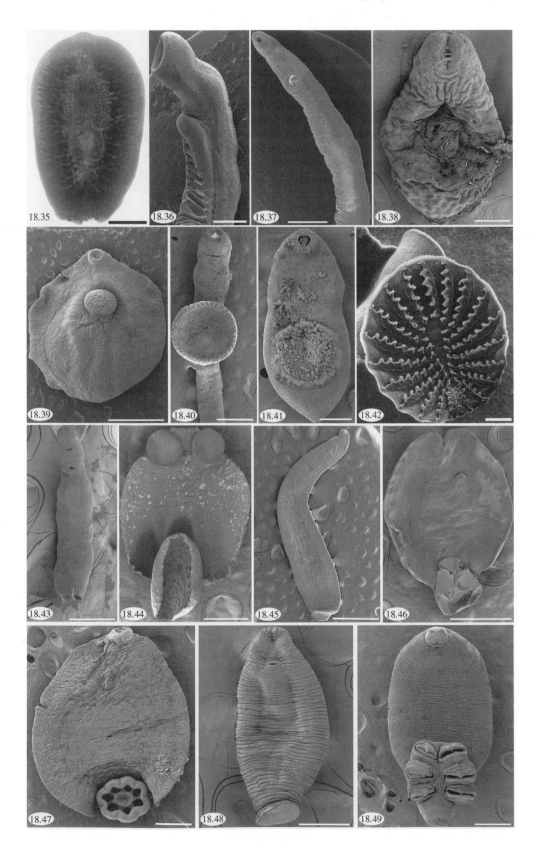

records are of species of *Argulus* (family Argulidae) from the dorsal surfaces of either dasyatid (e.g., Cressey, 1976) or potamotrygonid (e.g., Ross, 1999) stingrays. Marques (2000) reported finding members of a second genus, *Dolops*, on freshwater stingrays in South America.

Of the thousands of species of isopods known worldwide, only about 500 are known to associate with fishes (Bunkley-Williams and Williams, 1998), only approximately 42 of these associate with elasmo-branchs (see Moreira and Sadowsky, 1978), and only a very small subset of these attach to the skin. Bunkley-Williams and Williams (1998) provided a useful overview of the isopods infecting fishes. Of the five families of isopods infecting elasmobranchs, members of three have been found on the skin. Gnathids are unique among these isopods in that it is the larval stage, or praniza, rather than the adult stage, that is parasitic. These relatively small isopods feed on the blood of their hosts, becoming more conspicuous as they feed and their body swells with host blood. Knowledge of the taxonomy and host specificity of gnathids is limited by the fact that pranizae cannot be identified to species; the taxonomy of the family is currently based on the morphology of adult males. Although much more commonly associated with the gills, pranizae of gnathids have been reported from the skin (Heupel and Bennett, 1999). Representatives of two additional families of isopods occupy sites on the external surfaces of elasmobranchs. The cymothoid *Nerocila acuminata*, for example, occurs on the skin of several species of sharks and rays (Brusca, 1981); several different species of Aegidae have also been reported from the skin of elasmobranchs (see Moreira and Sadowsky, 1978).

The copepods are both the most diverse arthropod group parasitizing elasmobranchs and the most diverse group of ectoparasites of elasmobranchs. In 1993, Cheung listed 221 species of copepods from elasmobranchs. We estimate this number is now approaching 275 species. The copepods of elasmo-branchs belong to two of the eight known copepod orders (the Poecilostomatoida and Siphonostoma-toida). Elasmobranchs host members of 4 families of peocilostomatoid and 12 families of siphonostomatoid copepods. Species in 8 of these 16 families have been reported from the skin. These are the Taeniacanthidae (e.g., Braswell et al., 2002) in the Poecilostomatoidea and the siphonostomatoid families Caligidae (e.g., Bere, 1936), Dissonidae (e.g., Deets and Dojiri, 1990), Euryphoridae (e.g., Lewis, 1966), Koryeridae (e.g., Cheung, 1993), Lernaeopodidae (e.g., Pearse, 1953), Pandaridae (e.g., Lewis, 1966), Trebiidae (e.g., Pearse, 1953), and possibly the Pennellidae (but see Benz and Bullard, in review). The pennellids differ from the eight other skin-dwelling copepod families in that they exhibit a mesoparasitic, rather than ectoparasitic, lifestyle.

Among the ectoparasites of elasmobranchs, the platyhelminth subclass Monogenea is second in diversity only to the copepods. Cheung's (1993) estimate of 150 species parasitizing elasmobranchs consists of species in eight monogenean families. These species represent both superorders (Monopis-thocotylea and Polyopisthocotylea) of the subclass. Given the numerous descriptions of new taxa that have appeared over the past decade, Whittington and Chisholm (2003) estimated that monogeneans known from chondrichthyans now number 201 species; approximately 4% of these have been reported from chimaeras. Thus, their data suggest that 193 species of monogeneans are known from elasmobranchs at this time. Five families include species that parasitize the skin of their elasmobranch hosts. Among them, the Acanthocotylidae (e.g., Kearn, 1963), Capsalidae (e.g., Kearn, 1979), and Microbothriidae (e.g., Kearn, 1979) are primarily parasites of the skin. Although they are much more commonly found

---

FIGURES 18.35 TO 18.49 (See figure facing page.) Micrographs of non-cestode Platyhelminthes. 18.35. Tricladida: Procerodidae: *Micropharynx parasitica* (HWML No. 1904) ex *Dipturus laevis*. 18.36. Aspidogastrea: Multicalycidae: *Multicalyx cristata* ex *Dasyatis* sp. 18.37. Digenea: Azygiidae: *Otodistomum* sp. ex *Raja rhina*. 18.38. Digenea: Derogenidae: *Thometrema overstreeti* ex *Potamotrygon magdalenae*. 18.39. Digenea: Gorgoderidae: *Anaporrhutum* sp. ex *Rhinoptera* sp. 18.40. Digenea: Syncoeliidae: *Syncoelium vermilionensis* ex *Mobula japanica*. 18.41. Digenea: Zoogonidae: *Diphterosto-mum* sp. ex *Leptocharias smithii*. 18.42. Monogenea: Acanthocotylidae: posterior attachment structure (haptor) of *Acan-thocotyle* sp. ex. *Raja* sp. 18.43. Monogenea: Amphibdellidae: *Amphidelloides* sp. ex *Narcine tasmaniensis*. 18.44. Monogenea: Capsalidae ex *Dasyatis akajei*. 18.45. Monogenea: Loimoidae: *Loimopapillosum* sp. ex *Eusphyra blochii*. 18.46. Monogenea: Microbothriidae: *Dermopthirius penneri* ex *Carcharhinus limbatus*. 18.47. Monogenea: Monocotylidae: *Calicotyle* sp. ex *Rhizoprionodon terraenovae*. 18.48. Monogenea: Udonellidae: *Udonella* sp. ex caligid copepod on *Urogymnus asperrimus*. 18.49. Monogenea: Hexabothriidae: *Erpocotyle* sp. ex *Bathyraja magellanica*. Scale bars: Figure 18.35, 1.6 mm; Figures 18.36, 18.45, and 18.49, 1 mm; Figure 18.37 and 18.39, 2 mm; Figure 18.38, 50 μm; Figure 18.43 and 18.48, 200 μm; Figure 18.41, Figure 18.42, 100 μm; Figures 18.40, 18.44, 18.46, and 18.47, 500 μm.

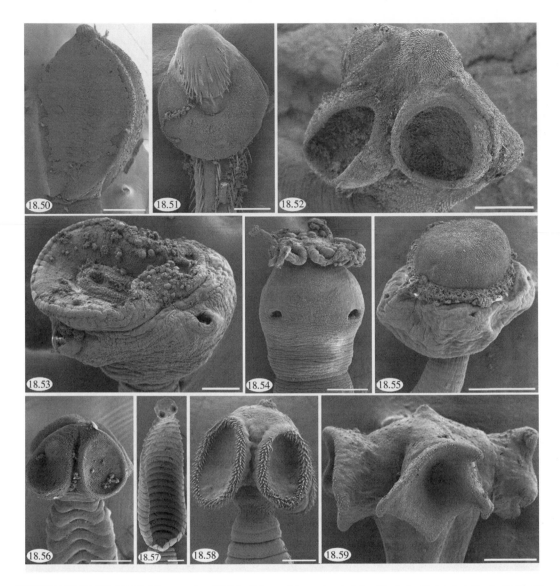

FIGURES 18.50 TO 18.59 Micrographs of Diphyllidea and Lecanicephalidea (Cestoda: Platyhelminthes). 18.50. Diphyllidea: Ditrachybothridiidae: *Ditrachybothridium macrocephalum* ex *Leucoraja fullonica.* 18.51. Diphyllidea: Echinobothriidae: *Echinobothrium helmymohamedi* ex *Taeniura lymma.* 18.52. Lecanicephalidea: Anteroporidae: *Anteropora* sp. ex *Hemiscyllium ocellatum.* 18.53. Lecanicephalidea: Lecanicephalidae: *Lecanicephalum* sp. ex *Dasyatis centroura.* 18.54. Lecanicephalidea: Polypocephalidae: *Polypocephalus* sp. ex *Rhinoptera* sp. 18.55. Lecanicephalidea: Tetragonocephalidae: *Tetragonocephalum* sp. ex *Himantura undulata.* 18.56. Lecanicephalidea: *Aberrapex senticosus* ex *Myliobatis californicus.* 18.57. Lecanicephalidea: *Eniochobothrium* sp. ex *Rhinoptera* sp. 18.58. Lecanicephalidea: *Hornellobothrium* sp. ex *Aetobatus narinari.* 18.59. Lecanicephalidea: *Quadcuspibothrium francisi* ex *Mobula japanica.* Scale bars: Figures 18.50 and 18.55, 200 μm; Figures 18.51, 18.53, 18.54, 18.56, 18.57, and 18.59, 50 μm; Figure 18.52, 100 μm; Figure 18.58, 20 μm.

on the gills, loimoids have also been reported from this site (Benz and Bullard, in review). As a group, the monocotylids are by far the least site specific of the monogenean families, despite the fact that individual species and genera are often very site specific. Monocotylids are found in or on the elasmobranch body from a wide diversity of sites. At present, the only monocotylid genus reported from the skin is *Dendromonocotyle* (see Kearn, 1979). Species in several of these skin-dwelling families have developed an interesting mode of camouflage, whereby they sequester pigment, which appears to be derived from host skin, in their digestive tract. This pigment renders them almost invisible against the

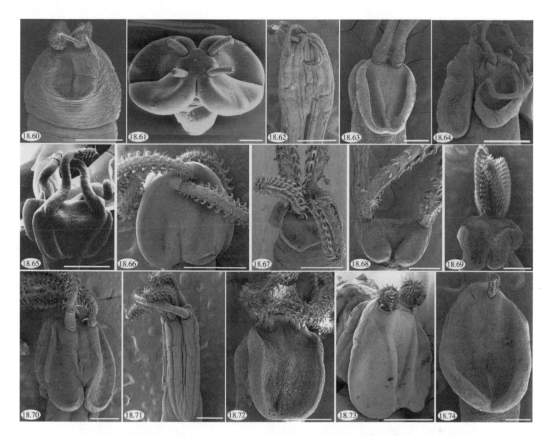

FIGURES 18.60 TO 18.74 Micrographs of Trypanorhyncha (Cestoda: Platyhelminthes). 18.60. Hepatoxylidae: *Hepatoxylon* sp. ex *Prionace glauca*. 18.61. Pseudotobothriidae: *Pseudotobothrium* sp. 18.62. Tentaculariidae: *Tentacularia* sp. ex *Prionace glauca*. 18.63. Eutetrarhynchidae: *Eutetrarhynchus lineatus* ex *Ginglymostoma cirratum*. 18.64. Gilquiniidae: *Gilquinia squali* ex *Squalus suckleyi*. 18.65. Shirleyrhynchidae: *Cetorhinocola acanthocapax* ex *Cetorhinus maximus*. 18.66. Grillotiidae: *Grillotia similis* ex *Ginglymostoma cirratum*. 18.67. Otobothriidae: *Otobothrium* sp. ex *Negaprion acutidens*. 18.68. Rhinoptericolidae: *Rhinoptericola* sp. ex *Rhinoptera bonasus*. 18.69. Dasyrhynchidae: *Dasyrhynchus* sp. ex *Carcharhinus plumbeus*. 18.70. Gymnorhynchidae: *Gymnorhynchus isuri* ex *Isurus oxyrhinchus*. 18.71. Hornelliellidae: *Hornelliella* sp. ex *Stegastoma fasciatum*. 18.72. Lacistorhynchidae: *Lacistorhynchus tenuis* ex *Mustelus canis*. 18.73. Mixodigmatidae: *Mixodigma leptaleum* ex *Megachasma pelagios*. 18.74. Mustelicolidae: *Diesingium* sp. ex *Mustelus mustelus*. Scale bars: Figure 18.60, 2 mm; Figures 18.61, 18.64, 18.67, 18.68, 18.73, and 18.74, 200 μm; Figures 18.62, 18.69, 18.70, and 18.71, 500 μm; Figures 18.63 and 18.72, 100 μm; Figure 18.65, 400 μm; Figure 18.66, 1 mm.

pigmented dorsal surfaces of their elasmobranch hosts (Kearn, 1979). The Udonellidae includes species that attach to copepods (primarily caligids) parasitizing the body of elasmobranchs and thus can be considered at least indirect inhabitants of this site (e.g., Price, 1938a). Evidence suggests that udonellids are particular about the site they inhabit on their copepod host (Causey, 1961). The possibility that these monogeneans occasionally feed directly on the fish hosting their copepod host has been suggested (Kearn, 1998). Bullard et al. (2000) recently reported the postoncomiracidial (i.e., juvenile) stages of a member of the family Dioncidae on the skin of *Carcharhinus limbatus*. This is the ninth monogenean family reported from elasmobranchs.

Species in the remaining platyhelminth groups rarely occupy the skin, but a few exceptions exist. For example, the triclad *Micropharyx parasitica* (family Procerodidae) has been reported with some regularity from skates in the Atlantic Ocean (see Beverley-Burton, 1984). This species is found on the dorsal surfaces of its host where, according to Ball and Khan (1976), it feeds on host epidermal tissue. Thus, unlike the majority of the other non-neodermatan platyhelminths, this species parasitizes vertebrates, rather than invertebrates. Cestodes and digeneans are also found, albeit rarely, on the skin of elasmobranchs. Plerocerci (i.e., larvae) of the trypanorhynch family Grillotiidae were reported by Guiart (1935)

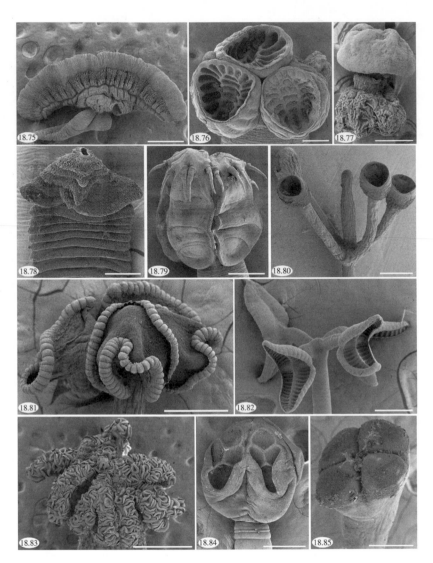

FIGURES 18.75 TO 18.85 Micrographs of Tetraphyllidea (Cestoda: Platyhelminthes). 18.75. Cathetocephalidae: *Cathetocephalus* sp. ex *Carcharhinus leucas*. 18.76. Dioecotaeniidae: *Dioecotaenia* sp. ex *Rhinoptera bonasus*. 18.77. Disculicepitidae: *Disculiceps* sp. ex *Carcharhinus brevipinna*. 18.78. Litobothriidae: *Litobothrium daileyi* ex *Alopias pelagicus*. 18.79. Onchobothriidae: *Acanthobothrium* sp. ex *Aetobatus narinari*. 18.80. Phyllobothriidae: Echeneibothriinae: *Pseudanthobothrium* sp. ex *Leucoraja erinacea*. 18.81. Phyllobothriidae: Phyllobothriinae: *Anthocephalum* sp. ex *Dasyatis longus*. 18.82. Phyllobothriidae: Rhinebothriinae: *Rhinebothrium* sp. ex *Dasyatis longus*. 18.83. Phyllobothriidae: Thysanocephalinae: *Thysanocephalum* sp. ex *Galeocerdo cuvier*. 18.84. Phyllobothriidae: Triloculariinae: *Zyxibothrium kamienae* ex *Malacoraja senta*. 18.85. Prosobothriidae: *Prosobothrium* sp. ex *Prionace glauca*. Scale bars: Figure 18.75, 500 μm; Figures 18.76, 18.77, 18.80, 18.81, 18.82, 18.84, and 18.85, 200 μm; Figure 18.78, 50 μm; Figure 18.79, 100 μm; Figure 18.83, 1 mm.

encysted in the skin of several species of sharks. *Paronatrema mantae*, a digenean belonging to the family Syncoeliidae, was reported from the skin of *Manta birostris* by Manter (1940).

18.2.2 Sensory Systems

18.2.2.1 Eyes — Arthropods are the primary associates of the eyes of elasmobranchs. Benz and his co-workers (e.g., Benz et al., 1998, 2002; Borucinska et al., 1998) have recently done much to document the interesting association between lernaeopodid copepods of the genus *Ommatokoita* and the eyes of their shark hosts. This work suggests that these copepods cause severe corneal displasia, resulting in at least partial blindness in their squaliform hosts. Newbound and Knott (1999) reported finding a member of a second family of copepods, the Caligidae, exclusively on the surface of the eyes of elasmobranchs off Western Australia. Although branchiurans have been observed on the eyes of elasmobranchs on occasion, there is no evidence to suggest that these vagile arthropods are doing anything more than traversing this site. Russo (1975) reported finding the piscicoline leech *Branchellion lobata* on the eyes of spiny dogfish. There exists a record of an adult didymozoid digenean from the back of the eye of the shark *Carcharhinus longimanus* (see Pozdnyakov, 1989).

18.2.2.2   *Olfactory Bulbs* — The olfactory bulbs (also known as olfactory sacs, olfactory capsules, or nasal fossae), and in particular the lamellae of these organs, are the sites of attachment of a diversity of metazoans, including arthropods such as ostracods, isopods, and copepods, as well as nematodes, leeches, and members of the platyhelminth subclass Monogenea. Benz (1993) suggested that the conspicuous overlap between the fauna of the olfactory bulbs and that of the gills, at least at higher taxonomic levels, is perhaps not surprising, given the remarkable similarity between the morphology and configuration of the lamellae of the olfactory bulbs and those of the gills. He hypothesized that the olfactory bulbs represent modified branchial chambers that were originally derived from gills.

Ostracods, isopods, leeches, and nematodes are, at best, occasional associates of the olfactory bulbs of elasmobranchs. For example, the ostracod *Vargula parasitica* has been reported from the olfactory bulbs of smooth hammerheads, *Sphyrna zygaena*, in the West Indies (Wilson, 1913; Williams and Bunkley-Williams, 1996). However, this species more commonly occupies the gills of elasmobranchs (e.g., Williams and Bunkley-Williams, 1996). Praniza larvae of the isopod family Gnathiidae have been reported from the olfactory bulbs of a diversity of sharks and rays (e.g., Smit and Basson, 2002). On occasion, leeches of the subfamily Piscicolinae have been found on the olfactory bulbs (e.g., Sawyer et al., 1975) or oronasal grooves (Llewellyn and Knight-Jones, 1984) of elasmobranchs. In such cases, however, this site appears to be one of many on which these species are found. The anisakid nematode *Terranova brevicapitata* has been reported from the olfactory bulbs of tiger and dusky sharks (Cheung, 1993).

The olfactory bulbs are also parasitized by representatives of at least seven families of copepods. These include, for example, chondracanthids such as *Acanthochondrites annulatus* (e.g., Kabata, 1970), ergasilids (e.g., Wilson, 1913), all species of the kroyerid genus *Kroyerina* (see Benz, 1993) and species in several other kroyerid genera (e.g., Rokicki and Bychawska, 1991), all species of the eudactylinid genus *Eudactylinella* (e.g., Benz, 1993), as well as several species of pandarids (e.g., Lewis, 1966) and lernaeopodids (Benz, 1991). The family Sphyriidae is also represented in this site. For example, Diebakate et al. (1997) described the mesoparasite *Thamnocephalus* from the olfactory bulbs of the shark *Leptocharias smithii* in Senegal. The relationship between this species and its host is complicated by the fact that it simultaneously parasitizes the brain of its host; while the posterior region of its body extends from the olfactory bulbs, the anterior regions of the body attach to the olfactory lobes of the brain (see below).

Members of the monogenean family Monocotylidae inhabit the olfactory bulbs of sharks (e.g., Kearn and Green, 1983) and rays (e.g., Kearn and Beverley-Burton, 1990; Whittington, 1990) with some regularity. In fact, the olfactory bulbs appear to represent the primary site of attachment for many species and even genera of monocotylids (Kearn, 1998), particularly the merizocotylines. Several species in the monogenean family Acanthocotylidae have also been reported from the olfactory bulbs of elasmobranchs, as have at least some species of microbothriids (see Price, 1963). However, Price's records of both of these families from this site are particularly unusual and should be verified.

18.2.2.3   *Acousticolateralis System* — In their diagnosis of the philichthyid copepod genus *Colobomatus*, Deboutteville and Nunes (1952, p. 599) noted that species "vivant dans les canaux muqueux de la tête de Téléostéens (rarement des Sélaciens)." Indeed, *C. lamnae* was described by Hesse (1873) from *Lamna nasus* (as *L. cornubica*). This site represents one of the most poorly known regions of the elasmobranch body because it is so infrequently examined for parasites. We suspect that efforts spent examining the pores and ducts of this system in a diversity of elasmobanchs might yield additional members of this copepod family.

18.2.3 Respiratory System

18.2.3.1 Spiracles — Although an uncommon site of attachment for parasites, the spiracles of elasmobranchs have been reported to host copepods of the families Lernaeopodidae, Taeniacanthidae, and Caligidae. For example, Kabata (1979) noted that the lernaeopodid *Pseudocharopinus bicaudatus* is commonly found within the spiracles of *Squalus acanthias* in British waters. Braswell et al. (2002) found a small percentage of the individuals of the taeniacanthid *Taeniacanthodes dojirii* from *Narcine entemedor* attached in the vicinity of the spiracles. Caligid copepods have been reported from the spiracles of several species of elasmobranchs in the Gulf of Mexico (e.g., Bere, 1936).

On rare occasions, piscicolid leeches of the subfamily Piscicolinae occur, among other sites, in the spiracular valves of elasmobranchs (Russo, 1975).

18.2.3.2 Gills and Branchial Chamber — The gills and branchial chamber of elasmobranchs offer a protected, oxygen-rich, compact living space for parasites (Kearn, 1998). The gill filaments are a rich source of blood for blood-feeding parasite taxa. In combination with their accessibility, these factors may explain the very high diversity of metazoan parasites found in this region of the elasmobranch body (Kearn, 1998). Although we treat the gills and branchial chamber together, in nature parasites often occupy much more specific sites within the gills and/or branchial chamber. This fact was nicely illustrated by Benz (1986) who noted that, among blue shark copepods, the interbranchial septa are inhabited by the pandarid *Phyllothyreus cornutus*, the secondary lamellae of the gill filaments by the pandarid *Gangliopus pyriformis*, the excurrent water channels between gill filaments by the kroyerid *Kroyeria carchariaeglauci*, and the efferent arterioles of gill filaments by eudactylinids of the genus *Nemesis*. Space limitations prevent us from presenting more specific data on attachment sites for the parasites of gills and branchial chamber here.

Somewhat surprisingly, leeches are only rarely found associated with the gills of elasmobranchs. In most cases, even species reported from this site (e.g., the pontobdelline *Stibarobdella tasmanica* and the piscicoline *Branchellion ravenelli*) have also been reported from a diversity of other sites, most notably the skin (Burreson, pers. comm.), suggesting that the gills are not a preferred site of attachment for these taxa.

At least four families of nematodes include species that have been found associated with the gills and/or branchial chamber of elasmobranchs. Such occurrences, however, are relatively uncommon. The distinctive eggs of the trichosomoidid nematode *Huffmanela carcharhini*, discussed in more detail in the section on skin above, have been observed in the mucosa covering the connective tissue of the gill arches of certain carcharhinid sharks (see Moravec, 2001). The distinctive vesicular females of the philometrid nematode *Phlyctainophora lamnae* have also been reported from this region by Steiner (1921), specifically between the hyomandibula arch and skull of the shark *Lamna nasus* (as *L. cornubica*). The gills were included among the sites from which Adamson et al. (1987) collected adults of the philometrid *P. squali* and from which Aragort et al. (2002) collected adults of the guyanemid dracunculoid *Histodytes microocellatus*. Microfilariae of a species of dracunculoid belonging to a fourth nematode group, but one that has not yet been assigned to family, were found in gill squashes from a spotted eagle ray by Adamson et al. (pers. comm.). Cheung (1993) noted that the anisakid *Contraceacum plagiostomum* has been found on the gills of the basking shark and thorny skate.

Copepods are among the most commonly encountered parasites of the gills and branchial chamber of elasmobranchs. Of the 16 families of copepods parasitizing elasmobranchs, 11 include species that parasitize these sites. The gills and/or branchial chamber are the primary site of attachment for nine families of siphonostomes. For example, all eight genera of Eudactylinidae that parasitize elasmobranchs include species that associate with the gills; six of these eight genera are restricted to the gills (Benz, 1993). Two of the three genera of kroyerids include species that are found on the gill lamellae or, in the case of the unusual mesoparasite *Kroyeria caseyi*, interbranchial septa of their elasmobranch hosts (see Benz and Deets, 1986). Benz (1993) considered the gill filaments and branchial chamber to be among the primary sites of attachment of caligid copepods. Although not the primary site of attachment, the gills and branchial chamber are included among the sites occupied by the following additional siphonostome families: Cecropidae (e.g., Benz and Deets, 1988), Dichelesthiidae (e.g., Benz, 1993), Dissonidae (e.g., Benz, 1993), Pandaridae (e.g., Lewis, 1966), Lernaeopodidae (e.g., Wilson, 1913), and a number of species of the mesoparasitic family Sphyriidae (e.g., Kabata, 1979). In addition, members of the poecilostome copepod families Taeniacanthidae (e.g., Wilson, 1913) and Chondracanthidae (Cheung, 1993) also occur on this site.

Three additional arthropod groups have been reported from the gills of elasmobranchs, albeit infrequently. This is the primary site of attachment of cypridinid ostracods when they have been found associated with elasmobranchs (e.g., Williams and Bunkley-Williams, 1996; Bennett et al., 1997). Records of members of the branchiuran genus *Argulus* from the branchial cavity of dasyatid stingrays exist, but this does not appear to be a primary site of attachment for these arthropods (see Cressey, 1976). Members of four families of isopods have been reported from the gills of elasmobranchs. Among these, the praniza larvae of the Gnathiidae are most commonly encountered in this site (e.g., Newbound and Knott, 1999). These arthropods are known to cause injury to the epithelium and are often associated with inflammation and severe tissue hypertrophy at these sites, particularly in heavy infections (e.g., Honma et al., 1991). Delaney (1984) found corallanids attached to the gills of *Aetobatus narinari*. The gills were among the sites occupied by the cirolanid *Cirolana borealis* (see Bird, 1981). At least four genera of aegids have been reported from the gills (see Moreira and Sadowsky, 1978). In addition, the cymothid *Lironeca ovalis* has been reported from the gills of sawfish (Cheung, 1993).

Although the vast majority of the approximately 18,000 species of platyhelminths belonging to the subclass Digenea live as endoparasites (see Cribb et al., 2001), there are some exceptions, and many of these can be found among the digeneans of elasmobranchs. In fact, as noted by Bray and Cribb (2003), the digeneans of elasmobranchs tend to be found in a diversity of relatively unusual sites, many of which are more typical of ectoparasitic taxa. For example, members of the family Syncoeliidae often associate with the gills of their hosts (see Curran and Overstreet, 2000). Didymozoids, such as *Tricharrhen okenii*, have also been reported from the gill arches and branchial chamber of elasmobranchs (Yamaguti, 1971). Bray and Cribb (2003) noted that a lepocreadiid digenean has also been reported from the gills of a porbeagle shark, but this was likely the result of an accidental infection.

On very rare occasions cestodes have been found associated with the gills of elasmobranchs. For example, Dollfus (1960) reported plerocerci of a species of the trypanorhynch genus *Nybelina* from the gills of *Mustelus canis*.

Representatives of all eight families of monogeneans have been reported from elasmobranch gills. Hexabothriids are known only from the gills of elasmobranchs, where they exhibit feeding habits that are relatively unusual for monogeneans in that they feed exclusively on blood (Kearn, 1998). Approximately 50 species of hexabothriids have been reported from elasmobranchs. With only a few exceptions, the Amphibdellidae have been reported from the gills of rays (e.g., Llewellyn, 1960). The loimoids are found primarily on the gills of their elasmobranch hosts (e.g., Cheung, 1993). The gills represent one of several sites occupied by capsalids (e.g., Cheung, 1993) and monocotylids (see Chisholm et al., 1997). In addition, acanthocotylids and microbothriids have been reported from this site (Bonham and Guberlet, 1938 and Price, 1963, respectively), but the latter reports are so rare that they should be considered suspect until verified.

18.2.4 Digestive System

Parasite site specificity within the digestive system is marked, reflecting the fact that, although the digestive system proper represents a single continuous tube beginning with the mouth and ending with the rectum and cloaca, it contains a diversity of physically and physiologically distinct environments.

18.2.4.1 Buccal Cavity and Esophagus — There exist records of at least one species in each of the three subfamilies of piscicolid leeches from the buccal cavity and/or esophagus of elasmobranchs. For example, the pontobdelline *Stibarobdella macrothela* and the piscicoline *Branchellion lobata* have been reported from, among other sites, the buccal cavity (Sawyer et al., 1975 and Llewellyn and Knight-Jones, 1984, respectively). The platybdelline leech *Rhopalobdella japonica*, however, may be specific to the buccal cavity; this species is known only from the buccal cavity of *Dasyatis akajei*, where it has been found dangling between the upper lip and tooth plates (see Burreson and Kearn, 2000). We have

found relatively large numbers of leeches occupying a similar site in *Pastinachus sephen* taken from Stradbrook Island, Australia. The unusual leech discovered by Curran et al. (pers. comm.) from *Zapteryx exasperata* in the Gulf of California, Mexico, was found in the buccal cavity in addition to the external body surfaces of this host. As noted above, to date these authors have been unable to assign this species to one of the three subfamilies of piscicolids.

Copepods are less commonly encountered in the buccal cavity than they are on other external body sites of elasmobranchs, but a few species occupy this site with some regularity. The only species of dichelesthiid that parasitizes elasmobranchs, *Anthosoma crassum*, has been reported from the buccal cavity of large pelagic sharks (e.g., Lewis, 1966). Species of the eudactylinid genus *Carniforssorus* are known from the oral chamber of elasmobranchs (Benz, 1993). Bere (1936) reported caligids from the mouth of several batoids and pandarids from the mouth of several sharks in the Gulf of Mexico. Euryphorids, lernaeopodids, and sphyriids have also been reported from the buccal cavity of elasmobranchs (Cheung, 1993). In addition, the pranizae larvae of gnathid isopods and adult cirolanids have been reported from the buccal cavity of elasmobranchs (Heupel and Bennett, 1999 and Moreira and Sadowsky, 1978, respectively).

Digeneans, monogeneans, nematodes, and barnacles have also occasionally been found parasitizing this site. Digeneans of the families Syncoeliidae and Ptychogonimidae are known to occur in the buccal cavity of sharks (e.g., Curran and Overstreet, 2000 and Cheung, 1993, respectively), as is the monogenean monocotylid *Tritestis ijime* (e.g., Price, 1938b). The buccal cavity was among the sites in which Adamson et al. (1987) found adults of the philometrid nematode *Phlyctainophora squali* in several species of sharks off California. The cystidicolid *Parascophorus galeata* was reported from the esophagus of *Sphyrna tiburo* in North Carolina (Cheung, 1993). The mouth was among the sites reported infected with the barnacle *Anelasma squalicola* by Yano and Musick (2000).

18.2.4.2 *Stomach* — Despite the potentially inhospitable nature of the stomach as an environment, it is home to a number of platyhelminth, nematode, and, to a lesser extent, acanthocephalan and arthropod species. Most of the species that inhabit the stomach do so to the exclusion of other sites in the elasmobranch host. In some cases this site specificity extends to higher taxonomic categories as well.

The presence of acanthocephalans in elasmobranchs is unusual. Adults of most of the 1200 or so species in this phylum are much more commonly encountered in teleosts, birds, turtles, and even mammals (see Crompton and Nickol, 1985). As a consequence, records of acanthocephalans from elasmobranchs are generally considered to represent accidental infections (Knoff et al., 2001). Williams et al. (1970) presented a possible explanation for this absence, suggesting that acanthocephalans may be unable to tolerate the high levels of urea found in elasmobranchs. Two of the eight species of acanthocephalans reported from elasmobranchs have been found in the stomach; both species belong to the polymorphid genus *Corynosoma* (see Knoff et al., 2001).

At least 12 species in four families of nematodes parasitize the stomach of elasmobranchs. These include members of the Anisakidae (see Olsen, 1952), Ascaridae (see McVicar, 1977), and, most commonly, the Acanthocheilidae (see Diaz, 1972) and Physalopteridae (see Moravec and Nagasawa, 2000). Many of the nematodes found in the stomach appear to be restricted to this site, or at least have not generally been reported from other sites in their hosts.

Although it is not clear if they represent parasites or food, members of three families of isopods have been reported from the stomach of elasmobranchs. These include the cirolanid *Cirolana borealis*, the aegid *Aega psora*, and the cymothid *Lironeca raynaudi* (see Moreira and Sadowsky, 1978).

Bray and Cribb (2003) suggested that members of as many as 20 of the 150 families of the platyhelminths in the subclass Digenea have been reported parasitizing elasmobranchs. However, they note that the records for at least 4 of these families are erroneous, and that while 8 of the remaining families include legitimate elasmobranch parasites, another 8 are likely to represent accidental infections. Among the 16 families found in elasmobranchs (as accidental or normal infections), species in 6 have been reported from the stomachs of these hosts with some regularity. One of the most commonly encountered groups is the Azygiidae. The large, muscular *Otodistomum* species are conspicuous inhabitants of the stomachs of a diversity of sharks and rays (Gibson and Bray, 1977); elasmobranchs appear to be the primary hosts of these taxa. The more delicate-bodied ptychogonimids and zoogonids have also been reported from the stomach of elasmobranchs (e.g., Bray et al., 1995 and Bray and Gibson, 1986, respectively). However, many of the zoogonid species also occur in teleosts (see Bray and Gibson, 1986). Bucephalids, opecoelids, and derogenids have also been reported from the stomachs of their hosts, the latter family with some regularity (see Bray and Cribb, 2003).

By far the majority of members of the platyhelminth subclass Cestoda reported from the stomach of elasmobranchs are trypanorhynchs. Site data provided by Bates (1990) and Cheung (1993) suggest that at least 6 of the 21 families of trypanorhynchs are able to live in the stomach either as larvae or adults. For example, adults of the tentaculariid genus *Tentacularia* are routinely found attached in the pyloric stomach of a diversity of sharks (e.g., Guiart, 1935; Dollfus, 1942). Adult hepatoxylids (e.g., *Hepatoxylon trichiri*; see Williams, 1960) and larval grillotiids (e.g., *Grillotia*; see Klimpel et al., 2001) have been reported from the lumen and wall of the stomach of elasmobranchs, respectively. Otobothriids, sphyriocephalids, and pterobothriids have also been reported from the stomach of elasmobranchs (e.g., Dollfus, 1942, Guiart, 1935, and Cheung, 1993, respectively). The report of the diphyllidean cestode *Echinobothrium benedeni* from the stomach of skates in the Mediterranean (Cheung, 1993) requires confirmation.

18.2.4.3 Spiral Intestine

18.2.4.3 Spiral Intestine — The spiral intestine is home to acanthocephalans, digeneans, nematodes, cestodes, and infrequently monogeneans. Five families of acanthocephalans have been reported from the spiral intestines of elasmobranchs. These families are the Cavisomidae (e.g., Golvan et al., 1964), Echinorhynchidae (e.g., Arai, 1989), Illiosentidae (e.g., Buckner et al., 1978), Rhadinorhynchidae, and Polymorphidae (e.g., Knoff et al., 2001). In most of these families, however, records exist for a total of only one or two species. Like the acanthocephalans found in the stomach, most records are thought to represent accidental, rather than normal infections, because in most cases these species are also known to parasitize other species of vertebrates. The cavisomid genus *Megapriapus* may, however, represent an interesting exception, for it is known only from potamotrygonid stingrays (see Golvan et al., 1964). The work of Marques (2000) suggests that this genus may be more widely distributed among potamotrygonids than currently thought.

The spiral intestine of elasmobranchs hosts a greater diversity of nematodes than any other site within the elasmobranch body. Approximately 50 of the 80 or so species in 10 of the 14 families of nematodes known to parasitize elasmobranchs have been reported from this site. These records include members of both classes of nematodes, the Rhabditea and Enoplea. The spiral intestine is the primary site of attachment of the Gnathostomatidae (e.g., Deardorff and Ko, 1983) and the Capillariidae (see Moravec, 2001). The following families also occur in this organ with some regularity: ascarids and anisakids (see McVicar, 1977), acanthocheilids (see Williams et al., 1970), physalopterids (see Moravec et al., 2002), cystidicolids (see Campana-Rouget, 1955), philometrids (see Adamson et al., 1987), and cucullanids (see Johnston and Mawson, 1943). In addition, Moravec et al. (1998) reported the dracunculoid *Mexiconema cichlasomae* from the spiral intestine of *Ginglymostoma cirratum*. However, these authors considered this an accidental infection because, although rare in elasmobranchs, this nematode commonly parasitizes teleosts. Most of the above nematodes occur in the spiral intestine as adults. It is uncommon to encounter more than a single species of nematode in the spiral intestine, and, in general, infections consist of a small number of individuals (i.e., infections are of low intensity). On occasion, however, we have encountered remarkably large concentrations of individuals of some of the larger nematodes (e.g., gnathostomatids) in the spiral intestines of, for example, *Aetomylaeus nichofii* and *Heterodontus mexicanus*.

In stark contrast to the incredible diversity of digeneans found parasitizing the intestinal tract of teleosts, only a handful of species in only 8 of the 150 known families of digeneans parasitize the spiral intestine of elasmobranchs (see Cribb et al., 2001; Bray and Cribb, 2003). In the cases of most families, records from the spiral intestine are rare. With the exception of the robustly muscular azygiids such as *Otodistomum veliporum* (see Gibson and Bray, 1977), most of these species are small and thus easy to overlook. Brooks (1979) described the derogenid *Thometrema overstreeti* (as *Paravitellotrema overstreeti*) from the spiral intestine of the freshwater stingray *Potamotrygon magdalenae*. Although this species also parasitizes teleosts, it appears to be a regular component of the intestinal fauna of these rays. Reports of bucephalids such as *Prosorhynchus squamatus* and *Bucephalopsis arcuatum* (e.g., Cheung, 1993) and zoogonids such as *Diptherostomum betencourti* (e.g., Bray and Gibson, 1986) from the spiral intestine also exist. Bray and Cribb (2003) also cite reports of didymozoids, hemiurids, opecoelids, and syncoelids from the spiral intestine of elasmobranchs. On rare occasion, monocotylid monogeneans have been reported parasitizing this organ (e.g., Chisholm et al., 1997).

The spiral intestine is by far the most heavily parasitized internal organ of elasmobranchs. This is because it is the primary site occupied by cestodes. Cestoda is unquestionably the most diverse group of elasmobranch parasites. It is very rare to encounter an elasmobranch in nature that does not host at least one species of cestode in its spiral intestine. In our experience, it is fairly routine for certain elasmobranchs, particularly batoids, to host ten or more species of cestodes in their spiral intestines. Of the 14 orders of cestodes recognized by Khalil et al. (1994), 4 parasitize elasmobranchs as adults. The Diphyllidea, Lecanicephalidea, and Trypanorhyncha are exclusive to elasmobranchs. With the exception of the Chimaerocestidae, which parasitizes ratfish (see Williams and Bray, 1984), this is also true of the Tetraphyllidea. Collectively, these four orders include 34 families and well over 800 species. Representatives of all 34 families have been reported from the spiral intestine of elasmobranchs. The spiral intestine is the only site occupied by most of these species.

These four orders of tapeworms differ conspicuously in diversity. In the most recent monographic treatment of the Diphyllidea, Tyler (2001) recognized two families, the Ditrachybothridiidae and the Echinobothriidae, each consisting of a single genus and, collectively, 34 species. Jensen (2001) considered the Lecanicephalidea to include as many as 100 valid species and 12 valid genera. Euzet (1994a) recognized four families of lecanicephalideans: Polypocephalidae, Anteroporidae, Tetragonocephalidae, and Lecanicephalidae. There are however, a number of forms that do not conform to the diagnoses of any of the families recognized by Euzet (1994a). We have included electron micrographs of some of these taxa (e.g., *Aberrapex*, *Eniochobothrium*, *Hornellobothrium*, and *Quadcuspibothrium*) to make their presence known to readers. Although most lecanicephalideans have been reported from batoids, records exist of members of this order parasitizing sharks (e.g., Caira et al., 1997b). Trypanorhynch cestodes are easy to recognize by their possession of four tentacles bearing hooks. The taxonomy of the over 200 described species is based largely on the shape and arrangement of these tentacular hooks. In the most recent treatment of the order, Campbell and Beveridge (1994) recognized 21 families in four superfamilies. Trypanorhynchs in general parasitize both sharks and rays; however, many families appear to be restricted to one or the other of these clades (see Campbell and Beveridge, 1994). Although most trypanorhynchs parasitize the spiral intestine, almost all cestode species that have been found in sites of the elasmobranch body other than the spiral intestine belong to this order.

The tetraphyllideans are the most diverse of the four cestode orders parasitizing elasmobranchs. The order includes approximately 50 genera and well over 500 species. Euzet (1994b) recognized eight families, seven of which are found in elasmobranchs: Cathetocephalidae, Dioecotaeniidae, Disculicepitidae, Litobothriidae, Onchobothriidae, Phyllobothriidae, and Prosobothriidae. Collectively, the species in this order exhibit morphological diversity of their attachment structure, or scolex, that is unparalleled in any other cestode group. Most systematically problematic among tetraphyllidean families is the Phyllobothriidae, which includes a suite of potentially unrelated groups (see Caira et al., 1999, 2001). We have included illustrations of the subfamilies of phyllobothriids recognized by Euzet (1994b) (Figure 18.81 through Figure 18.85) as a reminder of the diversity represented by what is currently considered to be a single family. With the exception of *Acanthobothrium*, genera of tetraphyllideans are generally restricted in distribution to either sharks or rays.

Although we have not attempted to present detailed data here, the spiral intestine is a complex environment that consists of a diverse suite of microhabitats (Williams et al., 1970). Many cestode species, including trypanorhynchs (e.g., Caira and Gavarrino, 1990) and tetraphyllideans (e.g., Williams et al., 1970; Cislo and Caira, 1993; Curran and Caira, 1995) exhibit site specificity for particular regions within this organ.

18.2.4.4   Rectum — Platyhelminths appear to be the sole occupants of the elasmobranch rectum. However, the diversity of platyhelminths parasitizing this site is limited. Members of the digenean families Gorgorderidae, Syncoeliidae, and Zoogonidae have, on occasion, been found inhabiting the rectum of sharks (e.g., Cheung, 1993, Curran and Overstreet, 2000, and Bray and Gibson, 1986, respectively). In addition, records of monogeneans of the family Monocotylidae from the rectum of elasmobranchs are not uncommon (e.g., Chisholm et al., 1997; Bullard and Overstreet, 2000).

18.2.4.5 Cloaca — The cloaca is home to a number of platyhelminths and copepods and, on occasion, leeches and isopods. The cloacal region of some batoids is the site of attachment of certain members of the monogenean family Monocotylidae (e.g., Chisholm et al., 1997; Bullard and Overstreet, 2000). Digeneans of the family Syncoeliidae have been reported from the cloaca of several large pelagic shark species (e.g., Curran and Overstreet, 2000). Species in three families of copepods have been found in the cloaca of elasmobranchs. These include, for example, a species of chondracanthid in the genus *Acanthochondrites*, the caligid *Caligus rabidus*, and several species of lernaeopodids belonging to the genus *Lernaeopoda* (see Cheung, 1993). The cloaca is among the numerous sites of the body parasitized by the piscicoline leech *Branchellion lobata* (see Cheung, 1993) and also by the leech Curran et al. (pers. comm.) found on *Zapteryx exasperata*. Hale (1940) reported the aegid *Aega antillensis* from the cloaca of a tiger shark in Australia. Praniza larvae of gnathids have also been reported from this site (Benz and Bullard, in review).

18.2.4.6 Gallbladder and Bile Ducts — In elasmobranchs, the gallbladder and bile ducts are the primary sites of infection of members of the platyhelminth subclass Aspidogastrea. This taxon currently consists of four families (Rohde, 2002), two of which include species that parasitize elasmobranchs. In the Multicalycidae, *Multicalyx cristata* has been reported from the gallbladder of sharks and rays (e.g., Thoney and Burreson, 1986). The only known species of Stichocotylidae, *Stichocotyle nephropis*, was found in the bile ducts of rays (e.g., MacKenzie, 1963).

We recently were surprised to discover that the gallbladders of *Mobula japonica* in the Gulf of California, Mexico were routinely occupied by adults of a relatively large, as of yet unidentified, species of trypanorhynch cestode belonging to the family Tentaculariidae.

18.2.4.7 Pancreatic Duct — Nematodes of the family Rhabdochonidae have been reported on several occasions from the pancreatic ducts of sharks and rays. For example, McVicar and Gibson (1975) reported adults of *Pancreatonema torriensis* from the pancreatic ducts of *Raja naevus*. Recently, Moravec et al. (2001) reported a new species of rhabdochonid nematode from the pancreatic duct of the dogfish shark, *Squalus acanthias*, off coastal Massachusetts.

18.2.4.8 Liver — The majority of our current knowledge about the parasites of the elasmobranch liver comes from samples taken from the outer surfaces of this organ. Little is known about the organisms that may inhabit the parenchyma. Plerocercoids of trypanorhynchs of the hepatoxylid genus *Hepatoxylon* are commonly found attached to the exterior surfaces of the liver in some of the larger species of pelagic sharks (e.g., Waterman and Sin, 1991). Adams et al. (1998) provided an interesting account of a single gravid campulid from the liver of *Alopias vulpinus*, a shark they justifiably consider to be an atypical host for this digenean species.

We suspect that careful examination of liver parenchyma is likely to reveal the presence of nematodes in at least some species of elasmobranchs, but we have been unable to find a verified report of this phylum from this organ.

18.2.5 Circulatory System

18.2.5.1 Heart and Vasculature — Representatives of several different groups of invertebrate metazoans have been reported, some fairly infrequently, from the heart and/or vasculature of elasmobranchs. Adult specimens of a new genus and species of dracunculoid nematode (*Lockenloia sanguinis*) were reported by Adamson and Caira (1991) from the heart of a nurse shark, *Ginglymostoma cirratum*, in the Florida Keys. Adamson et al. (pers. comm.) found microfilariae (i.e., larval stages) of a nematode that appears to be related to *Lockenloia* in the gill vasculature of a spotted eagle ray, *Aetobatus narinari*.

Aragort et al. (2002) recently reported a guyanemid dracunculoid from the heart, among many other sites, of *Raja microocellata*.

The heart and vasculature are the primary sites occupied by digeneans of the family Sanguinicolidae. At least three monotypic genera of sanguinicolids are known from sharks and rays. *Selachohemecus olsoni* was reported by Short (1954) from the heart of *Rhizoprionodon terraenovae* as (as *Scoliodon*). *Hyperandrotrema cetorhini* was described from the heart and blood vessels of basking sharks by Maillard and Ktari (1978). Madhavi and Hanumantha Rao (1970) described *Orchispirium heterovitellatum* from the mesenteric vessels of *Dasyatis imbricatus*.

Even less frequently encountered in the elasmobranch heart are mites, isopods, and monogeneans. Benz (see Benz and Bullard, in review) found what appeared to be either a deutonymph or adult mite in the lumen of the heart of a nurse shark in Florida Bay. Bird (1981) described an interesting "outbreak" of the isopod *Cirolana borealis* in the Cape Canaveral, FL shark fishery in 1977–1978. During that period of time, this cirolanid isopod was observed to cause extensive pathology in the heart of several carcharhiniform shark species. Although it was found in a number of other sites, it showed a distinct preference for the heart. Llewellyn (1960) described the unusual occurrence of the amphibdellid monogenean *Amphibdella flavolineata* from the heart of electric rays. Euzet and Combes (1998) provided an interesting account of the sites occupied by individuals of this monogenean species in different stages of maturity as they move toward the heart to mate.

18.2.5.2   Spleen — The spleen was one of the sites from which Aragort et al. (2002) reported finding the guyanemid nematode *Histodytes microocellatus* in *Raja microocellata*. To our knowledge, this is currently the only record of a metazoan parasite from the spleen of an elasmobranch.

18.2.6   Reproductive System

18.2.6.1   Gonads — Limited work has been done examining the gonads of elasmobranchs for parasites. We know of only three reports of metazoans from these organs; two of these are of nematodes, one is of a cestode. Rosa-Molinar et al. (1983) described larval philometrid nematodes in granulomas associated with the ovaries of blacktip sharks. In addition, Aragort et al. (2002) found specimens of the guyanemid nematode *Histodytes microocellatus* in the gonads of *Raja microocellata*. This was only one of a number of sites parasitized by this nematode. Tandon (1972) reported plerocerci of the pterobothriid trypanorhynch *Pterobothrium* sp. from the ovary of *Dasyatis uarnak*. Although there appear to be no published records of parasites of the male gonads, we recently encountered unidentified larval trypanorhynchs from the testes of a species of *Himantura* in Borneo.

18.2.6.2   Oviducts — The oviducts appear among the sites reported occupied by several species of monogeneans of the family Monocotylidae parasitizing elasmobranchs (e.g., Woolcock, 1936). The digenean syncoeliid *Paranotrema vaginacola* has also been reported from this site in a species of *Squalus* from Papua New Guinea (Dollfus, 1937).

18.2.6.3   Uterus — An unusual array of parasites, specifically leeches, copepods, isopods, and nematodes, have been reported from the elasmobranch uterus. All but the latter group are somewhat unexpected inhabitants of the uterus because they are typically considered to be ectoparasitic. Moser and Anderson (1977) found a species of piscicoline leech inhabiting the external surfaces of embryos of the Pacific angel shark, *Squatina californica, in utero*. Similarly, Nagasawa et al. (1998) found copepods of the family Trebiidae associated with the external surfaces of embryos in the uteri of several species of angel sharks. Benz et al. (1987) reported larval philometrid nematodes in the uterus of a specimen of the shark *Carcharhinus plumbeus* from the northeastern coast of the United States. In addition, the cirolanid *Cirolanis borealis* was reported from the uterus, among other sites, by Bird (1981), and we recently encountered multiple individuals of an unidentified corollanid isopod in the uteri of a specimen of the Atlantic weasel shark, *Paragaleus pectoralis*, taken off Senegal. Reports such as these are, however, uncommon. We suspect that the presence of at least some of these taxa in the uterus may be the result of accidental forays into this organ facilitated by the fact that the uterine pores open directly into the cloaca.

18.2.7   Nervous System

18.2.7.1   Brain — Like the liver, the elasmobranch brain and the other elements of the nervous system have been poorly sampled for parasites. Those records that do exist, however, are particularly interesting. Adamson et al. (pers. comm.) found adult individuals of a dracunculoid nematode in the brain of a spotted eagle ray. These worms were found wrapped around the optic nerves and surrounded much of the brain. As noted above, the unusual mesoparasitic sphyriid copepod described by Diebakate et al. (1997) is appropriately considered a parasite of the brain. While the posterior-most portions of its body extend from the capsule of the olfactory bulb into the external environment, the more anterior regions of the body of this animal penetrate the olfactory lobes of the brain of its shark host.

18.2.8   Body Cavities

18.2.8.1   Pericardial Cavity — Two groups of platyhelminths (Digenea and Monogenea) have been reported from the pericardial cavity of elasmobranchs. Several species of digeneans of the gorgoderid genus *Anaporrhutum* parasitize the pericardial chamber of batoids. This phenomenon was first described by Ofenheim (1900), who reported *A. albidum* from the pericardial chamber of a spotted eagle ray from the Pacific Ocean. More recently, Curran et al. (2003) reported finding one to three individuals of species of *Anaporrhutum* in the pericardial cavity of members of five different genera of batoids in the Gulf of California, Mexico.

There exists one record of a monogenean from the pericardial cavity of elasmobranchs; Bullard and Overstreet (2000) reported monocotylids that may have come from the pericardial cavity of rays in Mexico. However, this record requires verification before it is considered a valid site occupied by these platyhelminths.

18.2.8.2   Peritoneal Cavity — The inhabitants of the peritoneal cavity include species in each of the three major groups of parasitic platyhelminths, the Monogenea, Cestoda, and Digenea. Although most monogeneans are ectoparasitic, several genera of monocotylids have been reported from this site (e.g., Kearn, 1970). The peritoneal cavity is the site most commonly parasitized by larvae of trypanorhynch cestodes such as the hepatoxylid *Hepatoxylon trichuri* (e.g., Waterman and Sin, 1991). In most cases, these larvae are generally not lying free in the body cavity itself; rather, they are attached to either the serosa surrounding organs such as the liver, or to the various mesenteries of the body cavity. Plerocerci (i.e., larvae) of the trypanorhynch family Grillotiidae have been reported from the peritoneal cavity of elasmobranchs with some regularity (e.g., Dollfus, 1942; Williams, 1960). Plerocerci of the tetraphyllidean cestode *Phyllobothrium radioductum* have been reported from the body cavity of skates (Cheung, 1993). Pappas (1970) reported finding adults of the trypanorhynch *Lacistorhynchus tenuis* in the body cavity of *Triakis semifasciata*. Representatives of at least four genera of gorgoderid digeneans have been found in the peritoneal cavity of a diversity of sharks and rays, suggesting that this may be a common site for these platyhelminths (e.g., Markell, 1953; Curran et al., 2003). In addition, on several occasions, species belonging to the azygiid genus *Otodistomum* have been reported from the peritoneal cavity of elasmobranchs (e.g., Gibson and Bray, 1977).

Given the typical life cycle of most monogeneans and digeneans, the occurrence of adult members of these taxa in the peritoneal cavity of elasmobranchs leads to speculation about the portals of entry and exit that might be utilized by inhabitants of this site. Gibson and Bray (1977) suggested that the abdominal pore, which opens directly into the cloaca, would provide an appropriate portal both into and out of this site, making life in the peritoneal cavity possible.

The peritoneal cavity is only rarely parasitized by nematodes. Moravec and Little (1988) reported two species of micropleurid nematodes of the genus *Granulinema* that are likely to have come from the peritoneal cavity of bull sharks in Louisiana. We have occasionally encountered what we believe to be larval ascarid nematodes in the body cavities of a diversity of elasmobranchs in the Gulf of California.

18.2.9 Excretory System

18.2.9.1   Kidneys — The kidneys were one of the sites from which the guyanemid nematode *Histodytes microocellatus* was collected by Aragort et al. (2002).

18.2.9.2   Rectal Gland — The lumen of the rectal gland can be parasitized by monogeneans. Kearn (1987), for example, reported adults of the monogenean monocotylid genus *Calicotyle* from the lumen of the rectal gland of rays; Euzet and Williams (1960) reported a different species of *Calicotyle* from the lumen of the rectal gland of a shark. Plerocerci (i.e., larvae) of the trypanorhynch family Lacistorhynchidae have been reported from the external surface of the rectal gland (e.g., Pappas, 1970).

18.2.10   Body Musculature

Reports of metazoans from the musculature of the body are limited. Guiart (1935) found plerocerci (i.e., larvae) of a grillotiid trypanorhynch in the musculature of the shark *Pseudotriakis microdon*. On the few occasions that we examined musculature of elasmobranchs, such as, for example, *Rhinobatos* in the Gulf of California, Mexico, unidentified larval nematodes were found. Clearly this site represents another poorly studied region of the elasmobranch body and effort spent examining these tissues is likely to yield additional parasite data.

18.3   General Observations

The skin is home to a greater assortment of metazoan higher taxa than any of the other regions of the elasmobranch body. In fact, five of the six phyla and 12 of the 15 major lower taxa of metazoans considered here (see Figure 18.1) have been reported from the skin of the fins, claspers, head, torso, and/or tail of elasmobranchs. The spiral intestine hosts the greatest number of species of metazoan parasites, essentially because this is the primary site occupied by cestodes.

The sites occupied by the major groups of metazoan parasites are relatively predictable. However, each major group includes both families that exhibit fidelity for a particular site or suite of sites, and families that exhibit more relaxed site specificity. For example, cestodes found in sites other than the spiral intestine are generally tentaculariids, lacistorhynchids, or larval hepatoxylids. Among monogeneans, the species most often found in sites other than the skin or gills are monocotylids. Copepods found in sites other than the skin, gills, or olfactory bulbs are generally sphyriids. Digeneans found in unusual sites are usually gorgoderids or syncoeliids. However, in general, elasmobranch digeneans occupy unconventional sites relative to their counterparts in teleosts, the majority of which occur in the digestive system of their hosts. Nematodes, as a phylum, occupy the greatest diversity of sites within elasmobranchs; among nematodes, the Philometridae, Guyanemidae, and Dracunculoidea are the groups most commonly found in sites other than the digestive system.

Studies aimed at conducting complete necropsies of elasmobranchs are rare. It is much more common for investigators to target a particular site of the elasmobranch based on their taxonomic expertise, often to the exclusion of all other regions of the elasmobranch body. For example, researchers interested in copepods generally focus their necropsy efforts on the gills, olfactory bulbs, and outer body surfaces, while researchers interested in cestodes target the spiral intestine. As a consequence, some regions of the body are neglected, and the picture of the total parasite fauna of an elasmobranch species remains incomplete. Although they are unlikely ever to rank with the skin and digestive system in terms of the diversity they host, sites such as the brain, circulatory system, musculature, liver, and gonads are likely to yield a greater diversity of metazoans than is currently recognized, should it become more routine to include these sites in necropsies.

Despite the relatively extensive literature on elasmobranch parasites, a total of only approximately 1500 species of metazoan parasites have been reported from the 900 or so known species of elasmobranchs. More than 50% of these parasite species are cestodes. Given that individual species of elasmobranchs generally host from 3 to 14 unique species of cestodes, the metazoan parasite diversity of elasmobranchs is currently wildly underestimated. Hundreds of species of elasmobranchs have yet to be examined for parasites. The complete parasite faunas of hundreds more await description. As noted by Caira and Jensen (2001), the Scyliorhinidae, Dalatiidae, Urolophidae, Narcinidae, Rhinobatidae, and, in particular, Rajidae are especially poorly sampled for parasites.

Acknowledgments

We are grateful to George Benz and Stephen Bullard for allowing us to examine a version of their manuscript on metazoan parasites and associates of chondrichthyans prior to its publication. Stephen Bullard also provided microbothriid specimens (Figure 18.46). We are very much indebted to Eugene Burreson, who generated an up-to-date list of leeches of elasmobranchs and the sites within elasmobranchs they occupy. Ian Whittington and Leslie Chisholm provided extensive, thoughtful comments on an earlier version of this manuscript. They also supplied the image of the acanthocotylid (Figure 18.42), as well as specimens of amphibdellids and loimoids (Figures 18.43 and 18.45). The figure of the pseudotobothriid (Figure 18.61) was provided by Harry Palm, who also offered helpful advice about some of the more interesting sites occupied by trypanorhynchs. Frantisek Moravec was of enormous assistance with the classification of the nematodes of elasmobranchs; he also assisted with the identification of nematode specimens for illustration of the families and provided copies of much of the nematode literature. Many of the images presented here were made possible through the generosity of Clinton Duffy who provided us with specimens of the piscicoline (Figure 18.5), gnathiid (Figure 18.20), cercropid (Figure 18.24), dichelesthiid (Figure 18.25), kroyerid (Figure 18.29), lernaeopodid (Figure 18.30), and pandarid (Figure 18.31). He also provided the image of the shirleyrhynchid (Figure 18.65). Stephen Curran provided the specimens of the cavisomid (Figure 18.3), aspidogastrid (Figure 18.36), and syncoeliid (Figure 18.40), and also graciously allowed us to examine copies of several manuscripts describing parasites of elasmobranchs of the Gulf of California prior to their publication. Gaines Tyler provided the specimens of the illiosentid (Figure 18.4) and the echinobothriid (Figure 18.51). The corallanid specimen (Figure 18.19) was supplied by Burt Williams who also provided useful information on the isopods of elasmobranchs. Patricia Charvet-Almeida provided the specimen of the ascarid (Figure 18.10). We appreciate the timely assistance and access to material provided by the Invertebrate Zoology Section of the Smithsonian Institution (USNM) and the Harold W. Manter Parasitology Laboratory (HWML), Lincoln, Nebraska. Kirsten Jensen collected the euryphorid (Figure 18.28), monocotylid (Figure 18.47), and all but one of the lecanicephalidean (Figures 18.53 through 18.59) specimens. The azygiid (Figure 18.37) and gilquiniid (Figure 18.64) specimens came from the collection of Nathan Riser; the hexabothriid (Figure 18.49), hepatoxylid (Figure 18.60), and dasyrhynchid (Figure 18.69) specimens came from the collection of Ronald Campbell. The gymnorhynchid (Figure 18.70) specimen came from the collection of Murray Dailey. Kirsten Jensen and Florian Reyda provided very helpful comments on earlier versions of the manuscript. This work was supported in part by NSF Grants 9532943, 0103640, and 0118882.

References

Adams, A. M., E. P. Hoberg, D. F. McAlpine, and S. L. Clayden. 1998. Occurrence and morphological comparisons of *Campula oblonga* (Digenea: Campulidae), including a report from an atypical host, the thresher shark, *Alopias vulpinus*. *J. Parasitol.* 84:435–438.

Adamson, M. L. and J. N. Caira. 1991. *Lockenloia sanguinis* n. gen., n. sp. (Nematoda: Dracunculoidea) from the heart of a nurse shark, *Ginglymostoma cirratum*, in Florida. *J. Parasitol.* 77:663–665.

Adamson, M. L., G. B. Deets, and G. W. Benz. 1987. Description of male and redescription of female *Phlyctainophora squali* Mudry and Dailey, 1969 (Nematoda; Dracunculoidea) from elasmobranchs. *Can. J. Zool.* 65:3006–3010.

Aragort, W. S., F. Alvarez, R. Iglesias, J. Leiro, and M. L. Sanmartin. 2002. *Histodytes microocellatus* gen. et sp. nov. (Dracunculoidea: Guyanemidae), a parasite of *Raja microocellata* on the European Atlantic coast (north-western Spain). *Parasitol. Res.* 88:932–940.

Arai, J. P. 1989. Guide to the parasites of fishes of Canada. Part III. Acanthocephala and Cnidaria. *Can. Spec. Publ. Fish. Aquat. Sci.* 107.

Ball, I. R. and R. A Khan. 1976. On *Micropharynx parasitica* Jägerskiöld, a marine planarian ectoparasitic on the thorny skate, *Raja radiata* Donovan, from the North Atlantic Ocean. *J. Fish Biol.* 8:419–426.

Bates, R. M. 1990. *A Checklist of the Trypanorhyncha (Platyhelminthes: Cestoda) of the World (1935–1985).* National Museum of Wales, Zoological Series 1, Cardiff.

Bennett, M. B., M. R. Heupel, S. M. Bennett, and A. R. Parker. 1997. *Sheina orri* (Myodocopa: Cypridinidae) an ostracod parasitic on the gills of the epaulette shark, *Hemiscyllium ocellatum* (Bonnaterre, 1788) (Elasmobranchii: Hemiscyllidae). *Int. J. Parasitol.* 27:275–281.

Benz, G. W. 1984. Association of the pedunculate barnacle, *Conchoderma virgatum* (Spengler, 1790), with pandarid copepods (Siphonostomatoida: Pandaridae). *Can. J. Zool.* 62:741– 742.

Benz, G. W. 1986. Distributions of siphonostomatoid copepods parasitic upon large pelagic sharks in the western North Atlantic. *Syllogeus* 58:211–219.

Benz, G. W. 1991. Description of some larval stages and augmented description of adult stages of *Albionella etmopteri* (Copepoda: Lernaeopodidae), a parasite of deep-water lanternsharks (*Etmopterus*: Squalidae). *J. Parasitol.* 77:666–674.

Benz, G. W. 1993. Evolutionary Biology of Siphonostomatoida (Copepoda) Parasitic on Vertebrates. Ph.D. dissertation, University of British Columbia, Vancouver, B.C., Canada.

Benz, G. W. and S. A. Bullard. In review. Metazoan parasites and associates of chondricthyans with emphasis on taxa harmful to captive hosts, in *Elasmobranch Husbandry Manual.*

Benz, G. W. and G. B. Deets. 1986. *Kroyeria caseyi* sp. nov. (Kroyeriidae: Siphonostomatoida), a parasitic copepod infesting gills of night sharks (*Carcharhinus signatus* (Poey, 1868)) in the western North Atlantic. *Can. J. Zool.* 64:2492–2498.

Benz, G. W. and G. B. Deets. 1988. Fifty-one years later: an update on *Entepherus*, with a phylogenetic analysis of Cecropidae Dana, 1849 (Copepoda: Siphonostomatoida). *Can. J. Zool.* 66:856–865.

Benz, G. W., H. L. Pratt, and M. L. Adamson. 1987. Larval philometrid nematodes (Philometridae) from the uterus of a sandbar shark, *Carcharhinus plumbeus. Proc. Helm. Soc. Wash.* 54:154–155.

Benz, G. W., Z. Lucas, and L. F. Lowry. 1998. New host and ocean records for the copepod *Ommatokoita elongata* (Siphonostomatoida: Lernaeopodidae), a parasite of the eyes of sleeper sharks. *J. Parasitol.* 84:1271–1274.

Benz, G. W., J. D. Borucinska, L. F. Lowry, and H. E. Whiteley. 2002. Ocular lesions associated with attachment of the copepod *Ommatokoita elongata* (Lernaeopodidae: Siphonostomatoida) to corneas of Pacific sleeper sharks, *Somniosus pacificus*, captured off Alaska in Prince William Sound. *J. Parasitol.* 88:474–481.

Bere, R. 1936. Parasitic copepods from Gulf of Mexico fish. *Am. Midl. Nat.* 17:577–625.

Beverley-Burton, M. 1984. Guide to the parasites of fishes of Canada. Part I. Monogenea and Turbellaria. *Can. Spec. Publ. Fish. Aquat. Sci.* 74.

Bird, P. M. 1981. The occurrence of *Cirolana borealis* (Isopoda) in the hearts of sharks from Atlantic coastal waters of Florida. *Fish. Bull.* 79:376–383.

Bonham, K. and J. E. Guberlet. 1938. Ectoparasitic trematodes of Puget Sound fishes. *Am. Midl. Nat.* 20:590–602.

Borucinska, J. D., G. W. Benz, and H. E. Whiteley. 1998. Ocular lesions associated with attachment of the parasitic copepod *Ommatokoita elongata* (Grant) to corneas of Greenland sharks, *Somniosus microcephalus* (Bloch & Schneider). *J. Fish Dis.* 21:415–422.

Bousfield, E. L. 1987. Amphipod parasites of fishes of Canada. *Can. Bull. Fish. and Aquat. Sci.,* 217.

Braswell, J. S., G. W. Benz, and G. B. Deets. 2002. *Taeniacanthodes dojirii* n. sp. (Copepoda: Poecilostomatoia: Taeniacanthidae) from Cortez electric rays (*Narcine entemedor*: Torpediformes: Narcinidae) captured in the Gulf of California, and a phylogenetic analysis of and key to the species of *Taeniacanthodes. J. Parasitol.* 88:28–35.

Bray, R. A. and T. H. Cribb. 2003. The digeneans of elasmobranchs — distribution and evolutionary significance, in *Taxonomy, Ecology and Evolution of Metazoan Parasites.* (Livre hommage à Louis Euzet). Vol. 1. Combes, C. and Jourdane, J., Eds., PUP, Perpignan, France, 67–96.

Bray, R. A. and D. I. Gibson. 1986. The Zoogonidae (Digenea) of fishes from the north-east Atlantic. *Bull. Br. Mus. Nat. Hist. (Zool.)* 51:127–206.

Bray, R. A., A. Brockerhoff, and T. H. Cribb. 1995. *Melogonimus rhodanometra* n. g. n. sp. (Digenea, Ptychogonimidae) from the elasmobranch *Rhina ancylostoma* Bloch and Schneider (Rhinobatidae) from the Southeastern Coastal waters of Queensland, Australia. *Syst. Parasitol.* 30:11–18.

Brooks, D. R. 1979. *Paravitellotrema overstreeti* sp. n. (Digenea: Hemiuridae) from the Colombian freshwater stingray *Potamotrygon magdalenae* Dumeril. *Proc. Helm. Soc. Wash.* 46:52–54.

Brusca, R. C. 1981. A monograph on the Isopoda Cymothoidae (Crustacea) of the eastern Pacific. *Zool. J. Linn. Soc.* 73:117–199.

Buckner, R. L., R. M. Overstreet, and R. W. Heard. 1978. Intermediate hosts for *Tegorhynchus furcatus* and *Dollfusentis chandleri* (Acanthocephala). *Proc. Helm. Soc. Wash.* 45:195–201.

Bullard, S. A. and R. M. Overstreet. 2000. *Calicotyle californiensis* n. sp. and *Calicotyle urobati* n. sp. (Monogenea: Calicotylinae) from elasmobranchs in the Gulf of California. *J. Parasitol.* 86:939–944.

Bullard, S. A., G. W. Benz, and J. S. Braswell. 2000. *Dioncus postonchomiracidia* (Monogenea: Dioncidae) from the skin of blacktip sharks, *Carcharhinus limbatus* (Carcharhinidae). *J. Parasitol.* 86:245–250.

Bunkley-Williams, L. and E. H. Williams, Jr. 1998. Isopods associated with fishes: a synopsis and corrections. *J. Parasitol.* 84:893–896.

Burreson, E. M. and G. C. Kearn. 2000. *Rhopalobdella japonica* n. gen., n. sp. (Hirudinea, Piscicolidae) from *Dasyatis akajei* (Chondrichthyes: Dasyatididae) in the Northwestern Pacific. *J. Parasitol.* 86:696–699.

Caira, J. N. 1990. Metazoan parasites as indicators of elasmobranch biology, in Elasmobranchs as Living Resources: Advances in the Biology, Ecology, Systematics, and the Status of the Fisheries. H. L. Pratt, Jr., S. H. Gruber, and T. Taniuchi, Eds., NOAA Tech. Rep. 90, 71–96.

Caira, J. N. and M. M. Gavarrino. 1990. *Grillotia similis* (Linton, 1908) comb. n. (Cestoda: Trypanorhyncha) from nurse sharks in the Florida Keys. *J. Helm. Soc. Wash.* 57:15–20.

Caira, J. N. and K. Jensen. 2001. An investigation of the coevolutionary relationships between onchobothriid tapeworms and their elasmobranch hosts. *Int. J. Parasitol.* 31:959–974.

Caira, J. N., G. W. Benz, J. Borucinska, and N. E. Kohler. 1997a. Pugnose eels, *Simenchelys parasiticus* (Synaphobranchidae) from the heart of a shortfin mako, *Isurus oxyrinchus* (Lamnidae). *Environ. Biol. Fish.* 49:139–144.

Caira, J. N., K. Jensen, Y. Yamane, A. Isobe, and K. Nagasawa. 1997b. On the tapeworms of *Megachasma pelagios*: description of a new genus and species of lecanicephalidean and additional information on the trypanorhynch *Mixodigma leptaleum*, in *Biology of the Megamouth Shark*. K. Yano, J. F. Morrissey, Y. Yabumoto, and K. Naayam, Eds., Tokai University Press, Tokyo, 181–191.

Caira, J. N., K. Jensen, and C. J. Healy. 1999. On the phylogenetic relationships among the tetraphyllidean, lecanicephalidean and diphyllidean tapeworm genera. *Syst. Parasitol.* 42:77–151.

Caira, J. N., K. Jensen, and C. J. Healy. 2001. Interrelationships among tetraphyllidean and lecanicephalidean cestodes, in *Interrelationships of the Platyhelminthes*. D. T. J. Littlewood and R. Bray, Eds., Taylor & Francis, London, 135–158.

Campana-Rouget, Y. 1955. Parasites de poissons de mer ouest-africains récoltés per J. Cadenat. IV. Nematodes (1re note). Parasites de sélaciens. *Bull. Inst. Fr. A. Sci. Nat.* 17:818–839.

Campbell, R. A. and I. Beveridge. 1994. Order Trypanorhyncha, in *Keys to Cestode Parasites of Vertebrates*. L. F. Khalil, A. Jones, and R. A. Bray, Eds., CAB Int., Wallingford, U.K.,51–148.

Causey, D. 1961. The site of *Udonella califorum* (Trematoda) upon parasitic copepod hosts. *Am. Midl. Nat.* 66:314–318.

Cheung, P. 1993. Parasitic diseases of elasmobranchs, in *Fish Medicine*. M. K. Stoskopf, Ed., W. B. Saunders, Philadelphia, 782–807.

Chisholm, L. A., T. J. Hansknecht, I. D. Whittington, and R. M. Overstreet. 1997. A revision of the Calicotylinae Monticelli, 1903 (Monogenea: Monocotylidae). *Syst. Parasitol.* 38:159–183.

Cislo, P. R. and J. N. Caira. 1993. The parasite assemblage in the spiral intestine of the shark *Mustelus canis*. *J. Parasitol.* 79:886–899.

Cressey, R. F. 1976. The genus *Argulus* (Crustacea: Branchiura) of the United States. Water Pollution Control Research Series 18050 ELDO2/72, U.S. Environmental Protection Agency, Washington, D.C.

Cribb, T. H., R. A. Bray, D. T. J. Littlewood, S. P. Pichelin, and E. A. Herniou. 2001. The Digenea, in *Interrelationships of the Platyhelminthes*. D. T. J. Littlewood and R. A. Bray, Eds., Taylor & Francis, London, 168–185.

Crompton, D. W. T. and B. B. Nickol. 1985. *Biology of the Acanthocephala*. Cambridge University Press, Cambridge, U.K.

Curran, S. S. and J. N. Caira. 1995. Attachment site specificity and the tapeworm assemblage in the spiral intestine of the blue shark (*Prionace glauca*). *J. Parasitol.* 81:149–157.

Curran, S. S. and R. M. Overstreet. 2000. *Syncoelium vermilionensis* sp. n. (Hemiuroidea: Syncoeliidae) and new records for members of Azygiidae, Ptychogonimidae, and Syncoeliidae parasitizing elasmobranchs in the Gulf of California, in *Metazoan Parasites in the Neotropics: A Systematic and Ecological Perspective.* G. Salgado-Maldonado, A. N. Garcia Aldrete, and V. M. Vidal-Martinez, Eds., Instituto de Biologia, U.N.A.M., Mexico, 117–133.

Curran, S. S., C. K. Blend, and R. M. Overstreet. 2003. *Anaporrhutum euzeti* sp. n. (Gorgoderidae: Anaporrhutinae) from rays in the Gulf of California, Mexico, in *Taxonomy, Ecology and Evolution of Metazoan Parasites.* (Livre hommage à Louis Euzet). Vol. 1. C. Combes and J. Jourdane, Eds., PUP, Perpignan, France, 225–234.

Deardorff, T. L. and R. C. Ko. 1983. *Echinocephalus overstreeti* sp. n. (Nematoda: Gnathostomatidae) in the stingray, *Taeniura melanopilos* Bleeker, from the Marquesas Islands, with comments on *E. sinensis* Ko, 1975. *Proc. Helm. Soc. Wash.* 50:285–293.

Deboutteville, D. C. and L. P. Nunes. 1952. Copépodes Philichthyidae nouveaux, parasites de poissons Européens. *Ann. Parasitol.* 27:598–609.

Deets, G. B. and M. Dojiri. 1990. *Dissonus pastinum* n. sp. (Siphonostomatoida: Dissonidae) a copepod parasitic on a horn shark from Japan. *Beaufortia* 41:49–54.

Delaney, P. M. 1984. Isopods of the genus *Excorallana* Stebbing, 1904, in the Gulf of California, Mexico, with descriptions of two new species and a key to the known species (Crustacea, Isopoda, Corallanidae). *Bull. Mar. Sci.* 34:1–20.

Diaz, J. P. 1972. Cycle évolutif d'*Acanthoceilus quadridentatus* Molin, 1988 (Nematoda). *Vie Milieu* 22:289–304.

Diebakate, C., A. Raibaut, and Z. Kabata. 1997. *Thamnocephalus cerebrinoxius* n. g., n. sp. (Copepoda: Sphyriidae), a parasite in the nasal capsules of *Leptocharias smithii* (Müller and Henle, 1839) (Pisces: Leptochariidae) off the coast of Senegal. *Syst. Parasitol.* 38:231–235.

Dollfus, R. P. 1937. Les trématodes digénea des selaciens (Plagiostomes). Catalogue par hotes. Distribution géographique. *Ann. Parasitol. Hum. Comp.* 15:259–281.

Dollfus, R. P. 1942. Etudes critiques sur les tetrarhynques du Museum de Paris. *Arch. Mus. Nat. Hist. Nat.* 19:7–466.

Dollfus, R. P. 1960. Sur une collection de tetrarhynques homeacanthes de la famille des Tentaculariidae, recoltes principalement dans las region de Dakar. *Bull. Inst. Fr. Afr. Noire* 22:788–852.

Euzet, L. 1994a. Order Lecanicephalidea, in *Keys to Cestode Parasites of Vertebrates.* L. F. Khalil, A. Jones, and R. A. Bray, Eds., CAB Int., Wallingford, U.K., 195–204.

Euzet, L. 1994b. Order Tetraphyllidea, in *Keys to Cestode Parasites of Vertebrates.* L. F. Khalil, A. Jones, and R. A. Bray, Eds., CAB Int., Wallingford, U.K., 149–194.

Euzet, L. and C. Combes. 1998. The selection of habitatas among the monogenea. *Int. J. Parasitol.* 28:1645–1652.

Euzet, L. and H. H. Williams. 1960. A re-description of the trematode *Calicotyle stossichi* Braun, 1899, with an account of *Calicotyle palombi* sp. nov. *Parasitology* 50:21–30.

Gibson, D.I. and R. A. Bray. 1977. The Azygiidae, Hirudinellidae, Ptychogonimidae, Sclerodistomidae and Syncoeliidae of fishes form the north-east Atlantic. *Bull. Br. Mus. Nat. Hist. (Zool.)* 32:167–245.

Golvan, Y. I., A. Garcia-Rodrigo, and C. Díaz-Ungría. 1964. *Megapriapus ungriai* (Garcia- Rodrigo, 1960) n. gen. (Palaeacanthocephala) parasite d'une Pastenague d'eau douce du Vénézuéla (*Pomatotrygon hystrix*). *Ann. Parasitol. Hum. Comp.* 39:53–59.

Guiart, J. 1935. Cestodes parasites provenant des campagnes scientifiques du Prince Albert ler de Monaco. Resultats des Campagnes scientifiques accomplies sur son yacht par Albert Ier. Imprimerie de Monaco.

Hale, H. M. 1940. Report on the Cymothoid Isopoda obtained by the F.I.S. "Endeavour" on the coasts of Queensland, New South Wales, Victoria, Tasmania, and South Australia. *Trans. R. Soc. S. Aust.* 64:288–304.

Hesse, M. 1873. Mémoire sur des Crustacés rares ou nouveaux des côtes de France. *Ann. Sci. Nat.* 17–18.

Heupel, M. R. and M. B. Bennett. 1999. The occurrence, distribution and pathology associated with gnathiid isopod larvae infecting the epaulette shark, *Hemiscyllium ocellatum. Int. J. Parasitol.* 29:321–330.

Honma, Y., S. Tsunaki, and A. Chiba. 1991. Histological studies on the juvenile gnathiid (Isopoda, Crustacea) parasitic on the branchial chamber wall of the stingray, *Dasyatis akajei*, in the Sea of Japan. *Rep. Sado Mar. Biol. Stat. Niigata Univ.* 21:37–47.

Jensen, K. 2001. A Monograph of the Order Lecanicephalidea (Platyhelminthes: Cestoda). Ph.D. dissertation, University of Connecticut, Storrs.

Johnston, T. H. and P. M. Mawson. 1943. Some nematodes from Australian elasmobranchs. *Trans. R. Soc. S. Aust.* 67:187–190.

Kabata, Z. 1970. Crustacea as enemies of fishes, in *Diseases of Fishes*, Book 1. S. F. Sniexzko and H. R. Axelrod, Eds., Tropical Fish Hobbiest Publications, Neptune City, NJ.

Kabata, Z. 1979. *Parasitic Copepoda of British Fishes*. Ray Society, London.

Kabata, Z. 1988. Copepoda and Branchiura, in *Guide to the Parasites of Fishes of Canada*, Part II. *Crustacea*. L. Margolis and Z. Kabata, Eds., Department of Fisheries and Oceans, Ottawa, Canada, 3–127.

Kearn, G. C. 1963. Feeding in some monogenean skin parasites: *Entobdella soleae* on *Solea solea* and *Acanthocotyle* sp. on *Raja clavata. J. Mar. Biol. Assoc. U.K.* 42:93–104.

Kearn, G. C. 1970. The oncomiracidia of the monocotylid monogenans *Dictocotyle coeliaca* and *Calicotyle kroyeri. Parasitology* 61:153–160.

Kearn, G. C. 1979. Studies on gut pigment in skin-parasitic monogeneans, with special reference to the monocotylid *Dendromonocotyle kuhlii. Int. J. Parasitol.* 9:545–552.

Kearn, G. C. 1987. The site of development of the monogenan *Calicotyle kroyeri*, a parasite of rays. *J. Mar. Biol. Assoc. U.K.* 67:77–87.

Kearn, G. C. 1998. *Parasitism and the Platyhelminthes*. Chapman & Hall, London.

Kearn, G. C. and M. Beverley-Burton. 1990. *Mycteronastes undulatae* gen. nov., sp. nov. (Monongenea: Monotcotylidae) from the nasal cavities of *Raja undulata* in the Eastern Atlantic. *J. Mar. Biol. Assoc. U.K.* 70:747–753.

Kearn, G. C. and J. E. Green. 1983. *Squalotrema llewellyni* gen. nov., sp. nov. a monocotylid monogenean from the nasal fossae of the spur-dog, *Squalus acanthias*, at Plymouth. *J. Mar. Biol. Assoc. U.K.* 63:17–25.

Khalil, L. F., A. Jones, and R. A. Bray, Eds., 1994. *Keys to Cestode Parasites of Vertebrates*. CAB Int., Wallingford, U.K.

Klimpel, A., A. Seehagen, H. W. Palm, and H. Rosenthal. 2001. *Deep-Water Metazoan Fish Parasites of the World*. Lagos-Verlag, Berlin.

Knoff, M., S. C. Clemente, R. M. Pinto, and D. C. Gomes. 2001. Digenea and Acanthocephala of elasmobranch fishes from the southern coast of Brazil. *Mem. Inst. Oswaldo Cruz* 96:1095–1101.

Lewis, A. G. 1966. Copepod crustaceans parasitic on elasmobranch fishes of the Hawaiian islands. *Proc. U.S. Natl. Mus.* 118(No. 3524):57–154.

Llewellyn, J. 1960. Amphibdellid (Monogenean) parasites of electric rays (Torpedinidae). *J. Mar. Biol. Assoc. U.K.* 39:561–589.

Llewellyn, L. C. and W. Knight-Jones. 1984. A new genus and species of marine leech from British coastal waters. *J. Mar. Biol. Assoc. U.K.* 64:919–934.

Long, D. J. and B. M. Waggoner. 1993. The ectoparasitic barnacle *Anelasma* (Cirripedia, Thoracica, Lepado-morpha) on the shark *Centroscyllium nigrum* (Chondrichthyes, Squalidae) from the Pacific sub-Antarctic. *Syst. Parasitol.* 26:133–136.

MacCallum, G. A. 1925. Eggs of a new species of nematoid worm from a shark. *Proc. U.S. Natl. Mus.* 67(No. 2588):1–2.

MacDonald, S. and J. Llewellyn. 1980. Reproduction in *Acanthocotyle greeni* n. sp. (Monogenea) from the skin of *Raia* spp. at Plymouth. *J. Mar. Biol. Assoc. U.K.* 60:81–88.

MacKenzie, K. 1964. *Stichocotyle nephropis* Cunningham, 1887 (Trematoda) in Scottish waters. *Ann. Mag. Nat. Hist.* 6:505–506.

Madhavi, R. and K. Hanumantha Rao. 1970. *Orchispirium heterovitellatum* gen. et sp. n. (Trematoda: Sanguinicolidae) from the ray fish, *Dasyatis imbricatus* Day, from Bay of Bengal. *J. Parasitol.* 56:41–43.

Maillard, C. and M.-H. Ktari. 1978. *Hyperandrotrema cetorhini* n.g., n.sp. (Trematoda: Sanguinicolidae) parasite du systeme circulatoire de *C. maximus* (Sel.). *Ann. Parasitol. Hum. Comp.* 53:359–365.

Manter, H. W. 1940. Digenetic trematodes of fishes from the Galapagos Islands and the neighboring Pacific. *Rep. Allan Hanc. Pac. Exped.* 2:325–497.

Markell, E. K. 1953. *Nagmia floridensis* n. sp., an anaporrhutine trematode form the coelom of the sting ray *Amphotistius sabinus. J. Parasitol.* 39:45–51.

Marques, F. P. L. 2000. Evolution of Neotropical Freshwater Stingrays and Their Parasites: Taking into Account Space and Time. Ph.D. dissertation, University of Toronto, Toronto.

McVicar, A. H. 1977. Intestinal helminth parasites of the ray *Raja naevus* in British waters. *J. Helm.* 51:11–21.

McVicar, A. H. and D. I. Gibson. 1975. *Pancreatonema torriensis* gen. nov., sp. nov. (Nematoda: Rhabdochonidae) from the pancreatic duct of *Raja naevus*. *Int. J. Parasitol.* 5:529–535.

Moravec, F. 2001. *Trichinelloid Nematodes Parasitic in Cold-Blooded Vertebrates*. Academy of Sciences of the Czech Republic, Ceské Budejovice.

Moravec, F. and M. D. Little. 1988. *Granulinema* gen. n. a new dracunculoid genus with two new species (*G. carcharhini* sp. n. and *G. simile* sp. n.) from the bull shark, *Carcharhinus leucas* (Valenciennes), from Louisiana, USA. *Fol. Parasitol.* 35:113–120.

Moravec, F. and K. Nagasawa. 2000. Two remarkable nematodes from sharks in Japan. *J. Nat. Hist.* 34:1–13.

Moravec, F., M. I. Jimenez-Garcia, and G. Salgado Maldonado. 1998. New observations on *Mexiconema cichlasomae* (Nematoda: Dracunculoidea) from fishes in Mexico. *Parasite* 5:289–293.

Moravec, F., J. D. Borucinska, and S. Frasca. 2001. *Pancreatonema americanum* sp. nov. (Nematoda, Rhabdochonidae) from the pancreatic duct of the dogfish shark, *Squalus acanthias*, from the coast of Massachusetts, USA. *Acta Parasitol.* 46:293–298.

Moravec, F., J. G. Van As, and I. Dykova. 2002. *Proleptus obtusus* Dujardin, 1845 (Nematoda: Physalopteridae) from the puffadder shyshark *Haploblepharus edwardsii* (Scyliiorhinidae) from off South Africa. *Syst. Parasitol.* 53:169–173.

Moreira, P. S. and V. Sadowsky. 1978. An annotated bibliography of parasitic Isopoda (Crustacea) of Chondrichthyes. *Bolm. Inst. Oceanogr.* (São Paulo) 27:95–152.

Moser, M. and S. Anderson. 1977. An intrauterine leech infection: *Branchellion lobata* Moore, 1952 (Piscicolidae) in the Pacific angel shark (*Squatina californica*) from California. *Can. J. Zool.* 55:759–760.

Nagasawa, K., S. Tanaka, and G. W. Benz. 1998. *Trebius shiinoi* n. sp. (Trebiidae: Siphonostomatoida: Copepoda) from uteri and embryos of the Japanese angelshark (*Squatina japonica*) and the clouded angelshark (*Squatina nebulosa*), and redescription of *T. longicaudatus*. *J. Parasitol.* 84:1218–1230.

Newbound, D. R. and B. Knott. 1999. Parasitic copepods from pelagic sharks in western Australia. *Bull. Mar. Sci.* 65:715–724.

Ofenheim, E. von. 1900. Über eine neue Distomidengattung. *Z. Naturwiss.* 73:145–186, pl. III.

Olsen, L. W. 1952. Some nematodes parasitic in marine fishes. *Pub. Inst. Mar. Sci. Univ. Tex.* 11:173–215.

O'Sullivan, J. B., R. R. McConnaughey, and M. E. Huber. 1987. A blood-sucking snail: the Cooper's nutmeg, *Cancellaria cooperi* Gabb, parasitizes the California electric ray, *Torpedo californica* Ayres. *Biol. Bull.* 172:362–366.

Pappas, P. W. 1970. The trypanorhynchid cestods from Humboldt Bay and Pacific Ocean sharks. *J. Parasitol.* 56:1034.

Pearse, A. S. 1953. Parasitic crustaceans from Alligator Harbor Florida. *Q. J. Fla. Acad. Sci.* 15:187–243.

Pozdnyakov, S. E. 1989. On the finding of didymozoids in pelagic sharks of the Pacific. *Parazitologiya* 23:529–532 [in Russian].

Price, E. W. 1938a. North American monogenetic trematodes. II. The families Monocotylidae, Microbothriidae, Acanthocotylidae, and Udonellidae (Capsaloidea). *J. Wash. Acad. Sci.* 28:183–198.

Price, E. W. 1938b. North American monogenetic trematodes. II. The families Monocotylidae, Microbothriidae, Acanthocotylidae, and Udonellidae (Capsaloidea). *J. Wash. Acad. Sci.* 28:109–126.

Price, E. W. 1963. A new genus and species of monogenetic trematode from a shark, with a review of the family Microbothriidae Price, 1936. *Proc. Helm. Soc. Wash.* 30:213–218.

Rohde, K. 2002. Subclass Aspidogastrea Faust & Tang, 1936, in *Keys to the Trematoda*, Vol. 1. D. I. Gibson, A. Jones, and R. A. Bray, Eds., CAB Int./Natural History Museum, London, 5–14.

Rokicki, J. and D. Bychawska. 1991. Parasitic copepods of Carcharhinidae and Sphyridae [sic] (Elasmobranchia) from the Atlantic Ocean. *J. Nat. Hist.* 25:1439–1448.

Rosa-Molinar, E., C. S. Williams, and J. R. Lichtenfelds. 1983. Larval nematodes (Philometridae) in granulomas in ovaries of black-tip sharks, *Carcharhinus limbatus* (Valenciennes). *J. Wildlf. Dis.* 19:275–277.

Ross, R. 1999. *Freshwater Stingrays from South America (Aqualog Special)*. Aqualog Verlag, Mörfelden-Walldorf, Germany.

Russo, R. A. 1975. Notes on the external parasites of California inshore sharks. *Calif. Fish Game* 61:228–232.

Ruyck, R. and A. G. Chabaud. 1960. Un cas de parasitisme attribuable a des larves de *Phlyctainophora lamnae* Steiner chez un sélacien, et cycle évolutif probable de ce nematode. *Vie Milieu* 11:386–389.

Sawyer, R. T., A. R. Lawler, and R. M. Overstreet. 1975. Marine leeches of the eastern United States and the Gulf of Mexico with a key to the species. *J. Nat. Hist.* 9:633–667.

Short, R. B. 1954. A new blood fluke, *Selachohemecus olsoni*, n. g., n. sp. (Aporocotylidae) from the sharp-nosed shark, *Scoliodon terra-novae*. *Proc. Helm. Soc. Wash.* 21:78–82.

Smit, N. J. and L. Basson. 2002. *Gnathia pantherina* sp. n. (Crustacea: Isopoda: Gnathiidae), a temporary ectoparasite of some elasmobranch species from southern Africa. *Fol. Parasitol.* 49:137–151.

Steiner, G. 1921. *Phlyctainophora lamnae* n. g., n. sp., eine neue parasitische Nematodenform aus *Lamna conrubica* (Heringshai). *Centralbl. Bakt. Jena 1 Abth.* 86:91–595.

Tandon, R. S. 1972. Some observations on the plerocercus larva of a trypanorhynchid cestode obtained from the ovary of the ray *Dasyatis uarnak*. *Proc. Natl. Acad. Sci. India* 42:431–435.

Thoney, D. A. and E. M. Burreson. 1986. Ecological aspects of *Multicalyx cristata* (Aspidocotylea) infections in northwest Atlantic elasmobranchs. *Proc. Helm. Soc. Wash.* 53:162–165.

Tyler, G. A. 2001. A Monograph on the Diphyllidea (Platyhelminthes: Cestoda). Ph.D. dissertation, University of Connecticut, Storrs.

Waterman, P. B. and F. Y. T. Sin. 1991. Occurrence of the marine tapeworms, *Hepatoxylon trichirui* and *Hepatoxylon megacephalum*, in fishes from Kaikoura, New Zealand. *N. Z. Nat. Sci.* 18:71–73.

Whittington, I. D. 1990. *Empruthotrema kearni* n. sp. and observations on *Thaumatocotyle pseudodasybatis* Hargis, 1955 (Monogenea: Monocotylidae) from the nasal fossae of *Aetobatus narinari* (Batiformes: Myliobtatidae) from Moreton Bay, Queensland. *Syst. Parasitol.* 15:23–31.

Whittington, I. D. and L. Chisholm. 2003. Diversity of Monogenea from Chondrichthyes: do monogeneans fear sharks? in *Taxonomy, Ecology and Evolution of Metazoan Parasites*. (Livre hommage à Louis Euzet). Vol. 2. C. Combes and J. Jourdane, Eds., PUP, Perpignan, France, 339–364.

Williams, E. H., Jr. 1978. *Conchoderma virgatum* (Spengler) (Cirripedia, Thoracica) in association with *Dimemoura latifolia* (Steenstrup & Lutken) (Copepoda, Caligidae), a parasite of the shortfin mako, *Isurus oxyrhynchus* Rafinesque (Pisces, Chondrichthyes). *Crustaceana* 34:109–110.

Williams, E. H., Jr. and L. Bunkley-Williams. 1996. *Parasites of Offshore Big Game Fishes of Puerto Rico and the Western Atlantic*. Puerto Rico Dept. Nat. Environ. Res., San Juan, Puerto Rico and the University Puerto Rico, Mayaguez, Puerto Rico.

Williams, H. H. 1960. A list of parasitic worms, including 22 new records from marine fishes caught off the British Isles. *Ann. Mag. Nat. Hist.* 2:705–715.

Williams, H. H. and R. A. Bray. 1984. *Chimaerocestos prudhoei* gen. et sp. nov. representing a new family of tetraphyllideans and the first record of strobilate tapeworms from a holocephalan. *Parasitology* 88:105–116.

Williams, H. H., A. V. McVicar, and R. Ralph. 1970. The alimentary canal of fishes as an environment for helminth parasites. *Symp. Br. Soc. Parasitol.* 8:43–77.

Wilson, C. B. 1913. Crustacean parasites of West Indian fishes and land crabs, with descriptions of new genera and species. *Proc. U.S. Natl. Mus.* 44:189–277.

Woolcock, B. 1936. Monogenetic trematodes from some Australian fishes. *Parasitology* 28:79–91.

Yamaguti, S. 1971. *Synopsis of the Digenetic Trematodes of Vertebrates*, Vol. I. Keigaku, Tokyo.

Yano, K. and J. A. Musick. 2000. The effect of the mesoparasitic barnacle *Anelasma* on the development of reproductive organs of deep-sea squaloid sharks, *Centroscyllium* and *Etmopterus*. *Environ. Biol. Fish.* 59:329–339.

19

Assessing Habitat Use and Movement

Colin A. Simpfendorfer and Michelle R. Heupel

CONTENTS

19.1 Introduction

Sharks occur in all of the world's oceans and in waters that include the deep-sea, oceanic, neritic, and estuarine habitats. In addition, a few specialized species also occur in rivers and lakes connected to the ocean. The occurrence of sharks within these broad regions is well understood for most species. For example, the gummy shark, *Mustelus antarcticus*, is known to occur in the neritic waters of southern Australia, or the salmon shark, *Lamna ditropis*, is known to inhabit the boreal waters of the north Pacific. (Chapter 2 provides a detailed consideration of the zoogeography of the sharks, skates, and rays.) However, a shark will not occur in all of the habitats within its range; instead, it is more likely to have specific habitats in which it spends most of its time. It is this detailed analysis of the habitats that a species uses that is discussed here.

It is intuitive that information on the habitat use of sharks would be important for management and conservation. For example, Olsen (1954) identified that newborn school sharks, *Galeorhinus galeus*, occur in protected coastal bays around Tasmania and proposed that these areas be protected (Figure

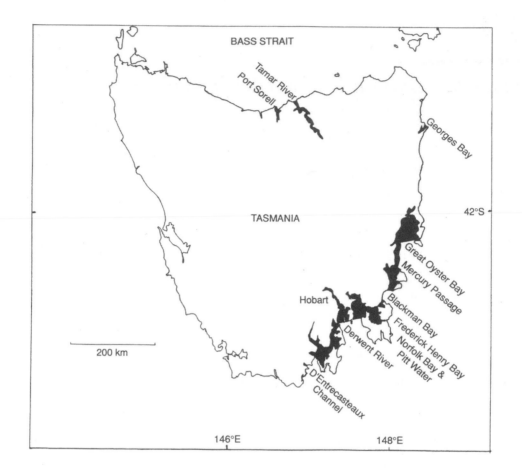

FIGURE 19.1 Nursery areas (black areas) of the school shark, *Galeorhinus galeus*, in Tasmania that are protected by state law. (From Williams, H. and Schaap, A.H. 1992. *Aust. J. Mar. Freshwater Res.* 43:237–250. With permission.)

19.1). Despite some early recognition of the importance of habitat use information it was only in the 1990s that resource managers and researchers began focusing research on *essential fish habitat, critical habitat,* and *marine protected areas.*

There have been relatively few studies of habitat use in the sharks, skates, and rays. By far the most widely investigated topic is that of nursery areas (i.e., habitat use patterns of juvenile sharks). There have been a variety of studies that have defined nursery areas and identified their importance (e.g., Springer, 1967; Clarke, 1971; Bass, 1978; Branstetter, 1990; Williams and Schaap, 1992; Simpfendorfer and Milward, 1993; Castro, 1993; Holland et al., 1993; Morrissey and Gruber, 1993a), and there is much ongoing research to better understand nursery area use and importance (e.g., Heupel and Hueter, 2001, 2002). Only a handful of studies have directly addressed questions of habitat preference. For example, Morrissey and Gruber (1993b) describe the habitat selection of juvenile lemon sharks, *Negaprion brevirostris,* in the Bahamas, and Heithaus et al. (2002) described the habitat selection of tiger sharks, *Galeocerdo cuvier,* in Shark Bay, Australia. We return to these examples later in the chapter.

In this chapter we consider *habitat use* to be the observed pattern of the habitats in which an individual or species occurs. This term has been used synonymously with *habitat selection* and *habitat utilization* in the ecological literature. When studying habitat use of a species a researcher aims to identify the species' *habitat preferences* and the factors underlying these preferences.

In this chapter we first briefly consider information requirements for measuring and testing habitat use; we then describe the different methods by which habitat use in the sharks skates and rays has been investigated and give examples of the results from some of these studies. The final section of the chapter discusses the importance of scale in habitat use studies and examines some of the mechanisms that drive habitat preferences.

19.2 Measuring Habitat Use and Habitat Preference

There are many approaches to describing habitat use and quantifying habitat preferences. Habitat use is most commonly determined by overlaying movement or location information on habitat data. Modern geographic information systems (GIS) have made this process relatively straightforward. Habitat preference, however, is a matter of determining if one habitat type is used more frequently than another, relative to the abundance of each habitat.

19.2.1 Habitat Use

One of the important concepts in describing habitat use is an individual's home range. The definition of home range has been refined over time. Burt (1943) originally defined home range as the area around the established home, which is traversed by an animal in its normal activities of food gathering, mating, and caring for its young. Many authors have felt that Burt's original definition was too general and did not apply to animals that do not care for their young or maintain specific home or nest sites. Cooper (1978) points out that the home range of an animal should not be treated as an inclusive area because an animal may use a small portion of the area intensively, other areas moderately, and may not use some areas at all. This type of observation led several authors to define home range as the smallest subregion of an area that accounts for a specific portion (often 95%) of the space an animal utilizes (e.g., Jennrich and Turner, 1969; Anderson, 1982; Worton, 1987). This type of mathematical approach to defining the home range is often referred to as a utilization distribution. The most widely used form of the utilization distribution is the kernel distribution (Worton, 1987). Several studies of shark movements have defined home range patterns and described habitat utilization patterns (e.g., McKibben and Nelson, 1986; Holland et al., 1993; Morrissey and Gruber, 1993a). These shark studies have typically included all areas and individual uses, and do not provide detailed information on whether specific habitats are selected preferentially.

19.2.2 Habitat Preference

The problem of determining if an individual or species shows a habitat preference can be broken down into two parts. The first is a test to determine if habitats are used in proportion to their availability (i.e., are there habitat preferences?), which is most often achieved using a chi-squared goodness-of-fit test. If there are differences, then the second part of the problem is to identify which habitats are preferred and which are avoided. There is a range of indices available. Krebs (1999) described a number of these, including the simple forage ratio, the rank preference, and more complex indices such as Manly's α. In one of the few studies of habitat selection in sharks Morrissey and Gruber (1993b) used a simple chi-squared goodness-of-fit test to compare habitat use to habitat availability and then Strauss's index of selection (Strauss, 1979) to investigate which habitats were preferred.

The simple comparison of use to availability can often lead to difficulties, especially with wide-ranging mobile species such as sharks. The question is often how much habitat should be considered available. In enclosed systems, such as the lagoon studied by Morrissey and Gruber (1993b), the area is well defined. However, in more open areas the limits are less clear, and depend much more on the temporal and spatial scales that are being considered. To account for this it is better to assume that not all habitats are equally available, and instead to generate randomized tracks of animals and measure the expected proportions of habitats used. These randomized habitat use patterns can then be compared to the observed pattern of habitat use using chi-squared tests. Heithaus et al. (2002) used two methods of generating randomized habitat use patterns for tiger sharks: correlated random walk and track randomization. These methods produced expected habitat use patterns that differed from those based simply on habitat availability.

19.3 Approaches to Assessing Habitat Use and Habitat Preferences

To study the habitat use and selection of a shark species requires that the movements of individuals, and the habitats that they are occurring in, be determined over sufficiently long time periods to obtain meaningful data. Over time a wide variety of approaches to this problem have been taken in elasmobranch species. Below we examine these approaches, provide a brief example from the literature, and discuss the advantages and disadvantages of these approaches. Finally, we summarize the constraints of each approach in a table to allow for easy comparison (Table 19.1).

19.3.1 Direct Observation

The simplest method of examining habitat use by sharks is to directly observe individuals and record the habitats that they use over time. This technique is effective only in areas where the water clarity is sufficient to enable direct observation. This approach was used by Economakis and Lobel (1998) who studied the daily aggregations of gray reef sharks (*Carcharhinus amblyrhynchos*) at Johnson Atoll. In this study the numbers of sharks present at a small island were counted to provide information on how often aggregations occurred and how many animals were present. The authors combined these sightings data with water temperature, tidal cycle, and habitat descriptions to determine why this habitat use pattern occurs.

There are a number of disadvantages to direct observation in shark studies. First, observations can normally only be made during the day, leaving any nocturnal changes in habitat use undetected. Second, the act of observing may change the behavior of individuals depending on the method (a disadvantage that must be addressed in any study of habitat use). It has been observed that the presence of divers can cause a response in some shark species (Johnson and Nelson, 1973; Nelson et al., 1986). Finally, it is not usually possible to identify individual animals, and so information about individual habitat use is not available.

Medved and Marshall (1983) took the direct observation method one step farther, and overcame the problem of water clarity. Working on small sandbar sharks, *C. plumbeus*, in Chincoteague Bay — an area with poor water clarity — they attached small Styrofoam floats to the dorsal fin and followed their movements. Using this approach they were able to describe sandbar shark movements and habitat use, identifying tidal flow as an important controlling factor.

19.3.2 Relative Catch Rates

Rather than directly observe sharks to determine their habitat use, it is also possible to use relative catch rates in different habitats to draw conclusions about habitat use. With this approach a sampling gear is set in all available habitats and the catch rates between them compared using selectivity or preference index values. Michel (2002) used this approach to examine the habitat use of four species of sharks in the Ten Thousand Islands region of Florida. Gillnets were set in three habitats (gulf edge, transition, and backwater) and a preference index was used to show that bonnetheads, *Sphyrna tiburo*, had no habitat preference, blacktip sharks, *C. limbatus*, preferred gulf edge habitats, bull sharks, *C. leucas*, preferred backwater habitats, and lemon sharks, *N. brevirostris*, avoided gulf edge habitats (Figure 19.2).

Catch rate comparisons are good for investigating population-level habitat use patterns. However, they have several major drawbacks. First, they are unable to resolve detailed individual movements that can help understand why the habitat use patterns occur. Second, if there are habitat-specific movement rates this can affect catch rates, which can be misidentified as habitat preference. Finally, sampling gears may be more effective in specific habitats and so also bias catch rates.

19.3.3 Acoustic Tracking

The most widely used approach in studies of shark habitat use is acoustic telemetry. With this technique an acoustic tag that generates a series of "pings" is attached to an individual. The acoustic signal is then

TABLE 19.1

Constraints of Different Approaches to Investigating Habitat Use in Sharks, Skates, and Rays

| Approach | Size of Animals | Accuracy of Positions | Constraints | | | | Best Use |
			Temporal Coverage	Geographic Coverage	Equipment Costs	Other	
Direct observation	Any	±10 m	Short, day-time	Limited	Low	Requires good water clarity Observer effects	Coral reef species
Relative catch rates	Any	n/a	Any	Any	Low	Biased by habitat-specific catchability or movement rates	Commercially fished species
Acoustic tracking	Any	±50 m	Short (days)	Any	Moderate	Only one animal tracked at a time Chasing effects	Detailed short-term studies of habitat use
Acoustic monitoring	Any	±225 m (omnidirectional) ±1 m (triangulating)	Any	Moderate Small	High High	Only good if animals stay in range of receivers	Long-term studies in confined environments
Satellite tracking	>1.5 m	±250 m to 10 km	Any	Global	High	Animal must surface to give location	Large species that surface regularly
Archival tags	Any	±0.5° (best)	Any	Global	High	Must recover animal or use pop-up satellite tags	Wide ranging species
Animal-borne video systems	>1.5 m	±10 m (if acoustically tracked)	Short (hours)	Limited	High	Size of equipment	Large species in clear water

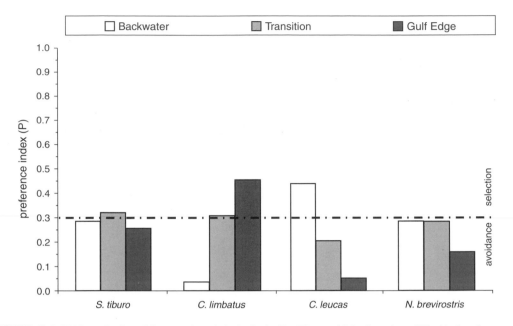

FIGURE 19.2 Habitat selection of four species of sharks in the Ten Thousand Islands region of Florida based on catch rates in gillnet and longlines. Bars above the dashed line indicate habitat preference, bars below the line indicate habitat avoidance. (From Michel, M. 2002. Ph.D. thesis, University of Basel, Switzerland.)

located using a receiver and hydrophone, and the movement of the shark followed and its position regularly recorded. The movement data can then be overlaid on habitat information to determine habitat use. We make a distinction between acoustic tracking (where an individual is followed and locations determined) and acoustic monitoring (where data-logging acoustic receivers are used to gather data remotely). Acoustic monitoring is covered in the next section.

There are many good examples of using acoustic tracking studies in sharks (e.g., Sciarotta and Nelson, 1977; Holland et al., 1993; Morrissey and Gruber, 1993a); however, few have directly addressed the issue of habitat use. Probably the best example is a series of studies conducted on lemon sharks, *N. brevirostris*, at Bimini, Bahamas, by Gruber and his students. Gruber et al. (1988) provided an initial glimpse into the behavior and habitat use of this species, and provided the groundwork for more detailed study. Sharks were captured in the lagoon at Bimini, fitted with acoustic tags, and tracked for up to 101 h at a time, with some individuals being reacquired and retracked for as long as 8 days. The results showed evidence of site attachment to various regions within the lagoon, and showed differences in activity space between night and day. Following this study, Morrissey and Gruber (1993a,b) used acoustic tracking over extended periods with 38 individuals, often reacquiring individuals many times, to generate a long time series of locations. The authors used these location data, and data on the habitat within the study site, to demonstrate that juvenile lemon sharks preferred shallower, warmer waters with rocky or sandy substrates (Figure 19.3) and that sharks did not show any preference based on salinity (Morrissey and Gruber, 1993b).

Acoustic tracking is a method that can provide detailed spatial data over a relatively large area, depending on the range of the tracking vessel. As such, it has been the most widely used method for investigating habitat use patterns in sharks, skates, and rays to date. However, it does have several disadvantages in habitat use studies. First, individual sharks can only be tracked for short periods (usually less than 48 h) because of the human resources required. This limits the technique to investigations of short-term habitat use and temporal shifts in habitat use (e.g., diurnal changes). Second, only one individual can normally be tracked at a time. Thus, population-level changes in habitat use are difficult to identify. Third, the need to follow a shark, normally in a boat, can possibly result in changes in behavior. This leads to the concern that the researcher in some way is chasing the shark, and so not observing natural behavior. The need to capture and handle the shark adds to concerns that following release normal behavior is not observed.

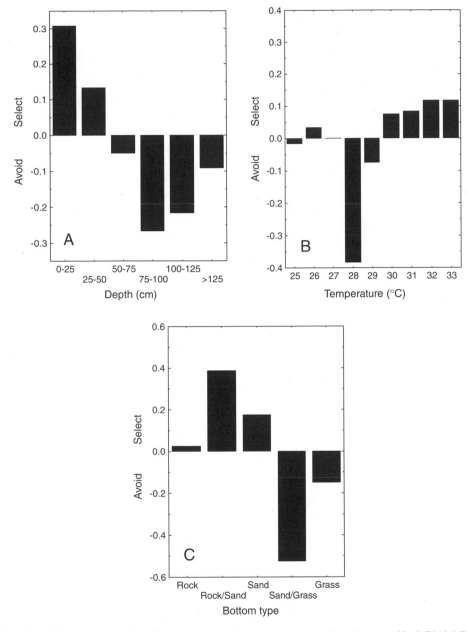

FIGURE 19.3 Habitat selection of juvenile lemon sharks, *Negaprion brevirostris*, in the lagoon at North Bimini, Bahamas, based on (A) water depth, (B) temperature, and (C) bottom type. Bars above the line indicate selection, bars below the line indicate avoidance. (From Morrissey, J.F. and Gruber, S.H. 1993. *Environ. Biol. Fish.* 38:311–319. With permission.)

19.3.4   Acoustic Monitoring

The results of acoustic tracking studies have been very important in defining short-term movement and habitat use patterns of sharks. However, understanding longer-term patterns of individuals and population-level factors is also important. Neither of these issues can be adequately tackled using acoustic tracking because of the resources required to continuously follow an individual for long periods (e.g., >1 week) or following more than one individual at a time. The development of underwater data-logging acoustic receivers opened up new possibilities in long-term population-level studies. Early equipment was large and relatively expensive and researchers were able to use only a limited number of receivers, covering small high-use areas. For example, Klimley et al. (1988) used two acoustic monitors to define the movements and habitat use of scalloped hammerheads, *Sphyrna lewini*, at a seamount over a 10-day period. They found that during the day sharks remained in a group at the seamount, but at night dispersed into the surrounding area.

As technology progressed, receivers became smaller and more affordable providing the opportunity to cover much larger areas for longer periods (Voegeli et al., 2001). Recent research on young blacktip sharks, *C. limbatus*, in Terra Ceia Bay, FL (Heupel and Hueter, 2001, 2002) has shown that a large array of omnidirectional acoustic receivers can be used to continuously monitor the movements of a population of sharks in a confined region for long periods. In this study up to 40 individuals per year were monitored within the study site for periods as great as 167 days. The use of an algorithm for taking the presence–absence data provided by the omnidirectional receivers and converting them to averaged positions (Simpfendorfer et al., 2002) enabled relatively fine-scale movement and habitat use data to be generated. Error estimates based on the averaged positions were 209 to 223 m, depending on the time period over which the averages were calculated. The results provided detailed long-term data on movements and habitat use (Figure 19.4), how they vary over time (Figure 19.5), and the synchronicity in habitat use changes across the population.

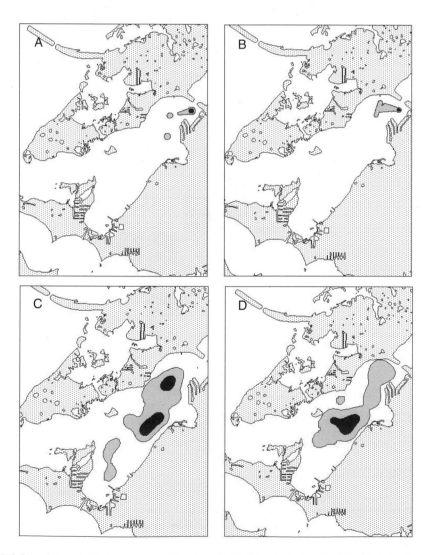

FIGURE 19.4 Increasing monthly home ranges of two juvenile blacktip sharks, *Carcharhinus limbatus*, from June (A and B) to October (C and D) in Terra Ceia Bay, FL. Home ranges were calculated using 50% (black areas) and 95% (gray areas) fixed kernels. (From Heupel et al., unpubl.)

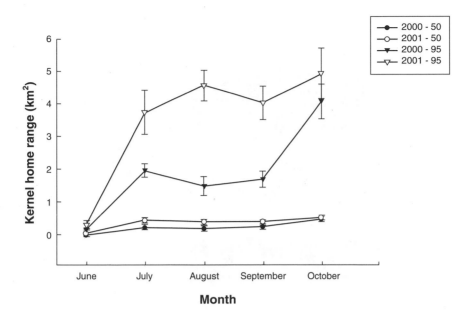

FIGURE 19.5 Increases in monthly home range size (50 and 95% kernels) of juvenile blacktip sharks, *Carcharhinus limbatus*, in Terra Ceia Bay, FL, in 2 years. (From Heupel et al., unpubl.)

The data-logging receivers used by Heupel and Hueter (2001) were small, inexpensive units that provide presence–absence data. An alternative type of data-logging receiver is one that provides triangulation of acoustic signals to give submeter accuracy in position (Voegeli et al., 2001). This type of equipment is much more expensive per unit and can be used to cover only a small area. However, for animals that have small activity spaces, or that a researcher wishes to study in detail in a certain area, this equipment can provide very good results. Klimley et al. (2001) used this type of system to study white sharks, *Carcharodon carchiaras*, at Año Nuevo Island off California and showed that sharks concentrated their movements to specific areas (Figure 19.6). These high-use areas were close to the islands in areas where seals haul out and provided the white sharks with the best opportunity for locating potential prey.

Acoustic monitoring is beginning to provide a greater understanding of habitat use in sharks, especially on longer temporal scales. As an emerging field, there still remains a significant amount of technical and analytical development to be undertaken. As with all approaches, however, it does have disadvantages. The largest drawback is that only sharks within the array of data-logging receivers can be studied. If an individual leaves the range of the array, then no data can be gathered. Second, with the use of the simpler data-logging receivers that do not triangulate positions, the accuracy of the positions may be relatively low, depending on the level of analysis applied (Simpfendorfer et al., 2002).

19.3.5 Satellite Telemetry

The biggest limitation of acoustic monitoring systems is that sharks must remain in the receiver array to be studied. Once they leave this area, habitat use data can no longer be collected. One technique that can address the issue of large spatial coverage is satellite telemetry. With this method a tag that transmits a signal to the ARGOS system is attached to a shark, and every time the shark comes to the surface the tag transmits and the ARGOS system estimates its location. As the ARGOS system uses polar-orbiting satellites, no matter where in the world's oceans the shark is, if it is at the surface when a satellite is overhead, its location can be determined. For sharks this approach is ideal for wide-ranging pelagic species that regularly come to the surface.

This approach was used by Eckert and Stewart (2001) to study whale sharks, *Rhincodon typus*, in the Sea of Cortez. They attached satellite transmitters to sharks using a towed float on a long tether, so that

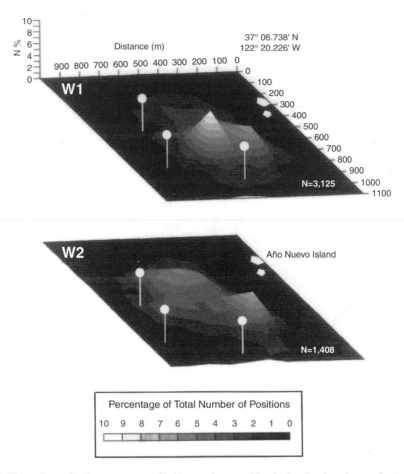

FIGURE 19.6 Three-dimensional contour maps of habitat use by two white sharks, *Carcharodon carcharias*, at Año Nuevo Island, CA. (From Klimley, A.P. et al. 2001. *Mar. Biol.* 138:429–446. With permission.)

the shark did not have to fully surface for the tag to be able to transmit. Using this technique they tagged 15 animals and tracked them from 1 to 1144 days and over distances of thousands of kilometers (Figure 19.7), including one animal that moved to the western north Pacific. In addition to location information the tags stored depth and temperature data that were also transmitted to the satellite. These data provided information on movements, migrations, and habitat use of whale sharks within the pelagic realm. The data indicated that they range widely, but preferred waters with sea surface temperatures between 28 and 32°C, and occupied depths mostly less than 10 m.

Satellite telemetry remains a relatively new approach in shark research, largely because of the need for animals to be at the surface (or at least very close to it) to transmit the required signals for positions to be generated. In addition to whale sharks, this approach has successfully been used with white sharks (Bruce, pers. comm.) and tiger sharks (Heithaus, pers. comm.). Another limitation of the approach is the accuracy of the position estimates. Many factors influence the accuracy of the positions, including the number of signals received per satellite pass, the temporal distribution of signals over the period of the satellite pass, the signal strength, and the relative movement of the tag. The best possible accuracy is to within approximately 250 m of the true position, and in most marine animal studies many position estimates will only be within 10 km (Hays et al., 2001).

19.3.6   Archival Tags

The use of archival tags — tags that store data on light level (for estimation of geographic position), depth, and temperature — were developed to overcome the problems of collecting long-term data on

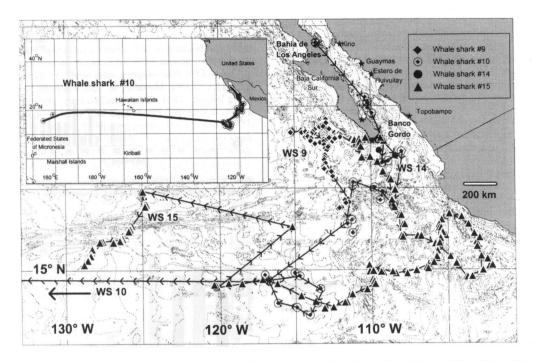

FIGURE 19.7 Long-term tracks of four whale sharks, *Rhincodon typus*, released in the Sea of Cortez. (From Eckert, S.A. and Stewart, B.S. 2001. *Environ. Biol. Fish.* 60:299–308. With permission.)

animals that rarely, if ever, come to the surface. These tags were originally developed for use on tuna and other pelagic teleosts, but have become relatively popular for use on sharks. The use of light levels to estimate location (a process known as light-base geolocation) relies on the ability to accurately estimate sunrise and sunset times, relative to Greenwich Mean Time (longitude), and day length (latitude). The accuracy of location estimates using light-based geolocation is low (Welch and Eveson, 1999, 2001; Musyl et al., 2001), and so only useful in habitat use studies with broad spatial scales (e.g., a species that migrates long distances) or in situations where location is of secondary importance (e.g., pelagic species where habitat use can be best defined using depth and temperature).

West and Stevens (2001) used archival tags to study school shark, *Galeorhinus galeus*, movements in southern Australia. As a heavily fished species the stock assessment process required information on habitat use, specifically how often, and for how long, did individuals enter pelagic habitats as opposed to neritic habitats where they were fished? In the study, 30 individuals were released, and at the time of publication 9 had been recaptured. The depth data stored by the tags showed that school sharks used pelagic habitats for variable periods that lasted as long as several months (Figure 19.8). This behavior of switching between pelagic and neritic habitats is unusual in sharks, and this study provided a good understanding of this phenomenon so that the stock assessment process could take into account periods when school sharks were not susceptible to particular fishing gears.

A large drawback of traditional archival tags is the need to recapture tagged animals to retrieve the data. This restricts work to heavily exploited species that have high rates of recapture. Recently, two manufacturers have developed archival tags that can be programmed to detach from an animal at a specific time, float to the surface, and transmit data via the ARGOS system (pop-up tags). These tags eliminate the need to recapture animals, opening the way for work on species that are not heavily fished. As these tags have only been recently developed, there are few published studies. Boustany et al. (2002) reported the results from six white sharks, *C. carcharias*, fitted with pop-up satellite tags off the coast of California. Four of these animals moved into open ocean habitats, including one that moved to Hawaii. The results showed a much greater use of pelagic habitats than previously believed for this species, and the use of archiving tags allowed information on patterns of depth and temperature utilization to be collected.

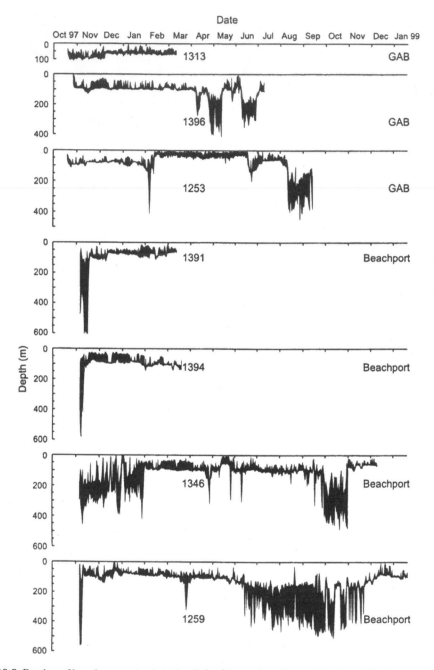

FIGURE 19.8 Depth profiles of seven school sharks, *Galeorhinus galeus*, from archival tags. Sharks were released in the Great Australian Bight (GAB) or off Beechport, South Australia. Sections of the depth profiles that show large variation between 150 and 500 m indicate periods of pelagic habitat use, while the remaining sections indicate neritic habitat use. (From West, G.J. and Stevens, J.D. 2001. *Environ. Biol. Fish.* 60:283–298. With permission.)

FIGURE 19.9 National Geographic's CritterCam, an animal-borne video and data telemetry recorder, placed on a male nurse shark, *Ginglymostoma cirratum*, in a long-term study of mating behavior in this species. (Photo © Harold L. Pratt, Jr. With permission.)

The disadvantages of pop-up tags are that they are relatively large (restricting them to use on larger species), that a relatively small amount of information can be downloaded via satellite from a relatively small platform, and that they are high cost. Some of these disadvantages will be overcome as the technology develops further, and they promise to be of great use in some studies of shark habitat use.

19.3.7 Animal-Borne Video Systems

Another recent addition to research tools available for studying habitat use in sharks is the animal-borne video camera and telemetry system (National Geographic's "CritterCam™"; Heithaus et al., 2001; Figure 19.9). Using this system the habitats in which a shark is swimming can be directly observed once the CritterCam is retrieved and the video tape viewed. In addition, telemetry data (depth, water temperature, and swimming speed) are also collected, providing a broad range of information on the shark's behavior. In studies using other telemetry methods, there is a level of error in the assignment of habitat type due to uncertainty in position information. Such errors are reduced by CritterCam because the habitats can be identified from the video. Heithaus et al. (2002) used this approach (combined with acoustic tracking) to investigate the habitat use of tiger sharks, *Galeocerdo cuvier*, in a seagrass ecosystem in Shark Bay, Australia. They fitted CritterCams to 37 individuals and recorded a total of 75 h of video. They used the habitat use data generated to compare with track randomization, correlated random walk models, and habitat availability to determine habitat preference. They found tiger sharks preferred shallow seagrass habitats (Figure 19.10) where their prey was most abundant.

Animal-borne video systems are a large piece of equipment and so can only be used with relatively large sharks (~1.5 m total length and larger). Continued development should see the size decrease. Other disadvantages of this approach are that they need to be used in areas with relatively high water clarity, are expensive to produce, and require a high degree of technical skill to use. As visual systems, they are also best used during the day, although some results can be obtained at night (Heithaus, pers. comm.). Systems are also limited by the amount of video that can be stored on the tape. To overcome this problem, these units can be programmed for a variety of recording schedules.

19.4 The Importance of Scale in Habitat Use Studies

How habitat use data are interpreted is often dependent on the scale at which it is collected. For example, if a study collects data only during the day, how do we understand how habitat use changes at night?

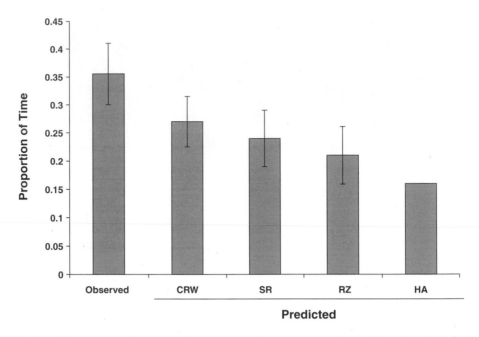

FIGURE 19.10 Habitat selection between shallow seagrass and deep sandy habitats in Shark Bay, Australia, by tiger sharks, *Galeocerdo cuvier*. Tiger sharks used shallow habitats significantly more than suggested by correlated random walk models (CRW), sample randomization (SR), randomization (RZ), and habitat availability (HA). Error bars indicate 95% confidence intervals. (From Heithaus, M.R. et al. 2002. *Mar. Biol.* 140:237–248. With permission.)

Similarly, we can only understand how habitat use changes as animals grow by collecting data at all ages, and preferably for the same individuals. Spatial scale is also important. If a study is limited to a specific area and individuals move in and out of the area regularly, then an incomplete understanding of habitat use will be gained. Below we consider examples of temporal and spatial scale in shark habitat use to demonstrate how they can influence interpretation of results.

19.4.1 Temporal Factors

Two levels of temporal scale are considered. The first are diel effects. Many studies of shark movements from which habitat use patterns can be inferred are based on short-term acoustic tracking. In some situations these tracks do not even last 24 h and so diel changes in habitat use may not be fully resolved. Second, we consider longer-term changes in habitat use that occur as sharks grow. Little is known of how habitat use changes for individual sharks over longer time frames, as most studies are based on the results of short-term acoustic tracks. However, we know that habitat use must change over time and it is important to understand how and why these changes occur.

19.4.1.1 Diel Effects — It has been commonly observed in sharks that there are diel changes in behavior and habitat use. Thus, it is important for researchers to collect data both during the day and at night to provide a full understanding of habitat use. Holland et al. (1993) demonstrated this for juvenile scalloped hammerheads in Hawaii using acoustic tracking. They found that activity space was larger at night than during the day, and that the center of activity shifted between day and night (Figure 19.11). The scalloped hammerheads used a small core daytime area, but ranged more widely at night as they hunted around patch reefs. If nighttime tracking was not carried out, then the importance of dispersed feeding in patch reef areas may never have been determined. There are many other examples of diel changes in shark behavior and habitat use (e.g., Nelson and Johnson, 1970; Klimley et al., 1988; Gruber et al., 1988; Carey and Scharold, 1990) showing that this is an important factor in this type of study.

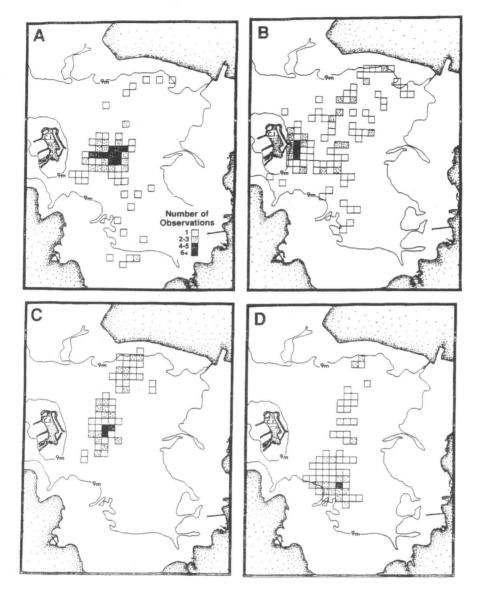

FIGURE 19.11 Diurnal habitat use of two juvenile scalloped hammerheads, *Sphyrna lewini*, in Kanahoe Bay, Hawaii. Daytime habitat use was smaller and more concentrated (A and C), while nighttime habitat use was larger and more dispersed (B and D). (From Holland, K.N. et al. 1993. *Copeia* 1993:495–502. With permission.)

19.4.1.2 Longer-Term Effects — Longer-term studies of movement and habitat use in sharks are rare, but the use of acoustic monitoring systems, archival tags, or satellite tracking can overcome these limitations. Studies of juvenile blacktip sharks in Terra Ceia Bay, FL using acoustic monitoring (Heupel and Hueter, 2001; Heupel, unpubl. data) have provided some of the best understanding of how habitat use can change over time. The young blacktips spend the first couple of months after birth (May to July) in a very small area at the north end of the bay, but this is followed by periods when they change, and rapidly expand, their home range (Figure 19.12). Interestingly, this expansion of home range occurred with a relatively high level of synchronicity within the population (Figure 19.12), suggesting that sharks may be responding to an environmental cue that triggers these changes in behavior. Thus, short-term studies (such as most acoustic tracking) do not necessarily reveal changes in habitat use patterns over time.

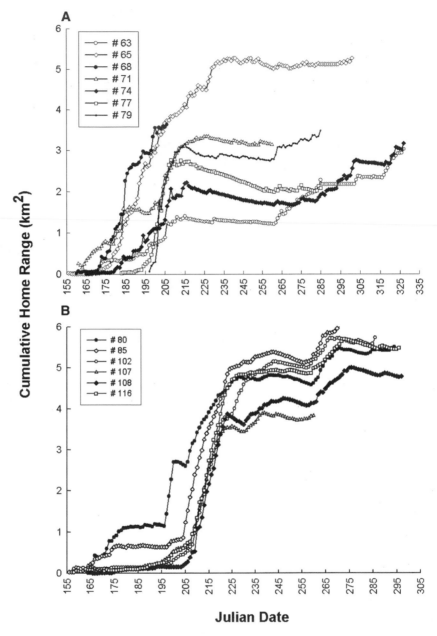

FIGURE 19.12 Increase in cumulative daily home range size for juvenile blacktip sharks, *Carcharhinus limbatus*, in Terra Ceia Bay, FL, from birth through the first summer of life; 2 years of data are presented: (A) 2000 and (B) 2001. Time is represented by days since January 1 of each year (Julian date). (From Heupel et al., unpubl.)

19.4.2   Spatial Factors

The spatial effects of animal movement patterns are critical to accurately defining habitat use. This is particularly true when examining large, highly migratory, pelagic species. Early studies of habitat use by pelagic sharks involved acoustic tracking of individuals to define their daily movements. In these studies horizontal and vertical movements of individuals were examined to define habitat use within the open ocean (Carey and Scharold, 1990; Holts and Bedford, 1993). These studies characterized the short-term movements of two common pelagic species: mako shark, *Isurus oxyrinchus* (Holts and Bedford, 1993), and blue shark, *Prionace glauca* (Carey and Scharold, 1990). Information was obtained regarding the use of the water column for thermal regulation and prey capture. More advanced spatial examination of habitat use was possible when archival and satellite tags were developed. These technologies have provided information on broad-scale habitat use by wide-ranging species such as white sharks, *C. carcharias* (Boustany et al., 2002), whale sharks, *R. typus* (Eckert and Stewart, 2001), and school sharks, *Galeorhinus galeus* (West and Stevens, 2001). These three studies showed that these species can at times undertake transoceanic scale movements, and that to fully understand habitat use a technique that enables information to be gathered on a broad spatial scale is required.

19.5 Factors Influencing Habitat Selection by Sharks

Understanding why sharks display the habitat use patterns that they do is a much more difficult task than simply describing them. A number of factors must be taken into account, and may work at different levels for different species, or in different locations for the same species. Johnson (1980) recognized that habitat selection is a hierarchical process, with different factors acting at different scales, including geographic range, home range, and use of habitats within the home range. Both physical and biotic factors may shape habitat use at all spatial scales. Physical factors include temperature, salinity, depth, and bottom sediment characteristics. Biotic factors include benthic vegetation (e.g., seagrasses or mangroves), prey distribution and availability, predator distribution, social organization, and reproductive activity.

19.5.1   Physical Factors

Habitat use is often bound by the physical parameters of the environment and the tolerance levels of the study species. Temperature is an important physical factor affecting shark habitat use on a broad scale. There are few species that can survive in the full range of temperatures that occur in the world's oceans, and so there are physical limits to the habitats that are available to a species. Within a species it is also common to see seasonal changes in distribution due to migrations. Although many factors may help drive these migrations, the inability of a species to tolerate the seasonal changes in temperature is an important factor in many cases.

Physical factors also act at finer spatial scales. Salinity may be an important factor for coastal species that enter estuaries. Some species such as bull sharks, *Carcharhinus leucas*, show a preference for lower salinity areas, particularly as juveniles (Compagno, 1984). Thus, they may actively seek these areas for their nurseries. Tidal flow is also an important factor for species that use shallow near-shore habitats. Ackerman et al. (2000) showed that habitat use in leopard sharks, *Triakis semifasciata*, in a coastal bay in California were directly influenced by tidal flow. Tidal flow can act in several different ways to affect habitat use. Decreasing depth can force animals to move to other habitats as shallow areas are exposed at low tide. Alternatively, changing physical parameters within the water column can provide constraints on the type and amount of habitat available for use by a species. Temperature is also an important physical factor at fine spatial scales. In a study of bat rays, *Myliobatis californica*, Matern et al. (2000) hypothesized that in Tomales Bay, CA this species selected areas that enabled it to behaviorally thermoregulate.

19.5.2   Biotic Factors

Intuitively, biotic factors should play an important role in habitat selection by sharks. The needs to feed, avoid predators, and reproduce are important parts of a shark's life. Despite this, their importance in habitat use studies has rarely been considered for sharks. Detailed behavioral studies by Heithaus et al. (2002) on tiger sharks in Shark Bay, Australia, are probably the best example of research involving prey distribution. In this study information on shark movements and habitat use was defined by acoustic tracking, animal-borne video systems, and prey distribution surveys. Tiger sharks showed a preference for shallow seagrass areas where their main prey (fish, turtles, sea snakes, and birds) were more commonly found.

In a study of juvenile blacktip sharks in Terra Ceia Bay, FL, Heupel and Hueter (2002) found that there was no correlation between prey availability (small fish density) and blacktip occurrence. On the basis of these data they concluded that blacktip shark habitat selection was not based on food availability, and suggested that the risk of predation by larger sharks was more likely to be driving habitat selection. Further research is needed to explore the importance of predation risk in relation to habitat use, especially in smaller sharks.

19.6 The Future of Habitat Use Studies on Sharks, Skates, and Rays

The importance of habitat use studies has been increasing over the past decade. The recognition of the need to define and understand essential fish habitat for commercially fished species, and critical habitat for endangered species, has provided the impetus for much of this work. This increase in research, combined with the development of technologies that are making it possible for researchers to answer relevant habitat use and preference questions (e.g., satellite tags, acoustic monitoring systems, archival tags, and animal-borne video systems), has made this type of work much more accessible. As such, this type of research should continue to grow in importance. To date, however, habitat use studies have been largely qualitative. Only a handful of studies have provided quantitative evidence of habitat selection in sharks (e.g., Morrissey and Gruber, 1993b; Heithaus et al., 2002; Michel, 2002). In the future, researchers will need to design and implement studies that specifically address questions of habitat use and preference.

References

Ackerman, J.T., Kondratieff, M.C., Matern, S.A. and Cech, J.J., Jr. 2000. Tidal influence on spatial dynamics of leopard sharks, *Triakis semifasciata*, in Tomales Bay, California. *Environ. Biol. Fish.* 58:33–43.

Anderson, D.J. 1982. The home range: a new nonparametric estimation technique. *Ecology* 63:103–112.

Bass, A.J. 1978. Problems in studies of sharks in the southwest Indian Ocean, in Sensory Biology of Sharks, Skates and Rays, E.S. Hodgson and R.F. Mathewson, Eds., Office of Naval Research, Department of the Navy, Arlington, VA, 545–594.

Boustany, A.M., Davis, S.F., Pyle, P., Anderson, S.D., Le Boef, B.J. and Block, B.A. 2002. Expanded niche for white sharks. *Nature* 415:35–36.

Branstetter, S. 1990. Early life history implications of selected carcharhinoid and lamnoid sharks of the northwest Atlantic. NOAA Tech. Rep., NMFS 90:17–28.

Burt, W.H. 1943. Territoriality and home range concepts as applied to mammals. *J. Mammal.* 24:346–352.

Carey, F.G. and Scharold, J.V. 1990. Movements of blue sharks (*Prionace glauca*) in depth and course. *Mar. Biol.* 106:329–342.

Castro, J.I. 1993. The shark nursery of Bulls Bay, South Carolina, with a review of the shark nurseries of the southeastern coast of the United States. *Environ. Biol. Fish.* 38:37–48.

Clarke, T.A. 1971. The ecology of the scalloped hammerhead shark, *Sphyrna lewini*, in Hawaii. *Pac. Sci.* 25:133–144.

Compagno, L.J.V. 1984. FAO Species Catalogue. Vol. 4. *Sharks of the World, an Annotated and Illustrated Catalogue of Shark Species Known to Date*. Part 2. *Carcharhiniformes*. FAO Fish. Synop. 125:251–655.

Cooper, W.E., Jr. 1978. Home range criteria based on temporal stability of area occupation. *J. Theor. Biol.* 73:687–695.

Eckert, S.A. and Stewart, B.S. 2001. Telemetry and satellite tracking of whale sharks, *Rhincodon typus*, in the Sea of Cortez, Mexico and the north Pacific Ocean. *Environ. Biol. Fish.* 60:299–308.

Economakis, A.E. and Lobel, P.S. 1998. Aggregation behavior of the grey reef shark, *Carcharhinus amblyrhynchos*, at Johnson Atoll, Central Pacific Ocean. *Environ. Biol. Fish.* 51:129–139.

Gruber, S.H., Nelson, D.R., and Morrissey, J.F. 1988. Patterns of activity and space utilization of lemon sharks, *Negaprion brevirostris*, in a shallow Bahamian lagoon. *Bull. Mar. Sci.* 43:61–76.

Hays, G.C., Akesson, S., Godley, B.J., Luschi, P., and Santidrian, P. 2001. The implications of location accuracy for the interpretation of satellite-tracking data. *Anim. Behav.* 61:1035–1040.

Heithaus, M.R., Marshall, G.J., Buheler, B.M., and Dill, L.M. 2001. Employing CritterCam to study habitat use and behavior of large sharks. *Mar. Ecol. Prog. Ser.* 209:307–310.

Heithaus, M.R., Dill, L.M., Marshall, G.J., and Buhleier, B. 2002. Habitat use and foraging behavior of tiger sharks (*Galeocerdo cuvier*) in a seagrass ecosystem. *Mar. Biol.* 140:237–248.

Heupel, M.R. and Hueter, R.E. 2001. Use of an automated acoustic telemetry system to passively track juvenile blacktip shark movements, in *Electronic Tagging and Tracking in Marine Fisheries,* J.R. Sibert and J.L. Nielsen, Eds., Kluwer Academic, Amsterdam, the Netherlands, 217–236.

Heupel, M.R. and Hueter, R.E. 2002. The importance of prey density in relation to the movement patterns of juvenile sharks within a coastal nursery area. *Mar. Freshwater Res.* 53:543–550.

Holland, K.N., Wetherbee, B.M., Peterson, J.D., and Lowe, C.G. 1993. Movements and distribution of hammerhead shark pups on their natal grounds. *Copeia* 1993:495–502.

Holts, D.B. and Bedford, D.W. 1993. Horizontal and vertical movements of the shortfin mako shark, *Isurus oxyrinchus*, in the Southern California Bight. *Aust. J. Mar. Freshwater Res.* 44:901–909.

Jennrich, R.I. and Turner, F.B. 1969. Measurement of non-circular home range. *J. Theor. Biol.* 22:227–237.

Johnson, D.H. 1980. The comparison of usage and availability measurements for evaluating resource preference. *Ecology* 61:65–71.

Johnson, R.H. and Nelson, D.R. 1973. Agonistic display in the gray reef shark, *Carcharhinus menisorrah*, and its relationship to attacks on man. *Copeia* 1973:76–84.

Krebs, C.J. 1999. *Ecological Methodology*, 2nd ed., Addison-Wesley/Longman, Menlo Park, CA.

Klimley, A.P., Butler, S.B., Nelson, D.R., and Stull, T. 1988. Diel movements of the scalloped hammerhead shark, *Sphyrna lewini* Griffith and Smith, to and from a seamount in the Gulf of California. *J. Fish Biol.* 33:751–761.

Klimley, A.P., Le Boeuf, B.J., Cantara, K.M., Richert, J.E., Davis, S.F., and Van Sommeran, S. 2001. Radio-acoustic positioning as a tool for studying site-specific behavior of the white shark and other large marine species. *Mar. Biol.* 138:429–446.

Matern, S.A., Cech, J.J., and Hopkins, T.E. 2000. Diel movements of bat rays, *Myliobatis californica*, in Tomales Bay, California: evidence for behavioral thermoregulation? *Environ. Biol. Fish.* 58:173–193.

McKibben, J.N. and Nelson, D.R. 1986. Patterns of movement and grouping of gray reef sharks, *Carcharhinus amblyrhynchos*, at Enewetak, Marshall Islands. *Bull. Mar. Sci.* 38:89–110.

Medved, R.J. and Marshall, J.A. 1983. Short-term movements of young sandbar sharks, *Carcharhinus plumbeus* (Pisces, Carcharhinidae). *Bull. Mar. Sci.* 33:87–93.

Michel, M. 2002. Environmental Factors Affecting the Distribution and Abundance of Sharks in the Mangrove Estuary of the Ten Thousand Islands, Florida, USA. Ph.D. thesis, University of Basel, Switzerland, 134 pp.

Morrissey, J.F. and Gruber, S.H. 1993a. Home range of juvenile lemon sharks. *Copeia* 1993:425–434.

Morrissey, J.F. and Gruber, S.H. 1993b. Habitat selection by juvenile lemon sharks, *Negaprion brevirostris*. *Environ. Biol. Fish.* 38:311–319.

Musyl, M.K., Brill, R.W., Curran, D.S., Gunn, J.S., Hartog, J.R., Hill, R.D., Welch, D.W., Eveson, J.P., Boggs, C.H., and Brainard, R.E. 2001. Ability of archival tags to provide estimates of geographical position based on light intensity, in *Electronic Tagging and Tracking in Marine Fisheries,* J.R. Sibert and J.L. Nielsen, Eds., Kluwer Academic, Amsterdam, the Netherlands, 343–367.

Nelson, D.R. and Johnson, R.H. 1970. Diel activity rhythms in the nocturnal, bottom-dwelling sharks, *Heterodontus francisci* and *Cephaloscyllium ventriosum*. *Copeia* 1970:732–739.

Nelson, D.R., Johnson, J.N., McKibben, J.N., and Pittenger, G.G. 1986. Agonistic attacks on divers and submersibles by gray reef sharks, *Carcharhinus amblyrhynchos*: antipredatory or competitive? *Bull. Mar. Sci.* 38:68–88.

Olsen, A.M. 1954. The biology, migration and growth rate of the school shark *Galeorhinus australis* (Macleay) (Carcharhinidae) in south-eastern Australian waters. *Aust. J. Mar. Freshwater Res.* 5:353–410.

Sciarotta, T.C. and Nelson, D.R. 1977. Diel behavior of the blue shark, *Prionace glauca*, near Santa Catalina Island, California. *U.S. Fish. Bull.* 75:519–528.

Simpfendorfer, C.A. and Milward, N.E. 1993. Utilisation of a tropical bay as a nursery area by sharks of the families Carcharhinidae and Sphyrnidae. *Environ. Biol. Fish.* 37:337–345.

Simpfendorfer, C.A., Heupel, M.R., and Hueter, R.E. 2002. Estimation of short-term centers of activity from an array of omnidirectional hydrophones, and its use in studying animal movements. *Can. J. Fish. Aquat. Sci.* 59:23–32.

Springer, S. 1967. Social organisation of shark populations, in *Sharks, Skates, and Rays,* P.W. Gilbert, R.F. Mathewson, and D.P. Rall, Eds., Johns Hopkins University Press, Baltimore, 149–174.

Strauss, R.E. 1979. Reliability for estimates for Ivlev's electivity index, the forage ratio, and a proposed linear index of food selection. *Trans. Am. Fish. Soc.* 108:344–352.

Voegeli, F.A., Smale, M.J., Webber, D.M., Andrade, Y., and O'Dor, R.K. 2001. Sonic telemetry, tracking and automated monitoring technology. *Environ. Biol. Fish.* 60:267–281.

Welch, D.W. and Eveson, J.P. 1999. An assessment of light-based geoposition estimates from archival tags. *Can. J. Fish. Aquat. Sci.* 56:1317–1327.

Welch, D.W. and Eveson, J.P. 2001. Recent progress in estimating geoposition using daylight, in *Electronic Tagging and Tracking in Marine Fisheries,* J.R. Sibert and J.L. Nielsen, Eds., Kluwer Academic, Amsterdam, the Netherlands, 369–383.

West, G.J. and Stevens, J.D. 2001. Archival tagging of school sharks, *Galeorhinus galeus,* in Australia: initial results. *Environ. Biol. Fish.* 60:283–298.

Williams, H. and Schaap, A.H. 1992. Preliminary results of a study into the incidental mortality of sharks in gill-nets in two Tasmanian shark nursery areas. *Aust. J. Mar. Freshwater Res.* 43:237–250.

Worton, B.J. 1987. A review of models of home range for animal movement. *Ecol. Model.* 38:277–298.

Subject Index

A

Acid-base regulation, 248
  digestive processes, 288
  gill function, 259, 261–263
  kidney function, 260–261
  rectal gland function, 260
Acoustic monitoring, 559–561, *See also* Telemetry
Acoustic tracking, 556–559
ACTH, 295
Adaptive immunity, 380–382
Adrenocorticotropic hormone (ACTH), 295
Age determination, 399–400, *See also* Growth
  back-calculation, 409
  banding patterns, 400, 405–407
  bomb carbon dating, 414
  chemical markers, 411–412
  edge and relative marginal increment analysis, 412–413
  implications for fisheries management, 438–439
  precision and bias, 408
  process, 400
  sample preparation, 403–404
  sampling, 402–403
  staining, 407–408
  structures used
    caudal thorns, 402
    jaw, 402
    neural arches, 401–402
    spines, 401
    vertebrae, 400
  study review, 414–437
  summary of studies (table), 426–437
  tag-recapture, 410–411
  verification and validation, 409–410
Age of maturity, 425, 438, 457–458
  sex differences, 452
Age-structured demographic models, 459–459
Alar scales or plates, fossil evidence, 5
Albinism, 473
Allozymes, 473–474
Ambushing behavior, 167, 498, 500
Ammonia excretion, 261–262
Amphistylic jaw support, 173
Ampullae of Lorenzini, 341–343
Anaerobic metabolism, 211
Angiotensin, 297
Animal-borne video systems ("CritterCam"), 565
Aplacental viviparity, 271
Apoptosis, 385–386
Apparent competition, 492

Aquarium trade, 413
Archival tags, 562–565
Arginine vasotocin (AVT), 298
ARGOS, 561, 563
Aristotle, 270, 273
Audiograms, 333–335
Autodiastyly, 173
Average percent error (APE) technique, 408

B

Back-calculation, 409
Basophils, 371
Batch feeding, 500
Bathymetric zones, 35–36
Batoid evolution, 55, 90
Batoid feeding behaviors, 169
Batoid feeding mechanics, functional morphology of, 183–186
Batoid locomotion mechanics, 141, 142, 157–159
Batoid phylogeny, 79–108
  analyses, 80–81, 83, 86, 88, 89
  annotated classification, 93–96
  basal clades, 80
  biogeography, 96–100
  cephalic and branchial musculature, 108
  character matrix, 81, 102
  claspers, 81–83, 93
  consensus tree, 83, 86, 88, 89
  DNA evidence, 80
  evolutionary implications, 90
  implications of analyses, 86–92
  monophyly, 79–80, 103
    external morphology, 103
    lateral line canals, 104
    squamation, 103
    tooth root vascularization, 103–104
  rajidae relationships, 93
  skeletal structures, 104–108
  specimens examined, 101
  swimming velocity, 90
Bear Gulch fossils, 7–16
  age, sex, and reproductive stage, 11, 15
  community structure and population dynamics, 9–11
  reproductive strategies, 15–16
  taxonomic composition and diversity, 8
Benthic predation, 500
  effect on invertebrate communities, 508–510
Bile duct parasites, 541

R

Animal Index

A

Acanthobothrium, 540
Acanthocephalans, 524, 538, 539
Acanthochondrites, 541
Acanthochondrites annulatus, 535
Acanthocotyle greeni, 529
Acanthorhina, 17
Acroteriobatus, 97
Aculeola nigra, 58, 64
Aega antillensis, 541
Aega psora, 538
Aetobatus, 80, 82, 89, 91, 96, 99, 103–108
Aetobatus narinari, 88, 89, 102, 186, 191, 282, 283, 356, 537, 541
Aetomylaeus, 96, 99, 103
Aetomylaeus nichofi, 539
Aetoplatea, 96, 99
Agnathan thelodonts, 6
Alligators, 356
Alopias, 46, 142, 478
Alopias pelagicus, 48, 67, 191, 272, 327, 419, 431, 438, 459, 478
Alopias vulpinus (thresher shark), 48, 67, 419, 431, 475, 500, 541
Alopiidae, 48, 67
Amblyraja, 95, 98
Amphibdella flavolineata, 542
Amphipods, 524, 529
Anacanthobatis, 95, 97–100, 103
Anaporrhutum, 543
Anelasma squalicola, 527, 529, 538
Angel shark, Pacific, See Squatina californica
Annelids, 524, 529, See also Leeches
Anoxypristis, 94
Antarctilamna, 6
Anthosoma crassum, 538
Apristurus, 49, 58, 142
Apristurus acanutus, 68
Apristurus aphyodes, 68
Apristurus atlanticus, 68
Apristurus brunneus, 68
Apristurus canutus, 68
Apristurus fedorvi, 68
Apristurus gibbosus, 68
Apristurus herklotsi, 68
Apristurus indicus, 68
Apristurus investigatoris, 68
Apristurus japonicus, 68
Apristurus kampae, 68
Apristurus laurussoni, 68

Apristurus longicephalus, 68
Apristurus macrorhynchus, 68
Apristurus macrostomus, 68
Apristurus manis, 68
Apristurus microps, 68
Apristurus micropterygeus, 68
Apristurus nasutus, 68
Apristurus parvipinnis, 68
Apristurus pinguis, 68
Apristurus platyrhynchus, 68
Apristurus profundorum, 68
Apristurus riveri, 68
Apristurus saldanha, 69
Apristurus sibogae, 69
Apristurus sinensis, 69
Apristurus sp. A, 68, 69
Apristurus sp. B, 68, 69
Apristurus sp. C, 68, 69
Apristurus sp. D, 68, 69
Apristurus sp. E, 68, 69
Apristurus sp. F, 68, 69
Apristurus spongiceps, 69
Apristurus stenseni, 69
Apristurus verwyi, 69
Aptychotrema, 91, 95
Argopecten irradians concentricus, 508
Argulus, 531, 537
Arhynchobatinae, 93, 95, 97, 100
Arhynchobatis, 95
Arthropods, 524, 528, 529, 538
Astropecten verrilli, 509
Asymbolus, 49
Asymbolus analis, 69
Asymbolus funebris, 69
Asymbolus occiduus, 69
Asymbolus pallidus, 69
Asymbolus parvus, 69
Asymbolus rubiginosus, 69
Asymbolus submaculatus, 69
Asymbolus vincenti, 69
Atelomycterus fasciatus, 69
Atelomycterus macleayi, 69
Atelomycterus marmoratus, 69
Atelomycterus sp. A, 69
Atlantic sharpnose shark, See Rhizoprionodon terraenovae
Atlantoraja, 95, 99, 103
Aulohalaelurus, 49
Aulohalaelurus kanakorum, 69
Aulohalaelurus labiosus, 69
Aztecodus, 6